Springer Collected Works in Mathematics

More information about this series at http://www.springer.com/series/11104

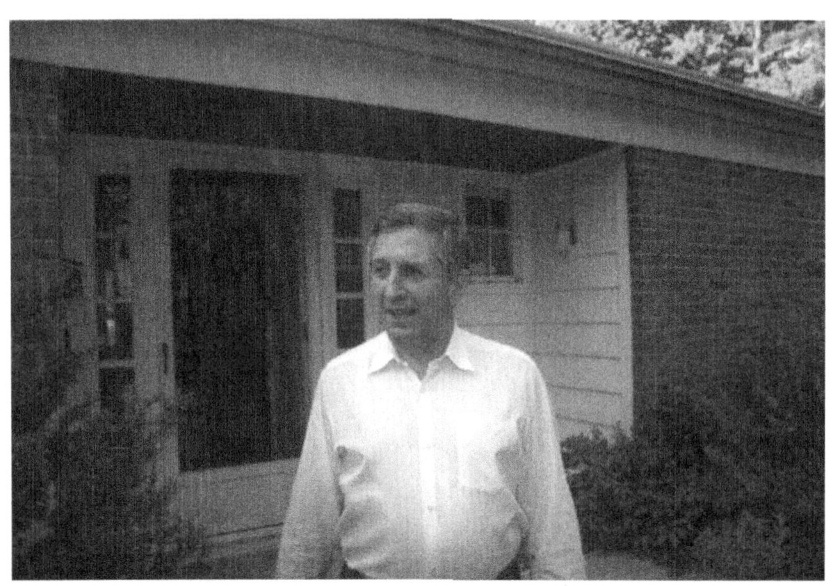

Princeton, 2005, Photo by D. Ridberg.

Yakov G. Sinai

Selecta II

Probability Theory, Statistical Mechanics, Mathematical Physics and Mathematical Fluid Dynamics

Reprint of the 2010 Edition

Yakov G. Sinai
Department of Mathematics
Princeton University
Princeton, NJ, USA

ISSN 2194-9875
Springer Collected Works in Mathematics
ISBN 978-1-4939-9788-6

Mathematics Subject Classification (2010): 37A35, 37A60, 37Dxx

This Springer imprint is published by the registered company Springer Science+Business Media, LLC, part of Springer Nature.
The registered company address is: 233 Spring Street, New York, NY 10013, U.S.A.

Preface

Two volumes of Selecta contain many of Yakov Sinai's papers spanning more than half of a century. Some of these papers became classics a long time ago, some later, and others were published only recently.

Sinai has pioneered building bridges between the theory of dynamical systems and statistical mechanics. Often switching his main research interests from one field to another, he has demonstrated how the ideas and approaches in one area can enrich and bring new understanding to the other one. This Selecta clearly demonstrates how Sinai succeeded in transforming these two areas essentially into one with a unified vision and a wealth of tools and ideas. This singular vision can be traced in throughout these papers, including those on mathematical physics, fluid dynamics, and Partial Differential Equations.

Sinai is universally considered as the major architect of the modern theory of dynamical systems. Naturally, the first volume is dedicated to ergodic theory and dynamical systems. The part on entropy theory demonstrates the dramatic beginning of one of the most revolutionary discoveries in mathematics of the twentieth century (which allowed to build a unified theory of probabilistic and deterministic systems). The last paper in this part was published 40 years ago. Chaotic billiards is another flourishing area whose foundations were laid in a pioneering work of Sinai.

The first three papers in the "Dynamical Systems" section have dramatically changed this field by bringing together the concepts and approaches from statistical mechanics and dynamics. These ideas are at the heart of thermodynamic formalism, which has since become the basic approach to the studio of strongly chaotic (hyperbolic) systems. Other papers in this volume include an influential paper on Feigenbaum universality and more recent papers related to number-theoretic [OIC??] problems.

The selection of the papers for this edition was made by Sinai himself. He has also provided commentary for each paper. (It was a tough selection, and in our opinion quite a few of Sinai's classical papers were not included here.) The reader of this Selecta will be impressed by the variety of brilliant and unconventional ideas which revolutionized so many areas of mathematics and created so many exciting new directions. Sinai's enthusiasm and infinite optimism, multiplied by brilliance and consistent hard work, are responsible for that. Very often his intuition, taste, and enthusiasm have led to a goal visible at the time only by him. Sinai created new mathematical machineries, which were later refined and polished by others while he was busy with other discoveries.

This collection of papers of one of the giants of modern mathematics will serve as an inspiration for students, as well as for senior researchers, demonstrating that an exciting scientific journey through the various disciplines never ends if one is truly fascinated by mathematics. A piece of advice to the readers: Try to borrow some of Sinai's optimism while reading this Selecta and keep that optimism with you in your research.

Dmitry Dolgopyat

Leonid A. Bunimovich
Konstantin M. Khanin

Contents

VOLUME 2

Probability Theory, Statistical Mechanics, Mathematics Physics and Mathematical Fluid Dynamics

Part I. Probability Theory

1. *Simple Random Walks on Tori*, J. Statist. Phys., **94** (1999), no. 3/4, 695–708 3

2. *The limiting behavior of a one-dimensional random walk in a random medium*, Teor. Veroyatnost. i Primenen., 27 (1982), no. 2, 247–258; translated in Theory Probab., **27** (1982), no. 2, 256–268 18

3. *A remark concerning random walks with random potentials*, Fund. Math., **147** (1995), no. 2, 173–180 32

4. *A Random Walk with Random Potential*, Teor. Veroyatnost. i Primenen., **38** (1993), no. 2, 457–460; translated in Theory Probab. Appl., **38** (1993), no. 2, 382–385 41

5. *Distribution of some functionals of the integral of a random walk*, Teoret. Mat. Fiz., 90 (1992), no. 3, 323–353, translated in Theoret. Math. Phys., **90** (1992), no. 3, 219–241 46

II. Statistical Mechanics

1. *Construction of dynamics in one-dimensional systems of statistical mechanics*, Teoret. Mat. Fiz., 11 (1972), no. 2, 248–258, translated in Theoret. Math. Phys., **11** (1972), no. 2, 487–494 73

2. (with S. A. Pirogov), *Phase diagrams of classical lattice systems*, Teoret. Mat. Fiz., **25** (1975), no. 3, 358–369; translated in Theoret. Math. Phys., **25** (1975), no. 3, 1185–1192. (with S. A. Pirogov), Phase diagrams of classical lattice systems (continuation), Teoret. Mat. Fiz., **26** (1976), no. 1, 61–76; translated in Theoret. Math. Phys., **26** (1976), no. 1, 39–49 82

3. (with E. I. Dinaburg), *An analysis of ANNNI model by Peierl's contour method*, Comm. Math. Phys., **98** (1985), no. 1, 119–144 102

4. (with P. M. Bleher), *Critical indices for Dyson's asymptotically-hierarchical models*, Comm. Math. Phys., **45** (1975), no. 3, 247–278 129

5. (with P. M. Bleher) *Investigation of the critical point in models of the type of Dyson's hierarchical models*, Comm. Math. Phys, **33** (1973), no. 1, 23–42 162

III. Mathematical Physics

1. (with E. I. Dinaburg), *The one-dimensional Schrödinger equation with a quasiperiodic potential*, Funkcional. Anal. i Priložen., 9 (1975), no. 4, 8–21; translated in Funct. Anal. Appl., **9** (1975), no. 4, 279–289 191

2. (with P. Bleher and D. Kosygin), *Distribution of Energy Levels of Quantum Free Particle on the Liouville Surface and Trace Formulae*, Comm. Math. Phys., **170** (1995), no. 2, 375–403 203

3. *Mathematical problems in* the *theory of quantum chaos*, Geometric aspects of functional analysis (1989–90), 41–59, Lecture Notes in Math., 1469, Springer, Berlin, 1991 233

4. *Dynamics of a heavy particle surrounded by a finite number of light particles*, Teoret. Mat. Fiz., 121 (1999), No. 1, 110–116; translated as Theoret. and Math. Phys., **121** (1999), no. 1, 1351–1357 253

5. (with A. I. Neishtadt), *Adiabatic piston as a dynamical system*, J. Statist. Phys. **116** (2004), no. 1-4, 815–820 261

6. *Poisson distribution in a geometric problem*, Dynamical systems and statistical mechanics (Moscow, 1991), 199–214, Adv. Soviet Math., **3**, Amer. Math. Soc., Providence, RI, 1991 268

IV. Mathematical Fluid Dynamics

1. (with Weinan E, K. Khanin, and A. Mazel), *Invariant Measures for Burgers Equation with Stochastic Forcing*, Ann. of Math. (2), **151** (2000), no. 3, 877–960 289

2. (with J. Mattingly), *An Elementary Proof of the Existence and Uniqueness Theorem for Navier-Stokes Equations*, Commun. in Contemp. Math., 1 (1999), no. 4, 497–516 375

3. *Statistics of shocks in solutions of Inviscid Burgers equation*, Comm. Math. Phys., **148** (1992), no. 3, 601–621 396

4. (with Weinan E and J. Mattingly), *Gibbsian dynamics and ergodicity for the stochastically forced Navier–Stokes equation*, Comm. Math. Phys., **224** (2001), no. 1, 83–106 418

5. (with D. Li), *Blow ups of complex solutions of the 3D-Navier-Stokes system and renormalization group method*, J. Eur. Math. Soc. (JEMS), **10** (2008), no. 2, 267–313 443

6. *Two results concerning asymptotic behavior of solutions of the Burgers equation with force*, J. Statist. Phys., **64** (1991), no. 1-2, 1–12 501

Acknowledgments

I thank L. A. Bunimovich, F. Cellarosi, D. Dolgopyat, E. Giaccaglia, K. Khanin, Ya. Pesin, and I. Vinogradov for their help in preparing this volume.

Ya. G. Sinai

Publisher Acknowledgments

We thank the publishers and copyright holders of Yakov G. Sinai's papers in Volume 1 of his Selected Papers for their kind permission to reprint his articles.

Doklady Akademii Nauk SSSR

2. *On the notion of entropy for a dynamic system* (Russian), Dokl. Akad. Nauk SSSR, 124 (1959), no. 4, 768–771.

Soviet mathematics - Doklady

10. (with V. A Rokhlin), *The structure and properties of invariant measurable partitions*, Dokl. Akad. Nauk SSSR, 141 (1961), no. 5, 1038–1041; translated in Soviet Math. Dokl., 2 (1961), no. 4-6, 1611-1614.

American Mathematical Society

10. (with V. A Rokhlin), *The structure and properties of invariant measurable partitions*, Dokl. Akad. Nauk SSSR, 141 (1961), no. 5, 1038–1041; translated in Soviet Math. Dokl., 2 (1961), no. 4-6, 1611-1614.

9. *Dynamical systems with countable Lebesque spectrum. I.*, Izv. Akad. Nauk SSSR Ser. Mat., 25 (1961), 899–924; translated in Amer. Math. Soc. transl. (2), 39 (1964), 83–110.

20. *Classical dynamical systems with countably-multiple Lebesque spectrum. II.*, Izv. Akad. Nauk SSSR Ser. Mat., 30 (1966), 15–68; translated in Amer. Math. Soc. transl. (2), 68 (1968), 34–88.

Cambridge University Press

195. (with C. Ulcigrai), *Renewal-type limit theorem for the Gauss map and continued fractions*, Ergodic Theory Dynam. Systems, 28 (2008), no. 2, 643–655.

71. (with Ya. B. Pesin), *Gibbs measures for partially hyperbolic attractors*, Ergodic Theory and Dynam. Systems, 2 (1982), no. 3-4, 417–438.

Springer

126. (with K. M. Khanin), *Mixing of some classes of special flows over rotations of the circle*, Funktsional. Anal. i Prilozhen., 26 (1992), no. 3, 1–21, translated in Funct. Anal. Appl., 26 (1992), no. 3, 155–169.

25. *Markov Partitions and U-diffeomorphisms*, Funkcional. Anal. i Priloˇzen., 2 (1968), no. 1, 64–89; translated in Funct. Anal. Appl., 2 (1968), no. 1, 61–82.

130. (with N. I. Chernov, G. Eyink, and J. Lebowitz), *Steady-state electrical conduction in the periodic Lorentz gas*, Comm. Math. Phys., 154 (1993), no. 3, 569–601.

London Mathematical Society, Turpion Ltd, and the Russian Academy of Sciences

105. (with K. M. Khanin), *Smoothness of conjugacies of diffeomorphisms of the circle with rotations*, Uspekhi Mat. Nauk, 44 (1989), no. 1, 57–82, 247; translated in Russian Math. Surveys, 44 (1989), no. 1, 66–99.

78. (with E. B. Vul and K. M. Khanin), *Feigenbaum universality and thermodynamic formalism*, Uspekhi Mat. Nauk, 39 (1984), no. 3, 3–37; translated in Russian Math. Surveys, 39 (1984), no. 3, 1–40.

37. *Gibbs measures in ergodic theory*, Uspehi Mat. Nauk, 27 (1972), no. 4(166), 21–64; translated in Russian Math. Surveys, 27 (1972), no. 4, 21–69.

29. *Dynamical systems with elastic reflections*, Uspehi Mat. Nauk, 25 (1970), no. 2(152), 141–192; translated in Russian Math. Surveys, 25 (1970), no. 2, 137–189.

38. (with L. A. Bunimovich), *On a fundamental theorem in the theory of dispersing billiards*, Mat. Sb. (N.S.), 90(132) (1973), 415–431; translated in Math. USSR-Sb., 19 (1973), no. 3, 407–423.

94. (with N. I. Chernov), *Ergodic properties of some systems of two-dimensional disks and three-dimensional balls*, Uspekhi Mat. Nauk, 42 (1987), no. 3, 153–174; translated in Russian Math. Surveys, 42 (1987), no. 3, 181–207.

51. *Billiard trajectories in a polyhedral angle*, Uspehi Mat. Nauk, 33 (1978), no. 1 (199), 229–230; translated in Russian Math. Surveys, 33 (1978), no. 1, 219–220.

Bibliography

1 *On the distribution of the first positive sum for the sequence of independent random variables*, Teor. Veroyatnost. i Primenen., **2** (1957), 126–135; translated in Theory Probab. Appl., **2** (1957), no. 1, 122–129.

2 *On the notion of entropy for a dynamic system* (Russian), Dokl. Akad. Nauk SSSR, **124** (1959), no. 4, 768–771.

3 *Flows with finite entropy* (Russian), Dokl. Akad. Nauk SSSR, **125** (1959), no. 6, 1200–1202.

4 *Geodesic flows on manifolds of negative constant curvature*, Dokl. Akac. Nauk SSSR, **131** (1960), no. 4, 752–755; translated in Soviet Math. Dokl., **1** (1960), no. 2, 335–339.

5 *The central limit theorem for geodesic flows on manifolds of constant negative curvature*, Dokl. Akad. Nauk SSSR, **133** (1960), no. 6, 1303–1306; translated in Soviet Math. Dokl., **1** (1960), no. 4, 983–987.

6 *Dynamical systems and stationary Markov processes*, Teor. Veroyatnost. i Primenen., **5** (1960), 335–338; translated in Theory Probab. Appl., **5** (1960), no. 3, 305–308.

7 (with L. D. Meshalkin), *Investigation of the stability of a stationary solution of a system of equations for the plane movement of an incompressible viscous liquid*, Prikl. Mat. Meh., **25** (1961), 1140–1143; translated in J. Appl. Math. Mech., **25** (1961), 1700–1705.

8 *Geodesic flows on compact surfaces of negative curvature*, Dokl. Akad. Nauk SSSR, **136** (1961), no. 3, 549–552; translated in Soviet Math. Dokl., **2** (1961), no. 1, 106–109.

9 *Dynamical systems with countable Lebesque spectrum. I.*, Izv. Akad. Nauk SSSR Ser. Mat., **25** (1961), 899–924; translated in Amer. Math. Soc. transl. (2), **39** (1964), 83–110.

10 (with V. A. Rokhlin), *The structure and properties of invariant measurable partitions*, Dokl. Akad. Nauk SSSR, **141** (1961), no. 5, 1038–1041; translated in Soviet Math. Dokl., **2** (1961), no. 4–6, 1611–1614.

11 (with V. I. Arnold), *Small perturbations of the automorphisms of the torus*, Dokl. Akad. Nauk SSSR, **144** (1962), no. 4, 695–698; translated in Soviet Math. Dokl., **3** (1962), no. 3, 783–786.

12 *A weak isomorphism of transformations with invariant measure*, Dokl. Akad. Nauk SSSR, **147** (1962), no. 4, 797–800; translated in Soviet Math. Dokl., **3** (1962), no. 6, 1725–1729.

13 *On limit theorems for stationary processes*, Teor. Veroyatnost. i Primenen., **7** (1962), 213–219; translated in Theory Probab. Appl., **7** (1962), no. 2, 205–210.

14 *Probabilistic ideas in ergodic theory*, Proc. Internat. Congr. Mathematicians (Stockholm, 1962), 540–559. Inst. Mittag-Leffler, Djursholm. Translated in Amer. Math. Soc. transl. (2), **31** (1963), 62–80.

15 *On the foundations of the ergodic hypothesis for a dynamical system of statistical mechanics*, Dokl. Akad. Nauk SSSR, **153** (1963), no. 6, 1261–1264; translated in Soviet Math. Dokl., **4** (1963), no. 6, 1818–1822.

16 *Some remarks on the spectral properties of ergodic dynamical systems*, Uspehi Mat. Nauk, **18** (1963), no. 5(113), 41–54; translated in Russian Math. Surv., **18** (1963), no. 5, 37–50.

17 *On a physical system with positive entropy* (Russian), Vestnik Moskov. Univ. Ser. I Mat. Meh., **1963** (1963), no. 5, 6–12.

18 *On higher order spectral measures of ergodic stationary processes*, Teor. Veroyatnost. i Primenen., **8** (1964), 463–470; translated in Theory Probab. Appl., **8** (1964), no. 4, 429–436.

19 *On a weak isomorphism of transformations with invariant measure*, Mat. Sb. (N.S.), **63** (105) (1964), 23–42; translated in Amer. Math. Soc. trasl. (2), **57** (1966), 123–149.

20 *Classical dynamical systems with countably-multiple Lebesque spectrum. II.*, Izv. Akad. Nauk SSSR Ser. Mat., **30** (1966), 15–68; translated in Amer. Math. Soc. transl. (2), **68** (1968), 34–88.

21 *The asymptotic behavior of the number of closed geodesics on a compact manifold of negative curvature*, Izv. Akad. Nauk SSSR Ser. Mat., **30** (1966), 1275–1296; translated in Amer. Math. Soc. transl. (2), **73** (1968), 227–250.

22 (with R. A. Minlos), *The phenomenon of "separation of phases" at low temperatures in certain lattice models of a gas. I.*, Mat. Sb. (N.S.), **73** (115) (1967), 375–448; translated in Math. USSR-Sb., **2** (1967), no. 3, 335–395.

23 *A lemma from measure theory*, Mat. Zametki, **2** (1967), 373–378; translated in Math. Notes, **2** (1967), no. 4, 715–718.

24 (with D. V. Anosov), *Certain smooth ergodic systems*, Uspehi Mat. Nauk, **22** (1967), no. 5, 107–172; translated in Russian Math. Surv., **22** (1967), no. 5, 103–167.

25 *Markov Partitions and U-diffeomorphisms*, Funkcional. Anal. i Priložen., **2** (1968), no. 1, 64–89; translated in Funct. Anal. Appl., **2** (1968), no. 1, 61–82.

26 *Construction of Markov Partitions*, Funkcional. Anal. i Priložen., **2** (1968), no. 3, 70–80; translated in Funct. Anal. Appl., **2** (1968), no. 3, 245–253.

27 (with R. A. Minlos), *The phenomenon of "separation of phases" at low temperatures*, II, Trudy Moscov. Mat. Obš., **19** (1968), 113–178; translated in Trans. Moscow Math. Soc., **19** (1968), 121–196.

28 *Theory of dynamical systems. Part I: Ergodic theory*, Lectures held in Warsaw, Spring, 1967. English edition by A. Szankowski, based on notes in Polish prepared by J. Strelcyn and issued by the University of Warsaw, 1969. Lecture Notes Series, No. 23 Matematisk Institut, Aarhus Universitet, Aarhus 1970 a–c+viii+122 pp.

29 *Dynamical systems with elastic reflections*, Uspehi Mat. Nauk, **25** (1970), no. 2(152), 141–192; translated in Russian Math. Surv., **25** (1970), no. 2, 137–189.
30 (with R. A. Minlos), *Spectra of stochastic arising operators in lattice models of a gas*, Teoret. Mat. Fiz., **2** (1970), 230–243; translated in Theoret. Math. Phys., **2** (1970), no. 2, 167–176.
31 *Measures Invariantes des Y -systems*, Actes du Congrès International des Mathématiciens (Nice, 1970), **2**, 929–940. Gauthier-Villars, Paris, 1971.
32 (with K. Volkovysski), *Ergodic properties of an ideal gas with an infinite number of degrees of freedom*, Funkcional. Anal. i Priložen., **5** (1971), no. 3, 19–21; translated in Funct. Anal. Appl., **5** (1971), no. 3, 185–187.
33 (with A. N. Livšic), *Invariant measures that are compatible with smoothness for transitive U-systems*, Dokl. Akad. Nauk SSSR, **207** (1972), no. 5, 1039–1041; translated in Soviet Math. Dokl., **13** (1972), no. 6, 1656–1659.
34 *Ergodic properties of a gas of one-dimensional hard rods with an infinite number of degrees of freedom*, Funkcional. Anal. i Priložen., **6** (1972), no. 1, 41–50; translated in Funct. Anal. Appl., **6** (1972), no. 1, 35–43.
35 (with A. Ya. Khelemski), *A description of differentiations in algebras of local observables*, Funkcional. Anal. i Priložen., **6** (1972), 99–100; translated in Funct. Anal. Appl., **6** (1972), no. 4, 343–344.
36 *Construction of the dynamics for one-dimensional systems of statistical mechanics*, Teoret. Mat. Fiz., **11** (1972), no. 2, 248–258, translated in Theoret. Math. Phys., **11** (1972), no. 2, 487–494.
37 *Gibbs measures in ergodic theory*, Uspehi Mat. Nauk, **27** (1972), no. 4(166), 21–64; translated in Russian Math. Surv., **27** (1972), no. 4, 21–69.
38 (with P. M. Bleher) *Investigation of the critical point in models of the type of Dyson's hierarchical models*, Comm. Math. Phys., **33** (1973), no. 1, 23–42.
39 (with L. A. Bunimovich), *On a fundamental theorem in the theory of dispersing billiards*, Mat. Sb. (N.S.), **90**(132) (1973), 415–431; translated in Math. USSR-Sb., **19** (1973), no. 3, 407–423.
40 (with B. M. Gurevich and Yu. M. Sukhov), *Invariant measures of dynamical systems of one-dimensional statistical mechanics*, Uspehi Mat. Nauk, **28** (1973), no. 5(173), 45–82; translated in Russian Math. Surv., **28** (1973), 49–86.
41 (with S. A. Pirogov), *Phase transitions of the first kind for small perturbations of the Ising Model*, Funkcional. Anal. i Priložen., **8** (1974), no. 1, 25–30; translated in Funct. Anal. Appl., **8** (1974), no. 1, 21–25.
42 (with Yu. M. Sukhov), *On an existence theorem for the solutions of Bogoljubov's chain of equations*, Teoret. Mat. Fiz., **19** (1974), 344–363; translated in Theoret. Math. Phys., **19** (1974), no. 3, 560–573.
43 *Construction of a cluster dynamics for dynamical systems of statistical mechanics*, Vestnik Moskov. Univ. Ser. I Mat. Meh., **29** (1974), no. 1, 152–158; translated in Moscow Univ. Math. Bull., **29** (1974), no. 1, 124–129.
44 (with P. M. Bleher), *Critical indices for Dyson's asymptotically-hierarchal models*, Comm. Math. Phys., **45** (1975), no. 3, 247–278.
45 (with E. I. Dinaburg), *The one-dimensional Schrödinger operator with quasi-periodic potential*, Funkcional. Anal. i Priložen., **9** (1975), no. 4, 8–21; translated in Funct. Anal. Appl., **9** (1975), no. 4, 279–289.
46 (with S. A. Pirogov), *Phase diagrams of classical lattice systems*, Teoret. Mat. Fiz., **25** (1975), no. 3, 358–369; translated in Theoret. Math. Phys., **25** (1975), no. 3, 1185–1192.
47 *Introduction to Ergodic Theory*, Princeton University Press, Princeton, New Jersey USA, 1976, pp. 144.
48 *Self-similar probability distribution*, Teor. Veroyatnost i Primenen., **21** (1976), no. 1, 63–80; translated in Theory Probab. Appl., **21** (1976), no. 1, 64–80.
49 (with S. A. Pirogov), *Phase diagrams of classical lattice systems (continuation)*, Teoret. Mat. Fiz., **26** (1976), no. 1, 61–76; translated in Theoret. Math. Phys., **26** (1976), no. 1, 39–49.
50 (with V. V. Anshelevich), *Some questions on noncommutative ergodic theory*, Uspehi Mat. Nauk, **31** (1976), no. 4(190), 151–167; translated in Russian Math. Surv., **31** (1976), no. 4, 157–174.
51 (with S. A. Pirogov), *Ground states in the two-dimensional boson quantum field-theory*, Ann. Phys., **109** (1977), no. 2, 393–400.
52 *Billiard trajectories in a polyhedral angle*, Uspehi Mat. Nauk, **33** (1978), no. 1 (199), 229–230; translated in Russian Math. Surv., **33** (1978), no. 1, 219–220.
53 *Ergodic properties of a Lorentz gas*, Funktsional. Anal. i Prilozhen., **13** (1979), no. 3, 46–59; translated in Funct. Anal. Appl., **13** (1979), no. 3, 192–202.
54 (with K. M. Khanin), *Existence of free energy for models with long-range random Hamiltonians*, J. Stat. Phys., **20** (1979), no. 6, 573–584.
55 *Ergodic and kinetic properties of the Lorentz gas. Nonlinear dynamics*, Annals of the New York Academy of Sciences, **357**, 143–149. New York Academy of Sciences, New York, 1980.
56 (with E. B. Vul) *Discovery of closed orbits of dynamical systems with the use of computers*, J. Stat. Phys., **23** (1980), no. 1, 27–47.
57 (with E. B. Vul) *A structurally stable mechanism of appearance of invariant hyperbolic sets*, Multicomponent random systems, pp. 595–606, Adv. Probab. Related Topics, **6**, Dekker, New York, 1980.
58 *Theory of Phase Transitions. Rigorous results* (Russian), Nauka, Moscow, 1980, p. 208.
59 (with I. P. Cornfeld and S. V. Fomin) *Ergodic Theory* (Russian), Nauka, Moscow, 1980, p. 384.
60 (with L. A. Bunimovich), *Markov partitions for dispersed billiards*, Comm. Math. Phys., **78** (1980/81), no. 2, 247–280.

61 (with L. A. Bunimovich), *Statistical properties of Lorentz gas with a periodic configuration of scatterers*, Comm. Math. Phys., **78** (1980/81), no. 4, 479–497.

62 (with Ya. B. Pesin), *Hyperbolicity and stochasticity of dynamical systems*, Math. Phys. Rev. 2, 53–115, Soviet Sci. Rev. Sect. C: Math. Phys. Rev., **2**, Harwood Academic, Chur, 1981.

63 (with E. B. Vul), *Hyperbolicity conditions for the Lorenz model*. Nonlinear and turbulent processes in physics (Proc. Internat. Workshop, Kiev, 1979). Phys. D, **2** (1981), no. 1, 3–7.

64 *The stochasticity of dynamical systems*, Selecta Math. Soviet, **1** (1981), no. 1, 100–119.

65 *Theory of Phase Transitions. Rigorous Results*. N. Holland, 1982.

66 (with I. P. Cornfeld and S. V. Fomin), *Ergodic Theory*, Springer-Verlag, 1982.

67 *Lorentz Gas and Random Walks*, Mathematical problems in theoretical physics, Lecture Notes in Phys., **153**, 12–14, Springer-Verlag, Berlin-New York, 1982.

68 (with V. V. Anshelevich and K. M. Khanin), *Symmetric random walks in random environments*, Comm. Math. Phys., **85** (1982), no. 3, 449–470.

69 *Commensurate-incommensurate phase transition in one-dimensional chains*, J. Stat. Phys., **29** (1982), no. 3, 401–425.

70 *The limiting behavior of a one-dimensional random walk in random medium*, Teor. Veroyatnost. i Primenen., **27** (1982), no. 2, 247–258; translated in Theory Probab., **27** (1982), no. 2, 256–268.

71 (with N. I. Chernov), *Entropy of a gas of hard spheres with respect to the group of space – time shifts*, Trudy Sem. Petrovsk., (1982), no. 8, 218–238; translated in J. Math. Sci., **32** (1986), no. 4, 389–406.

72 (with Ya. B. Pesin), *Gibbs measures for partially hyperbolic attractors*, Ergodic Theory Dynam. Syst., **2** (1982), no. 3–4, 417–438.

73 (with A. Goldberg and K. M. Khanin), *Universal properties of sequences of period-tripling bifurcations*, Uspehi Mat. Nauk, **38** (1983), no. 1(229), 159–160; translated in Russian Math. Surv., **38** (1983), no. 1, 187–188.

74 (with C. Boldrighini and L. A. Bunimovich), *On the Boltzmann equation for the Lorentz gas*, J. Stat. Phys., **32** (1983), no. 3, 477–501.

75 *Ground states of the Dyson antiferromagnetic model*, Teoret. Mat. Fiz., **57** (1983), no. 1, 97–104; translated in Theoret. Math. Phys., **57** (1983), no. 1, 1014–1019.

76 (with S. Burkov), *Phase diagrams of the one-dimensional lattice models with long-range antiferromagnetic interaction*, Uspekhi Mat. Nauk, **38** (1983), no. 4(232), 205–225; translated in Russian Math. Surv., **38** (1983), no. 4, 235–257.

77 (with S. Burkov), *Rotation numbers that have nothing to do with KAM theory*, Funktsional. Anal. i Prilozhen., **18** (1984), no. 4, 73–74; translated in Funct. Anal. Appl., **18** (1984), no. 4, 327–328.

78 (with N. Khimchenko), *On the description of classical reflectionless potentials*, Rep. Math. Phys., **20** (1984), no. 1, 53–63.

79 (with E. B. Vul and K. M. Khanin), *Feigenbaum universality and thermodynamic formalism*, Uspekhi Mat. Nauk, **39** (1984), no. 3, 3–37; translated in Russian Math. Surv., **39** (1984), no. 3, 1–40.

80 (with K. M. Efimov), *Hydrodynamic modes for a Lorentz gas with a periodic configuration of scatters* (Russian), Some Problems in Modern Analysis, 102–119, Moskov. Gos. Univ. Mekh.-Mat. Fak., Moscow, 1984.

81 (with E. I. Dinaburg), *An analysis of ANNNI model by Peierls' contour method*, Comm. Math. Phys., **98** (1985), no. 1, 119–144.

82 (with I. M. Khalatnikov, E. M. Lifshitz, K. M. Khanin, and L. N. Shchur) *On the stochasticity in relativistic cosmology*, J. Stat. Phys., **38** (1985), no. 1–2, 97–114.

83 *An answer to a question by J. Milnor*, Comment. Math. Helv., **60** (1985), no. 2, 173–178.

84 *Lectures in Probability Theory. Part 1* (Russian), Moskov. Gos. Univ. Mekh.-Mat. Fak., Moscow, 1985, p. 128.

85 *Lectures in Probability Theory. Part 2* (Russian), Moskov. Gos. Univ. Mekh.-Mat. Fak., Moscow, 1986, p. 106.

86 (with E. Presutti and M. Soloveichik), *Hyperbolity and Möller-morphism for a model of classical statistical mechanics*, Statistical physics and dynamical systems (Kszeg, 1984), 253–284, Progr. Phys., **10**, Birkhäuser Boston, Boston, MA, 1985.

87 (with C. Boldrighini, A. Pellegrinotti, E. Presutti, and M. Soloveichik), *Ergodic properties of semi-infinite one-dimensional system of statistical mechanics*, Comm. Math. Phys., **101** (1985), no. 3, 363–382.

88 *Structure of the spectrum of the Schrödinger operator with almost-periodic potential in the vicinity of its left edge*, Funktsional. Anal. i Prilozhen., **19** (1985), no. 1, 42–48, 96; translated in Funct. Anal. Appl., **19** (1985), no. 1, 34–39.

89 (with L. A. Bunimovich), *Erratum: "Markov partition for dispersed billiards"* [Comm. Math. Phys., **78** (1980/81), no. 2, 247–280]. Comm. Math. Phys., **107** (1986), no. 2, 357–358.

90 (with K. M. Khanin and L. N. Shchur), *A new approach to the construction of fixed points of the renormalization group in dynamical systems*, Izv. Vyssh. Uchebn. Zaved. Radiofiz., **29** (1986), no. 9, 1061–1066; translated in Radiophys. Quantum Electron., **29** (1986), no. 9, 1061–1066.

91 (with M. Soloveichik), *One-dimensional classical massive particle in the ideal gas*, Comm. Math. Phys., **104** (1986), no. 3, 423–443.

92 (with E. I. Dinaburg and A. E. Mazel), *The ANNNI model and contour models with interaction*, Math. Phys. Rev., **6**, 113–168, Soviet Sci. Rev. Sect. C Math. Phys. Rev., **6**, Harwood Academic Publ., Chur, 1987.

93 (with K. Khanin), *A new proof of Herman theorem*, Comm. Math. Phys., **112** (1987), no. 1, 89–101.

94 *Anderson localization for one-dimensional difference Schrödinger operator with quasi-periodic potential*, J. Stat. Phys., **46** (1987), no. 5, 861–909.

95 (with N. I. Chernov), *Ergodic properties of some systems of two-dimensional disks and three-dimensional balls*, Uspekhi Mat. Nauk, **42** (1987), no. 3, 153–174; translated in Russian Math. Surv., **42** (1987), no. 3, 181–207.
96 (with K. M. Khanin), *The renormalization group method in the theory of dynamical systems*, Conference "Renormalization Group—86" (Dubna, 1986), 203–222, 397–398, 407, Joint Inst. Nuclear Res., Dubna, 1987.
97 (with E. I. Dinaburg), *Contour models with interaction and their applications*, Selecta Math. Soviet, **7** (1988), no. 3, 291–315.
98 (with Ya. B. Pesin), *On stable manifolds for a class of two-dimensional diffeomorphisms*, Topology and Geometry-Rohlin Seminar, Lecture Notes in Math., Springer-Verlag, **1346** (1988), 113–125.
99 *About A. N. Kolmogorov's work on the entropy of dynamical systems*, Ergodic Theory Dynam. Syst., **8** (1988), no. 4, 501–502.
100 *The absence of the Poisson distribution for spacings between quasi-energies in the quantum kicked-rotator model*, Progress in chaotic dynamics. Phys. D, **33** (1988), no. 1–3, 314–316.
101 (with K. M. Khanin), *Renormalization group method in the theory of dynamical systems*, Internat. J. Modern Phys. B, **2** (1988), no. 2, 147–165.
102 (with L. Bunimovich), *Spacetime chaos of coupled map lattices*, Nonlinearity, **1** (1988), no. 4, 491–516.
103 (with Ya. Pesin), *Space-time chaos in the system of weakly interactive hyperbolic systems*, J. Geom. Phys., **5** (1988), no. 3, 483–492.
104 (with V. A. Chulaevski), *Anderson localization for the 1D-discrete Schrödinger operator with two-frequency potential*, Comm. Math. Phys., **125** (1989), no. 1, 91–112.
105 *Limit theorems for multiple trigonometric Weyl sums*, Trudy Mat. Inst. Steklov., **191** (1989), 118–129; translated in Proc. Steklov Inst. Math., (1992), no. 2, 131–143. Statistical mechanics and the theory of dynamical systems.
106 (with K. M. Khanin), *Smoothness of conjugacies of diffeomorphisms of the circle with rotations*, Uspekhi Mat. Nauk, **44** (1989), no. 1, 57–82, 247; translated in Russian Math. Surv., **44** (1989), no. 1, 66–99.
107 *Kolmogorov's works on ergodic theory*, Ann. Probab., **17** (1989), no. 3, 833–839.
108 *Hyperbolic billiards*. A plenary address presented at the International Congress of Mathematicians held in Kyoto, August 1990. ICM-90. Mathematical Society of Japan, Tokyo; distributed outside Asia by the American Mathematical Society, Providence, RI, 1990.
109 *Mathematical problems in the theory of quantum chaos*, Chaos (Woods Hole, MA, 1989), 395–414, Amer. Inst. Phys., New York, 1990.
110 (with C. Series), *Ising models on the Lobachevsky plane*, Comm. Math. Phys., **128** (1990), no. 1, 63–76.
111 (with L. Bunimovich and N. Chernov), *Markov partitions for the two-dimensional hyperbolic billiards*, Uspekhi Mat. Nauk, **45** (1990), no. 3(273), 97–134, 221; translated in Russian Math. Surveys, **45** (1990), no. 3, 105–152.
112 (with S. K. Nechaev), *Scaling behavior of random walks with topological constraints*, New trends in probability and statistics, **1** (Bakuriani, 1990), 683–693, VSP, Utrecht, 1991.
113 *Erratum: "Some mathematical problems in the theory of quantumchaos"*, [Phys. A 163 (1990), no. 1, 197–204]. Phys. A, **173** (1991), no. 1–2, 302.
114 *Mathematical problems in the theory of quantum chaos*, Geometric aspects of functional analysis (1989–90), 41–59, Lecture Notes in Math., **1469**, Springer, Berlin, 1991.
115 *Hyperbolic billiards*, Proceedings of the International Congress of Mathematicians, Vol. I, II (Kyoto, 1990), 249–260, Math. Soc. Japan, Tokyo, 1991.
116 (with E. I. Dinaburg) *The statistics of the solutions of the integer equation* $ax - by = \pm 1$, Funktsional. Anal. i Prilozhen. **24** (1990), no. 3, 1–8, 96; translated in Funct. Anal. Appl. **24** (1990), no. 3, 165–171.
117 *Two results concerning asymptotic behavior of solutions of the Burgers equations with force*, J. Stat. Phys., **64** (1991), no. 1–2, 1–12.
118 (with L. Bunimovich and N. I. Chernov), *Statistical properties of two-dimensional hyperbolic billiards*, Uspekhi Mat. Nauk, **46** (1991), no. 4(280), 43–92, 192; translated in Russian Math. Surv., **46** (1991), no. 4, 47–106.
119 *Finite-dimensional randomness*, Uspekhi Mat. Nauk, **46** (1991), no. 3(279), 147–159; translated in Russian Math. Surv., **46** (1991), no. 3, 177–190.
120 *Poisson distribution in a geometric problem*, Dynamical systems and statistical mechanics (Moscow, 1991), 199–214, Adv. Soviet Math., **3**, Amer. Math. Soc., Providence, RI, 1991.
121 (with Ya. B. Pesin), *Space-time chaos in chains of weakly interacting hyperbolic maps*, Dynamical systems and statistical mechanics (Moscow, 1991), 165–198, Adv. Soviet Math., **3**, Amer. Math. Soc., Providence, RI, 1991.
122 (with V. A. Chulaevski), *The exponential localization and structure of the spectrum for 1D quasi-periodic discrete Schrödinger operator*, Rev. in Math. Phys., **3** (1991), no. 3, 241–284.
123 *Probability theory. An introductory course*. Translated from the Russian and with a preface by D. Haughton. Springer Textbook. Springer-Verlag, Berlin, 1992. vi+138 pp.
124 (with A. E. Mazel), *A limiting distribution connected with fractional parts of linear forms*, Ideas and methods in mathematical analysis, stochastics, and applications (Oslo, 1988), 220–229, Cambridge University Press, Cambridge, 1992.
125 *Statistics of shocks in solutions of the Inviscid Burgers equation*, Comm. Math. Phys., **148** (1992), no. 3, 601–621.
126 *Distribution of some functionals of the integral of a random walk*, Teoret. Mat. Fiz., **90** (1992), no. 3, 323–353; translated in Theoret. Math. Phys., **90** (1992), no. 3, 219–241.

127 (with K. M. Khanin), *Mixing of some classes of special flows over rotations of the circle*, Funktsional. Anal. i Prilozhen., **26** (1992), no. 3, 1–21, translated in Funct. Anal. Appl., **26** (1992), no. 3, 155–169.

128 (with L. B. Koralov and S. K. Nechaev), *Limit behavior of a two-dimensional random walk with topological constraints*, Teor. Veroyatnost. i Primenen., **38** (1993), no. 2, 331–344; translated in Theory Probab. Appl., **38** (1993), no. 2, 296–306.

129 *Random Walk with Random Potential*, Teor. Veroyatnost. i Primenen., **38** (1993), no. 2, 457–460; translated in Theory Probab. Appl., **38** (1993), no. 2, 382–385.

130 (with D. V. Kosygin and A. A. Minasov), *Statistical properties of the spectra of Laplace-Beltrami operators on Liouville surfaces*, Uspekhi Mat. Nauk, **48** (1993), no. 4(292), 3–130; translated in Russian Math. Surv., **48** (1993), no. 4, 1–142.

131 (with N. I. Chernov, G. Eyink, and J. Lebowitz), *Steady-state electrical conduction in the periodic Lorentz gas*, Comm. Math. Phys., **154** (1993), no. 3, 569–601.

132 (with K. M. Khanin, A. E. Mazel, and S. B. Shlosman), *Loop condensation effects in the behavior of random walks*, The Dynkin Festschrift, 167–184, Progr. Probab., **34**, Birkhäuser Boston, Boston, MA, 1994.

133 *Topics in Ergodic Theory*, Princeton Mathematical Series, **44**. Princeton University Press, Princeton, NJ, 1994.

134 *A probabilistic approach to the analysis of the statistics of convex polygonal lines*, Funktsional. Anal. i Prilozhen., **28** (1994), no. 2, 41–48, 96; translated in Funct. Anal. Appl., **28** (1994), no. 2, 108–113.

135 (with P. Bleher and D. Kosygin), *Distribution of Energy Levels of a Quantum Free Particle on a Liouville Surface and Trace Formulae*, Comm. Math. Phys., **170** (1995), no. 2, 375–403.

136 *Geodesic flows on manifolds of negative curvature*, Algorithms, fractals, and dynamics (Okayama/Kyoto, 1992), 201–215, Plenum, New York, 1995.

137 *A remark concerning random walks with random potentials*, Fund. Math., **147** (1995), no. 2, 173–180.

138 (with E. Weinan and Yu. G. Rykov), *The Lax-Oleinik variational principle for some one-dimensional systems of quasi-linear equations*, Uspekhi Mat. Nauk, **50** (1995), no. 1(301), 193–194; translated in Russian Math. Surv., **50** (1995), no. 1, 220–222.

139 (with K. Khanin, J. Lebowitz, and A. Mazel), *Self-avoiding random walks in five or more dimensions: an approach using polymer expansions*, Uspekhi Mat. Nauk, **50**, (1995), no. 2(302), 175–206; translated in Russian Math. Surv., **50** (1995), no. 2, 403–434.

140 (with E. Weinan and Yu. Rykov), *Generalized Variational Principles, Global Weak Solutions and Behaviour with Random Initial Data for Systems of Conservation Laws Arising in the Adhesion Particle Dynamics*, Comm. Math. Phys., **177** (1996), no. 2, 349–380.

141 (with B. R. Hunt, K. M. Khanin, and J. A. Yorke), *Fractal properties of critical invariant curves*, J. Stat. Phys., **85** (1996), no. 1–2, 261–276.

142 (with H. Spohn), *Remarks on the delocalization transition for heteropolymers*, Topics in statistical and theoretical physics, 219–223, Amer. Math. Soc. Transl. Ser. 2, **177**, Amer. Math. Soc., Providence, RI, 1996.

143 *Burgers system driven by a periodic stochastic flow*, Itô's stochastic calculus and probability theory, 347–353, Springer, Tokyo, 1996.

144 *A Remark Concerning the Thermodynamic Limit of the Lyapunov Spectrum*, Internat. J. Bifur. Chaos Appl. Sci. Engrg., **6** (1996), no. 6, 1137–1142.

145 (with E. I. Dinaburg and A. B. Soshnikov), *Splitting of the low Landau levels into a set of positive Lebesgue measure under small periodic perturbations*, Comm. Math. Phys., **189** (1997), no. 2, 559–575.

146 (with E. Weinan, K. M. Khanin, and A. Mazel), *Probability Distribution Functions for the Randomly Forced Burgers Equation*, Phys. Rev. Lett., **78** (1997), no. 10, 1904–1907.

147 *On the distribution of the maximum of fractional Brownian motion*, Uspekhi Mat. Nauk, **52** (1997), no. 2(314), 119–138; translated in Russian Math. Surv., **52** (1997), no. 2, 359–378.

148 (with K. M. Khanin), *Hyperbolicity of minimizing trajectories for two-dimensional Hamiltonian systems with random forcing*, Tr. Mat. Inst. Steklova, **216** (1997), Din. Sist. i Smezhnye Vopr., 176–180; translated in Proc. Steklov Inst. Math., (1997), no. 1 (216), 169–173.

149 (with S. Goldstein and J. L. Lebowitz), *Remark on the (non)convergence of ensemble densities in dynamical systems*, Chaos and irreversibility (Budapest, 1997). Chaos, **8** (1998), no. 2, 393–395.

150 *Parabolic Perturbations of Hamilton-Jacobi Equations*, Fund. Math. **157**, (1998), no. 2–3, 299–303.

151 (with A. B. Soshnikov), *A refinement of Wigner's semicircle law in a neighborhood of the spectrum edge for random symmetric matrices*, Funktsional. Anal. i Prilozhen., **32** (1998), no. 2, 56–79, 96; translated in Funct. Anal. Appl., **32** (1998), no. 2, 114–131.

152 (with A. B. Soshnikov), *Central Limit Theorem for Traces of Large Random Symmetric Matrices with Independent Matrix Elements*, Bol. Soc. Brasil. Mat. (N.S.), **29** (1998), no. 1, 1–24.

153 *Asymptotic Behavior of Solutions of -Burgers Equation with Quasi-periodic Forcing*, Topol. Methods Nonlinear Anal., **11** (1998), no. 2, 219–226.

154 *Convex Hulls of Random Processes*, Geometry of differential equations, 153–159, Amer. Math. Soc. Transl. Ser. 2, **186**, Amer. Math. Soc., Providence, RI, 1998.

155 *Simple Random Walk on Tori*, J. Stat. Phys., **94** (1999), no. 3/4, 695–708.

156 (with J. Mattingly), *An Elementary Proof of the Existence and Uniqueness Theorem for Navier–Stokes Equations*, Commun. in Contemp. Math., **1** (1999), no. 4, 497–516.

157 *Navier–Stokes Equation with Periodic Boundary Conditions*, Regul. Chaotic Dyn., **4** (1999), 1–15.
158 *On some problems of the theory of dynamical systems and mathematical physics*, GAFA 2000 (Tel Aviv, 1999). Geom. Funct. Anal. 2000, Special Volume, Part I, 425–433.
159 *Anomalous transport in almost periodic media*, Uspekhi Mat. Nauk, **54** (1999), no. 1(325), 181–208; translated in Russian Math. Surv. **54** (1999), no. 1, 181–208.
160 *Dynamics of a heavy particle surrounded by a finite number of light particles*, Teoret. Mat. Fiz., **121** (1999), no. 1, 110–116; translated as Theoret. Math. Phys., **121** (1999), no. 1, 1351–1357.
161 (with E. Weinan, K. Khanin, and A. Mazel), *Invariant Measures for Burgers Equation with Random Forcing*, Ann. Math. (2), **151** (2000), no. 3, 877–960.
162 (with J. Piasecki), *A Model of Non-Equilibrium Statistical Mechanics*, Dynamics: models and kinetic methods for non-equilibrium many body systems (Leiden, 1998), 191–199, NATO Sci. Ser. E Appl. Sci., 371, Kluwer Acad. Publ., Dordrecht, 2000.
163 (with J. Lebowitz and J. Piazecki), *Dynamics of an adiabatic piston*, Dokl. Akad. Nauk, **375** (2000), no. 6, 734–736; translated in Dokl. Math., **62** (2000), no. 3, 398–401.
164 (with E. Weinan), *New results in mathematical and statistical hydrodynamics*, Uspekhi Mat. Nauk, **55** (2000), no. 4(334), 25–58; translated in Russian Math. Surv., **55** (2000), no. 4, 635–666.
165 (with V. Yu. Kaloshin), *Nonsymmetric simple random walks along orbits of ergodic automorphisms*, On Dobrushin's way. From probability theory to statistical physics, 109–115, Amer. Math. Soc. Transl. Ser. 2, **198**, Amer. Math. Soc., Providence, RI, 2000.
166 (with V. Yu. Kaloshin), *Simple random walks along orbits of Anosov diffeomorphisms*, Tr. Mat. Inst. Steklova, **228** (2000), Probl. Sovrem. Mat. Fiz., 236–245; translated in Proc. Steklov Inst. Math., (2000), no. 1 (228), 224–233.
167 (with J. Lebowitz and J. Piasecki), *Scaling Dynamics of a Massive Piston in an Ideal Gas*, Hard ball systems and the Lorentz gas, 217–227, Encyclopaedia Math. Sci., **101**, Springer, Berlin, 2000.
168 *Remembrances of A. N. Kolmogorov*, Kolmogorov in perspective, 177–230, Hist. Math., **20**, Amer. Math. Soc., Providence, RI, 2000.
169 (with E. I. Dinaburg), *A quasilinear approximation for the three-dimensional Navier–Stokes system*, Mosc. Math. J., **1** (2001), no. 3, 381–388, 471.
170 (with E. Weinan and J. Mattingly), *Gibbsian dynamics and ergodicity for the stochastically forced Navier–Stokes equation*, Comm. Math. Phys., **224** (2001), no. 1, 83–106.
171 (with N. Chernov and J. Lebowitz), *Scaling dynamics of a massive piston in a cube filled with ideal gas: exact results*, J. Stat. Phys., **109** (2002), no. 3–4, 529–548.
172 (with A. V. Kontorovich), *Structure theorem for (d,g,h)-maps*, Bull. Braz. Math. Soc. (N.S.), **33** (2002), no. 2, 213–224.
173 (with N. Chernov and J. Lebowitz), *Dynamics of a massive piston immersed in an ideal gas*, Uspekhi Mat. Nauk, **57** (2002), no. 6(348), 3–86; translated in Russian Math. Surv., **57** (2002), no. 6, 1045–1125.
174 *Statistical (3x + 1)-problem*, Comm. Pure Appl. Math., **56** (2003), no. 7, 1016–1028.
175 (with E. I. Dinaburg and V. S. Posvyanski), *On some approximations of the quasi-geostrophic equation*, Geometric methods in dynamics. II. Astérisque No. 287 (2003), xvii, 19–32.
176 (with E. I. Dinaburg), *Existence and uniqueness of solutions of a quasilinear approximation of the three-dimensional Navier–Stokes system*, Problemy Peredachi Informatsii, **39** (2003), no. 1, 53–57; translated in Probl. Inf. Transm., **39** (2003), no. 1, 47–50.
177 (with A. I. Neishtadt), *Adiabatic piston as a dynamical system*, J. Stat. Phys. **116** (2004), no. 1–4, 815–820.
178 *What is ... a billiard?*, Notices Amer. Math. Soc., **51** (2004), no. 4, 412–413.
179 (with D. Kosygin), *From Kolmogorov's work on entropy of dynamical systems to non-uniformly hyperbolic dynamics*, Kolmogorov's heritage in mathematics, 239–252, Springer, Berlin, 2007.
180 (with Yu. Yu. Bakhtin and E. I. Dinaburg), *On solutions with infinite energy and enstrophy of the Navier–Stokes system*, Uspekhi Mat. Nauk, **59** (2004), no. 6(360), 55–72; translated in Russian Math. Surv., **59** (2004), no. 6, 1061–1078
181 *On local and global existence and uniqueness of solutions of the 3D Navier–Stokes system on R^3*, Perspectives in analysis, 269–281, Math. Phys. Stud., **27**, Springer, Berlin, 2005.
182 *Power series for solutions of the 3D-Navier–Stokes system on R^3*, J. Stat. Phys., **121** (2005), no. 5–6, 779–803.
183 *A diagrammatic approach to the 3D Navier–Stokes system*, Uspekhi Mat. Nauk, **60** (2005), no. 5(365), 47–70; translated in Russian Math. Surv., **60** (2005), no. 5, 849–873.
184 (with C. Ulcigrai), *Weak mixing in interval exchange transformations of periodic type*, Lett. Math. Phys., **74** (2005), no. 2, 111–133.
185 (with A. Bufetov and C. Ulcigrai), *A condition for continuous spectrum of an interval exchange transformation*, Representation theory, dynamical systems, and asymptotic combinatorics, 23–35, Amer. Math. Soc. Transl. Ser. 2, **217**, Amer. Math. Soc., Providence, RI, 2006.
186 *How mathematicians and physicists found each other in the theory of dynamical systems and in statistical mechanics*, Mathematical events of the twentieth century, 399–407, Springer, Berlin, 2006.
187 *Mathematicians and physicists = cats and dogs?*, Bull. Amer. Math. Soc. (N.S.), **43** (2006), no. 4, 563–565.
188 *On a Separating Solution of a Recurrent Equation*, Regul. Chaotic Dyn., **12** (2007), no. 5, 490–501.

189 (with J. Bourgain), *Limit behavior of large Frobenius numbers*, Uspekhi Mat. Nauk, **62** (2007), no. 4(376), 77–90; translated in Russian Math. Surv., **62** (2007), no. 4, 713–725.

190 (with L. Koralov), *Theory of probability and random processes*. Second edition. Universitext. Springer, Berlin, 2007.

191 *Congratulations to Professor K. Itô*, Jpn. J. Math., **2** (2007), no. 1, 129–131.

192 (with E. I. Dinaburg), *Existence theorems for the 3D-Navier–Stokes system having as initial conditions sums of plane waves*, Instability in models connected with fluid flows. I, 289–300, Int. Math. Ser. (N. Y.), **6**, Springer, New York, 2008.

193 *Two limit theorems*, Nonlinearity, **21** (2008), no. 12, T253–T254.

194 *Mathematical results related to the Navier–Stokes system*, SPDE in hydrodynamic: recent progress and prospects, 151–164, Lecture Notes in Math., **1942**, Springer, Berlin, 2008.

195 (with D. Li), *Blow ups of complex solutions of the 3D Navier–Stokes system and renormalization group method*, J. Eur. Math. Soc. (JEMS), **10** (2008), no. 2, 267–313.

196 (with C. Ulcigrai), *Renewal-type limit theorem for the Gauss map and continued fractions*, Ergodic Theory Dynam. Syst., **28** (2008), no. 2, 643–655.

197 (with C. Ulcigrai), *A Limit Theorem for Birkhoff sums of non-integrable functions over rotations*, Contemporary Mathematics, **469** (2008), 317–340.

198 (with S. Albeverio and F. Flandoli), *SPDE in hydrodynamic: recent progress and prospects*, Lectures given at the C.I.M.E. Summer School held in Cetraro, August 29–September 3, 2005. Edited by Giuseppe Da Prato and Michael Rekner. Lecture Notes in Math., 1942. Springer-Verlag, Berlin; Fondazione C.I.M.E., Florence, 2008. viii+166 pp.

199 (with M. Arnold), *Global existence and uniqueness theorem for 3D-Navier–Stokes system on T^3 for small initial conditions in the spaces $\Phi(\alpha)$*, Pure Appl. Math. Q., **4** (2008), no. 1, part 2, 71–79.

200 (with D. Li), *Complex singularities of solutions of some 1D hydrodynamic models*, Phys. D, **237** (2008), no. 14–17, 1945–1950.

201 (with D. Li), *Complex singularities of the Burgers system and renormalization group method*, Current developments in mathematics, 2006, 181–210, Int. Press, Somerville, MA, 2008.

202 (with V. Shur and A. Ustinov), *Limiting Distribution of Frobenius Number for $n = 3$*, "Journal of Number Theory", 13 pp., submitted for publication.

203 (with I. Vinogradov), *Critical Solution of a Quadratic Recurrent Equation*, J. Stat. Phys., submitted for publication.

204 (with E. I. Dinaburg and D. Li), *A New Boundary Problem for the Two-dimensional Navier-Stokes System*, 16 pp., J. Stat. Phys., submitted for publication.

205 (with C. Ulcigrai), *Estimates from above of certain double trigomometric sums*, Topological Methods Nonlinear Anal., 15 pp., submitted for publication.

206 (with E. I. Dinaburg and D. Li), *Navier–Stokes System on the Flat Cylinder and Unit Square with Slip Boundary Condition*, 25 pp., Commun. Contemp. Math., submitted for publication.

Other works

1 N. S. Krylov, *Works on the foundations of statistical physics*. Translated from the Russian by A. B. Migdal, Ya. G. Sinai, and Yu. L. Zeeman. With a preface by A. S. Wightman. With a biography of Krylov by V. A. Fock. With an introductory article "The views of N. S. Krylov on the foundations of statistical physics" by Migdal and Fok. With a supplementary article "Development of Krylov's ideas" by Sinai. Princeton Series in Physics. Princeton University Press, Princeton, N.J., 1979. xxviii+283 pp.

2 J. von Neumann, *Selected works on functional analysis. I.* (Russian). Translated by L. A. Bunimovich, A. M. Vershik, and P. M. Blekher. With commentaries by L. A. Bunimovich, A. M. Vershik, P. M. Blekher, and A. V. Marchenko. Compiled by Ya. G. Sinai. With von Neumann's biography by A. M. Vershik, A. N. Kolmogorov, and Ya. G. Sinai. Classics of Science. "Nauka", Moscow, 1987. 376 pp.

3 G. A. Galperin, N. I. Chernov, *Billiards and chaos* (Russian). With a preface by Ya. G. Sinai. Current Life, Science and Technology: Series "Mathematics and Cybernetics", 91-5. "Znanie", Moscow, 1991. 48 pp.

4 V. A. Rokhlin, *Selected works* (Russian). *Appendix: Reminiscences of V. A. Rokhlin* (Russian). Translated by V. I. Arnold, A. M. Vershik, S. P. Novikov, and Ya. G. Sinai. With commentaries on Rokhlin's work by N. Yu. Netsvetaev and A. M. Vershik. Edited and with a preface by A. M. Vershik. Moskovski Tsentr Nepreryvnogo Matematicheskogo Obrazovaniya, Moscow, 1999. 496 pp.

5 *Russian mathematicians in the 20th century*. Edited by Yakov Sinai. World Scientific Publishing Co., Inc., River Edge, NJ, 2003. xii+700 pp.

6 *The Kolmogorov legacy in physics*. Translated from the 2003 French original. Edited by R. Livi and A. Vulpiani. Lecture Notes in Physics, 636. Springer-Verlag, Berlin, 2003. xvi+246 pp. "About A. N. Kolmogorov", introduction by Ya. G. Sinai.

Part I
Probability Theory

Journal of Statistical Physics, Vol. 94, Nos. 3/4, 1999

Simple Random Walks on Tori

Ya. G. Sinai[1]

Received March 26, 1998

We consider a Markov chain whose phase space is a d-dimensional torus. A point x jumps to $x+\omega$ with probability $p(x)$ and to $x-\omega$ with probability $1-p(x)$. For Diophantine ω and smooth p we prove that this Markov chain has an absolutely continuous invariant measure and the distribution of any point after n steps converges to this measure.

KEY WORDS: Markov chain; homological equation; Levy excursion; stable law.

1. INTRODUCTION

Consider Markov Chain on the d-dimensional torus Tor^d where a moving point jumps from $x \in \mathrm{Tor}^d$ to $x \pm \omega$ with probabilities $p(x)$ and $(1-p(x))$, respectively. Here $\omega = (\omega_1, \omega_2, ..., \omega_d) \in \mathrm{Tor}^d$ is fixed. We shall call such Markov chains simple random walks on tori. In the one-dimensional case a point wanders along the unit circle jumping from x to $x \pm \omega$.

We shall impose the following conditions:

1° The point ω is Diophantine, i.e. for some positive K, γ

$$\inf_{m \in \mathbb{Z}^1} |(\omega, n) - m| \geqslant \frac{K}{|n|^\gamma}$$

where $n = (n_1, ..., n_d) \in \mathbb{Z}^d$, $|n| = \sum_{i=1}^d |n_i| \neq 0$. The shift on Tor^d by ω is denoted by T, i.e. $Tx = x + \omega$ (the addition mod 1).

2° Let $H^r(\mathrm{Tor}^d) = H^r$ be the space of continuous functions on Tor^d such that for their Fourier expansions $f = \sum_{s \in \mathbb{Z}^d} f_s e^{2\pi i(s, x)}$ the series $\sum_{s \in \mathbb{Z}^d} |f_s| \cdot |s|^r < \infty$. Then $0 < p(x) < 1$, $x \in \mathrm{Tor}^d$ and $p \in H^r$ for some $r > 2\gamma$.

[1] Department of Mathematics, Princeton University, Princeton, New Jersey, and Landau Institute of Theoretical Physics, Moscow, Russia.

Y.G. Sinai, *Selecta II: Probability Theory, Statistical Mechanics, Mathematical Physics and Mathematical Fluid Dynamics*, DOI 10.1007/978-1-4419-6205-8_1, © Springer Science+Business Media, LLC 2010

We shall prove for this Markov chain that under formulated above conditions it has a unique invariant measure which is absolutely continuous with respect to the Lebesgue measure. The density of this measure π satisfies the equation

$$\pi(x) = p(T^{-1}x)\,\pi(T^{-1}x) + (1 - p(Tx))\,\pi(Tx) \tag{1}$$

Let us show that (1) always has a solution.

Definition. Simple random walk is symmetric if

$$\int_{\mathrm{Tor}^d} \ln p(x)\,dx = \int_{\mathrm{Tor}^d} \ln(1 - p(x))\,dx.$$

Otherwise it is called non-symmetric.

Consider first symmetric case. Here the homological equation

$$\frac{p(x)}{1 - p(x)} = \frac{h(x)}{h(T^{-1}x)} \tag{2}$$

has a positive solution $h \in H^{r-\gamma}$. Let us check that the function $\pi_0(x) = h(x)/p(x)$ satisfies (1). We have

$$\begin{aligned} &p(T^{-1}x)\,\pi_0(T^{-1}x) + (1 - p(Tx))\,\pi_0(Tx) \\ &\quad = h(T^{-1}x) + \frac{1 - p(Tx)}{p(Tx)} \cdot h(Tx) = h(T^{-1}x) + h(x) \end{aligned}$$

But $h(T^{-1}x) + h(x) = h(x)/p(x)$. Indeed, $h(x)(1/p(x) - 1) = h(T^{-1}x)$ or

$$\frac{1 - p(x)}{p(x)} = \frac{h(T^{-1}x)}{h(x)}$$

which is our homological equation (2).

In the non-symmetric case we write

$$\frac{p(x)}{1 - p(x)} = \lambda \frac{h(x)}{h(T^{-1}x)} \tag{3}$$

where $\ln \lambda = \int \ln p(x)\,dx - \int \ln(1 - p(x))\,dx$ and $h \in H^{r-\gamma}$. Assume that $\lambda > 1$. Consider the equation

$$\lambda^{-1} r^{(+)}(Tx) - r^{(+)}(x) = \frac{1}{h(x)} \tag{4}$$

For $\lambda \neq 1$ it always has a solution since no small denominators arise. We shall see later that $r^{+}(x) > 0$. Let us show that $\pi_0(x) = r^{(+)}(x)\, h(x)/p(x)$ satisfies (1). We have

$$\frac{r^{(+)}(x)\, h(x)}{p(x)} = r^{(+)}(T^{-1}x)\, h(T^{-1}x) + \frac{(1-p(Tx))\, r^{+}(Tx)\, h(Tx)}{p(Tx)}$$

Using (3) we can write

$$\frac{r^{(+)}(x)}{p(x)} = r^{(+)}(T^{-1}x) \cdot \frac{h(T^{-1}x)}{h(x)} + \lambda^{-1} r^{+}(Tx)$$

The right-hand side can be rewritten using (4) as

$$\begin{aligned}
&\frac{r^{(+)}(T^{-1}x)\, h(T^{-1}x)}{h(x)} + \lambda^{-1} r^{+}(Tx) \\
&= \left(\lambda^{-1} r^{(+)}(x) - \frac{1}{h(T^{-1}x)}\right) \frac{h(T^{-1}x)}{h(x)} + \lambda^{-1} r^{(+)}(Tx) \\
&= \lambda^{-1} r^{(+)}(x) \frac{h(T^{-1}x)}{h(x)} - \frac{1}{h(x)} + \left(\frac{1}{h(x)} + r^{(+)}(x)\right) \\
&= r^{(+)}(x) \left(\lambda^{-1} \frac{h(T^{-1}x)}{h(x)} + 1\right) \\
&= r^{(+)}(x) \left(\frac{1-p(x)}{p(x)} + 1\right) \\
&= \frac{r^{(+)}(x)}{p(x)}
\end{aligned}$$

The case $\lambda < 1$ is considered in the same way by replacing T by T^{-1}. The probabilistic meaning of the functions h, $r^{(+)}$ will become clear later.

The main result of this paper is the following theorem.

Theorem 1. Under the conditions formulated above the invariant measure of simple random walk on Tor^d is unique. Even more, if $P_k(T^m x)$ is the probability that after k steps the moving point is at $T^m x$, $|m| \leqslant k$, then for $k \to \infty$ the probability measures $P_k = \{P_k(T^m x)\}$ converge weakly to $\pi_x(x)\, dx$.

This theorem is proven in Sections 2, 3, and 4 for the symmetric case. In Section 5 we explain the needed changes in the non-symmetric case. In Sections 6 we formulate some extensions of Theorem 1 and mention several open problems.

2. RECURRENCE OF SYMMETRIC RANDOM WALKS

Fix $x \in \mathrm{Tor}^d$ and consider simple random walk $b = \{b(k), k \geqslant 0\}$ on $\mathbb{Z}^1$ where $b(0) = 0$ and the probabilities to go from m to $m \pm 1$ are $p(T^m x)$, $1 - p(T^m x)$, respectively. We shall show in this section that in the symmetric case random walk b is recurrent.

Take any segment $[k_1, k_2]$, $k_1 < k_2$ and introduce the probabilities $R_k^{(\pm)}$ of random walks b which go out of k and reach k_2, respectively k_1, earlier than k_1, respectively k_2. These probabilities satisfy the following set of recurrent relations and boundary conditions

$$R_k^{(+)} = p(T^\kappa x)\, R_{k+1}^{(+)} + (1 - p(T^\kappa x))\, R_{k-1}^{(+)}, \qquad k_1 < k < k_2, \qquad R_{k_1}^{(+)} = 0, \qquad R_{k_2}^{(+)} = 1 \tag{5}$$

$$R_k^{(-)} = p(T^\kappa x)\, R_{k+1}^{(-)} + (1 - p(T^\kappa x))\, R_{k-1}^{(-)}, \qquad k_1 < k < k_2, \qquad R_{k_1}^{(-)} = 1, \qquad R_{k_2}^{(-)} = 0 \tag{6}$$

To construct needed solutions we have to find two linearly independent solutions. One is trivial: $R \equiv 1$. Let us check that the other one is given by the formulas

$$R_{k_1} = 0,\ R_k = \sum_{i=k_1+1}^{k} h(T^i x), \qquad k > k_1$$

We must check that for $k > k_1$

$$R_k = p(T^k x)\, R_{k+1} + (1 - p(T^\kappa x)\, R_{k-1}$$

which in our case is reduced to

$$0 = p(T^k x)\, h(T^{k+1} x) - (1 - p(T^k x))\, h(T^k x)$$

But this is equivalent to (2). Now it is easy to see that

$$R_k^{(+)} = \frac{\sum_{i=k_1+1}^{k} h(T^i x)}{\sum_{k_1 < i \leqslant k_2} h(T^i x)}, \qquad k > k_1$$

and

$$R_k^{(-)} = \frac{\sum_{i=k}^{k_2-1} h(T^i x)}{\sum_{k_1 \leqslant i < k_2} h(T^i x)}$$

are needed solutions of (2), (3). Take $k = 0, k_2 = 1$. We immediately see that $\lim_{k_1 \to -\infty} R_0^{(+)} = 1$. In the same way for $k_1 = -1$ the limit

$\lim_{k_2 \to \infty} R_0^{(-)} = 1$. This shows that random walk b with probability 1 returns to the origin and infinitely many times, i.e. random walk b is recurrent.

3. LEVY EXCURSIONS AND THEIR DISTRIBUTIONS

Consider the probabilities $p_{2n}^{(+)}(x)$ of random walks b such that $b(0) = 0$, $b(k) > 0$ for $1 \leqslant k < 2n$, $b(2n) = 0$. In view of recurrence $\sum_{n \geqslant 1} p_{2n}^{(+)}(x) = p(x)$. We shall use generating functions $\varphi^{(+)}(\theta, x) = \sum_{n \geqslant 1} \theta^{2n} p_{2n}^{(+)}(x)$, $|\theta| \leqslant 1$. We have

$$
\begin{aligned}
p_2^{(+)}(x) &= p(x)(1 - p(Tx)),\\
p_{2n}^{(+)}(x) &= p(x) \sum_{s \geqslant 1} \sum_{n_1 + n_2 + \cdots n_s = n-1} p_{2n_1}^{(+)}(Tx)\, p_{2n_2}^{(+)}(Tx) \cdot \cdots\\
&\qquad \cdot p_{2n_s}^{(+)}(Tx)(1 - p(Tx)), \qquad n > 1
\end{aligned}
$$

Multiplying both sides by θ^{2n} and adding over n we arrive at the equation (see also [S2])

$$
\begin{aligned}
\varphi^{(+)}(\theta; x) &= p(x)(1 - p(Tx))\, \theta^2 \left(1 + \sum_{k \geqslant 1} (\varphi^{(+)}(\theta; Tx))^k \right)\\
&= \frac{p(x)(1 - p(Tx))\, \theta^2}{1 - \varphi^{(+)}(\theta; Tx)} \qquad (7)
\end{aligned}
$$

Put $\theta_1 = \sqrt{1 - \theta^2}$, $\psi^{(+)}(\theta_1; x) = h(x)(1 - \varphi^{(+)}(\theta; x)/p(x))$. We have from (7) the following equation for $\psi^{(+)}(\theta_1; x)$

$$
\psi^{(+)}(\theta_1; x) = \frac{\theta_1^2(x) + \psi^{(+)}(\theta_1; Tx)}{1 + \dfrac{\psi^{(+)}(\theta_1; Tx)}{h(x)}}
$$

which we rewrite in the form

$$
\begin{aligned}
\psi^{(+)}(\theta_1; x) &= h(x) - (1 - \theta_1^2)\, h(x) \left(1 - \frac{\psi^{(+)}(\theta_1; Tx)}{h(x)} \cdot \frac{1}{1 + \dfrac{\psi^{(+)}(\theta_1; Tx)}{h(x)}} \right)\\
&= \theta_1^2 h(x) + (1 - \theta_1^2) \frac{1}{\dfrac{1}{h(x)} + \dfrac{1}{\psi^{(+)}(\theta_1; Tx)}} \qquad (8)
\end{aligned}
$$

Introduce the one-dimensional map $Q_x(z) = \theta_1^2 h(x) + (1-\theta_1^2)\, 1/(1/h(x) + 1/z)$, $z \geqslant 0$. The dependence on θ_1 is not mentioned explicitly. From (8) for any $k > 0$

$$\psi^{(+)}(\theta_1; x) = Q_x(\psi^{(+)}(\theta_1; Tx)) = Q_x \circ Q_{Tx} \circ \cdots \circ Q_{T^\kappa x}(\psi^{(+)}(\theta_1; T^{\kappa+1}x))$$

The maps Q_x have the following properties:

1° Any Q_x is a contraction:

$$|Q_x(z') - Q_x(z'')| = (1-\theta_1^2)\left[\frac{1}{\dfrac{1}{h(x)} + \dfrac{1}{z'}} - \frac{1}{\dfrac{1}{h(x)} + \dfrac{1}{z''}}\right]$$

$$= (1-\theta_1^2)\,\frac{z'' - z'}{\left(1 + \dfrac{z'}{h(x)}\right)\left(1 + \dfrac{z''}{h(x)}\right)}$$

and the needed statement follows from positivity of z and h.

The property implies, in particular, the uniqueness of solutions of (8) and (7).

2° There exists a positive constant K_1 such that for all sufficiently small θ_1

$$0 \leqslant Q_x(z) \leqslant K_1\theta_1 \qquad \text{if} \quad 0 \leqslant z \leqslant K_1\theta_1$$

Indeed, put $H = \max_x h(x)$, $H_1 = \min h(x)$. We have

$$Q_x(z) = \theta_1^2 h(x) + (1-\theta_1^2)\,\frac{z}{1 + \dfrac{z}{h(x)}}$$

$$= \theta_1^2 h(x) + (1-\theta_1^2)\left(z - \frac{\dfrac{z^2}{h(x)}}{1 + \dfrac{z}{h(x)}}\right)$$

$$\leqslant \theta_1^2 H + K_1\theta_1 - \frac{(1-\theta_1^2)\, K_1^2\theta_1^2}{H_1 + K_1\theta_1} \leqslant K_1\theta_1$$

provided that K_1 is large enough.

Using 1° and 2° we can write $\psi^{(+)}(\theta_1; x)$ as an infinite fraction

$$\psi^{(+)}(\theta_1; x) = \theta_1^2(x) + (1-\theta_1^2)\cfrac{1}{\cfrac{1}{h(x)} + \cfrac{1}{\theta_1^2 h(Tx) + (1-\theta_1^2)\cfrac{1}{\cfrac{1}{h(Tx)}} + \cdots}}$$

which should be understood as the limit of finite fractions. It can be rewritten as a continued fraction consisting of the infinite number of fragments

$$(1-\theta_1^2)^k\, \theta_1^2 h(T^\kappa x) + \cfrac{1}{\cfrac{1}{(1-\theta_1^2)^{\kappa+1}\, h(T^\kappa x)}}$$

3° If $h \in H^{r-\gamma}$ then $\psi^{(+)}(\theta_1; x) \in C^{r-2\gamma}(\mathrm{Tor}^d)$ as a function of x.

Indeed, if $h \in H^{r-\gamma}$ then $h \in C^{r-\gamma}(\mathrm{Tor}^d)$ and the statement follows by direct differentiation of the infinite fraction giving $\psi^{(+)}(\theta_1; x)$.

4° Let ω be Diophantine. Then $\psi^{(+)}(\theta_1; x)$ has the following representation

$$\psi^{(+)}(\theta_1; x) = a^{(+)}\theta_1 + \theta_1^2 f^{(+)}(\theta_1; x)$$

where $a^{(+)}$ is a constant, $a^{(+)} \leqslant K_1$ (see 2°), $f^{(+)} \in C^{r-\gamma}(\mathrm{Tor}^d)$.

Proof. We can write $\psi^{(+)}(\theta_1; x) = \theta_1 a^{(+)} + \theta_1 F^{(+)}(\theta_1; x)$ where

$$\theta_1 a^{(+)} = \int_{\mathrm{Tor}^d} \psi^{(+)}(\theta_1; x)\, dx \leqslant \theta_1 K_1$$

in view of 2°, $|F(\theta_1; x)| \leqslant 2K_1$. Rewrite (8) as follows

$$\begin{aligned} &F^{(+)}(\theta_1; x) - F^{(+)}(\theta_1; Tx) \\ &\quad = \theta_1 h(x) - \theta_1^2 a^{(+)} - \theta_1^2 F^{(+)}(\theta_1; Tx) \\ &\quad = \frac{\dfrac{\theta_1(1-\theta_1^2)(a^{(+)} + F^{(+)}(\theta_1; Tx))}{h(x)}}{1 + \dfrac{\theta_1 a^{(+)} + \theta_1 F^{(+)}(\theta_1; Tx)}{h(x)}} = \theta_1 G^{(+)}(\theta_1; x) \end{aligned} \tag{9}$$

where $G^{(+)}(\theta_1; x) \in C^{r-\gamma}(\mathrm{Tor}^d)$. Since ω is Diophantine the solution of (9) can be written as $F^{(+)}(\theta_1; x) = \theta_1 f^{(+)}(\theta_1; x)$, $f^{(+)}(\theta_1; x) \in C^{r-\gamma}(\mathrm{Tor}^d)$.

Return back to $\psi^{(+)}(\theta; x)$. Our previous analysis gives the following representation for $\psi^{(+)}$:

$$\varphi^{(+)}(\theta; x) = p(x)\left(1 - \frac{\psi^{(+)}(\sqrt{1-\theta^2}; x)}{h(x)}\right)$$

$$= p(x)\left(1 - \frac{a^{(+)}\sqrt{1-\theta^2}}{h(x)} + (1-\theta^2)\, f^{(+)}\left(\sqrt{1-\theta^2}; x\right)\right) \quad (10)$$

The Tauberian theorem for generating functions (see [F]) implies

$$p_{2n}^{(+)}(x) = \frac{p(x)}{\text{const}}\frac{1}{n^{3/2}}\left(1 + \mathcal{O}\left(\frac{1}{n}\right)\right)$$

In other words, the probabilities $p_{2n}^{(+)}(x)$ decay in the same way as the similar probabilities for the usual symmetric simple random walk.

The asymptotics of probabilities $p_{2n}^{(-)}(x)$ of such b that $b(1) = -1$, $b(k) < 0$ for $1 \leqslant k \leqslant 2n$, $b(2n) = 0$ is investigated in the same way. For the corresponding generating function $\varphi^{(-)}(\theta; x)$ we can write

$$\varphi^{(-)}(\theta; x) = (1 - p(x))\left(1 - \frac{a^{(-)}\sqrt{1-\theta^2}}{h(x)} + (1-\theta^2)\, f^{(-)}(\sqrt{1-\theta^2}; x)\right) \quad (11)$$

For the generating function of the moment of the first return to the origin $\varphi(\theta; x) = \varphi^{(+)}(\theta; x) + \varphi^{(-)}(\theta; x)$ we have

$$\varphi(\theta; x) = 1 - \frac{a\sqrt{1-\theta^2}}{h(x)} + (1-\theta^2)\, f(\sqrt{1-\theta^2}; x) \quad (12)$$

where $a = a^{(+)}p(x) + a^{(-)}(1-p(x))$, $f(\sqrt{1-\theta^2}; x) = f^{(+)}(\sqrt{1-\theta^2}; x)\, p(x) + f^{(-)}(\sqrt{1-\theta^2}; x)(1-p(x))$.

4. PROOF OF THEOREM 1 IN THE SYMMETRIC CASE

Return back to our random walk on Tor^d. After k steps the moving point can be in any point $T^m x$, $|m| \leqslant k$ and the probability to be at $T^m x$ is the probability $P_k(m)$ that $b(k) = m$. We shall study the asymptotics of $P_\kappa(m)$ as $k \to \infty$, $m \to \infty$ so that m^2/k tends to a constant $z \neq 0$. Assume for definiteness that $z > 0$. Introduce the following random moments $\mathcal{T}_{i,k}(b) = \max\{n \mid b(n) = i, n \leqslant k\}$, $i = 0, 1, ..., m$. It is clear that $b(\mathcal{T}_{0,k}(b)) = 0$,

$b(n)>0$ for $n>\mathcal{T}_{0,k}(b)$. The difference $\mathcal{T}_{i+1,k}(b)-\mathcal{T}_{i,k}(b)=\tau_{i+1}$ consists of several positive excursions starting at i and of the last step from i to $i+1$. Let $v_i\geqslant 0$ be the number of these excursions, $\xi_{i1},...,\xi_{iv_1}$ are their lengths. If $v_i=0$ then the particle jumps from i to $i+1$ and after that does not return to i before k. We can write (see [S2])

$$P_k(m)=\sum_{v_0,v_1,...,v_m}\sum_{\sum_{i=1}^{m}(2k_1^{(i)}+\cdots+2k_{v_i}^{(i)})=k-m} p(x)\prod_{s=1}^{v_0} p_{2k_s^{(0)}}(x)\cdot\prod_{i=1}^{m-1}\left[p(T^ix)\cdot\prod_{i=1}^{v_i} p^{(+)}_{2k_i^{(i)}}(T^ix)\right] \tag{13}$$

Introduce the generating function $\phi(\theta;m)=\sum P_k(m)\,\theta^k$, $|\theta|<1$. We have from (13)

$$\phi(\theta;m)=(p(T^mx)\,\theta)^{-1}\frac{p(x)\,\theta}{1-\varphi(\theta;x)}\cdot\prod_{i=1}^{m}(p(T^ix)\,\theta)\cdot\sum_{v_1,...,v_m,\,v_i\geqslant 0}\prod_{i=1}^{m}\varphi^{(+)}(\theta;T^ix)^{v_i}$$
$$=(p(T^mx)\,\theta)^{-1}\,\theta^m\frac{\theta p(x)}{1-\varphi(\theta;x)}\prod_{i=1}^{m}\frac{(p(T^ix))}{1-p(T^ix)}\cdot\prod_{i=1}^{m}\frac{(1-p(T^ix))}{1-\varphi^{(+)}(\theta;t^ix)} \tag{14}$$

Now we use again the symmetry of our random walk (see Section 1), which yields

$$\prod_{i=1}^{m}\frac{p(T^ix)}{1-p(T^ix)}=\frac{h(T^mx)}{h(x)}$$

Each of the functions $1-p(T^ix)/1-\varphi^{(+)}(\theta;t^ix)$ is the generating function of the distribution of $\sum_{j=1}^{v_i}\xi_{ij}=\xi_i$ belonging to the domain of attraction of the stable one-sided law with exponent $\alpha=\frac{1}{2}$ (see [GK] and [F]). This follows easily from the results in Section 3. Therefore the distribution of the normed sum $1/m^2\sum_{i=1}^{m}\zeta_i=\zeta$ converges to this law. One can see this also from the expression for the characteristics function of ζ equal to

$$
\begin{aligned}
f(t) &= Ee^{it\zeta} = Ee^{i\,t/m^2 \cdot \sum_{i=1}^m \zeta_i} \\
&= \prod_{i=1}^{m} \frac{1-p(T^i x)}{1-\varphi^{(+)}(e^i\, t/m^2;\, T^i x)} \\
&= \prod_{i=1}^{m} \frac{1}{1+\dfrac{p(T^i x)\, a^{(+)} \sqrt{2}\sqrt{it}\,(1+O(1/m^2))}{(1-p(T^i x))\, h(T^i x)\cdot m}} \\
&\quad \times \exp\left\{ it\frac{1}{m}\sum_{i=1}^{m} \frac{p(T^i x)\, a^{(+)}\sqrt{2}}{(1-p(T^i x))\, h(T^i x)} \cdot (1+o(1))\right\}
\end{aligned}
$$

The average $1/m \sum_{i=1}^m p(T^i x)\, a^{(+)} \sqrt{2}/(1-p(T^i x))\, h(T^i x) = a^{(+)}\sqrt{2} 1/m \sum_{i=1}^m 1/h(T^{i-1}x)$ in view of (2) converges for every x and $m \to \infty$ to $\sigma_0 = a^{(+)}\sqrt{2}\int dx/h(x)$. Thus $f(t)$ converges to $\exp\{\sqrt{it\sigma_0}\}$ which is the characteristic function of the above-mentioned law.

In (14) we have also the factor $1/1-\varphi(\theta;x) = \sum_{n=0}^{\infty} \varphi^n(\theta;x)$. Our arguments below imply easily the "arcsin"-law (see [F]) for random walk b but we shall not discuss this in detail. Each $\varphi^n(\theta;x)$ is the generating function for the sum of n independent identically distributed random variables whose distribution belongs to the domain of attraction of the same stable law. Again putting $\exp\{i\,t/m^2\}$ we can write using (12)

$$
\begin{aligned}
&\varphi^n(e^{i\,t/m^2};x)\prod_{i=1}^{m}\frac{1-p(T^i x)}{1-\varphi^{(+)}(e^{i\,t/m^2};\,T^i x)} \\
&\qquad = \left(1-\frac{a\sqrt{it}}{h(x)\,m}\right)^n \cdot e^{\sqrt{it\sigma_0}} \cdot (1+o(1)) \\
&\qquad = e^{-an\sqrt{it}/h(x)\,m} \cdot e^{\sqrt{it\sigma_0}}(1+o(1))
\end{aligned}
$$

and

$$
\begin{aligned}
&\sum_{n\geqslant 0} \varphi^n(e^{i\,t/m^2};x)\prod_{i=1}^{m}\frac{1-p(T^i x)}{1-\varphi^{(+)}(e^{i\,t/m^2};\,T^i x)} \\
&\quad = \sum_{n\geqslant 0} e^{-an\sqrt{it}/h(x)\,m}\prod_{i=1}^{m}\frac{1-p(T^i x)}{1-\varphi^{(+)}(e^{it/m^2};\,T^i x)}(1+o(1)) \\
&\quad = \frac{mh(x)}{a\sqrt{it}}\, e^{\sqrt{it\sigma_0}}(1+o(1)) \qquad (15)
\end{aligned}
$$

Now we can use the local limit theorem of probability theory (see [GK]) which says that

$$P\left\{\frac{1}{m^2}\sum_{i=1}^{m}\zeta_i=z\right\}=\frac{g(\sigma_1^{-1}z)\,\sigma_1^{-1}}{m^2}(1+\varepsilon(m,z))$$

where g is the standard Gaussian density and $\varepsilon(m,z)$ tends to zero as $m\to\infty$ uniformly for all z in any fixed interval $[z_1,z_2]$, $0<z_1<z_2<\infty$. It is easy to check that (15) implies

$$P_\kappa(m)=\frac{h(T^m x)}{p(T^m x)}\cdot e^{-m^2/2k\sigma_1}\cdot\frac{1}{\sqrt{2\pi\sigma_1}}(1+o(1))$$

for some $\sigma_1>0$ where $o(1)$ is uniformly small in any finite interval of values of $z=m/\sqrt{k}$. This implies Theorem 1 in the symmetric case.

5. NON-SYMMETRIC RANDOM WALKS

In the non-symmetric case we begin with the definition of the mean drift. We use the same systems of equations (5) and (6) and again want to find two linearly independent solutions. As before, one is $R_k\equiv 1$. We shall try to find another one in the form $R_k=\mu^\kappa r(T^k x)$ for some unknown μ and r. We have for them the equation

$$\mu r(Tx)-r(x)=\mu^{-1}\frac{1-p(x)}{p(x)}(\mu r(x)-r(T^{-1}x)) \tag{16}$$

Choose $\mu=\lambda^{-1}$ (see (3)) so that

$$\mu\frac{1-p(x)}{p(x)}=\frac{h(T^{-1}x)}{h(x)}$$

If $p\in H^r$ and ω is Diophantine, the function $h\in H^{r-\gamma}$ and (16) leads to

$$\lambda^{-1}r(Tx)-r(x)=\frac{1}{h(x)} \tag{17}$$

The equation (17) always has a solution $h\in H^{r-\gamma}$ since no small denominators arise. We can write explicit expressions for $R_k^{(\pm)}$:

$$R_k^{(+)}=\frac{\lambda^{-k}r(T^k x)-\lambda^{-k_1}r(T^{k_1}x)}{\lambda^{-k_2}r(T^{k_2}x)-\lambda^{-k_1}r(T^{k_1}r(T^{k_1}x)},\qquad k\geqslant k_1 \tag{18'}$$

$$R_k^{(-)}=\frac{\lambda^{-k_2}r(T^{k_2}x)-\lambda^{-k}r(T^k x)}{\lambda^{-k_2}r(T^{k_2}x)-\lambda^{-k_1}r(T^{k_1}r(T^{k_1}x)},\qquad k\leqslant k_2 \tag{18''}$$

Let $\lambda < 1$. Take $k=0, k_2=1, k_1 \to -\infty$. From (18′), (18″) it follows that $R^{(+)} \to \lambda^{-1} r(x)/r(Tx) < 1$. This means that we have a mean drift to the left. If $\lambda > 1$ a mean drift is to the right.

Consider for definiteness the case $\lambda > 1$. It is easy to see that with probability 1 the limit $\lim_{k\to\infty} b(k) = \infty$. A positive excursion of length $2n$ is a part of b such that $b(m)=0$, $b(2n+m)=0$, $b(k)>0$ for $m<k<2n+m$. The probabilities $p_{2n}^{(+)}(x)$ of the positive excursions of the length $2n$ satisfy the same equations as in the symmetric case (see Section 3) and for the corresponding generating function $\varphi^{(+)}(\theta; x)$ we have the same equation (7). However in this case $\varphi^{(+)}(1, x) = \sum_{n=1}^{\infty} p_{2n}^{(+)}(x) = p(x)\, q(x)$ where $0 < q(x) = \lambda^{-1} r(Tx)/r(x)$ is the conditional probability that random walk b goes out of 1 and eventually comes back to 0. It follows from the last inequality that $r(x) > 0$.

We shall not switch from θ to θ_1 as we did in Section 3 and shall consider $\psi^{(+)}(\theta; x) = h(x)(1 - \varphi^{(+)}(\theta; x)/p(x))$. It satisfies the equation

$$\psi^{(+)}(\theta; x) = \frac{(1-\theta^2)\, h(x) + \psi^{(+)}(\theta; Tx)\, \lambda}{1 + \dfrac{\psi^{(+)}(\theta; Tx)\, \lambda}{h(x)}}$$

or

$$\psi^{(+)}(\theta; x) = (1-\theta^2)\, h(x) + \cfrac{1}{\cfrac{1}{h(x)} + \cfrac{1}{\psi^{(+)}(\theta; Tx)\, \lambda}} \tag{19}$$

From this equation it follows that

$$\psi^{(+)}(1; x) = \cfrac{1}{\cfrac{1}{h(x)} + \cfrac{1}{\lambda h(Tx)} + \cfrac{1}{\lambda^2 h(T^2 x)} + \cdots}$$

and $0 < \psi^{(+)}(1; x) < \infty$ since $\lambda > 1$.

The analysis of this equation is quite similar to the symmetric case. We write $\psi^{(+)}(\theta; x) = \psi^{(+)}(1; x)(1 + \delta(\theta; Tx))$ and for $\delta(\theta; x)$ we have from (19)

$$\delta(\theta; x) = \frac{1}{1 + \dfrac{\psi_0(Tx)\, \lambda}{h(x)}}\, \delta(\theta; Tx) + \cdots$$

where dots mean terms of higher order of smallness. This shows that in a small neighborhood of $\psi^{(+)}(1, x)$ the mapping $Q_z(\theta; z) = (1-\theta^2)\, h(x) + \theta^2/1/h(x) + 1/z\lambda$ is a contraction and in some neighborhood $U = (1-\varepsilon, 1+\varepsilon)$ the solution $\psi^{(+)}(\theta; x)$, $\theta \in U$ of (19) is a real analytic function on U. It implies in particular that the probabilities $p_{2n}(x)$ decay in this case exponentially.

Again we study the asymptotic behavior of probabilities $P_\kappa(m)$. We can write the same expression (13) and (14) takes the form

$$\psi(\theta; m) = p(T^m x) \cdot \theta)^{-1} \cdot \theta^m \prod_{i=1}^{m} \frac{1-\varphi^{(+)}(1; T^i x)}{1-\varphi^{(+)}(\theta; T^i x)}$$

$$\cdot \prod_{i=1}^{m} \cdot \frac{p(T^i x)}{1-\varphi^{(+)}(1, T^i x)} \cdot \frac{\theta p(x)}{1-\varphi(\theta; x)}$$

We already saw (see above) that $\varphi^{(+)}(1, x) = p(x)\, \lambda^{-1} r(Tx)/r(x)$ and

$$\begin{aligned}
1-\varphi^{(+)}(1, x) &= 1 - \frac{p(x)\, \lambda^{-1} r(Tx)}{r(x)} \\
&= 1 - \frac{p(x)(r(x) + h^{-1}(x))}{r(x)} \\
&= (1-p(x)) \left(1 - \frac{p(x)}{1-p(x)} \cdot \frac{1}{h(x)\, r(x)}\right) \\
&= (1-p(x)) \cdot \left(1 - \frac{\lambda}{(r(x)\, h(T^{-1}x)}\right) \\
&= (1-p(x)) \left(\lambda^{-1} r(x) - \frac{1}{h(T^{-1}x)}\right) \cdot \frac{\lambda}{r(x)} \\
&= (1-p(x))\, r(T^{-1}x) \cdot \frac{\lambda}{r(x)}
\end{aligned}$$

Thus

$$\begin{aligned}
\prod_{i=1}^{m} \frac{p(T^i x)}{1-\varphi^{(+)}(1, x)} &= \prod_{i=1}^{m} \frac{p(T^i x)\, r(T^i x)}{(1-p(T^i x))\, r(T^{i-1}x)^{-1}} \\
&= \prod_{k=1}^{m} \frac{h(T^i x)\, r(T^i x)}{h(T^{i-1}x)\, r(T^{i-1}x)} = \frac{h(T^m x)\, r(T^m x)}{h(x)\, r(x)}
\end{aligned}$$

The ratio $1-\varphi^{(+)}(1, T^i x)/1-\varphi^{(+)}(\theta; T^i x)$ is a generating function of a positive random variable with an exponentially decaying distribution. So the product $\theta^m \prod_{i=1}^m 1-\varphi^{(+)}(1; T^i x)/1-\varphi^{(+)}(\theta; T^i x)$ has the usual Gaussian asymptotics and we can write

$$P_\kappa(m) = \frac{h(T^m x)\, r(T^m x)}{p(T^m x)} e^{-1/2\,(m-ak)^2/\sigma} C(x)(1+\varepsilon(m,k)) \qquad (20)$$

Here a is the mean drift, $\sigma > 0$ is a constant, $C(x)$ is a number depending on x, $\varepsilon(m,k)$ tend to zero as $k \to \infty$ and m remains in $\mathcal{O}(\sqrt{k})$-neighborhood of $a \cdot k$. It is clear that (20) implies Theorem 1.

6. SOME GENERALIZATIONS AND OPEN PROBLEMS

The same methods give the existence and uniqueness of stationary measures for one-dimensional diffusion processes with smooth local characteristics taking place on orbits of Diophantine groups of shifts on Tor^d.

The main problem considered in this paper is a particular case of the following more general problem. Suppose that we have a measure-preserving automorphism T acting on a measure space $(M, \mathcal{M}, \mu)$ and $p < 1$ is positive a.e.. Consider Markov chain where a point $x \in M$ jumps to Tx with probability $p(x)$ and to $T^{-1}x$ with probability $1-p(x)$. Problem: does this Markov chain have an invariant measure equivalent to μ. We believe that in the case of T with strong mixing properties like Anosov transitive diffeomorphisms the answer is negative. Probably this case is connected with random walks in random environments (see [S1]). It would be interesting to extend the results of this paper to groups $\mathbb{Z}^k$, $k < d$ acting on Tor^d and to Markov chains where a point can jump from x to $T^i x$, $|i| \leqslant i_0$.

ACKNOWLEDGMENTS

The financial support from NSF, Grant No. DMS-9706794 and RFFI, Grant No. 96-01-0037 are acknowledged. I thank V. Kaloshin and W. Schlag for useful remarks.

REFERENCES

[F] Feller, W., *Introduction to Probability Theory and Its Applications*, Vol. 1 & 2 (Wiley, New York, 1971).

[GK] Gnedenko, B. V., and Kolmogorov, A. N., *Limit Distributions for Sums of Independent Random Variables* (Addison-Wesley, Reading, MA, 1968).

[S1] Sinai, Ya. G., The limiting behavior of one-dimensional random walks in random media, *Probability Theory and Its Applications* **27**:247–258 (1982).

[S2] Sinai, Ya. G., Distribution of some functionals of the integral of the brownian motion, *Theor. and Math. Physics* (in Russian) **90**:323–353 (1992).

Communicated by J. L. Lebowitz

Comments

The problem which is considered in this paper is a part of a more general problem. Let (M, μ) be a probability space and T be an automorphism of this space. Take a function $p(x)$ defined on M such that $0 \leq p(x) \leq 1$ and consider a Markov chain whose phase space is M and the moving particle goes to Tx with probability $p(x)$ and to $T^{-1}x$ with probability $1 - p(x)$.

In this way, we get random walk along trajectories of the dynamical system (M, μ, T). The main question is to find conditions under which this random walk has an invariant measure equivalent to μ. In this paper, the case of the shift on the torus was considered and simple sufficient conditions when this is true were given.

In our paper with V. Kaloshin (see [KS]), we proved that random walks along trajectories of hyperbolic (Anosov) diffeomorphisms have no absolutely continuous invariant measures, even for smooth f.

THEORY OF PROBABILITY
Volume XXVII AND ITS APPLICATIONS Number 2
1982

THE LIMITING BEHAVIOR OF A ONE-DIMENSIONAL RANDOM WALK IN A RANDOM MEDIUM

YA. G. SINAI

(*Translated by A. R. Kraiman*)

1. Statement of the Problem and Formulation of Results

We consider the simplest random walk on the set of integer points of the straight line Z^1, under which a randomly moving point passes from $x \in Z^1$ to $x \pm 1$ with probabilities $p(x)$ and $q(x) = 1 - p(x)$, respectively. The position of the point after n transitions is denoted by $x(n)$. It is assumed that $x(0) = 0$.

A knowledge of all the $p(x)$ determines the probability distribution of the random variables $x(n)$, $n > 0$. All the probabilities relating to events connected with the behavior of $x(n)$, $n > 0$, will be denoted below by P.

It is said that a random walk occurs in a *random medium* if the probabilities $p(x)$ themselves are realizations of some random process. The simplest version arises when $p(x)$ forms a sequence of independent random variables; for example, $p(x) = \frac{1}{2} + \varepsilon\xi(x)$, where $\xi(x) = \pm 1$ is a random sign, and the signs $\xi(x)$ for different x are mutually independent, $0 < \varepsilon < \frac{1}{2}$.

In this paper we mainly discuss the situation when the $p(x)$ are independent. Certain generalizations are indicated at the end. The probabilities relating to events depending on the realization of $p(x)$ are denoted by $\mathbf{P}$.

Problems connected with the recurrence and non-recurrence properties of a random walk in a random medium were examined in [1]–[3]. In this paper we investigate the behaviour of $x(n)$ as $n \to \infty$. The basic assumption is that $p(x)$, $q(x) \geqq \text{const} > 0$ and

$$\mathbf{E}_{\mathbf{P}} \log \frac{q(x)}{p(x)} = 0.$$

It will be shown that, under these conditions, in contrast to the ordinary random walk, for large n the variable $x(n)$ takes on values of order $\log^2 n$. But if $x(n)$ is normalized, i.e., the variable $x(n)/\log^2 n$ is considered, then as $n \to \infty$ the probability distribution for $x(n)/\log^2 n$ becomes localized, i.e., concentrated in an arbitrarily small neighborhood of some point depending on the realization $p = \{p(x)\}$. The basic result can be formulated more precisely in the following manner.

Let $\alpha > 0$, $\delta > 0$ be given. For all sufficiently large n there exist a set C_n in the space of realizations p and a point $m^{(n)} = m^{(n)}(p)$ for each $p \in C_n$ such that

Y.G. Sinai, *Selecta II: Probability Theory, Statistical Mechanics, Mathematical Physics and Mathematical Fluid Dynamics*, DOI 10.1007/978-1-4419-6205-8_2, © Springer Science+Business Media, LLC 2010

$\mathbf{P}(C_n) \geqq 1-\alpha$, *and for* $p \in C_n$

$$P\left(\left|\frac{x(n)}{\log^2 n} - m^{(n)}\right| < \delta\right) \to 1$$

as $n \to \infty$ *uniformly in* $p \in C_n$. *As* $n \to \infty$ *the probability distributions for* $m^{(n)}$ *converge weakly to some limit distribution.*

The function $w^{(n)}(t)$ is basic for the analysis; it is formed from the transition probabilities $p(x)$ in the following way:

$$w^{(n)}(t) = \begin{cases} \dfrac{1}{\log n} \sum\limits_{0 \leqq x \leqq k} \log \dfrac{q(x)}{p(x)} & \text{for } t = \dfrac{k}{\log^2 n}, \quad k = 1, 2, \ldots, \\ \dfrac{1}{\log n} \sum\limits_{k \leqq x \leqq 0} \log \dfrac{q(x)}{p(x)} & \text{for } t = \dfrac{k}{\log^2 n}, \quad k = -1, -2, \cdots. \end{cases}$$

For the remaining t the function $w^{(n)}(t)$ is defined with the help of linear interpolation, $w^{(n)}(0) = 0$. In what follows it turns out that $m^{(n)}(p)$ is one of the local minima of the function $w^{(n)}$.

The proof of the basic result is carried out in Sections 4 and 5. In Sections 2 and 3, we shall conduct auxiliary constructions, and concluding remarks are made in Section 6.

I thank S. A. Molchanov for stimulating discussions on the questions considered in this article, and K. Kipnis for acquainting me with bibliographies concerning the question of random walks in random media. I also thank L. A. Bunimovich and K. M. Khanin for useful discussions.

2. Exit Probabilities and Their Estimates

Let $[a, b]$ be a segment with end-points belonging to Z^1. We shall denote by $k^+_{[a,b]}(x)$ the P-probability of paths starting at x which hit b before a; $k^-_{[a,b]}(x) = 1 - k^+_{[a,b]}(x)$ is the P-probability of paths leaving x and hitting a before b. Then

$$k^+_{[a,b]}(x) = p(x) k^+_{[a,b]}(x+1) + q(x) k^+_{[a,b]}(x-1),$$
$$k^+_{[a,b]}(a) = 0, \qquad k^+_{[a,b]}(b) = 1$$

and

$$k^-_{[a,b]}(x) = p(x) k^-_{[a,b]}(x+1) + q(x) k^-_{[a,b]}(x-1),$$
$$k^-_{[a,b]}(a) = 1, \qquad k^+_{[a,b]}(b) = 0.$$

The solutions of these equations are easily found. Let us cite the corresponding results in a form suitable for what follows.

Lemma 1. *The following equations hold*:

$$k^+_{[a,b]}(x) = \left(\sum_{y=a+1}^{x} \exp\{\log n[w^{(n)}(y \log^{-2} n) - w^{(n)}(a \log^{-2} n)]\}\right)$$
$$\cdot \left(\sum_{y=a+1}^{b} \exp\{\log n[w^{(n)}(y \log^{-2} n) - w^{(n)}(a \log^{-2} n)]\}\right)^{-1},$$

$$k^-_{[a,b]}(x)=\Big(\sum_{y=x+1}^{b}\exp\{\log n[w^{(n)}(y\log^{-2}n)-w^{(n)}(b\log^{-2}n)]\}\Big)$$
$$\cdot\Big(\sum_{y=a+1}^{b}\exp\{\log n[w^{(n)}(y\log^{-2}n)-w^{(n)}(b\log^{-2}n)]\}\Big)^{-1}.$$

We shall state two consequences of these relations. Let us introduce the notation $T_{r,s}=[r\log^{-2}n,\, s\log^{-2}n]$.

Corollary 1. *Let the variable x be such that*

$$w^{(n)}(x\log^{-2}n)=\max_{t\in T_{a,b}}w^{(n)}(t),\qquad w^{(n)}(a\log^{-2}n)=\min_{t\in T_{a,x}}w^{(n)}(t),$$

$$w^{(n)}(b\log^{-2}n)=\min_{t\in T_{x,b}}w^{(n)}(t).$$

Then $k^+_{[a,b]}(x)\geqq(b-a)^{-1}$, $k^-_{[a,b]}(x)\geqq(b-a)^{-1}$.

In fact,

$$\max_{y\in[a,b]}[w^{(n)}(y\log^{-2}n)-w^{(n)}(a\log^{-2}n)]=w^{(n)}(x\log^{-2}n)-w^{(n)}(a\log^{-2}n).$$

Therefore,

$$k^+_{[a,b]}(x)\geqq\exp\{\log n[w^{(n)}(x\log^{-2}n)-w^{(n)}(a\log^{-2}n)]\}$$
$$\cdot[(b-a)\max_{y\in[a,b]}\exp\{\log n[w^{(n)}(y\log^{-2}n)-w^{(n)}(a\log^{-2}n)]\}]^{-1}$$
$$=(b-a)^{-1}.$$

The estimate for $k^-_{[a,b]}(x)$ is obtained similarly.

Corollary 2. (a) *Let* $x=a+1$, $w^{(n)}(a\log^{-2}n)=\min_{t\in T_{a,b}}w^{(n)}(t)$, $w^{(n)}(b)=\max_{t\in T_{a,b}}w^{(n)}(t)$. *Then*

$$\text{const}\,(b-a)^{-1}\exp\{-\log n[w^{(n)}(b\log^{-2}n)-w^{(n)}(a\log^{-2}n)]\}$$
$$\leqq k^+_{[a,b]}(x)\leqq\text{const}\exp\{-\log n[w^{(n)}(b\log^{-2}n)-w^{(n)}(a\log^{-2}n)]\}.$$

(b) *Let* $x=b-1$, $w^{(n)}(b\log^{-2}n)=\min_{t\in T_{a,b}}w^{(n)}(t)$, $w^{(n)}(a\log^{-2}n)=\max_{t\in T_{a,b}}w^{(n)}(t)$. *Then*

$$\text{const}\,(b-a)^{-1}\exp\{-\log n[w^{(n)}(a\log^{-2}n)-w^{(n)}(b\log^{-2}n)]\}$$
$$\leqq k^-_{[a,b]}(x)\leqq\text{const}\exp\{-\log n[w^{(n)}(a\log^{-2}n)-w^{(n)}(b\log^{-2}n)]\}.$$

Let us prove, for example, assertion (a); assertion (b) is obtained similarly. We have

$$k^+_{[a,b]}(x)=q(a+1)(p(a+1))^{-1}$$
$$\cdot\Big(\sum_{a+1}^{b}\exp\{\log n[w^{(n)}(y\log^{-2}n)-w^{(n)}(a\log^{-2}n)]\}\Big)^{-1}.$$

According to the condition the numerator is contained between two constants, while

$$\max_{a\leqq y\leqq b}[w^{(n)}(y\log^{-2}n)-w^{(n)}(a\log^{-2}n)]=w^{(n)}(b\log^{-2}n)-w^{(n)}(a\log^{-2}n).$$

Therefore,

$$k^+_{[a,b]}(x)\leqq \text{const}\exp\{-\log n[w^{(n)}(b\log^{-2}n)-w^{(n)}(a\log^{-2}n)]\},$$

$$k^+_{[a,b]}(x)\geqq \frac{\text{const}}{b-a}\exp\{-\log n[w^{(n)}(b\log^{-2}n)-w^{(n)}(a\log^{-2}n)]\},$$

which it was required to show.

Now we shall consider the behavior for large n of the expectation $h_{[a,b]}(x)$ of the time a path, exiting from x, will reach the end-points of the segment $[a, b]$. It is clear that $h_{[a,b]}(a)=h_{[a,b]}(b)=0$ and, for $a<x<b$,

$$h_{[a,b]}(x)=p(x)h_{[a,b]}(x+1)+q(x)h_{[a,b]}(x-1)+1.$$

It is also not difficult to write out explicitly the solution of this chain of equations. That is,

$$\begin{aligned}h_{[a,b]}(x)=&\sum_{y=a+1}^{x}\left(\prod_{z=a+1}^{y-1}q(z)p^{-1}(z)\right)\\&\cdot\left(\sum_{a+1\leqq y_1<y_2\leqq b}(p(y_1))^{-1}\prod_{z=y_1+1}^{y_2-1}q(z)(p(z))^{-1}\right)\\&\cdot\left(\sum_{a+2\leqq y\leqq b}\prod_{z=a+1}^{y-1}q(z)(p(z))^{-1}\right)^{-1}\\&-\sum_{a+1\leqq y_1<y_2<x}(p(y_1))^{-1}\prod_{z=y_1+1}^{y_2-1}q(z)(p(z))^{-1}.\end{aligned}$$

With the aid of the function $w^{(n)}$ the latter expression can be rewritten in the following way:

$$\begin{aligned}h_{[a,b]}(x)=&\sum_{y=a+1}^{x}\exp\{\log n[w^{(n)}((y-1)\log^{-2}n)-w^{(n)}(a\log^{-2}n)]\}\\&\cdot\Bigg(\sum_{a+1\leqq y_1<y_2\leqq b}(p(y_1))^{-1}\\&\qquad\cdot\exp\{\log n[w^{(n)}((y_2-1)\log^{-2}n)-w^{(n)}(y_1\log^{-2}n)]\}\Bigg)\\&\cdot\left(\sum_{a+2\leqq y\leqq b}\exp\{\log n[w^{(n)}((y-1)\log^{-2}n)-w^{(n)}(a\log^{-2}n)]\}\right)^{-1}\\&-\sum_{a+1\leqq y_1<y_2<x}(p(y_1))^{-1}\\&\qquad\cdot\exp\{\log n[w^{(n)}((y_2-1)\log^{-2}n)-w^{(n)}(y_1\log^{-2}n)]\}.\end{aligned}$$

We shall now deduce a series of corollaries. Let the segment $[a, b]$ and the point x be such that $x = a+1$,

$$w^{(n)}(a \log^{-2} n) = \min_{t \in T_{a,b}} w^{(n)}(t), \qquad w^{(n)}(b \log^{-2} n) = \max_{t \in T_{a,b}} w^{(n)}(t).$$

Then

$$h_{[a,b]}(x) = \Big(\sum_{a+1 \leqq y_1 < y_2 \leqq b} (p(y_1))^{-1} \exp \{\log n [w^{(n)}((y_2-1) \log^{-2} n)$$

$$- w^{(n)}(y_1 \log^{-2} n)]\}\Big)$$

$$\cdot \Big(\sum_{a+2 \leqq y \leqq b} \exp \{\log n [w^{(n)}((y-1) \log^{-2} n) - w^{(n)}(a \log^{-2})]\}\Big)^{-1}$$

$$\leqq \text{const } (b-a)^2,$$

since

$$\max_{a+1 \leqq y_1 \leqq y_2 \leqq b} [w^{(n)}(y_2 \log^{-2} n) - w^{(n)}(y_1 \log^{-2} n)]$$

$$= w^{(n)}(b \log^{-2} n) - w^{(n)}((a+1) \log^{-2} n).$$

A similar inequality is also valid in the case when $x = b-1$,

$$w^{(n)}\Big(\frac{b}{\log^2 n}\Big) = \min_{a \leqq t \leqq b} w^{(n)}\Big(\frac{t}{\log^2 n}\Big), \qquad w^{(n)}\Big(\frac{a}{\log^2 n}\Big) = \max_{a \leqq t \leqq b} w^{(n)}\Big(\frac{t}{\log^2 n}\Big).$$

Let us now consider the situation when

$$w^{(n)}\Big(\frac{a}{\log^2 n}\Big) = \max_{a \leqq t \leqq b} w^{(n)}\Big(\frac{t}{\log^2 n}\Big), \qquad w^{(n)}\Big(\frac{b}{\log^2 n}\Big) = \min_{a \leqq t \leqq x} w^{(n)}\Big(\frac{t}{\log^2 n}\Big).$$

In this case,

$$h_{[a,b]}(x) \leqq \text{const } (b-a)^2 \exp \Big\{ \log n \max_{a < y_1 \leqq y_2 < b} \Big[w^{(n)}\Big(\frac{y_2}{\log^2 n}\Big) - w^{(n)}\Big(\frac{y_1}{\log^2 n}\Big)\Big]\Big\}.$$

In what follows the difference $(b-a)$ takes on, as $n \to \infty$, values of order $\log n$, and the difference $w^{(n)}(y_2/\log^2 n) - w^{(n)}(y_1/\log^2 n)$ takes on values of order 1.

The variance of the time of reaching the boundary can be estimated similarly.

3. Local Maxima and Minima of the Function $w^{(n)}(t)$

In this section we shall introduce the maxima and minima of the curve $w^{(n)}(t)$ of interest to us. Our construction is carried out in a manner such that it also has meaning as $n \to \infty$, i.e., for typical realizations of a Wiener measure.

Let the curve $w^{(n)}(t)$, $-\infty < t < \infty$, (see Section 1) be given. We shall call a segment $[t_1, t_2]$ a *depression* if for $\bar{t}$ such that $w^{(n)}(\bar{t}) = \min_{t_1 \leqq t \leqq t_2} w^{(n)}(t)$,

$$w^{(n)}(t_1) = \max_{t_1 \leqq t \leqq \bar{t}} w^{(n)}(t), \qquad w^{(n)}(t_2) = \max_{\bar{t} \leqq t \leqq t_2} w^{(n)}(t).$$

We shall call $w^{(n)}(t_1)-w^{(n)}(\bar{t})(w^{(n)}(t_2)-w^{(n)}(\bar{t}))$ the left (right) depth of the depression. We shall call $\min(w^{(n)}(t_1)-w^{(n)}(\bar{t}),\ w^{(n)}(t_2)-w^{(n)}(\bar{t}))$ the depth of the depression $d([t_1,t_2])$. We shall call a set of points $\mathfrak{M}=\{M_0, m_1, M_1, m_2, \cdots, M_r, m_{r+1}, M_{r+1}\}$ such that each segment $[M_i, M_{i+1}]$, $0\leqq i\leqq r$, is a depression, $w^{(n)}(m_i)=\min_{t\in[M_i,M_{i+1}]} w^{(n)}(t)$, the *set* of maxima and minima of the function $w^{(n)}$. The set $\mathfrak{M}'\in\mathfrak{M}''$ if all points of $\mathfrak{M}'$ are contained among the points of $\mathfrak{M}''$.

We shall now define the operation of refining a depression. Let $\mathfrak{M}=\{M_0, m_1, M_1, \cdots, M_r, m_{r+1}, M_{r+1}\}$. We consider a segment $[m_i, M_i]$. We find t_1, t_2 such that $m_i\leqq t_1<t_2\leqq M_i$ and

$$w^{(n)}(t_1)-w^{(n)}(t_2)=\max_{m_i\leqq t'\leqq t''\leqq M_i}(w^{(n)}(t')-w^{(n)}(t'')).$$

It is not difficult to see that $[M_{i-1}, t_1]$ and $[t_1, M_i]$ will be depressions; therefore, adding the points t_1, t_2 to the set $\mathfrak{M}$ again reduces to a set of maxima and minima. We shall call the operation described a *right refinement operation*. In a similar manner let us take the segment $[M_i, m_{i+1}]$ and consider t_1, t_2, $M_i\leqq t_1\leqq t_2\leqq m_{i+1}$, such that $w^{(n)}(t_2)-w^{(n)}(t_1)=\max_{M_i\leqq t'\leqq t''\leqq m_{i+1}}(w^{(n)}(t'')-w^{(n)}(t'))$. It is directly verified that $[M_i, t_2]$ and $[t_2, M_{i+1}]$ are depressions, and adding the points t_1, t_2 to the set $\mathfrak{M}$ likewise leads us to a set of maxima and minima. We shall call this operation a *left refinement operation*.

Let M^+ (M^-) be the smallest (largest) $t>0$ ($t<0$) for which $w^{(n)}(t)=1$, and let m_0 be such that $w^{(n)}(m_0)=\min_{M^-\leqq t\leqq M^+} w^{(n)}(t)$. Then $[M^+, M^+]$ is a depression and $\{M^-, m_0, M^+\}$ is a set of maxima and minima.

Lemma 2. *Let $\alpha>0$, $\delta>0$ be given. Then there exist, for all sufficiently large n, a set C_n and numbers r and δ_1 such that $\mathbf{P}(C_n)\geqq 1-\alpha$, and for any choice of $p\in C_n$ from the set $\{M^-, m_0, M^+\}$ a set of maxima and minima $\mathfrak{M}=\{M^-=M_0, m_1, M_1, \cdots, M_r, m_r, M_{r+1}=M^+\}$ can be obtained using not more than r left and right refinement operations, such that*

$$\max_{0\leqq j\leqq r}(M_{j+1}-M_j)\leqq\delta,\qquad \min_{\substack{t',t''\in\mathfrak{M}\\ t'\neq t''}}|w^{(n)}(t')-w^{(n)}(t'')|\geqq\delta_1,$$

$$\min_{\substack{t\in\mathfrak{M}\\ t\neq M_0M_{r+1}}}|w^{(n)}(t)-1|\geqq\delta_1,\qquad \min_{t\in\mathfrak{M}}|t|\geqq\delta_1,\qquad M_{r+1}-M_0\leqq\delta_1^{-1}.$$

The proof of the lemma follows from the fact that the assertion of the lemma is valid for a Wiener measure, and a measure in the space of paths $w^{(n)}$ converges weakly to a Wiener measure.

Lemma 3. *Let $M_{i_1}\in\mathfrak{M}$, $M_{i_2}\in\mathfrak{M}$ and $[M_{i_1}, M_{i_2}]$ be a depression, $d([M_{i_1}, M_{i_2}])>1$ and $0\in[M_{i_1}, M_{i_2}]$. Then*

$$P\{x(m)\in[M_{i_1}\log^2 n, M_{i_2}\log^2 n]\text{ for all }0\leqq m\leqq n\}\to 1\quad \text{as } n\to\infty.$$

PROOF. Let m_j be a minimum of the depression under consideration, $m_j\in\mathfrak{M}$, and for definiteness $m_j<0$. Also let the point 0 be contained in the depression

$[M_{i_0}, M_{i_0+1}]$. We take a segment $[m_j \log^2 n, M_{i_2} \log^2 n]$. From Lemma 1 for $p \in C_n$

$$k^+_{[m_j\log^2 n, M_{i_2}\log^2 n]}(0) \leqq \exp\{-\log n[w^{(n)}(M_{i_2}) - w^{(n)}(m_j)]\} \cdot |M_{i_2} - m_j| \log^2 n \exp\{\log n[w^{(n)}(M_{i_0}) - w^{(n)}(m_j)]\} \leqq \delta_1^{-1} \log^2 n \exp\{-(\log n)\delta_1\} \to 0$$

as $n \to \infty$. Thus, as $n \to \infty$, a random point hits $m_j \log^2 n$ earlier than $M_{i_2} \log^2 n$ with probability tending to 1.

Let us introduce the P-probability $\pi_{[a,b]}(x)$ that a random point, leaving x, hits one of the end-points of $[a, b]$ without getting to x on the way. It is clear that

$$\pi_{[a,b]}(x) = p(x)k^+_{[x,b]}(x+1) + q(x)k^-_{[a,x]}(x-1).$$

We shall now estimate $\pi_{[M_{i_1}\log^2 n, M_{i_2}\log^2 n]}(m_j \log^2 n)$. By virtue of Lemma 1,

$$k^+_{[m_j\log^2 n, M_{i_2}\log^2 n]}(m_j \log^2 n + 1) \leqq \text{const} \exp\{-\log n[w^{(n)}(M_{i_2}) - w^{(n)}(m_j)]\},$$

$$k^-_{[M_{i_1}\log^2 n, m_j\log^2 n]}(m_j \log^2 n - 1) \leqq \text{const} \exp\{-\log n[w^{(n)}(M_{i_1}) - w^{(n)}(m_j)]\}.$$

Hence it follows that $\pi_{[M_{i_1}\log^2 n, M_{i_2}\log^2 n]}(m_j \log^2 n) \leqq \text{const}/n^{1+\delta_1}$. From the latter inequality it follows that the P-probability that the random point, leaving m_j, returns to m_j at least n times before it hits M_{i_1} or M_{i_2} is equal to

$$(1 - \pi_{[M_{i_1}\log^2 n, M_{i_2}\log^2 n]}(m_j \log^2 n))^n \geqq \left(1 - \frac{\text{const}}{n^{1+\delta_1}}\right)^n \to 1$$

as $n \to \infty$. The lemma is proved.

Let $[M_{i'_1}, M_{i'_2}]$, $[M_{i''_1}, M_{i''_2}]$ be two depressions satisfying the conditions of Lemma 3. We shall show that their intersection $D = [M_{i'_1}, M_{i'_2}] \cap [M_{i''_1}, M_{i''_2}]$ is again a depression satisfying the conditions of Lemma 3. We shall consider only the case when $D = [M_{i''_1}, M_{i'_1}]$, since the remaining cases are simpler.

Let us denote by $m_{j'}$, $m_{j''}$ the minima corresponding to the initial depressions. We shall show that D must contain at least one of the points $m_{j'}$, $m_{j''}$. Let us assume that this is not so, and let $w^{(n)}(M_{i'_2}) > w^{(n)}(M_{i''_2})$. But $M_{i''_1} < M_{i'_2} < m_{j''}$, and we obtain a contradiction with the fact that $[M_{i''_1}, M_{i''_2}]$ is a depression. If $w^{(n)}(M_{i''_1}) > w^{(n)}(M_{i'_2})$, then we obtain a contradiction with the fact that $[M_{i'_1}, M_{i'_2}]$ is a depression.

Assume that $m_{j'} \in D$ and $M_{i'_1} < M_{i''_1} < m_{j'}$. If $w^{(n)}(M_{i''_1}) - w^{(n)}(m_{j'}) < 1$, then $w^{(n)}(m_{j'}) > w^{(n)}(m_{j''})$, since $w^{(n)}(M_{i''_1}) - w^{(n)}(m_{j''}) > 1$. Hence $M_{i'_2} > m_{j''}$, since otherwise $w^{(n)}(M_{i''_1}) < w^{(n)}(M_{i'_2})$, and we again obtain a contradiction with the fact that $[M_{i''_1}, M_{i''_2}]$ is a depression. Hence $M_{i'_2} > m_{j''}$ and both minima belong to D. Then $m_{j''}$ is a minimum of D and $w^{(n)}(M_{i'_1}) - w^{(n)}(m_{j''}) \geqq w^{(n)}(M_{i'_1}) - w^{(n)}(m_{j'}) > 1$. Thus we see that D is a depression satisfying the conditions of Lemma 3.

We can now derive the smallest depression $[M_-^{(0)} = M_{i_1}, M_+^{(0)} = M_{i_2}] \subset \mathfrak{M}$ satisfying the conditions of Lemma 3. We shall call this depression *basic*, and the set of maxima and minima $\mathfrak{M}_1 = \{M_{i_1}, m_{i_1+1}, M_{i_1+1}, \cdots, M_{i_2-1}, m_{i_2}, M_{i_2}\}$ the *basic set*. Let $m^{(0)} = m_{j_0} \in \mathfrak{M}_1$ be a minimum of the basic depression. We can now formulate the basic result of this paper more precisely in the form of the following theorem.

Theorem 1. *For all $p \in C_n$ (see Lemma 2) the relation*

$$P\left(\frac{x(n)}{\log^2 n} \in [M_{j_0-1}, M_{j_0}]\right) \to 1 \quad as\ n \to \infty$$

holds uniformly in $p \in C_n$.

4. Auxiliary Process

Let $\mathfrak{M}_1 = \{M_-^{(0)} = M_{i_1}, m_{i_1+1}, M_{i_1+1}, \cdots, m^{(0)} = m_j, M_j, \cdots, m_{i_2}, M_{i_2} = M_+^{(0)}\}$ be the basic set and $m^{(0)}$ the absolute minimum for $\mathfrak{M}_1$. Let us fix an increasing sequence of sets of maxima and minima $\mathfrak{M}^{(1)} \subset \mathfrak{M}^{(2)} \subset \cdots \subset \mathfrak{M}^{(r)} = \mathfrak{M}_1$, where $\mathfrak{M}^{(1)} = \{M_-^{(0)}, m^{(0)}, M_+^{(0)}\}$ and $\mathfrak{M}^{(i+1)}$ is obtained from $\mathfrak{M}^{(i)}$ with the aid of the (right or left) refinement operation.

Let n be fixed and let $x = \{x(i), i \geqq 0\}$ be the path of the random walk under consideration. During the walk the point $x(i)/\log^2 n$ passes through points of any set $\mathfrak{M}^{(l)}$, $1 \leqq l \leqq r$, at certain times. Let us write out these points in succession until the path of $x/\log^2 n$ has passed through $m^{(0)}$ n times: $\omega = (\omega_1, \cdots, \omega_t)$, $\omega_i \in \mathfrak{M}^{(l)}$, $\omega_t = m^{(0)}$, and the symbol $m^{(0)}$ is encountered n times in the *message* ω. It is clear that $t \geqq n$. We shall denote by $\Omega_{\mathfrak{M}^{(l)}}$ the space of such messages. The probability distribution on paths of x induces a probability distribution on messages $\omega \in \Omega_{\mathfrak{M}^{(l)}}$. The form of this distribution will be described shortly. Let $\mathfrak{M}^{(i)}$ be one of the sets and let $\mathfrak{M}^{(i+1)}$ be obtained with the aid of the refinement operation, for definiteness the right one. This means that for some $m_k, M_k \subset \mathfrak{M}^{(i)}$ the points M'', m'' are added in order to obtain $\mathfrak{M}^{(i+1)}$.

We take $\omega' \in \Omega_{\mathfrak{M}^{(i)}}$ and consider all possible pairs of the form (m_k, m_k), (m_k, M_k), (M_k, m_k), (M_k, M_k). Then the message $\omega'' \in \Omega_{\mathfrak{M}^{(i+1)}}$ is obtained from ω' by a refinement inside each pair of a certain number of symbols m'', M''. The refinement m'', M'' inside the pair m_k, m_k means that the path of x, leaving m_k, will be in the states m'', M'' up to the following return to m_k in accordance with the refined set. The refinement inside the other pairs has a similar meaning.

Lemma 4. *The conditional distribution on messages ω'' under the condition ω' is such that sets of m'', M'', refined inside separate pairs, are independent of one another.*

The proof follows directly from the definitions. We investigate the probability distribution for the refined set in more detail. We shall dwell on the case of the pair (m_k, m_k). The probability that not a single symbol is refined is equal to

$$q(m_k \log^2 n) k^+_{[M_{k-1}\log^2 n, m_k \log^2 n]}(m_k \log^2 n + 1)$$
$$+ p(m_k \log^2 n) k^-_{[m_k \log^2 n, M'' \log^2 n]}(m_k \log^2 n + 1).$$

The probability that exactly one symbol M'' is refined is equal to

$$p(m_k \log^2 n) k^+_{[m_k \log^2 n, M'' \log^2 n]}(m_k \log^2 n + 1) q(M'' \log^2 n)$$
$$\cdot k^-_{[m_k \log^2 n, M'' \log^2 n]}(M'' \log^2 n - 1).$$

From Lemma 1 and Corollary 1 it follows that this probability is contained between

$$\text{const}\exp\{-\log n[w^{(n)}(M'')-w^{(n)}(m_k)]\}=\text{const}\, n^{-(w^{(n)}(M'')-w^{(n)}(m_k))}$$

and

$$\text{const}\,(\log n)^{-2} n^{-(w^{(n)}(M'')-w^{(n)}(m_k))}.$$

Further, any refined set consists of a series of symbols M'' followed by a series of symbols m'', then again by a series of M'', and so on. At the end one has invariably a series of symbols M''. The lengths of various series are mutually independent. An important remark concerns the length of the series of m''. We shall find a conditional probability π of passing from m'' to m'' without getting to M'' and M_k. Obviously we can write

$$\begin{aligned}1-\pi &= q(m''\log^2 n)k^-_{[M''\log^2 n, m''\log^2 n]}(m''\log^2 n-1)\\ &\quad + p(m''\log^2 n)k^+_{[m''\log^2 n, M_k\log^2 n]}(m''\log^2 n+1),\end{aligned}$$

from which we obtain, with the aid of Lemma 1 and its corollaries;

$$\frac{\text{const}}{\log^2 n} n^{-(w^{(n)}(M'')-w^{(n)}(m''))}\leqq 1-\pi\leqq \text{const}\, n^{-(w^{(n)}(M'')-w^{(n)}(m''))}.$$

This inequality together with the preceding ones shows that the symbols M'', m'' are sparsely refined, with a probability approximately equal to $n^{-(w^{(n)}(M'')-w^{(n)}(m_k))}$; but if the refinement has occurred, then the length of the refined series of m'' is approximately equal to $n^{(w(M'')-w(m''))}$. Here the inequality $w^{(n)}(M'')-w^{(n)}(m_k)>w^{(n)}(M'')-w^{(n)}(m'')$ is of substantial importance, showing that on the whole a relatively small number of symbols m'', M'' is refined. In essence it is this circumstance which is basic for the validity of Theorem 1. We point out also that the length of each series of m'' or M'' has an exponential distribution and the lengths of various series are mutually independent. A similar analysis is carried out for the other pairs (m_k, M_k), (M_k, M_k), (M_k, m_k).

Let us introduce the set $\Xi^{(i+1)}$ of pairs (M'', m_k), (m_k, M''), (M'', M''), (M'', m''), (m'', m''), (m'', M''), (m'', M_k), (M_k, m''), and for $\omega''\in\Omega_{\mathfrak{M}^{(i+1)}}$ and $\xi\in \Xi^{(i+1)}$ denote by $\nu_\xi(r;\omega'')$ the number of places of the pair ξ appearing up to the r-th appearance of $m^{(0)}$. If $\omega'\in\Omega_{\mathfrak{M}^{(i)}}$ is a message from which ω'' was produced with the aid of a refinement of symbols, then for fixed ω' each $\nu_\xi(r;\omega'')$ is representable in the form of a sum of independent random variables. Therefore the conditional expectation and variance can be represented in the form of linear combinations of $\nu_\zeta(r;\omega')$, where $\zeta\in Z^{(i)}$ and $Z^{(i)}$ is the set of all possible pairs of symbols for $\mathfrak{M}^{(i)}$.

We fix the chain $\omega^{(1)}, \omega^{(2)}, \cdots, \omega^{(i)}=\omega'$, where the message $\omega^{(l+1)}$ has been obtained from $\omega^{(l)}$ by a refinement of the corresponding pairs of symbols, and $Z^{(l)}$ is the set of pairs of symbols for the set $\mathfrak{M}^{(l)}$ encountered in the messages $\omega\in\Omega_{\mathfrak{M}^{(l)}}$. Let us consider the conditional variance $\mathbf{D}(\nu_\xi(r;\omega'')|\omega')$. As was already stated, it is a linear combination with positive coefficients $\nu_\zeta(r;\omega^{(i)})$, $\zeta\in Z^{(i)}$.

From Chebyshev's inequality it follows that, with a probability tending to 1, as $n\to\infty$ and for $\frac{1}{2}n\leqq r\leqq n$, each $\nu_\zeta(r;\omega^{(i)})$ is equivalent (in the sense of the ratio tending to 1) to its conditional expectation under the condition $\omega^{(i-1)}$, which is a linear combination of $\nu_\xi(r;\omega^{(i-1)})$, $\zeta\in Z^{(i-1)}$. The coefficients of these linear combinations do not depend on the conditions. Moving farther on from $(i-1)$ to $(i-2)$, $(i-3)$ and so on, we see that with a probability tending to 1, as $n\to\infty$, the conditional variance $\mathbf{D}(\nu_\xi(r;\omega'')|\omega')$ is equivalent to some expression $D_\xi^{(n)}$ depending on n, but not depending on ω'.

The following assertion can be obtained by the usual methods of proof of the local limit theorem for sums of independent random variables. Let $\mathbf{E}(\nu_\xi(r;\omega''|\omega'))$ be the conditional expectation of $\nu_\xi(r;\omega'')$ under the condition ω', and A a constant number. Then for any set of integers z_ξ, $\xi\in\Xi^{(i+1)}$ such that $|z_\xi-\mathbf{E}(\nu_\xi(r;\omega'')|\omega')|\leqq A\sqrt{\mathbf{D}(\nu_\xi(r;\omega'')|\omega')}$

$$P(\nu_\xi(r;\omega'')=z_\xi|\omega')\sim\Big(\prod_{\xi\in\Xi^{(i+1)}}D_\xi^{(n)}\Big)^{-1/2}g\Big(\frac{z-E}{\sqrt{D^{(n)}}}\Big),$$

where z, E, $D^{(n)}$ are vectors with components z_ξ, $\mathbf{E}(\nu_\xi)$, $D_\xi^{(n)}$, respectively, and g is a non-singular eight-dimensional Gaussian distribution.

Let us consider the pair $\zeta=(t',t'')\in Z^{(i)}$. It can be one of six types: 1) $t'=t''$ and $t'=m_j$; 2) $t'=m_j$ and $t''=M_j$; 3) $t'=m_j$, $t''=M_{j-1}$; 4) $t'=t''=M_j$; 5) $t'=M_j$, $t''=m_j$; 6) $t'=M_j$, $t''=m_{j+1}$. The following assertion is easily obtained by induction on m: for any $\varepsilon>0$ as $n\to\infty$,

in the 1st case

$$P(rn^{-(w^{(n)}(m_j)-w^{(n)}(m^{(0)}))-\varepsilon}\leqq\nu_\zeta(r;\omega)\leqq rn^{-(w^{(n)}(m_j)-w^{(n)}(m^{(0)}))+\varepsilon})\to 1,$$

in the 2nd case

$$P(rn^{-(w^{(n)}(M_{j-1})-w^{(n)}(m^{(0)}))-\varepsilon}\leqq\nu_\zeta(r;w)\leqq rn^{-(w^{(n)}(m_j)-w^{(n)}(m^{(0)}))+\varepsilon})\to 1,$$

in the 3rd case

$$P(rn^{-(w^{(n)}(M_{j-1})-w^{(n)}(m^{(0)}))-\varepsilon}\leqq\nu_\zeta(r;\omega)\leqq rn^{-(w^{(n)}(M_{j-1})-w^{(n)}(m^{(0)}))+\varepsilon})\to 1,$$

in the 4th case

$$P(rn^{-(w^{(n)}(M_j)-w^{(n)}(m^{(0)}))-\varepsilon}\leqq\nu_\zeta(r;\omega)\leqq rn^{-(w^{(n)}(M_j)-w^{(n)}(m^{(0)}))+\varepsilon})\to 1,$$

in the 5th case

$$P(rn^{-(w^{(n)}(M_j)-w^{(n)}(m^{(0)}))-\varepsilon}\leqq\nu_\zeta(r;\omega)\leqq rn^{-(w^{(n)}(M_j)-w^{(n)}(m^{(0)}))+\varepsilon})\to 1,$$

in the 6th case

$$P(rn^{-(w^{(n)}(M_j)-w^{(n)}(m^{(0)}))-\varepsilon}\leqq\nu_\xi(r;\omega)\leqq rn^{-(w^{(n)}(M_j)-w^{(n)}(m^{(0)}))+\varepsilon})\to 1.$$

Since $\frac{1}{2}n\leqq r\leqq n$, while $w^{(n)}(t)-w^{(n)}(m^{(0)})>\delta_1$ for $t\in\mathfrak{M}^{(i)}\backslash m^{(0)}$, these relations mean that the probable number of appearances of any symbol different from $m^{(0)}$ is small compared to n. We shall now derive the basic result of this section.

Let $\omega\in\Omega_{\mathfrak{M}_1}$ and ω_i be the i-th coordinate of the message ω.

Theorem 2. *Let $p \in C_n$. Then for any r with $\frac{1}{2}n \leqq r \leqq n$*

$$P(\omega_r = m^{(0)}) \to 1 \quad \text{as } n \to \infty$$

uniformly in $p \in C_n$.

PROOF. For any r_1 we denote by $\nu(r_1)$ the total number of inserted symbols up to the r_1-th appearance of the symbol $m^{(0)}$. Then $\nu(r_1)$ is also a sum of independent random variables satisfying the conditions of the local central limit theorem. If $\mathbf{E}(r_1)$, $\mathbf{D}(r_1)$ are the expectation and variance, then there is a $\gamma > 0$ such that $r_1 n^{1-\gamma-\varepsilon} \leqq \mathbf{E}(r_1) \leqq r_1 n^{1-\gamma+\varepsilon}$ for any $\varepsilon > 0$ and $n \to \infty$, and

$$\text{const}\, r_1^{-1}\mathbf{D}(r_1) \leqq \left(\frac{1}{r_1}\mathbf{E}r_1\right)^2 \leqq \text{const}\, r_1^{-1}\mathbf{D}(r_1).$$

We can now write

$$P(\omega_r = m^{(0)}) = \sum_{r_1} P(\nu(r_1) = r - r_1) \sim \sum_{r_1} \frac{1}{\sqrt{2\pi\mathbf{D}(r_1)}} \exp\left\{-\frac{(r - r_1 - \mathbf{E}(r_1))^2}{2\mathbf{D}(r_1)}\right\}.$$

In the latter sum one can restrict oneself to those r_1 for which the local limit theorem is valid. For such r_1 we obviously have $\mathbf{D}(r_1) \sim \mathbf{D}(r)$ and

$$\exp\left\{-\frac{(r - r_1 - \mathbf{E}(r_1))^2}{2\mathbf{D}(r_1)}\right\} \sim \exp\left\{-\frac{(r - r_1 - \mathbf{E}(r))^2}{2\mathbf{D}(r)}\right\},$$

since

$$\begin{aligned} \frac{|r - r_1|\,|\mathbf{E}(r_1) - \mathbf{E}(r)|}{\mathbf{D}(r)} &\leqq \text{const}\, \frac{(r - r_1)^2 |\mathbf{E}(r_1) - \mathbf{E}(r)|}{\mathbf{D}(r)|r - r_1|} \\ &\leqq A\, \text{const}\, \frac{|\mathbf{E}(r) - \mathbf{E}(r_1)|}{|r - r_1|} \to 0 \end{aligned}$$

as $n \to \infty$. Therefore the latter sum is equivalent to

$$\sum_{r_1} \frac{1}{\sqrt{2\pi\mathbf{D}(r)}} \exp\left\{-\frac{(r - r_1 - \mathbf{E}(r))^2}{2\mathbf{D}(r)}\right\} \to 1$$

as $r \to \infty$. Thus Theorem 2 is proved.

REMARK. The proof of Theorem 2 recalls the equivalence proofs of various canonical ensembles in statistical mechanics (for example, see [4]). A similar remark relates to the arguments of the following section.

5. Proof of Theorem 1

In this section we shall establish that, for any $\delta > 0$ and $p \in C_n$,

$$P\left(\left|\frac{x(n)}{\log^2 n} - m^{(0)}\right| \leqq \delta\right) \to 1 \quad \text{as } n \to \infty$$

uniformly in $p \in C_n$.

Let us consider the path of the random walk $x=\{x(t), t \geqq 0\}$ and the normalized path

$$x' = \frac{1}{\log^2 n} x = \left\{\frac{1}{\log^2 n} x(t), t \geqq 0\right\}.$$

To it corresponds the message $\omega \in \Omega_{\mathfrak{M}_1}$, $\omega=(\omega_1, \cdots, \omega_s)$. We denote by τ_i, $1 \leqq i \leqq s$, the time from the i-th to the $(i+1)$st hitting of one of the points of $\mathfrak{M}_1$, and by τ_0 the time until first hitting of the path x' in $\mathfrak{M}_1$, having left from 0. Obviously, for a fixed ω the τ_i form a sequence of independent random variables. We shall be interested in a value of r such that $\tau_0+\tau_1+\cdots+\tau_r<n$, $\tau_0+\tau_1+\cdots+\tau_r+\tau_{r+1}>n$, and the probability corresponding to this event:

$$P=\sum_r \sum_{l_1<n} \sum_{l_2 \geqq n-l} P(\tau_0+\cdots+\tau_r=l_1, \tau_{r+1}=l_2, \omega_r=m^{(0)}).$$

The assertion we need is equivalent to the fact that $P \to 1$ as $n \to \infty$.

For an assigned ω the probability distribution of τ_i, $i>0$, depends only on ω_i and ω_{i+1}. Let Z be the set of admissible pairs of symbols of $\mathfrak{M}_1$, and for $\zeta \in Z$, let $\nu_\zeta(r)$ be the number of appearances of the pair ζ up to the i-th appearance of the symbol $m^{(0)}$. Then

$$E_r = \mathbf{E}(\tau_1+\cdots+\tau_r|\omega) = \sum_{\zeta \in Z} h_\zeta \nu_\zeta(r),$$

$$D_r = \mathbf{D}(\tau_1+\cdots+\tau_r|\omega) = \sum_{\zeta \in Z} d_\zeta \nu_\zeta(r).$$

Here h_ζ and d_ζ are the expectation and variance of the transition time between states of the pair ζ. If $\zeta^{(0)}=(m^{(0)}, m^{(0)})$, then $E_r \sim h_{\zeta^{(0)}} r$, $D_r \sim d_{\zeta^{(0)}} r$. Moreover, $|E_r - h_{\zeta^{(0)}} r| \leqq r^{1-\gamma}$ for some $\gamma>0$.

On the basis of the local limit theorem, the applicability of which is established here by the usual methods (for example, see [5]), we can now write, for typical ω,

$$P(\tau_0+\tau_1+\cdots+\tau_r=l|\omega, \omega_r=m^{(0)}) \sim (2\pi n d_{\zeta^{(0)}})^{-1/2} \exp\left\{-\frac{(l-E_r)^2}{2D_r}\right\}$$

as $n \to \infty$, for $l=E_r+z\sqrt{D_r}$, $|z| \leqq A$, where A is an arbitrary fixed number. We have used the fact that the basic contribution to the variance of the sum for typical ω and $r \sim n$ is a term corresponding to the pair $\zeta^{(0)}$, since $\nu_{\zeta^{(0)}}(r) \sim r$. Further, we have,

$$P \sim \sum_r \sum_{l_1<n} (2\pi n d_{\xi^{(0)}})^{-1/2} \exp\left\{-\frac{(l_1-E_r)^2}{2nd}\right\} \sum_{l_1 \geqq n-l} p_{l_1}, \tag{1}$$

where p_l is the probability distribution for the random variable τ_r corresponding to the pair $\xi^{(0)}$. Here $\sum_k \sum_{l_1 \geqq k} p_{l_1} = h_{\zeta^{(0)}}$. We interchange the order of summation over l_1 and r. Let us use the fact that

$$\sum_r (2\pi n d)^{-1/2} \exp\left\{-\frac{(l_1-E_r)^2}{2nd}\right\} \sim \frac{1}{h_{\xi^{(0)}}}.$$

Therefore we obtain

$$P \sim \sum_{l_1<n} \sum_{r} (2\pi n d_{\xi^{(0)}})^{-1/2} \exp\left\{-\frac{(l_1-E_r)^2}{2nd}\right\} \sum_{l_2 \geqq n-l_1} p_{l_2}$$

$$\sim \frac{1}{h_{\xi^{(0)}}} \sum_{n-l_1 \geqq 1} \sum_{l_2 \geqq n-l_1} p_{l_2} = 1.$$

Our assertion is proved.

6. Concluding Remarks

1. The existence of a limit distribution for the point $m^{(0)}$ is easily derived from our constructions. The assertion of the recurrence of the random walk under consideration follows.

2. The basic result of this paper concerns the limiting behavior of $x(n)$ in probability. It would be interesting to clarify the nature of the behavior of $x(n)$ with probability 1. Apparently the random process $x(n)$ has a step-wise character. For a given n the path passes a great part of the time in the deepest minimum corresponding to this n. Thereupon, with an increase in n this minimum becomes insufficiently deep, and a deeper minimum appears into which the particle jumps, and so on.

3. The methods of this paper have been adapted for the analysis of random excitation of the simplest random walk. It would be interesting to extend them in order to include the more general case.

4. The condition of independence of probabilities $p(x)$ can be weakened appreciably. It suffices to require that the measure in the space of paths $w^{(n)}(t)$ converge weakly to a Wiener measure.

5. The problem considered above for random walks also arises naturally for diffusion processes. We are then led to the problem of the limiting behavior of diffusion processes for which the drift coefficient is a random time function. The case considered corresponds to the situation when the diffusion coefficient is constant or varies within restricted limits, while the drift coefficient is white noise, i.e., an arbitrary (of course, generalized) Wiener process. Our methods permit one to obtain results similar to those presented above for this case also.

Received by the editors
July 25, 1980

REFERENCES

[1] H. KESTEN, M. W. KOSLOW AND F. SPITZER, *A limit law for random walk in a random environment*, Compos. Math., 30 (1975), pp. 145–168.
[2] F. SOLOMON, *Random walks in a random environment*, Ann. Probab., 3, 1 (1975), pp. 1–31.
[3] G. RITTER, *Random walk in a random environment, critical case*, thesis, Cornell University, Ithaca, NY, 1976.
[4] R. A. MINLÖS AND A. M. KHALFINA, *A two-dimensional limit theorem for the number of particles and the energy in a large canonical ensemble*, Izv. Akad. Nauk SSSR, Ser. Mat., 34, 5 (1970), pp. 1173–1191. (In Russian.)
[5] V. V. PETROV, *Sums of Independent Random Variables*, Springer-Verlag, New York, 1975.

Comments

While Bunimovich, Chernov, and I were working on various properties of Lorentz gases it became clear that the dynamics of Lorentz gas particles resembles trajectories of random walks in random environment.

At that time, the famous paper by Kesten, Koslow, and Spitzer became already known. It showed that random walks in random environments can be trapped. I studied the symmetric case and showed that in this case the displacements can grow as $\log^2 t$. The numerous discussions with S. A. Molchanov at that time were very useful.

This paper had many continuations. The extension to continuous time was obtained in the paper by T. Brox (see [B]).

Some other results can be found in the paper by K. Kawazu, Y. Tamura, and H. Tanaka (see [KTT]) and in O. Zeitouni (see [Z]).

In the independent papers by O. Golosov (see [G]) and H. Kesten (see [K]), the distribution of the difference between the coordinate of the moving point and the position of the related minimum was derived.

FUNDAMENTA
MATHEMATICAE
147 (1995)

A remark concerning random walks with random potentials

by

Yakov G. Sinai (Princeton, N.J., and Moscow)

Abstract. We consider random walks where each path is equipped with a random weight which is stationary and independent in space and time. We show that under some assumptions the arising probability distributions are in a sense uniformly absolutely continuous with respect to the usual probability distribution for symmetric random walks.

We consider random walks on the d-dimensional lattice $\mathbb{Z}^d$ with each path having a random statistical weight. Paths starting at (x,k) and ending at (y,n) will be denoted by $\omega_{x,k}^{y,n}$, i.e. $\omega_{x,k}^{y,n} = \{\omega(t) \in \mathbb{Z}^d,\ k \le t \le n,\ \omega(k) = x,\ \omega(n) = y,\ \|\omega(t+1) - \omega(t)\| = 1\}$. To define a random weight introduce a sequence of iid rv $F = \{F(x,t)\}$, $x \in \mathbb{Z}^d$, $t \in \mathbb{Z}$. Without any loss of generality we may assume that the $F(x,t)$ are given for all $x \in \mathbb{Z}^d$, $t \in \mathbb{Z}$. The space of all possible realizations of F is denoted by Φ. The measure corresponding to F is denoted by Q, the expectation with respect to Q is denoted by M. We do not use any special notation for the natural σ-algebra in Φ. Our main assumption concerning the distribution of $F(x,t)$ is

$$M\exp(2F(x,t)) < \infty.$$

The natural group of space-time translations acting in Φ is denoted by $\{T^{x,t}\}$. It preserves the measure Q.

We shall consider the statistical weight of $\omega_{x,k}^{y,n}$ equal to

$$\pi(\omega_{x,k}^{y,n}) = \exp\Big\{\sum_{t=k}^{n} F(t,\omega(t))\Big\}\frac{1}{(2d)^{n-k}}.$$

Introduce partition functions

$$Z_{x,k}^{y,n} = \sum_{\omega_{x,k}^{y,n}} \pi(\omega_{x,k}^{y,n}), \qquad Z_{x,k}^{n} = \sum_{y} Z_{x,k}^{y,n}.$$

1991 *Mathematics Subject Classification*: Primary 34F05.

Y.G. Sinai, *Selecta II: Probability Theory, Statistical Mechanics, Mathematical Physics and Mathematical Fluid Dynamics*, DOI 10.1007/978-1-4419-6205-8_2,

Now we may define the "random" probability distribution $P^n_{F;\,x,k}$ defined on paths $\omega^{y,n}_{x,k}$ by the formula

$$p(\omega^{y,n}_{x,k}) = \frac{\pi(\omega^{y,n}_{x,k})}{Z^n_{x,k}}.$$

The induced probability distribution of $y = \omega(n)$ is

$$p^{y,n}_{x,k} = \sum_{\omega^{y,n}_{x,k}} p(\omega^{y,n}_{x,k}) = \frac{Z^{y,n}_{x,k}}{Z^n_{x,k}}.$$

We shall also need the usual transition probabilities

$$q^{(n-k)}(y-x) = \sum_{\omega^{y,n}_{x,k}} \frac{1}{(2d)^{n-k}}.$$

It is well known that for any $A > 0$ and y for which $\|y - x\| \le A\sqrt{n}$,

$$q^{(n-k)}(y-x) = \frac{1}{(2\pi(n-k)/d)^{d/2}} \exp\left\{-\frac{d\|y-x\|^2}{2(n-k)}\right\}(1 + \gamma^{(n-k)}(y-x)),$$

where $\|\cdot\|$ is the Euclidean norm and $\gamma^{(n-k)}(z)$ tends to zero uniformly in z satisfying the above-mentioned restrictions.

Our purpose in this note is to study the behavior of the distribution of the normalized displacement

$$\frac{\omega(n) - \omega(k)}{\sqrt{(n-k)/d}} = \frac{y-x}{\sqrt{(n-k)/d}}$$

with respect to $P^n_{F;\,x,k}$ as $n \to \infty$. The problem was considered by J. Imbrie and T. Spencer [3] and later by E. Bolthausen [1]. In [1] and [3], it was shown that if the $F(x,t)$ are small enough in appropriate sense, and $d \ge 3$, then the limiting distribution of the displacement is Gaussian and for typical F the mean of the square of displacement grows proportionally to time. Recently these results were extended to some random processes with continuous time by J. Conlon and P. Olsen [2]. All these results can be formulated also in terms of diffusion of directed polymers in random environments.

We show below that some of the results of [1] and [3] remain valid under weaker assumptions on the distribution of F and the dimension d. Define

$$\alpha_d = \sum_{n>0} \sum_z (q^{(n)}(z))^2.$$

This is finite if $d \ge 3$. Put

$$\Lambda = M \exp\{F(x,t)\} \quad \text{and} \quad \lambda = \frac{M \exp\{2F(x,t)\} - \Lambda^2}{\Lambda^2}.$$

Our main assumption is

$$\lambda\alpha_d < 1. \tag{1}$$

It is easy to see that (1) is valid for $d \geq 3$ if λ is small enough. If $F(x,t)$ takes two values $\pm c$ with probability $1/2$, then (1) is valid for those d for which $\alpha_d < 1$, and does not require the smallness of c. Indeed, in this case always $\lambda < 1$, i.e.,

$$M\exp\{2F(x,t)\} \leq 2\Lambda^2,$$

because this is equivalent to the obvious inequality

$$\frac{1}{2}(e^{2c}+e^{-2c}) \leq 2\left(\frac{e^c+e^{-c}}{2}\right)^2.$$

If the $F(x,t)$ have Gaussian distribution $N(0,\sigma)$, then (1) is valid for small enough σ.

Put

$$h(x,t) = \frac{\exp\{F(x,t)\}-\Lambda}{\Lambda}$$

and introduce the series

$$\varphi(x,k) = \sum_{r\geq 1}\sum_{k\leq k_1<\ldots<k_r}\sum_{z_1,\ldots,z_r} q^{(k_1-k)}(z_1-x)q^{(k_2-k_1)}(z_2-z_1)\ldots$$
$$\ldots q^{(k_r-k_{r-1})}(z_r-z_{r-1})h(z_1,k_1)h(z_2,k_2)\ldots h(z_r,k_r),$$
$$\psi(y,n) = \sum_{r\geq 1}\sum_{k_1<\ldots<k_r\leq n}\sum_{z_1,\ldots,z_r} q^{(k_2-k_1)}(z_2-z_1)\ldots$$
$$\ldots q^{(k_r-k_{r-1})}(z_r-z_{r-1})q^{(n-k_r)}(y-z_r)h(z_1,k_1)\ldots h(z_r,k_r).$$

It is clear that $\varphi(x,t)$ and $\psi(y,t)$ constitute stationary (with respect to space-time translations) random fields, i.e. $\varphi(x,k) = T^{x,k}\varphi(0,0)$ and $\psi(y,n) = T^{y,n}\psi(0,0)$. Also they are transformed into each other by reversal of time in random walks. This implies, in particular, that the distributions of $\varphi(x,t)$ and $\psi(x,t)$ coincide.

Below we prove the following theorems.

THEOREM 1. *If* (1) *is valid then the series giving* $\varphi(x,k)$ *and* $\psi(y,n)$ *converge in the space* $L^2(\Phi,Q)$.

THEOREM 2. *If* (1) *is valid and* $\|y-x\| \leq A\sqrt{n-k}$ *where* A *is any constant, then the partition function* $Z^{y,n}_{x,k}$ *has the representation*

$$Z^{y,n}_{x,k} = \Lambda^{n-k+1}q^{(n-k)}(y-x)[(1+\varphi(x,k))(1+\psi(y,n))+\delta^{(y,n)}_{(x,k)}],$$

where $M\delta^{y,n}_{x,k} = 0$ *and* $M(\delta^{y,n}_{x,k})^2 \to 0$ *as* $n\to\infty$, x,k *remain fixed and* y *satisfies the above-mentioned restriction.*

P r o o f o f T h e o r e m 1. It is clear that φ and ψ are represented as sums of orthogonal vectors in the space $L^2(\Phi, Q)$. Therefore

$$\begin{aligned} M\varphi^2(x,k) &= \sum_{r\geq 1} \lambda^r \sum_{k<k_1<\ldots<k_r} \sum_{z_1,\ldots,z_r} (q^{(k_1-k)}(z_1-x))^2 \\ &\quad \times (q^{(k_2-k_1)}(z_2-z_1))^2 \ldots (q^{(k_r-k_{r-1})}(z_r-z_{r-1}))^2 \\ &= \sum_{r\geq 1} (\lambda\alpha_d)^r < \infty. \end{aligned}$$

The same is true for $\psi(x,t)$. We also have $M\varphi(x,k) = M\psi(y,n) = 0$. ∎

Theorem 2 is proven in Appendix 1.

THEOREM 3. *If* (1) *holds then* $1+\varphi(x,t) > 0$ *and* $1+\psi(y,t) > 0$ *for* Q*-a.e.* F.

P r o o f. We already showed that $M\varphi(x,t) = M\psi(x,t) = 0$, $M\varphi^2(x,t) > 0$ and $M\psi^2(x,t) > 0$. It is enough to consider $\varphi(x,k)$ since $\varphi(x,k)$ and $\psi(y,n)$ have the same distribution. By Theorem 2,

$$\frac{Z^{y,n}_{x,k}}{\Lambda^{n-k+1}} - q^{(n-k)}(y-x)[(1+\varphi(x,k))(1+\psi(y,n))] = \delta^{(y,n)}_{(x,k)} q^{(n-k)}(y-x).$$

Take a continuous non-negative function f with compact support on $\mathbb{R}^d$, and write

$$\begin{aligned} &\sum_y \frac{Z^{y,n}_{x,k}}{\Lambda^{n-k+1}} f\left(\frac{x-y}{\sqrt{n-k}}\sqrt{d}\right) \\ &\qquad = (1+\varphi(x,k)) \sum_y q^{(n-k)}(y-x) f\left(\frac{y-x}{\sqrt{n-k}}\sqrt{d}\right)(1+\psi(y,n)) \\ &\qquad\quad + \sum_y q^{(n-k)}(y-x) f\left(\frac{y-x}{\sqrt{n-k}}\sqrt{d}\right) \delta^{(y,n)}_{(x,k)}. \end{aligned}$$

Theorem 2 immediately implies that the last term tends to zero in $L^2(\Phi, Q)$ for any fixed x, k and $n \to \infty$. Since $M\psi(y,n) = 0$ the sum

$$\sum_y q^{(n-k)}(y-x) f\left(\frac{y-x}{\sqrt{n-k}}\sqrt{d}\right)(1+\psi(y,n))$$

converges in $L^2(\Phi, Q)$ to $C = \int e^{-\|z\|^2/2} f(z)\, dz/(2\pi)^{d/2} > 0$. Thus

$$\operatorname*{l.i.m.}_{n\to\infty} \frac{1}{C} \sum_y \frac{Z^{y,n}_{x,k}}{\Lambda^{n-k+1}} f\left(\frac{y-x}{\sqrt{n-k}}\sqrt{d}\right) = 1 + \varphi(x,k).$$

Now we can use the obvious inequality

$$Z_{x,k-2}^{y,n} \geq \sum_{\langle x,x'\rangle} \left(\frac{1}{2d}\right)^2 e^{F(x,k-2)+F(x',k-1)} Z_{x,k}^{y,n} = g(x,k-2) Z_{x,k}^{y,n},$$

where the last expression gives also the definition of $g(x,k-2)$ which is positive a.e., and the sum is taken over x' such that $\|x-x'\|=1$. We use the notation $\langle x,x'\rangle$ for the nearest neighbors on the lattice. Thus we have

$$1+\varphi(x,k-2) \geq g(x,k-2)(1+\varphi(x,k)). \tag{2}$$

Assume that $1+\varphi(x,k-2)=0$ with positive probability. Take x and consider the set $\mathcal{H}^+$ of those numbers $2k$ such that $1+\varphi(x,2k)>0$. It follows from (2) that if $2k \in \mathcal{H}^+$ then $2k-2 \in \mathcal{H}^+$. Therefore $\mathcal{H}^+ = 2\mathbb{Z}^1$ for a.e. F. The ergodicity of $T^{0,2}$ implies that $Q(\{F : 1+\varphi(x,k)=0\})=0$. ∎

Let the conditions of Theorem 2 be valid. As in the proof of Theorem 3 take a continuous function f on $\mathbb{R}^d$ with compact support. Using Theorem 2 we can write

$$\begin{aligned}(3)\quad & \sum_y f\left(\frac{y-x}{\sqrt{n-k}}\sqrt{d}\right)\frac{Z_{x,k}^{y,n}}{Z_{x,k}^n} \\ &= \frac{(1+\varphi(x,k))\Lambda^{n-k+1}}{Z_{x,k}^n} \\ &\quad \times \left[\sum_y f\left(\frac{y-x}{\sqrt{n-k}}\sqrt{d}\right) q^{(n-k)}(y-x)(1+\psi(y,n))\right. \\ &\quad \left. + \sum_y f\left(\frac{y-x}{\sqrt{n-k}}\sqrt{d}\right)\delta_{(x,k)}^{(y,n)} q^{(n-k)}(y-x)\right].\end{aligned}$$

Our estimations during the proof of Theorem 2 in the Appendix give

$$\underset{n\to\infty}{\text{l.i.m.}} \frac{Z_{x,k}^n}{\Lambda^{n-k+1}} = 1+\varphi(x,k).$$

Also the last sum in (3) tends to zero in $L^2(\Phi,Q)$ as $n\to\infty$. Therefore, we have the following theorem.

THEOREM 4.

$$\underset{n\to\infty}{\text{l.i.m.}} \frac{1}{Z_{x,k}^n}\sum_y f\left(\frac{y-x}{\sqrt{n-k}}\sqrt{d}\right) Z_{x,k}^{y,n} = \int f(z) e^{-\|z\|^2/2}\frac{dz}{(2\pi)^{d/2}}.$$

This theorem shows in what sense the normalized displacement $(\omega(n)-\omega(k))\sqrt{d}/\sqrt{n-k}$ has the limiting Gaussian distribution. Its variance is the same as for the usual random walk.

Appendix

Proof of Theorem 2. We have

$$\begin{aligned}
Z_{x,k}^{y,n} &= \sum_{\omega_{x,k}^{y,n}} \exp\Big\{\sum_{t=k}^{n} F(t,\omega(t))\Big\}\frac{1}{(2d)^{n-k}}\\
&= \sum_{\omega_{x,k}^{y,n}} \prod_{t=k}^{n} (\Lambda + \exp\{F(t,\omega(t))\} - \Lambda)\frac{1}{(2d)^{n-k}}\\
&= \Lambda^{n-k+1}\sum_{\omega_{x,k}^{y,n}} \prod_{t=k}^{n} (1 + h(\omega(t),t))\frac{1}{(2d)^{n-k}}\\
&= \Lambda^{n-k+1}\Big[q^{(n-k)}(y-x) + \sum_{r\geq 1}\sum_{k\leq k_1<\ldots<k_r<n}\sum_{z_1,\ldots,z_r} q^{(k_1-k)}(z_1-x)\\
&\quad\times q^{(k_2-k_1)}(z_2-z_1)\ldots q^{(k_r-k_{r-1})}(z_r-z_{r-1})\\
&\quad\times q^{(n-k_r)}(y-z_r)h(z_1,k_1)\ldots h(z_r,k_r)h(y,n)\Big].
\end{aligned}$$

In what follows we only deal with the finite sum

$$\begin{aligned}
\widetilde{Z}_{x,k}^{y,n} = \sum_{r\geq 1}\sum_{k\leq k_1<\ldots<k_r\leq n}\sum_{z_1,\ldots,z_r} q^{(k_1-k)}(z_1-x)q^{(k_2-k_1)}(z_2-z_1)\ldots\\
\ldots q^{(k_r-k_{r-1})}(z_r-z_{r-1})q^{(n-k_r)}(y-z_r)h(z_1,k_1)\ldots h(z_r,k_r).
\end{aligned}$$

It is clear that $M\widetilde{Z}_{x,k}^{y,n}=0$ and

$$\begin{aligned}
M(\widetilde{Z}_{x,k}^{y,n})^2 = \sum_{r\geq 1}\lambda^r \sum_{k\leq k_1<\ldots<k_r\leq n}\sum_{z_1,\ldots,z_r} (q^{(k_1-k)}(z_1-x))^2\\
\times (q^{(k_2-k_1)}(z_2-z_1))^2\ldots(q^{(n-k_r)}(y-z_r))^2.
\end{aligned}$$

Fix some constant B whose value will be chosen later and consider

$$\begin{aligned}
\widetilde{Z}_{x,k}^{y,n}(1) = \sum_{r\leq B\ln n}\sum_{k\leq k_1<\ldots<k_r\leq n}\sum_{z_1,\ldots,z_r} q^{(k_1)}(z_1-x)\\
\times q^{(k_2-k_1)}(z_2-z_1)\ldots q^{(k_r-k_{r-1})}(z_r-z_{r-1})\\
\times q^{(n-k_r)}(y-z_r)h(z_1,k_1)\ldots h(z_r,k_r).
\end{aligned}$$

Let $\widetilde{Z}_{x,k}^{y,n}(2)$ be a similar sum where $r>B\ln n$. Then the trivial estimation gives

$$M(\widetilde{Z}_{x,k}^{y,n}(2))^2 \leq \sum_{r>B\ln n}(\lambda\alpha_d)^r = \frac{(\lambda\alpha_d)^{B\ln n}}{1-\lambda\alpha_d}.$$

Take B so large that

$$\frac{(\lambda\alpha_d)^{B\ln n}}{1-\lambda\alpha_d} \leq \frac{1}{n^{2d}} \qquad \text{for all large enough } n.$$

We can write

$$\frac{Z_{x,k}^{y,n}}{\Lambda^{n-k+1}} = q^{(n-k)}(y-x)(1+\widetilde{Z}_{x,k}^{y,n}(1)+\widetilde{Z}_{x,k}^{y,n}(2)).$$

From our estimations it follows that

(i) for all y with $\|y-x\| \leq A\sqrt{n-k}$ the ratio $\widetilde{Z}_{x,k}^{y,n}(2)/q^{(n-k)}(y-x)$ tends to zero in $L^2(\Phi, Q)$ uniformly in y;

(ii) for any continuous function f with compact support, the sum

$$\sum_y f\left(\frac{y-x}{\sqrt{(n-k)/d}}\right)\widetilde{Z}_{x,k}^{y,n}(2)$$

converges to zero in $L^2(\Phi, Q)$.

Thus it remains to study $\widetilde{Z}_{x,k}^{y,n}(1)$ assuming $\|y-x\| \leq A\sqrt{n-k}$. Let us call an interval (k_{j-1}, k_j) *large* if $k_j - k_{j-1} \geq n^\beta$ for some β with $1/2 < \beta < 1$. Here $k_0 = k$, $k_{r+1} = n$. If $r \leq B\ln n$ then at least one large interval in the sequence $(0, k_1, k_2, \ldots, k_r, n)$ is present. We shall show that the main contribution to $\widetilde{Z}_{x,k}^{y,n}(1)$ comes from r-tuples $(k_1, k_2, \ldots, k_r)$ with only one large interval. Write

$$\begin{aligned}
&\widetilde{Z}_{x,k}^{y,n}(1,1) \\
&= \sum_{\substack{0\leq r_1 \leq B\ln n \\ 0\leq r_2\leq B\ln n \\ 1\leq r=r_1+r_2\leq B\ln n}} \ \sum_{\substack{k\leq k_1<\ldots<k_r\leq n \\ (k_{r_1},k_{r_1+1}) \text{ is the unique} \\ \text{large interval}}} \ \sum_{z_1,\ldots,z_r} q^{(k_1)}(z_1-x) \\
&\quad \times q^{(k_2-k_1)}(z_2-z_1)\ldots q^{(k_{r_1+1}-k_{r_1})}(z_{r_1+1}-z_{r_1}) \\
&\quad \times q^{(k_{r_1+2}-k_{r_1+1})}(z_{r_1+2}-z_{r_1+1})\ldots q^{(k_r-k_{r-1})}(z_r-z_{r-1})q^{(n-k_r)}(y-z_r) \\
&\quad \times [h(z_1,k_1)\ldots h(z_{r_1},k_{r_1})]\cdot[h(z_{r_1+1},k_{r_1+1})\ldots h(z_r,k_r)].
\end{aligned}$$

We can write

$$\frac{\widetilde{Z}_{x,k}^{y,n}(1,1)}{q^{(n)}(y-x)} = (1+\varphi(x,k))(1+\psi(y,n)) - 1 + \delta_{(x,k)}^{(y,n)}(2).$$

The last formula also implies the definition of $\delta_{(x,k)}^{(y,n)}(2)$. Since we can restrict ourselves by summation over those $(z_1,\ldots,z_r)$ where $\|z_{r_1}-x\| \leq n^{2\beta}$, $\|z_{r_1+1}-y\| \leq n^{2\beta}$, the summation over all other z is exceedingly small. Thus $M(\delta_{x,k}^{y,n}(2))^2 \to 0$ as $n \to \infty$ uniformly over all y under consideration.

The rest of our argument is to show that the contribution of r-tuples where the number of large intervals is greater than 1 is relatively small. Again we write down the square of the norm of the corresponding sum:

$$S_{x,k}^{y,n} = \alpha_d \sum_{r\geq 1} (\alpha_d \lambda)^r \sum_{k\leq k_1<\ldots<k_r\leq n} \sum_{z_1,\ldots,z_r} p^{(k_1-k)}(z_1-x)$$
$$\times p^{(k_2-k_1)}(z_2-z_1)\ldots p^{(k_r-k_{r-1})}(z_2-z_1)\ldots$$
$$\ldots p^{(k_r-k_{r-1})}(z_r-z_{r-1})p^{(n-k_r)}(y-z_r),$$

where $p^{(i)}(z) = (q^{(i)}(z))^2/\alpha_d$. The last double sum can again be considered as the probability that the sum $\vec{\eta}_1+\ldots+\vec{\eta}_r$ takes the values $y-x$, $n-k$, where $\vec{\eta}_j = (z_j - z_{j-1}, k_j - k_{j-1})$. It is easy to show that the distribution of the time component of η_j decays as $\text{const}/t^{d/2}$. Direct probabilistic arguments show that the probability to have at least two values of j for which the value of the "time" component is greater than n^β decays as $1/n^{(\beta+1)d}$. This shows that the contribution of terms with two large increments $(k_j - k_{j-1})$ to $S_{x,k}^{y,n}(1)$ is small in $L^2(\Phi, Q)$ compared with the norm of $q^{(n-k)}(y-x)$. We omit the details.

I thank K. M. Khanin and Yu. I. Kifer for useful discussions.

The financial support from NSF (grant DMS-9404437) and from the Russian Foundation of Fundamental Research (grant N93-01-16090) are highly appreciated.

References

[1] E. Bolthausen, *A note on the diffusion of directed polymers in a random environment*, Comm. Math. Phys. 123 (1989), 529–534.

[2] J. G. Conlon and P. A. Olson, *A Brownian motion version of directed polymer problem*, preprint, University of Michigan, 1994.

[3] J. Z. Imbrie and T. Spencer, *Diffusion of directed polymers in a random environment*, J. Statist. Phys. 52 (1988), 609–626.

MATHEMATICS DEPARTMENT
PRINCETON UNIVERSITY
PRINCETON, NEW JERSEY 08540
U.S.A.

LANDAU INSTITUTE OF THEORETICAL PHYSICS
MOSCOW, RUSSIA

Received 5 December 1994;
in revised form 30 January 1995

Comments

This article does not require comments.

A RANDOM WALK WITH RANDOM POTENTIAL*

YA. G. SINAI†

(*Translated by E. E. Dyakonova*)

Abstract. Recurrence properties are investigated for a model of one-dimensional random walk with random potential arising in polymer physics [1]. Two theorems giving the distinctive localization of probability distributions of the walk are presented.

Key words. random walks with random potential, statistical weights, statistical sums, the distribution localization

A beautiful model of one-dimensional random walk was offered in [1], [2] for some physics polymer problems and the question about its properties such as recurrence was posed. The problem is to study the trajectories of a simple random walk $\omega^{(n)} = \{\omega(k),\ 0 \leqq k \leqq n\}$, $\omega(0) = 0$, $\omega(k+1) - \omega(k) = \pm 1$ supplied with a statistical weight

$$\pi_n\left(\omega^{(n)}\right) = \exp\left\{\varepsilon \sum b(k)U\left(\omega(k)\right)\right\}p(\omega^{(n)}) = \frac{1}{2^n}\exp\left\{\varepsilon \sum_{k=0}^{n} b(k)U\left(\omega(k)\right)\right\}.$$

Here $b(k)$ is a sequence of independent random variables taking on values ± 1 with probability $\frac{1}{2}$; the function $U(x)$ takes on three values: $U(x) = -1$ for $x < 0$ and $U(0) = 0$, $U(x) = 1$ for $x > 0$; ε is a nonzero parameter. We introduce the statistical sum $Z^{(0,n)} = \sum_{\omega^{(n)}} \pi_n(\omega^{(n)})$, where

*Received by the editors October 30, 1992.

†Landau Institute of Theoretical Physics of the Russian Academy of Sciences, Kosygin str. 2, Moscow, 117940, Russia.

Y.G. Sinai, *Selecta II: Probability Theory, Statistical Mechanics, Mathematical Physics and Mathematical Fluid Dynamics*, DOI 10.1007/978-1-4419-6205-8_4,

$p_n(\omega^{(n)}) = (Z^{(0,n)})^{-1}\pi_n(\omega^{(n)})$ and the corresponding probability distribution P_n. Of course, P_n depends on the properties of the sequence $\{b(k)\}$ and the problem is to investigate the asymptotic behavior (as $n \to \infty$) of the distributions of the random variables $\omega(k)$ for typical sequences $\{b(k)\}$. The main result of this note are two theorems describing the distinctive localization of these distributions for $\varepsilon = 0$. We observe that [1] solved the case of the periodic sequence $\{b(k)\} = \{+1; -1; +1, -1, \ldots\}$ explicitly by means of the transfer-matrix and also found the localization for $\varepsilon = 0$.

It is convenient to assume that $\{b(k),\ k \geqq 0\}$ is a part of a two-sided infinite sequence $b = \{b(k),\ -\infty < k < \infty\}$. We introduce the Bernoulli probability distribution P on the natural σ-algebra of subsets of the space B of such sequences. The distribution P is invariant under translation T. Recall that $Tb = b'$, where $b'(k) = b(k+1)$.

THEOREM 1. *There exist an integer-valued random variable $\nu(b) \geqq 0$, random variables $\nu_n(b)$ depending on the random variables $b(k)$, $0 \leqq k \leqq n$, and, for almost all b, a number $n_0(b)$ such that, for all $n \geqq n_0(b)$,*

a$_1$) $\nu_n(b) = \nu(T^n b)$;

a$_2$) $P_n(\omega(n) = s) \leqq \text{const}\ \exp\{-\delta|s|\}$ *for* $|s| \geqq \nu(T^n b) = \nu_n(b)$.

Here const *is some absolute constant and $\delta = \delta(\varepsilon) > 0$.*

Proof. First we note that, for almost all b, there is an $m'(b)$ such that for all $k > m'(b)$

$$\left|b(-1) + \cdots + b(-k)\right| \leqq k^{2/3}.$$

In what follows we denote by $Z_0^{(m_1,m_2)}(b)$ the statistical sum

$$Z_0^{(m_1,m_2)}(b) = \sum_{\omega}^{(m_1,m_2)} \exp\left\{\varepsilon \sum_{k=m_1}^{m_2} b(k)U\big(\omega(k)\big)\right\},$$

where $\sum_{\omega}^{(m_1,m_2)}$ stands for summation over all the trajectories of simple random walk within the time interval $m_1 \leqq k \leqq m_2$ such that $\omega(m_1) = \omega(m_2) = 0$. Of course, both m_1, m_2 should be either even or odd.

LEMMA 1. *There exist $\delta_1(\varepsilon) > 0$ and $m''(b) > 0$ such that $Z_0^{(-m,0)}(b) \geqq \exp\{\delta_1(\varepsilon)m\}$ for all $m \geqq m''(b)$.*

We prove the lemma a little bit later.

Let $m' = m'(n;b) < n$ be such that for all $m > m'$

$$\left|b(n) + b(n-1) + \cdots + b(n-m)\right| \leqq m^{2/3}.$$

If there is no $m'(n,b) < n$ satisfying the previous condition, we put $m'(n;b) = n$. Let $m'' = m''(n;b) < n$ be such that $Z_0^{(n-m,n)} \geqq \exp\{\delta_1(\varepsilon)m\}$ for all $m'' \leqq m \leqq n$. We put $m'' = n$ if the last condition does not hold.

We set now $\nu(b) = \max(m'(b), m''(b))$ and $\nu_n(b) = \max(m'(n;b), m''(b;n))$.

LEMMA 2. *There exists an $n_0(b)$ such that $\nu_n(b) = \nu(T^n b)$ for all $n \geqq n_0(b)$.*

In fact, there exists an $n_0(b)$ such that $m'(n;b) = m(T^n b)$ and $m''(n;b) = m^n(T^n b)$ for all $n > n_0(b)$. For m' this statement is elementary and for m'' it follows from Theorem 1 below.

Assuming that Lemmas 1, 2 have been established, we prove Theorem 1.

Let $n \geqq n_0(b)$ and $|s| \geqq \nu(T^n b)$. By definition

$$p_n(s) = P_n(\omega(n) = s)\frac{1}{Z^{(0,n)}} \sum^{(s)} \pi_n(\omega^{(n)}).$$

Here $\sum^{(s)}$ stands for summation over all the trajectories $\omega^{(n)}$ such that $\omega(n) = s$. For every trajectory in $\sum^{(s)}$ we denote by t the latest time $t = t(\omega^{(n)}) < n$ such that $\omega(t) = 0$. It is clear that $n - t \geqq |s|$. Then

$$p_n(s) = \sum_{t=0}^{n-s} \frac{Z_t^{(s)}}{Z^{(0,n)}},$$

where $Z_t^{(s)}$ is the sum over the trajectories $\omega^{(n)}$ with a given t.

Evidently, n and s may be assumed to be even. In this case $Z^{(0,t)} \geqq Z_0^{(0,t)} Z_0^{(t,n)}$, and $Z_t^{(s)} = Z_0^{(0,t)} Z_1^{(t,n)}$, where $Z_1^{(t,n)}$ is the corresponding sum over the trajectories of simple random walk that leave 0 at time t, hit point s at time n, and do not cross 0 within the time interval (t,n). For this reason $p_n(s) \leqq \sum_{t=0}^{n-s} Z_1^{(t,n)} (Z_0^{(t,n)})^{-1}$. It is clear that

$$Z_1^{(t,n)} = \exp\left\{\varepsilon \sum_{k=t+1}^{n} b(k) \operatorname{sign} s\right\} q^{(t,n)}(s),$$

where $q^{(t,n)}$ is the probability that the trajectories of simple random walk that leave 0 at time t, hit s at time n and do not cross 0 within the interval (t,n). The rough estimate $q^{(t,n)} \leqq \text{const}\,(n-t)^{-1/2}$ is sufficient for us. Now we can write

$$p_n(s) \leqq \sum_{t=0}^{n-s} \exp\left\{\varepsilon |n-t|^{2/3} + \text{const} + \frac{1}{2}\lg(n-t) - \delta_1(\varepsilon)|n-t|\right\}.$$

We may always assume that $n - t \geqq \text{const}$. Then the right-hand side of the last inequality does not exceed $\exp\{-\delta(\varepsilon)|s|\}$. The theorem is proved.

Proof of Lemma 1. We fix a number $\Delta = \Delta(\varepsilon)$, whose value will be specified below. It is sufficient to consider only m proportional to Δ, i.e., $m = r\Delta$. For such m we have

$$Z_0^{(-m,0)}(b) \geqq \prod_{j=1}^{r} Z_{00}^{(-j\Delta,-(j-1)\Delta)},$$

where $Z_{00}^{(-j\Delta,-(j-1)\Delta)}$ is the statistical sum over the trajectories such that $\omega(-j\Delta) = \omega(-(j-1)\Delta) = 0$, $\omega(m) = 0$, for $-j\Delta < m < -(j-1)\Delta$. Then

$$Z_{00}^{(-j\Delta,-(j-1)\Delta)} = \left(\exp\left\{\varepsilon \sum_{k=-j\Delta+1}^{-(j-1)\Delta-1} b(k)\right\} + \exp\left\{-\varepsilon \sum_{k=-j\Delta+1}^{-(j-1)\Delta-1} b(k)\right\}\right) q(\Delta),$$

where $q(\Delta) \sim \text{const}\,\Delta^{-3/2}$. Therefore,

$$\log Z_{00}^{(-j\Delta,-(j-1)\Delta)} = \left|\varepsilon \sum_{k=-j\Delta}^{-(j-1)\Delta} b(k)\right| - \frac{3}{2}\log\Delta + O(1).$$

It is easy to see that $\log Z_{00}^{(-j\Delta,-(j-1)\Delta)}$ constitute a sequence of independent (with respect to j) and identically distributed random variables and their logarithms have expectations of the form $\delta_2(\varepsilon)\sqrt{\Delta} - \frac{3}{2}\log\Delta + O(1)$ for some $\delta_2(\varepsilon) > 0$. For Δ large enough this quantity exceeds some $\delta_3(\varepsilon) > 0$. Now the lemma follows by the strong law of large numbers. Lemma 1 is proved.

Remark 1. From the proof of Lemma 1 some explicit estimates of the distribution of $\nu(b)$ can be obtained.

Remark 2. It is natural to treat $\nu(T^n b)$ as the radius of localization at time n. By the Birkhoff–Khinchin ergodic theorem, for almost all b, the frequency has a limit with which $\nu(T^n b)$ takes on any given value m.

Improving the reasoning and estimates somewhat used in the proving Theorem 1, one can obtain estimates of the distribution of random variables $\omega(k)$ for $0 < k < n$. The following theorem holds.

THEOREM 2. *There exist a random variable $\bar{\nu}(b)$, constants $\gamma > 0, \delta(\varepsilon) > 0$, and a number $n_1(b)$ such that*

$$P_n\left\{\omega(k) = s\right\} \leqq \exp\left\{-\delta(\varepsilon)|s|\right\} \qquad \textit{for} \quad |s| > \bar{\nu}(T^k b)$$

for almost all b, $n \geqq n_1(b)$ and $\lg^\gamma n \leqq k \leqq n - \lg^\gamma n$.

We omit the proof of the theorem.

It would be interesting to show that $\lim_{n\to\infty} P_n\big(\omega(k) = s\big)$ exists for any fixed k.

Acknowledgments. I am grateful to S. K. Nechaev, D. E. Mazel, K. M. Khanin, and H. Spohn for numerous discussions of the problem considered.

REFERENCES

[1] A. Yu. Grosberg, S. F. Izrailev, and S. K. Nechaev, J. Stat. Phys. (to appear).

[2] T. Garel, D. A. Huse, S. Leibler, and H. Orland, *Localization transition of random chains at interfaces*, Europhys. Lett., 8 (1989), pp. 9–13.

Comments

The model considered in this paper was proposed by S. K. Nechaev. He also formulated a hypothesis according to which this random walk becomes localized. The basic result of this paper is the proof of his hypothesis.

DISTRIBUTION OF SOME FUNCTIONALS OF THE INTEGRAL OF A RANDOM WALK*

Ya. G. Sinai

The asymptotic behavior of the probabilities associated with the first crossing of a straight line or parabola by the integral of a Brownian curve is studied.

1. FORMULATION OF THE PROBLEM AND MAIN RESULTS

The analysis of some problems relating to solutions of the Burgers equation with random initial data has given rise to some beautiful problems in probability theory that can be formulated rather simply but, so far as we know, have not hitherto been studied in detail. Some of them are investigated in the present paper. The connections with the Burgers equation will be discussed elsewhere.

We consider a sequence of independent random variables ξ_i taking the values ± 1 with probabilities ½, $i \geq 1$. We set $\eta_0 = 0$, $\eta_n = \xi_1 + \ldots + \xi_n$ for $n > 0$ and $\zeta_0 = 0$, $\zeta_n = \eta_1 + \ldots + \eta_n$ for $n > 0$. We introduce the random variable τ, which is equal to the least positive n for which $\eta_n = 0$. Clearly, τ takes only even values. We set $\zeta = \zeta_\tau$ and consider the generating function $\varphi(s, t)$ of the joint distribution of the random variables ζ and τ, i.e.,

$$\varphi(s, t) = \sum_{m>0, j>0} s^m t^{2j} p_{m,2j}, \quad \text{where} \quad p_{m,2j} = P\{\zeta = m, \ \tau = 2j\}, \ |s| \leq 1, \ |t| \leq 1.$$

Then $\varphi(s, t)$ corresponds to random-walk trajectories for which $\xi_1 = \eta_1 = 1$. Therefore $\varphi(1, 1) = ½$.

Lemma 1.

$$\varphi(s, t) = \frac{st^2}{4} \frac{1}{1 - \varphi(s, st)}.$$

Proof. The minimum possible value of ζ is 1 and corresponds to $\tau = 2$. For $\tau > 2$, we denote by k the number of returns to 1 of the η_n random-walk trajectory that occur before the first return to 0. Then $1 \leq k \leq (\tau - 2)/2$, and we can write

$$p_{m,2j} = \sum_{k=1}^{j-1} \sum_{\substack{(m_1, j_1), \ldots, (m_k, j_k) \\ j_1 + j_2 + \ldots + j_k = j-1 \\ (m_1 + 2j_1) + (m_2 + 2j_2) + \ldots + (m_k + 2j_k) + 1 = m}} \frac{1}{4} p_{m_1, 2j_1} p_{m_2, 2j_2} p_{m_k, 2j_k}.$$

Here, $2j_1, 2j_2, \ldots, 2j_k$ are the lengths of the successive k cycles corresponding to the returns to 1, and

$$\frac{1}{4} = \frac{1}{2} \cdot \frac{1}{2}$$

corresponds to the first and last steps of the random walk. Hence

$$\sum_{m>0, j>0} s^m t^{2j} p_{m,2j} = \frac{st^2}{4} + \sum_{k=1}^{\infty} \frac{st^2}{4} \left(\sum_{m,j} p_{m,2j} s^m (st)^{2j} \right)^k = \frac{st^2}{4} + \frac{st^2}{4} \sum_{k=1}^{\infty} (\varphi(s, st))^k = \frac{st^2}{4} \frac{1}{1 - \varphi(s, st)}.$$

The lemma is proved.

Lemma 1 makes it possible to obtain an "explicit" expression for $\varphi(s, t)$.

*Dedicated to the memory of Mikhail Konstantinovich Polivanov.

L. D. Landau Institute of Theoretical Physics, USSR Academy of Sciences. Translated from Teoreticheskaya i Matematicheskaya Fizika, Vol. 90, No. 3, pp. 323-353, March, 1992. Original article submitted August 22, 1991

Y.G. Sinai, *Selecta II: Probability Theory, Statistical Mechanics, Mathematical Physics and Mathematical Fluid Dynamics*, DOI 10.1007/978-1-4419-6205-8_5, © Springer Science+Business Media, LLC 2010

Lemma 2.

$$\varphi(s,t)=\cfrac{\frac{st^2}{4}}{1-\cfrac{\frac{s^3t^2}{4}}{1-\cfrac{\frac{s^5t^2}{4}}{1-\cfrac{\frac{s^7t^2}{4}}{1-\cdots}}}} \tag{1}$$

Proof. Suppose $\varphi_k(s,t)=\varphi(s,s^kt)$. Then $\varphi(s,t)=\varphi_0(s,t)$, and from Lemma 1

$$\varphi_k(s,t)=\frac{s^{2k+1}t^2}{4}\,\frac{1}{1-\varphi_{k+1}(s,t)}$$

Integrating the last relation, we obtain the proposition of the lemma.

We investigate the fraction (1) by the methods of the theory of continued fractions in Sec. 2. The main result of Sec. 2 is the proof of the following theorem.

THEOREM. 1. *Let*

$$s=\exp\left\{i\frac{\lambda}{n^{3/2}}\right\},\qquad t=\exp\left\{i\frac{\mu}{n}\right\}.$$

Then for fixed λ *and* μ *and* $n\to\infty$, *we have the asymptotic behavior*

$$\varphi\left(e^{i\frac{\lambda}{n^{3/2}}},e^{i\frac{\mu}{n}}\right)=\frac{1}{2}\exp\left\{\frac{1}{n^{1/2}}\left(f(\lambda,\mu)+o(1)\right)\right\},$$

where

$$f(\lambda,\mu)=\theta\int_0^\infty\frac{dv}{v^2}\left\{\exp\left[-2\int_v^\infty\left(2\sqrt{-2\mu i-2u\lambda i}+Y_u(\lambda,\mu)-\frac{1}{v}\right)du\right]=1\right\}.$$

and $Y_u(\lambda,\mu)$ *is a solution of the differential equation*

$$\frac{dY_u(\lambda,\mu)}{du}=-4\sqrt{-2\mu i-2u\lambda i}\,Y_u(\lambda,\mu)-Y_u^2(\lambda,\mu).$$

for which $Y_u(\lambda,\mu)\sim 1/\mu$ *as* $u\to\infty$. *The form of the constant factor* θ *will be established during the proof.*

We note that the asymptotic behavior of Y_u as $u\to 0$ is such that the expression in the curly brackets behaves as v^2 as $v\to 0$ and therefore the integrand in the curly brackets does not have a singularity in the neighborhood of $v=0$. In addition, the function f possesses the homogeneity property $\gamma^{1/2}f(\gamma^{3/2}\lambda,\gamma\mu)=f(\lambda,\mu)$, which, in fact, also follows from its definition and can be verified analytically independently. In Theorem 1 it is not assumed that n is an integer.

We now drop the assumption that $\zeta>0$. By virtue of the symmetry, the joint characteristic function of the random variables (ζ,τ) has the form $\varphi(e^{i\lambda},e^{i\mu})+\varphi(e^{-i\lambda},e^{i\mu})$. We consider the sequence of independent random variables $(\zeta^{(1)},\tau^{(1)})$, $(\zeta^{(2)},\tau^{(2)}),\ldots,(\zeta^{(k)},\tau^{(k)})$ each of which has the same distribution as the pair (ζ,τ).

THEOREM 2. *In the limit* $k\to\infty$, *there exists a joint limit distribution of normalized sums*

$$\frac{\zeta^{(1)}+\ldots+\zeta^{(k)}}{k^3},\ \frac{\tau^{(1)}+\ldots+\tau^{(k)}}{k^2}.$$

The joint characteristic function of this limit distribution has the form $\exp\{\frac{1}{2}(f(\lambda,\mu)+f(-\lambda,\mu))\}$.

Corollary 1. The distribution of the random variable ζ belongs to the region of attraction of the symmetric stable law with exponent $\alpha={}^1/_3$.

Also of interest is the conditional distribution of the random variable $\zeta/m^{3/2}$ calculated under the condition that $\tau=2m$.

Corollary 2. In the limit $m\to\infty$, this conditional distribution converges to a limit whose characteristic function has the form

$$C\int_{-\infty}^{\infty} e^{-it\mu}(f(\lambda,\mu)+f(-\lambda,\mu))d\mu,$$

where C is a constant.

Corollaries 1 and 2 are also proved in Sec. 2. It appears that the result of Corollary 2 has a bearing on the problem of the distribution of the integral of Brownian motion between two successive zeros, but this question requires further analysis. In Sec. 3, we study the asymptotic behavior of the probability $P_N^{(0)}=P\{\zeta_n\geq 0 \text{ for all } 0\leq n\leq N\}$ and prove the following theorem.

THEOREM 3. *There exist positive c_1 and c_2 such that*

$$c_1/N^{1/4}\leq P_N^{(0)}\leq c_2/N^{1/4}.$$

Further, in Sec. 4 we shall consider analogous questions for the integral of a Wiener process. Let $b(t)$ be a Wiener process, $0\leq t<\infty$, and $r<0$ be a fixed negative number.

THEOREM 4. *Let*

$$P_r(T)=P\left\{\int_0^t b(u)du\geq r \quad \text{for all} \quad 0\leq t\leq T\right\}.$$

Then

$$\frac{A_1(r)}{T^{1/4}}\leq P_T\leq\frac{A_2(r)}{T^{1/4}},$$

where $A_1(r)$ and $A_2(r)$ are positive constants.

Let $r<0$ be a fixed negative number, and a be an arbitrary number.

THEOREM 5. *Let*

$$P_{r,a}(T)=P\left\{\int_0^t b(u)du\geq r+a\cdot t \quad \text{for all} \quad t,\ 0\leq t\leq T\right\}.$$

Then there exist positive constants $A_1(r, a)$ and $A_2(r, a)$ such that

$$\frac{A_1(r,a)}{T^{1/4}}\leq P_{r,a}(T)\leq\frac{A_2(r,a)}{T^{1/4}}.$$

Let r and a be the same as in Theorem 4, $\sigma>0$. We introduce the probability

$$P_{r,a,\sigma}=P\left\{\int_0^t b(u)du>r+at-\sigma t^2 \quad \text{for all} \quad t\geq 0\right\}.$$

THEOREM 6. *There exist positive $B_1(r, a)$ and $B_2(r, a)$ such that for all $\sigma>0$*

$$B_1(r,a)\sigma^{1/2}\leq P_{r,a,\sigma}\leq B_2(r,a)\sigma^{1/2}.$$

In Theorems 4–6 there evidently exist corresponding asymptotic expressions and not merely inequalities.

I thank M. Aisenmann, H. Kesten and H. Spohn for discussing some problems associated with the material of this paper. In particular, suggestions made by G. Kesten were very important for obtaining the results of Sec. 3.

2. PROOF OF THEOREMS 1 AND 2 AND OF COROLLARIES 1 AND 2

Let $\varphi(s, t)$ be the continued fraction (1). We set

$$s=\exp\left\{i\frac{\lambda}{n^{3/2}}\right\},\qquad t=\exp\left\{i\frac{\mu}{n}\right\}.$$

Proof of Theorem 1. Note first that the theory of continued fractions is constructed for expressions of the type (see [1])

(2)

$$[a_1, a_2, a_3, \dots] = \cfrac{1}{a_1 + \cfrac{1}{a_2 + \cfrac{1}{a_3 + \cfrac{1}{\dots}}}}.$$

The infinite fraction (1) reduces to the following continued fraction of the type (2):

$$\varphi(s, t) = \cfrac{1}{\cfrac{4}{st^2} + \cfrac{1}{-\cfrac{1}{s^2} + \cfrac{1}{\cfrac{4}{s^3t^2} + \cfrac{1}{-\cfrac{1}{s^4} + \ddots}}}}. \tag{3}$$

In other words

$$a_{2k-1} = \frac{4}{t^2 s^{2k-1}}, \quad a_{2k} = -\frac{1}{s^{2k}}, \quad k=1, 2, \dots .$$

Let p_m/q_m be the m-th convergent for (3). Then the numbers p_m and q_m satisfy the chain of difference equations (see [1])

$$p_{m+1} = a_{m+1} p_m + p_{m-1}, \qquad q_{m+1} = a_{m+1} q_m + q_{m-1}$$

with initial conditions $p_0 = 0$, $p_1 = 1$, $q_0 = 1$, $q_1 = a_1 = 4t^{-2}s^{-1}$. One can say that if

$$u_{m+1} = a_{m+1} u_m + u_{m-1}, \tag{4}$$

then p_m and q_m are particular solutions of (4) with different initial data. We set $r_m = u_m/u_{m-1}$. Then from (4)

$$r_{m+1} = a_{m+1} + \frac{1}{r_m}. \tag{5}$$

Because of the form of the coefficients a_n, it is more convenient to go over to the expression of r_{m+1} in terms of r_{m-1}:

$$r_{m+1} = a_{m+1} + \frac{1}{a_m} - \frac{a_m^{-2}}{a_m^{-1} + r_{m-1}}. \tag{6}$$

In what follows, wishing to emphasize the dependence on λ and μ, we shall write $a_m(\lambda, \mu)$, $r_m(\lambda, \mu)$, etc., assuming that s and t are replaced by their expressions in terms of λ and μ. In addition, the solutions corresponding to $p_m(\lambda, \mu)$ and $q(\lambda, \mu)$ will be denoted, respectively, by

$$\bar{r}_m(\lambda, \mu), \bar{\bar{r}}_m(\lambda, \mu).$$

For our analysis, an important role is played by the equation

$$\varphi\left(e^{i\frac{\lambda}{n^{3/2}}}, e^{i\frac{\mu}{n}}\right) = \lim_{s\to\infty} \frac{p_s(\lambda, \mu)}{q_s(\lambda, \mu)} = \frac{p_3(\lambda, \mu)}{q_3(\lambda, \mu)} \prod_{m=2}^{m_0} \frac{\bar{r}_{2m+1}(\lambda, \mu)}{\bar{\bar{r}}_{2m+1}(\lambda, \mu)} \prod_{m=2}^{m_0} \frac{\bar{r}_{2m}(\lambda, \mu)}{\bar{\bar{r}}_{2m}(\lambda, \mu)} \times$$

$$\prod_{m>m_0} \frac{\bar{r}_{2m+1}(\lambda, \mu)}{\bar{\bar{r}}_{2m+1}(\lambda, \mu)} \prod_{m>m_0} \frac{\bar{r}_{2m}(\lambda, \mu)}{\bar{\bar{r}}_{2m}(\lambda, \mu)}. \tag{7}$$

In the last relation, m_0 may be arbitrary. We choose $m_0 \sim a\sqrt{n}$ as $n \to \infty$, where $a > 0$ is arbitrary, and we then go to the limit $a \to 0$ in the final expression.

The investigation of (7) breaks up into two parts. In the first part, we study the products

$$\prod_{m=2}^{m_0} \frac{\bar{r}_{2m+1}(\lambda, \mu)}{\bar{\bar{r}}_{2m+1}(\lambda, \mu)}, \quad \prod_{m=2}^{m_0} \frac{\bar{r}_{2m}(\lambda, \mu)}{\bar{\bar{r}}_{2m}(\lambda, \mu)}$$

and show that each of them is close, respectively, to

$$\prod_{m=2}^{m_0} \frac{\bar{r}_{2m+1}(0, 0)}{\bar{\bar{r}}_{2m+1}(0, 0)} \quad \text{and} \quad \prod_{m=2}^{m_0} \frac{\bar{r}_{2m}(0, 0)}{\bar{\bar{r}}_{2m}(0, 0)}.$$

At the same time,

$$\bar{r}_m(\lambda, \mu), \ \bar{\bar{r}}_m(\lambda, \mu)$$

approach each other and for $m=2m_0$ the differences

$$\bar{r}_m(\lambda, \mu)-\bar{\bar{r}}_m(\lambda, \mu)$$

become small and subsequently indeed tend to zero.

1. Analysis of $\prod_{m=2}^{m_0} \frac{\bar{r}_{2m+1}(\lambda, \mu)}{\bar{\bar{r}}_{2m+1}(\lambda, \mu)}, \quad \prod_{m=2}^{m_0} \frac{\bar{r}_{2m}(\lambda, \mu)}{\bar{\bar{r}}_{2m}(\lambda, \mu)}$.

We consider only the first product, the second is treated similarly.

For $\lambda=\mu=0$, the relations (6) take the form

$$r_{2m+1}(0,0)=3+\frac{1}{1-r_{2m-1}(0,0)}. \tag{8}$$

It is easy to show that the solutions of (8) in which we are interested admit the asymptotic expansion

$$r_{2m-1}(0,0)=2+\frac{1}{m}+\frac{\theta}{m^2}+O\left(\frac{1}{m^3}\right). \tag{9}$$

We denote by $\bar{\theta}, \bar{\bar{\theta}}$ the values of the constant θ corresponding to $\bar{r}_{2m-1}(0,0)$, $\bar{\bar{r}}_{2m-1}(0,0)$, respectively. For each of the solutions

$$\bar{r}_{2k-1}(\lambda, \mu), \ \bar{\bar{r}}_{2k-1}(\lambda, \mu),$$

we now write

$$r_{2k-1}(\lambda, \mu)=2+r_{2k-1}^{(1)}(0,0)+h_{2k-1}(\lambda, \mu),$$

where $r_{2k-1}^{(1)}(0, 0)$ satisfy the relations

$$r_{2k+1}^{(1)}=1-\frac{1}{1-r_{2k-1}^{(1)}}$$

Then for $h_{2m+1}(\lambda, \mu)$ we obtain the expression

$$h_{2m+1}(\lambda, \mu)=\delta_m(\lambda, \mu)+\frac{s^{4m}}{(1+r_{2m-1}^{(1)}+(1-t))^2}h_{2m-1}(\lambda, \mu)+h_{2m-1}^{(1)}(\lambda, \mu), \tag{10}$$

where, we recall

$$s=\exp\left\{i\frac{\lambda}{n^{3/2}}\right\}, \quad t=\exp\left\{i\frac{\mu}{n}\right\}$$

and

$$\delta_m(\lambda, \mu)=4(t^{-2}s^{-(2m+1)}-1)-(s^{2m}-1)-\frac{1}{(1-r_{2m-1}^{(1)})(1+r_{2m-1}^{(1)}+(1-s^{2m})h_{2m-1}(\lambda, \mu))},$$

and $h_{2m-1}^{(1)}(\lambda, \mu)$ contains terms that are small and of higher order in h. It is easy to show that for $|h_{2m-1}| \leq 1/2$ the estimate

$$|h_{2m-1}^{(1)}(\lambda, \mu)| \leqslant B_1 |h_{2m-1}(\lambda, \mu)|^2$$

holds. Here, B_1 is an absolute constant whose exact value is not of interest to us. In addition,

$$\delta_m(\lambda, \mu)=-\frac{8i\mu}{m}+O\left(\frac{1}{m^{3/2}}\right)$$

for the considered m and in the leading order does not depend on λ.

Suppose we have proved that

$$|h_{2m-1}(\lambda, \mu)| \leqslant A\frac{m}{n}$$

for some constant A. Then

$$|h_{2m+1}(\lambda,\mu)| \leqslant |\delta_m(\lambda,\mu)| + |h_{2m-1}(\lambda,\mu)| + B_1 A^2 \frac{m^2}{n^2}. \tag{11}$$

It can be assumed that

$$|\delta_m(\lambda,\mu)| \leqslant \frac{8(1+|\mu|)}{n}.$$

Then from (11)

$$|h_{2m+1}(\lambda,\mu)| \leqslant \frac{8(1+|\mu|)}{n} + \frac{Am}{n^2} + \frac{m^2}{n^2} A^2 B_1.$$

We choose $A = 10(1 + |\mu|)$ and then take a sufficiently small that for $m^2/n \leq a^2$ the right-hand side does not exceed $A(m+1)/n$. Then the required estimate of $h_{2m+1}^{(1)}(\lambda, \mu)$ will be obtained for all the considered m. Further, from (10)

$$h_{2m_0+1}(\lambda,\mu) = \sum_{m=2}^{m_0} \sum_{k=m}^{m_0} \frac{s^{4k}}{1 + r_{2k-1}^{(1)} + (1 - s^{2k})} (\delta_m(\lambda,\mu) + h_{2m-1}^{(1)}(\lambda,\mu)).$$

It follows from our estimates that for the considered m the term $h_{2m-1}^{(1)}(\lambda, \mu)$ is small compared with $\delta_m(\lambda, \mu)$. In addition

$$\frac{s^{4k}}{1 + r_{2k-1}^{(1)} + (1 - s^{2k})} = \frac{1}{1 + \dfrac{1}{k} + O\left(\dfrac{1}{n}\right)},$$

and therefore

$$\prod_{k=m}^{m_0} \frac{s^{4k}}{1 + r_{2k-1}^{(1)} + (1 - s^{2k})} = \frac{m}{m_0}\left(1 + O\left(\frac{1}{m}\right)\right).$$

This gives

$$h_{2m-1}(\lambda,\mu) = B_2 \frac{m}{n} + h_{2m-1}^{(2)}(\lambda,\mu), \tag{12}$$

where B_2 is a constant and

$$|h_{2m-1}^{(2)}(\lambda,\mu)| \leqslant \frac{m^2}{n^2} \,\text{const}.$$

It is also important that B_2 depends on μ but in the same way for $\bar{r}_{2m-1}^{(1)}$ and $\bar{\bar{r}}_{2m-1}^{(1)}$. We can now write

$$\prod_{m=3}^{m_0} \frac{\bar{r}_{2m-1}(\lambda,\mu)}{\bar{\bar{r}}_{2m-1}(\lambda,\mu)} = \prod_{m=3}^{m_0} \frac{\bar{r}_{2m-1}(0,0)}{\bar{\bar{r}}_{2m-1}(0,0)} \cdot \prod_{m=3}^{m_0} \frac{\bar{r}_{2m-1}(\lambda,\mu)}{\bar{r}_{2m-1}(0,0)} \left(\prod_{m=3}^{m_0} \frac{\bar{\bar{r}}_{2m-1}(\lambda,\mu)}{\bar{\bar{r}}_{2m-1}(0,0)}\right)^{-1} \tag{13}$$

and

$$\prod_{m=3}^{m_0} \frac{\bar{r}_{2m-1}(\lambda,\mu)}{\bar{r}_{2m-1}(0,0)} = \prod_{m=3}^{m_0}\left(1 + \frac{\bar{r}_{2m-1}(\lambda,\mu) - \bar{r}_{2m-1}(0,0)}{\bar{r}_{2m-1}(0,0)}\right) =$$

$$\prod_{m=3}^{m_0}\left(1 + \frac{\bar{h}_{2m-1}(\lambda,\mu)}{\bar{r}_{2m-1}(0,0)}\right) = \exp\left\{\sum_{m=3}^{m_0} \ln\left(1 + \frac{\bar{h}_{2m-1}(\lambda,\mu)}{\bar{r}_{2m-1}(0,0)}\right)\right\} = \exp\left\{\sum_{m=3}^{m_0} \frac{\bar{h}_{2m-1}(\lambda,\mu)}{\bar{r}_{2m-1}(0,0)} + \bar{\rho}_0(\lambda,\mu)\right\}.$$

We recall that $\bar{h}_{2m-1}(\lambda, \mu)$ is the sequence (10) corresponding to $\bar{r}_{2m-1}(0, 0)$; the term $\bar{\rho}_0(\lambda, \mu)$ is the error, for which we obtain from (12) the estimate

$$|\bar{\rho}_0(\lambda,\mu)| \leqslant \frac{\text{const}\, a^3}{\sqrt{n}}.$$

Using (12) and (9), we obtain

$$\sum_{m=3}^{m_0} \frac{\bar{h}_{2m-1}(\lambda, \mu)}{\bar{r}_{2m-1}(0,0)} = B_2 \sum_{m=3}^{m_0} \frac{\frac{m}{n} + \bar{h}^{(2)}_{2m-1}(\lambda, \mu)}{2 + \frac{1}{m} + \frac{\theta}{m^2} + O\left(\frac{1}{m^3}\right)} = B_2 \sum_{m=3}^{m_0} \frac{\frac{m}{n}}{2 + \frac{1}{m}} + \bar{\rho}_1(\lambda, \mu), \tag{14}$$

where $\bar{\rho}_1(\lambda, \mu)$ is one further error, satisfying $|\bar{\rho}_1(\lambda, \mu)| \leq a/\sqrt{n^{1/2}}$. Exactly the same argument can be made for the product

$$\prod_{m=3}^{m_0} \frac{\bar{\bar{r}}_{2m-1}(\lambda, \mu)}{\bar{\bar{r}}_{2m-1}(0,0)}.$$

Then in (13) the first terms from (14) cancel. We obtain

$$\prod_{m=3}^{m_0} \frac{\bar{r}_{2m-1}(\lambda, \mu)}{\bar{\bar{r}}_{2m-1}(\lambda, \mu)} = \prod_{m=3}^{m_0} \frac{\bar{r}_{2m-1}(0,0)}{\bar{\bar{r}}_{2m-1}(0,0)} \exp\{\rho_2(\lambda, \mu)\}, \tag{15}$$

where

$$|\rho_2(\lambda, \mu)| \leq \frac{\text{const}\, a}{\sqrt{n}}.$$

A similar relation is derived in the same manner for even indices:

$$\prod_{m=2}^{m_0} \frac{\bar{r}_{2m}(\lambda, \mu)}{\bar{\bar{r}}_{2m}(\lambda, \mu)} = \prod_{m=2}^{m_0} \frac{\bar{r}_{2m}(0,0)}{\bar{\bar{r}}_{2m}(0,0)} \exp\{\rho_3(\lambda, \mu)\},$$

where

$$|\rho_3(\lambda, \mu)| \leq \frac{\text{const}\, a}{\sqrt{n}}.$$

Note that

$$\frac{1}{2} = \varphi(1,1) = \frac{p_3(0,0)}{q_3(0,0)} \cdot \prod_{m=3}^{m_0} \frac{\bar{r}_{2m-1}(0,0)}{\bar{\bar{r}}_{2m-1}(0,0)} \cdot \prod_{m=2}^{m_0} \frac{\bar{r}_{2m}(0,0)}{\bar{\bar{r}}_{2m}(0,0)} \prod_{m>m_0} \frac{\bar{r}_{2m-1}(0,0)}{\bar{\bar{r}}_{2m-1}(0,0)} \cdot \prod_{m>m_0} \frac{\bar{r}_{2m}(0,0)}{\bar{\bar{r}}_{2m}(0,0)}$$

It is readily deduced from (9) that

$$\prod_{m>m_0} \frac{\bar{r}_{2m-1}(0,0)}{\bar{\bar{r}}_{2m-1}(0,0)} = \exp\left\{\frac{1}{2}\, \frac{\theta - \bar{\bar{\theta}}}{m_0} + O\left(\frac{1}{m_0^2}\right)\right\}.$$

Using (5), we can write

$$\bar{r}_{2m}(0,0) = -1 + \frac{1}{2 + \frac{1}{m} + \frac{\theta}{m^2} + O\left(\frac{1}{m^3}\right)} = -\frac{1}{2} - \frac{1}{4m} - \frac{\theta}{4m^2} + \frac{1}{4m^2} + O\left(\frac{1}{m^3}\right).$$

$$\bar{\bar{r}}_{2m}(0,0) = -1 + \frac{1}{2 + \frac{1}{m} + \frac{\bar{\bar{\theta}}}{m^2} + O\left(\frac{1}{m^3}\right)} = -\frac{1}{2} - \frac{1}{4m} - \frac{\bar{\bar{\theta}}}{4m^2} + \frac{1}{4m^2} + O\left(\frac{1}{m^3}\right).$$

Hence

$$\prod_{m>m_0} \frac{\bar{r}_{2m}(0,0)}{\bar{\bar{r}}_{2m}(0,0)} = \exp\left\{\frac{1}{2}\, \frac{\theta - \bar{\bar{\theta}}}{m_0} + O\left(\frac{1}{m^2}\right)\right\}.$$

Therefore

$$\frac{p_3(\lambda, \mu)}{q_3(\lambda, \mu)} \cdot \prod_{m=3}^{m_0} \frac{\bar{r}_{2m-1}(\lambda, \mu)}{\bar{\bar{r}}_{2m-1}(\lambda, \mu)} \cdot \prod_{m=2}^{m_0} \frac{\bar{r}_{2m}(\lambda, \mu)}{\bar{\bar{r}}_{2m}(\lambda, \mu)} = \frac{p_3(\lambda, \mu)}{q_3(\lambda, \mu)} \prod_{m=3}^{m_0} \frac{\bar{r}_{2m-1}(0,0)}{\bar{\bar{r}}_{2m-1}(0,0)} \cdot \prod_{m=2}^{m_0} \frac{\bar{r}_{2m}(0,0)}{\bar{\bar{r}}_{2m}(0,0)} \cdot e^{\rho_4(\lambda, \mu)} =$$

$$\frac{1}{2} \exp\left\{-\frac{\theta - \bar{\bar{\theta}}}{a\sqrt{n}} + O\left(\frac{1}{m_0^2}\right) + \rho_5(\lambda, \mu)\right\}, \tag{16}$$

where

$$|\rho_5(\lambda,\mu)|\leqslant\frac{\text{const}\,a}{\sqrt{n}}.$$

2. Analysis of the products $\prod_{m>m_0}\frac{\bar r_{2m-1}(\lambda,\mu)}{\bar{\bar r}_{2m-1}(\lambda,\mu)}$, $\prod_{m>m_0}\frac{\bar r_{2m}(\lambda,\mu)}{\bar{\bar r}_{2m}(\lambda,\mu)}$.

We again investigate in detail only the first product, the second being treated similarly.

For such m the numbers $\bar r_{2m}(\lambda,\mu)$ and $r_{2m}(\lambda,\mu)$ are already close to each other, and the proximity is only increased with increasing m. Therefore, it is natural to write

$$\prod_{m>m_0}\frac{\bar r_{2m-1}(\lambda,\mu)}{\bar{\bar r}_{2m-1}(\lambda,\mu)}=\prod_{m>m_0}\left(1+\frac{\bar r_{2m-1}(\lambda,\mu)-\bar{\bar r}_{2m-1}(\lambda,\mu)}{\bar{\bar r}_{2m-1}(\lambda,\mu)}\right)\sim\exp\left\{\sum_{m>m_0}\frac{\bar r_{2m-1}(\lambda,\mu)-\bar{\bar r}_{2m-1}(\lambda,\mu)}{\bar{\bar r}_{2m-1}(\lambda,\mu)}\right\}. \tag{17}$$

In the last relation, the error of each term is the order of the square of the numerator.

Further, on the basis of (12)

$$\bar r_{2m_0-1}(\lambda,\mu)-\bar{\bar r}_{2m_0-1}(\lambda,\mu)=\frac{\theta-\bar\theta}{m_0^2}(1+O(a)). \tag{18}$$

It follows from (6) that

$$\bar r_{2m+1}(\lambda,\mu)-\bar{\bar r}_{2m+1}(\lambda,\mu)=\frac{(\bar r_{2m-1}(\lambda,\mu)-\bar{\bar r}_{2m-1}(\lambda,\mu))s^{4m}}{(\bar r_{2m-1}(\lambda,\mu)-s^{2m})(\bar{\bar r}_{2m-1}(\lambda,\mu)-s^{2m})}. \tag{19}$$

We denote by $x_m(\lambda,\mu)$ the stable fixed point of the mapping

$$x\to g_m(x)=a_{2m+1}+\frac{1}{a_{2m}}-\frac{a_{2m}^{-2}}{a_{2m}^{-1}+x}.$$

It is readily verified that

$$x_m(\lambda,\mu)=2t^{-2}s^{-2m-1}(1+\sqrt{1-t^2s^{2m+1}}).$$

For $m\gg n^{3/2}$, this point is close to 2, then increases slowly, and ultimately tends to infinity. For $m=m_0$, both

$$\bar r_{2m_0-1}(\lambda,\mu),\ \bar{\bar r}_{2m_0-1}(\lambda,\mu)$$

are near 2, and then, with increasing m, approach $x_m(\lambda,\mu)$. The main approach occurs at times $m=O(\sqrt{n})$, when

$$\bar r_{2m-1}(\lambda,\mu)-x_m(\lambda,\mu)=O\left(\frac{1}{\sqrt{n}}\right),\quad \bar{\bar r}_{2m-1}(\lambda,\mu)-x_m(\lambda,\mu)=O\left(\frac{1}{\sqrt{n}}\right).$$

Subsequently, both these differences become negligibly small, and the corresponding part of the sum on the right-hand side of (17) will also be negligibly small.

We now study the behavior of the differences

$$\bar r_{2m-1}(\lambda,\mu)-x_m(\lambda,\mu),\ \bar{\bar r}_{2m-1}(\lambda,\mu)-x_m(\lambda,\mu)$$

at times $O(\sqrt{n})$ when both differences have the order $O(1/n^{1/2})$. Below, by $r_{2m-1}(\lambda,\mu)$ we understand any of the sequences

$$\bar r_{2m-1}(\lambda,\mu),\ \bar{\bar r}_{2m-1}(\lambda,\mu).$$

We set

$$r_{2m+1}(\lambda,\mu)-x_m(\lambda,\mu)=\frac{1}{\sqrt{n}}Y^{(n)}_{\frac{m}{\sqrt{n}}}(\lambda,\mu).$$

Then from (6)

$$x_m(\lambda,\mu)+\frac{1}{\sqrt{n}}Y^{(n)}_{\frac{m}{\sqrt{n}}}(\lambda,\mu)=a_{2m+1}+\frac{1}{a_{2m}}-$$

$$\frac{a_{2m}^{-2}}{a_{2m}^{-1} + x_{m-1}(\lambda, \mu) + \frac{1}{\sqrt{n}} Y_{\frac{m-1}{\sqrt{n}}}^{(n)}(\lambda, \mu)} = a_{2m+1} + \frac{1}{a_{2m}} -$$

$$\frac{a_{2m}^{-2}}{a_{2m}^{-1} + x_m(\lambda, \mu) + (x_{m-1}(\lambda, \mu) - x_m(\lambda, \mu)) + \frac{1}{\sqrt{n}} Y_{\frac{m-1}{\sqrt{n}}}^{(n)}(\lambda, \mu)}. \tag{20}$$

We now note that if $|\lambda| + |\mu| \neq 0$ then

$$x_{m-1}(\lambda, \mu) - x_m(\lambda, \mu) = O\left(\frac{1}{n^{3/4}}\right)$$

and this quantity is small compared with

$$\frac{1}{\sqrt{n}} Y_{\frac{m-1}{\sqrt{n}}}^{(n)}(\lambda, \mu).$$

Further

$$x_m(\lambda, \mu) + a_{2m}^{-1} = 1 + 2\sqrt{-\frac{2\mu i}{n} - \frac{(2m+1)\lambda i}{n^{3/2}}} - \frac{2\mu i}{n} - \frac{(2m+1)\lambda i}{n^{3/2}} + O\left(\frac{1}{n^{3/2}}\right).$$

Therefore

$$\frac{a_{2m}^{-2}}{a_{2m}^{-1} + x_m(\lambda, \mu) + (x_{m-1}(\lambda, \mu) - x_m(\lambda, \mu)) + \frac{1}{\sqrt{n}} Y_{\frac{m-1}{\sqrt{n}}}^{(n)}(\lambda, \mu)} =$$

$$\frac{a_{2m}^{-2}}{(a_{2m}^{-1} + x_m(\lambda, \mu))}\left(1 - \frac{1}{\sqrt{n}} Y_{\frac{m-1}{\sqrt{n}}}^{(n)}(\lambda, \mu) \frac{1}{a_{2m}^{-1} + x_m(\lambda, \mu)} + \frac{1}{n}\left(Y_{\frac{m-1}{\sqrt{n}}}^{(n)}(\lambda, \mu)\right)^2 + O\left(\frac{1}{n^{3/2}}\right)\right).$$

Hence and from (19) we obtain

$$\frac{1}{\sqrt{n}}\left(Y_{\frac{m}{\sqrt{n}}}^{(n)} - Y_{\frac{m-1}{\sqrt{n}}}^{(n)}\right) = -4\sqrt{-2\mu i - \frac{2(m+1)\lambda i}{\sqrt{n}}} \times$$

$$\frac{Y_{m-1}^{(n)}(\lambda, \mu)}{n} - \frac{1}{n}\left(Y_{\frac{m-1}{\sqrt{n}}}^{(n)}(\lambda, \mu)\right)^2 + O\left(\frac{1}{n^{3/2}}\right). \tag{21}$$

Multiplying both sides by n, we see that (21) is a difference scheme with step $\Delta u = 1/n^{1/2}$ for the differential equation

$$\frac{dY_u(\lambda, \mu)}{du} = -4\sqrt{-2\mu i - 2u\lambda i}\, Y_u(\lambda, \mu) - Y_u^2(\lambda, \mu). \tag{22}$$

Since

$$\bar{r}_{2m_0+1}(0, 0) = 2 + \frac{1}{m_0} + O\left(\frac{1}{m_0^2}\right), \quad \bar{\bar{r}}_{2m_0+1}(0, 0) = 2 + \frac{1}{m_0} + O\left(\frac{1}{m_0^2}\right),$$

it follows that

$$\bar{r}_{2m_0+1}(\lambda, \mu), \quad \bar{\bar{r}}_{2m_0+1}(\lambda, \mu)$$

differ by $O(1/n)$, and therefore the solutions of (22) corresponding to them are identical. Since

$$x_{m_0}(\lambda, \mu) = 2 + 2\sqrt{-\frac{2\mu i + 2a\lambda i}{n}} + O\left(\frac{1}{n^{3/2}}\right),$$

the initial data for (22) are

$$Y_a(\lambda, \mu) = \frac{1}{a} - 2\sqrt{-2\mu i - 2a\lambda i}.$$

In what follows, we shall denote this solution by $Y_u(\lambda, \mu; a)$. Using (17) and (19), we have

$$\sum_{m>m_0} \frac{\bar{r}_{2m-1}(\lambda,\mu)-\bar{\bar{r}}_{2m-1}(\lambda,\mu)}{\bar{\bar{r}}_{2m-1}(\lambda,\mu)} = (\bar{r}_{2m_0-1}(\lambda,\mu)-\bar{\bar{r}}_{2m_0-1}(\lambda,\mu))\sum_{m>m_0}\frac{1}{\bar{\bar{r}}_{2m-1}(\lambda,\mu)}\times$$

$$\prod_{k=m_0+1}^{m}\frac{1}{(\bar{r}_{2k-1}(\lambda,\mu)-s^{2k})}\,\frac{1}{(\bar{\bar{r}}_{2k-1}(\lambda,\mu)-s^{2k})} \tag{23}$$

We now note that in the last sum it can be assumed that $m \le n^{1/2+\varepsilon}$ for any $\varepsilon>0$, since, as can be shown, the sums

$$\sum_{m>n^{1/2+\varepsilon}}$$

are negligibly small. For the indicated m, we can replace $\bar{\bar{r}}_{2m-1}(\lambda, \mu)$ by 2, and include the error in the remainder term. To the same accuracy

$$\prod_{k=m_0+1}^{m}\frac{1}{(\bar{r}_{2k-1}(\lambda,\mu)-s^{2k})(\bar{\bar{r}}_{2k-1}(\lambda,\mu)-s^{2k})} \sim \exp\left\{-2\int_{0}^{\frac{m}{\sqrt{n}}}[2\sqrt{-2\mu i-2u\lambda i}+Y_u(\lambda,\mu;a)]\,du\right\}.$$

The error in this relation tends to 0 as $n\to\infty$. Therefore, returning to (23) and using the fact that

$$\bar{r}_{2m_0-1}(\lambda,\mu)-\bar{\bar{r}}_{2m_0-1}(\lambda,\mu)=\frac{\theta-\bar{\theta}}{a^2 n}(1+O(a)),$$

we obtain

$$\prod_{m>m_0}\frac{\bar{r}_{2m+1}(\lambda,\mu)}{\bar{\bar{r}}_{2m+1}(\lambda,\mu)}=\exp\left\{\frac{\bar{\theta}-\bar{\bar{\theta}}}{2a^2\sqrt{n}}\left[\int_0^\infty dv\exp\left\{-2\int_0^v[2\sqrt{-2\mu i-2u\lambda i}+Y_u(\lambda,\mu;a)]\right\}du+o(1)\right]\right\}. \tag{24}$$

We show that $Y_u(\lambda, \mu; a)$ admits the representation

$$Y_u(\lambda,\mu;\quad a)=\frac{1}{u}-2\sqrt{-2\mu i}+g(u),$$

where $g(u)=O(u)$ as $u\to 0$. Assuming that this representation holds, we substitute it in the right-hand side of (24). We obtain

$$\exp\left\{\frac{\bar{\theta}-\bar{\bar{\theta}}}{2a^2\sqrt{n}}\int_0^\infty dv\left[\exp\left\{\left[-2\int_0^v\left(2\sqrt{-2\mu i-2u\lambda i}+\frac{1}{u}-2\sqrt{-2\mu i}+g(u)\right)du\right]\right\}\right]\right\}=$$

$$\exp\left\{\frac{(\theta-\bar{\theta})}{2a^2\sqrt{n}}\int_0^\infty dv\frac{a^2}{v^2}\left[\exp\left\{-2\int_0^v[2\sqrt{-2\mu i-2u\lambda i}-2\sqrt{-2\mu i}+g(u)]\,du\right\}\right]\right\}=$$

$$\exp\left[\frac{(\theta-\bar{\bar{\theta}})}{2a^2\sqrt{n}}\left[a+\int_a^\infty\frac{a^2}{v^2}\left[\exp\left\{-2\int_a^v[2\sqrt{-2\mu i-2\lambda ui}-2\sqrt{-2\mu i}+g(u)]\,du\right\}-1\right]dv\right]\right]=$$

$$\exp\left\{\frac{\bar{\theta}-\bar{\bar{\theta}}}{2a\sqrt{n}}+\frac{(\bar{\theta}-\bar{\bar{\theta}})}{2\sqrt{n}}\int_a^\infty\frac{dv}{v^2}\left[\exp\left\{-2\int_a^v[2\sqrt{-2\mu i-2\lambda ui}-2\sqrt{-2\mu i}+g(u)]\,du\right\}-1\right]\right\}.$$

In the last integral, we can, by virtue of the asymptotic behavior of $g(u)$, go to the limit $a\to 0$ and write the limit in the form

$$\int_0^\infty\frac{dv}{v^2}\left[\exp\left\{-2\int_0^v\left[2\sqrt{-2\mu i-2u\lambda i}+Y_u(\lambda,\mu)-\frac{1}{u}\right]du\right\}-1\right].$$

Now here $Y_u(\lambda, \mu)$ is a solution of the differential equation

$$\frac{dY_u(\lambda,\mu)}{du}=-4\sqrt{-2\mu i-2u\lambda i}\,Y_u(\lambda,\mu)-Y_u^2(\lambda,\mu) \tag{25}$$

that depends on λ and μ as on parameters and for which $Y_u(\lambda, \mu)\sim 1/u$ as $u\to 0$, i.e., $Y_u(\lambda, \mu)u\to 1$ as $u\to 0$.

Further, from (5)

$$\bar{r}_{2m}(\lambda,\mu)-\bar{\bar{r}}_{2m}(\lambda,\mu)=\frac{1}{\bar{r}_{2m-1}(\lambda,\mu)}-\frac{1}{\bar{\bar{r}}_{2m-1}(\lambda,\mu)}=\frac{\bar{\bar{r}}_{2m-1}(\lambda,\mu)-\bar{r}_{2m-1}(\lambda,\mu)}{\bar{r}_{2m-1}(\lambda,\mu)\bar{\bar{r}}_{2m-1}(\lambda,\mu)}, \tag{26}$$

whence

$$\prod_{m>m_0}\frac{\bar{r}_{2m}(\lambda,\mu)}{\bar{\bar{r}}_{2m}(\lambda,\mu)}=\prod_{m>m_0}\left(1+\frac{\bar{r}_{2m}(\lambda,\mu)-\bar{\bar{r}}_{2m}(\lambda,\mu)}{\bar{\bar{r}}_{2m}(\lambda,\mu)}\right).$$

The last expression differs from

$$\exp\left\{\sum_{m>m_0}\frac{\bar{r}_{2m}(\lambda,\mu)-\bar{\bar{r}}_{2m}(\lambda,\mu)}{\bar{\bar{r}}_{2m}(\lambda,\mu)}\right\}$$

by a quantity that is of a higher order of smallness. In addition, on the basis of (5) we can replace $\bar{r}_{2m}(\lambda,\mu)$ by $-\frac{1}{2}$ for $m\gg n^{1/2+\varepsilon}$, and, as is readily shown, the part of the sum with larger m is negligibly small. Further, from (26), using the fact that

$$\bar{r}_{2m-1}(\lambda,\mu),\ \bar{\bar{r}}_{2m-1}(\lambda,\mu)$$

for the considered m is close to 2, we have

$$\sum_{m>m_0}\frac{\bar{r}_{2m}(\lambda,\mu)-\bar{\bar{r}}_{2m}(\lambda,\mu)}{\bar{\bar{r}}_{2m}(\lambda,\mu)}\sim-2\sum_{m>m_0}(\bar{r}_{2m}(\lambda,\mu)-\bar{\bar{r}}_{2m}(\lambda,\mu))\sim\frac{1}{2}\sum_{m>m_0}(\bar{r}_{2m-1}(\lambda,\mu)-\bar{\bar{r}}_{2m-1}(\lambda,\mu)).$$

We have already estimated the last sum. As a result

$$\prod_{m>m_0}\frac{\bar{r}_{2m}(\lambda,\mu)}{\bar{\bar{r}}_{2m}(\lambda,\mu)}=\exp\left\{\frac{1}{\sqrt{n}}\left[\frac{\theta-\bar{\theta}}{2a}+\frac{\bar{\theta}-\bar{\bar{\theta}}}{2}\int_a^\infty\frac{dv}{v^2}\left[\exp\left\{-2\int_a^v\left[2\sqrt{-2\mu i-2u\lambda i}+\right.\right.\right.\right.$$

$$\left.\left.\left.\left.Y_u(\lambda,\mu)-\frac{1}{u}\right]du\right\}-1\right]\right]\right\}.$$

and from (7), (15), and (16), going to the limit $a\to 0$, we obtain

$$\varphi\left(e^{i\frac{\lambda}{n^{3/2}}},e^{i\frac{\mu}{n}}\right)=\frac{1}{2}\exp\left\{\frac{1}{\sqrt{n}}(f(\lambda,\mu)+o(1))\right\},$$

where

$$f(\lambda,\mu)=(\bar{\theta}-\theta)\int_0^\infty\frac{dv}{v^2}\left[\exp\left\{-2\int_0^v\left[2\sqrt{-2\mu i-2u\lambda i}+Y_u(\lambda,\mu)-\frac{1}{u}\right]du\right\}\right].$$

To complete the proof of the theorem, it remains to investigate the properties of the function $Y_u(\lambda,\mu)$ in the neighborhood of $u=0$.

We first assume that $\lambda=0$. Then the differential equation (22) takes the form

$$\frac{dY_u(0,\mu)}{du}=-4\sqrt{-2\mu i}\,Y_u(0,\mu)-Y_u^2(0,\mu).$$

Apart from unimportant corrections, the solution $Y_u(0,\mu;a)$ in which we are interested has the form

$$Y_u(0,\mu;a)=\frac{4\sqrt{-2\mu i}\,e^{-4u\sqrt{-2\mu i}}}{1-e^{-4u\sqrt{-2\mu i}}},$$

and for small u

$$Y_u(0,\mu;a)=\frac{4\sqrt{-2\mu i}(1-4\sqrt{-2\mu i}\,u+O(u^2))}{u4\sqrt{-2\mu i}-\frac{1}{2}(4\sqrt{-2\mu i})^2u^2+O(u^3)}=$$

$$\frac{1}{u}\frac{1-4\sqrt{-2\mu i}\,u+O(u^2)}{1-2\sqrt{-2\mu i}\,u+O(u^2)}=\frac{1}{u}-2\sqrt{-2\mu i}+O(u).$$

This is the required asymptotic expansion in the neighborhood of $u=0$. It is readily seen on the basis of the form of the equation that for $\lambda\neq 0$ the first two terms of this expansion remain the same as for $\lambda=0$. In fact, an explicit expression for the solution $Y_\mu(\lambda, \mu)$ can be given [S. M. Zarbaniev, Mechanics and Mathematics Faculty of the Moscow State University] and the validity of the expansion can be established directly.

With this, the proof of Theorem 1 is completed.

Proof of Theorem 2. Theorem 2 is proved by the method of characteristic functions on the basis of the expansion obtained in Theorem 1.

We denote by $g(u, v)$ the two-dimensional density of the limit distribution considered in Theorem 2. Then

$$g_1(u) = \int_{-\infty}^{\infty} g(u, v)\,dv$$

is the density corresponding to ζ, and

$$g_2(v) = \int_{-\infty}^{\infty} g(u, v)\,du$$

is the analogous density for the random variable τ.

Corollary 1 follows directly from Theorem 2. It can also be obtained from Theorem 1 by setting $\mu=0$ and using the homogeneity of the function $f(\lambda, \mu)$ noted in Sec. 1.

Proof of Corollary 2. It is well known (see [2]) that

$$P\{\tau=2n\} = 2\int_{-1/2}^{1/2} \exp\{-4\pi i n\mu\}\varphi(1, e^{2\pi i\mu})\,d\mu$$

decreases as $n^{-3/2}$, i.e., $P\{\tau=2n\} \sim C_0 n^{-3/2}$ for $n\to\infty$, where C_0 is a positive constant. We consider the characteristic function of the conditional distribution of the random variable ζ under the condition that $\tau=2n$. By definition, it is

$$\frac{1}{P\{\tau=2n\}}\sum_m P\{\zeta=m, \tau=2n\}e^{2\pi i\lambda m} = \frac{1}{P\{\tau=2n\}}\int_{-1/2}^{1/2}(\varphi(e^{2\pi i\lambda}, e^{2\pi i\mu})+\varphi(e^{-2\pi i\lambda}, e^{2\pi i\mu}))e^{-4\pi i n\mu}\,d\mu.$$

We are interested in the characteristic function of the random variable $\zeta/n^{3/2}$ and its limit as $n\to\infty$, i.e., the $n\to\infty$ limit of the expression

$$\psi_n(\lambda) = \frac{1}{P\{\tau=2n\}}\int_{-1/2}^{1/2}(\varphi(e^{2\pi i\frac{\lambda}{n^{3/2}}}, e^{2\pi i\mu}) + \varphi(e^{-2\pi i\frac{\lambda}{n^{3/2}}}, e^{2\pi i\mu}))e^{-4\pi i\mu n}\,d\mu =$$

$$\frac{1}{P\{\tau=2n\}\,n}\int_{-\frac{1}{2}n}^{\frac{1}{2}n}(\varphi(e^{2\pi i\frac{\lambda}{n^{3/2}}}, e^{2\pi i\frac{\mu}{n}}) + \varphi(e^{-2\pi i\frac{\lambda}{n^{3/2}}}, e^{2\pi i\frac{\mu}{n}}) - 1)e^{-4\pi i\mu}\,d\mu.$$

We represent the last expression as a sum of three terms:

$$\psi_n(\lambda) = \frac{1}{P\{\tau=2n\}\,n}\int_{|\mu|\leqslant R}(\varphi(e^{2\pi i\frac{\lambda}{n^{3/2}}}, e^{2\pi i\frac{\mu}{n}}) + \varphi(e^{-2\pi i\frac{\lambda}{n^{3/2}}}, e^{2\pi i\frac{\mu}{n}}) - 1)e^{-4\pi i\mu}d\mu +$$

$$\frac{2}{P\{\tau=2n\}\,n}\int_{|\mu|\geqslant R}\left(\varphi(1, e^{2\pi i\frac{\mu}{n}}) - \frac{1}{2}\right)e^{-4\pi i\mu}\,d\mu +$$

$$\frac{1}{P\{\tau=2n\}\,n}\int_{|\mu|\geqslant R}(\varphi(e^{2\pi i\frac{\lambda}{n^{3/2}}}, e^{2\pi i\frac{\mu}{n}}) + \varphi(e^{-2\pi i\frac{\lambda}{n^{3/2}}}, e^{2\pi i\frac{\mu}{n}}) -$$

$$2\varphi(1, e^{2\pi i\frac{\mu}{n}}))e^{-4\pi i\mu}\,d\mu = \psi_n^{(1)}(\lambda) + \psi_n^{(2)}(\lambda) + \psi_n^{(3)}(\lambda).$$

It follows from Theorem 1 that, choosing R, we can make $\psi_n^{(1)}(\lambda)$ arbitrarily close to the integral

$$C\int_{-\infty}^{\infty}\exp\{-4\pi i\mu\}(f(\lambda, \mu)+f(-\lambda, \mu))\,d\mu,$$

where C can be regarded as a normalizing constant. It is this integral that occurs in the statement of Corollary 2.

Further, choosing R, we can make the term $\psi_n^{(2)}(\lambda)$ arbitrarily small. This is well known, since we have at our disposal the explicit form of $\varphi(1, t)$. It therefore remains to discuss the smallness of $\psi_n^{(3)}(\lambda)$. In this paper, we only describe in general terms how it is to be obtained. Since $|\mu|$ is much larger than $|\lambda|$ and λ can be assumed fixed, we can, in investigating the continued fraction that determines $\varphi(s, t)$, construct a perturbation theory about the fixed point corresponding to $s=1$, μ. Then the same arguments as above show that

$$\varphi(e^{i\lambda/n^{3/2}}, t) = \varphi(1, t) + \frac{1}{\sqrt{n}} r(\lambda, t)$$

and the function $r(\lambda, t)$ admits an integrable estimate. We omit the details.

3. PROOF OF THEOREM 3

We return to the original situation. We consider the sequence of independent random variables ξ_n, which take the values ± 1 with probabilities ½ and set $\eta_n=\xi_1+\ldots+\xi_n$, $\zeta_n=\eta_1+\ldots+\eta_n$. We are now interested in the nature of the decrease of the probabilities $P\{\zeta_n \geq 0 \text{ for all } 0 \leq n \leq N\}$.

We denote by $0<\tau_1<\tau_2<\ldots\tau_k<\ldots$ the sequence of times of the returns of η_n to 0. In other words, τ_k are all times for which $\eta_{\tau_k}=0$. We set

$$\tau^{(k)} = \tau_k - \tau_{k-1}, \quad \zeta^{(k)} = \zeta_{\tau_k} - \zeta_{\tau_{k-1}}.$$

An important observation, which I owe to G. Kesten, is that

$$(\zeta^{(1)}, \tau^{(1)}), (\zeta^{(2)}, \tau^{(2)}), \ldots, (\zeta^{(k)}, \tau^{(k)}), \ldots$$

form a sequence of independent random variables. In fact, it is convenient to regard the entire process ζ_n as consisting of independent fragments $(\zeta^{(j)}, \tau^{(j)})$. We introduce the random variable γ_1, which is equal to a k such that

$$\zeta^{(1)} \geq 0, \quad \zeta^{(1)}+\zeta^{(2)} \geq 0, \quad \zeta^{(1)}+\ldots+\zeta^{(k-1)} \geq 0, \quad \zeta^{(1)}+\ldots+\zeta^{(k)} < 0.$$

Then the least n for which $\zeta_n<0$ satisfies the inequalities

$$\tau^{(1)}+\ldots+\tau^{(\gamma_1-1)} < n \leq \tau^{(1)}+\ldots+\tau^{(\gamma_1)}.$$

We now study the joint distribution of the random variables γ_1 and $\tau=\tau^{(1)}+\ldots+\tau^{(\gamma_1)}$, using the methods of recovery theory (see [3]). We introduce the generating function

$$\chi(s, t) = \sum_{k,m} P\{\gamma_1=k, \bar{\tau}=2m\} s^k \cdot t^{2m}.$$

A small strengthening of the well-known Sparre-Andersen theorem (see [3,4]) is the following lemma.

Lemma 3.

$$\ln \frac{1}{1-\chi(s, t)} = \sum_{n=1}^{\infty}\sum_{m=1}^{\infty} \frac{s^n t^{2m}}{n} P\{\zeta^{(1)}+\ldots+\zeta^{(n)}<0, \; \tau^{(1)}+\ldots+\tau^{(n)}=2m\}.$$

Proof. We follow the method in Feller's book [3]. We consider the realization $Z^{(0)}=\{(\zeta^{(1)}, \tau^{(1)}), \ldots, (\xi^{(n)}, \tau^{(n)})\}$ and also the $n-1$ realizations $Z^{(\nu)}$ obtained from $Z^{(0)}$ by applying the group of cyclic permutations. We fix r and define n random variables $X^{(\nu)}$, where $X^{(\nu)}=1$ if n is the r-th lowest ladder moment for the sequence of independent random variables $\zeta^{(1)}, \zeta^{(2)}, \ldots, \zeta^{(n)}, \ldots$ (see [3], Chap. 12, for the definition of ladder moments) and 0 otherwise. If $X^{(0)} = 1$, this means that n is the lowest ladder moment for the original sequence $Z^{(0)}$. Therefore

$$P\{X^{(0)}=1, \; \tau^{(1)}+\ldots+\tau^{(n)}=2m\} = q^{(r)}_{n,m}$$

is the distribution of the sum of r independent two-dimensional random variables each of which has the distribution $P\{\gamma_1 = k, \bar{\tau} = 2m\}$. Further, by virtue of the symmetry, the random variables $X^{(\nu)}$, $\tau^{(1)}+\ldots+\tau^{(n)}$ have the same distribution. Therefore, as in the ordinary case (see [3]),

$$\frac{1}{r} q^{(r)}_{n,m} = \frac{1}{n} P\{X^{(0)}+\ldots+X^{(n-1)}=r, \; \tau^{(1)}+\ldots+\tau^{(n)}=2m\}.$$

Summing the last expression over r, we obtain

$$\sum_{r=1}^{\infty} \frac{1}{r} q^{(r)}_{n,m} = \frac{1}{n} P\{\zeta^{(1)}+\ldots+\zeta^{(n)}<0, \; \tau^{(1)}+\ldots+\tau^{(n)}=2m\}.$$

We multiply both sides of the last relation by $s^n t^{2m}$ and sum over all n and m. We obtain

$$\sum_{r=1}^{\infty}\frac{1}{r}\chi^r(s,t)=\sum_{n=1}^{\infty}\sum_{m=1}^{\infty}\frac{s^n t^{2m}}{n}P\{\zeta^{(1)}+\ldots+\zeta^{(n)}<0,\ \tau^{(1)}+\ldots+\tau^{(n)}=2m\}.$$

The last relation is equivalent to the proposition of the lemma. Lemma 3 is proved.

Lemma 3 also gives an expression for the generating function $\chi(1, t)$ of the random variable $\bar{\tau}$:

$$\ln\frac{1}{1-\chi(1,t)}=\sum_{n=1}^{\infty}\frac{1}{n}\sum_{m}P\{\zeta^{(1)}+\ldots+\zeta^{(n)}<0,\ \tau^{(1)}+\ldots+\tau^{(n)}=2m\}t^{2m}.$$

By virtue of the symmetry,

$$P\{\zeta^{(1)}+\ldots+\zeta^{(n)}>0,\ \tau^{(1)}+\ldots+\tau^{(n)}=2m\}=P\{\zeta^{(1)}+\ldots+\zeta^{(n)}<0,\ \tau^{(1)}+\ldots+\tau^{(n)}=2m\}.$$

In addition, we readily deduce from the local limit theorem for the two-dimensional distribution of the random variables $\zeta^{(1)}$, $\tau^{(1)}$ that

$$P\{\zeta^{(1)}+\ldots+\zeta^{(n)}=0,\ \tau^{(1)}+\ldots+\tau^{(n)}=2m\}\sim\frac{\text{const}\ g(0,2m/n^2)}{n^3}.$$

where $g(u, v)$ is the density of the two-dimensional stable limit distribution of Theorem 2 (see Corollary 1 of this theorem). Hence

$$P\{\zeta^{(1)}+\ldots+\zeta^{(n)}<0,\ \tau^{(1)}+\ldots+\tau^{(n)}=2m\}={}^1\!/_2P\{\tau^{(1)}+\ldots+\tau^{(n)}=2m\}+\delta_{n,m},$$

where

$$\sum_{m}|\delta_{n,m}|\leqslant\frac{\text{const}}{n^{3}}.$$

Therefore, for

$$\omega(t)=\sum_{m=1}^{\infty}P\{\tau^{(1)}=2m\}t^{2m},\qquad \delta(t)=\sum_{n}\frac{1}{n}\sum_{n,m}\delta_{n,m}t^{2m}$$

we have

$$\sum_{n=1}^{\infty}\frac{1}{n}\sum_{m}P\{\zeta^{(1)}+\ldots+\zeta^{(n)}<0,\ \tau^{(1)}+\ldots+\tau^{(n)}=2m\}t^{2m}=$$

$$\frac{1}{2}\sum_{n=1}^{\infty}\frac{1}{n}\sum_{m}P\{\tau^{(1)}+\ldots+\tau^{(n)}=2m\}t^{2m}+$$

$$\delta(t)=\frac{1}{2}\sum_{n=1}^{\infty}\frac{1}{n}(\omega(t))^n+\delta(t)=-{}^1\!/_2\ln(1-\omega(t))+\delta(t).$$

From the above estimates there follows the existence of the limit

$$\lim_{t\to1}\delta(t)=\delta.$$

Since $1-\omega(t)\sim\text{const}(1-t)^{1/2}$ as $t\to0$, for the generating function $\chi(1, t)$ of the random variable $\bar{\tau}$ we obtain in the limit $t\to1$

$$1-\chi(1,t)\sim\text{const}\cdot(1-t)^{1/4}.$$

Then on the basis of the Tauber theorem for generating functions (see [3]) $P\{\bar{\tau}\geq 2m\}\sim\text{const}\ m^{-5/4}$. Since the random variable $\bar{\nu}$, which is equal to the least n for which $\zeta_n<0$, satisfies the inequality $\bar{\nu}\leq\bar{\tau}$, we have

$$P\{\bar{\nu}>T\}\leqslant P\{\bar{\tau}>T\}\sim\frac{\text{const}}{T^{1/4}}.$$

We now show how to obtain a similar estimate from the other side. We can strengthen the assertion of the lemma and show that for any $j<0$ and generating function

$$\chi_j(s,\ t)=\Sigma P\{\gamma_1=k,\ \bar{\tau}=2m,\ \zeta^{(1)}+\ldots+\zeta^{(\bar{\tau})}=j\}s^k t^{2m}$$

the following relation holds:

$$\ln \frac{1}{1-\chi_j(s,t)} = \sum_{n=1}^{\infty} \frac{s^n \cdot t^{2m}}{n} P\{\zeta^{(1)}+\ldots+\zeta^{(n)}=j,\ \tau^{(1)}+\ldots+\tau^{(n)}=2m\}.$$

From this we obtain the possibility, using the Tauber theorems, of obtaining asymptotic expressions for the probabilities

$$P\{\gamma_1=k,\ \bar{\tau}=2m,\ \zeta^{(1)}+\ldots+\zeta^{(\tau)}=j\}.$$

Further, on the basis of the duality principle (see [3], Chap. 12, §7), this gives asymptotic expressions for the probabilities

$$P\{\zeta^{(1)}+\ldots+\zeta^{(r)}\geqslant 0 \text{ for } 0\leqslant r\leqslant n,\ \zeta^{(1)}+\ldots+\zeta^{(n)}=-j,\ \tau^{(1)}+\ldots+\tau^{(n)}=2m\},$$

from which there also follow asymptotic behaviors for the probabilities

$$P\{\zeta^{(1)}+\ldots+\zeta^{(r)}\geqslant 0 \text{ for } 0\leqslant r<n,\quad \zeta^{(1)}+\ldots+\zeta^{(n)}=-j,\ \tau^{(1)}+\ldots+\tau^{(n)}=2m,$$

$$\zeta^{(1)}+\ldots+\zeta^{(n)}+\zeta^{(n+1)}<0,\ \tau^{(1)}+\ldots+\tau^{(n)}+\tau^{(n+1)}=2m_1\}.$$

The existence of this asymptotic behavior means, in particular, the existence of a limit distribution for the hop $\tau^{(\gamma_1)}$ and for the value of the sum $\zeta^{(1)}+\ldots+\zeta^{(\gamma_1-1)}$ directly before the hop. Summing over j, we obtain the joint distribution of $\bar{\tau}$ and $\bar{\tau}-\tau_{\gamma_1}$. Since

$$P\{\bar{\nu}\geqslant T\}\geqslant P\{\bar{\tau}-\tau_{\gamma_1}\geqslant T\},$$

we directly obtain the result we require.

Note that τ_{γ_1} and $\bar{\tau}$ take values of the same order and that there exists a limit distribution of the ratio $\bar{\tau}/T$, and also a joint limit distribution τ_{γ_1}/T as $T\to\infty$.

4. DISTRIBUTION OF THE TIME OF INTERSECTION OF THE STRAIGHT LINE $w=r+at$ BY THE INTEGRAL OF THE WIENER PROCESS

It is readily seen that for $r<0$ the probability P_T that

$$\int_0^t b(u)\,du>r+at$$

for all $0\leq t\leq T$ tends to zero as $T\to\infty$. Here, $b(u)$ is a Wiener process that starts from zero. In this section, we first investigate the rate of this limiting behavior. Suppose $a=0$. We prove Theorem 4 (we repeat its formulation given in Sec. 1).

THEOREM 4. *Let $r<0$ be a fixed number. Then*

$$P_r(T)=P\{\int_0^t b(u)\,du>r \quad \text{for all} \quad t,\ 0\leqslant t\leqslant T\}$$

satisfies the inequalities

$$\frac{A_1(r)}{T^{1/4}}\leqslant P_r(T)\leqslant\frac{A_2(r)}{T^{1/4}}$$

where $A_1(r)$ and $A_2(r)$ are positive constants.

Proof of Theorem 4. We first derive an explicit expression for the moment of first attainment of the level r, and we then deduce from it the required asymptotic behavior. Regarding r as fixed, we take the parameter R, which subsequently will tend to infinity, and we consider the probabilities $Q_R(n, 2m)$, where

$$Q_R(n,2m)=P\{\zeta^{(1)}+\ldots+\zeta^{(n-1)}\geqslant rR^{3/2},\quad \zeta^{(1)}+\ldots+\zeta^{(n)}<rR^{3/2},\ \bar{\tau}=\tau_1+\ldots+\tau_n=2m\}.$$

We introduce the generating functions

$$\chi_R(s,t)=\sum_n\sum_m s^n t^{2m} Q_n(n,2m).$$

Then, arguing exactly as in the previous section, we obtain the relation

$$\ln\frac{1}{1-\chi_R(s,t)}=\sum_{n=1}^{\infty}\sum_{m=1}^{\infty}\frac{s^n t^{2m}}{n}P\{\zeta^{(1)}+\ldots+\zeta^{(n)}<rR^{3/2},\ \bar\tau=2m\}.$$

Setting $t=e^{-\lambda}$, $s=e^{-\mu}$, we can write down for the Laplace transform

$$h_R(\lambda,\mu)=M\exp\left\{-\left(\frac{\lambda\gamma_1}{R^{1/2}}+\frac{\mu\bar\tau}{R}\right)\right\}=\chi_R\left(\exp\left\{-\frac{\lambda}{R^{1/2}}\right\},\ \exp\left\{-\frac{\mu}{R}\right\}\right)$$

the expression

$$\ln\frac{1}{1-h_R(\lambda,\mu)}=\sum_{n,m}\frac{1}{n}\exp\left\{-\frac{\lambda n}{R^{1/2}}-\frac{\mu\bar\tau}{R}\right\}P\{\zeta^{(1)}+\ldots+\zeta^{(n)}<rR^{3/2},\ \tau_1+\ldots+\tau_n=2m\}.$$

Using the method of characteristic functions, we can readily show that

$$P\{\zeta^{(1)}+\ldots+\zeta^{(n)}<rR^{3/2},\tau_1+\ldots+\tau_n=2m\}\sim\frac{1}{n^2}\int_{-\infty}^{rR^{3/2}/n^3}g\left(u,\frac{2m}{n^2}\right)du=G\left(\frac{rR^{3/2}}{n^3},\frac{2m}{n^2}\right)\frac{1}{n^2},$$

where the density $g(u,v)$ was introduced in the proof of Corollary 1 of Theorem 1 and

$$G(u,v)=\int_{-\infty}^{u}g(w,v)\,dw.$$

As always, this asymptotic behavior is valid in any finite range of variation of the variables $rR^{3/2}/n^3$ and $2m/n^2$. Further, for $n=uR^{1/2}$, $2m=vR$ the sum

$$\sum_{n,m}\frac{1}{n}\exp\left\{-\frac{\lambda n}{R^{1/2}}-\frac{\mu 2m}{R}\right\}P\{\zeta^{(1)}+\ldots+\zeta^{(n)}<rR^{3/2},\ \tau_1+\ldots+\tau_n=2m\}\sim$$

$$\sum\frac{1}{n}\exp\left\{-\frac{\lambda n}{R^{1/2}}-\frac{\mu 2m}{R}\right\}G\left(\frac{rR^{3/2}}{n^3},\frac{2m}{n^2}\right)\frac{1}{n^2}=\frac{1}{R^{1/2}}\sum_{u,v}\frac{1}{u}\exp\{-\lambda u-\mu v\}G\left(\frac{r}{u^3},\frac{v}{u^2}\right)\frac{1}{n^2}\sim$$

$$\sum\frac{1}{R^{1/2}}\frac{1}{u}e^{-\lambda u}\int_0^{\infty}e^{-\mu v_1}G\left(\frac{r}{u^3},v_1\right)dv_1$$

for $v_1=v/u^2$. In the limit $R\to\infty$, the last expression converges to the limit

$$\Phi(\lambda,\mu)=\int_0^{\infty}\frac{du}{u}e^{-\lambda u}\int_0^{\infty}e^{-\mu v_1}G\left(\frac{r}{u^3},v_1\right)dv_1.$$

Note that in the neighborhood of $u=0$ there are no singularities in the integrand, since

$$\int_0^{\infty}e^{-\mu v_1 u^2}G\left(\frac{r}{u^3},v_1\right)dv_1\leq G\left(\frac{r}{u^3}\right)=\int_0^{\infty}g\left(\frac{r}{u^3},v\right)dv.$$

$G(t)\leq\mathrm{const}/|t|^{1/3}$ as $t\to-\infty$. Thus, we find that in the limit $R\to\infty$ there exists a joint limit distribution of the random variables $\gamma_1/R^{1/2}$ and $\bar\tau/R$ as $R\to\infty$.

Setting $\lambda=0$, we find the Laplace transform of the limit distribution of the random variable $\bar\tau/R$ as $R\to\infty$. It has the form

$$\varphi(\mu)=1-\exp\left\{-\int_0^{\infty}\frac{du}{u}\int_0^{\infty}\exp\{-\mu v u^2\}G\left(\frac{r}{u^3},v\right)dv\right\}.\tag{27}$$

We now investigate the expression in the argument of the exponential in (27) in more detail. After simple manipulations, we obtain

$$\Phi(0,\mu)=\int_0^{\infty}\frac{du}{u}\int_0^{\infty}\exp\{-\mu v u^2\}G\left(\frac{r}{u^3},v\right)dv=\int_0^{1}\frac{du}{u}\int_0^{\infty}\exp\{-\mu v u^2\}G\left(\frac{r}{u^3},v\right)dv+$$

$$\int_1^{\infty}\frac{du}{u}\int_0^{\infty}\left(G\left(\frac{r}{u^3},v\right)-G(0,v)\right)\exp\{-\mu u^2 v\}dv+\frac{1}{2}\int_{\mu}^{1}\frac{h(w)-h(0)}{w}dw+\frac{1}{2}\int_1^{\infty}\frac{h(w)dw}{w}-$$

$$\tfrac{1}{4}\ln\mu=\Phi_1(\mu)+\Phi_2(\mu)+\Phi_3(\mu)+\Phi_4(\mu)-\tfrac{1}{4}\ln\mu.$$

We have here denoted

$$h(w)=\int_0^\infty G(0,v)e^{-wv}\,dv$$

and used the fact that $h(0)=\tfrac{1}{2}$. We shall now show that each of the terms $\Phi_i(\mu)$, $1\le i\le 4$, has a finite limit as $\mu\to 0$. Indeed,

$$\Phi_1(\mu)=\int_0^1\frac{du}{u}\int_0^\infty e^{-\mu u^2 v}G\left(\frac{r}{u^3},v\right)dv\leqslant\int_0^1\frac{du}{u}G\left(\frac{r}{u^3}\right)\leqslant\text{const},$$

since $G(t)$ is the distribution function of a stable law with exponent $\alpha=1/3$, and therefore

$$G(t)\leqslant\frac{\text{const}}{1+|t|^{1/3}}$$

for $t<0$. Note that for $r=0$ our estimates are not valid and $\Phi_1(\mu)\to\infty$ as $\mu\to 0$, which is in fact a natural result. Further,

$$\Phi_2(\mu)=\int_1^\infty\frac{du}{u}\int_0^\infty\exp\{-\mu u^2 v\}\left(G\left(\frac{r}{u^3},v\right)-G(0,v)\right)dv=$$

$$-\int_1^\infty\frac{du}{u}\int_0^\infty dv\exp\{-\mu u^2 v\}\int_{-\frac{r}{u^3}}^0 g(z,v)\,dz=-\int_0^\infty dv\int_r^0 g(z,v)\,dz\int_1^{\left(\frac{r}{z}\right)^{1/3}}\frac{du}{u}\exp\{-\mu u^2 v\}$$

and as $\mu\to 0$ the last expression converges to a finite limit:

$$-\frac{1}{3}\int_0^\infty dv\int_r^\infty g(z,v)\ln\left(\frac{r}{z}\right)dz=-\frac{1}{3}\int_r^0 g_1(z)\ln\left(\frac{r}{z}\right)dz.$$

The convergence of the last expression to a limit follows from the relation $G(0,v)=\tfrac{1}{2}g_2(v)$, the asymptotic behavior $g_2(v)\sim\text{const}/v^{3/2}$ as $v\to\infty$, and the inequality $h(0)-h(w)\le\text{const}\,w^{1/2}$ for $0<w<1$, which follows from it. In reality, the function $\Phi_4(\mu)$ does not depend on μ. Thus, in the limit $R\to\infty$ the random variable $\bar\tau/R$ has a limit distribution whose Laplace transform $\varphi(\mu)$ is given in (27) and has the asymptotic behavior $\varphi(\mu)=1-\text{const}\,\mu^{1/4}$ for $\mu\to+0$.

From this we obtain the result we need. We consider the random processes

$$\eta_t^{(R)}=\frac{1}{R^{1/2}}\eta_{[tR]},\quad\zeta_t^{(R)}=\frac{1}{R^{3/2}}\zeta_{[tR]}.$$

It is readily seen that in the limit $R\to\infty$ the probability measures corresponding to these processes converge, respectively, to the Wiener measure of the process $b(t)$ and to the measure corresponding to the process

$$\int_0^t b(u)\,du.$$

Further, if the random variable $\lambda^{(R)}$ is the first time at which the random process $\zeta_t^{(R)}$ reaches the level r, then, obviously,

$$\frac{\tau_1+\tau_2+\ldots+\tau_{n-1}}{R}\leqslant\lambda^{(R)}\leqslant\frac{\tau_1+\ldots+\tau_n}{R}=\frac{\bar\tau}{R}.$$

Hence

$$P_r(T)=P\left\{\int_0^t b(u)\,du>r\quad\text{for all}\quad 0\leqslant t\leqslant T\right\}=$$

$$\lim_{R\to\infty}P\{\zeta_t^{(R)}>r\quad\text{for all}\quad 0\leqslant t\leqslant T\}\leqslant\lim_{R\to\infty}P\{\lambda^{(R)}>T\}\leqslant\lim_{R\to\infty}P\left\{\frac{\bar\tau}{R}\geqslant T\right\}.$$

The Laplace transform of the limit distribution for the random variable $\bar\tau/R$ is given in (27). From the asymptotic behavior of it given above we obtain the required upper bound for $P_r(T)$. To obtain the lower bound, we note that, as in the discrete case, we can prove the existence of a nondegenerate joint limit distribution for τ_n/R and $\bar\tau/R$. We denote this distribution by F. Then, arguing as in Sec. 3, we can write

$$P_r(T)\geqslant \lim_{R\to\infty} P\left\{\frac{\tau_1+\ldots+\tau_{n-1}}{R}\geqslant T\right\}=\lim_{R\to\infty} P\left\{\frac{\tau_1+\ldots+\tau_n}{R}-\frac{\tau_n}{R}\geqslant T\right\},$$

from which the required asymptotic behavior follows.

We now prove Theorem 5 (we repeat the formulation of it given in Sec. 1), which strengthens Theorem 4.

THEOREM 5. *Suppose $r<0$, a are fixed numbers. Then*

$$P_{r,a}(T)=P\left\{\int_0^t b(u)du>r+at \quad \text{for all} \quad t,\ 0\leqslant t\leqslant T\right\}$$

satisfies the inequalities

$$\frac{A_1(r,a)}{T^{1/4}}\leqslant P_{r,a}(T)\leqslant \frac{A_2(r,a)}{T^{1/4}},$$

where $A_1(r, a)$ and $A_2(r, a)$ are positive constants.

Proof of Theorem 5. We rewrite the inequality

$$\int_0^t b(u)du>r+at$$

in the form

$$\int_0^t (b(u)-a)du>r.$$

As in the case of the previous theorem, we introduce the parameter R, which we subsequently allow to tend to infinity. It can also be assumed that a is a multiple of $1/\sqrt{R}$ and that $\sqrt{R}$ is an integer. Having the random variables ξ_i (see Secs. 1 and 3), we introduce the random variable $\tau_1(a)$, at which η_n first becomes equal to $R^{1/2}a$ and we then consider the random variables $\tau_1(a) <\tau_2(a)<\ldots<\tau_k(a)<\ldots$, which are equal to the successive times at which the trajectory of η_n reaches the point $[R^{1/2}a]$. As above, we set

$$\tau^{(k)}(a)=\tau_k(a)-\tau_{k-1}(a),\qquad \zeta^{(1)}(a)=\zeta_{\tau_1(a)}-\tau_1(a)R^{1/2},\qquad \zeta^{(k)}(a)=\zeta_{\tau_k(a)}-\zeta_{\tau_{k-1}(a)}-a\sqrt{R}\tau^{(k)}(a).$$

Then $\zeta^{(k)}(a)$, $k=2, 3, \ldots$, form a sequence of independent equally distributed random variables and

$$\zeta^{(1)}(a)+\ldots+\zeta^{(m)}(a)=\zeta_{\tau_m(a)}-a\sqrt{R}\tau_m(a).$$

Since $\zeta^{(1)}(a)$ has a distribution that differs slightly from the distribution of the remaining $\zeta^{(j)}(a)$, $j>1$, it is convenient to fix $\zeta^{(1)}(a)=w_1>rR^{3/2}$ and $\tau_1(a)=u_1$ and under this condition introduce the random variable $\gamma_1(a)$ which is equal to the k such that

$$\zeta^{(1)}(a)+\zeta^{(2)}(a)\geqslant rR^{3/2},\ \ldots,\ \zeta^{(1)}(a)+\zeta^{(2)}(a)+\ldots$$
$$+\zeta^{(k-1)}(a)\geqslant rR^{3/2},\ \zeta^{(1)}(a)+\zeta^{(2)}(a)+\ldots+\zeta^{(k)}(a)<rR^{3/2}$$

or

$$\zeta^{(2)}(a)\geqslant rR^{3/2}-w_1,\ \zeta^{(2)}(a)+\zeta^{(3)}(a)\geqslant rR^{3/2}-w_1,\ldots$$
$$\zeta^{(2)}(a)+\zeta^{(3)}(a)+\ldots+\zeta^{(k-1)}(a)\geqslant rR^{3/2}-w_1,\ \zeta^{(2)}(a)+\ldots+\zeta^{(k)}(a)<rR^{3/2}.$$

Then the least n for which $\zeta_n-aR^{1/2}n<0$ satisfies the inequalities

$$\tau^{(1)}(a)+\ldots+\tau^{(\gamma_1(a))-1}<n\leqslant\tau^{(1)}(a)+\ldots+\tau^{(\gamma_1(a))}$$

or

$$\tau^{(2)}(a)+\ldots+\tau^{(\gamma_1(a)-1)}<n-u_1<\tau^{(2)}(a)+\ldots+\tau^{(\gamma_1(a))}.$$

As in the proof of the previous theorem, we study the joint distribution of $\gamma_1(a)$ and $\bar{\tau}(a)=\tau^{(2)}(a)+\ldots+\tau^{(\gamma_1)}(a)$. We consider the probabilities

$$Q_R^{(a)}(n,2m)=P\left\{\sum_{j=2}^{i}\zeta^{(j)}(a)\geqslant rR^{3/2}-w_1 \quad \text{for all} \quad i<n,\right.$$

$$\sum_{j=2}^{n} \zeta^{(j)}(a) < rR^{3/2} + w_1,\ \bar{\tau}(a) = 2m \mid \zeta^{(1)}(a) = w,\ \tau_1(a) = u_1 \Big\}$$

and the corresponding generating functions

$$\chi_R^{(a)}(s,t) = \sum_{n,m} s^n t^{2m} Q_R^{(a)}(n, 2m).$$

Then, using the same arguments as in the previous section, we obtain

$$\ln \frac{1}{1-\chi_R^{(a)}(s,t)} = \sum_{n=1}^{\infty} \sum_{m=1}^{\infty} \frac{s^n t^{2m}}{n} P\Big\{ \sum_{j=2}^{n+1} (\zeta^{(j)} - a\tau^{(j)} R^{1/2}) < rR^{3/2},$$

$$\tau_2(a) + \ldots + \tau_{n+1}(a) = 2m \mid \zeta^{(1)}(a) = w,\ \tau_1(a) = u_1 \Big\}.$$

For the Laplace transform we can now write

$$h_R^{(1)}(\lambda, \mu) = M \exp\Big\{ -\Big(\frac{\lambda \bar{\gamma}_1}{R^{1/2}} + \frac{\mu \bar{\tau}}{R} \Big) \Big\} = \chi_R^{(1)}\Big(\exp\Big\{ -\frac{\lambda}{R^{1/2}} \Big\}, \exp\Big\{ -\frac{\mu}{R} \Big\} \Big) =$$

$$\sum_{n,m} \frac{1}{n} \exp\Big\{ -\frac{\lambda n}{R^{1/2}} - \frac{\mu \cdot 2m}{R} \Big\} P\{ \zeta^{(1)} + \ldots + \zeta^{(n)} -$$

$$a(\tau^{(1)}(a) + \ldots + \tau^{(n)}(a)) < rR^{3/2},\ \tau^{(1)}(a) + \ldots + \tau^{(n)}(a) = 2m \}. \tag{28}$$

It was noted above that the pair (ζ, τ) belongs to the region of attraction of a certain two-dimensional limit distribution which with respect to one variable is a stable law with exponent $\alpha = {}^1/_3$ and with respect to the other variable is a stable law with exponent $\alpha = 1/2$. Therefore, the probabilities

$$P\{ \zeta^{(1)} + \ldots + \zeta^{(n)} - a(\tau^{(1)}(a) + \ldots + \tau^{(n)}(a)) R^{1/2} < rR^{3/2},$$

$$\tau^{(1)}(a) + \ldots + \tau^{(n)}(a) = 2m \} = P\{ \zeta^{(1)} + \ldots + \zeta^{(n)} < rR^{3/2} + a \cdot 2m R^{1/2},$$

$$\tau^{(1)}(a) + \ldots + \tau^{(n)}(a) = 2m \} = P\Big\{ \frac{\zeta^{(1)} + \ldots + \zeta^{(n)}}{n^3} < \frac{rR^{3/2}}{n^3} + a \frac{2m}{n^2} \frac{R^{1/2}}{n},$$

$$\sim \frac{1}{n^2} \int_{-\infty}^{r\left(\frac{R^{1/2}}{n}\right)^3 + a \frac{2m}{n^2} \frac{R^{1/2}}{n}} g\Big(u, \frac{2m}{n^2}\Big) du =$$

$$\tau^{(1)}(a) + \ldots + \tau^{(n)}(a) = 2m \} \sim$$

$$\frac{1}{n^2} G\Big(r\Big(\frac{R^{1/2}}{n}\Big)^3 + a \frac{2m}{n^2} \frac{R^{1/2}}{n} \Big).$$

As before, this asymptotic behavior holds in any finite range of variation of the variables $2m/n^2$, $R^{1/2}/n$. Similarly, as in the proof of the previous theorem, we have

$$\lim_{R\to\infty} h_R^{(1)}(\lambda, \mu) = \Phi^{(1)}(\lambda, \mu) = \int_0^{\infty} \frac{du}{u} e^{-\lambda u} \int_0^{\infty} G\Big(\frac{r}{u^3} + \frac{av}{u}, v \Big) dv.$$

The last relation means the existence in the limit $R \to \infty$ of a joint distribution of the random variables $\bar{\gamma}_1/R^{1/2}$ and $\bar{\tau}/R$ as $R \to \infty$. Setting $\lambda = 0$, we obtain the limit distribution as $R \to \infty$ for the random variable $\bar{\tau}/R$, which has the form

$$\Phi^{(1)}(\mu) = 1 - \exp\Big\{ -\int_0^{\infty} \frac{du}{u} \int_0^{\infty} e^{-\mu u^2 v} G\Big(\frac{r}{u^3} + \frac{av}{u}, v \Big) dv \Big\}.$$

The behavior of $\Phi^{(1)}(\mu)$ in the limit $\mu \to +0$ is investigated in the same way as we investigated $\Phi(\mu)$ in the proof of the previous theorem. The end of the proof is the same. We omit the details.

We investigate similarly the behavior of the probabilities

$$P\left\{\int_0^t b(u)\,du<r+at \quad \text{for all} \quad 0\leq t\leq T\right\}$$

for $r>0$. Further, it is readily seen that for $r>0$ and any $\sigma>0$ the probability

$$P_{r,a,\sigma}=P\left\{\int_0^t b(u)\,du\geq r+at-\sigma t^2 \quad \text{for all} \quad t>0\right\}$$

is positive. We now investigate the behavior of $P_{r,a,\sigma}$ as $\sigma\to 0$, assuming $r<0$, a to be fixed.

We prove Theorem 6 (we repeat the formulation of it given in Sec. 1).

THEOREM 6. *There exist positive constants $B_1(r, a)$ and $B_2(r, a)$ such that*

$$B_1(r,a)\sigma^{1/2}\leq P_{r,a,\sigma}\leq B_2(r,a)\sigma^{1/2}.$$

Proof of Theorem 6. We introduce the random variable $\nu_1(b)$, which is equal to the first time $u\geq 1/\sigma^2$ at which $b(u)-2\sigma u=a$ and we denote by D the set of trajectories b such that $1/\sigma^2\leq\nu_1(b)\leq 2/\sigma^2$ and

$$\left|\int_0^{\nu_1(b)} b(u)\,du\right|\leq\frac{1}{2}\sigma^3.$$

Then

$$P_{\sigma,r,a}=P\left\{\int_0^t (b(u)-2\sigma u)\,du<r+at \quad \text{for all} \quad t\right\}\geq$$

$$P\left\{\int_0^t (b(u)-2\sigma u)\,du<r+at \quad \text{for all} \quad t,\ b\in D\right\}=$$

$$\int_{\frac{1}{\sigma^2}<x<\frac{2}{\sigma^2}}\int_{|y|<\frac{1}{2\sigma^3}} P\left\{\int_0^t (b(u)-2\sigma u)\,du<r+at \quad \text{for all} \quad t\,\Big|\,\nu_1(b)=x,\ \int_0^x b(u)\,du=y\right\}dP(x,y),$$

where $P(x, y)$ is the joint distribution function of the random variables ν_1 and

$$\int_0^{\nu_1} b(u)\,du.$$

Further, b being strictly Markov, we have

$$P\left\{\int_0^t (b(u)-2\sigma u)\,du<r+at \quad \text{for all} \quad t\,\Big|\,\nu_1(b)=x,\ \int_0^{\nu_1} b(u)\,du=y\right\}=P\left\{\int_0^t (b(u)-2\sigma u)\,du<\right.$$

$$\left.r+at \quad \text{for all} \quad t\geq x\,\Big|\,\nu_1(b)=x,\ \int_0^x b(u)\,du=y\right\}\times$$

$$P\left\{\int_0^t (b(u)-2\sigma u)\,du<r+at \quad \text{for all} \quad 0\leq t\leq x\,\Big|\,\nu_1(b)=x,\ \int_0^x b(u)\,du=y\right\}.$$

Since the process $b(u)$ increases, roughly speaking, as $u^{1/2}$, it is easy to show that for all considered b and under conditions satisfying the inequalities given above,

$$P\left\{\int_0^t (b(u)-2\sigma u)\,du<r+at \quad \text{for all} \quad t\geq x\,\Big|\right.$$

$$\left.\nu_1(b)=x,\ \int_0^x b(u)\,du=y\right\}=P\left\{\int_0^t b(u)\,du<r+at+\sigma t^2 \quad \text{for all} \quad t\geq x\,\Big|\right.$$

$$\left.\nu_1(b)=x,\ \int_0^x b(u)\,du=y\right\}\geq\text{const}.$$

Therefore, denoting by $d\Pi(b)$ the Wiener measure, we can write

$$P_{\sigma,r,a} \geqslant \text{const} \int_F d\Pi(b), \tag{29}$$

where F is the set of realizations of b such that

$$\frac{1}{\sigma^2} < \nu_1(b) < \frac{2}{\sigma^2}, \qquad \left| \int_0^{\nu_1(b)} b(u)\,du \right| < \frac{1}{2\sigma^3}$$

and

$$\int_0^t (b(u) - 2\sigma u)\,du < r + at \quad \text{for all} \quad t,\ 0 \leqslant t \leqslant \nu_1(b).$$

In the functional integral (29) we now make a change of variables, setting $b_1(u) = b(u) - 2\sigma u$. Then in accordance with Girsanov's formula

$$\int_F d\Pi(b) = \int_{F_1} e^{-2\sigma b_1(\nu_1'(b_1)) - 2\sigma^2\nu_1(b)}\, d\Pi(b_1) \geqslant e^{-2\sigma a - 4} \int_{F_1} d\Pi(b_1),$$

where F_1 is the set of functions $b_1(u)$ distributed with respect to the usual Wiener measure for which

$$\int_0^t b_1(u)\,du < r + at$$

for all t, $0 < t < \nu_1'(b_1)$, where $\nu_1'(b_1) = x > 1/\sigma^2$ is the first instant of time after $1/\sigma^2$ at which $b_1(u) = a$, $1/\sigma^2 < x < 2/\sigma^2$ and

$$-\frac{3}{2}\frac{1}{\sigma^3} < \int_0^x b_1(u)\,du = \int_0^x b_1(u)\,du - \sigma x^2 < -\frac{1}{2}\frac{1}{\sigma^3}.$$

The calculation of the probability

$$\int_{F_1} d\Pi(b_1)$$

again reduces to calculation of the distribution of sums of a random number of random variables and is done by the same methods as above. As a result, we find that this probability behaves in the same way as the probability that

$$\int_0^t b_1(u)\,du < r + at$$

for all t, $0 \leq t \leq 2/\sigma^2$, which we have already estimated.

To obtain the upper bound, we write

$$P_{\sigma,r,a} \leqslant P\left\{ \int_0^t (b(u) - 2\sigma u)\,du < r + at \quad \text{for all} \quad t,\ 0 \leqslant t \leqslant \frac{1}{\sigma^2}: \right.$$

$$\left. b\left(\frac{1}{\sigma^2}\right) - \frac{2}{\sigma} < a \right\} + P\left\{ \int_0^t (b(u) - 2\sigma u)\,du < r + at \quad \text{for all} \quad t,\ 0 \leqslant t \leqslant \frac{1}{\sigma^2}:\ b\left(\frac{1}{\sigma^2}\right) - \frac{2}{\sigma} > a \right\} = P_1 + P_2.$$

It is easier to estimate P_2. We set $b_1(u) = b(u) - 2\sigma u$. Then by means of Girsanov's formula,

$$P_2 = \int_{M_2} e^{-2\sigma b_1\left(\frac{1}{\sigma^2}\right) - 2}\, d\Pi(b_1) \leqslant e^{-2\sigma a - 2} \int_{M_2} d\Pi(b_1),$$

where M_2 is the set of b_1 such that

$$\int_0^t b_1(u)\,du \leqslant r + at$$

for all t, $0 \leq t \leq 1/\sigma^2$. We have already estimated this probability. As a result $P_2 \leq \text{const}\ \sigma^{1/2}$.

We now estimate P_1. As before, we make the change of variables $b(u) \to b_1(u)$. Then

$$P_1 = \int_{M_1} e^{-2\sigma b_1\left(\frac{1}{\sigma^2}\right) - 2}\, d\Pi(b_1),$$

where M_1 is the set of b_1 such that

$$\int_0^t b_1(u)\,du \leq r+at \quad \text{for all} \quad t,\ 0\leq t\leq \frac{1}{\sigma^2},$$

Π is the Wiener measure,

$$d\Pi(b_1) \sim \exp\left\{-\frac{1}{2}\int_0^t \left(\frac{db_1}{du}\right)^2 du\right\} db_1.$$

We introduce the random variable $\nu_2(b_1)$, which is equal to the largest $u^* \leq 1/\sigma^2$ at which $b_1(u^*)=a$. Since $b_1(u)-a<0$ for $u^* < u \leq 1/\sigma^2$, the inequality

$$\int_0^t b_1(u)\,du \leq r+at \quad \text{for all} \quad t,\ 0\leq t\leq \frac{1}{\sigma^2}$$

holds if and only if

$$\int_0^t b_1(u)\,du \leq r+at \quad \text{for all} \quad t,\ 0\leq t\leq \nu_2(b_1).$$

Therefore

$$P_1 \leq \int_0^{1/\sigma^2} dP_{\nu_2}(x)\, P(M_1 \mid \nu_2(b_1)=x)\, E\left(e^{-2\sigma b_1\left(\frac{1}{\sigma^2}\right)} \mid \nu_2(b_1)=x\right).$$

The conditional mathematical expectation

$$E\left(e^{-2\sigma b_1\left(\frac{1}{\sigma^2}\right)} \mid \nu_2(b_1)=x\right) = e^{-2\sigma a} E\left(e^{-2\sigma\left(b_1\left(\frac{1}{\sigma^2}\right)-a\right)} \mid \nu_2(b_1)=x\right)$$

can be calculated explicitly, since the conditional distribution

$$b_1\left(\frac{1}{\sigma^2}\right)-a = b_1\left(\frac{1}{\sigma^2}\right)-b_1(x)$$

is well known. The result does not exceed

$$\text{const}\, e^{\text{const}\,\sigma^2\left(\frac{1}{\sigma^2}-x\right)} \leq \text{const}.$$

Further, the density $dP_{\nu_2}(x)/dx$ is readily calculated by going over to discrete random variables, calculating the corresponding probabilities explicitly, and then going to the limit in which the discretization step tends to zero. The density does not exceed

$$\frac{\text{const}}{\sqrt{x\left(\frac{1}{\sigma^2}-x\right)}}.$$

We actually estimated the probability $P\{M_1 \mid \nu_2(b_1)=x\}$ above, and it does not exceed $x^{-1/4}$. As a result

$$P_1 \leq \text{const} \int_0^{1/\sigma^2} \frac{dx}{\sqrt{x\left(\frac{1}{\sigma^2}-x\right)}}\, x^{1/4} = \text{const}\, \sigma^{1/2}.$$

The theorem is proved.

This theorem admits some stronger forms. We give one of them.

THEOREM 7. *Let $t_0=C/\sigma^2$, $y\in(-A, C)$, where $A, C>0$ are fixed. Consider the conditional Wiener process under the condition that $b(t_0)=y/\sigma^2$. Then the probability that*

$$\int_0^t b(u)\,du < r+at+\sigma t^2 \quad \text{for all} \quad t,\ 0\leq t\leq t_0,$$

is bounded above and below by const $\sigma^{1/2}$, *where* const *depends, of course, on r, a, A, C.*

This theorem is proved in essentially the same manner as Theorem 6.

REFERENCES

1. A. Ya. Khinchin, *Continued Fractions* [in Russian], Nauka, Moscow (1978).
2. W. Feller, *An Introduction to Probability Theory and its Applications*, Vol. 1, Wiley, New York (1950).
3. W. Feller, *An Introduction to Probability Theory and its Applications*, Vol. 2, Wiley, New York (1966).
4. E. Sparre-Andersen, "On the fluctuations of sums of random variables," *Math. Scand.*, **1**, 263 (1953); **2**, 195 (1954).

Comments

The problem appeared in connection with the question formulated by U. Frisch about fractal properties of solutions of the one-dimensional Burgers equation whose initial conditions are Brownian curves. The answer was obtained in my paper [S]. The derivation of the main result was obtained with the help of the following statement.

Let $b(t)$, $T \geq 0$ be a standard Brownian motion. Then for every $C > 0$ one can find $C_1 > 0$ such that

$$P\left\{\int_0^t b(s)\mathrm{d}s \geq -C, 0 \leq t \leq T\right\} \geq \frac{C_1}{T^{1/4}}. \tag{1}$$

The distribution of the minimum of $\int_0^t b(s)\mathrm{d}s$, $0 \leq t \leq T$, was derived in several papers. However, it seems that the asymptotic (1) was new.

Various analytic results related to (1) can be found in the book by Lachal ([L]).

References

[B] T. Brox, A one-dimensional diffusion process in a Wiener medium, *Ann. Prob.*, **14** (1986), 1206–1218.

[G] O. Golosov, Localization of random walks in one-dimensional random environments, *Comm. Math. Phys.*, **92** (1984), 491–506.

[KS] V. Kaloshin, Ya. G. Sinai, Simple random walks along orbits of Anosov diffeomorphisms, *Proc. Steklov Inst. Math.*, **228** (2000), no. 1, 224–233.

[KTT] K. Kawazu, Y. Tamura, H. Tanaka, One-dimensional diffusions and random walks in random environments, *Lecture Notes in Math.*, **1299** (1988), Springer Verlag, 170–184.

[K] H. Kesten, The limit distribution of Sinai's random walk in random environment, *Physica*, **138 A** (1986), 299–309.

[L] A. Lachal, *Etudes probabiliste et analytique d'une classe de foncionelles rattachees à la primitive du movement brownien*, Lyon, 1995.

[S] Ya. G. Sinai, Statistics of shocks in solutions of the inviscid Burgers equation, *Comm. Math. Phys.*, **148** (1992), no. 3, 601–622.

[Z] O. Zeitouni, Random walks in random environments, *Lecture Notes in Math.*, **1837** (2004), Springer Verlag, 190–312.

Part II
Statistical Mechanics

CONSTRUCTION OF DYNAMICS IN ONE-DIMENSIONAL SYSTEMS OF STATISTICAL MECHANICS

Ya. G. Sinai

It is well known that in one-dimensional systems the microcanonical, small canonical, and grand canonical distributions have the same thermodynamic limit. This limit can be regarded as a measure on the phase space of an infinite system of particles. Under the assumption that the binary interaction potential has compact support, it is shown that one can find a one-parametric group of transformations in the phase space that preserve this measure and are related in a natural manner to the infinite system of Hamiltonian equations that describe the motion of the particles. This result has been previously proved by Lanford under the assumption that the potential has bounded modulus and finite range.

The concept of the thermodynamic transition to the limit has proved very helpful in equilibrium statistical physics (see [1-4]). On the other hand, this concept has not been developed very far in problems of nonequilibrium statistical physics. The first difficulty that is encountered in such a case consists of constructing a dynamics for systems with infinitely many degrees of freedom obtained after the thermodynamic passage to the limit.

As is customary in ergodic theory, we shall understand by a dynamical system a triplet $(M, \mu, \{S_t\})$, where M is a phase space, μ is a normalized measure defined on the natural σ-algebra of subsets of M, and $\{S_t\}$ is one-parameter group of measure-preserving transformations of M. To the best of our knowledge, the desired dynamical systems corresponding to a one-dimensional gas with infinitely many particles were first constructed by Lanford [5, 6]. He assumed that the binary interaction potential of the particles has finite range and that its absolute magnitude is bounded.

The second requirement is rather restrictive since it is more natural to consider potentials that are only bounded below in problems of statistical physics.

In this note we construct dynamical systems of one-dimensional statistical physics for a large class of finite-range potentials that are bounded below. Our method differs from that of Lanford.

1. Construction of the Phase Space

The reader is referred to [4, 7-9] for the constructions of this and the following sections.

We assume that the binary interaction potential U(r) satisfies these conditions:

1) $U(r) \equiv \infty$ for $0 \le r \le r_0$, $U(r) < \infty$ for $r > r_0$;

2) U(r) is continuously differentiable for $r > r_0$ and $U(r) \ge A > -\infty$;

3) $U(r) \equiv 0$ for $r \ge r_1 > r_0$.

For each s consider the interval $\Delta_s = [-s, s]$. By a configuration of molecules in Δ_s we mean an arbitrary finite subset $\hat{q}_s$ of Δ_s such that $\min_{q', q'' \in \hat{q}_s} |q' - q''| > r_0$. Let $\hat{Q}_s$ be the space of all such configurations; let $n(\hat{q}_s)$ be the number of points in $\hat{q}_s$ and $\hat{Q}_{s,k}$ be the space of configurations with $n(\hat{q}_s) = k$. The space $\hat{Q}_s$ contains a smallest σ-algebra $\hat{\mathfrak{S}}_s$, which contains all possible sets of the form $\{\hat{q}_s : n(\hat{q}_s \cap \Delta) = l\}$, where Δ is a Borel subset of Δ_s.

Moscow State University. Translated from Teoreticheskaya i Matematicheskaya Fizika, Vol. 11, No. 2, pp. 248-258, May, 1972. Original article submitted July 9, 1971.

Y.G. Sinai, *Selecta II: Probability Theory, Statistical Mechanics, Mathematical Physics and Mathematical Fluid Dynamics*, DOI 10.1007/978-1-4419-6205-8_6, © Springer Science+Business Media, LLC 2010

We assume that the points of the configuration $\hat{q}_s$ are labelled: $\hat{q}_s = \{q_1 < q_2 < \ldots < q_{n(\hat{q}_s)}\}$. A point of the phase space is a pair $x_s = (\hat{q}_s, \hat{v}_s)$, where $\hat{v}_s = \{v_1, \ldots, v_{n(\hat{q}_s)}\}$ is the vector of the molecule velocities. Let M_s be the space of such pairs x_s. Let $M_{s,k}$ be the part of M_s for which $n(\hat{q}_s) = k$. The space M_s contains a natural σ-algebra of the subsets $\mathfrak{S}_s$. Let $\mathfrak{S}_{s,k}$ be its restriction to $M_{s,k}$. Let π_s be the natural projection of M_s onto $\hat{Q}_s$.

For $s_1 < s_2$ we have the natural embedding

$$\varphi_{s_1}^{s_2} : M_{s_2} \to M_{s_1},$$

generated by the mapping of the restriction. Consider the inductive limit of the spaces M_s as $s \to \infty$. A point x of this space is a pair $x = (\hat{q}, \hat{v})$, where $\hat{q}$ is an infinite subset of the straight line R^1 for which $\inf_{q', q'' \in q} |q' - q''| \geq r_0$ and $\hat{v}$ is an R^1-valued function defined on $\hat{q}$. The subset $\hat{q}$ is a configuration of molecules of the gas with infinitely many particles and $\hat{v}$ is the vector of the molecules velocities. A pair of numbers (q', v'), $q' \in \hat{q}$, $v' = \hat{v}(q')$ gives the coordinate and velocity of one molecule in x.

Let $\hat{Q}$ be the space of infinite configurations; M be the phase space of the points x; and π be the natural projection of M onto Q.

The inclusion $\Delta_s \subset R^1$ generates a mapping φ_s of M onto M_s. Without danger of confusion, we shall denote by the same letters $\hat{\mathfrak{S}}_s$ and $\hat{\mathfrak{S}}_s$ the σ-algebras of subsets of M of the form $\varphi_s^{-1}(A)$, $A \in \mathfrak{S}_s$, $\hat{\mathfrak{S}}_s$. We shall denote the σ-algebra containing all the σ-algebras $\mathfrak{S}_s(\hat{\mathfrak{S}}_s)$ by $\mathfrak{S}(\hat{\mathfrak{S}})$.

Thus, we have constructed the phase, M, and configuration, $\hat{Q}$, spaces for a system of infinitely many molecules and we have defined a natural σ-algebra of subsets of these spaces.

2. Construction of a Measure. Equivalence of the Microcanonical, Small Canonical, and Grand Canonical Ensembles in One-Dimensional Systems

For all $s > 0$, $k > 0$, and integral $2s > r_0 k$ consider the Hamiltonian system with Hamiltonian

$$H_{s,k} = \sum_{i=1}^{k} \frac{p_i^2}{2m} + \sum_{i,j} U(|q_i - q_j|) + W_s(q),$$

where $p_i = mv_i$, m is the molecule mass, and

$$W_s(q) = \begin{cases} 0 & \text{for } |q| < s, \\ \infty & \text{for } |q| \geq s. \end{cases}$$

The phase space of this system is $M_{s,k}$. The introduction of the function W_s is equivalent to the specification of boundary conditions in the form of elastic reflections from the ends of Δ_s.

By Liouville's theorem, $\prod_{i=1}^{k} dp_i dq_i$ defines the density of an invariant measure in $M_{s,k}$; of course, it is not normalized.

Since $H_{s,k}$ does not depend explicitly on the time, it is a first integral of the system. On any constant-energy manifold, $H_{s,k} = \mathcal{H}^{(0)}_{s,k} = \text{const}$, the measure $\prod dp_i dq_i$ induces a finite conditional measure $\mu(\cdot | s, k, \mathcal{H}^{(0)}_{s,k})$, which can be assumed normalized. This measure is frequently called the microcanonical ensemble.

Suppose $s \to \infty$, $k \to \infty$, $\mathcal{H}^{(0)}_{s,k} \to \infty$. Then we have a family of measures $\mu(\cdot | s, k\ \mathcal{H}^{(0)}_{s,k})$, each of which is defined on the σ-algebra $\mathfrak{S}_{s,k}$. For any fixed s_0, the mapping $\varphi_{s_0}^{s}$ for $s > s_0$ generates a family of measures $\mu_{s_0}(\cdot | s, k\ \mathcal{H}^{(0)}_{s,k})$ on the σ-algebra $\mathfrak{S}_{s_0}$.

The following theorem actually follows from the results of [7, 8, 10]. It is given in [9] in the form that we require.

THEOREM 1. Suppose $s \to \infty$, $k \to \infty$ and $\mathscr{H}^{(1)}_{s,k} \to \infty$ so that

1) $\lim 2s/k = v > 0$;

2) $\lim \mathscr{H}^{(0)}_{s,k}/k = h$.

Then for any fixed s_0 there exists a limit of the measures $\mu_{s_0}(\cdot\,|s, k, \mathscr{H}^{(0)}_{s,k})$ in the sense that for any set $A \in \mathfrak{S}_{s_0}$ there exists $\lim \mu_{s_0}(A\,|s, k, \mathscr{H}^{(0)}_{s,k})$.

The limit measures μ_{s_0} are naturally consistent for different s_0 and, by Kolmogorov's theorem, generate a single measure μ on the complete σ-algebra $\mathfrak{S}$. The limit measure μ can be described explicitly, which we now do.

Consider the measure ν on the configuration space $\hat{Q}$ generated by μ under the mapping π. It is a Markov measure in this sense: suppose the molecules are labelled in such a way that the molecule with the smallest nonnegative coordinates has the label 0; then for fixed coordinates $\ldots q_{-n}, \ldots, q_i$ of molecules with numbers (labels) that do not exceed i the conditional distribution of the interval $q_{i+1} - q_i$ depends only on the molecules that are not situated at a distance greater than r_1 from q. For us it will be important that for $s > r_1$,

$$\nu\{q : q_{i+1} - q_i > s\,|\,q_i, \ldots\} = \text{const}\, e^{-\alpha s},$$

where the constant depends on $q_i, q_{i-1}, \ldots$, but is uniformly bounded below, and α is a constant that depends only on v.

Further, the conditional distribution of velocities for fixed configuration $\hat{q}$ generated by μ is such that the velocity of each molecule has a Gaussian distribution with mean 0 and variance β^{-1}, where the parameter β is chosen such that the mean value of the energy of one molecule is h. Then the velocities of different molecules are mutually independent. This assertion is a precise form of the assertion that the velocities of the molecules have a Maxwellian distribution. In the one-dimensional case the thermodynamic limits of the microcanonical, small canonical, and grand canonical ensembles are equal (see [9, 10]). This means that if for the same values of β and v and for every k we consider the measure on M_s whose restriction to $M_{s,k}$ is defined by the density $\Xi_s^{-1}(\beta,\mu) \times \exp[-\beta H(p_1, \ldots, p_k; q_1, \ldots, q_k) + \mu k]$, where $\Xi_s(\beta, \mu)$ is a normalizing factor, μ is the chemical potential, and β and μ are chosen from the condition that the limit of the mean energy per particle is equal to h when $s \to \infty$ and the limit of the mean density k/2s is equal to $\rho = v^{-1}$, all these measures converge to the measure μ of Theorem 1 as $s \to \infty$.

This has a corollary which we shall use in the following section.

COROLLARY 1. Suppose the molecules are labelled as above and suppose the coordinates and the velocities of all the molecules with numbers i, $|i| > n$ are fixed. Then the conditional distribution of the molecules with numbers i, $|i| \leq n$, is given by the density

$$\tilde{\Xi}^{-1}(\beta, \mu) \exp\Big[-\beta\Big(\sum_{|i|\leqslant n} \frac{p_i^2}{2m} + \sum_{|i_1|,\,|i_2|\leqslant n} U(|q_{i_1} - q_{i_2}|) + \sum_{\substack{|j|>n,\\ |i|\leqslant n}} U(|q_i - q_j|)\Big)\Big],$$

where $\tilde{\Xi}(\beta, \mu)$ is a normalizing factor that depends on the positions of the molecules with numbers i, $|i| > n$.

To prove this we must write down the formula for such conditional probabilities calculated in the small canonical ensemble for finite s and then make the passage to the limit $s \to \infty$ in this formula (in this connection see the papers [2, 11] of Dobrushin on Gibbs distributions).

COROLLARY 2.

$$\int \sum_{|i_1|\leqslant n} \Big[\sum_{i_2} \frac{\partial U(|q_{i_1} - q_{i_2}|)}{\partial r}\Big]^2 d\mu \leqslant \text{const}\, n,$$

where the constant depends only on v or h.

This corollary is proved in [9].

3. Construction of Dynamics

Thus, in §§ 1 and 2, we have constructed the phase space M of our system of infinitely many particles and a measure μ on M which is the limit of the microcanonical ensembles. We now wish to define the motion of our infinite system of molecules. Formally, this motion must be described by the following infinite system of differential equations:

$$\frac{dq_i}{dt}=\frac{p_i}{m}, \quad \frac{dp_i}{dt}=-\sum_j \frac{\partial U(|q_i-q_j|)}{\partial r}. \tag{1}$$

However, such systems do not have existence and uniqueness theorems for the solutions. Physically, this is because when there is a rapid growth of velocities at infinity there can be instantaneous collapses when the initial data are such that the molecules fly instantaneously to the minimal possible distances (see also [5]).

It will be shown later that such effects occur if the initial data belong to a set of measures 0. We shall prove the following result: in M one can chose a subset $\tilde{M}$, $\mu(\tilde{M}) = 1$, in which for every point $x_0 \in \tilde{M}$ all the molecules admit a partitioning into finite groups $G_1, G_2, \ldots$ such that if we write down a finite system of equations of the type (1) for each group and integrate the system over t from 0 to 1, then for all such t the molecules of the different groups G_i will not interact with one another. Thus, the system (1) splits up into an infinite number of finite systems. Solving these finite systems of equations with respect to t from 0 to 1, we obtain a new point $x_1 \in \tilde{M}$, for which one can perform a similar, but different partitioning of all the molecules into groups, etc. Thus, on the set $\tilde{M}$ we can construct a continuous one-parametric group of transformations, the system of equations (1) being satisfied along the trajectory of each point $x_0 \in \tilde{M}$. A similar notion concerning the dynamics of systems with infinitely many particles has been frequently expressed by Bogolyubov [12, 13].

We now proceed to a more formal description.

Suppose we are given a point $x_0 = (\hat{q}_0, \hat{v}_0) \in M$. We label the molecules that belong in x_0 with numbers from $-\infty$ to $+\infty$ in such a way that the molecule with the smallest nonnegative coordinate has the label 0. For every $n > 0$ we fix the molecules for which the modulus of their number (label) is greater than n and we consider the motion of molecules with numbers i, $|i| \le n$, described by the system of equations

$$\frac{dq_i}{dt}=\frac{p_i}{m}, \quad \frac{dp_i}{dt}=-\sum_{|j|\le n} \frac{\partial U(|q_i-q_j|)}{\partial r}-\sum_{|j|>n} \frac{\partial U(|q_i-q_j^0|)}{\partial r},$$

where $q_j^{(0)}$ is the coordinate of a fixed molecule with number j, $|j| > n$, in x_0. In other words, we assume that the molecules with numbers j, $|j| > n$, are fixed and generate an external fixed for the molecules with i, $|i| \le n$. Suppose $P_n(x_0) = \max_{|t|\le 1, |i|\le n} |p_i(t)|$, i.e., $P_n(x_0)$ is the maximal modulus of the momentum of the moving molecules during the time from -1 to $+1$.

Let $M(c_1, c_2)$ be the subset of M distinguished by these two conditions: for $x_0 \in M(c_1, c_2)$ there exists $n_0(x_0)$ such that for all $n \ge n_0(x_0)$: 1) $P_n(x_0) \le c_1\sqrt{\ln n}$; 2) there exist i_n' and i_n'' such that $-n \le i_n' < -n/2$, $n/2 \le i_n'' < n$, and

$$q^{(0)}_{i_n'+1} - q^{(0)}_{i_n'} \geq c_2 \ln n, \quad q^{(0)}_{i_n''+1} - q^{(0)}_{i_n''} \geq c_2 \ln n;$$

here, as above, $q_k^{(0)}$ is the coordinate of a molecule with number k corresponding to x_0.

PRINCIPAL LEMMA. There exist numbers $c_1 > 0$ and $c_2 > 0$ such that $\mu(M(c_1, c_2)) = 1$.

We defer the proof of this lemma for a while in order to construct the dynamics first. Suppose numbers c_1 and c_2 have been chosen for which the lemma holds. Take a point $x_0 \in M(c_1, c_2)$ and a number $n_1(x_0) = \max(n_0(x_0), \bar{n})$, where $\bar{n}$ is such that $c_2 \ln \bar{n} - (c_1/m)\sqrt{\ln \bar{n}} - (c_1/m)\ln 2 > r_0$.

Set $n_k(x_0) = 2^{k-1} n_1(x_0)$, $k = 1, 2, \ldots$, $m_k'(x_0) = i'_{n_k(x_0)} + 1$, $m_k''(x_0) = i''_{n_k(x_0)}$. For each k consider the motion of the molecules when the molecules with numbers j, $|j| > n_k(x_0)$ (see above), are fixed. It then follows from conditions 1 and 2 that for all t, $-1 \le t \le 1$, the molecules with numbers f_1, $m_k'(x_0) \le f_1 \le m_k''(x_0)$, are situated at a distance from the molecules with the numbers $j_2 - n_k(x_0) \le j_2 < m_k'(x_0)$, $m_k''(x_0) < j_2 \le n_k(x_0)$, greater than $c_2 \ln n_k(x_0) - (c/m)\sqrt{\ln n_k(x_0)} > r_0$; this is because of the choice of $n_1(x_0)$. Therefore, the inner group of molecules $m_k'(x_0) \le j \le m_k''(x_0)$ does not interact with the remainder. We go over from k to k + 1. For the same reasons, the molecules with numbers j, $m_k'(x_0) \le j \le m_k''(x_0)$, are situated during the time from -1 to $+1$ at a distance greater than

$$c_2 \ln n_k(x_0) - \frac{c_1}{m}\sqrt{\ln n_{k+1}(x_0)} \geq c_2 \ln n_k - \frac{c_1}{m}\sqrt{\ln n_k} - \frac{c_1}{m}\ln 2 > r_0,$$

from the molecules with the numbers j, $-n_{k+1}(x_0) \le j < m_k'(x_0)$, $m_k''(x_0) < j \le n_{k+1}(x_0)$, for the motion of all molecules with numbers f, $|f| \le n_{k+1}(x_0)$. Therefore, the group of molecules $m_k'(x_0) \le j \le m_k''(x_0)$ does not interact with the remaining molecules. This had an important consequence. For all $k \ge 1$, $l \ge k$, consider the motion of molecules with numbers j, $m_k'(x_0) \le j \le m_k''(x_0)$ for $-1 \le t \le 1$ as we have described above, i.e., molecules with numbers j, $|j| > n_l(x_0)$, $l \ge k$, are frozen. Then for all such l the motion of these molecules does not depend on l!

For all t, $-1 \le t \le 1$, we now determine the motion of the point $x^{(0)}$; for every k for molecules with numbers j, $m_k'(x_0) \le j \le m_k''(x_0)$, this motion is identical to the motion that arises if all the molecules with j, $|j| > n_k(x_0)$, are fixed. Let S_t, $-1 \le t \le 1$, be the shift of the point x_0 during the time t along its trajectory. If $-1 \le t_1, t_2, t_1 + t_2 \le 1$, it is readily seen that $S_{t_1}S_{t_2} = S_{t_1+t_2}$.

We now show that μ is invariant under these transformations. Let f(x) be a bounded function that is measurable with respect to some σ-algebra $\mathfrak{S}_s$. If the molecules with numbers i, $|i| > n$, are fixed, the conditional distribution of the internal molecules with numbers i, $|i| \le n$, has the form specified in Corollary 1 of § 2, and is identical to the invariant distribution for the motion of the internal molecules when the external molecules are fixed. Let $\{S_t^{(n)}\}$ be the group of shifts along the trajectories that arises for this motion. Then

$$I = \int_M f(x)\, d\mu(x) = \int d\mu \int f(x)\, d\mu(x\,|$$

(the coordinates and momenta of the molecules with i, $|i| > n$ are fixed) where the external integration is over the coordinates and the momenta of the molecules with i, $|i| > n$. Further, for any t, $|t| \le 1$,

$$I = \int d\mu \int f(S_t^{(n)}x)\, d\mu\ (x\,|$$

(the coordinates and momenta of the molecules with i, $|i| > n$ are fixed) and if $B_m^{(s)}$ is the set x_0 for which the motion of the molecules in the interval Δ_s under the action of the group $\{S_t(m)\}$ is identical with the limit motion and $\chi_m(s)$ is the indicator of this set, then

$$I = \int d\mu \int f(S_t x)\, \chi_n^{(s)}(x)\, d\mu\,(x\,|$$

(the coordinates and momenta of the molecules with i, $|i| > n$ are fixed)

$$+ \int d\mu \int f(S_t^{(n)}x)(1 - \chi_n^{(s)}(x))\, d\mu\,(x\,|$$

(the coordinates and momenta of the molecules with i, $|i| > n$ are fixed)

$$= \int_M f(S_t x)\, \chi_n^{(s)}(x)\, d\mu(x) + \int_M f(S_t^{(n)}x)(1 - \chi_n^{(s)}(x))\, d\mu(x).$$

As $n \to \infty$, the first term tends to $\int_M f(S_t x) d\mu(x)$, and the second to 0. Therefore,

$$\int_M f(S_t x)\, d\mu(x) = \int_M f(x)\, d\mu(x),$$

and the measure is invariant.

Thus, S_t is a measure-preserving, one-to-one mapping $M(c_1, c_2) \to M$, $|t| \le 1$. Set $\tilde{M} = \bigcap_{n=-\infty}^{\infty} S_1^n M(c_1, c_2)$. Clearly, $\mu(\tilde{M}) = 1$. For any point $x \in \tilde{M}$, t, $m \le t \le m+1$, set $S_t x = S_{t-m}(S_1^m x)$. By the foregoing, the transformations S_t form a group and preserve the measure. Thus, we have constructed the desired dynamical system $(M, \mu, \{S_t\})$.

4. Proof of the Main Lemma

Let N_n' be the set of $x \in M$ for which conditions 2 of § 3 for given n is not satisfied; let $N_n''(N_n''')$ be the set of $x \in M$ for which

$$P_n^+(x) = \max_{\substack{0 \le t \le 1 \\ |i| \le n}} |p_i(t)| > c_1\sqrt{\ln n},$$

$$P_n^-(x) = \max_{\substack{-1 \le t \le 0 \\ |i| \le n}} |p_i(t)| > c_1\sqrt{\ln n}.$$

Here $p_i(t)$ is the velocity at the time t of the molecule that has the label i at the initial instant of time. We shall show that if c_i is sufficiently small and c_2 is sufficiently large, then

$$\sum \mu(N_n') < \infty, \quad \sum \mu(N_n'') < \infty. \tag{2}$$

The assertion of the main lemma will then follow from the Borel–Cantelli lemma of probability theory (see [14]).

As we said in §2, the conditional distribution of the $q_{i+1} - q_i$ between the molecules satisfies, if the disposition of the molecules $\ldots q_{-n}, \ldots, q_i$ is given, the following relation for sufficiently large r:

$$\mu(q_{i+1} - q_i > r \mid q_i, q_{i-1}, \ldots) = \text{const}\, e^{-\alpha r},$$

where α is a positive constant. Therefore

$$\mu(q_{i+1} - q_i > c_2 \ln n \mid q_i, q_{i-1}, \ldots) = \frac{\text{const}}{n^{c_2\alpha}}$$

for sufficiently large n. It follows that

$$\mu_n = \mu(\max_i (q_{i+1} - q_i) \le c_2 \ln n \quad \text{for all} \quad i, i = -n, \ldots, [-n/2])$$
$$= \int \nu(q_{-[n/2]} - q_{-[n/2]-1} \le c_2 \ln n \mid q_{-[n/2]-1}, \ldots, q_{-n})\, d\mu(q_{-[n/2]-1} \ldots q_{-n}),$$

where the integration is extended over the q, where

$$\max(q_{i+1} - q_i) \le c_2 \ln n, \quad -n \le i \le -[n/2] - 1,$$

$$\mu_n \le \left(1 - \frac{\text{const}}{n^{\alpha c_2}}\right)\mu_{n-1} \le \ldots \le \left(1 - \frac{\text{const}}{n^{\alpha c_2}}\right)^{n/2-1}.$$

If $\alpha c_2 < 1$, a term of a convergent series appears on the right. One can similarly estimate the probability that $\max(q_{i+1} - q_i) \le c_2 \ln n$ for $i = [n/2], \ldots, n$. This proves the first of the inequalities (2).

We now proceed to prove the second. Our arguments are based on the next lemma.

LEMMA. Let $\mathfrak{M}$ be an open Riemannian C^∞ manifold and α be a C^∞ vector field on $\mathfrak{M}$ that preserves the finite measure $ds = \rho(x)\, d\sigma(x)$, where $d\sigma$ is the element of Riemannian volume on $\mathfrak{M}$. Let $\{T_t\}$ be the one-parametric group of shifts along the trajectories of the vector field α. Then, if Γ is an open C^∞ submanifold of codimensionality 1 in $\mathfrak{M}$ and $d\sigma_\Gamma$ is the element of Riemannian volume on Γ induced by the metric in $\mathfrak{M}$,

$$s\Big(\coprod_{-1 \le \tau \le 0} T_\tau \Gamma\Big) \le \int_\Gamma \rho(x)\, d\sigma_\Gamma(x)\, \|\alpha(x)\|.$$

Proof. The s measure for an infinitesimally small tube surrounding a section of the trajectory $\{T_{-\tau}x\}$, $0 \le \tau \le 1$, $x \in \Gamma$ is equal to $\rho(x)\, d\sigma_\Gamma(x)\, (n_\Gamma(x), \alpha(x))$, where $n_\Gamma(x)$ is the normal to Γ at the point x. The assertion of the lemma follows because $|(n_\Gamma(x), \alpha(x))| \le \|\alpha(x)\|$.

We now apply the lemma to our case. Having the point $x_0 = (\hat{q}_0, \hat{v}_0)$, we consider $q^{(0)}_{-n-1}, q^{(0)}_{n+1} \in \hat{q}_0$ and take $\mathfrak{M}$ to be the phase space of the molecules with numbers i, $|i| \le n$ contained within the interval $q^{(0)}_{-n-1} + r_0$, $q^{(0)}_{n+1} - r_0$, with the condition $q_{i+1} - q_i > r_0$. We take the Euclidean metric $\sum_{i=-n}^{n} dq_i^2 + \sum_{i=-n}^{n} dp_i^2$, as the metric. On the phase space $\mathfrak{M}$ there is defined a vector field $\alpha(x)$ generated by the Hamiltonian system of equations with the Hamiltonian

$$H=\sum_{i=-n}^{n}\frac{p_i^2}{2m}+\sum_{-n\leqslant i,j\leqslant n}U(|q_i-q_j|)+\sum_{\substack{|i|\leqslant n\\|j|>n}}U(|q_i-q_j^{(0)}|).$$

This Hamiltonian does not depend explicitly on the time and is therefore a first integral of the system. It follows that we can take the Gibbs distribution in the form $ds = \Xi_n^{-1}\times\exp(-\beta H)\cdot\prod_{|i|\leqslant n} dp_i dq_i$ as the invariant measure ds; here the parameter β is chosen in accordance with § 2 in such a way that we have convergence to the measure μ as $n\to\infty$, and Ξ_n^{-1} is a normalizing factor.

By what we have said in Corollary in §2 this measure ds is equal to the conditional measure induced by μ if the coordinates and velocities of all molecules with numbers j, $|j|>n$, are fixed. Thus $\rho(x)=\Xi_n^{-1}\exp(-\beta H)$.

Let $\Gamma_i^+=\{x\in\mathfrak{M}\colon\ p_i=c_1\sqrt{\ln n},\ p_j\leqslant c_1\sqrt{\ln n},\ j\neq i\}$, $\Gamma_i^-=\{x\notin\mathfrak{M}\colon\ p_i=-c_1\sqrt{\ln n},\ p_j\geqslant -c_1\sqrt{\ln n},\ j\neq i\}$, $\Gamma=\bigcup_{i=-n}^{n}\Gamma_i^+\cup\bigcup_{i=-n}^{n}\Gamma_i^-$. It is readily seen that the set

$$\{x\colon \max|p_i(0)|<c_1\sqrt{\ln n},\quad P_n^{-1}(x_0)\geqslant c_1\sqrt{\ln n}\}$$

to within the set of measure 0 of the points x_0 for which the velocities of at least two molecules simultaneously take the value $c_1\sqrt{\ln n}$ is contained in the set $\bigcup_{-1\leqslant\tau\leqslant} T_\tau\Gamma$. Applying this lemma, we obtain

$$s\Big(\bigcup_{-1\leqslant\tau\leqslant 0}T_\tau\Gamma\Big)\leqslant\int_\Gamma\rho(x)\,d\sigma_\Gamma(x)\,\|\alpha(x)\|$$

$$\leqslant\sum_i\int_{\Gamma_i^+}\rho(x)\,\|\alpha(x)\|\,d\sigma_{\Gamma_i^+}(x)+\sum_i\int_{\Gamma_i^-}\rho(x)\,\|\alpha(x)\|\,d\sigma_{\Gamma_i^-}(x)$$

$$\leqslant\sum_i e^{-\beta c_2^2\ln n}\Big(\frac{1}{\sqrt{2\pi\beta^{-1}}}\Big)^{2n+1}\int_{\Gamma_i^+}\exp\Big[-\beta\Big(\frac{p_{-n}^2}{2}+\ldots+\frac{p_{i-1}^2}{2}$$

$$+\frac{p_{i+1}^2}{2}+\ldots+\frac{p_n^2}{2}\Big)\Big]\prod_{s\neq i}dp_s\,\frac{\exp\big[-\beta\sum_{i_1 i_2}U(|q_{i_1}-q_{i_2}|)\big]}{Z(n,\beta)}\prod dq_s\,\|\alpha(x)\|$$

$$+\text{ a similar sum for the integral }\int_{\Gamma_i^-}=\Sigma^+ +\Sigma^-, \tag{3}$$

where $Z(n,\beta)$ is a normalizing factor with respect to q. The factor is $\exp[-\beta c_2^2\ln n]=1/n^{\beta c_2^2}$. The number to terms of Σ^+ is 2n. The norm is

$$\|\alpha(x)\|=\sqrt{\sum_{s\neq i}\frac{p_s^2}{2m}+\sum_{i_1}\Big[\sum_{i_2}\frac{\partial U(|q_{i_1}-q_{i_2}|)}{\partial r}\Big]^2}.$$

By the Cauchy–Schwarz–Bunyakovskii inequality

$$(\Sigma^+)^2\leqslant\sqrt{\frac{\beta}{2\pi}\frac{1}{n^{\beta c_2^2}}}\sum_i\int\Big(\sum_{\substack{s\neq i\\|s|\leqslant n}}\frac{p_s^2}{2m}+\sum_{|s|\leqslant n}\Big(\sum_j\frac{\partial U(|q_s-q_j|)}{\partial r}\Big)^2\Big)$$

$$\times\exp\Big(-\beta\sum_{s\neq i}\frac{p_s^2}{2}\Big)\exp\Big(-\beta\sum_{j_1,j_2}U(|q_{j_1}-q_{j_2}|)\Big)$$

$$\times\frac{1}{(2\pi\beta^{-1})^n}\frac{1}{Z(n,\beta)}\prod_{s\neq i}dp_s\prod_{|r|\leqslant n}dq_r\leqslant\sqrt{\frac{\beta}{2\pi}}\frac{\beta^{-1}}{mn^{\beta c_2^2-1}}$$

$$+\frac{1}{\sqrt{2\pi\beta^{-1}}\,n^{\beta c_2^2}}\int\sum_{|s|\leqslant n}\Big(\sum_j\frac{\partial U(|q_s-q_j|)}{\partial r}\Big)^2$$

$$\times\exp\Big[-\beta\sum_{-n\leqslant i_1,i_2\leqslant n}U(|q_{i_1}-q_{i_2}|)\Big]Z^{-1}(n,\beta)\prod_{|r|\leqslant n}dq_r. \tag{4}$$

We can now write

$$\mu(N_n'') \leqslant \mu(\max_{|i|\leqslant n} |p_i(0)| > c_1 \sqrt{\ln n})$$
$$+ \mu(\max_{|i|\leqslant n} |p_i(0)| \leqslant c_1 \sqrt{\ln n}, \max_{\substack{|i|\leqslant n \\ 0\leqslant t\leqslant 1}} |p_i(t)| > c_1 \sqrt{\ln n})$$
$$\leqslant \left[1 - \left(1 - \sqrt{\frac{\beta}{2\pi}} \int_{|u|>c_1\sqrt{\ln n}} \exp\left(-\frac{\beta u^2}{2}\right) du\right)^{2n+1}\right]$$
$$+ \int d\mu\mu(\max_{|i|\leqslant n} |p_i(0)| < c_1 \sqrt{\ln n}, \max_{\substack{|i|\leqslant n \\ 0\leqslant t\leqslant 1}} |p_i(t)| > c_1 \sqrt{\ln n} |$$

(the coordinates and momenta of the molecules with i, |i| > n are fixed). For sufficiently large c_2 the first term is a term of a convergent series. As regards the second term, it follows from Corollary 1 of §2 and the inequalities (3) and (4) that

$$\int d\mu\mu(\max_{|i|\leqslant n} |p_i(0)| < c_1 \sqrt{\ln n}, \max_{\substack{|i|\leqslant n \\ 0\leqslant t\leqslant 1}} |p_i(t)| > c_1 \sqrt{\ln n} |$$

(the coordinates and momenta of the molecules with i, |i| > n are fixed)

$$\leqslant \int d\mu_S(\bigcup_{-1\leqslant\tau\leqslant 0} T_\tau \Gamma) \leqslant \int d\mu(\Sigma^+ + \Sigma^-) \leqslant \frac{\text{const}}{n^{\frac{\beta c_2^2}{2} - \frac{1}{2}}} \left\{1 + \int d\mu \left[\int L \sum_{|i_1|\leqslant n} \left(\sum_{i_2} \frac{\partial U(|q_{i_1} - q_{i_2}|)}{\partial r}\right)^2 d\mu(x|\right.\right.$$

(the coordinates and momenta of the molecules with i, |i| > n are fixed)

$$)^{1/2}\} \leqslant \frac{\text{const}}{n^{\frac{1}{2}\beta c_2^2 - \frac{1}{2}}} \left\{1 + \left[\int d\mu \int \sum_{|i_1|\leqslant n} \left(\sum \frac{\partial U(|q_{i_1} - q_{i_2}|)}{\partial r}\right)^2 d\mu(x|\right.\right.$$

(the coordinates and momenta of the molecules with i, |i| > n are fixed)

$$\left.\left.\right]^{1/2}\right\} \leqslant \frac{\text{const}}{n^{\frac{1}{2}\beta c_2^2 - \frac{1}{2}}} \left\{1 + \left[\int d\mu \sum_{|i_1|\leqslant n} \left(\sum_{i_2} \frac{\partial U(|q_{i_1} - q_{i_2}|)}{\partial r}\right)^2\right]^{1/2}\right\}.$$

The last integral does not exceed const n by virtue of Corollary 2 of §2. Consequently, for sufficiently large c_2 the last expression is a term of a convergent series. The main lemma is proved.

I am grateful to A. Zemlyakov and R. A. Minlos for valuable comments.

LITERATURE CITED

1. N. N. Bogolyubov, D. Ya. Petrina, and V. I. Khatset, Teor. Mat. Fiz., 1, 251 (1969).
2. R. L. Dobrushin, Teoriya Veroyatnostei i Ee Primeneniya, 2, 201 (1968).
3. R. A. Minlos, Funktsional'nyi Analiz i Ego Prilozheniya, 2, 60 (1967).
4. D. Ruelle, Statistical Mechanics, New York (1969).
5. O. Lanford, Commun. Math. Phys., 9, 179 (1968).
6. O. Lanford, Commun. Math. Phys., 11, 257 (1969).
7. R. A. Minlos and A. M. Khalfina, Izv. Akad. Nauk SSSR, Ser. Matem., 34, 1173 (1970).
8. A. M. Khalfina, Matem. Sb., 80, 3 (1969).
9. A. Khaitov, Tr. Mosk. Matem. Ob-va (1973) (in press).
10. D. Ruelle, Commun. Math. Phys., 9, 267 (1968).
11. R. L. Dobrushin, Funktsional'nyi Analiz i Ego Prilozheniya, 1, 27 (1969).
12. N. N. Bogolyubov, "Problems of a dynamical theory in statistical physics," in: Studies in Statistical Mechanics, Vol. 1 (ed. J. de Boer and G. E. Uhlenbeck), North-Holland, Amsterdam (1962).
13. N. N. Bogolyubov, Zh. Éksp. Teor. Fiz., 16, 691 (1946).
14. W. Feller, An Introduction to Probability Theory and Its Applications, Vol. 1, Wiley, New York (1950).

Comments

First results concerning the existence and uniqueness of solutions of the infinite dimensional system of ODE describing the dynamics of the infinite set of particles in statistical mechanics were proven by O. Lanford (see [L1]).

But the results were unsatisfactory because many restrictions on the dynamics were imposed.

In this paper, I construct for one-dimensional particles the so-called cluster dynamics in which for each value of time t_0 the whole ensemble of particles can be decomposed into finite sets (clusters) so that each set moves independently of the other sets during the time interval $[0, t_0]$.

The idea of cluster dynamics in the gas phase can be found in the works of Bogolyubov. Some time later C. Marchioro, A. Pellegrinotti and E. Presutti (see [MPP]) proved a general theorem giving even in the multi-dimensional case the existence and uniqueness of dynamics preserving every equilibrium Gibbs measure.

The problem was solved more or less completely in the paper by R. L. Dobrushin and J. Fritz (see [DF]) in which they proved the existence of dynamics even in nonequilibrium ensembles.

Some numerical studies of cluster dynamics can be found in the paper [GKSZ].

PHASE DIAGRAMS OF CLASSICAL LATTICE SYSTEMS

S.A. Pirogov and Ya. G. Sinai

A method is developed for constructing phase diagrams for a large number of classical lattice systems. For these systems, a description is obtained of the stable thermodynamic phases in terms of contour models. The general theory is applied to the investigation of spontaneous symmetry breaking and the construction of phase diagrams for a number of concrete lattice systems.

I. Basic Definitions. Formulation of the Basic Theorem

This paper is devoted to an investigation of the phase diagrams of classical lattice systems, and it consists of six sections. For technical reasons, the material is divided into two parts published in two issues of this journal. This, the first part, contains Secs. I and II, in which we set forth in detail the main problem, formulate the basic theorem, and derive consequences from it. The second part will be published in the next issue. Sections III-V contain the complete proof of the basic theorem. The proof is based on the method of contour functionals proposed and developed in the earlier papers [1-3]. In these sections, we formulate without proof some of the results of [1], which we use to prove the basic theorem. In Sec. VI we derive results on spontaneous symmetry breaking.

We now discuss the problems. Suppose $H(p, q) = p^2/2 + U(q)$ is the Hamilton function of some classical dynamical system with bounded configuration space. Then for large β the main contribution to the integral

$$\int f(p, q) \exp(-\beta H(p, q))\, dp\, dq \tag{1}$$

is made by integrals over small neighborhoods of the points where the function H attains an absolute minimum. These points are called the ground states of the Hamiltonian H. Rigorous assertions of this kind are usually proved under "stability" assumptions on the ground states. What follows can, in a certain sense, be regarded as a generalization of the above fact to the infinite-dimensional case. Of course, a direct infinite-dimensional generalization without specification of the types of the Hamiltonians H is rather meaningless. We shall study infinite-dimensional Hamiltonians H corresponding to classical lattice models of statistical physics. The phase spaces of these systems are constructed from discrete objects and are therefore somewhat simpler than in the continuous case. For the lattice systems considered we define what is a ground state of the Hamiltonian (in the class of periodic configurations). Further, we introduce an assumption about the Hamiltonian, satisfied for a large class of models, that the set of ground states is finite. Under this assumption, we formulate a condition on the Hamiltonian, which we call the Peierls condition, and which can be naturally regarded as an analog of the condition of "stability" of ground states. With regard to the neighborhood of a ground state, on the transition to the infinite-dimensional case it is replaced, as is well known to specialists in probability theory and quantum field theory, by the set of "typical realizations" of the infinite-dimensional probability distribution concentrated around this ground state.

We now turn to more precise definitions. Since we consider lattice models, we assume that a lattice is given in the space R^d. For what follows, only the case $d \geq 2$ is of interest. In contrast, the form of the lattice is not of significance to us, and for simplicity we shall deal only with an ordinary integral lattice Z^d. Let X be a finite set. We define a configuration of the system as a function $s(i)$, $i \in Z^d$, with values in X, i.e., $s(i) \in X$. We denote the set of all configurations by $\mathfrak{M}$. This set is the

Institute of Information Transmission Problems, Academy of Sciences of the USSR; L.D. Landau Institute of Theoretical Physics, Academy of Sciences of the USSR. Translated from Teoreticheskaya i Matematicheskaya Fizika, Vol. 25, No. 3, pp. 358-369, December, 1975. Original article submitted April 21, 1975.

Y.G. Sinai, *Selecta II: Probability Theory, Statistical Mechanics, Mathematical Physics and Mathematical Fluid Dynamics*, DOI 10.1007/978-1-4419-6205-8_7, © Springer Science+Business Media, LLC 2010

phase space of our infinite-dimensional system and the group Z^d of spatial shifts acts on it. We shall say that a periodic configuration is a configuration $s(i)$ that is invariant under a certain subgroup $\hat{Z} \subset Z^d$ of finite index. We shall denote the set of all periodic configurations by $\mathfrak{M}^{per}$.

We now turn to the description of the Hamiltonian H. We assume that for every $i \in Z^d$ there is given a function $U_i(s(j), |j-i| \leq R)$ of the variables $s(j)$, $|j-i| \leq R$, and this describes the interaction of the variable $s(i)$ with the other variables $s(j)$ in an R neighborhood of it. The existence of the number R, which is the same for all i, means that we consider systems in which the interaction range of the particles is finite (and equal to R).

In many cases one can assume that U_i does not depend on i, i.e., the interaction is invariant under the group of spatial shifts Z^d. We shall consider the hardly more general situation when U_i is periodic, i.e., invariant under some subgroup $\hat{Z} \subset Z^d$ of finite index. For a number of models this generalization is helpful. In practice, the assumption of periodicity means that there is a finite number of functions U_i and that the function which describes the interaction $s(i)$ with its neighbors in the neighborhood of radius R is determined by the coset that contains the point $i \in Z^d$.

One would like to write the Hamiltonian of the system as the infinite sum

$$H(s) = \sum_{i \in Z^d} U_i(s(j), |j-i| \leqslant R). \tag{2}$$

But obviously this would be meaningless. Fortunately, in none of the cases does the function H itself appear, but only the difference $H(s') - H(s'') = H(s', s'')$ for a pair of configurations s' and s'' which coincide almost everywhere, i.e., everywhere except for a finite number of places. And these differences are meaningful. Namely,

$$H(s', s'') = \sum_{i \in Z^d} (U_i(s') - U_i(s'')), \tag{2'}$$

where for brevity we have set $U_i(s) = U_i(s(j), |j-i| \leqslant R)$.

In the last sum only a finite number of terms are nonzero.

We shall call $H(s', s'')$ the relative Hamiltonian. In all that follows, only relative Hamiltonians are important. Below, we shall drop the word "relative". We shall say that an expression of the type (2) is a formal Hamiltonian.

Having the Hamiltonian $H = H(s', s'')$, we can introduce the definition of a ground state.

DEFINITION 1.1. A periodic configuration a is called a ground state of the Hamiltonian H if for any configuration s that coincides with a almost everywhere

$$H(s, a) \geqslant 0. \tag{3}$$

A ground state can be defined differently as a periodic configuration with smallest specific energy. The specific energy is defined as a function on the space of periodic configurations defined to within an additive constant as follows. Suppose $s, s' \in \mathfrak{M}^{per}$ and V is a cube in the space R^d with edges parallel to the coordinate axes. We set

$$s_V(i) = \begin{cases} s(i), & i \in V, \\ s'(i), & i \notin V. \end{cases}$$

Then by definition the difference of the specific energies of the configurations s and s' is equal to the limit

$$\lim_{|V| \to \infty} \frac{H(s_V, s')}{|V|} = \lambda_H(s, s'), \tag{4}$$

where $|V|$ is the number of points $i \in V$. The existence of a limit follows from the periodicity of the configurations s and s' and the interaction $U_i(s)$. It is easy to show that on $\mathfrak{M}^{per}$ one can define a function $e_H(s)$ such that $\lambda_H(s, s') = e_H(s) - e_H(s')$. The function $e_H(s)$ is defined uniquely to within an additive constant.

PROPOSITION 1.1. A periodic function a is called a ground state of the Hamiltonian H if and only if

$$e_H(a) \leqslant e_H(s) \tag{5}$$

for any $s \in \mathfrak{M}^{per}$.

(This is proved in the Appendix).

For any Hamiltonian H, we denote the set of its ground states by S(H). We now turn to the formulation of the requirement of stability of ground states. We impose on H a condition which will mean that any deviation from one of the ground states has energy proportional to the area of the surface that separates the regions of space occupied by the different ground states. We shall it the Peierls condition because Peierls as long ago as 1936 pointed out its importance for the ferromagnetic Ising model [12].

Suppose we are given a Hamiltonian H_0 with finite set of ground states $S(H_0) = \{s_1, \ldots, s_r\}$. Then there exists a common subgroup $\hat{Z} \subseteq Z^d$ of finite index with respect to which all the configurations $s_1, \ldots, s_r$ are invariant. We set $N = (Z^d : \hat{Z})$ and introduce the following notation. We denote the restriction of the configuration s to the subset $V \subset Z^d$ by pr(s, V). We define the distance between two points i, $j \in Z^d$ as $|i-j| = \max_{1 \leqslant k \leqslant d} |i_k - j_k|$. For a given point $i \in Z^d$ we denote by $U_N(i)$ the cube $\{j : |j - i| \leq N\}$.

DEFINITION 1.2. Suppose $s \in \mathfrak{M}$. The cube $U_N(i)$ is said to be an irregular cube of the configuration s if $\mathrm{pr}(s, U_N(i)) \neq \mathrm{pr}(s_q, U_N(i))$ for all $q = 1, \ldots, r$. The union of the irregular cubes of the configuration s is called the boundary of this configuration and denoted by B(s).

DEFINITION 1.3. The Hamiltonian H_0 with ground states $s_1, \ldots, s_r$ satisfies the Peierls condition if for any $q = 1, \ldots, r$ and any configuration s that coincides with s_q almost everywhere

$$H_0(s, s_q) \geqslant \rho |B(s)|, \tag{6}$$

where ρ is a positive constant. By $|B|$, $B \subset Z^d$, we shall, as above, denote the number of points $i \in B$.

From the intuitive point of view, Hamiltonians satisfying the Peierls condition must be regarded as Hamiltonians in which the ground states are separated from the excited states by a finite energy gap.

The subject of our investigation is the set of pure thermodynamic phases for the family of Hamiltonians $H_0 + \mu_1 H_1 + \ldots + \mu_{r-1} H_{r-1}$, where $H_1, \ldots, H_{r-1}$ are external fields applied to lift the degeneracy of the ground state.

If the external fields $H_1, \ldots, H_{r-1}$ really are to lift the r-fold degeneracy of the ground state, they must satisfy a certain condition of the type of linear independence. We shall now give an appropriate exact definition.

Having the Hamiltonian $L(\mu) = \mu_1 H_1 + \ldots + \mu_{r-1} H_{r-1}$, we define to within an additive constant the specific energies $e_\mu(s_q)$, $q = 1, \ldots, r$. We set

$$t_\mu(s_q) = e_\mu(s_q) - \min_q e_\mu(s_q), \quad q = 1, \ldots, r. \tag{7}$$

The set of numbers $t_\mu(s_q)$, $q = 1, \ldots, r$, can be regarded as a point of the boundary O_r of the positive octant R^r_+ of the space R^r, i.e., $O_r = \{a = (a_1, \ldots, a_r) : \min_q a_q = 0\}$.

DEFINITION 1.4. The set of Hamiltonians $H_1, \ldots, H_{r-1}$ lifts the degeneracy of the ground state of the Hamiltonian H_0 if the mapping

$$\varphi : \mu = (\mu_1, \ldots, \mu_{r-1}) \to t_\mu = (t_\mu(s_1), \ldots, t_\mu(s_r)) \tag{8}$$

maps the space of parameters $(\mu_1, \ldots, \mu_{r-1})$ onto the complete boundary O_r of the positive octant.

We now give a clear formulation of our basic theorem.

Let H_0 be a Hamiltonian with ground states $s_1, \ldots, s_r$ satisfying the Peierls condition and suppose the set of Hamiltonians $H_1, \ldots, H_{r-1}$ lifts the degeneracy of a ground state of H_0. Consider the family of Hamiltonians

$$H = H_0 + \mu_1 H_1 + \ldots + \mu_{r-1} H_{r-1}. \tag{9}$$

THEOREM. For all $\beta \geqslant \beta_0(H_0; H_1, \ldots, H_{r-1}; d)$

1) there exists a point $\bar{\mu}(\beta)$ in the space of parameters $(\mu_1, \ldots, \mu_{r-1})$ at which there exist r pure thermodynamic phases at the reciprocal temperature β;

2) there exist $r = C^1_r$ curves $\gamma_1^1(\beta), \ldots, \gamma_r^1(\beta)$ beginning at the point $\bar{\mu}(\beta)$ in the parameter space on each of which there exist $r - 1$ pure thermodynamic phases at the reciprocal temperature β;

3) there exist $C^2_r = r(r-1)/2$ bounded curves $\gamma_1^1(\beta), \ldots, \gamma_r^1(\beta)$ of two-dimensional surfaces $\gamma_{ij}^2(\beta)$, $1 \leqslant i < j \leqslant r$, in the parameter space on each of which there exist $r - 2$ pure thermodynamic phases

at the reciprocal temperature β; and so forth, right up to

4) there exist $r = C_r^{r-1}$ bounded $r - 2$-dimensional surfaces constructed at the point $r - 1$ of open domains $\gamma_1^{r-1}(\beta), \ldots, \gamma_r^{r-1}(\beta)$ in the parameter space on each of which there exists one pure thermodynamic phase at the reciprocal temperature β.

To formulate the theorem more accurately, it remains to define what we mean by pure thermodynamic phases. The idea of Dobrushin and Lanford and Ruelle is to define thermodynamic phases by means of probability distributions on $\mathfrak{M}$ that are limits of Gibbs distributions in finite volumes $V \subset Z^d$ with definite boundary conditions (see below).

These limits are called limit Gibbs distributions. According to Dobrushin [4] and Lanford and Ruelle [10], limit Gibbs distributions can be defined in a purely internal manner without recourse to a limiting process. Namely, suppose we are given a finite volume $V \subset Z^d$ and a fixed configuration $\bar{s} \in \mathfrak{M}$. Consider the finite set $\mathfrak{M}(\bar{s}, V) \subset \mathfrak{M}$, consisting of the configurations s' for which $s'(i) = \bar{s}(i)$ for $i \notin V$. We define a probability distribution $P_{\bar{s}, V}$ on the set $\mathfrak{M}(\bar{s}, V)$ in such a way that

$$P(s')/P(s'') = \exp(-\beta H(s', s'')) \tag{10}$$

for all $s', s'' \in \mathfrak{M}(\bar{s}, V)$.

The probability distribution $P_{\bar{s}, V}$ is called a Gibbs distribution in the finite volume V with the boundary conditions $\bar{s}$.

DEFINITION 1.5 ([4]). A limit Gibbs distribution for a Hamiltonian H and parameter $\beta > 0$ is defined as a probability distribution P on $\mathfrak{M}$ for which the conditional probabilities

$$P\{s(i), i \in V \mid s(i) = \bar{s}(i), i \notin V\} \tag{11}$$

P-almost certainly coincide with $P_{\bar{s}, V}\{s(i), i \in V\}$.

The main problem of equilibrium statistical physics in relation to the models considered here is to describe, for every Hamiltonian H and every $\beta > 0$, the convex set of limit Gibbs distributions. The limit Gibbs distribution P is called a pure thermodynamic phase if it is spatially periodic and an extreme point of the convex set of all spatially periodic limit Gibbs distributions.

Our aim is to study the set of pure thermodynamic phases for the Hamiltonian $H = H_0 + \sum \mu_i H_i$ as a function of the values of the parameters μ_i, $1 \leq i \leq r - 1$, for a fixed, sufficiently large parameter $\beta > 0$.

THEOREM 1. Let H_0 be a Hamiltonian with ground states $s_1, \ldots, s_r$ satisfying the Peierls condition, and suppose the set of Hamiltonians $H_1, \ldots, H_{r-1}$ lift the degeneracy of a ground state of H_0. Then for all $\beta > \beta_0(H_0; H_1, \ldots, H_{r-1}; d)$ there is defined a homeomorphism of the parameters $(\mu_1, \ldots, \mu_{r-1})$ onto a certain neighborhood of the point $0 = (0, \ldots, 0)$ on O_r which describes the phase diagram as follows. For the Hamiltonian $H = H_0 + \sum \mu_i H_i$, $\mu = (\mu_1, \ldots, \mu_{r-1}) \in U$ and given $\beta > \beta_0$ there exist different pure thermodynamic phases P_q for the q values for which $a_q = 0$ for $a = (a_1, \ldots, a_r) = \mathcal{F}(\beta)\mu$.

This theorem encompasses all the previously obtained special results on phase diagrams of classical lattice systems with finite interaction range (see [5], [6], [7], [8], [9], [16]).

From the proof of the theorem there follows additional information about the structure of the typical configurations for each of the constructed distributions. Namely, for the distribution P_q almost every configuration s is such that on a connected set whose density on the lattice Z^d tends to unity as $\beta \to \infty$ the configuration s coincides with s_q, and all the connected components of the set $\{i: s(i) \neq s_q(i)\}$ are finite.

Theorem 1 can be extended to the case of slightly more general lattice models for which the potential has a hard core. We introduce the corresponding definitions. Suppose now that the function $U_i(s(j), |j - i| \leq R)$ takes finite values or $+\infty$ and is spatially periodic in the same sense as before. We introduce the formal Hamiltonian with hard core

$$H = \sum_{i \in Z^d} U_i(s(j), |j-i| \leqslant R). \tag{12}$$

DEFINITION 1.6. A configuration $s \in \mathfrak{M}$ is said to be admissible if $U_i(s(j), |j-i| \leqslant R) < \infty$ for all i.

We denote the set of admissible configurations of the formal Hamiltonian H by $\mathfrak{M}_H$. If U_i is periodic with respect to the subgroup $\hat{Z} \subset Z^d$, then so is the set $\mathfrak{M}_H$. The formal Hamiltonians considered above can be characterized by the fact that for them $\mathfrak{M}_H = \mathfrak{M}$. We shall call them bounded formal Hamiltonians. We shall also say that the corresponding relative Hamiltonians are bounded.

The relative Hamiltonian $H(s', s'')$ is defined for any pair of configurations $s', s'' \in \mathfrak{M}_H$ that coincide almost everywhere. It is convenient to augment the definition in the case when the configurations s' and s'' coincide almost everywhere and $s' \notin \mathfrak{M}_H$, $s'' \in \mathfrak{M}_H$ by setting $H(s', s'') = +\infty$.

For these Hamiltonians, without any modifications, one can define ground states and formulate the Peierls condition.

Suppose we are given a Hamiltonian H_0 with hard core satisfying the Peierls condition. We include it in the family of Hamiltonians $H_0 + \mu_1 H_1 + \ldots + \mu_{r-1} H_{r-1}$, where $H_1, \ldots, H_{r-1}$ are bounded Hamiltonians that lift the degeneracy of the ground state of the Hamiltonian H_0.

The formulation of Theorem 1 is not changed.

For all the Hamiltonians $H = H_0 + \sum \mu_i H_i$ the sets of admissible configurations $\mathfrak{M}_H$ are identical.

Appendix

Proof of Proposition 1.1. It follows from the definition of a ground state that if a is a ground state of the bounded Hamiltonian H and s is an arbitrary periodic configuration, then $e_H(a) \le e_H(s)$.

We show that if $a \in \mathfrak{M}^{per}$ is not a ground state of H, then there exists $b \in \mathfrak{M}^{per}$ for which $e_H(b) < e_H(a)$.

Since a is not a ground state, there exists a configuration s that coincides with a almost everywhere and for which $H(s, a) < 0$.

We denote by A the R neighborhood of the set $\{i: s(i) \ne a(i)\}$ (where R is the range of the interaction Hamiltonian H).

Suppose that $\hat{Z} \subset Z^d$ is a subgroup of finite index with respect to which the Hamiltonian H and the configuration a are invariant and that all points of the set A belong to different cosets $Z^d/\hat{Z}$. We introduce a finite set $K \subset Z^d$ for which 1) $K \supset A$; 2) K contains one element from each coset $Z^d/\hat{Z}$.

Then there exists a unique configuration $b \in \mathfrak{M}^{per}$ which is invariant under the group $\hat{Z}$ and coincides with the configuration s on K. It is easy to see that $e_H(b) < e_H(a)$.

II Examples

1. Ising Model with Attraction. The simplest example of a Hamiltonian satisfying the Peierls condition is that of the Ising model. There exist several equivalent descriptions of this model. We shall choose the one that enables one to verify the Peierls condition in the most direct manner. Namely, for the Ising model (with attraction) in this interpretation $X = \{0, 1\}$, (i.e., the Ising model is regarded as a model of a lattice gas ([11]), and the Hamiltonian is written in the form

$$H_0 = \frac{1}{2} \sum_{d(i,j)=1} U(s(i), s(j)), \tag{13}$$

where

$$d(i, j) = \sqrt{\sum (i_k - j_k)^2},$$

and

$$U(x, y) = \begin{cases} 1, \text{ if } x \ne y; \\ 0, \text{ if } x = y. \end{cases}$$

The Hamiltonian H_0 satisfies the Peierls condition with two ground states $s_0 = 0$ and $s_1 = 1$. The particular interest in the Ising model is due primarily to the fact that for the Ising model with $d = 2$ the thermodynamic functions can be calculated explicitly. In [3] it was shown that for the Hamiltonian H_0 for $d \ge 2$ and at sufficiently low temperatures there exist two different limit Gibbs distributions.

The proof of the existence of two different limit Gibbs distributions is due basically to the early work of Peierls [12], who introduced for the first time configuration contours. In [1], [2] a contour model is defined, the properties of contour models are investigated, and they are used to study the behavior of an isotherm near a phase transition point. In [3], detailed studies are made of the properties of typical configurations at a phase transition point. In [14] it is shown that for Hamiltonian $H_h = H_0 + hN$, where $N=\sum_i s(i)$ for $h \neq 0$, there exists exactly one limit Gibbs distribution, and its properties are studied. This result is generalized by Martirosyan [15] to the case of small perturbations of the Ising model (see below).

2. Small Perturbations. A large class of Hamiltonians satisfying the Peierls condition can be obtained from one given Hamiltonian by means of small perturbations. The following proposition holds.

PROPOSITION 2.1. If the Hamiltonian H_0 satisfies the Peierls condition with the ground states $s_1, \ldots, s_r$ and constant ρ, and $L_1, \ldots, L_k$ are bounded Hamiltonians such that for each of them the difference of the specific energies satisfies $\lambda_{L_i}(s_p, s_q)=0,\ 1\leqslant p,\ q\leqslant r$, the Hamiltonian $H=H_0+\sum\alpha_iL_i$ for sufficiently small $|\alpha_i|$, $i = 1, \ldots, k$, satisfies the Peierls condition with the ground states $s_1, \ldots, s_r$ and constant $\rho - C\cdot\max_i|\alpha_i|$ (where C is a number that depends on $L_1, \ldots, L_k$).

Proof. Since $\lambda_L(s_p, s_q) = 0$, for any configuration s that almost everywhere coincides with s_q, $L_i(s, s_q) \leq C_i|B(s)|$, where $C_i > 0$ is a constant. Therefore

$$H(s, s_q)\geqslant(\rho - C\cdot\max_i|\alpha_i|)|B(s)|, \tag{14}$$

where $C=\sum C_i$.

If H_0 is the Hamiltonian of the Ising model with attraction and L is an arbitrary bounded Hamiltonian such that $\lambda_L(s_0, s_1) = 0$ (s_0 and s_1 are ground states of the Hamiltonian H_0), then in accordance with Proposition 2.1 the Hamiltonian $H = H_0 + \varepsilon L$ for sufficiently small $|\varepsilon|$ satisfies the Peierls condition with the ground states s_0 and s_1.

Applying Theorem 1, we obtain

COROLLARY. There exist positive constants ε_0 and β_0 and a function $h = h(\beta, \varepsilon)$ defined for $\beta > \beta_0$, $|\varepsilon| < \varepsilon_0$ such that for the Hamiltonian $H_h=H_0+\varepsilon L+h(\beta, \varepsilon)\cdot N$ and the parameter β there exist two different limit Gibbs distributions.

Martirosyan showed that for the Hamiltonian $H_h=H_0+\varepsilon L+hN$, $h\neq h(\beta, \varepsilon)$ and the parameter β there exists exactly one limit Gibbs distribution [15].

3. Ising Model with Several States. The Ising model with several states differs from the ordinary Ising model with attraction in that for it $X = \{1, \ldots, r\}$, and the function $U(x, y)$ has the following property: $U(x, x) = 0$ for $x = 1, 2, \ldots, r$ and $U(x, y) > 0$ if $x \neq y$.

The Hamiltonian H_0 of the Ising model with several states satisfies the Peierls condition with r ground states.

As Hamiltonians $H_1, \ldots, H_{r-1}$ that lift the degeneracy of the ground state of H_0 we use the Hamiltonians N_q, $1\leqslant q\leqslant r-1$,

$$N_q=\sum_i U_q(s(i)), \tag{15}$$

where

$$U_q(x)=\begin{cases}1, & \text{if } x=q,\\ 0, & \text{if } x\neq q.\end{cases}$$

The Ising model with several states is the simplest model that enables one to illustrate Theorem 1. For this model, one can slightly strengthen the formulation of Theorem 1 by noting that as neighborhood of the origin in the parameter space one can take the entire space, and $\mathcal{J}(\beta)$ is the mapping of the parameter space onto the complete boundary O_r of the positive octant.

4. Model with Repulsion. Models with repulsion between particles differ from the ones considered above in that their ground states are not translationally invariant but are periodic with period that depends on the model. The simplest of these models is the Ising model with repulsion, which was investigated by Dobrushin in [5]. For the Ising model with repulsion, $X = \{0, 1\}$, and this model differs from the one with attraction by the fact that

$$U(x,y)=\begin{cases}1, & \text{if } x=y;\\ 0, & \text{if } x\neq y.\end{cases}$$

It is easy to show that not only the Hamiltonian H_0 of the Ising model with repulsion but also the entire family of Hamiltonians $H_h = H_0 + h\cdot N$ $\left(N=\sum_i s(i)\right)$ for $|h| < 2d$ satisfy the Peierls condition with the two ground states s_1 and s_2, which can be described as follows.

We shall say that a point $i \in Z^d$ is even if $\sum_k i_k$ is even and odd otherwise. The division of Z^d (say for $d = 2$) into even and odd points is analogous to the division of a chessboard into white and black squares. Then

$$s_1(i)=\begin{cases}0, & \text{if i is even;}\\ 1, & \text{if i is odd.}\end{cases}\quad s_2(i)=\begin{cases}1, & \text{if i is even;}\\ 0, & \text{if i is odd.}\end{cases}$$

The constant ρ_h with which the Hamiltonian H_h satisfies the Peierls condition depends on h; namely, ρ_h is proportional to $2d - |h|$. In [5] it is shown that for the Hamiltonian H_h, $|h| < 2d$, for $\beta > \beta_0(h)$ there exist two different limit Gibbs distributions. These Gibbs distributions are not translationally invariant, and go over into one another under a shift of the integral lattice Z^d through 1 along any of the coordinate axes.

Another model with repulsion between particles is the model of hard spheres. For this model, $X = \{0, 1\}$, and the Hamiltonian H_0 is determined by a function $U(x, y)$ of the form

$$U(x,y)=\begin{cases}+\infty, & \text{if } xy=1;\\ 0, & \text{if } xy=0.\end{cases}$$

All the Hamiltonians $H_h = H_0 + h\cdot N$, $h < 0$, satisfy the Peierls condition with ground states s_1 and s_2 that are the same as for the Ising model with repulsion and constant ρ_h proportional to $|h|$. In [5] it is shown that for the Hamiltonians H_h, $h < 0$, for $\beta > \beta_0(h)$ there exist two different limit Gibbs distributions, these being analogous to the two for the Ising model with repulsion.

5. Fisher Antiferromagnet. The Fisher antiferromagnet belongs to the large class of so-called spin models that are usually used to describe magnetic systems. For these models, $X = \{-1, +1\}$. The Ising model (with attraction and with repulsion) can also be regarded as a spin system if one goes over from the variables $s(i) = 0, 1$ to the variables $2s(i) - 1$, which take the values +1 and −1 and are called spin variables. The Hamiltonian H_0 is written in the form

$$H_0=\sigma\frac{1}{4}\sum_{d(i,j)=1} s(i)s(j), \tag{16}$$

and by $s(i)$, $i \in Z^d$ we denote spin variables which take the values +1 and −1, and $\sigma = -1$ for the Ising model with attraction and $\sigma = +1$ for the model with repulsion. The Hamiltonian N can be expressed in terms of the spin variables $s(i)$ in the form $N={}^1/_2\sum_i s(i)$ and interpreted as the total magnetic moment of the system, so that the term $h\cdot N$ in the Hamiltonian $H_h = H_0 + h\cdot N$ now corresponds to an external magnetic field of strength h. The Ising model with attraction (in the spin interpretation) is used to describe ferromagnetism, and the Ising model with repulsion (in the spin interpretation) to describe antiferromagnetism. In the Fisher model, nearest spins interact in the same way as in the Ising antiferromagnet, while spins that are neighbors along a diagonal interact as in an Ising ferromagnet; more precisely,

$$H_h=\frac{1}{4}\sum_{d(i,j)=1} s(i)s(j)-\frac{\mathcal{J}}{4}\sum_{d(i,j)=\sqrt{2}} s(i)s(j)+\frac{h}{2}\sum_i s(i). \tag{17}$$

Thus, the Hamiltonian of the Fisher model contains two parameters, $\mathcal{J}>0$ (attraction between diagonal neighbors) and h (external magnetic field). From the results of Sec. VI of our paper, using some of the specific features of the Fisher model, one can obtain

<u>THEOREM.</u> There exists a positive constant $\mathcal{D}$ and a positive function $h(\beta, \mathcal{J})$, which is defined for $\mathcal{J}>0$, $\beta>\mathcal{D}/\mathcal{J}$, such that for the Fisher model with given $\mathcal{J}>0$ and parameter $\beta>\mathcal{D}/\mathcal{J}$

a) for $|h|<h(\beta, \mathcal{J})$ there exist two different limit Gibbs distributions;

b) for $|h|<h(\beta, \mathcal{J})$ there exist three limit Gibbs distributions, no one of which is a convex combination of the other two.

The two Gibbs distributions that exist for $|h|<h(\beta, \mathscr{J})$ have similar properties to the two for the Ising model with repulsion, and the third Gibbs distribution, which arises for $h=\pm h(\beta, \mathscr{J})$ and continues its existence for $|h|>h(\beta, \mathscr{J})$, when the first two disappear, is transaltionally invariant and analogous to the Gibbs distribution for the Ising model with attraction in nonzero external field.

Sections III-VI of our paper will appear in the next issue of this journal.

LITERATURE CITED

1. R. A. Minlos and Ya. G. Sinai, Tr. Mosk. Mat. Obshch., 19, 113 (1968).
2. R. A. Minlos and Ya. G. Sinai, Tr. Mosk. Mat. Obshch., 17, 213 (1967).
3. R. A. Minlos and Ya. G. Sinai, Mat. Sb., 73, No. 3, 375 (1967).
4. R. L. Dobrushin, Funktsional. Analiz i Ego Prilozhen., 2, No. 4, 31 (1968).
5. R. L. Dobrushin, Funktsional. Analiz i Ego Prilozhen., 2, No. 4, 44 (1968).
6. V. M. Gertsik and R. L. Dobrushin, Funktsional. Analiz i Ego Prilozhen., 8, No. 3, 12 (1974).
7. S. A. Pirogov and Ya. G. Sinai, Funktsional. Analiz i Ego Prilozhen., 8, No. 1, 25 (1974).
8. S. A. Pirogov, Dokl. Akad. Nauk SSSR, 214, No. 6, 1273 (1974).
9. S. A. Pirogov, Usp. Mat. Nauk, 30, No. 2, 223 (1975).
10. O. E. Lanford and D. Ruelle, Commun. Math. Phys., 13, No. 3, 194 (1969).
11. D. Ruelle, Statistical Mechanics, New York (1969).
12. R. E. Peierls, Proc. Cambridge Philos. Soc., 32, 477 (1936).
13. R. L. Dobrushin, Teor. Mat. Fiz., 12, No. 1, 115 (1972).
14. A. Nasr, Usp. Mat. Nauk, 26, No. 5, 222 (1971).
15. D. G. Martirosyan, Teor. Mat. Fiz., 22, No. 3, 335 (1975).
16. V. M. Gertsik, Usp. Mat. Nauk, 30, No. 3, 159 (1975).

PHASE DIAGRAMS OF CLASSICAL LATTICE SYSTEMS

CONTINUATION*

S.A. Pirogov and Ya.G. Sinai

III. Contour Models

In this section we introduce contour models and study some of their properties. In Sec. V we shall show that every pure thermodynamic phase which we construct can be described by means of a suitably chosen contour model. In the language of contour models one can clearly elucidate what is meant by the expression "typical configurations of every phase are small perturbations of a corresponding ground state".

Definition of Contours

Suppose we have a fixed Hamiltonian H_0 with ground states $s_1, \ldots s_r$ satisfying the Peierls condition and a fixed set of Hamiltonians $H_1, \ldots, H_{r-1}$ that lift the degeneracy of the ground state of H_0. As in Sec. I, we denote by N the index of the largest subgroup $Z \subset Z^d$ with respect to which all the configurations $s_1, \ldots, s_r$ are invariant. We choose a number $\bar{R}$ sufficiently large for the following conditions to hold: 1) $\bar{R}$ is greater than the range of each of the Hamiltonians $H_0, H_1, \ldots, H_{r-1}$; 2) $\bar{R} > N$.

We denote by $U_{\bar{R}}(i)$ the cube $\{j : |j - i| \leq \bar{R}\}$. The following definition is a modification of Definition (1.2).

DEFINITION (3.1). Let s be an arbitrary configuration. The cube $U_{\bar{R}}(i)$ is said to be an irregular cube of the configuration s if $\mathrm{pr}(s, U_{\bar{R}}(i)) \neq \mathrm{pr}(s_q, U_{\bar{R}}(i))$ for all $q = 1, \ldots, r$.

In what follows, the union of the irregular cubes $U_{\bar{R}}(i)$ of the configuration s will be called the boundary of the configuration s. This definition of the boundary of the configuration differs from the one given in Sec. I, but the Peierls condition is retained with this change in the definition, and only the constant ρ changes. The word boundary will not henceforth be used in the previous meaning, so that the notation B(s) is retained for the boundary in the new sense.

Two subsets $K, L \subset Z^d$ are said to be far apart if $\mathrm{dist}(K, L) > 1$. (We emphasize that dist is understood in the sense of the metric defined on Russian page 361 of Part I.) A subset $K \subset Z^d$ is said to be connected if it cannot be represented as the union of two subsets that are far apart. The last two concepts can be made geometrically more perspicuous if with each point of the lattice $i \in Z^d$ we associate a closed unit cube with center at this point, and with every subset $K \subset Z^d$ the union of all these cubes with respect to the points $i \in K$. We denote this union by [K]. It is easy to see that: 1) $K, L \subset Z^d$ are far apart if and only if [K] and [L] do not intersect; 2) K is connected if and only if [K] is connected in the usual sense.

Any set $K \subset Z^d$ can be uniquely decomposed into connected components that are far apart.

DEFINITION (3.2). Let s be an arbitrary configuration $i \in Z^d$. A cube $U_N(i)$ is said to be a q-regular cube of the configuration s, $1 \leq q \leq r$, if $\mathrm{pr}(s, U_N(i)) = \mathrm{pr}(s_q, U_N(i))$. If $|i - j| = 1$ and $U_N(i)$, $U_N(j)$ – q, q′ are regular cubes of the configuration s, respectively, then $q = q'$. Therefore, if K is connected set and for every point $i \in K$ the cube $U_N(i)$ is q(i)-regular, then $q(i) = q$ is constant on K.

* Sections I and II of this investigation were published in the previous issue (Teor. Mat. Fiz., 25, 358 (1975)).

L. D. Landau Institute of Theoretical Physics, Academy of Sciences of the USSR; Institute of the Problems of Information Transmission, Academy of Sciences of the USSR. Translated from Teoreticheskaya i Matematicheskaya Fizika, Vol. 26, No. 1, pp. 61-76, January, 1976. Original article submitted April 21, 1975.

DEFINITION (3.3). The pair $(M, pr(s, M))$, where M is the connected component of $B(s)$ and $pr(s, M)$ is the restriction of the configuration s to the set M is called a contour of s.

DEFINITION (3.4). The pair $\Gamma = (M, s_M)$ consisting of the connected subset $M \subset Z^d$ and the function $s_M(i)$ defined on M with values in X is called a contour; it can be a contour of only one configuration s.

The set M is called the support of Γ and denoted $\operatorname{supp}\Gamma$.

If we were dealing with only the Ising model (or a small perturbation of it), contours could be defined more geometrically, and they would then be closed nonself-intersecting curves in the case $d = 2$ (see [1-4]). We shall show below that many properties of these simplest contours can be transferred to our more complicated objects.

2. Interior of a Contour

Let $\Gamma = (M, s_M)$ be a contour with support M. The set $Z^d \backslash M$ decomposes into a finite number of connected components A_α, and for every α the set $\partial A_\alpha = \{i \in A_\alpha : \operatorname{dist}(i, M) = 1\}$ is connected.

If Γ is a contour with finite support M, $|M| < \infty$, and dimension $d \geq 2$, then among the connected components A_α all except one are finite.

We shall say that the finite connected components A_α are the components of the interior of the contour Γ and that the infinite connected component is the exterior of the contour Γ.

If Γ is a contour of the configuration s, then for every point $i \in \partial A_\alpha$ the cube $U_N(i)$ is $q(i)$-regular. It then follows from the foregoing that $q(i) = q$ is constant and can be regarded as a characteristic of the connected component A_α. We denote it by $q(A_\alpha)$. It is clear that $q(A_\alpha)$ depends only on the contour $\Gamma = (M, s_M)$ itself and not on the configuration s.

We define the m-interior, $1 \leq m \leq r$, of the contour Γ as the union of all the components A_α of the interior for which $q(A_\alpha) = m$. We denote the m-interior of Γ by $\operatorname{Int}_m \Gamma$, and the exterior of Γ by $\operatorname{Ext}\Gamma$. If $q(\operatorname{Ext}\Gamma) = q$, we shall say that Γ is a contour with boundary conditions s_q, and to emphasize this we shall denote it by Γ^q.

With every contour $\Gamma = (M, s_M)$ we can associate a special configuration, which we denote s_Γ. This configuration is uniquely determined by the fact that it coincides with s_M on M, while on A_α it coincides with $s_{q(A_\alpha)}$. The configuration s_Γ can also be uniquely characterized by the fact that the set of its contours consists of the unique contour Γ. If Γ is a contour with the boundary conditions s_q, then the configuration s_Γ almost everywhere coincides with s_q.

We define the interior of the contour Γ as the union of all the components of the interior, $\operatorname{Int}\Gamma = \bigcup_m \operatorname{Int}_m \Gamma$.

In what follows we shall use the notation* $V_m(\Gamma^q) = |\operatorname{Int}_m \Gamma^q|$, $V(\Gamma^q) = |\operatorname{Int}\Gamma^q|$, $|\Gamma^q| = |\operatorname{supp}\Gamma^q|$.

We consider only contours with finite support.

3. Exterior Contours

Suppose we are given a finite set of contours $\Gamma_1, \ldots, \Gamma_n$ whose supports are pairwise far apart. Then there exists a subset of these contours $\Gamma_{i_1}, \ldots, \Gamma_{i_k}$ such that

1) $\operatorname{supp}\Gamma_{i_t} \subset \operatorname{Ext}\Gamma_j$ for any $j \neq i_t$;
2) for any $j \neq i_1, \ldots, i_k$, $\operatorname{supp}\Gamma_j \subset \operatorname{Int}\Gamma_{i_t}$, for some t.

The contours $\Gamma_{i_1}, \ldots, \Gamma_{i_k}$ are called exterior contours of the set of contours $\Gamma_1, \ldots, \Gamma_n$. Note that if a given set of contours $\Gamma_1, \ldots, \Gamma_n$ is infinite, then from it one cannot always separate exterior contours. If Γ and Γ' are contours for which $\operatorname{supp}\Gamma \subset \operatorname{Ext}\Gamma'$ and $\operatorname{supp}\Gamma' \subset \operatorname{Ext}\Gamma$, then $\operatorname{Int}\Gamma \subset \operatorname{Ext}\Gamma'$, $\operatorname{Int}\Gamma' \subset \operatorname{Ext}\Gamma$.

Suppose we are given a configuration s that coincides almost everywhere with s_q. Consider the set of all contours of the configuration s. By definition, these contours have supports that are pairwise far apart, their number is finite, and their supports are finite. Therefore, among them we can

* By $|M|$, $M \subset Z^d$, we everywhere denote the number of points $i \in M$.

distinguish exterior contours. These exterior contours, which we shall henceforth call exterior contours of the configuration s, have boundary conditions s_q.

Having the Hamiltonian H, we consider for any contour Γ^q the configuration s^q_Γ corresponding to it and the relative energy $H(\Gamma^q)=H(s_{\Gamma^q}, s_q)$. We shall call this the energy of the contour Γ^q. Thus, there arises a function that is defined on the contours Γ^q (with fixed q). We shall say that such functions are contour functionals.

Let F_q be an arbitrary functional defined on the contours Γ^q. We define the norm $\|\cdot\|$ of this functional by

$$\|F_q\|=\sup_{\Gamma^q}\frac{|F_q(\Gamma^q)|}{|\Gamma^q|}. \tag{18}$$

For an arbitrary bounded Hamiltonian L we have the equation

$$L(\Gamma^q)=\sigma(L)(\Gamma^q)+\sum_m (h_m-h_q)V_m(\Gamma^q), \tag{19}$$

where $h_m = e_L(s_m)$ is the specific energy of s_m, and $\|\sigma(L)\| < \infty$.

For the contour Γ^q we denote by $\Lambda(\Gamma^q)$ the set of configurations s coinciding almost everywhere with s_q for which Γ^q is a unique exterior contour.

DEFINITION (3.5). The crystal partition function for the contour Γ^q of the Hamiltonian H and parameter $\beta > 0$ is defined by

$$\Xi(\Gamma^q|\beta H)=\sum_{s\in\Lambda(\Gamma^q)}\exp(-\beta H(s,s_q)). \tag{20}$$

Let V be a finite subset of the lattice Z^d. By $\Re_q(V)$ we denote the set of configurations s such that $s(i) = s_q(i)$ for $i\in Z^d\setminus V$, and also

1) $\operatorname{dist}(B(s), Z^d\setminus V)>1$,
2) for any exterior contour Γ^q of the configuration s, $\operatorname{Int}\Gamma^q\subset V$.

We shall say that the configurations $s\in\Re_q(V)$ are "rarefied" configurations in the volume V with boundary conditions s_q.

DEFINITION (3.6). The rarefied partition function for the volume V with boundary conditions s_q for the Hamiltonian H and parameter β is defined as

$$\Xi^q_{\text{rar}}(V|\beta H)=\sum_{s\in\Re_q(V)}\exp(-\beta H(s,s_q)). \tag{21}$$

For Hamiltonian H with range not exceeding $\bar{R}$ the rarefied and the crystal partition function are directly related to one another. This can be seen from the following lemma.

LEMMA 3.1. For any finite volume $V\subset Z^d$

$$\Xi^q_{\text{rar}}(V|\beta H)=\sum\prod_{i=1}^{n}\Xi(\Gamma_i^q|\beta H), \tag{22}$$

where the summation is over all possible sets of exterior contours $\{\Gamma_1^q,\dots,\Gamma_n^q\}$ with supports that are pairwise far apart and such that $\operatorname{supp}\Gamma_i^q\subset V$, $\operatorname{dist}(\operatorname{supp}\Gamma_i^q, Z^d\setminus V)>1$, $\operatorname{Int}\Gamma_i^q\subset V$ (we shall write these last three conditions in abbreviated form as $\Gamma_i^q \subset V$). The empty set of contours corresponds to a term equal to 1.

For any contour Γ^q

$$\Xi(\Gamma^q|\beta H)=\exp(-\beta H(\Gamma^q))\prod_m \Xi^m_{\text{rar}}(\operatorname{Int}_m\Gamma^q|\beta H). \tag{23}$$

The proofs of both assertions of the lemma follow directly from the definitions.

An important role in the proof of the fundamental Lemma 4.1 which follows below is played by the following obvious remark.

Suppose that for every contour Γ^q there is defined the quantity $\Xi(\Gamma^q)$ and for every finite volume $V\subset Z^d$ the quantity $\Xi_{\text{rar}}^q(V)$ is defined by Eq. (22) in terms of $\Xi(\Gamma^q)$.

LEMMA 3.2. If $\Xi(\Gamma^q)$, $\Xi_{\text{rar}}^q(V)$ satisfy Eqs. (23), then $\Xi(\Gamma^q)=\Xi(\Gamma^q|\beta H)$. Thus, Eqs. (23) augmented by the definition of the $\Xi_{\text{rar}}^q(V)$'s in accordance with Eq. (22) can be regarded as recursion relations for

$\Xi(\Gamma^q|\beta H)$.

We emphasize that in (22) and (23) the dependence on the original Hamiltonian H is expressed by the factor $\exp(-\beta H(\Gamma^q))$.

4. Contour Models

We now construct an auxiliary ensemble of sets of contours and consider on them special probability distributions. This ensemble of sets of contours will not be an ensemble of sets of contours of configurations, but certain statistical properties of the latter can be conveniently formulated using an auxiliary ensemble. This method of describing the statistical properties of an ensemble of sets of contours of a configuration was first proposed by Minlos and Sinai [3] and has been widely used in recent years in a number of investigations of phase transitions [1, 2, 5-7, 8-10]. For any finite set $V \subset Z^d$ we denote by $\mathfrak{C}(V)$ the ensemble whose element is the finite set of contours $\{\Gamma_1^q, \ldots, \Gamma_n^q\}$, $\Gamma_i^q \subset V$, with supports that are pairwise far apart. A set that contains no contour at all also occurs in $\mathfrak{C}(V)$ (the contours $\Gamma_1^q, \ldots, \Gamma_n^q$ are not in general contours of a single configuration s).

Suppose there is given a nonnegative contour functional F_q defined on the contours Γ^q. We consider the probability distribution on the ensemble $\mathfrak{C}(V)$ for which

$$P_V(\{\Gamma_1^q, \ldots, \Gamma_n^q\}|F_q) = \frac{\exp(-\sum_i F_q(\Gamma_i^q))}{\Xi_{rar}(V|F_q)}, \tag{24}$$

where the normalizing factor (contour rarefied partition function) is

$$\Xi_{rar}(V|F_q) = \sum \exp\Big(-\sum_i F_q(\Gamma_i^q)\Big), \tag{25}$$

and the summation is over all possible dispositions $\{\Gamma_1^q, \ldots, \Gamma_n^q\} \in \mathfrak{C}(V)$.

DEFINITION (3.7). The set of probability distributions P_V for all volumes V is called a contour model.

For given contour Γ^q we consider the set of collections of contours $\{\Gamma^q, \Gamma_1^q, \ldots, \Gamma_n^q\}$ with supports that are pairwise far apart and whose unique exterior contour is Γ^q. We call the sum of the statistical weights $\exp\Big(-F_q(\Gamma^q) - \sum_i F_q(\Gamma_i^q)\Big)$ of all these collections the contour crystal partition function $\Xi(\Gamma^q|F_q)$.

The contour rarefied partition function can be expressed in terms of the crystal partition functions by the same formula as in Lemma 3.1:

$$\Xi_{rar}(V|F_q) = \sum \prod_{i=1}^{n} \Xi(\Gamma_i^q|F_q), \tag{26}$$

where the summation is over all possible sets of exterior contours $\{\Gamma_1^q, \ldots, \Gamma_n^q\}$, $\Gamma_i^q \subset V$ with supports that are pairwise far apart. In addition, the crystal partition function $\Xi(\Gamma^q|F_q)$ can be written in the form

$$\Xi(\Gamma^q|F_q) = \exp(-F_q(\Gamma^q)) \prod_m \Xi_{rar}(\mathrm{Int}_m \Gamma^q|F_q). \tag{27}$$

In all of the following analysis, the contour models play a decisive role. As we shall see, the pure thermodynamic phases that we construct can be described by means of contour models, i.e., by means of appropriately chosen contour functionals.

In what follows, we require only periodic contour functionals. We give the corresponding definition.

Let $\hat{Z} \subset Z^d$ be some subgroup of finite index with respect to which the configuration s_q is invariant and the remaining configurations $\{s_m, m \neq q\}$ form an invariant set, i.e., are transformed by the group $\hat{Z}$ into each other. Then if Γ^q is a contour with boundary conditions s_q and $t \in \hat{Z}$, then $T_t\Gamma^q$ is also a contour with boundary conditions s_q.

DEFINITION (3.8). The functional F_q is said to be periodic if it is invariant under a subgroup $\hat{Z} \subset Z^d$ of the above form, i.e., $F_q(T_t\Gamma^q) = F_q(\Gamma^q)$ for $t \in \hat{Z}$.

5. Properties of Contour Models

For every contour model Peierls's inequality holds: For any contour Γ^q the probability of its belonging to the set of contours does not exceed $\exp(-F_q(\Gamma^q))$ (see [1]). More subtle properties of contour models can be established under the following additional condition: $F_q(\Gamma^q)\geqslant\tau|\Gamma^q|$, where $\tau > 0$ is a sufficiently large constant. We shall say that a functional which satisfies this last inequality is a τ functional. In [1, 2] the following important lemma is actually proved.

LEMMA 3.3. If τ is sufficiently large, i.e., $\tau \geq \tau_0$, and $V_i\uparrow Z^d$ as $i\to\infty$, i.e., $V_i\subset V_{i+1}$ and $\bigcup_i V_i=Z^d$, then the probability distributions $P_V(\cdot\,|\,F_q)$ converge weakly to the limit distribution $P(\cdot\,|\,F_q)$ on the ensemble $\mathfrak{C}_\infty$ of infinite sets of contours $\xi=\{\Gamma_i^q,\ i=1,2,\ldots,\}$ with supports that are pairwise far apart. With probability 1, each infinite set of contours $\xi\in\mathfrak{C}_\infty$ corresponds to an infinite set of exterior contours $\theta(\xi)$ lying one within the other. The density of the set of points of the lattice lying in the exterior of all the exterior contours in $\theta(\xi)$ tends to 1 as $\tau\to\infty$.

If F_q is a periodic τ functional, then there exists the limit

$$s(F_q)=\lim_{i\to\infty}\frac{\log\Xi_{\mathrm{rar}}(V_i|F_q)}{|V_i|} \tag{28}$$

with respect to any sequence of volumes V_i for which $\dfrac{|\partial V_i|}{|V_i|}\to 0$. At the same time $s(F_q)\leqslant\varepsilon(\tau)\to 0$ as $\tau\to\infty$ (here, by ∂V we understand the set of points $j\in V$ such that $\mathrm{dist}(j, Z^d\setminus V)=1$).

The following estimate of the remainder term in the expansion of the logarithm of the rarefied partition function holds:

$$\log\Xi_{\mathrm{rar}}(V|F_q)=s(F_q)\,|V|+\Delta(V|F_q),\quad |\Delta(V|F_q)|<\varepsilon(\hat{Z};\tau)\,|\partial V|, \tag{29}$$

where $\varepsilon(\hat{Z};\tau)$ depends not only on τ but also on the subgroup $\hat{Z}\subset Z^d$ with respect to which the functional F_q is invariant. For fixed subgroup $\hat{Z}$, $\varepsilon(\hat{Z};\tau)\to 0$ as $\tau\to\infty$.

It follows from (29) and also (27) that

$$\log\Xi(\Gamma^q|F_q)=s(F_q)\,V(\Gamma^q)-F_q(\Gamma^q)+\nabla(\Gamma^q|F_q), \tag{30}$$

where $\nabla(\Gamma^q|F_q)=\sum_m\Delta(\mathrm{Int}_m\,\Gamma^q|F_q)$.

In what follows, we consider contour functionals that satisfy the condition of Lemma 3.3. The constant τ_0 that occurs in the lemma depends on the ground states $s_1,\ldots,s_r$ and also on $\bar{R}$.

From the results of [1] we obtain more complete information about the distribution $P(\cdot\,|\,F)$ as well and also about the rate of convergence to the limit distribution.

6. Dependence of s(F) on F and Contour Models with a Parameter

We define a new norm in the space of functionals F_q:

$$|||F_q|||_c=\sup_{\Gamma^q}\frac{|F_q(\Gamma^q)|}{(|\Gamma^q|+V(\Gamma^q))\,c^{\delta(\Gamma^q)}}. \tag{31}$$

Here, c is a fized constant greater than 1 and $\delta(\Gamma^q)=\delta(\mathrm{supp}\,\Gamma^q)$, where $\delta(K)$ is the diameter of the subset $K\subset Z^d$ in the sense of the metric introduced on Russian page 361 of Part I. It follows from Peierls's inequality that

$$|s(F_q)-s(F_q')|<\varepsilon(\tau;c)\,|||F_q-F_q'|||_c \tag{32}$$

for any two periodic τ functionals F_q and F_q'. Here $\varepsilon(\tau;c)\to 0$ for $\tau\to\infty$ and fixed $c>1$.

It is sufficient to show that

$$|\log\Xi_{\mathrm{rar}}(V|F_q)-\log\Xi_{\mathrm{rar}}(V|F_q')|<\varepsilon(\tau;c)\,|||F_q-F_q'|||_c\,|V|. \tag{33}$$

By Peierls's inequality

$$\left|\frac{\partial\log\Xi_{\mathrm{rar}}(V|F_q)}{\partial F_q(\Gamma^q)}\right|=P(\Gamma^q)\leqslant\exp(-F_q(\Gamma^q)), \tag{34}$$

where $P(\Gamma^q)$ is the probability that the contour Γ^q belongs to the set of contours. Estimating the left-hand side of (33) in accordance with Lagrange's formula and remembering that $F_q(\Gamma^q)\geqslant\tau|\Gamma^q|$, we find

$$|\log \Xi_{rar}(V|F_q)-\log \Xi_{rar}(V|F_q')|\leqslant \sum_{\Gamma^q\subset V} \exp(-\tau|\Gamma^q|)\,|F_q(\Gamma^q)-F_q'(\Gamma^q)|, \tag{35}$$

from which we readily obtain (33) by assuming that τ is sufficiently large (it is important that $\delta(\Gamma^q)\leqslant|\Gamma^q|$).

In the proof of the main theorem we shall also use contour models with a parameter, which are designed to describe thermodynamically unstable phases. Models like this were first introduced in [10].

Suppose that, besides the functional F_q, we are given a number $a \geq 0$. We consider the probafility distribution on the ensemble $\mathfrak{C}(V)$ for which

$$P_V(\{\Gamma_1^q,\dots,\Gamma_n^q\}|F_q,a)=\frac{\exp(-\sum_i F_q(\Gamma_i^q)+a|\bigcup_i \operatorname{Int}\Gamma_i^q|)}{\Xi_{rar}(V|F_q,a)}, \tag{36}$$

where the normalizing factor (contour partition function with parameter) is

$$\Xi_{rar}(V|F_q,a)=\sum \exp\Big(-\sum_i F_q(\Gamma_i^q)+a|\bigcup \operatorname{Int}\Gamma_i^q|\Big), \tag{37}$$

and the summation is over all possible sets $\{\Gamma_1^q,\dots,\Gamma_n^q\}\in\mathfrak{C}(V)$.

DEFINITION (3.9). The set of probability distributions P_V for all volumes V is called a contour model with the parameter a. If $a = 0$, we have an ordinary contour model. The contour partition function with parameter can be expressed in terms of the crystal contour partition functions as follows:

$$\Xi_{rar}(V|F_q,a)=\sum\prod_{i=1}^{n}\exp(aV(\Gamma_i^q))\,\Xi(\Gamma_i^q|F_q), \tag{38}$$

where the summation is over all possible sets of exterior contours $\{\Gamma_1^q,\dots,\Gamma_n^q\}$, $\Gamma_i^q\subset V$, with supports that are pairwise far apart. For the periodic τ functional F_q we set

$$\log \Xi_{rar}(V|F_q,a)=(s(F_q)+a)\,|V|+\Delta(V|F_q,a). \tag{39}$$

In what follows, we require some rough estimates of the contour partition function with parameter; namely, it follows from the inequality

$$1\leqslant\Xi_{rar}(V|F_q,\ a)\leqslant\exp(a|V|)\,\Xi_{rar}(V|F_q) \tag{40}$$

that

$$-(a+\varepsilon(\tau))\,|V|\leqslant\Delta(V|F_q,a)\leqslant\varepsilon(\hat{Z};\tau)\,|\partial V|. \tag{41}$$

LEMMA 3.4. For any two periodic τ functionals F_q and F_q' and parameters a, $a' \geq 0$ we have the inequality

$$|\Delta(V|F_q,a)-\Delta(V|F_q',a')|\leqslant\Big(\frac{1}{c-1}+\varepsilon(\tau;c)\Big)c^{\delta(V)}|V|\,|||F_q-F_q'|||_c+|V|\,|a-a'|. \tag{42}$$

It is obviously sufficient to prove this lemma separately for the case $a = a'$ and for the case $F_q = F_q'$. In the second case, the proof is obtained in the same way as in the first case but it is much simpler. We shall therefore consider only the case $a = a'$.

Since

$$|\Delta(V|F_q,a)-\Delta(V|F_q',a)|\leqslant|\log \Xi_{rar}(V|F_q,a)-\log \Xi_{rar}(V|F_q',a)|+|s(F_q)-s(F_q')|\cdot|V|, \tag{43}$$

and (see above) $|s(F_q)-s(F_q')|<\varepsilon(\tau;c)|||F_q-F_q'|||_c$, it is sufficient to prove the estimate

$$|\log \Xi_{rar}(V|F_q,a)-\log \Xi_{rar}(V|F_q',a)|\leqslant\frac{c^{\delta(V)}}{c-1}|V|\,|||F_q-F_q'|||_c. \tag{44}$$

We use the fact that

$$\left|\frac{\partial}{\partial F_q(\Gamma^q)}\log \Xi_{rar}(V|F_q,a)\right|=P_V(\Gamma^q|F_q,a) \tag{45}$$

is the probability of the contour Γ^q belonging to the set of contours $\{\Gamma_1^q,\dots,\Gamma_n^q\}\in\mathfrak{C}(V)$. Estimating the left-hand side of (44) with respect to the multidimensional Lagrange formula, we obtain

$$|\log \Xi_{rar}(V|F_q,a)-\log \Xi_{rar}(V|F_q',a)|\leqslant\sum_{\Gamma^q\subset V}P_V(\Gamma^q|\bar{F}_q,a)\,|F_q(\Gamma^q)-F_q'(\Gamma^q)|, \tag{46}$$

where $\bar{F}_q=t\cdot F_q+(1-t)F_q'$ for some t, $0\leqslant t\leqslant 1$. Further

$$|\log \Xi_{rar}(V|F_q, a) - \log \Xi_{rar}(V|F_q', a)| \leqslant$$

$$\leqslant \sum_{\Gamma^q \subset V} P_V(\Gamma^q|F_q, a) c^{\delta(\Gamma^q)} (|\Gamma^q| + V(\Gamma^q)) \cdot |||F_q - F_q'|||_c = E_V\left(\sum_{i=1}^{n} (|\Gamma_i^q| + V(\Gamma_i^q)) c^{\delta(\Gamma_i)} |F_q, a\right) \cdot |||F_q - F_q'|||_c \tag{47}$$

(the symbol $E_V(\cdot|F_q, a)$ denotes the mathematical expectation in the corresponding ensemble). The inequality we want to prove now follows from an elementary lemma.

LEMMA 3.5. For any finite volume $V \subset Z^d$ and any set $\{\Gamma_1^q, \ldots, \Gamma_n^q\} \in \mathfrak{C}(V)$

$$\sum_{i=1}^{n} (|\Gamma_i^q| + V(\Gamma_i^q)) c^{\delta(\Gamma_i^q)} < \frac{c^{\delta(V)}}{c-1} \cdot |V|. \tag{48}$$

Proof. For every finite volume $V \subset Z^d$ we set

$$\varphi(V) = \frac{1}{|V|} \max \sum_{i=1}^{n} (|\Gamma_i| + V(\Gamma_i)) c^{\delta(\Gamma_i)}, \tag{49}$$

where the maximum is taken over all sets of contours $\{\Gamma_1, \ldots, \Gamma_n\} \in \mathfrak{C}(V)$ (we omit the index q). We separate from the set $\{\Gamma_1, \ldots, \Gamma_n\}$ all the exterior contours $\Gamma_{i_1}, \ldots, \Gamma_{i_k}$. Then

$$\sum_{i=1}^{n} (|\Gamma_i| + V(\Gamma_i)) c^{\delta(\Gamma_i)} \leqslant \sum_{t=1}^{k} (|\Gamma_{i_t}| + V(\Gamma_{i_t})) c^{\delta(\Gamma_{i_t})} + \sum_{t=1}^{k} V(\Gamma_{i_t}) \varphi(\mathrm{Int}\, \Gamma_{i_t}). \tag{50}$$

For every natural number d we set $\psi(d) = \max\limits_{V: \delta(V) \leqslant d} \varphi(V)$. Since

$$\sum_{t=1}^{k} (|\Gamma_{i_t}| + V(\Gamma_{i_t})) \leqslant |V|, \delta(\Gamma_{i_t}) \leqslant \delta(V) - 1$$

and $\delta(\mathrm{Int}\, \Gamma_{i_t}) \leqslant \delta(V) - 1$, it follows from (50) that $\varphi(V) \leqslant c^{\delta(V)-1} + \psi(\delta(V) - 1)$, i.e., $\psi(d) \leqslant c^{d-1} + \psi(d-1)$, and hence

$$\psi(d) \leqslant c^{d-1} + c^{d-2} + \ldots + 1 = \frac{c^d - 1}{c-1}. \tag{51}$$

IV. Fundamental Lemma

Let H be a Hamiltonian whose range does not exceed $\bar{R}$. We choose a subgroup of finite index $\hat{Z} \subset Z^d$ with respect to which the Hamiltonian H and all the configurations $s_1, \ldots, s_r$ are invariant. Suppose we have the expansion

$$H(\Gamma^q) = \Phi(\Gamma^q) + \sum_m (h_m - h_q) V_m(\Gamma^q), \tag{52}$$

where $|h_q| < \infty$, $1 \leq q \leq r$, and $\Phi(\Gamma^q) \geqslant \rho|\Gamma^q|$, where ρ is a positive constant.

LEMMA 4.1. There exists $\bar{\tau} = \bar{\tau}(s_1, \ldots, s_r; \bar{R}; \hat{Z})$ such that for $\beta \cdot \rho > \bar{\tau} + 1$ for the Hamiltonian H there exist and are unique a point $a = (a_1, \ldots, a_r) \in O_r$ and periodic $\bar{\tau}$ functionals F_q, $1 \leq q \leq r$, for which the following equations hold:

$$\Xi(\Gamma^q|\beta H) = \exp(a_q V(\Gamma^q)) \Xi(\Gamma^q|F_q), \tag{53}$$

$$a_q = \beta h_q - s(F_q) + \mathrm{const}, \tag{54}$$

where the constant term const, which does not depend on q, is determined by the condition that $\min a_q = 0$, i.e., $a \in O_r$.

Proof of Lemma 4.1. It follows from Eqs. (53) that the functionals F_q, $1 \leq q \leq r$, belong to the class of $\hat{Z}$-invariant functionals. Substituting (53) into the chain of recursion relations (22) and (23) of Lemma 3.1, we obtain

$$\exp(a_q V(\Gamma^q)) \Xi(\Gamma^q|F_q) = \exp(-\beta H(\Gamma^q)) \prod_m \Xi_{rar}(\mathrm{Int}_m \Gamma^q|F_m, a_m). \tag{55}$$

In accordance with Lemma 3.2, Eqs. (55) are equivalent to Eqs. (53).

Taking the logarithms of Eqs. (55) and substituting them into the expansion (52) and Eqs. (54), we obtain

$$F_q(\Gamma^q) = \beta \Phi(\Gamma^q) - \sum_m \Delta(\mathrm{Int}_m \Gamma^q|F_m, a_m) + \nabla(\Gamma^q|F_q), \tag{56}$$

where we have already assumed that a_q have the form (54). Equations (56) and (54) form a closed system of equations for the contour functionals F_q, $1 \leq q \leq r$.

We now show that for $\beta\rho > \bar{\tau} + 1$ these equations have a unique solution in the class of $\hat{Z}$-periodic $\bar{\tau}$ functionals. Lower bounds on $\bar{\tau}$ will be given during the proof. First, we require that $4^d\varepsilon(\hat{Z};\bar{\tau})<1$. Then the right-hand side of Eqs. (56) [with a_m found from Eqs. (54)] can be regarded as a mapping that associates the set of $\hat{Z}$-invariant $\bar{\tau}$ functionals $\{F_q,\ 1 \le q \le r\}$ with a new set of these functionals. We denote by $\mathfrak{B}(\bar{\tau})$ the space of the sets $\hat{F} = \{F_q,\ 1 \le q \le r\}$ of $\hat{Z}$-invariant $\bar{\tau}$ functionals equipped with the metric

$$|||\hat{F}-\hat{F}'|||_c = \max_{1\le q\le r} |||F_q-F_q'|||_c. \tag{57}$$

We write the system of equations (56) in abridged form as

$$\hat{F}=S(\hat{F}\,|\,\beta\hat{\Phi},\beta h), \tag{58}$$

where

$$S(\hat{F}\,|\,\beta\hat{\Phi},\beta h)=\beta\hat{\Phi}+T(\hat{F}\,|\,\beta h) \tag{59}$$

and $T(\hat{F}\,|\,\beta h)$ is determined by the right-hand side of Eqs. (56) with the values of a_m substituted from Eqs. (54).

Then by virtue of Lemma 3.4

$$|||T(\hat{F}\,|\,\beta h)-T(\hat{F}'\,|\,\beta h')|||_c\le 2\left(\frac{1}{c-1}+\varepsilon(\bar{\tau};c)\right)|||\hat{F}-\hat{F}'|||_c+2\beta|h-h'|+2\varepsilon(\bar{\tau};c)\,|||\hat{F}-\hat{F}'|||_c. \tag{60}$$

We set $c = 13$ and require $\varepsilon(\bar{\tau};c) < 1/12$. Then the mapping $T(\hat{F}\,|\,\beta h)$ for fixed h satisfies a Lipschitz condition with respect to $\hat{F}$ with constant $\frac{1}{2}$. In addition, $|||T(\hat{F}\,|\,\beta h)|||_c<\infty$. It follows from this that the mapping $S(\hat{F}\,|\,\beta\hat{\Phi},\beta h)$ has a unique fixed point $\hat{F}$. Lemma 4.1 is proved.

From the estimate (60) used in the proof of Lemma 4.1 it follows that the fixed point $\hat{F}$ of the mapping $S(\hat{F}\,|\,\beta\hat{\Phi},\beta h)$ depends continuously on $\beta\hat{\Phi}$ and βh. More precisely,

$$\tfrac{1}{2}|||\hat{F}-\hat{F}'|||_c\le\beta|||\hat{\Phi}-\hat{\Phi}'|||_c+2\beta|h-h'|. \tag{61}$$

We now turn to the situation of Theorem 1. We suppose that $H = H_0 + L$, where the Hamiltonian H_0 with ground states $s_1, \ldots, s_r$ satisfies Peierls's condition with constant ρ, and L is a bounded Hamiltonian. Then if $\|\sigma(L)\| < \rho/2$,

$$H(\Gamma^q)=\Phi(\Gamma^q)+\sum_m (h_m-h_q)V_m(\Gamma^q), \tag{62}$$

where

$$h_m=e_L(s_m),\ \Phi(\Gamma^q)=H_0(\Gamma^q)+\sigma(L)(\Gamma^q)\ge\frac{\rho}{2}|\Gamma^q|.$$

Thus, Lemma 4.1 applies to H.

Suppose we are given a Hamiltonian H_0 with ground states $s_1, \ldots, s_r$ that satisfies Peierls's condition with constant ρ and a set of Hamiltonians $H_1, \ldots, H_{r-1}$ which lift the degeneracy of the ground state of H_0. We fix a neighborhood U of the origin in the space of parameters $(\mu_1, \ldots, \mu_{r-1})$ such that for any Hamiltonian $L=\mu_1H_1+\ldots+\mu_{r-1}H_{r-1}$, $\mu\in U$, the inequality $\|\sigma(L)\|<\rho/2$ holds. It follows from what we have said above that Lemma 4.1 applies to the Hamiltonian $H=H_0+\mu_1H_1+\ldots+\mu_{r-1}H_{r-1}$ for $\mu \in U$ and parameter $\beta>\beta_0=2(\bar{\tau}+1)/\rho$ (where $\bar{\tau}=\bar{\tau}(s_1,\ldots,s_r;\bar{R};\hat{Z})$); this enables us to find numbers $a_1, \ldots, a_r$ which serve as coordinates of the point $a\in O_r$. In this case we set $\mathcal{J}(\beta)\mu=a$ and show that the mapping $\mathcal{J}(\beta): U\to O_r$ has the properties indicated in Theorem 1. The coordinates a_q, $1\le q\le r$, of the point $a=\mathcal{J}(\beta)\mu$ can be found from the equation

$$a_q=\beta h_q-s(F_q)+\text{const}, \tag{63}$$

where $h_q=e_{L(\mu)}(s_q)$ for $L(\mu)=\mu_1H_1+\ldots+\mu_{r-1}H_{r-1}$, and the functionals F_q, $1 \le q \le r$, can be found from the system of equations (56) and (54) of Lemma 4.1 applied to $H = H_0 + L(\mu)$.

Since $\sigma(L(\mu))(\Gamma^q)$ and $h_q=e_{L(\mu)}(s_q)$ are linear functions of μ, it follows from the inequality (61) that the functionals F_q, $1 \le q \le r$, satisfy a Lipschitz condition with respect to $\beta\mu$ with a constant that depends on the set of Hamiltonians $H_1, \ldots, H_{r-1}$ but not on β. The quantity $s(F_q)$ satisfies a Lipschitz condition with respect to F_q with the constant $\varepsilon(\bar{\tau};c)$. Increasing $\bar{\tau}$ if necessary, we can assume that $s(F_q)$ for $\beta\rho > 2(\bar{\tau} + 1)$ satisfies a Lipschitz condition with respect to $\beta\mu$ with arbitrarily fixed small constant.

In particular, $\mathcal{J}(\beta)\mu$ is a continuous function of $\mu \in U$. We show that the implicit equation $a=\mathcal{J}(\beta)\mu$ defines μ as a unique and continuous function of $a\in O_r$. We write this equation in the form

$$\beta h_q = a_q + s(F_q) + \text{const}, \tag{64}$$

where $h_q = e_{L(\mu)}(s_q)$, and the functionals F_q, $1 \leq q \leq r$, are determined from β and μ in accordance with Lemma 4.1 applied to $H = H_0 + L(\mu)$. As we have noted above, $s(F_q)$ satisfies a Lipschitz condition with respect to $\beta\mu$ with arbitrarily small constant.

From the fact that the set of Hamiltonians $H_1, \ldots, H_{r-1}$ lifts the degeneracy of the ground state of H_0 it follows that $\mu_1, \ldots, \mu_{r-1}$ can be expressed linearly in terms of the differences $h_q - h_{q'}$, $1 \leqslant q, q' \leqslant r$. Therefore, from Eq. (64) we can find μ by the method of successive approximations provided $\max_q a_q$ is sufficiently small for the iterations not to leave the set U. Thus, $\mathcal{J}(\beta)$ is a one-to-one mapping of U onto some neighborhood of the point 0 on O_r that is continuous in both directions.

V. Construction of Limit Gibbs Distributions

Let V be a finite subset of the lattice Z^d. We recall the definition of "rarefied" configurations in the volume V with boundary conditions s_q (see subsection 3). We define "rarefied" configurations in the volume V with boundary conditions s_q as configurations s for which $s(i) = s_q(i)$ for $i \in Z^d \setminus V$, and also

1) $\text{dist}(B(s), Z^d \setminus V) > 1$,
2) for any exterior contour Γ^q of the configuration s, Int $\Gamma^q \subset V$.

The probability distribution on the set $\mathfrak{R}_q(V)$ of rarefied configurations in the volume V with boundary conditions s_q induced by the Gibbs distribution in the volume V with boundary conditions s_q will be called the rarefied Gibbs distribution in the volume V with boundary conditions s_q. For a Hamiltonian H whose range does not exceed $\overline{R}$ the rarefied Gibbs distribution in the volume V with boundary conditions s_q can be described as follows. Consider all possible sets of exterior contours $\{\Gamma_1^q, \ldots, \Gamma_n^q\}$, $\Gamma_i^q \subset V$, with supports that are pairwise far apart. The set $\{\Gamma_1^q, \ldots, \Gamma_n^q\}$ is ascribed a probability proportional to $\Xi(\Gamma_1^q|\beta H) \ldots \Xi(\Gamma_n^q|\beta H)$. The random variables $\{s(i), i \in \text{Int}_m \Gamma_j^q\}$ for different j or m for fixed set of exterior contours are conditionally independent, and their conditional distributions are identical to the rarefied Gibbs distributions in the volume $\text{Int}_m \Gamma_j^q$ with boundary conditions s_m. In the common exterior of the exterior contours $\{\Gamma_j^q, j=1, \ldots, n\}$ we have $s(i) = s_q(i)$. Suppose that $\Xi(\Gamma^q|\beta H) = \Xi(\Gamma^q|F_q)$ for some τ functional F_q and τ is sufficiently large. It then follows from Lemma 3.3 that if $V_i \uparrow Z^d$ as $i \to \infty$ (i.e., $V_i \subset V_{i+1}$ $\bigcup V_i = Z^d$), then the rarefied Gibbs distribution in the volume V_i with boundary conditions s_q converges weakly as $i \to \infty$ to the limit Gibbs distribution P_q, which can be described as follows. In accordance with Lemma 3.3, we define a probability distribution on the ensemble of infinite sets of exterior contours $\{\Gamma_j^q, j=1, 2, \ldots\}$ with supports that are pairwise far apart. The conditional probability distribution on the set of configurations $\mathfrak{M}$ for fixed exterior contours $\{\Gamma_j^q, j=1, 2, \ldots\}$ is constructed as follows. In the common exterior of the exterior contours $\{\Gamma_j^q, j=1, 2, \ldots\}$ we have $s(i) = s_q(i)$. The random variables $\{s(i), i \in \text{Int}_m \Gamma_j^q\}$ for different j or m are conditionally independent and their conditional distributions coincide with the rarefied Gibbs distributions in the volumes $\text{Int}_m \Gamma_j^q$ with boundary conditions s_m.

Description of the typical (i.e., that form a set of total probability) configurations for the distribution P_q shows in what sense they can be regarded as small deviations from the ground state s_q; there is an infinite connected "sea" of a high density of points in the q-th ground state in which islands are precipitated seldom but statistically uniformly, these islands being bounded by finite exterior contours. Within these islands the remaining contours of the configuration are arranged.

For the Hamiltonian $H = H_0 + \mu_1 H_1 + \ldots + \mu_{r-1} H_{r-1}$ there exist limit Gibbs distributions P_q for the q's for which $a_q = 0$, where $(a_1, \ldots, a_r) = a = \mathcal{J}(\beta)\mu$. The distribution P_q is periodic, and as follows readily from the estimates of [1], has the property of correlation weakening (mixing) and is a pure thermodynamic phase (one can even show that the distribution P_q is an extreme point of the convex set of all limit Gibbs distributions (see [11])). For the case when the Hamiltonian H_0 is bounded, the natural conjecture arises that any periodic limit Gibbs distribution of the Hamiltonian $H = H_0 + \mu_1 H_1 + \ldots + \mu_{r-1} H_{r-1}$ is a mixture of the pure thermodynamic phases P_q which we have constructed. This conjecture was recently proved by D. G. Martirosyan for the values of the parameters $\mu_1, \ldots, \mu_{r-1}$ for which r or r − 1 pure thermodynamic phases are constructed.

VI. Spontaneous Symmetry Breaking

The group of all motions D of the space R^d that carry the integral lattice Z^d into itself acts naturally on the phase space $\mathfrak{M}$ of the lattice system. In addition, on $\mathfrak{M}$ there acts the group K of transformations of the form $(ks)(i) = k(i)s(i)$, where $k(i)$ is a permutation of the set X which depends periodically on i. By K·D we denote the group of transformations of $\mathfrak{M}$ generated by K and D.

We define the action of the group K·D on the space of bounded Hamiltonians by the formula

$$(gL)(s, s') = L(gs, gs'), \quad g \in K \cdot D. \tag{65}$$

Let G be a subgroup of K·D.

DEFINITION (6.1). The group G is called a symmetry group of the Hamiltonian H if:
1) any transformation $g \in G$ maps the set of admissible configurations of H onto itself;
2) for any pair of admissible configurations s and s' that coincide almost everywhere and any $g \in G$

$$H(gs, gs') = H(s, s'). \tag{66}$$

We consider a Hamiltonian H_0 with ground states $s_1, \ldots, s_r$ which satisfies Peierls's condition and has symmetry group G. The group G maps the set $\{s_1, \ldots, s_r\}$ onto itself. We denote the configuration gs_m, $g \in G$, by s_{gm}.

We take the set of Hamiltonians $H_1, \ldots, H_{r-1}$, which lifts the degeneracy of the ground state of H_0.

We assume that the family of Hamiltonians $L(\mu) = \mu_1 H_1 + \ldots + \mu_{r-1} H_{r-1}$ is invariant under the action of G. Then the group G acts on the space of the parameters μ in accordance with the formula

$$L(g\mu) = gL(\mu). \tag{67}$$

We define, in addition, the action of G on the boundary of the octant O_r as $(ga)_q = a_{gq}$.

Under these assumptions, it follows directly from Lemma 4.1 that the mapping $\mathcal{I}(\beta)$ constructed from the Hamiltonian H_0 and the set of Hamiltonians $H_1, \ldots, H_{r-1}$ commutes with the action of G.

Let us consider the following frequently encountered situation. Suppose, as above, that we are given a Hamiltonian H_0 with ground states $s_1, \ldots, s_r$ satisfying Peierls's condition and having the symmetry group G. Suppose that the set $\{s_1, \ldots, s_r\}$ consists of $\varkappa$ orbits of the group G. We denote by $O_r(G)$ the subset of O_r consisting of the fixed points $a = (a_1, \ldots, a_r)$ of the action of G. Now, instead of the set of Hamiltonians $H_1, \ldots, H_{r-1}$ that lifts the degeneracy of the ground state of H_0 we have the set of bounded Hamiltonians $H_1, \ldots, H_{\varkappa-1}$ c with symmetry group G. Thus, any Hamiltonian $H = H_0 + \mu_1 H_1 + \ldots + \mu_{\varkappa-1} H_{\varkappa-1}$ has the same symmetry group G. Therefore, the external fields $H_1, \ldots, H_{\varkappa-1}$ can lift only the "fortuitous" degeneracy of the ground state of H_0 not associated with the presence of the symmetry group G for this Hamiltonian. Namely, for this reason we assume that the number of external fields is equal to $\varkappa - 1$, where $\varkappa$ can be naturally called the multiplicity of the "fortuitous" degeneracy. If the external fields $H_1, \ldots, H_{\varkappa-1}$ are completely to lift the $\varkappa$-fold "fortuitous" degeneracy of the ground state, they must satisfy a condition of the type of linear independence, analogous to Definition 1.4 in the situation of Theorem 1.

Having the Hamiltonian $L(\mu) = \mu_1 H_1 + \ldots + \mu_{\varkappa-1} H_{\varkappa-1}$, $\mu = (\mu_1, \ldots, \mu_{\varkappa-1})$, we define to within an additive constant the specific energies $e_\mu(s_q) = e_{L(\mu)}(s_q)$, $1 \leq q \leq r$. It is clear that for any $g \in G$

$$e_{L(\mu)}(gs_q) = e_{L(\mu)}(s_q). \tag{68}$$

Therefore the set of numbers

$$t_\mu(s_q) = e_\mu(s_q) - \min e_\mu(s_q), \quad 1 \leq q \leq r, \tag{69}$$

can be regarded as a point $t_\mu \in O_r(G) \subset O_r$.

DEFINITION (6.2). The set of Hamiltonians $H_1, \ldots, H_{\varkappa-1}$ with symmetry group G lifts the fortuitous degeneracy of the ground state of H_0 if the mapping

$$\varphi: \mu = (\mu_1, \ldots, \mu_{\varkappa-1}) \to t_\mu = (t_\mu(s_1), \ldots, t_\mu(s_r)) \tag{70}$$

maps the space of parameters $(\mu_1, \ldots, \mu_{\varkappa-1})$ onto the complete set $O_r(G)$.

PROPOSITION 6.1. Under the above assumptions there exist $\beta_0 = \beta_0(H_0; H_1, \ldots, H_{\varkappa-1}; d)$ and a neighborhood U of the origin in the space of parameters $(\mu_1, \ldots, \mu_{\varkappa-1})$ such that for any $\beta > \beta_0$ there is defined a homeomorphism $\mathcal{I}(\beta)$ of the neighborhood U onto some neighborhood of the point 0 on $O_r(G)$ for which for the point $\mu = (\mu_1, \ldots, \mu_{\varkappa-1}) \in U$ the Hamiltonian $H = H_0 + \mu_1 H_1 + \ldots + \mu_{\varkappa-1} H_{\varkappa-1}$ has different pure thermodynamical phases P_q with given β for the q's for which $a_q = 0$ for $a = (a_1, \ldots, a_r) = \mathcal{I}(\beta)\mu$. All that we said in Sec. I about the properties of typical configurations of the distribution P_q apply to the pure thermodynamic phases P_q. Since the Hamiltonian H has the symmetry group G, the transformation $g \in G$ of the phase space $\mathfrak{M}$ carries the Gibbs distribution P_q into the Gibbs distribution P_{gq}.

The proof of Proposition 6.1 is obtained from Lemma 4.1 in the same way as Theorem 1 with O_r in Eq. (64) replaced by $O_r(G)$ and with allowance for Definition (6.2). In the special case when there is no fortuitous degeneracy, i.e., $\varkappa = 1$, $O_r(G) = 0$, we obtain from Proposition 6.1 the Dobrushin–Gertsik theorem on spontaneous symmetry breaking [12-14].

LITERATURE CITED

1. R. A. Minlos and Ya. G. Sinai, Tr. Mosk. Matem. Ob-va, 19, 113 (1968).
2. R. A. Minlos and Ya. G. Sinai, Tr. Mosk. Matem. Ob-va, 17, 213 (1967).
3. R. A. Minlos and Ya. G. Sinai, Mat. Sb., 73, No. 3, 375 (1967).
4. R. L. Dobrushin, Funkts. Analiz, 2, No. 4, 44 (1968).
5. S. A. Pirogov and Ya. G. Sinai, Funkts. Analiz, 8, No. 1, 25 (1974).
6. S. A. Pirogov, Dokl. Akad. Nauk SSSR, 214, No. 6, 1273 (1974).
7. S. A. Pirogov, Usp. Mat. Nauk, 30, No. 2, 223 (1975).
8. N. Nasr, Usp. Mat. Nauk, 26, No. 5, 222 (1971).
9. D. G. Martirosyan, Teor. Mat. Fiz., 22, 335 (1975).
10. S. A. Pirogov, Izv. Akad. Nauk SSSR, Ser. Mat., 39, No. 6, 1403 (1975).
11. R. L. Dobrushin, Teor. Mat. Fiz., 12, 115 (1972).
12. V. M. Gertsik and R. L. Dobrushin, Funkts. Analiz., 8, No. 3, 12 (1974).
13. V. M. Gertsik, Usp. Mat. Nauk, 30, No. 3, 159 (1975).
14. V. M. Gertsik, Izv. Akad. Nauk SSSR, Ser. Mat. 40, No. 2 (1976) (in print).

Comments

C. Gruber proposed proving the existence of parameter h such that the classical lattice spin model with Hamiltonian

$$H = h \sum_{x \in \mathbb{Z}^d} \sigma(x) - J_1 \sum_{\|x' - x''\| = 1} \sigma(x')\sigma(x'') + J_2 \sum_{\substack{\|x' - x''\| \leq a \\ \|x' - x'''\| \leq a}} \sigma(x')\sigma(x'')\sigma(x''')$$

has for sufficiently low temperatures a phase transition of the first kind. In the last expression, $\sigma(x)$ are spin variables taking values ± 1 and $J_1 > 0$. Pirogov and I started to work on this problem. Quite soon it became clear that it is connected with the theory of contour models in statistical mechanics on which we were working earlier with R. A. Minlos. Pirogov and I realized that at the point of phase transition partition functions for all finite domains should be represented as partition functions of some contour models. After that it was not so difficult to write down the equations for the contour functionals and to prove the existence and uniqueness of the corresponding expressions. Our paper containing the solution of Gruber's problem was published in the journal *Functional Analysis and Its Applications* [PS1].

The next step was to extend the class of models to which our approach could be applied. At that time the so-called Peierls contour method was already rather well developed. The paper [DG] by V. M. Gertzik and R. L. Dobrushin was especially important. There the role of ground states and the Peierls condition were very clearly emphasized. Pirogov and I proposed a general scheme which could be applied to families of Hamiltonians depending on k parameters where k is the number of ground states. It is assumed that each family has some stratification in the parameter space describing the structure of the ground states. Under some additional conditions we proved that for small temperatures the set of periodic Gibbs states is a small perturbation of the set of the ground states. In a sense, this resembles the phenomenological Gibbs phase rule. The exposition of our theory can be found in the book [S1] and in the paper of M. Zahradnik [Z1]. Let me also mention our paper with S. A. Pigorov in which we analyzed the structure of the ground states in $: P(\phi) :_2$ models of quantum field theory (see [PS4]), and the recent book [Pr] by E. Presutti.

Commun. Math. Phys. 98, 119–144 (1985)

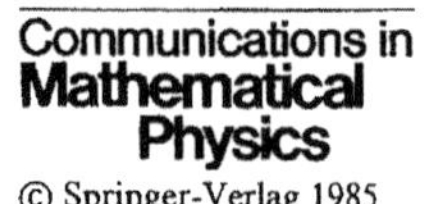

An Analysis of ANNNI Model by Peierl's Contour Method

E. I. Dinaburg and Ya. G. Sinai

Landau Institute of Theoretical Physics, Academy of Sciences, Moscow, USSR

Abstract. For the three-dimensional ANNNI model a converging expression for the curve of the coexistence of the (3.3)-phase and ferromagnetic phase is derived for low temperatures using a new extension of the Peierls contour method.

1. Description of ANNNI Model and Formulation of the Result

We consider a classical spin model on the lattice $\mathbb{Z}^3$, where the spin variables take the values ± 1 and the Hamiltonian has the form

$$H(\varphi(U)) = -J_0 \sum_{(x,x')\in U_{\mathrm{hor}}} \varphi(x)\varphi(x') - J_1 \sum_{(x,x')\in U^{(1)}_{\mathrm{ver}}} \varphi(x)\varphi(x') + J_2 \sum_{(x,x')\in U^{(2)}_{\mathrm{ver}}} \varphi(x)\varphi(x') . \quad (1)$$

Here $U \subset \mathbb{Z}^{(3)}$ is a finite set, U_{hor} is the set of horizontal bonds, $U^{(1)}_{\mathrm{ver}}(U^{(2)}_{\mathrm{ver}})$ is the set of vertical bonds of length 1 (2); in all cases the ends belong to U, $\varphi(x)$ is the spin variable at the point x, $\varphi(U)$ is the notation for a configuration on U. Parameters J_0, J_1, J_2 are positive coupling constants.

This model is called the axial next-nearest neighbor Ising model or, briefly, ANNNI model. It was introduced more than twenty years ago by Domb [1] and Elliott [2], and recently attracted much attention in connection with experimental results concerning compounds of rare-earth elements (see [3]).

This paper was motivated by the deep analysis of the phase diagram of the ANNNI model performed in the paper by Fisher and Selke (see [4]). Using a formal perturbation theory the authors have shown that for low temperatures T there are infinitely many separation-phase lines on the plane $(T, J_1/J_2)$. All these lines start in the point $(0, 1/2)$, where an infinite degeneracy of ground states takes place. From the mathematical point of view these lines determine the values of parameters where the number of periodic extreme limit Gibbs states is discontinuous.

The main idea of [4] which apparently is of more general importance is that for finite temperatures there appears a splitting of ground states if we take into account one-point spin flips of the least energy.

Y.G. Sinai, *Selecta II: Probability Theory, Statistical Mechanics, Mathematical Physics and Mathematical Fluid Dynamics*, DOI 10.1007/978-1-4419-6205-8_7, © Springer Science+Business Media, LLC 2010

We reproduce below the corresponding arguments. Assume that U is a cube of size l and we consider configurations $\varphi(U)$ with periodic boundary conditions. It is easy to see that it is sufficient to look for ground states for which $\varphi(x)$, $x=(x_1, x_2, x_3)$ depend only on x_3. All such configurations decompose onto series of horizontal planes $x_3=\text{const}$, where $\varphi(x)$ is a constant. We denote by $N_r = N_r(\varphi(U))$ the number of such series of the width r. Assuming that $\sum_{r\geqq 1} N_r > 1$ we can rewrite (1) as follows:

$$H(\varphi(U)) = -2J_0 l^3 + l^3\Big[J_1 \sum_{r\geqq 1} N_r - J_1 \sum_{r\geqq 1}(r-1)N_r + J_2 N_1 + J_2 \sum_{r\geqq 2}(r-2)N_r - 2J_2 \sum_{r\geqq 2} N_r\Big]$$

$$= -2J_0 l^3 - (J_1 - J_2)l^3 + 2(J_1 - 2J_2)\Big(\sum_{r\geqq 1} N_r - 1\Big)l^2 + 4J_2 N_1 l^2 .$$

The first two terms give the energy of the ferromagnetic configuration for which $\varphi(x)=\text{const}$, $x\in U$. For this configuration $N_l=1$, $N_r=0$ for $r<l$.

For $J_1 - 2J_2 = 0$ we have an infinite degeneracy of ground states because all configurations with $N_1=0$ are ground states. If $J_1 - 2J_2 > 0$ then the ground states (g.s.) are ferromagnetic configurations (f-g.s.). If $J_1 - 2J_2 < 0$ then the ground states are four configurations for which $N_r = 0$ for $r \neq 2$.

Assume that an inverse temperature β is chosen and fixed. In the usual picture of phase transitions of the first kind the limit Gibbs states corresponding to physical phases are concentrated on configurations which look like small islands of perturbations of an underlying "ground state sea." If this picture is valid then the main contribution to the free energy of the corresponding phase comes from one-point perturbations of the ground state. We shall list now all possible one-point perturbations and their energies in the case of ANNNI-model. In Table 1 the

Table 1

Configuration	Perturbed configuration	Change of energy
1) $++ + - -$ $- - - + +$	$+ + - - -$ $- - + + +$	$\varepsilon_0 = 8J_0$
2) $+ + + + +$ $- - - - -$	$+ + - + +$ $- - + - -$	$\varepsilon_1 = 8J_0 + 4J_1 - 4J_2$
3) $- + + - -$ $- - + + -$ $+ - - + +$ $+ + - - +$	$- + - - -$ $- - - + -$ $+ - + + +$ $+ + + - -$	$\varepsilon_2 = 8J_0 + 4J_2$
4) $+ + + + -$ $- + + + +$ $- - - - +$ $+ - - - -$	$+ + - + -$ $- + - + +$ $- - + - -$ $+ - + - -$	$\varepsilon_3 = 8J_0 + 4J_1$
5) $- + + + -$ $+ - - - +$	$- + - + -$ $+ - + - +$	$\varepsilon_4 = 8J_0 + 4J_1 + 8J_2$

configurations which are drawn horizontally are indeed vertical. Also we assume that J_0 is sufficiently large comparing with J_1, J_2. Otherwise one should take into account two-point perturbations (we are indebted to G. Uimin for this remark).

The table shows that the least energy ε_0 have perturbations on the boundaries between different series. Let us denote by U_i the set of points where spin flips of type i can happen, $i=0, 1, 2, 3, 4$. It is easy to check that $|U_0|=2l^3\sum_{r\geq 3} N_r$, $|U_1| = l^2 \sum_{r\geq 5}(r-4)N_r$, $|U_3|=2l^2\sum_{r\geq 4} N_r$, $|U_2|=l^2N_2$, $|U_4|=l^2N_3$.

Here and further the absolute value of a set means its cardinality. The number of configurations of k non-interacting spin-flips can be written up to terms of the next order as $C^k_{|U_i|}$. Using binomial coefficients we neglect an interaction of spin flips. Let us introduce for a ground state $\bar{\varphi}(U)$ the partition function

$$\Xi^{(a)}(\bar{\varphi}(U))=\sum_{\varphi(U)}\exp\{-\beta H(\varphi(U))\},$$

where the sum is taken over only those $\varphi(U)$ which arise after some number of non-interacting spin flips from $\bar{\varphi}(U)$. As was said before, in the main order

$$\begin{aligned}\Xi^{(a)}(\bar{\varphi}(U)) &\approx \exp\{-\beta H(\bar{\varphi}(U))\}\cdot\prod_{i=0}^{4}\sum_{k=0}^{|U_i|} C^k_{|U_i|}e^{-\beta\varepsilon_i k}\\ &=\exp\left\{-\beta H(\bar{\varphi}(U))+\sum_{i=0}^{4}|U_i|\ln(1+\exp(-\beta\varepsilon_i))\right\}.\end{aligned}$$

Thus the approximate value of the free energy corresponding to the ground state $\bar{\varphi}(U)$ is equal to

$$\begin{aligned}\ln\Xi^{(a)}(\bar{\varphi}(U)) &= (2J_0+J_1-J_2)\beta l^3 - 2(J_1-2J_2)\beta\sum_{r\geq 2}N_r\cdot l^2\\ &\quad +2l^2e^{-\beta\varepsilon_0}\sum_{r\geq 3}N_r + l^2e^{-\beta\varepsilon_1}\sum_{r\geq 4}(r-4)N_r\\ &\quad +2l^2e^{-\beta\varepsilon_3}\sum_{r\geq 4}N_r + l^2e^{-\beta\varepsilon_2}N_2 + l^2e^{-\beta\varepsilon_4}N_3.\end{aligned}$$

The difference $\varepsilon_1-\varepsilon_2=4(J_1-2J_2)=4\delta$. In the domain $\delta=J_1-2J_2>0$ we consider the sum

$$\begin{aligned}&-2\beta\delta l^2\sum_{r\geq 2}N_r + 2\exp\{-\beta\varepsilon_0\}l^2\sum_{r\geq 3}N_r\\ &+\exp\{-\beta\varepsilon_1\}\cdot l^2\cdot\sum_{r\geq 4}N_r(r-4)+l^2N_2e^{-\beta\varepsilon_2}\\ &\approx -2\beta\delta N_2l^2 + (-2\beta\delta+2\exp\{-\beta\varepsilon_0\})l^2\\ &\cdot\sum_{r\geq 3}N_r + \exp\{-\beta\varepsilon_1\}\cdot l^2\sum_{r\geq 4}(r-4)N_r + \exp\{-\beta\varepsilon_1\}N_2l^2\\ &\approx -(2\beta\delta+\exp\{-\beta\varepsilon_1\})l^2N_2 + [2(-\beta\delta+\exp\{-\beta\varepsilon_0\})\\ &-4\exp\{-\beta\varepsilon_1\}]l^2\cdot\sum_{r\geq 4}N_r + [2(-\beta\delta+\exp\{-\beta\varepsilon_0\})\\ &-3\exp\{-\beta\varepsilon_1\}]N_3l^2 + \exp\{-\beta\varepsilon_1\}l^3.\end{aligned}$$

If

$$\delta>0, \quad 2(-\beta\delta+\exp\{-\beta\varepsilon_0\})-3\exp\{-\beta\varepsilon_1\}>0,$$

then the sum takes the largest value for the ground state for which $N_2=N_r=0$ for $r\geqq 4$. This ground state consists of series of three horizontal planes of the same sign ((3,3)-g.s.) If

$$\delta>0, \quad 2(-\beta\delta+\exp\{-\beta\varepsilon_0\})-3\exp\{-\beta\varepsilon_1\}<0$$

then the sum takes the largest value when $\sum_{r\geqq 4} N_r$ takes the least value, i.e. for f-g.s. If

$$2(-\beta\delta+\exp\{-\beta\varepsilon_0\})-3\exp\{-\beta\varepsilon_1\}=0,$$

then

$$[2(-\beta\delta+\exp\{-\beta\varepsilon_0\})-4\exp\{-\beta\varepsilon_1\}]<0$$

and the sum takes the largest value for $N_2=N_r=0$, $r\geqq 4$. Thus if we neglect terms of smallness less than $\exp\{-\beta\varepsilon_1\}$ then for

$$\delta_0=\beta^{-1}\exp\{-\beta\varepsilon_0\}-3/2\beta^{-1}\exp\{-\beta\varepsilon_1\} \tag{2}$$

the free energies of f-g.s. and (3,3)-g.s. coincide. In other words the equality (2) gives an approximate equation for the co-existence of ferromagnetic and (3,3)-phases.

The arguments presented above give evidence for the following picture: there exist β_0 and a curve $\delta=\delta(\beta)$ defined for $\beta>\beta_0$ and close to $\delta_0(\beta)$ up to terms of order less than $\exp\{-\beta\varepsilon_1\}$ such that for all $\beta>\beta_0$ and $\delta>\delta(\beta)$ there are two limit Gibbs states which are small perturbations of f.g.s. (see [6]), for $\delta<\delta(\beta)$ and close enough to $\delta_0(\beta)$ there are six limit Gibbs states which are small perturbations of six (3, 3)-g.s. while for $\delta=\delta(\beta)$ we have eight limit Gibbs states corresponding to the coexistence of f.g.s. and (3,3)-g.s.

Now we can describe the main result of this paper. We develop a version of Peierls contour method suitable for the ANNNI model. We shall see that it has several peculiarities comparing with the usual situation (see [6]). With the help of the new technique we show the following theorem.

Main Theorem. *For large enough β there exists a continuous function $\delta=\delta(\beta)$ such that for $J_1-2J_2=2\delta(\beta)$ the* ANNNI *model has eight periodic limit Gibbs states.*

One can already see from the previous discussion a special role of isolated spin flips with the energies ε_0, ε_1 and ε_2. All functions of β which tend to zero faster than $\beta^{-1}\exp\{-\beta\varepsilon_1\}$ will be called very small (v.s.). In particular, we shall see that $\delta(\beta)-\delta_0(\beta)$ is v.s. Also we put

$$d(x,y)=\max(|x_1-y_1|+|x_2-y_2|, |x_3-y_3|)$$

for $x, y\in\mathbb{Z}^3$.

2. Boundaries for ANNNI Model

Assume that a configuration φ is given which coincides outside a finite set either with a f.g.s. or with a (3,3)-g.s. We shall say that φ at a point $x=(x_1, x_2, x_3)\in\mathbb{Z}^3$ is in a ferromagnetic phase (f.ph) if $\varphi(y)$ coincides with a f.g.s. for all y, $d(x, y) \leqq 9$. Also φ is in (3,3)-phase ((3,3)-ph) at $x \in \mathbb{Z}^3$ if there is a (3,3)-g.s. ψ and an interval (a, b), $b-a=18$ such that 1) $\psi(x_1, x_2, a-1) \neq \psi(x_1, x_2, a)$, $\psi(x_1, x_2, b-1) \neq \psi(x_1, x_2, b)$; 2) $a+6 \leqq x_3 < a+12$; 3) $\varphi(y)$ coincides with $\psi(y)$ for all $y=(y_1, y_2, y_3)$ such that $a \leqq y_3 < b$, $|y_1-x_1|+|y_2-x_2| \leqq 9$.

Due to this definition any boundary of the domain occupied by the (3,3)-phase lies between series of different signs.

All points where φ is not in a phase are called boundary points. The set of all boundary points is the boundary of φ and will be denoted by $B(\varphi)$.

Q(x), $x \in \mathbb{Z}^3$ is the closed unit cube with the centrum at x. A set of cubes is called connected if for any two cubes one can find a chain of cubes belonging to the set such that the first one and the last one coincide with the given cubes and every two neighboring cubes of the chain have a non-empty intersection.

We shall identify $B(\varphi)$ with the set of cubes $Q(x)$, $x \in B(\varphi)$. The boundary $B(\varphi)$ can be decomposed onto connected components $B_1, B_2, \ldots, B_r$. The boundary ∂B_s of B_s, $1 \leqq s \leqq r$, consists of faces which separate points of B_s and $\mathbb{Z}^3 \backslash B_s$. A set of faces is called connected if for any two faces of the set one can find a chain of faces belonging to the set such that the first one and the last one coincide with the given faces and every two neighboring faces of the chain have a non-empty intersection. Each ∂B_s is decomposed onto connected components one of which is the exteriour component $\partial B_s(\text{ext})$ while the others are interiour components $\partial B_{sk}(\text{int})$, $1 \leqq k \leqq K_s$. We denote O_{sk} the intrinsic connected components of cubes $Q(x)$, $x \in O_{sk}$, not belonging to B_s, whose boundaries are exactly $\partial B_{sk}(\text{int})$. For each $\partial B_s(\text{ext})$, $\partial B_{sk}(\text{int})$ there is a uniquely defined phase which is adjacent to the component.

A face is called vertical (horizontal) if it is parallel (orthogonal) to the axis x_3. Each component $\partial B_s(\text{ext})$, $\partial B_{sk}(\text{int})$ has several components of vertical faces $\partial B_{sm}^{(\text{ver})}(\text{ext})$, $m=1, \ldots, M_s$, $\partial B_{skm}^{(\text{ver})}(\text{int})$, $m=1, 2, \ldots, M_{sk}$ and several components of horizontal faces $\partial B_{sn}^{(\text{hor})}(\text{ext})$, $n=1, \ldots, N_s$, $\partial B_{skn}^{(\text{hor})}(\text{int})$, $n=1, \ldots, N_{sk}$. Due to our definition of points in the (3,3)-phase the horizontal components which separate the boundary from a (3,3)-g.s. cut an adjacent (3,3)-g.s. just exactly between ($\pm$)-series.

The least component b^0 of the boundary is the component which appears as a result of an isolated spin-flip inside the sea of a g.s. We write $b^0(x)$ if the spin-flip takes place at the point x.

Let us take a connected component B_p of the boundary $B(\varphi)$ for a configuration φ, and $\partial B_{pj}^{(\text{ver})}$ be one of components $\partial B_{pm}^{(\text{ver})}(\text{ext})$, $1 \leqq m \leqq M_p$ or $\partial B_{pkm}^{(\text{ver})}(\text{int})$, $1 \leqq m \leqq M_{pk}$. A component $\partial B_{pj}^{(\text{ver})}$ is called small if there is a point x such that $\partial B_{pj}^{(\text{ver})} \subset \partial(b^0(x))^{(\text{ver})}$.

A small component $\partial B_{pj}^{(\text{ver})}$ is called filled (empty) if there exists (does not exist) at least one point $y \in \mathbb{Z}^3$ such that $Q(y) \subset b^0(x) \cap B_p$. It is clear that we can make one or several spin-flips after which the small component $\partial B_{pj}^{(\text{ver})}$ disappears while other vertical components do not change. Certainly some components of the horizontal boundary also may change.

A connected component B_q is called smooth if none of the components of its vertical boundary is small. By definition any $b^0(x)$ is also a smooth component. Having a non-smooth component B_q we can make all spin-flips, described above, and get a new configuration φ' with one or several smooth components appearing from B_q.

3. Contours and Contour Models for ANNNI Model

Let φ coincide at infinity either with a f.g.s. or with a (3,3)-g.s. and $B(\varphi)$ be its boundary with the components $B_i(\varphi)$, $1 \leqq i \leqq r$. Two components $B_{i_1}(\varphi)$, $B_{i_2}(\varphi)$ of $B(\varphi)$ are called s-connected if the distance between $(\partial B_{i_1})^{(\mathrm{hor})}$ and $(\partial B_{i_2})^{(\mathrm{hor})}$ is not more than 38. This notion gives a possibility to decompose $B(\varphi)$ onto maximal components which we shall call s-components and denote by $b_1, b_2, \ldots, b_p$, $p \leqq r$. The components $B_i(\varphi) \subset b_j$ are called connected components belonging to the s-component b_j. For each b_j a function ph is defined on ∂b_j whose value on a face is a ground state which is adjacent to the face.

Definition 1. A contour γ is a pair $\gamma = (b, ph)$, where b is a maximal s-connected component of the boundary of a configuration and ph is the function defined on ∂b.

It is easy to see that ph takes a constant value on the exterior part of ∂b. We shall write $\gamma^{(f)}(\gamma^{(3,3)})$ if the value of ph on the exterior part is f.g.s. ((3,3)-g.s.). The set b is called the body of γ. If $\gamma = (b, ph)$ and $\varphi(b)$ are given one can complete $\varphi(b)$ till the configuration φ_γ on the whole lattice using the boundary conditions in such a way that $B(\varphi_\gamma) = b$.

For any contour $\gamma = (b, ph)$ we put $s\gamma$ to be the contour (sb, ph), where sb is the union of smooth components corresponding to contours of b. These smooth components appear after spin-flips destroying all small components. We denote

$$\partial_f(b) = \{z \mid z \in \partial b, ph(z) = \text{f.g.s.}\},$$

$$\partial_{(3,3)}(b) = \{z \mid z \in \partial b, ph(z) = (3,3)\text{-g.s.}\},$$

$$b_{(\mathrm{int})} = \{x \mid x \in b, d(x, (\partial_f(b))^{(\mathrm{hor})} > 9, d(x, (\partial_{(3,3)} b)^{(\mathrm{hor})}) > 6\}.$$

A contour $\gamma = (b, ph)$ is called smooth if $b = sb$. By definition the least contour $\gamma^0 = (b^0, ph)$ is smooth. For any smooth contour $\gamma = (b, ph)$ we put $Sm(\gamma)$ to be equal to the set of all contours $\gamma_1 = (b_1, ph)$ which after the spin-flips give the smooth contour γ. Also

$$\begin{aligned} V(\gamma) = V(b) = \{ & x = (x_1, x_2, x_3) \in \mathbb{Z}^3 \mid \partial Q^{(\mathrm{hor})}(y) \\ & \cap \partial b^{(\mathrm{hor})} \neq \emptyset \text{ for some } y = (x_1, x_2, z) \in b, \\ & ph(\partial Q^{(\mathrm{hor})}(y) \cap \partial b^{(\mathrm{hor})}) = \text{f.g.s.}, |x_3 - z| \leqq 10\} \\ & \cup \{x = (x_1, x_2, x_3) \in \mathbb{Z}^3 \mid \partial Q^{(\mathrm{hor})}(y) \cap \partial b^{(\mathrm{hor})} \neq \emptyset \\ & \text{for some } y = (x_1, x_2, z) \in b, ph(\partial Q^{(\mathrm{hor})}(y) \cap \partial b^{(\mathrm{hor})}) \\ & = (3,3) - \text{g.s.}, |x_3 - z| \leqq 12\} \cup b. \end{aligned}$$

We shall define now contour models appropriate for the ANNNI model (see the usual case in [6]). Let be given a finite set $U \subset \mathbb{Z}^3$. We introduce ensembles $\vartheta^{(f)}(U)$ $(\vartheta^{(3,3)}(U))$ whose elements are formal configurations $\{\gamma_i\}$ of mutually disjoint contours $\gamma_i^{(f)}(\gamma_i^{(3,3)})$ belonging to U. The distance between any pair of contours is more than 38.

Remark. Each contour can occupy only a position compatible with its boundary conditions ph. This is important for the (3,3)-phase.

Suppose that functions $F^{(f)}$, $F^{(3,3)}$ are defined on contours $\gamma^{(f)}$, $\gamma^{(3,3)}$ respectively taking the same values on congruent contours. We put the statistical weight of a configuration of contours $\{\gamma_i\}$ to be equal to

$$W(\{\gamma_i^{(f)}\})=\exp\left\{-\sum_i F^{(f)}(\gamma_i^{(f)})\right\},$$

$$W(\{\gamma_i^{(3,3)}\})=\exp\left\{-\sum_i F^{(3,3)}(\gamma_i^{(3,3)})\right\}.$$

The corresponding partition functions are

$$\Xi(U|F^{(f)})=\sum_{\{\gamma_i^{(f)}\}} W(\{\gamma_i^{(f)}\}), \qquad \Xi(U|F^{(3,3)})=\sum_{\{\gamma_i^{(3,3)}\}} W(\{\gamma_i^{(3,3)}\}).$$

In the usual cases one assumes that $F^{(f)}(\gamma^{(f)})\geqq \text{const}\,|b|$, $F^{(3,3)}(\gamma^{(3,3)})\geqq \text{const}\,|b|$ and const is sufficiently large (see [6]). This gives a possibility to present the logarithms of partition functions as sums of two terms where the first one is proportional to $|U|$ while the second one proportional to $|\partial U|$ is a remainder term with nice propoerties (see [6]). We shall see that this main property of contour models remains valid under much more mild assumptions concerning $F^{(f)}$, $F^{(3,3)}$.

Now we formulate these assumptions. We omit the indices "f" and "(3,3)." It means that the formulations are similar in both cases.

Assumptions Concerning F

Let γ be a contour, $\partial(sb)^{(\text{ver})}$ be a vertical boundary of sb, $s\gamma=(sb, ph)$, and $s\gamma$ is the smooth contour corresponding to γ. For any vertical face z we denote by $T_0(z)$ the set of contours $\gamma=(b, ph)$ such that $\partial b^{(\text{ver})}$ consists of a single connected component and $z\in\partial b^{(\text{ver})}$. It means that γ is completely defined by the boundary condition and $\partial b^{(\text{ver})}$. If the boundary condition is fixed we shall write $\partial b^{(\text{ver})}\in T_0(z)$.

The functional

$$F(\gamma)=F_1(\gamma)+F_2(\gamma),$$

where

1) $F_1(\gamma)>K_1|\partial(sb)^{(\text{ver})}|$ for a constant $K_1>0$;
2) $F_2(\gamma)>-d_1^{(1)}|b|$ for a constant $d_1^{(1)}>0$;
3) for any smooth $\gamma_1=(b_1, ph)$ and any

$$\ln\left[\sum_{\gamma=(b,ph)|\gamma\in Sm(\gamma_1),\, b(\text{int})\subset W}\exp\{-F_2(\gamma)\}\right]$$
$$\leqq d_1^{(1)}|b_1(\text{int})|-d_1^{(2)}(|b_1\backslash b_1(\text{int})|)$$
$$+\frac{1}{N}\exp\{-F(\gamma^{(0)})\}\cdot|(V(b_1)\backslash b_1(\text{int}))\cap W|$$

for the same constant $d_1^{(1)}$ and a constant $d_1^{(2)}>0$; (here N is a period of g.s. ($N=1$ if g.s. $=f$, $N=6$ if g.s. $=(3,3)$));

4) if $\gamma=(b, ph)$ and $b=\bigcup b_i$, where b_i are maximal connected components of b then

$$F(\gamma)=\sum_{\gamma_i=(b_i,ph)} F(\gamma_i).$$

Contrary to the usual cases $F(\gamma)$ may be negative. The property 3) is very important. In fact it requires some estimations of partition functions with summation over contours corresponding to any fixed smooth contour. In our case we shall get needed inequalities during the proof of Theorem B (Sect. 6). We shall use norms

$$\|F_1\| = \sup_{\gamma=(b,ph)} |F_1(\gamma)|/|\partial(sb)^{(\mathrm{ver})}|,$$

$$\|F_2\| = \sup_{\gamma=(b,ph)} |F_2(\gamma)|/|b|,$$

$$\|F\| = \|F_1\| + \|F_2\|.$$

Theorem A. *Assume that*

$$d_1^{(1)} \leqq \frac{1}{N}\exp\{-F(\gamma^0)\} - \exp\{-R_1K_1\}$$

$$\cdot \exp\{-F(\gamma^0)\} - R_2 \sum_{\substack{\gamma\in T_0(z)\\ \gamma\neq\gamma^0}} \exp\{-K_1|\partial\gamma^{(\mathrm{ver})}|\},$$

$$d_1^{(2)} > \exp\{-R_1K_1\}\exp\{-F(\gamma^0)\}$$

$$-R_2 \sum_{\substack{\gamma\in T_0(z)\\ \gamma\neq\gamma^0}} \exp\{-K_1|\partial\gamma^{(\mathrm{ver})}|\}.$$

There exist absolute constants K_0, $R_1 = R_1(K_0)$, $R_2 = R_2(K_0)$ such that for $K_1 > K_0$ one can find a number $a(F)$ for which

$$\ln \Xi(U|F) = a|U| + \Delta(U|F).$$

The remainder term $\Delta(U|F)$ satisfies the inequalities

a$_1$) $|\Delta(U|F)| \leqq \varrho\cdot|\partial U|$;

a$_2$) $|\Delta(U|\bar{F}) - \Delta(U|\bar{\bar{F}})| \leqq \varrho\,\|\bar{F} - \bar{\bar{F}}\|\cdot|\partial U|$.

Here

$$\varrho = \varrho(K_1, d_1^{(1)}) \leqq \tfrac{1}{2}\exp\{-R_1K_1 - F(\gamma^0)\}$$

is a constant.

Let a contour $\gamma^{(f)} = (b, ph)$ be given. O_s, $s = 1, 2, \ldots, r$ are interior domains corresponding to b, $\varkappa(O_s)$ are boundary conditions on $\partial b_s(\mathrm{int})$. We introduce another ensemble $\vartheta_{cr}(\gamma^{(f)})$ whose points are formal admissible configurations of non-intersecting contours inside O_s with the same indices "f" and

$$\Xi^{(cr)}(\gamma^{(f)}|F^{(f)}) = \exp\{-F(\gamma^{(f)})\}\cdot\prod_s\sum_{\{\gamma_i\}}\exp\left\{-\sum_i F^{(f)}(\gamma_i^{(f)})\right\},$$

where the last sum is taken over all configurations of contours $\{\gamma_i^{(f)}\}$ inside O_s. Then under the conditions of Theorem A

$$\ln\Xi^{(cr)}(\gamma^{(f)}|F^{(f)}) = -F^{(f)}(\gamma^{(f)}) + a\cdot\sum_s|O_s| + \Delta^{(cr)}(\gamma^{(f)}|F^{(f)}),$$

where

$$\Delta^{(cr)}(\gamma^{(f)}|F^{(f)}) = \sum_s \Delta(O_s|F^{(f)}).$$

Therefore $\Delta^{(cr)}$ satisfy the inequalities:

$$|\Delta^{(cr)}(\gamma^{(f)}|F^{(f)})| \leqq \varrho |\partial b|,$$
$$|\Delta^{(cr)}(\gamma^{(f)}|\bar{F}) - \Delta^{(cr)}(\gamma^{(f)}|\bar{\bar{F}})| \leqq \varrho[\|\bar{F}_1 - \bar{\bar{F}}_1\| + \|\bar{F}_2 - \bar{\bar{F}}_2\|] \cdot |\partial\gamma^{(f)}|.$$

In the same manner one can define $\Xi^{(cr)}(\gamma^{(3,3)}|F^{(3,3)})$, remainder terms $\Delta^{(cr)}(\gamma^{(3,3)}|F^{(3,3)})$ with the same properties.

4. Equations for Contour Functionals for the ANNNI Model

We shall proceed as in [6]. Let us take a configuration φ which coincides at infinity either with a f.g.s. or with a (3,3)-g.s. Its boundary consists of some number of maximal s-connected subsets $b_1, \ldots, b_r$ or of contours $\gamma_1 = (b_1, ph), \ldots, \gamma_r = (b_r, ph)$. For any set $V \subset \mathbb{Z}^3$ we put

$$-H(\varphi(V)) = \sum_{x \in V} \varphi(x)\Big[\tfrac{1}{2}J_0 \sum_{(x,x') \in (\mathbb{Z}^3)^{(1)}_{\text{hor}}} \varphi(x')$$
$$+ \tfrac{1}{2}J_1 \sum_{(x,x') \in (\mathbb{Z}^3)^{(1)}_{\text{ver}}} \varphi(x') - \tfrac{1}{2}J_2 \sum_{(x,x') \in (\mathbb{Z}^3)^{(2)}_{\text{ver}}} \varphi(x')\Big].$$

A contour γ_i is an outer contour if it is not contained inside any domain bounded by other contours. For any contour $\gamma^{(f)} = (b^{(f)}, ph)$ we denote by $\mathfrak{A}(\gamma^{(f)})$ the set of configurations for which the boundary consists of the single contour $b^{(f)}$ and put

$$\Xi_1(\gamma^{(f)}|\beta,\delta) = \sum_{\varphi \in \mathfrak{A}(\gamma^{(f)})} \exp\{-\beta H(\varphi(b^f))\}.$$

For any smooth contour $\gamma_1^{(f)} = (b_1^{(f)}, ph)$ and any set $W \supset b_1^{(f)}(\text{int})$, $(W \subset \mathbb{Z}^3)$ we put

$$\Xi_W^{(s)}(\gamma_1^{(f)}|\beta,\delta) = \sum_{\substack{\gamma^{(f)} = (b^{(f)}, ph) \in Sm(\gamma_1^{(f)}) \\ b^{(f)}(\text{int}) \subset W}} \exp\{\beta H^{(f)}(b^{(f)})\} \cdot \Xi_1(\gamma^{(f)}|\beta,\delta),$$

where $H^{(f)}(b)$ is the energy of the ferromagnetic configuration on b. In an analogous way we can introduce $\Xi_1(\gamma^{(3,3)}|\beta,\delta)$ and $\Xi_W^{(s)}(\gamma_1^{(3,3)}|\beta,\delta)$. Let $h^{(f)}$, $h^{(3,3)}$ be the energies per particles for f.g.s. and (3,3)-g.s. respectively which are functions of δ.

Theorem B. *Let $\delta = J_1 - 2J_2 > 0$, $\delta = \delta_0(\beta)(1 + o(1))$ as $\beta \to \infty$. There exist absolute constants $d_2^{(1)}$, $0 < d_2^{(1)} < 1$,* and $d_2^{(2)} > 0$ *not depending on β and such that*

$$\ln \Xi_1(\gamma^{(f)}|\beta,\delta) = -\beta h^{(f)} \cdot |b| - G_1^{(f)}(\gamma^{(f)}) - G_2^{(f)}(\gamma^{(f)}),$$
$$\ln \Xi_1(\gamma^{(3,3)}|\beta,\delta) = -\beta h^{(f)} \cdot |b| - G_1^{(3,3)}(\gamma^{(3,3)}) - G_2^{(3,3)}(\gamma^{(3,3)}),$$
$$\ln \Xi_W^{(s)}(\gamma^{(f)}|\beta,\delta) = -G_{1,W}^{(f)}(\gamma^{(f)}) - G_{2,W}^{(f)}(\gamma^{(f)}),$$
$$\ln \Xi_W^{(s)}(\gamma^{(3,3)}|\beta,\delta) = -G_{1,W}^{(3,3)}(\gamma^{(3,3)}) - G_{1,W}^{(3,3)}(\gamma^{(3,3)}),$$
$$\beta C_1|\partial(sb)^{(\text{ver})}| \leqq G_1(\gamma) \leqq \beta C_2|\partial(sb)^{(\text{ver})}|,$$
$$\beta C_1|\partial b^{(\text{ver})}| \leqq G_{1,W}(\gamma) \leqq \beta C_2|\partial b^{(\text{ver})}|,$$
$$G_2(\gamma) \geqq -d_2^{(1)} \exp\{-\beta\varepsilon_1\} \cdot |b|,$$
$$G_{2,W}(\gamma) \geqq -d_2^{(1)} \cdot e^{-\beta\varepsilon_1} \cdot |b(\text{int}) + d_2^{(2)} e^{-\beta\varepsilon_1}|b \backslash b(\text{int})|$$
$$- e^{-\beta\varepsilon_1}|(V(b) \backslash b(\text{int})) \cap W|$$

for some absolute positive constants, C_1, C_2. The absence of the index "f" or "(3,3)" means that the inequalities are valid for both types of contours.

Proof of Theorem B is given in Sect. 6. The presence of $d_2^{(1)}$ is very important. If the volume b were occupied by a phase then the corresponding partition function would satisfy a similar inequality with $d_2^{(1)}=1$, $d_2^{(2)}=0$. We shall take $d_2^{(1)}$, $d_2^{(2)}$ from Theorem B and use Theorem A with the constants

$$d_1^{(1)}=d_2^{(1)}\exp\{-\beta\varepsilon_1\}, \quad d_1^{(2)}=d_2^{(2)}\exp\{-\beta\varepsilon_1\}.$$

We recall the notation O_s for domains bounded by $\partial b_s(\text{int})$ of any smooth contour $\gamma=(b,ph)$, $O=\bigcup_s O_s$, $W(b)=\bigcup_s O_s\cup b$. Now we introduce the most important partition functions

$$\begin{aligned}&\Xi(\gamma|\beta,\delta)\\&=\sum\exp\Big\{\beta\sum_{x\in W(b)}\varphi(x)\Big[\tfrac{1}{2}J_0\sum_{(x,x')\in(\mathbf{Z}^3)_{\text{hor}}}\varphi(x')\\&\quad+\tfrac{1}{2}J_1\sum_{(x,x')\in(\mathbf{Z}^3)^{(1)}_{\text{ver}}}\varphi(x')-\tfrac{1}{2}J_2\sum_{(x,x')\in(\mathbf{Z}^3)^{(2)}_{\text{ver}}}\varphi(x')\Big]\Big\}.\end{aligned}$$

The exterior sum is taken over such configurations $\varphi(W(b))=(\varphi(b),\varphi(O))$ that $\varphi(b)\in\mathfrak{A}(\gamma)$ and $\varphi(O)$ is compatible with boundary conditions.

Main Hypothesis. *The needed curve $\delta=\delta(\beta)$ is uniquely defined by the following assumptions: there exist contour functionals $F^{(f)}$, $F^{(3,3)}$ satisfying the assumptions of Theorem A and such that*

1) $a^{(f)}-\beta h^{(f)}=a^{(3,3)}-\beta h^{(3,3)}$,
2) $\Xi(\gamma^{(f)}|\beta,\delta)=\Xi^{(cr)}(\gamma^{(f)}|F^{(f)})\exp\{h^{(f)}\cdot|W(b)|\}$.

$$\Xi(\gamma^{(3,3)}|\beta,\delta)=\Xi^{(cr)}(\gamma^{(3,3)}|F^{(3,3)})\exp\{h^{(3,3)}|W(b)|\}.$$

The relations 1), 2) are quite similar to analogous relations in [6]. Using them we derive first the equations for contour functionals and then solve 1) for finding δ.

Let us take a contour $\gamma^{(f)}$ with its inner components $O_s\subset O(\gamma^{(f)})$ and fix outer smooth contours $\gamma_{sl}^{(\kappa_s)}$ inside each O_s. The index "κ_s" takes the values "f" or "(3,3)" depending on the boundary condition on $\partial O_s=\partial b_s(\text{int})$. We can write

$$\begin{aligned}&\Xi(\gamma^{(f)}|\beta,\delta)\\&=\Xi_1(\gamma^{(f)}|\beta,\delta)\prod_s\sum_{\{\gamma_{sl}^{(\kappa_s)}\}}\exp\Big\{-\beta g^{(\kappa_s)}\Big(|O_s|-\sum_l|W(\gamma_{sl}^{(\kappa_s)})|\Big)\Big\}\cdot\prod_l\Xi(\gamma_{sl}^{(\kappa_s)}|\beta,\delta).\end{aligned}\tag{3}$$

Using the main hypothesis we rewrite (3) as follows:

$$\begin{aligned}&\Xi(\gamma^{(f)}|\beta,\delta)\\&=\Xi_1(\gamma^{(f)}|\beta,\delta)\cdot\prod_s\exp\{-\beta h^{(\kappa_s)}|O_s|\}\\&\quad\cdot\sum_{\{\gamma_{sl}^{(\kappa_s)}\}}\prod_l\Xi^{(cr)}(\gamma_{sl}^{(\kappa_s)}|\beta,\delta)=\Xi_1(\gamma^{(f)}|\beta,\delta)\cdot\prod_s\exp\{-\beta h^{(\kappa_s)}|O_s|\}\cdot\Xi(O_s|F^{(\kappa_s)})\\&=\Xi_1(\gamma^{(f)}|\beta,\delta)\cdot\prod_s\exp\{(-\beta h^{(\kappa_s)}+a^{(\kappa_s)})|O_s|+\Delta(O_s|F^{(\kappa_s)})\},\end{aligned}\tag{4}$$

or with the help of $-\beta h^{(f)}+a^{(f)}=-\beta h^{(3,3)}+a^{(3,3)}$

$$\begin{aligned}&\exp\{(-\beta h^{(f)}+a^{(f)})\cdot|W(b)|\\&\quad -F^{(f)}(\gamma^{(f)})-a^{(f)}|b|+\Delta^{(cr)}(\gamma^{(f)}|F^{(f)})\}\\&\quad=\Xi_1(\gamma^{(f)}|\beta,\delta)\exp\left\{(-\beta h^{(f)}+a^{(f)})\cdot\sum_s|O_s|+\sum_s\Delta(O_s|F^{(\kappa_s)})\right\}.\end{aligned}$$

Thus we get

$$-\ln\Xi_1(\gamma^{(f)}|\beta,\delta)-\beta h^{(f)}\cdot|b^{(f)}|=F^{(f)}(\gamma^{(f)})-\Delta^{(cr)}(\gamma^{(f)}|F^{(f)})+\sum_s\Delta(O_s|F^{(\kappa_s)}).$$

Using Theorem B we get a needed equation in its final form:

$$\begin{aligned}&G_1^{(f)}(\gamma^{(f)})+G_2^{(f)}(\gamma^{(f)})\\&\quad=F^{(f)}(\gamma^{(f)})-\Delta^{(cr)}(\gamma^{(f)}|F^{(f)})+\sum_s\Delta(O_s|F^{(\kappa_s)})\end{aligned}\tag{5}$$

In the same manner we get an analogous equation for $F^{(3,3)}$:

$$\begin{aligned}&G_1^{(3,3)}(\gamma^{(3,3)})+G_2^{(3,3)}(\gamma^{(3,3)})\\&\quad=F^{(3,3)}(\gamma^{(3,3)})-\beta(h^{(f)}-h^{(3,3)})|b|-\Delta^{(cr)}(\gamma^{(3,3)}|F^{(3,3)})+\sum_s\Delta(O_s|F^{(\kappa_s)}).\end{aligned}\tag{6}$$

The existence of solutions of (5), (6) is shown in Sect. 5. In Sect. 7 we discuss the final steps and make some conclusions.

5. Generalized Contour Models and Proof of Theorem A

We shall consider only ensembles $\vartheta^{(f)}(U)$; the case of $\vartheta^{(3,3)}(U)$ is treated in a similar way. The index "f" is therefore omitted. A contour γ is called an outer contour of a configuration $\{\gamma_i\}$ if it is not contained in any inner domain $O_s(\gamma_i)$ of another contour. As usual (see [5, 6]), we introduce correlation functions $\pi_s(\gamma_1,\dots,\gamma_s|U,F)=\pi_s(\gamma_1,\dots,\gamma_s)$ which are equal to probabilities of the presence of s outer contours $\gamma_1,\dots,\gamma_s$, $\gamma_i=(b_i,ph)$ in a random configuration of contours. We shall derive now correlation equations for π_s which differ slightly from the usual correlation equations for outer contours (see [5, 6]).

We use the notation $T(z)$ for the set of all contours $\gamma=(b,ph)$ such that $z\in\partial b^{(\mathrm{ver})}$. Also we put $r_1=19$ if $ph=f$ and $r_1=18$ if $ph=(3,3)$.

For any outer contour $\gamma=(b,ph)$ we denote by $W(\gamma)$ the set $\{x|x=(x_1,x_2,x_3)\notin b$; there exists a point $y=(y_1,y_2,y_3)\subset\partial b^{(\mathrm{hor})}$ for which $y_1=x_1$, $y_2=x_2$, $|x_3-y_3|\leqq 38\}$. Put

$$W_1(\gamma_1,\dots,\gamma_p)=\bigcup_{i=1}^{p}(W(\gamma_i)\cup b_i),$$

$$W_2(\gamma_1,\dots,\gamma_p)=W_1(\gamma_1,\dots,\gamma_p)\cup\Bigg(\bigcup_{x|b^{(0)}(x)\cap W_1(\gamma_1,\dots,\gamma_p)\neq\emptyset}b^{(0)}(x)\Bigg),$$

$$W_3(\gamma_1,\dots,\gamma_p)=W_1(\gamma_1,\dots,\gamma_p)\cup\{x|b^{(0)}(x)\cap W_1(\gamma_1,\dots,\gamma_p)\neq\emptyset\},$$

$$W_i^{(m)}(\gamma_1,\dots,\gamma_p)=\{x|x\in W_i(\gamma_1,\dots,\gamma_p),\,d(\partial W_i(\gamma_1,\dots,\gamma_p),x)\leqq m\},$$

$$i=1,2,3.$$

We have

$$\pi_p(\gamma_1,\ldots,\gamma_p|U,F)=\frac{\Sigma^{(1)}\prod_l w(\gamma_l)}{\Xi(U,F)}$$

$$=\frac{\Xi_1(\gamma_1,\ldots,\gamma_p)\cdot\Sigma^{(2)}\prod_l w(\gamma_l)}{\Xi(U,F)}\cdot\frac{\exp\left\{-\sum_{i=1}^{p}F(\gamma_i)\right\}}{\Xi_1(\gamma_1,\ldots,\gamma_p)}.$$

Here the sum $\sum^{(1)}$ is taken over all configurations containing outer contours $\gamma_1,\ldots,\gamma_p$; $\sum^{(2)}$ is the sum over all configurations not containing outer contours $\gamma_1,\ldots,\gamma_p$ and containing only contours which do not intersect and not encircle any of contours $\gamma_1,\ldots,\gamma_p$; $\Xi_1(\gamma_1,\ldots,\gamma_p)$ is the partition function over all configurations of contours $\gamma=(b,ph)$ such that the following two properties are valid:

a$_1$) $b\subset W_2(\gamma_1,\ldots,\gamma_p)$,

a$_2$) if $b\cap W_1^{(38)}(\gamma_1,\ldots,\gamma_p)\neq\emptyset$ then $\gamma=(b,ph)$ is the least contour.

Let us explain why we need the volume $W_1^{(38)}$. Assume for simplicity that $p=1$ and we are dealing with a single contour γ_1. Then in view of our definitions in the 38-neighbourhood of $\partial b_1^{(\mathrm{hor})}$ the configuration is fixed and the points where spin-flips produce the least contours s-connected with γ_1 must belong to $W_2(\gamma_1)$. We put now

$$\lambda(\gamma_1,\ldots,\gamma_p)=\Xi_1^{-1}(\gamma_1,\ldots,\gamma_p)\exp\left\{-\sum_{i=1}^{p}F(\gamma_i)\right\}. \tag{8}$$

Lemma 1. *There exists an absolute constant $R_0^{(1)}$ such that*

$$\lambda(\gamma_1,\ldots,\gamma_p)\leqq\exp\left\{-\sum_{i=1}^{p}F(\gamma_i)-\frac{1}{N}\exp\{-F(\gamma^{(0)})\}\right.$$
$$\left.\cdot|W_3(\gamma_1,\ldots,\gamma_p)|\,(1+\varrho(K_1;\gamma_1,\ldots,\gamma_p))\right\},$$

where

$$|\varrho(K_1;\gamma_1,\ldots,\gamma_p)|\leqq\exp\{-R_1^{(1)}K_1\}.$$

Proof. We restrict ourselves by summation in Ξ_1 over configurations consisting only of the least contours. The result of this summation can be written in the form

$$\exp\left\{\frac{1}{\mathrm{N}}\exp\{-F(\gamma^{(0)})\}\cdot|W_3(\gamma_1,\ldots,\gamma_p)|\,(1+\varrho(K_1;\gamma_1,\ldots,\gamma_p))\right\},$$

where $\varrho(K_1;\gamma_1,\ldots,\gamma_p)$ satisfies the needed estimation. Q.E.D.

Lemma 2. *Let $\gamma=(b,ph)$ be a smooth contour such that b is connected. Put*

$$\lambda_1(\gamma)=\lambda_1(b)=\exp\{-K|\partial b^{(\mathrm{ver})}|-d_2|b|\}, \tag{9}$$

where K, d_2 are constants.

There exist constant K_0, $R=R(K_0)$ such that if

$$K>K_0,\quad d_2>R\sum_{\gamma\in T_0(z),\,\gamma\neq\gamma^{(0)}}\exp\{-K|\partial b^{(\mathrm{ver})}|\}$$

then

$$\sum_{\substack{\gamma\in T(z)\\ \gamma\neq\gamma^{(0)}}} \lambda_1(\gamma)\leqq \exp\{-R_1K\}\cdot\exp\{-F(\gamma^{(0)})\},$$

where R_1 is an absolute constant.

Proof. Firstly we shall consider a simpler model where one can understand better the essence of the situation and of our arguments and then shall make necessary additional remarks. Twice during the proof we shall use a trick which we have learned from Zahradnik [7].

In the simplified model we consider Ising type contours which are connected volumes. The statistical weight of any such contour b is equal to

$$\lambda_1(b)=\exp\{-K|\partial b^{(\mathrm{ver})}|-D|b|\}, \tag{10}$$

where

$$D=R\sum_{b\in T_0(z),\, b\neq b_0(z)} \exp\{-K|\partial b^{(\mathrm{ver})}|\},$$

R is a constant. Here $b_0(z)$ is the least contour for which $|b_0(z)|=1$. A contour b is smooth if either it is the least contour or $|\partial b_i^{(\mathrm{ver})}|>4$ for all vertical components of the boundary.

We fix $\overline{\partial b}$ and B and consider first the sum $\sum\lambda_1(b)$ over such b for which $\partial b(\mathrm{ext}) =\overline{\partial b}$ and $|b|=B$. We denote by $\bar{b}$ the connected bounded set, for which $\partial(\bar{b})=\overline{\partial b}$, $B_1=|\bar{b}|$. We introduce an auxiliary model. For every finite set O we denote by $\mathfrak{L}\,(O)$ an ensemble whose points are formal configurations of disjoint sets a_i, where each a_i is a boundary of a simply connected set, $a_i=\partial\mathscr{A}_i$, $a_i\cap O\neq\emptyset$, and put

$$\lambda_1(\{a_i\})=\prod_i \exp\{-K|a_i^{(\mathrm{ver})}|\}, \qquad \Xi_{\mathfrak{L}}(O)=\sum_{\{a_i\}}\lambda_1(\{a_i\}).$$

Here $a_i^{(\mathrm{ver})}$ is the set of vertical faces of a_i which determines uniquely a_i. This model can be investigated by the usual methods of the theory of contour models. In particular, one can use Kirkwood-Salzburg equations for correlation functions of a_i and investigate the behaviour of $\Xi_{\mathfrak{L}}(O)$ for large O provided that K is large enough. For such K one can find $\alpha(K)$ for which

$$\Xi_{\mathfrak{L}}(O)=\exp\{\alpha(K)\cdot|O|+\alpha_1(K,O)\,|\partial O|\},$$

where

$$|\alpha_1(K,O)|\leqq\exp\{-Kr+\mathrm{const}\}, \tag{11}$$

and r is the cardinality of the least possible $|a_i^{(\mathrm{ver})}|$. Now we have

$$S=\sum_{\substack{b|\partial b^{(\mathrm{ver})}(\mathrm{ext})=\overline{\partial b},\\ |b|=B}} \lambda_1(b)=\exp\{-K|\overline{\partial b}^{(\mathrm{ver})}|-D\cdot B\}\cdot\sum_{\substack{\{a_i\}\cup\overline{\partial b}=\partial b,\\ |b|=B}}\exp\{-K|a_i^{(\mathrm{ver})}|\}.$$

The last sum is estimated in the following way:

$$\begin{aligned}\sum_{\{a_i\}\cup\overline{\partial b}=\partial b} \exp\{-K|a_i^{(\mathrm{ver})}|\} &=\exp\{-\alpha(K)\,(B_1-B)\}\\ &\cdot\sum_{\{a_i\}}\prod_i\exp\{-K|a_i^{(\mathrm{ver})}|\}\cdot\Xi_{\mathfrak{L}}(\mathscr{A}_i)\cdot\exp\{-\alpha_1(K,\mathscr{A}_i)\cdot|a_i|\}\\ &\leqq\exp\{-\alpha(K)\,(B_1-B)+\max_i|\alpha_1(K,\mathscr{A}_i)|B\}\cdot\Xi_{\mathfrak{L}}(\bar{b})\\ &=\exp\{\alpha(K)\cdot B+2\max_{\mathscr{A}}|\alpha_1(K,\mathscr{A})|\cdot B\}.\end{aligned}$$

Thus

$$S \leqq \exp\{-K|\overline{\partial b}^{(\mathrm{ver})}| - D_1 B\}, \tag{12}$$

where

$$D_1 = D - 2\alpha(K) - 2\max_{\mathscr{A}} |\alpha_1(K, \mathscr{A})|.$$

Now we have to consider in a more detail the structure of $\overline{\partial b}^{(\mathrm{ver})}(\mathrm{ext})$. Each $\overline{\partial b}^{(\mathrm{ver})}(\mathrm{ext})$ consists of several connected vertical components $\overline{\partial b}_i^{(\mathrm{ver})}(\mathrm{ext})$. It is an appropriate moment to introduce several notions. A vertical component a is a connected set of vertical faces such that each vertical edge belongs to an even number of faces. The space of all vertical components is denoted by A, the space of a belonging to a set O is denoted by $A(O)$. Each $\overline{\partial b}^{(\mathrm{ver})}$ is a union of a finite number of $a \in A$, $\overline{\partial b}^{(\mathrm{ver})} = \bigcup_i a_i$.

Having a we can complete it in a unique way to the boundary of a connected set $[a]$ adding some number of connected horizontal components $\eta_j = \eta_j(a)$ in such a way that $\partial \eta_j$ is contained in the set of horizontal edges of faces of a. Thus $a \cup \bigcup_j \eta_j = \partial[a]$, $\partial([a])^{(\mathrm{ver})} = a$. The set $[a]$ will be called an interior of the vertical component a.

Assume that z is a vertical face, $z \in a_0 \subset \overline{\partial b}^{(\mathrm{ver})}$. We shall call a_0 a vertical component of the zeroth level. By induction, suppose that vertical components of $\overline{\partial b}^{(\mathrm{ver})}$ of the ℓth level are defined. Denote them by $a_{\ell,i}$, $1 \leqq i \leqq I_\ell$. We construct $\eta_j(a_{\ell,i})$ and define the components of the $(\ell+1)^{\mathrm{th}}$ level as those components of $\overline{\partial b}^{(\mathrm{ver})}$ which intersect $\eta_j(a_{\ell,i})$ and are not components of the previous levels. The level of a component a_i is denoted by $\ell(a_i)$, $\ell(\overline{\partial b}^{(\mathrm{ver})}) = \max_{a_i \subset \overline{\partial b}^{(\mathrm{ver})}} \ell(\tilde{a}_i)$. If we are given vertical components $a_{\ell,i}$, $1 \leqq i \leqq I_\ell$, $\ell = \ell_0$, we can construct $\overline{\partial b}_{\ell_0} = \bigcup_{\ell=0}^{\ell_0} \bigcup_{i=1}^{I_\ell} a_{\ell,i}$ and consider b, $\partial b = \overline{\partial b}_{\ell_0}$. We have obviously $\ell(\overline{\partial b}_{\ell_0}^{(\mathrm{ver})}) = \ell_0$. We denote $\partial T_{\ell_0}(z)$ the set of $\overline{\partial b}$, $z \in \overline{\partial b}$ and $\ell(\overline{\partial b}) = \ell_0$. Now we fix a_0 and have to estimate the sum

$$\sum_{x \in a_0} \sum_{\ell=0}^{\infty} \sum_{\overline{\partial b}^{(\mathrm{ver})} | \ell(\overline{\partial b}^{(\mathrm{ver})}) = \ell} \cdot \sum_{B} \sum_{\substack{b | \partial b^{(\mathrm{ver})}(\mathrm{ext}) = \overline{\partial b}^{(\mathrm{ver})} \\ |b| = B}} \lambda_1(b).$$

The last two sums were in fact already estimated. Indeed, if we denote by $\mathscr{H}(\overline{\partial b}^{(\mathrm{ver})})$ the number of horizontal faces of $\overline{\partial b}$ which correspond uniquely to $\overline{\partial b}^{(\mathrm{ver})}$, then

$$\tfrac{1}{2}\mathscr{H}(\partial b^{(\mathrm{ver})}) \leqq B \leqq |\partial b^{(\mathrm{ver})}|^2,$$

and using (12) we get

$$\begin{aligned} &\sum_{B} \sum_{b | \partial b^{(\mathrm{ver})}(\mathrm{ext}) = \overline{\partial b}^{(\mathrm{ver})}, |b| = B} \lambda_1(b) \\ &\leqq \sum_{B} \exp\{-K|\overline{\partial b}^{(\mathrm{ver})}| - D_1 B\} \\ &\leqq |\overline{\partial b}^{(\mathrm{ver})}|^2 \cdot \exp\{-K|\overline{\partial b}^{(\mathrm{ver})}| - \tfrac{1}{2} D_1 \mathscr{H}(\overline{\partial b}^{(\mathrm{ver})})\} \\ &\leqq \exp\{-K^{(1)}|\overline{\partial b}^{(\mathrm{ver})}| - \tfrac{1}{2} D_1 \mathscr{H}(\overline{\partial b}^{(\mathrm{ver})})\}, \end{aligned}$$

where

$$K^{(1)}=K_1-2\max_{t\geqq 1} t^{-1}\ln t.$$

Now we have a simpler sum,

$$S=\sum_{x\in a_0}\sum_{\ell=0}^{\infty}\sum_{\overline{\partial b}^{(\mathrm{ver})}|\ell(\partial b^{(\mathrm{ver})})=\ell}\cdot\exp\{-K^{(1)}|\overline{\partial b}^{(\mathrm{ver})}|-\tfrac{1}{2}D_1\mathscr{H}(\overline{\partial b}^{(\mathrm{ver})})\}. \tag{13}$$

We shall estimate it by induction from large to small ℓ. Starting with an arbitrarily large $\mathscr{L}$ we assume that the summation over all contours b, for which $\ell_0<\ell(b)\leqq\mathscr{L}$ is already performed, and the result is

$$S_{\ell_0}=\sum_{\overline{\partial b}\in\partial T_{\ell_0}(z)}\cdot\exp\left\{-K^{(1)}\sum_{\ell=0}^{\ell_0-1}\sum_{i=1}^{I_\ell}|a_{\ell,i}|-K^{(2)}\sum_{i=1}^{I_{\ell_0}}|a_{\ell_0,i}|-\tfrac{1}{2}D_1\mathscr{H}(\overline{\partial b})\right\}.$$

Here

$$K^{(2)}=K^{(1)}-2\max_{t\geqq 1} t^{-1}\ln t.$$

We shall use again Zahradnik's trick. Let η_k be a horizontal component of $a_{\ell_0-1,j}$. We consider an ensemble $\mathfrak{B}_1(\eta_k)$ whose elements are mutually disjoint vertical components $a_{\ell_0,j}$ such that $[a_{\ell_0,j}]\cap\eta_k\neq\emptyset$. We shall estimate the sum

$$W=\sum_{\{a_{\ell_0,j}\}\in\mathfrak{L}_1(\eta_k)}\exp\left\{-K^{(2)}\sum_j|a_{\ell_0,j}|-\tfrac{1}{2}D_1M\right\}, \tag{14}$$

where M is the cardinality of the set of horizontal faces of η_k belonging to $\overline{\partial b}^{(\mathrm{hor})}$, and thus lying outside all $[a_{\ell_0,j}]$. We can write

$$W=\sum_M\exp\{-\tfrac{1}{2}D_1M\}\cdot\sum\exp\left\{-K^{(2)}\sum_j|a_{\ell_0,j}|\right\}, \tag{14'}$$

where the inner summation goes over all configurations $\{a_{\ell_0,j}\}$ with the fixed value of M.

For any set $O\subset\eta_k$ and $K>0$ we put

$$\Xi_{\mathfrak{L}_1}(O)=\sum_{\substack{\{a_{\ell_0,j}\}\subset\mathfrak{L}_1(\eta_k)\\ [a_{\ell_0,j}]\cap\eta_k\subset O}}\exp\left\{-K\sum_j|a_{\ell_0,j}|\right\}.$$

For sufficiently large K one can find $\bar{\alpha}(K)$ and $\bar{\alpha}_1(K,O)$ such that

$$\ln\Xi_{\mathfrak{L}_1}(O)=\bar{\alpha}(K)|O|+\bar{\alpha}_1(K,O)\cdot|\partial O|, \tag{15}$$

where $\bar{\alpha}(K)$, $\bar{\alpha}_1(K,O)\leqq\mathrm{const}\,e^{-Kr}$ as $K\to\infty$, and r is the least possible value of $|a_{\ell_0,j}|$. In all these arguments we assumed that all $a_{\ell_0,j}\subset\mathfrak{L}_1(\eta_k)$. But there can be also components $a_{\ell_0,j}$ which encircle connected components of η_k. The summation over such configurations only changes slightly the remainder term in (15). We shall assume later that this correction is already present in (15). Then in (14)

$$\begin{aligned}\sum_{\{a_{\ell_0,j}\}}\exp\left\{-K^{(2)}\sum_j|a_{\ell_0,j}|\right\}&\leqq\exp\{-\bar{\alpha}(K^{(2)})[|\eta_k|-M]\}\\ &\cdot\sum_{\{a_{\ell_0,j}\}}\prod_j\exp\{-K^{(2)}|a_{\ell_0,j}|\}\cdot\Xi_{\mathfrak{L}_1}([a_{\ell_0,j}]\cap\eta_k)\\ &\cdot\exp\{-\bar{\alpha}_1(K^{(2)},[a_{\ell_0,j}]\cap\eta_k)\cdot|\partial([a_{\ell_0,j}]\cap\eta_k)|\}\\ &\leqq\exp\{-\bar{\alpha}(K^{(2)})\cdot[|\eta_k|-M]+\max|\bar{\alpha}_1(K^{(2)},[a_{\ell_0,j}]\cap\eta_k)|\cdot M\}\\ &\cdot\Xi_{\mathfrak{L}_1}(\eta_k)=\exp\{\bar{\alpha}_2(K^{(2)})\cdot M\},\end{aligned}$$

where

$$\bar{\alpha}_2(K^{(2)}) = \bar{\alpha}(K^{(2)}) + 2\max|\bar{\alpha}_1(K^{(2)}, [a_{\ell_0,j}] \cap \eta_k)|.$$

Assuming that $\frac{1}{2}D_1 \geqq \bar{\alpha}_2(K^{(2)})$, we have

$$W \leqq \sum_M \exp\{(-\tfrac{1}{2}D_1 + \bar{\alpha}_2(K^{(2)})M\} \leqq \sum_i |a_{\ell_0-1,i}|^2.$$

Finally

$$S_{\ell_0-1} \leqq \sum_{\bar{\partial b} \in \partial T_{\ell_0-1}(z)} \cdot \exp\left\{-K^{(1)} \sum_{\ell=0}^{\ell_0-2} \sum_{i=1}^{I_\ell} |a_{\ell,i}| - K^{(2)} \sum_{i=1}^{I_{\ell_0-1}} |a_{\ell_0-1,i}| - \tfrac{1}{2}D_1 \mathscr{H}(\overline{\partial b})\right\}.$$

Thus we passed from ℓ_0 to $\ell_0 - 1$. Therefore for $\ell_0 = 0$, the sum $S_0 \leqq \sum_{a_0} \exp\{-K^{(2)}|a_0|\}$. The number of all possible components a_0 with $|a_0| = m$ is not more than C^m, where C is an absolute constant. This gives the final result provided that

$$D \geqq 2\alpha(K) + 2\max_O |\alpha_1(K, O)| + 2\left[\bar{\alpha}(K^{(2)}) + \max_O |\bar{\alpha}_1(K^{(2)}, O)|\right]. \qquad (16)$$

One easily finds that for sufficiently large K the numbers $D \sim R\exp\{-Kr\}$, $\alpha(K)$, $\max_O |\alpha_1(K, O)|$, $\bar{\alpha}(K^{(2)})$, $\max_O |\bar{\alpha}_1(K^{(2)}, O)| \leqq \text{const} \exp\{-Kr\}$, where r is the least possible value of $|a_0| \in T_0(z)$ and const is an absolute constant. Thus if R is sufficiently large the inequality (16) is valid.

Our arguments can be extended without any difficulties to more complicated situations such as contour models introduced in Sect. 3. The final conclusion and the result will be the same provided that K and R are sufficiently large. Q.E.D.

We shall say that a contour $\gamma = (b, ph)$ has the property A if the following is true: For the strip $S = W_1^{(38)}(\gamma_1, \ldots, \gamma_p)$ the intersection $b \cap S \neq \emptyset$, and there exists a configuration of the least contours

$$\{\gamma^{(0)}(x_i) = (b^{(0)}(x_i), ph)\}, \qquad i = 1, \ldots, q(\gamma),$$

such that

$$b = \bigcup_{i=1}^{q(\gamma)} b^{(0)}(x_i) \cup (b \backslash S), \qquad x_i \in W_3^{(38+2r_1)}(\gamma_1, \ldots, \gamma_p).$$

We denote by γ_x any contour containing $x \in \mathbb{Z}^3$.

Let us consider the fraction

$$Q = \frac{\Xi_1(\gamma_1, \ldots, \gamma_p) \cdot \sum^{(2)} \prod_i w(\gamma_i)}{\Xi(U|F)}.$$

It is easy to see that

$$Q = E\left\{\left[\sum_{M \subset S} \prod_{x \in M} \sum_{\gamma_x^A} \chi_{\gamma_x^A} \cdot f(\gamma_x^A) \prod_{x \in S \backslash M} \prod_{\gamma'_x} (1 - \chi_{\gamma'_x})\right] \prod_{\bar{\gamma}} (1 - \chi_{\bar{\gamma}})\right\}.$$

Here γ_x^A is the notation for the contour having the property A and containing x, γ is a contour encircling at least one of the contours $\gamma_1, \dots, \gamma_p$;

$$f(\gamma^A) = \exp\Bigg\{F(\gamma^A) - F(\hat{\gamma}^A) - \sum_{j|b^{(0)}(x_j)\cap\partial W_1 \neq \emptyset} F(\gamma^{(0)}(x_j))\Bigg\}, \quad \gamma^A = (b, ph),$$

$$\hat{\gamma}^A = (\hat{b}, ph), \quad \hat{b} = b \setminus \bigcup_j b^{(0)}(x_j) \cap S, \; \chi_\gamma$$

is the indicator for the set of configurations having an outer contour γ, γ_x' is a contour intersecting S and is not the least contour, E means the expectation with respect to the ensemble $\vartheta(U)$.

The last expression shows that Q can be written as a linear combination of correlation functions which gives us the desired system of correlation equations:

$$\begin{aligned}\pi_p(\gamma_1, \dots, \gamma_p) &= \lambda(\gamma_1, \dots, \gamma_p)\Bigg[1 + \sum_{\gamma_1^A, \dots, \gamma_r^A} \prod_{i=1}^{r} f(\gamma_i^A)\Bigg(\sum_{\gamma_1', \dots, \gamma_m'} (-1)^m \\ &\cdot \pi_{r+m}(\gamma_1^A, \dots, \gamma_r^A, \gamma_1', \dots, \gamma_m') \\ &+ \sum_{\bar{\gamma}_1, \dots, \bar{\gamma}_\ell} (-1)^\ell \pi_{r+m+\ell}(\gamma_1^A, \dots, \gamma_r^A, \gamma_1', \dots, \gamma_m', \bar{\gamma}_1, \dots, \bar{\gamma}_\ell)\Bigg)\Bigg].\end{aligned}$$

We introduce the following norm in the space of sequences $\pi = \{\pi_p(\gamma_1, \dots, \gamma_p)\}$:

$$\|\pi\| = \sup_{p, \{\gamma_1, \dots, \gamma_p\}} \frac{|\pi_p(\gamma_1, \dots, \gamma_p)|}{\bar{\lambda}(\gamma_1, \dots, \gamma_p)},$$

where

$$\begin{aligned}\bar{\lambda}(\gamma_1, \dots, \gamma_p) = \lambda(\gamma_1, \dots, \gamma_p) \exp\Bigg\{&\frac{1}{N}\exp\{-F(\gamma^{(0)}) - cK_1\} \\ &\cdot |(\partial W_3(\gamma_1, \dots, \gamma_p))^{(\mathrm{hor})}| \\ &+ \frac{D}{N}\exp\{-F(\gamma^{(0)})\} \cdot \sum_{i=1}^{p} |\partial(s\gamma_i)^{(\mathrm{ver})}|\Bigg\},\end{aligned}$$

c, D are constants.

The system of correlation equations can be written in the form $\pi = A\pi + \Phi$. Here A is a linear operator, $\Phi = \{\lambda(\gamma_1, \dots, \gamma_p)\}$.

Lemma 3. *There exist constants K_0, $R^{(1)} = R^{(1)}(K_0)$, $R^{(2)} = R^{(2)}(K_0)$ such that for $K_1 > K_0$,*

$$\begin{aligned}\Sigma^{r,m} = \Sigma^{r,m}(z_1, \dots, z_r, z_1', \dots, z_m') = &\sum_{\substack{\gamma_i^A \in T(z_i),\, 1 \leq i \leq r \\ \gamma_j' \neq \gamma^{(0)} \in T(z_j'),\, 1 \leq j \leq m}} \\ &\cdot \bar{\lambda}(\gamma_1^A, \dots, \gamma_r^A, \gamma_1', \dots, \gamma_m') \cdot f(\gamma_1^A, \dots, \gamma_r^A) \\ \leq &\Bigg(\exp\{-R^{(1)}K_1\} \cdot \frac{1}{N}\exp\{-F(\gamma^{(0)})\} + R^{(2)} \sum_{b \in T_0(z)} \exp\{-K_1 |\partial b^{(\mathrm{ver})}|\}\Bigg)^{r+m},\end{aligned}$$

where

$$z_i \in W_1^{(38)}(\gamma_1, \ldots, \gamma_p), \qquad z'_j \in W_1^{(38)}(\gamma_1, \ldots, \gamma_p)$$

are arbitrary vertical faces.

Proof. We shall write for brevity $f(\Gamma)$ instead of $f(\gamma_1, \ldots, \gamma_\ell)$ if f is a function of a configuration $\Gamma = \{\gamma_1, \ldots, \gamma_\ell\}$,

$$W(\Gamma, D, c) = \exp\{-cK_1\} \, |\partial W_3^{(\mathrm{hor})}(\Gamma)| + D \sum_{\gamma \in \Gamma} |\partial(s\gamma)^{(\mathrm{ver})}|,$$

$$\Gamma_A^{(r)} = \{\gamma_i^A\}_{i=1}^r, \qquad \Gamma^{(m)} = \{\gamma'_j\}_{j=1}^m,$$

$$\Gamma^{(r,m)} = \Gamma_A^{(r)} \cup \Gamma^{(m)}, \qquad \Gamma_{A,I}^{(r)} = \{\gamma_i^A\}_{i \in I},$$

I is a subset of the set $(1, \ldots, r)$,

$$\Gamma_{A,I}^{(r,m)} = \Gamma_{A,I}^{(r)} \cup \Gamma^{(m)}, \qquad \Gamma_p = \{\gamma_1, \ldots, \gamma_p\},$$

$$\hat{\Gamma}_{A,I}^{(r)} = \{\hat{\gamma}_i^A\}_{i \in I}, \qquad \hat{\Gamma}_{A,I}^{(r,m)} = \hat{\Gamma}_{A,I}^{(r)} \cup \Gamma^{(m)}.$$

It is easy to see that if $\hat{\gamma}_i^A = \gamma^{(0)}$, then

$$\sum_{\gamma_i^A \mid \gamma_i^A = \gamma^{(0)}} \exp\{-F(\gamma_i^A)\} \cdot f(\gamma_i^A) \cdot \exp\left\{\frac{1}{N} \exp\{-F(\gamma^{(0)})\} \cdot |W(\gamma_i^A; D, c)|\right\} \leqq R_3 \exp\{-2F(\gamma^{(0)})\}$$

for a constant $R_3 = R_3(D, c)$. We have

$$\Sigma^{(r,m)} = \sum_J \Sigma^{(J,m)},$$

where the sum $\Sigma^{(J,m)}$ has the same form as $\Sigma^{(r,m)}$ with the additional condition of fixing the indices $j \in J$ for which $\hat{\gamma}_j^A$ are the least contours. It corresponds to our general strategy to consider separately the contribution of least contours.

For $\Sigma^{(J,m)}$ we can write the estimate

$$\begin{aligned}
\Sigma^{(J,m)} \leqq {} & (R_3 \exp\{-2F(\gamma^{(0)})\})^{|J|} \sum_{\gamma_i^A \in T(z_i),\, i \in I = (1, \ldots, r) - J} \\
& \cdot \left[\exp\left\{-\sum_{\gamma \in \Gamma_{A,I}^{(r,m)}} F(\gamma)\right\} \cdot \prod_{i \in I} f(\gamma_i^A) \cdot \Xi_1(\Gamma_{A,I}^{(r,m)})^{-1} \right. \\
& \left. \cdot \exp\left\{\frac{1}{N} \cdot \exp\{-F(\gamma^{(0)})\} \cdot |W(\Gamma_{A,I}^{(r,m)}; D, c)|\right\}\right] \\
= {} & (R_3 \exp\{-2F(\gamma^{(0)})\})^{|J|} \cdot \sum_{\substack{\gamma_i^A \in T(z_i),\, i \in I \\ \gamma'_q \in T(z'_q),\, 1 \leqq q \leqq m}} \\
& \cdot \left[\exp\left\{-\sum_{\gamma \in \hat{\Gamma}_{A,I}^{(r,m)}} F(\gamma)\right\} \cdot \prod_{i \in I} \prod_{j=1}^{q(\gamma_i^A)} \exp\{-F(\gamma^{(0)}(x_j))\} \right. \\
& \left. \cdot (\Xi_1(\Gamma_{A,I}^{(r,m)}))^{-1} \cdot \exp\left\{\frac{1}{N} \exp\{-F(\gamma^{(0)})\} \, |W(\Gamma_{A,I}^{(r,m)}; D, c)|\right\}\right].
\end{aligned}$$

It is not difficult to see that

$$\prod_{i\in I}\prod_{j=1}^{q(\gamma_i^A)}\exp\{-F(\gamma_{x_j}^{(0)})\}\leqq\exp\left\{\frac{1}{N}\exp\{-F(\gamma^{(0)})\}\cdot|\tilde{W}(\hat{\Gamma}_{A,I}^{(r)})|\right.,$$

where

$$\tilde{W}(\hat{\Gamma}_{A,I}^{(r)})=\bigcup_{i\in I}(V(\tilde{b}_i)\backslash\tilde{b}_i(\text{int}))\cap W_3(\Gamma_p),\qquad \tilde{b}_i=b_i^A\backslash S.$$

Consequently

$$\begin{aligned}\Sigma^{(J,m)}\leqq{}&(R_3\exp\{-2F(\gamma^{(0)})\})^{|J|}\sum_{\substack{\gamma_i^A|d(z_i,\gamma_i^A)\leqq r_1\\ \gamma_q'\in T(z_q'),\,\gamma_q'\neq\gamma^{(0)}}}\\ &\cdot\exp\Big\{-\sum_{\gamma\in\Gamma_{A,I}^{(r,m)}}F(\gamma)\Big\}\cdot(\Xi_1(\hat{\Gamma}_{A,I}^{(r,m)}))^{-1}\\ &\cdot\exp\left\{\frac{1}{N}\exp\{-F(\gamma^{(0)})\}\cdot[|\tilde{W}(\hat{\Gamma}_{A,I}^{(r)})|+W(\Gamma_{A,I}^{(r,m)};D,c)]\right\}.\end{aligned}$$

This sum can be estimated as follows. First we can sum up over all contours γ_ℓ', $\ell=1,\dots,m$ corresponding to some smooth contours $\bar{\gamma}_\ell'$, $\ell=1,\dots,m$ and over all contours $\hat{\gamma}_i^A$, $i\in I$, which correspond to some smooth contours $\bar{\gamma}_i^A$, $i\in I$, and not intersect $\partial W_1(\Gamma_{A,I}^{(r,m)})$ (here we use the property 3) of contour functional F with the following choice of the set W: $W=V(\bar{\gamma}_\ell')$ for contours $\bar{\gamma}_\ell'$, $1\leqq\ell\leqq m$ and $W=V(\bar{\gamma}_i^A)\backslash\tilde{W}(\hat{\Gamma}_{A,I}^{(r)})$ for contours $\bar{\gamma}_i^A$, $i\in I$); then we estimate the sum over smooth contours using Lemma 2.

The result of the first summation gives us the estimate:

$$\begin{aligned}\Sigma^{(J,m)}\leqq{}&(R_3\exp\{-2F(\gamma^{(0)})\})^{|J|}(C(r_1))^{|I|+m}\\ &\cdot\max_{\substack{\bar{z}_i|d(\bar{z}_i,z_i)\leqq 2r_1,\,i\in I\\ \bar{z}_q'|d(\bar{z}_q',z_q')\leqq 2r_1,\,1\leqq q\leqq m}}\ \sum_{\substack{\bar{\gamma}_i^A\in T(\bar{z}_i)\\ \bar{\gamma}_q'\in T(\bar{z}_q')}}\ \prod_{\bar{\gamma}\in\bar{\Gamma}_{A,I}^{(r,m)}}D_1(\bar{\gamma})\\ &\cdot\exp\Big\{-\sum_{\gamma\in\bar{\Gamma}_{A,I}^{(r,m)}}F_1(\gamma)+d_1^{(1)}\sum_{\bar{\gamma}=(b,ph)\in\bar{\Gamma}_{A,I}^{(r,m)}}|b(\text{int})|\\ &-d_1^{(2)}\sum_{\bar{\gamma}=(b,ph)\in\bar{\Gamma}_{A,I}^{(r,m)}}|b\backslash b(\text{int})|+\frac{1}{N}\exp\{-F(\gamma^{(0)})\}\\ &\cdot\Big(\sum_{\bar{\gamma}=(b,ph)\in\bar{\Gamma}_{A,I}^{(r,m)}}|V(b)\backslash b(\text{int})|+D_2\Sigma|\partial(s\gamma^{(\text{ver})})|\Big)\cdot(\Xi_1(\bar{\Gamma}_{A,I}^{(r,m)}))^{-1}.\end{aligned}$$

Here $C(r_1)$ is a constant, $D_1(\bar{\gamma})=\max_{b_i\subset b}|\partial b_i|$, $\bar{\gamma}=(b,ph)$, b_i is the connected component of b. It is clear that $|D_1(\bar{\gamma})|<\max_i|\partial b_i^{(\text{ver})}|^2$, D_2 is a constant.

For any contour $\gamma=(b,ph)$ we denote by $n(\gamma)$ the number of connected components of the set b which differ from the least components. If $b=Ub_i$ is a decomposition on connected components, then we can assume that $b_i\neq b^{(0)}$ for $i\leqq n(\gamma)$ and $b_i=b^{(0)}$ for $i>n(\gamma)$.

Let $\mathcal{M}(\bar{b}_1,\dots,\bar{b}_n)$ be the set of all contours $\gamma=(b,ph)$ for which $n(\gamma)=n$ and let $b_i=\bar{b}_i$, $1\leqq i\leqq n(\gamma)$, $\mathcal{M}(\emptyset)$ be the set of contours with $n(\gamma)=0$.

Consider separately the sum over such configurations $\bar{\Gamma}_{A,I}^{(r,m)}$ for which

$$\bar{\gamma}_i^A \in \mathscr{M}(b_1^{(i)}, \ldots, b_{n_i}^{(i)}), \qquad \bar{\gamma}'_q \in \mathscr{M}(b_1^{\prime(q)}, \ldots, b_{n'_q}^{\prime(q)}),$$

$i \in I$, $1 \leqq q \leqq m$. This sum is not more than

$$\exp\left\{-\sum_{\gamma\in\bar{\Gamma}_{A,I}^{(r,m)}}\left[(K_1-D_3)\cdot\sum_{j=1}^{n_i}|(\partial b_j^{(i)})^{(\mathrm{ver})}|-d_1^{(1)}\sum_{j=1}^{n_i}|b_j^{(i)}(\mathrm{int})|\right.\right.$$
$$\left.\left.-d_1^{(2)}\sum_{j=1}^{n_i}|b_j^{(i)}\backslash b_j^{(i)}(\mathrm{int})|\right]\right\}\cdot\exp\left\{-\sum_{q=1}^{m}\left[(K_1-D_3)\right.\right.$$
$$\left.\left.\cdot\sum_{\ell=1}^{n'_q}|(\partial b_q^{\prime(\ell)})^{(\mathrm{ver})}|-d_1^{(1)}\sum_{\ell=1}^{n'_q}|b_q^{\prime(\ell)}(\mathrm{int})|-d_1^{(2)}\sum_{\ell=1}^{n'_q}|b_q^{\prime(\ell)}\backslash b_q^{(\ell)}(\mathrm{int})|\right]\right\}$$
$$\cdot\prod_{i\in I}(R_4\exp\{-F(\gamma^{(0)})\})^{d(b_1^{(1)},\ldots,b_{n_i}^{(1)})}\cdot\prod_{j=1}^{m}(R_4\exp\{-F(\gamma^{(0)})\})^{d(b_1^{\prime(J)},\ldots,b_{n_i}^{\prime(J)})}$$
$$\cdot\exp\left\{-\frac{1}{N}\exp\{-F(\gamma^{(0)})\}\cdot(1-\varrho)\cdot|W_0(\bar{\Gamma}_{A,I}^{(r,m)})|\right\},$$

where

$$W_0(\bar{\Gamma}_{A,I}^{(r,m)})=\left(\bigcup_{i\in I}\bigcup_{j=1}^{n_i}b_j^{(i)}\right)\cup\left(\bigcup_{i=1}^{m}\bigcup_{j=1}^{n'_j}V(b_j^{\prime(i)})\right),\qquad d(b_1,\ldots,b_n)$$

is the length of the minimal tree which connects the sets $W_3(b_i)$, $1\leqq i\leqq n$, $W_3(b_i) = W(\gamma_i)$, $\gamma_i=(b_i, ph)$, $\varrho\leqq\exp\{-C_1K_1\}$, D_3, R_4, C_2 are absolute constants. Now we see that the last sum is not more than

$$\prod_{\gamma_i\in\bar{\Gamma}_{A,I}^{(r,m)}\,|\,\gamma_i\in\mathscr{M}(b_1^{(i)},\ldots,b_{t_i}^{(i)})}$$
$$\cdot\left[\exp\left\{-(K_1-D_3)\sum_{j=1}^{t_i}|(\partial b_j^{(i)})^{(\mathrm{ver})}|-d_2\sum_{j=1}^{t_i}|b_j^{(i)}|\right\}\right.$$
$$\left.\cdot(R_4\exp\{-F(\gamma^{(0)})\})^{d(b_1^{(i)},\ldots,b_{t_i}^{(i)})}\right]$$
$$\cdot\prod_{\gamma\in\bar{\Gamma}_{A,I}^{(r,m)}\,|\,\gamma\in\mathscr{M}(\emptyset)}(R_5\exp\{-2F(\gamma^{(0)})\}).$$

Here

$$d_2=\min((\exp\{-F(\gamma^{(0)})\}-d_1^{(1)})(1-\varrho),d_1^{(2)}(r_2)^{-1}),$$

R_5 is an absolute constant, $r_2=9$ if $ph=$f.g.s. and $r_2=6$, if $ph=$(3.3)-g.s.

Now Lemma 3 follows from Lemma 2.

From the main estimate of Lemma 3 it is not difficult to see that the operator Λ is a contraction and $\|\Lambda\|\leqq\mathrm{const}(K_1)\to 0$ as $K_1\to\infty$. Therefore for sufficiently large K_1 the limit correlation functions can be written as $\pi=\sum_{i=0}^{\infty}\Lambda^i\Phi$. The estimation of the difference between the limit correlation functions and the correlation functions in a finite volume is done by a routine method (see [5, 6]).

6. Proof the Theorem B

We shall consider the case of contours $\gamma^{(f)}$ and $\Xi_W^{(s)}(\gamma^{(f)}|\beta,\delta)$. Other cases can be considered in the same manner. We have (see Sect. 4).

$$\begin{aligned}&\Xi_W^{(s)}(\gamma^{(f)}|\beta,\delta)\\&=\sum_{\substack{\gamma_1=(b_1,ph)\in Sm(\gamma^{(f)})\\ b_i(\mathrm{int})\in W}}\exp\{\beta\mathcal{H}^{(f)}(b_1^{(f)})\}\cdot\sum_{\varphi(b_1)\in\mathfrak{A}(\gamma_1^{(f)})}\\&\cdot\exp\Big\{\beta\Big[\sum_{x\in b_1}\varphi(x)\Big(\tfrac{1}{2}J_0\sum_{(x,x')\in(\mathbb{Z}^3)^{(1)}_{\mathrm{hor}}}\varphi(x')+\tfrac{1}{2}J_1\sum_{(x,x')\in(\mathbb{Z}^3)^{(1)}_{\mathrm{ver}}}\varphi(x')\\&-\tfrac{1}{2}J_2\sum_{(x,x')\in(\mathbb{Z}^3)^{(2)}_{\mathrm{ver}}}\varphi(x')\Big)\Big]\Big\}.\end{aligned}$$

Here $\mathcal{H}^{(f)}(b)$ is the energy of the ferromagnetic configuration in b.

For any $\varphi(b)=\varphi$ we denote

a$_1$) $G^{(\mathrm{ver})}(\varphi)$ is the set of ordered bonds $(x,x')\in(\mathbb{Z}^3)^{(1)}_{\mathrm{hor}}$, $x\in b$, for which $\varphi(x)\neq\varphi(x')$ and $d((x,x'),(\partial(sb))^{\mathrm{ver}})\leqq r_1$. We recall that $\partial(sb)^{(\mathrm{ver})}$ is the set of vertical faces of $\partial(sb)$ which can be also considered as the set of horizontal bonds perpendicular to these faces.

a$_2$) $G_1^{(\mathrm{ver})}(\varphi)$ is the set of ordered bonds $(x,x')\in(\mathbb{Z}^3)^{(1)}_{\mathrm{hor}}$, where $x\in b$, $\varphi(x)\neq\varphi(x')$ and $(x,x')\notin G^{(\mathrm{ver})}(\varphi)$.

a$_3$) $G_2^{(\mathrm{ver})}(\varphi)$ is the set of $x\in b$ for which $d(x,G^{(\mathrm{ver})}(\varphi)\cup G_1^{(\mathrm{ver})}(\varphi))\leqq 2r_1$. The values of $\varphi(x)$, $x\in G_2(\varphi)$, are defined uniquely by values of φ on $G^{(\mathrm{ver})}(\varphi)\cup G_1^{(\mathrm{ver})}(\varphi)$.

a$_4$) $W_1(\varphi)=b\backslash G_2^{(\mathrm{ver})}(\varphi)$.

a$_5$) $N_1(\varphi)$ is the set of $x=(x_1,x_2,x_3)\in W_1(b)$, where $\varphi(x_1,x_2,x_3+1)=\varphi(x_1,x_2,x_3-1)\neq\varphi(x_1,x_2,x_3)$.

a$_6$) $N_r(\varphi)$, $r>1$ is the number of series of points lying on the same vertical line, i.e. series of points $(x_1,x_2,x')\in W_1(b)$, $a<x'\leqq a+r$, and $\varphi(x_1,x_2,a)\neq\varphi(x_1,x_2,a+1)=\varphi(x_1,x_2,a+2)=\ldots=\varphi(x_1,x_2,a+r)\neq\varphi(x_1,x_2,a+r+1)$, $N_r^1(\varphi)$ is the set of points of the form $(x_1,x_2,a+1)$ or $(x_1,x_2,a+r)$, $N_r^{>2}(\varphi)$ is the set of points of the form $(x_1,x_2,a+\ell)$, $2<\ell<r-1$, $N_r^2(\varphi)$ is the set of points of the form $(x_1,x_2,a+2)$, $(x_1,x_2,a+r-1)$. An important remark is that $r\leqq 2r_1$, because otherwise there will be points in a ferromagnetic phase.

We rewrite the energy of $\varphi(b)$ as follows:

$$\begin{aligned}-H(\varphi(b))=&-\mathcal{H}^{(f)}(b)-\tfrac{1}{2}J_0|G^{(\mathrm{ver})}(\varphi)|-\tfrac{1}{2}J_0|G_1^{(\mathrm{ver})}(\varphi)|\\&-2\delta\sum_{r\geqq 2}N_r(\varphi)-4J_1N_1(\varphi)+\mathcal{H}_1.\end{aligned}$$

Here $\mathcal{H}_1$ is a correction term $|\mathcal{H}_1|\leqq\mathrm{const}\cdot(J_1+J_2)\cdot(|G_1(\varphi)|+|G_2(\varphi)|)$. We recall our assumption that J_1, J_2 are sufficiently small comparing with J_0.

Lemma 1. *There exists a constant* c_2 such that

$$|G^{(\mathrm{ver})}|>c_2|\partial(sb)^{(\mathrm{ver})}|.$$

Proof. Let a bond (x,x') be orthogonal to a face of $\partial(sb)^{(\mathrm{ver})}$ and $\varphi(x)$ is in a phase while $x'\in\partial(sb)^{(\mathrm{ver})}$ and $\varphi(x')$ is not in the phace. It means that in the neighbourhood of the radius r_1 there must be horizontal bond $(y, y') \subset b$, where $\varphi(y) \neq \varphi(y')$. Otherwise $\varphi(x')$ will be in a phase. Q.E.D.

The cardinality of the set of connected components $\partial b_i^{(\text{ver})}$, $|\partial b_i^{(\text{ver})}| = n$, passing through a fixed face is not more than c_3^n, where c_3 is a constant. We can choose c_3 so large that the cardinality of the set of all possible $\varphi(G^{(\text{ver})})$ is not more than $c_3^{|\partial b^{(\text{ver})}|}$.

A point $x = (x_1, x_2, x_3) \in b \backslash G^{(\text{ver})}(\varphi)$ is called an elementary defect (e.d.) if $\varphi(x) \neq \varphi(y)$ for all $y = (y_1, y_2, y_3)$, $|y_1 - x_1| + |y_2 - x_2| = 1$, $y_3 = x_3$, and there are no other horizontal bonds $(x', x'') \in b$ such that $(x', x'') \cap b^{(0)}(x) \neq \emptyset$ and $\varphi(x') \neq \varphi(x'')$.

A configuration $\varphi(b)$ is called ideal if there are no e.d. The set of ideal configurations in b is denoted by $\mathfrak{A}^{(i)}(b)$.

Let $\varphi_0(b) \in \mathfrak{A}^{(i)}(b)$. The set of all $\varphi(b)$ which differ from $\varphi_0(b)$ by some number of e.d. is denoted by $\mathfrak{A}(b \mid \varphi_0(b))$. Each $\varphi(b)$ corresponds for one and only one ideal configuration $\varphi_0(b)$.

We can write

$$
\begin{aligned}
&\Xi_W^{(s)}(\gamma^{(f)} \mid \beta, \delta) \\
&= \sum_{\substack{\gamma_1^{(f)} = (b_1^{(f)}, ph) \in Sm\gamma^{(f)} \\ b_1^{(f)}(\text{int}) \subset W}} \; \sum_{\varphi(b_1^{(f)}) \in \mathfrak{A}(\gamma_1^{(f)})} \\
&\quad \cdot \exp\left\{ -\beta\left[\tfrac{1}{2} J_0 |G^{(\text{ver})}(\varphi(b_1^{(f)}))| + \tfrac{1}{2} J_0 |G_1^{(\text{ver})}(\varphi(b_1^{(f)}))| \right.\right. \\
&\quad \left.\left. + 2\delta \sum_{r=2}^{2r_1} |N_r(\varphi(b_1^{(f)}))| + 4J_1 N_1(\varphi(b_1^{(f)})) + \mathscr{H}_1(\varphi(b_1^{(f)})) \right]\right\} \\
&= \sum_{\substack{\gamma_1^{(f)} = (b_1^{(f)}, ph) \in Sm\gamma^{(f)} \\ b_1^{(f)}(\text{int}) \subset W}} \; \sum_{\varphi(b_1^{(f)}) \in \mathfrak{A}(\gamma_1^{(f)})} \\
&\quad \cdot \exp\left\{ -\beta\left[\tfrac{1}{2} J_0 |G^{(\text{ver})}(\varphi(b_1^{(f)}))| + \tfrac{1}{2} J_0 |G_1^{(\text{ver})}(\varphi(b_1^{(f)}))| \right.\right. \\
&\quad \left.\left. + \delta \sum_{r=2}^{2r_1} |N_r^{(1)}(\varphi(b_1^{(f)}))| + 4J_1 N_1(\varphi(b_1^{(f)})) + \mathscr{H}_1(\varphi(b_1^{(f)})) \right]\right\}.
\end{aligned}
$$

Furthermore

$$
\begin{aligned}
&\Xi_W^{(s)}(\gamma^{(f)} \mid \beta, \delta) \\
&= \sum_{\substack{\gamma_1^{(f)} = (b_1^{(f)}, ph) \in Sm\gamma^{(f)} \\ b_1^{(f)}(\text{int}) \subset W}} \; \sum_{\substack{G \subset W \\ G_1 \subset W}} \cdot \exp\{-\beta[\tfrac{1}{2} J_0 - \beta^{-1} e^{-\beta\varepsilon_1}] \\
&\quad \cdot (|G| + |G_1|)\} \cdot \sum_{\substack{\varphi_0(b_1^{(f)}) \subset \mathfrak{A}^{(i)}(b_1^{(f)}) \\ G^{(\text{ver})}(\varphi_0) = G \\ G_1^{(\text{ver})}(\varphi_0) = G_1}} \cdot \exp\{-ce^{-\beta\varepsilon_1} \cdot |G_2|\} \\
&\quad \cdot \exp\left\{ -\beta\left[4J_1 N_1(\varphi_0(b_1^{(f)})) + \delta \sum_{r=2}^{2r_1} |N_r^{(1)}(\varphi_0(b_1^{(f)}))| + \mathscr{H}_1 \right]\right\} \\
&\quad \cdot \sum_{\varphi(b_1^{(f)}) \in \mathfrak{A}(b_1^{(f)} \mid \varphi_0(b_1^{(f)}))} \cdot \exp\left\{ -\beta \sum_{n=0}^{4} n_i(\varphi(b_1^{(f)}) \cdot \varepsilon_i \right\}.
\end{aligned}
$$

Here $n_i(\varphi(b_1^{(f)}))$ is the number of e.d. of type i of configuration $\varphi(b_1^{(f)})$ (see Sect. 1) and c is an absolute constant,

$$0<c<\min_{G,G_1}(|G_2|^{-1}(|G|+|G_1|).$$

Let $v_i(\varphi_0(b_1^{(f)}))$ be the set of points $x\in b_1^{(f)}(\mathrm{int})\backslash G_2$, where the spin-flip changes the energy ε_i.

Then

$$\begin{aligned}\Sigma\exp\left\{-\beta\sum_{i=0}^{4}n_i\varepsilon_i\right\} &\leqq \sum_{\{n_i\}}\prod_{i=0}^{4}C_{|v_i|}^{n_i}e^{-\beta\varepsilon_i n_i}\\ &\approx \exp\left[\sum_{i=0}^{4}|v_i|\ln(1+\exp(-\beta\varepsilon_i))\right]\\ &\approx \exp(|v_0|\exp(-\beta\varepsilon_0)+(|v_1|+|v_2|)\exp(-\beta\varepsilon_1)).\end{aligned}$$

It is easy to see that

$$|v_0(\varphi_0)|=\sum_{r\geqq 3}|N_r^{(1)}(\varphi_0)|,\qquad |v_1(\varphi_0)|=\sum_{r\geqq 4}^{2r_1}|N_r^{>2}(\varphi_0)|,$$
$$|v_2(\varphi_0)|=|N_2^{(1)}(\varphi_0)|.$$

We remark that every configuration $\varphi_0(b_1^{(f)})$ can be uniquely extended to $V(b)\backslash G\cup G_1$ and different configurations will have different extensions. As a result we can write

$$\begin{aligned}&\Xi_W^{(s)}(\gamma^{(f)}|\beta,\delta)\\ &\leqq \sum_{\substack{G\subset W\\ G_1\subset W}}\exp\{-(\tfrac{1}{2}\beta J_0-e^{-\beta\varepsilon_1})(|G|+|G_1|)\}\cdot\sum_{\substack{\varphi_0(V(b))\\ G^{(\mathrm{ver})}(\varphi_0)=G\\ G_1^{(\mathrm{ver})}(\varphi_0)=G_1}}\\ &\cdot\exp\{-ce^{-\beta\varepsilon_1}|G_2|\}\cdot\exp\left\{-\beta\left[4J_1|N_1(\varphi_0(V(b)\cap W)|\right.\right.\\ &+\delta\sum_{r=2}^{2r_1}|N_r^{(1)}(\varphi_0(V(b)))\cap b|-\beta^{-1}e^{-\beta\varepsilon_0}\sum_{r\geqq 3}|N_r^{(1)}(\varphi_0(V(b))\cap W|\\ &-\beta^{-1}e^{-\beta\varepsilon_1}\sum_{r=5}^{2r_1}|N_r^{>2}(\varphi_0(V(b)))\cap W|-\beta^{-1}e^{-\beta\varepsilon_1}\\ &\left.\left.\cdot|N_2^{(1)}(\varphi(V(b)))\cap W|+\text{v.s.n.}\,|W-G_2|+\mathscr{H}_1\right]\right\}.\end{aligned}$$

Remember that $2\beta\delta-2e^{-\beta\varepsilon_0}+3e^{-\beta\varepsilon_1}$ is v.s.n.. We have

$$\begin{aligned}&\beta\delta\sum_{r=2}^{2r_1}|N_r^{(1)}(\varphi_0(V(b)))\cap b|\\ &-e^{-\beta\varepsilon_0}\sum_{r\geqq 3}|N_r^{(1)}(\varphi_0(V(b)))\cap b|-e^{-\beta\varepsilon_1}\sum_{r=5}^{2r_1}|N_r^{>2}(\varphi_0(V(b)))\cap b|\\ &-e^{-\beta\varepsilon_1}|N_2^{(1)}(\varphi_0(V(b)))\cap b|=(\beta\delta-2e^{-\beta\varepsilon_1})|N_2^{(1)}(\varphi_0(V(b)))\cap b|\\ &+\tfrac{1}{2}e^{-\beta\varepsilon_1}\sum_{r=3}^{2r_1}|N_r^{(1)}(\varphi_0(V(b)))\cap b|\\ &-e^{-\beta\varepsilon_1}|b\backslash G_2|+e^{-\beta\varepsilon_1}\sum_{r>2}|N_r^{(2)}(\varphi_0(V(b)))\cap b|.\end{aligned}$$

As $\gamma=(b,ph)$ is the contour we have

$$\tfrac{1}{2}e^{-\beta\varepsilon_1}\sum_{r=3}^{2r_1}|N_r^{(1)}(\varphi_0(V(b)))\cap b(\mathrm{int})|+(\beta\delta-2e^{-\beta\varepsilon_1})$$
$$\cdot|N_2^{(1)}(\varphi_0(V(b)))\cap b(\mathrm{int})|>de^{-\beta\varepsilon_1}|b(\mathrm{int})\backslash G_2|$$

and

$$|N_r^{(2)}(\varphi_0(V(b)))\cap(b\backslash b(\mathrm{int})\cup G_2)|>d_1|b\backslash b(\mathrm{int})\cup G_2|,$$

where d, d_1 are absolute constants. Also

$$e^{-\beta\varepsilon_1}|N_r^{(1)}(\varphi_0(V(b)))\cap(V(b)\backslash b)\cap W)|$$
$$+e^{-\beta\varepsilon_1}|N_r^{>2}(\varphi_0(V(b)))\cap(V(b)\backslash b)\cap W)|$$
$$\leqq e^{-\beta\varepsilon_1}|(V(b)\backslash b)\cap W|.$$

If the sets G and G_1 are fixed, then the number of ideal configurations $\varphi_0(V(b))$ is not more than $\mathrm{const}^{|G|+|G_1|}$.

Thus

$$\Xi_W^{(s)}(\gamma^{(f)}|\beta,\delta)\leqq\sum_{\substack{G_1\subset W\\ G\subset W}}\exp\{-(\beta J_0-\mathrm{const})(|G|+|G_1|)\}$$
$$\cdot\exp\{e^{-\beta\varepsilon_1}|V(b)\cap W\backslash(b\cup G\cup G_1)|$$
$$+\bar{d}e^{-\beta\varepsilon_1}|b(\mathrm{int})\backslash(G\cup G_1)|-\bar{d}_1e^{-\beta\varepsilon_1}|b\backslash b(\mathrm{int})|\},$$

where d, $\bar{d}_1$ are positive constants and $0<\bar{d}<1$.

Lemma 2. *Put* $e^{-G_1^{(f)}(\gamma^{(f)})}=\sum_G\exp\left\{-\left(\frac{\beta}{2}J_0-\mathrm{const}\right)|G|\right\}$. *Then*

$$e^{-G_1^{(f)}(\gamma^{(f)})}\cdot\sum_{G,G_1}\exp\left\{-\left(\frac{\beta}{2}J_0-\mathrm{const}\right)(|G|+|G_1|)\right\}$$
$$\leqq\exp\{\mathrm{v.s.n.}\,|V(b)|\}.$$

We shall give only the sketch of the proof. We decompose $G\cup G_1$ onto connected components and get a usual contour model (see [6]). We remark that the statistical weights of the least components are v.s.n. because we excluded e.d. The needed estimate follows by a direct application of the method of correlation equations (see [6]).

Thus we get the desired result.

7. Solution of (5) and (6) and Final Remarks

Now we return to Eqs. (5), (6). We look for the solution $F^{(f)}=F_1^{(f)}+F_2^{(f)}$, $F^{(3,3)}=F_1^{(3,3)}+F_2^{(3,3)}$. From the very beginning we put $F_1^{(f)}(\gamma^{(f)})=G_1^{(f)}(\gamma^{(f)})$, $F_1^{(3,3)}(\gamma^{(f)})=G_1^{(3,3)}(\gamma^{(3,3)})$. Assumption 1 follows easily from Theorem B. Equations (5), (6) for F_2 can be rewritten in the following form:

$$G_2^{(f)}(\gamma^{(f)}) = F_2^{(f)}(\gamma^{(f)}) - \Delta^{(cr)}(\gamma^{(f)}|F^{(f)}) + \sum_s \Delta(O_s|F^{(\kappa_s)}),$$

$$\begin{aligned} G_2^{(3,3)}(\gamma^{(3,3)}) = F_2^{(3,3)}(\gamma^{(3,3)}) - \beta(h^{(f)} - h^{(3,3)}) \cdot |b| \\ - \Delta^{(cr)}(\gamma^{(3,3)}|F^{(3,3)}) + \sum_s \Delta(O_s|F^{(\kappa_s)}). \end{aligned} \tag{17}$$

Let us put

$$G_3^{(3,3)}(\gamma^{(3,3)}) = G_2^{(3,3)}(\gamma^{(3,3)}) + \beta(h^{(f)} - h^{(3,3)}) \cdot |b|.$$

Then (17) means that

$$(G_2^{(f)}, G_3^{(3,3)}) = (F_2^{(f)}, F_2^{(3,3)}) + T(F_2^{(f)}, F_2^{(3,3)}). \tag{18}$$

It follows from Theorem B that $G_2^{(f)} \geqq -d_2^{(1)} \exp\{-\beta\varepsilon_1\} \cdot |b|$, $d_2^{(1)}$ is some absolute positive constant independent of β, $d_2^{(1)} < 1$. For $G_3^{(3,3)}$ we have

$$G_3^{(3,3)} \geqq [-\tfrac{2}{3} e^{-\beta\varepsilon_0} + (1 - d_2^{(1)}) e^{-\beta\varepsilon_1}] \cdot |b|,$$

with the same $d_2^{(1)}$. From the other side

$$\sum_{\substack{\gamma \in T_0(z) \\ \gamma \neq \gamma^{(0)}}} \exp\{-F_1(\partial b^{(\mathrm{ver})})\}$$

is v.s.n. In the case of f-phase $N = 1$ and $\exp\{-F(\gamma^{(0)})\} = e^{-\beta\varepsilon_1}$, in (3,3)-phase, $N = 6$ and

$$\frac{1}{N} \exp\{-F(\gamma^{(0)})\} = \tfrac{2}{3} \exp\{-\beta\varepsilon_0\} + \text{v.s.n.}$$

Thus the conditions of Theorem A are valid for sufficiently large β.

It follows from the proof of Theorem B that if a contour consists of several components then the values of $G_2^{(f)}$, $G_2^{(3,3)}$ are sums of values of different components, i.e. Assumption 4 for contour models is true.

In the space of pairs $(F_2^{(f)}, F_2^{(3,3)})$ we introduce the norm

$$\|(F_2^{(f)}, F_2^{(3,3)})\| = \max(\|F_2^{(f)}\|, \|F_2^{(3,3)}\|).$$

Then in (18) the operator T is a contraction and the contraction coefficient is a v.s.n. as one can easily check. Thus we get the solution of (5), (6). It will satisfy also Assumptions 4, Sect. 3 because G satisfies it and it remains valid under iterations.

Assumption 3 of the contour functional follows from the estimates of the partition function $\Xi_W^{(s)}(\gamma|\beta, \delta)$ in Theorem B and from the fact that for any smooth contour $\gamma_1 = (b_1, ph)$ and contour $\gamma \in Sm(\gamma_1)$ we have

$$|\Delta^{(cr)}(\gamma|F)| + \left|\sum_s \Delta(O_s(\gamma)|F)\right| \leqq \text{v.s.n. } |b_1|.$$

Having $F^{(f)}$, $F^{(3,3)}$ we find δ using the equation $a^{(3,3)} - a^{(f)} = 2\beta\delta$.

Peierls's method in principle does not give any possibility to approach a critical point. However, it is apparently possible to estimate the number of values of δ where the set of limit Gibbs states is discontinuous as a function of β.

Acknowledgements. We express our sincere gratitude to V. Pokrovsky and G. Uimin for valuable discussions. Special thanks to A. Mazel who very carefully read the manuscript and gave us very many useful remarks which led to the substantial improvement of the text.

References

1. Domb, C.: On the theory of cooperative phenomena in crystals. Adv. Phys. **9**, No. 34, 149 (1960)
2. Elliott, R.J.: Phenomenological discussion of magnetic ordering in the heavy-earth metals. Phys. Rev. **124**, 346 (1961)
3. Bak, P., von Boehm, J.: Ising model with soliton, phasons and the "Devil's Staircase". Phys. Rev. B **21**, 5297 (1980)
4. Fisher, M.E., Selke, W.: Low temperature analyses of the axial next-nearest neighbour Ising model near its multiphase point. Phil. Trans. Royal Soc. **302**, 1 (1981)
5. Pirogov, S.A., Sinai, Ya.G.: Phase transitions of the first order for small perturbations of the Ising model. Funkt. Anal. Prilozh. **8**, 25 (1974)
6. Sinai, Ya.G.: Theory of phase transitions. Rigorous results. Budapest: Acad. Kiado 1982
7. Zahradnik, M.: An alternate version of Pirogov-Sinai theory. Commun. Math. Phys. **94**, 559–581 (1984)

Communicated by J. Fröhlich

Received July 23, 1984

Comments

The first results for ANNNI-model were obtained by M. Fisher and W. Selke in their paper [FS].

In this paper, we studied the phase diagram of ANNNI-model using the Peierls contour method. This model has infinitely many ground states. What is more, the entropy of the set of ground states is positive. Therefore, the usual PS theory (see [PS2, PS3]) cannot be applied. However, for finite temperatures only finitely many ground states are stable enough to produce Gibbs states at positive temperatures. Moreover, the number of Gibbs states tends to infinity as the temperature goes to zero. There are other models which display the same properties such as Ising models with power-like decay of interaction. Some results concerning phase diagrams in such models were obtained by A. Kerimov (see [K]).

Commun. math. Phys. 45, 247—278 (1975)

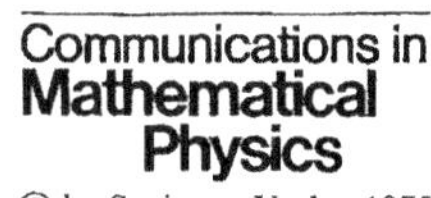

Critical Indices for Dyson's Asymptotically-Hierarchical Models

P. M. Bleher and Ya. G. Sinai
Landau, Theoretical Physics Institute, Moscow V334, URSS

Abstract. It is known that the investigation of the critical point for models of the type of Dyson's hierarchical models is reduced to the solution of some non-linear integral equation. In our previous publication the Gaussian solution was investigated. Here we construct non-Gaussian solutions of the equation and find the expressions for critical indices connected with them. Our procedure permits us to construct meaningful ε-expansions.

§ 1. Introduction

Dyson's hierarchical models or their generalization – asymptotically-hierarchical models – (a.h.m.) are of great interest because the renormalization group method in the theory of critical points by K. Wilson [3] and M. Fisher [4] becomes rigorous for such models (see [2] and the papers by Jona-Lasinio [5] and Gallavotti-Knops [6]). The investigation of critical points for a.h.m. is reduced to the solution of the corresponding nonlinear integral equation, which can be considered as an equation for the fixed point of the corresponding renormalization group. In [2, 8] a case with the Gaussian solution was investigated. It was shown that the critical indices in that case are precisely the same as predicted by the Landau semiphenomenological theory of phase transitions of the second kind. However, the Gaussian solution is stable only when the potential of interaction decreases sufficiently slowly.

In this paper we construct non-Gaussian solutions of our main integral equation. These solutions appear as bifurcations branches from the Gaussian solutions. The total number of the branches is infinite but only one of them has the necessary properties of stability to appear in general as a limit distribution for normed mean spin at the critical temperature. In the second part of this paper we find the values for critical indices corresponding to this branch. They coincide with the values found in the general theory by Wilson [3].

From the formal point of view the non-Gaussian solutions can be represented by a series of the parameter ε where ε is the deviation of the given value of the parameter from its bifurcation value. These series are always asymptotic because they describe the functions with different asymptotics at infinity. The method we apply can be regarded as a procedure which permits to make these ε-series meaningful. Roughly speaking at a given ε the formal series gives a good

Y.G. Sinai, *Selecta II: Probability Theory, Statistical Mechanics, Mathematical Physics and Mathematical Fluid Dynamics*, DOI 10.1007/978-1-4419-6205-8_9, © Springer Science+Business Media, LLC 2010

approximation for the solution only in the domain depending on ε. Beginning from a family of test functions we apply the transformations of the renormalization group and construct ε-expansion near every iteration. The values of ε rapidly decrease because the iterations rapidly converge to the solution which we are seeking. Therefore, ε-expansion becomes more and more exact on the increasing sequence of domains when the number of iterations tends to infinity.

Now we want to recall the definition of Dyson's hierarchical models, asymptotically-hierarchical models and to deduce the main integral equation (see [1, 2]). Let an integer $r>1$ and a positively-defined quadratic form $Q(t_1,\dots,t_r)=g((t_1+\dots+t_r)/r)^2+h(t_1^2+\dots+t_r^2)/r$ with g, h as parameters be fixed. Assume also that for any integer $n>1$, there is given a volume V_n consisting of r^n points divided into r equal subvolumes $V_{n-1,i}$, $i=1,\dots,r$. We consider a classical spin system, configurations of which can be represented as functions $u(x)$, $x\in V_n$, taking the values of ± 1. The Hamiltonian of Dyson's hierarchical model depends on a parameter c, $1<c<r$, and is defined by the following recurrence relation:

$$H_n(u)=\sum_{i=1}^r H_{n-1}(u_i)-c^r Q(s_1^{(n-1)},\dots,s_r^{(n-1)}). \tag{1.1}$$

Here $s_i^{(n-1)}=(1/r^{n-1})\sum_{x\in V_{n-1,i}}u(x)$ is the mean spin in the subvolume $V_{n-1,i}$ of the configuration u and u_i is the restriction of the whole configuration $u(x)$, $x\in V_n$ in the subvolume $V_{n-1,i}$.

Let us introduce $g_n(t;\beta)=\mathrm{Prob}_n\{s^{(n)}=t;\beta\}$, where Prob_n is the probability, calculated by the Gibbs distribution in the volume V_n with β as the inverse temperature, $s^{(n)}$ is the mean spin in the volume V_n, $s^{(n)}=(1/r^n)\sum_{x\in V_n}u(x)$. Then from (1.1) easily follows the system of recurrent equations for functions g_n:

$$g_n(t;\beta)=(\Xi_{n-1}^r(\beta)/\Xi_n(\beta))\sum_{(1/r)\sum_{i=1}^r t_i=t}g_{n-1}(t_1;\beta)\dots g_{n-1}(t_r;\beta)e^{\beta c^n Q(t_1,\dots,t_r)} \tag{1.2}$$

where $\Xi_\kappa(\beta)$ is the grand partition function in the volume V_k, $k\geqq 1$. The main assumption which is made at the investigation of hierarchical models, is that for $\beta=\beta_{cr}$ the typical values of the mean spin have the order $c^{-n/2}$. Making the change of coordinates $t=c^{-n/2}z$ and putting $\Delta_n=c^{n/2}r^{-n}$, $f_n(z;\beta)=g_n(z\cdot c^{-n/2};\beta)\Delta_n^{-1}$ we obtain from (1.2) the following system of recurrent equations for functions $f_n(z;\beta)$:

$$f_n(z;\beta)=L_n(\beta)\sum_{(z_1+\dots+z_r)/r=z/\sqrt{c}}f_{n-1}(z_1;\beta)\dots f_{n-1}(z_r;\beta)e^{\beta Q(z_1,\dots,z_r)}\Delta_n^{r-1} \tag{1.2'}$$

where $L_n(\beta)$ is a normed constant. From the mathematical point of view, the previous assumption is equivalent to the assumption that the functions $f_n(z;\beta)$ converge at $n\to\infty$ to a limit and the limit function $f(z;\beta)$ of continuous argument z satisfies the equation

$$f(z;\beta)=L(\beta)\int\dots\int\prod_{i=1}^r f(z_i;\beta)e^{\beta Q(z_1\dots z_r)}\delta(\Sigma z_i-rz/\sqrt{c})\prod_{i=1}^r dz_i. \tag{1.3}$$

The constant $L(\beta)$ is the normalization factor. Equation (1.3) is the main integral equation in the theory of hierarchical models.

It is easy to verify that (1.3) has the Gaussian solution $f(z;\beta)=\sqrt{a_0(\beta)/\pi}e^{-a_0(\beta)z^2}$ with $a_0(\beta)=((g+h)/(r-c))\beta$. General solutions of (1.3) for different β are related to each other via the equality $f(z;\beta_1)=\sqrt{\beta_2/\beta_1}f(z\sqrt{\beta_2/\beta_1};\beta_2)$. Therefore, it is sufficient to consider (1.3) with $\beta=1$.

If the relations (1.2) are valid for $n \geqq n_0$ and the family of initial probability distributions $g_{n_0}(t;\beta)$ is arbitrary then the corresponding model is called asymptotically-hierarchical model. Here t takes the values from $-r^{n_0}$ to r^{n_0} and all probabilities $g_{n_0}(t;\beta)$ are defined for some closed interval $[\beta^-,\beta^+]$ and are C^1-functions of β. For any fixed interval $[\beta^-,\beta^+]$ there is a natural topology in the space of such distributions $\{g_{n_0}(t;\beta), \beta\in[\beta^-,\beta^+]\}$.

Definition 1. The solution $f(z;\beta), 0<\beta<\infty$, of (1.3) is called thermodynamically-stable if there exists an integer n_0 and a closed interval $[\beta^-,\beta^+]$ for which one can find an open set Ω in the space of families of probability distributions $\{g_{n_0}(t;\beta), \beta\in[\beta^-,\beta^+]\}$ such that for any family $\{g_{n_0}(t;\beta), \beta\in[\beta^-,\beta^+]\}\in\Omega$ there exists one and only one $\beta^*\in[\beta^-,\beta^+]$ for which $f_{n_0}(z;\beta^*)$ converge weakly to $f(z;\beta^*)$.

One of the main results of [2, 8] is that the Gaussian solution is thermodynamically stable for $\sqrt{r}<c<r$ and for $c<\sqrt{r}$ it is unstable. Therefore, for $c<\sqrt{r}$ it is necessary to construct non-Gaussian solutions of (1.3).

Let $c_k=r^{1/(k+1)}, k=1,2,\ldots,\varepsilon=c_k-c$. The following theorem is the main result of this paper.

Theorem 1. *For any* $k=1,2,\ldots$ *one can find* $\delta_k>0$ *such that for any* ε, $0<\varepsilon\leqq\delta_k$ *there exists a normed solution* $f_\varepsilon(z)$ *of the equation*

$$f_\varepsilon(z)=L_\varepsilon\int\ldots\int f_\varepsilon(z_1)\ldots f_\varepsilon(z_r)e^{Q(z_1,\ldots,z_r)}\delta\left(\textstyle\sum_{i=1}^r z_i-rz/\sqrt{c}\right)\prod_{i=1}^r dz_i\,. \tag{1.4}$$

For this solution $0<f_\varepsilon(z)\leqq 2\sqrt{a_0/\pi}\exp[-(a_0z^2+A_0\varepsilon|z|^\alpha)]$, $a_0=(h+g)/(r-c)$, $A_0=A_0(k)$, α *is the root of the equation* $c^\alpha=r$. *These solutions* $f_\varepsilon(z)$ *continuously depend on* ε *for any fixed* z. ⌟

It is possible to show that the branches f_ε for $k>1$ are thermodynamically unstable. The branch f_ε for $k=1$ is thermodynamically stable (see § 8 of this paper).

Theorem 1 gives the existence of the solution of (1.3) for c sufficiently close to $r^{1/2}$. In [10] this branch was investigated on computers for $r=2$ (see also Appendix 2 below). The results of [10] doubtlessly show that there is no other bifurcations for $1<c<\sqrt{2}$.

During the proof we discuss in detail only the case $r=2$ and $Q=(t_1+t_2)^2$ which corresponds to Dyson's hierarchical model. The general case can be treated by obvious modifications. The reader can easily notice the similarity between the methods of this paper and papers [2, 8].

§ 2. The Idea of the Proof of Theorem 1

For $r=2$ and $Q(t_1,t_2)=(t_1+t_2)^2$ Eq. (1.3) takes the form:

$$f(z;\beta)=L_c e^{\beta z^2}\int_{-\infty}^{\infty} f(z/\sqrt{c}+u;\beta)f(z/\sqrt{c}-u;\beta)du\,.$$

The substitution $f(z;\beta)=f_1(z;\beta)\exp(-a_0(\beta)z^2)$, $a_0(\beta)=\beta c/(2-c)$ reduces the latter equation to the equation

$$f(z;\beta)=L_c\int_{-\infty}^{\infty} e^{-2a_0(\beta)u^2} f(z/\sqrt{c}+u;\beta)f(z/\sqrt{c}-u;\beta)du\,.$$

The next step is to give up the normalization condition and to consider the equation

$$f(z)=(1/\sqrt{\pi})\int_{-\infty}^{\infty} e^{-u^2} f(z/\sqrt{c}+u)f(z/\sqrt{c}-u)du = Af. \tag{2.1}$$

After normalization of the solution of (2.1) we shall obtain the solution of the initial equation (1.3) for $\beta=(2-c)/(2c)$. As was mentioned above, the solution of (1.3) for any β can be obtained from this one by a simple change of variables.

A depends on c and (2.1) defines a family of non-linear transformations when c changes in the interval $1<c<2$. It has an obvious solution $f\equiv 1$ which corresponds to the Gaussian solution of (1.3). It is very essential that it does not depend on c. When one has a smooth family of non-linear transformations of the finite-dimensional space with a fixed point which does not depend on the parameter of the family, one should consider the family of linearized transformation near this point and find such values of the parameter for which the spectrum of the corresponding linear transformation contains 1. If the second derivative in the direction, according to the eigenvalue 1, enters the Taylor series with non-zero coefficient, then through the fixed point there passes a new branch of fixed points of transformations of our family. One can say that the initial fixed point generates new fixed points.

The procedure which is applied below, can be considered as an adaptation of the methods of the finite-dimensional case to our transformation A, acting in the infinite-dimensional functional space. The linearized operator L_1 corresponding to $f(z)\equiv 1$ takes the form

$$L_1 g(z)=(2/\sqrt{\pi})\int_{-\infty}^{\infty} e^{-u^2} g(z/\sqrt{c}-u)du\,.$$

This operator is known as the Gauss integral operator (see [11]). We consider its action in the space of even functions f. Its eigenvalues are equal to $2, 2c^{-1}, 2c^{-2},\ldots,$ $2c^{-k},\ldots$. The corresponding eigenvectors are the Hermite polynomials, which are orthogonal with the weight $\exp(-\gamma z^2)$, $\gamma=1-c^{-1}$. Thus, the critical values of c near which one can expect the appearance of new solutions have the form $c_k=2^{1/(k+1)}$, $k=1,2,\ldots$. For $c<c_k$ and close to c_k the point $f\equiv 1$ has $(k+1)$-dimensional unstable eigenspace. Accordingly, the new solution must have k-dimensional unstable eigenspace for these values of c.

Our method of construction of new solutions of (2.1) has much in common with the widely-known Hadamard-Perron theorem in the theory of smooth dynamical systems (see [12, 13]). The direct method of contracting mappings cannot be applied because we are looking for unstable solutions. The construction must begin with the construction of the stable separatrice of the solution which we are seeking. The next step is the proof that the induced mapping on the separatrice is a contraction. The first step, i.e. the construction of the separatrice is usually taken in the following way. One takes a k-dimensional manifold which is close in a natural sense to the unstable subspace and finds its intersection with the separatrice. This intersection lies in one point. This point is determined by the property that all its images lie in a small region of the fixed point.

Our procedure is similar to the above process. However, we do not construct the whole separatrice but take a special k-dimensional family of test functions, find one point of this family which lies on the separatrice and prove that it converges to the solution which we are seeking.

§ 3. Properties of Operators L_f

The differential L_f of the non-linear transformation A in an arbitrary point f has the form

$$L_f g(z) = (2/\sqrt{\pi}) \int_{-\infty}^{\infty} e^{-u^2} f(z/\sqrt{c} - u) g(z/\sqrt{c} + u) du. \tag{3.1}$$

In this section we shall investigate several largest eigenvalues and eigenvectors of the linear operator L_f when the function f is sufficiently close to 1.

Roughly speaking we shall prove that in the case under consideration the formulae of the perturbation theory are applicable. One cannot hope that the series of the perturbation theory converge because the spectrum of non-perturbed operator when $f \equiv 1$ consists of numbers 2, $2c^{-1}$, $2c^{-2}, \ldots$ and tends to zero. However, we shall show that when the perturbation has the order ε in the appropriate norm the difference of n-th eigenvectors for perturbed and unperturbed operators is no more than $\varepsilon^{0,8}$ if ε is sufficiently small and n is fixed.

The consideration of this section will not be used below. The reader may acquaint himself with the formulation of Theorem 3.1. and proceed to the next section.

Now we are going to formulate the exact condition concerning a perturbation and to give the formulation of the theorem. Let ε, $0 < \varepsilon < 1$, be a certain number and

$$D_\varepsilon = [-d_0\sqrt{\ln(1/\varepsilon)}, d_0\sqrt{\ln(1/\varepsilon)}], \ d_0 = 10/(c-1).$$

Assume that there is given an even function $f(z) \in C^1(R^1)$ such that for $z \in D_\varepsilon$ the function f can be written in the form

$$f(z) = 1 - \varepsilon G_{2k}(\sqrt{\gamma} z) + R(z) \tag{3.2}$$

where $G_{2k}(z)$ is the $2k$-th Hermite polynomial (see [11]), $\gamma = 1 - c^{-1}$ and

$$|R(z)|, |dR(z)/dz| < \varepsilon^{5/3}. \tag{3.3}$$

For $z \notin D_\varepsilon$ the function f satisfies the estimates

$$f(z) < \exp(-(\varepsilon'/2)|z|^\alpha), \ \alpha = 2(\log_2 c)^{-1}, \tag{3.4}$$

$\varepsilon' = \varepsilon \cdot p_0$, where p_0 is the $2k$-th coefficient of $G_{2k}(\sqrt{\gamma} z)$,

$$|df(z)/dz| < |z|^\alpha \exp(-(\varepsilon'/2)|z|^\alpha). \tag{3.5}$$

Let us denote the Hilbert space of even functions on the line which have an integrable square with respect to the weight $\exp(-\gamma z^2)$ by $L^2_{ev}(R^1; \exp(-\gamma z^2))$.

Theorem 3.1. *Let N be fixed. Then there exists a number $\varepsilon_0 = \varepsilon_0(N)$ such that for any function f satisfying* (3.2)–(3.5) *with ε, $0 < \varepsilon < \varepsilon_0$, the operator L_f has $(N+1)$ eigenvectors $e_0(z; f), \ldots, e_N(z; f)$ and accordingly, eigenvalues $\lambda_0, \ldots, \lambda_N$ such that*

a$_1$) $|\lambda_i - 2/c^i| \leq \varepsilon^{4/5}, i = 0, \ldots, N$;

$|\lambda_k - 2/c^k - 2\varepsilon \int_{-\infty}^{\infty} \exp(-\gamma z^2) G_{2k}(\sqrt{\gamma} z) A G_{2k}(\sqrt{\gamma} z) dz| \leq \varepsilon^{\frac{3}{2}}$;

a$_2$) $\|e_i(z; f) - G_{2i}(\sqrt{\gamma} z)\|_{C^1(D_\varepsilon)} \leq \varepsilon^{4/5}$; $\quad i = 0, \ldots, N$;

a$_3$) $|e_i(z; f)| \leq |z|^{2i+1} \exp(-(\varepsilon'/2)|z|^\alpha)$,

$|de_i(z; f)/dz| \leq |z|^{2i} \exp(-(\varepsilon'/2)|z|^\alpha)$,

for $z \notin D_\varepsilon$; $i = 0, \ldots N$;

a$_4$) *in the Hilbert space* $L^2_{\mathrm{ev}}(R^1; \exp(-\gamma z^2))$ *there exists the closed subspace* $H_{f,N}$ *of the co-dimension* $(N+1)$ *invariant under* L_f *and such that*

$$\|L_f\|_{H_{f,N}} \leqq 2c^{-N-\frac{1}{2}}$$

$$\mathrm{dist}(H_{f,N}, L^2_N(R^1; \exp(-\gamma z^2))) \leqq \varepsilon^{4/5}$$

where $L^2_N(R^1; \exp(-\gamma z^2))$ *is the subspace of the Hilbert space* $L^2_{\mathrm{ev}}(R^1; \exp(-\gamma z^2))$ *generated by the Hermite polynomials* $G_{2i}(\sqrt{\gamma}z)$, $i > N$. ⌟

The proof of the theorem will be divided into several lemmas.

Lemma 3.1. *If the function* $f(z)$ *satisfies the condition* (3.2)–(3.5) *and* $\varepsilon > 0$ *is sufficiently small, then*

$$\|L_f - L_1\|_{L^2(R^1; \exp(-\gamma z^2))} \leqq \varepsilon^{31/32},$$

$$\|L_f - L_{1-\varepsilon G_{2k}}\|_{L^2(R^1; \exp(-\gamma z^2))} \leqq \varepsilon^{\frac{3}{2}}.$$

Proof. We have from (3.1) and (3.5)

$$(L_f - L_{1-\varepsilon G_{2k}})g(z) = (2/\sqrt{\pi})\int_{-\infty}^{\infty} \exp(-u^2) g(z/\sqrt{c} - u) R(z/\sqrt{c} + u) du$$
$$= (2/\sqrt{\pi})\int_{-\infty}^{\infty} \exp[-(u - z/\sqrt{c})^2] R(2z/\sqrt{c} - u) g(u) du = \int_{-\infty}^{\infty} K(z,u) g(u) du = Kg,$$

where

$$K(z,u) = (2/\sqrt{\pi})\exp[-(u - z/\sqrt{c})^2] R(2z/\sqrt{c} - u).$$

Moreover

$$\|K\|_{L^2(R^1; \exp(-\gamma z^2))} = \|K_0\|_{L^2(R^1)}, \tag{3.6}$$

where

$$K_0(z,u) = K(z,u)\exp[-(\gamma/2)(z^2 - u^2)] = (2/\sqrt{\pi})\exp[-Q(z,u)] R(2z/\sqrt{c} - u),$$
$$Q(z,u) = (z/\sqrt{c} - u)^2 + (\gamma/2)(z^2 - u^2) = (1/\sqrt{c})(z-u)^2 + (1/2 + 1/(2c) - 1/\sqrt{c})$$
$$\cdot (z^2 + u^2) \geqq \tfrac{1}{2}(1 - 1/\sqrt{c})^2 (z^2 + u^2) = \alpha_0 (z^2 + u^2) > 0. \tag{3.7}$$

We shall show that

$$\int_{-\infty}^{\infty}\int_{-\infty}^{\infty} |K_0(z,u)|^2 dz du < \varepsilon^3. \tag{3.8}$$

Let $\Omega_\varepsilon = \{\sqrt{z^2 + u^2} \leqq (d_0/3)\sqrt{\ln(1/\varepsilon)}\}$. For $(z,u) \in \Omega_\varepsilon$ the point $2z/\sqrt{c} - u \in D_\varepsilon$ and thus [see (3.3)]

$$\iint_{\Omega_\varepsilon} |K_0(z,u)|^2 dz du = \iint_{\Omega_\varepsilon} [(2/\sqrt{\pi})\exp(-Q(z,u)) R(2z/\sqrt{c} - u)]^2 dz du$$
$$\leqq (4/\pi)\iint_{\Omega_\varepsilon} \varepsilon^{10/3} \exp(-2\alpha_0(z^2 + u^2)) dz du \leqq \mathrm{const}\,\varepsilon^{10/3} < \tfrac{1}{2}\varepsilon^3 \tag{3.9}$$

for a sufficiently small ε. From (3.2)–(3.4) it follows that the inequality

$$|R(z)| < 1 + z^{4k} \tag{3.10}$$

is valid for all $z \in R^1$. Therefore $|R(2z/\sqrt{c} - u)| \leqq 1 + 4(z^2 + u^2)^{2k}$ and for sufficiently small ε

$$\iint_{R^2 \setminus \Omega_\varepsilon} |K_0(z,u)|^2 dz du \leqq (4/\pi)\iint_{R^2 \setminus \Omega_\varepsilon} \exp[-2\alpha_0(z^2 + u^2)]$$
$$(1 + 4(z^2 + u^2)^{2k})^2 dz du$$
$$\leqq \mathrm{const}\int_{(d_0/3)\sqrt{\ln(1/\varepsilon)}}^{\infty} \exp(-2\alpha_0 \varrho^2)(1 + 4\varrho^{4k})^2 d\varrho \leqq \mathrm{const}\,\varepsilon^{2\alpha(d_0/3)^2} (\ln^{4k}\varepsilon)$$
$$\leqq \mathrm{const}\,\varepsilon^4 \ln^{4k}\varepsilon < \tfrac{1}{2}\varepsilon^3.$$

Thus (3.8) is proved.

From the Schwartz inequality we have

$$\|K_0\|_{L^2(R^1)} \leqq \sqrt{\int_{-\infty}^{\infty}\int_{-\infty}^{\infty} |K_0(z,u)|^2 dz du}\,.$$

Then from this inequality, (3.6) and (3.8) one can easily derive

$$\|K\|_{L^2(R^1;\exp(-\gamma z^2))} = \|L_f - L_{1-\varepsilon G_{2k}}\|_{L^2(R^1;\exp(-\gamma z^2))} \leqq \varepsilon^{\frac{3}{4}}\,.$$

The second inequality of Lemma 3.1 is proved. The first one is obtained in a similar way. Thus Lemma 3.1 is proved.

Similar considerations lead to the proof of the following lemma.

Lemma 3.2. *Under the conditions of Lemma 3.1*

$$\|e^{-\gamma z^2}(L_f - L_1)g(z)\|_{C^1(R^1)} \leqq \varepsilon^{31/32}\|g(z)\|_{L^2(R^1;\exp(-\gamma z^2))}\,,$$

$$\|e^{-\gamma z^2}(L_f - L_{1-\varepsilon G_{2k}})g(z)\|_{C^1(R^1)} \leqq \varepsilon^{\frac{3}{4}}\|g(z)\|_{L^2(R^1;\exp(-\gamma z^2))}\,.$$

We shall omit the proof of Lemma 3.2. Up to the end of this section we shall write $\|g(z)\|$ instead of $\|g(z)\|_{L^2(R^1;\exp(-\gamma z^2))}$.

Lemma 3.3. *Under the conditions of Lemma 3.1 the operator L_f has the main eigenfunction $e_0(z;f)$ with eigenvalue $\lambda_0(f)$ such that*

$$|2-\lambda_0(f)| < \varepsilon^{15/16}\,;$$

$$\|e_0(z;f) - 1\| < \varepsilon^{15/16}\,.$$

Proof. We shall use the method of the contraction mappings. Let us denote

$$S = \{f(z) : \|f(z)\| = 1\}\,,$$

$$S_\delta = \{f(z) : f(z) \in S,\ \|f(z) - 1\| < \delta\}$$

and consider the non-linear mapping $U_f : g(z) \to \|L_f g\|^{-1} L_f g(z)$, $U_f : S \to S$, $\delta = \frac{1}{2}\varepsilon^{15/16}$ and ε be sufficiently small. We shall show that $U_f S_\delta \subseteqq S_\delta$ and $U_f|S_\delta$ is a contraction mapping.

Let U_1 be the mapping U_f corresponding to the function $f \equiv 1$ and D_g be the differential of this mapping at the point g.

It is easy to see that the spectrum of the operator D_1 consists of the numbers $c^{-1}, c^{-2}, c^{-3}, \ldots$ and D_1 is selfadjoint. So $\|D_1\| = c^{-1} < 1$ and D_1 is a contraction operator. Hence, we deduce that the differentials D_g of the operator U_1 at the points g close to 1, namely at the points $g \in S_\delta$, are contraction operators and then we deduce that the differentials $D_g U_f$, $g \in S_\delta$ are also contraction operators. That means that $U_f|_{S_\delta}$ is a non-linear contraction operator. Moreover, it follows from our considerations that for $g_1, g_2 \in S_\delta$

$$\|U_f(g_1) - U_f(g_2)\| \leqq \tfrac{1}{2}(1 + c^{-1})\|g_1 - g_2\|\,,$$

if ε is sufficiently small. Furthermore, due to the evident estimate

$$\|U_f(1) - 1\| < \varepsilon^{31/32}$$

the latter inequality implies $U_f : S_\delta \to S_\delta$.

Thus, the mapping $U_f : S_\delta \to S_\delta$ is contractive and therefore there exists a fixed point $e_0(z;f)$ of this mapping. It is evident that the function $e_0(z;f)$ is the eigenfunction of the operator L_f:

$$L_f e_0 = \lambda_0 e_0\,.$$

Then from the inequalities $\|L_f - L_1\| < \varepsilon^{31/32}$, $\|e_0 - 1\| < \frac{2}{3}\varepsilon^{15/16}$ we obtain the estimate $|\lambda_0 - 2| < \varepsilon^{15/16}$. Q.E.D.

Lemma 3.4. *For* $z, u \in R^1$ *and* $\alpha \geqq 2$

$$|z+u|^\alpha + |z-u|^\alpha \geqq 2|z|^\alpha + u^2|z|^{\alpha-2}\alpha(\alpha-1)/2 .$$

Proof. Let us divide the both sides of the inequality into $|z|^\alpha$ (for $z=0$ the inequality is evident) and denote $uz^{-1}=d$:

$$|1+d|^\alpha + |1-d|^\alpha \geqq 2 + (\alpha(\alpha-1)/2)d^2 .$$

Without losing generality we can consider $d \geqq 0, d \neq 1$. Let $F(d) = |1+d|^\alpha + |1-d|^\alpha - 2 - (\alpha(\alpha-1)/2)d^2$. Then

$$F''(d) = \alpha(\alpha-1)(|1+d|^{\alpha-2} + |1-d|^{\alpha-2} - 1) > 0$$

and $F'(0)=0$, therefore, $F'(d)>0$ for $d>0$. Then $F(0)=0$ and thus, the inequality $F'(d)>0$ for $d>0$ implies the inequality $F(d)>0$ for $d>0$. Q.E.D.

Proof of Theorem 3.1. We have

$$\|L_f^* - L_1\| = \|(L_f - L_1)^*\| = \|L_f - L_1\| < \varepsilon^{31/32} .$$

Therefore, from proof of Lemma 3.3 it follows that there exists the main eigenfunction $e_0^*(z;f)$ of the operator L_f^* and $\|e_0^*(z;f) - 1\| < \varepsilon^{15/16}$. The hyperplane H_0 which is orthogonal to the function $e_0^*(z;f)$, is invariant with respect to the operator L_f. Using the method of contraction mappings (see Lemma 3.3) in the hyperplane H_0 we shall prove the existence of the eigenfunction $e_1(z;f) \in H_0$ close to $e_1(z;1) = G_2(\sqrt{\gamma}z)$. Then we shall prove the existence $e_1^*(z;f)$ etc. As a result, $N+1$ eigenfunctions $e_0(z;f), \ldots, e_N(z;f)$ and eigenvalues $\lambda_0, \ldots, \lambda_N$ of the operator L_f will be constructed. Besides, the following inequalities are true for $i=0, 1, \ldots, N$

$$\|e_i(z;f) - G_{2i}(\sqrt{\gamma}z)\| < \varepsilon^{7/8} ,$$
$$|\lambda_i - 2c^{-i}| < \varepsilon^{7/8}$$

and at the end we shall construct the subspace $H_{f,N} \subset L^2_{\mathrm{ev}}(R^1; \exp(-\gamma z^2))$, satisfying the condition a_4) of Theorem 3.1.

Let us now prove that $|\lambda_k - 2/c^k - \varepsilon(G_{2k}, L_{G_{2k}}G_{2k})| < \varepsilon^{\frac{3}{2}}$. For this we must find the eigenfunction $e_k(z;f)$ using the perturbation theory up to the terms of order ε included. We have

$$L_f G_{2k} = (L_1 + L_{-\varepsilon G_{2k}} + L_R)G_{2k} = (2/c^k)G_{2k} - \varepsilon L_{G_{2k}}G_{2k} + O(\varepsilon^{8/5}) ;$$

let $e_k = G_{2k} + \varepsilon\varphi$, $\lambda_k = 2c^{-k} + \varepsilon l$. Then, in the formula $L_f e_k = \lambda_k e_k$ equating all the terms of order ε, we obtain

$$-L_{G_{2k}}G_{2k} + L_1\varphi = 2c^{-k}\varphi + lG_{2k} ,$$
$$\varphi(z) = \psi(z) + \alpha G_{2k}(\sqrt{\gamma}z),\ \psi(z) \perp G_{2k}(\sqrt{\gamma}z) ,$$
$$-L_{G_{2k}}G_{2k} = (2c^{-k} - L_1)\psi + lG_{2k},\ l = -(L_{G_{2k}}G_{2k}, G_{2k}) ,$$
$$\psi = (2c^{-k} - L_1)^{-1}(-L_{G_{2k}}G_{2k} - lG_{2k}) .$$

The function $L_{G_{2k}}G_{2k}$ is the polynomial of $4k$ degree and therefore it is easy to find ψ from the latter equality (it is also the polynomial of $4k$ degree).

Thus, the function φ is found and

$$\|L_f(G_{2k}+\varepsilon\varphi)-\lambda_k(G_{2k}+\varepsilon\varphi)\|=O(\varepsilon^{5/3})\,.$$

Then, using the method of contraction mappings we prove the estimates

$$\|e_k-G_{2k}-\varepsilon\varphi\|<\varepsilon^{\frac{3}{2}}$$

$$|\lambda_k-2c^{-k}-\varepsilon(G_{2k},L_{G_{2k}}G_{2k})|<\varepsilon^{\frac{3}{2}}$$

if ε is sufficiently small. Thus, the condition a_1) of Theorem 3.1 is proved.

Let us now prove a_3). It is necessary to point out that all the previous considerations were of a general character and they are applied to various problems of the perturbation theory. The proof of the conditions a_2) and a_3) is based on the nature of the perturbation of the main operator, reflected by conditions (3.2)–(3.5).

Let us consider the function $g_0(z)=\exp(-(\varepsilon'/2)|z|^\alpha)$ and the operator $T_f=\lambda_0^{-1}L_f$. The main eigenvalue of the operator T_g is equal to 1 and others do not exceed $\frac{1}{2}(c^{-1}+1)<1$, therefore, the iterations $g_n=T_f^n g_0$ tend to the function $\text{const}\, e_0(z;f)$ in the space $L^2(R^1;\exp(-\gamma z^2))$ where $\text{const}\approx 1$. In reality there takes place the convergence in C^1 on compacts because T_f is an integral operator with a smooth kernel.

More precisely from Lemma 3.2 it follows that

$$\|\exp(-\gamma z^2)(g_{n+1}(z)-g_n(z))\|_{C^1(R^1)}\leqq \text{const}\,\|g_n-g_{n-1}\|\,.$$

The following estimate is evident from the definition of the function g_0:

$$\|g_1-g_0\|\leqq\varepsilon^{15/16}\,.$$

Besides, due to the inequality

$$\|g_{n+1}-g_n\|\leqq\tfrac{1}{2}(c^{-1}+1)\|g_n-g_{n-1}\|\,,$$

we have

$$\|g_{n+1}-g_n\|\leqq[\tfrac{1}{2}(c^{-1}+1)]^n\varepsilon^{15/16}$$

and

$$\|\exp(-\gamma z^2)(g_{n+1}-g_n)\|_{C^1(R^1)}\leqq\text{const}[\tfrac{1}{2}(c^{-1}+1)]^n\varepsilon^{15/16}\,.$$

Let $D_\varepsilon^{(0)}=[-d_\varepsilon^{(0)},d_\varepsilon^{(0)}]$, where $d_\varepsilon^{(0)}=0,01(1-c^{-\frac{1}{2}})^{-1}\sqrt{\ln(1/\varepsilon)}$. It follows from the latter estimate that

$$\|g_n\|_{C^1(D_\varepsilon^{(0)})}\leqq 1+\text{const}\,\varepsilon^{15/16}\,.$$

Therefore the inequality

$$g_n(z)\leqq(2+|z|^{\frac{1}{2}})\exp(-(\varepsilon'/2)|z|^\alpha) \tag{3.11}$$

is fulfilled for all n in any case for $z\in D_\varepsilon^{(0)}$. It is evident that the function $g_0(z)$ satisfies this inequality for all $z\in R^1$.

Now let us assume that the function $g_n(z)$ satisfies inequality (3.11) for all $z\in R^1$ and prove that the function $g_{n+1}(z)$ satisfies this inequality for $z\in R^1\diagdown D_\varepsilon^{(0)}$.

The following inequality results from properties (3.2)–(3.4) of the function $f(z)$:

$$f(z)<(1+\chi_D(z))\exp(-(\varepsilon'/2)|z|^\alpha)\,,$$

where $\chi_D(z)$ is an indicator of the interval $D=[-d,d]$, containing all zeroes of the polynomial $G_{2k}(\sqrt{\gamma}z)$. It is very important to point out that D does not depend on ε. From here

$$g_{n+1}(z) \leqq (2/\lambda_0\sqrt{\pi}) \int_{-\infty}^{\infty} e^{-u^2}(1+\chi_D(z/\sqrt{c}-u))e^{-(\varepsilon'/2)|z/\sqrt{c}-u|^\alpha} \cdot (2+|z/\sqrt{c}+u|^{\frac{1}{2}})e^{-(\varepsilon'/2)|z/\sqrt{c}+u|^\alpha}du .$$

Let us now use Lemma 3.4. As $2=c^{\alpha/2}$

$$g_{n+1}(z) \leqq \exp(-(\varepsilon'/2)|z|^\alpha)\,(2/\lambda_0\sqrt{\pi}) \int_{-\infty}^{\infty} e^{-u^2}(1+\chi_D(z/\sqrt{c}-u))\cdot(2+|z/\sqrt{c}+u|^{\frac{1}{2}})du .$$

It is easy to verify that the main contribution into the right-hand side of this inequality is made by the item

$$S_0=\exp(-(\varepsilon'/2)|z|^\alpha)\,(2/\lambda_0\sqrt{\pi}) \int_{-\infty}^{\infty} e^{-u^2}|z/\sqrt{c}+u|^{\frac{1}{2}}du , \qquad z \notin D_\varepsilon^{(0)} .$$

From the inequality $|2-\lambda_0|<\varepsilon^{15/16}$ we have $2/(\lambda_0\sqrt{c})<1-(c-1)/8$ and thus for $z \notin D_\varepsilon^{(0)}$

$$S_0 \leqq |z|^{\frac{1}{2}}(1-(c-1)/10)\exp(-(\varepsilon'/2|z|^\alpha) .$$

So we have proved inequality (3.11) for the function g_{n+1} for $z \notin D_\varepsilon^{(0)}$. As we have established it for $z \in D_\varepsilon^{(0)}$ too, inequality (3.11) is valid for all $z \in R^1$. That means that the first inequality of condition a_3) of Theorem 3.1 is proved for the eigenfunction $e_0(z;f)$. The second one is deduced similarly and its proof is omitted.

Now we shall sketch the proof of conditions a_3) for all the other eigenfunctions. Let us consider such μ,

$$\mu < \text{const}\,\varepsilon^{15/16}$$

that the function $g_0(z)=G_2(\sqrt{\gamma}z)\exp(-(\varepsilon'/2)|z^\alpha|)+\mu e_0(z;f)$ belongs to the hyperplane H_0 which is orthogonal to $e_0^*(z;f)$. Then the iterations $g_n=T_f^n g_0$, where $T_f=\lambda_1^{-1}A_f$, converge with $e_1(z;f)$ and, besides, the function $g_0(z)$ satisfies the inequality

$$|g_0(z)|<(\text{const}+|z|^{5/2})\,(\exp(-(\varepsilon'/2)|z|^\alpha)) .$$

As above it is proved by induction that all the sequent functions g_n satisfy this inequality too, therefore it is fulfilled also for the function $e_1(z;f)$. Similar considerations are true for the sequent eigenfunctions.

Let us now prove a_2). Let $g_0(z)=\exp(-(\varepsilon'/2)|z|^\alpha)$. We have established already that the iterations $g_n=T_f^n g_0$, $T_f=\lambda_0^{-1}A_f$ converge to the eigenfunction $e_0(z;f)$ in C^1 on compacts and satisfy inequality (3.11). Let us show now that there exists a sequence of numbers $\{n_i\}_{i=0}^\infty$, $n_i \to \infty_{i\to\infty}$ such that

$$\|g_{n_i}-1\|_{C^1(D_\varepsilon)} \leqq \varepsilon^{7/8} . \tag{3.12}$$

It is evident that as a result, we shall prove a_2) for the eigenfunction $e_0(z;f)$.

Let $n_0=0$, $n_1=[0,0001\ln\varepsilon^{-1}]$. Let us expand the function $h_0=g_0-1$ in the Hermite polynomials up to the order M (the value M will be indicated below):

$$h_0(z)=\sum_{j=0}^M \delta_0^{(j)} G_{2j}(\sqrt{\gamma}z)+H_0(z) ,$$

where $(H_0(z), G_{2j}(\sqrt{\gamma}z))=0$, $j=0,1,\ldots,M$. Let us denote

$$r(z)=-\varepsilon G_{2k}(\sqrt{\gamma}z)+R(z),\ T_f=T_1+T_r$$

and write

$$g_1 = T_f g_0 = T_1 g_0 + T_r g_0 = T_1 1 + T_1 h_0 + T_r g_0 = 2/\lambda_0 + T_1 h_0 + T_r g_0 .$$

Hence,

$$h_1(z) = g_1(z) - 1 = (2/\lambda_0 - 1) + \sum_{j=0}^{M} (2/(\lambda_0 c^j))\delta_0^{(j)} G_{2j}(\sqrt{\gamma} z)$$
$$+ T_1 H_0^{(z)} + T_r g_0(z) = \sum_{j=0}^{M} \delta_1^{(j)} G_{2j}(\sqrt{\gamma} z) + H_1(z) + S_1(z)$$

where

$$\delta_1^{(0)} = (2/\lambda_0 - 1)\pi^{\frac{1}{4}}\gamma^{-\frac{1}{4}} + (2/\lambda_0)\delta_0^{(0)} , \tag{3.13}$$

$$\delta_1^{(j)} = (2/(\lambda_0 c^j))\delta_0^{(j)}, \; j = 1, \ldots, M , \tag{3.14}$$

$$H_1 = T_1 H_0 , \tag{3.15}$$

$$S_1 = T_r g_0 . \tag{3.16}$$

Analogous expansions are obtained for all the sequent functions $h_n(z) = g_n(z) - 1$ and the following estimates are true:

$$|\delta_n^{(0)}| \leqq 2^n \varepsilon^{15/16} , \tag{3.17}$$

$$|\delta_n^{(j)}| \leqq 2^n \lambda_0^{-n} c^{-jn} \varepsilon^{15/16} , \tag{3.18}$$

$$\|H_n(z)\| \leqq 2^n \lambda_0^{-n} c^{-n(M+1)} , \tag{3.19}$$

$$\|S_n(z)\| \leqq e^n \varepsilon^{31/32} . \tag{3.20}$$

It is very important that the validity of all these estimates is proved on the basis of the following properties of the function g_0:

1. g_0 satisfies the estimate (3.11),
2. $\|g_0(z) - 1\|_{C^1(D_\varepsilon)} \leqq \varepsilon^{15/16}$, (3.21)
3. $\|g_0(z) - 1\| \leqq \varepsilon^{31/32}$. (3.22)

Relations (3.17)–(3.20) are easily proved by induction. Let us now use estimates (3.17)–(3.20) for $n = n_1$. Then we receive for $z \in D_\varepsilon$ that

$$|h_{n_1}(z)| \leqq |\delta_{n_1}^{(0)}| \pi^{\frac{1}{4}} \gamma^{-\frac{1}{4}} + \sum_{j=0}^{M} |\delta_{n_1}^{(j)} G_{2j}(\sqrt{\gamma} z)| + |H_{n_1}(z)|$$
$$+ |S_{n_1}(z)| \leqq \varepsilon^{0.99} + \mathrm{const}(\ln \varepsilon^{-1})^M \varepsilon^{31/32} + \varepsilon^{31/32 - 0.001} \leqq \varepsilon^{15/16}$$

if ε is sufficiently small. We have used here the following considerations: the estimates $|\delta_{n_1}^{(0)}| \pi^{\frac{1}{4}} \gamma^{-\frac{1}{4}} \leqq \varepsilon^{0.99}$ and $\sum_{j=0}^{M} |\delta_{n_1}^{(j)} G_{2j}(\sqrt{\gamma} z)| \leqq \mathrm{const}\varepsilon^{31/32} [\ln \varepsilon^{-1})^M$ are deduced from (3.22), the estimate $|S_{n_1}(z)| < \varepsilon^{31/32 - 0.001}$ – from (3.20) and the estimate $|H_{n_1}(z)| < \varepsilon$ is proved in the following way

$$\|H_{n_1}(z)\|_{C^1(D_\varepsilon)} \leqq \varepsilon^{-100} \|\exp(-\gamma z^2) H_{n_1}\|_{C^1(R^1)}$$
$$\leqq \varepsilon^{-100} \|H_{n_1 - 1}\| \leqq \varepsilon^{-100} (2\lambda_0^{-1} c^{-M-1})^{n_1 - 1}$$

[see (3.19)]. Next $c^{-n_1} = \varepsilon^{0.0001 \ln c}$. Let us choose $M = 10^7 (\ln c)^{-1}$ (M does not depend on ε). Then it follows from the latter inequality that $\|H_{n_1}\|_{C^1(D_\varepsilon)} \leqq \varepsilon$. Q.E.D.

Thus it is proved that the function g_{n_1} satisfies the same conditions 1, 2, 3 as the function $g_0(z)$. Now in a similar way we prove that the function $g_{n_2}(z)$, $n_2 = 2n_1$ satisfies these conditions too and so on. As a result, we establish (3.12) Theorem 3.1 is proved.

§ 4. Inductive Assumptions, Formulations of Main Lemmas Proof of Theorem 1

We shall begin with some notations. Let

$$a=(G_{2k}, AG_{2k})=\int_{-\infty}^{\infty} e^{-\gamma z^2} G_{2k}(\sqrt{\gamma} z) A(G_{2k}(\sqrt{\gamma} z))dz$$

and $\varepsilon=(c_k-c)/a$. In Appendix 1 we show that $a \neq 0$. Below we consider only the case $\varepsilon>0$. All our assertions should begin with the phrase: "Let ε be sufficiently small". For this reason we shall omit it everywhere. Let us put $D_n=[-d_n, d_n]$ where $d_n=10(\ln\varepsilon^{-1}+\ln(1-\varepsilon)^{-n})^{\frac{1}{\alpha}}/(c-1)$. We shall take $\omega>1$ which is the root of the equation $5^{\omega-1}=(1-\varepsilon)^{-1/100}$, i.e. $\omega-1\approx\varepsilon/(100\ln 5)$. Furthermore, we shall consider the sequence of integers $n_i=[\omega n_{i-1}+n_0]$, $i=1,2,\ldots$, $n_0=2\log_5\varepsilon^{-1}$. Our procedure will be slightly different for $n=n_i$ and $n_i<n<n_{i+1}$. At each step we shall deal with a family of functions $f_n(z;a)=A^n f_0(z;a)$, where a is a parameter of the family, all the values of which form the k-dimensional parallelepiped: $a=\{a_0,\ldots,a_{k-1}\}$, $|a_i|\leqq A_i^{(n)}$, $i=1,\ldots,k$. All the functions of the family are even.

Inductive Assumptions for $n=n_i$. Conditions (U_{n_i})

For $n=n_i$ the k-dimensional parallelepiped $\mathfrak{B}_{n_i}=\{a=(a_0,a_1,\ldots,a_{k-1}): |a_s|\leqq \varepsilon^{5/3}(1-\varepsilon/2)^{n_i}, s=0,\ldots,k-1\}$, for each $a=(a_0,\ldots,a_{k-1})\in\mathfrak{B}_{n_i}$ the even function $f_{n_i}(z;a)$ is given so that

u$_1$) for some $a^{(0)}\in\mathfrak{B}_{n_i}$ the function $f_{n_i}(z;a^{(0)})=\bar{f}_i$ satisfies the conditions of Theorem 3.1; therefore the operator $L_{\bar{f}_i}$ has $N+1$ eigenfunctions $e_s(z;\bar{f}_i)=e_s^{(i)}$, $s=0,1,\ldots,N$ with eigenvalues $\lambda_s(\bar{f}_i)=\lambda_s^{(i)}$ and the invariant space $H_{\bar{f}_i,N}$; besides $|\lambda_s^{(i)}-2c^{-s}|<\varepsilon^{4/5}$, $s=0,1,\ldots,N$; the number N does not depend on ε and will be indicated below;

u$_2$) the function $g_{n_i}(z;a)=Af_{n_i}(z;a)-f_{n_i}(z;a)$, $a\in\mathfrak{B}_{n_i}$ can be represented in the form

$$g_{n_i}(z;a)=\sum_{j=0}^{k-1} a_j e_j^{(i)}(z)+\delta_{n_i}(a)e_k^{(i)}(z)+R_{n_i}(z;a)$$

here the function $R_{n_i}(z;a)$ being expanded on the subspace $H_{\bar{f}_i,K}$ and one-dimensional subspaces generated by $e_s^{(i)}$, $s=0,1,\ldots,K$ has zero projections on these one-dimensional subspaces; for $z\in D_{n_i}$

u$_{21}$) $|\delta_{n_i}(a)|<2\gamma^{-1/2}\varepsilon^{8/3}(1-2\varepsilon/3)^{n_i}$; $|\nabla_a\delta_{n_i}(a)|<2\gamma^{-1/2}\varepsilon^{5/2}(1-2\varepsilon/3)^{n_i}$;

u$_{22}$) $|R_{n_i}(z;a)|<\varepsilon^{8/3}(1-2\varepsilon/3)^{n_i}$;

$|\partial R_{n_i}(z;a)/\partial z|<\varepsilon^{5/2}(1-2\varepsilon/3)^{n_i}$;

$|\nabla_a R_{n_i}(z;a)|<\varepsilon^{5/2}(1-2\varepsilon/3)^{n_i}$;

u$_{23}$) $|\partial g_{n_i}(z;a)/\partial z|<\varepsilon^{5/2}(1-3\varepsilon/5)^{n_i}$;

$|\partial g_{n_i}(z;a)/\partial a_j|-e_j(z;\bar{f}_i)|<\varepsilon^{5/2}(1-3\varepsilon/5)^{n_i}$; $j=0,\ldots,k-1$;

u$_3$) for $z\notin D_{n_i}$

$$0\leqq f_{n_i}(z;a)\leqq\exp(-(\varepsilon'/2)|z|^\alpha),$$

where $\alpha=2(\log_2 c)^{-1}$ and is sufficiently close to $2k$ for small ε;

$$|\partial f_{n_i}(z;a)/\partial z|\leqq(1+|z|^{2k})\exp(-(\varepsilon'/2)|z|^{\alpha}),$$
$$|V_a f_{n_i}(z;a)|\leqq(1+|z|^{2k+1})\exp(-(\varepsilon'/2)|z|^{\alpha}),$$
$$|V_a\partial f_{n_i}(z;a)/\partial z|\leqq(1+|z|^{4k+1})\exp(-(\varepsilon'/2)|z|^{\alpha}).$$

Inductive Assumptions for $n_i<n\leqq n_{i+1}$. Conditions (V_n)

Let a k-dimensional parallelepiped $\mathfrak{A}_n=\{a=(a_0,\ldots,a_{k-1})\colon |a_i|\leqq\frac{1}{2}\varepsilon^{5/3}(1-\varepsilon/2)^n\cdot c^{(n-n_i)/10},\ i=0,1,\ldots,k-1\}$ and for each $a\in\mathfrak{A}_n$ an even function $f_n(z;a)$ be given. We put $g_n(z;a)=Af_n(z;a)-f_n(z,a)$ and denote

$$v_k=1-3\varepsilon/4\,,\qquad v_j=\lambda_j+\varepsilon^{\frac{1}{2}}<c^{-2(j-k)/3}\,,\qquad k<j\leqq N+1$$

where λ_j are eigenvalues of the operator $L_{\bar f_i}$ acting in the Hilbert space $L^2_{\mathrm{ev}}(R^1;\exp(-\gamma z^2))$ of even square-integrable functions with the weight $\exp(-\gamma z^2)$, $\gamma=1-c^{-1}$. Then the family $\{g_n(z;a),\ a\in\mathfrak{A}_n\}$ satisfies the conditions:

v$_1$) for $z\in D_n$

$$g_n(z;a)=\textstyle\sum_{s=0}^{k-1}a_s e_s^{(i)}(z)+\sum_{s=k}^{N}\delta_n^{(s)}(a)e_s^{(i)}(z)+h_n(z;a)+t_n(z;a)$$

where N is the same number as in the conditions (U_{n_i}) and will be indicated below and

v$_{11}$) $|\delta_n^{(s)}(a)|\leqq 2\gamma^{-\frac{1}{4}}\varepsilon^{8/3}(1-2\varepsilon/3)^{n_i}v_s^{n-n_i}\,,\qquad s=k,\ldots,N\,;$

$|V_a\delta_n^{(s)}(a)|\leqq 2\gamma^{-\frac{1}{4}}\varepsilon^{5/2}(1-2\varepsilon/3)^{n_i}v_s^{n-n_i}\,,\qquad s=k,\ldots,N\,;$

v$_{12}$) $h_n(z;a)=0$ for $z\notin D_n$; $\quad h_n(z;a)\in H_{\bar f_i}$,

$$\|h_n(z;a)\|_{L^2(R^1;\exp(-\gamma z^2))}<2\gamma^{-\frac{1}{4}}\varepsilon^{8/3}(1-2\varepsilon/3)^{n_i}v_{N+1}^{n-n_i}$$
$$\|\,|h_n(z;a)|+|V_a h_n(z;a)|\,\|_{L^2(R^1;\exp(-\gamma z^2))}\leqq\varepsilon^{7/3}(1-2\varepsilon/3)^{n_i}v_{N+1}^{n-n_i}\,,$$
$$\|h_n(z;a)\|_{C^m(D_n)}+\|V_a h_n(z;a)\|_{C^m(D_n)}<L_m^{(0)}\varepsilon^{7/3}(1-2\varepsilon/3)^{n_i}3^{n-n_i},$$

$m=0,1,\ L_m^{(0)}=\mathrm{const}$

v$_{13}$) $\|t_n(z;a)\|_{C(D_n)}\leqq\varepsilon^3(1-2\varepsilon/3)^{n_i+1}3^{n-n_i+1}\,,$

$$\left\|\,|V_a t_n(z;a)|+\left|\frac{\partial}{\partial z}t_n(z;a)\right|+\left|V_a\frac{\partial}{\partial z}t_n(z;a)\right|\right\|_{C(D_n)}<\varepsilon^{5/2}(1-3\varepsilon/5)^{n_i+1}3^{n-n_i}$$

v$_{14}$) $\|g_n(z;a)\|_{C(D_n)}<\varepsilon^{8/3}(1-2\varepsilon/3)^n 3^{n-n_i+1}$

$$\left\|\frac{\partial}{\partial z}g_n(z;a)\right\|_{C(D_n)},\ \left\|\frac{\partial}{\partial a_i}g_n(z;a)-e_j(z;\bar f_i)\right\|_{C(D_n)}<\varepsilon^{5/2}(1-3\varepsilon/5)^{k_i}3^{n-n_i}\,,$$

v$_2$) for $z\notin D_n$

v$_{21}$) $0\leqq f_n(z;a)<\exp(-\varepsilon'/2|z|^{\alpha})$;

v$_{22}$) $\left|\dfrac{\partial}{\partial z}f_n(z;a)\right|<(1+|z|^{2k})\exp(-\varepsilon'/2|z|^{\alpha})$;

v$_{23}$) $|V_a f_n(z;a)|<(1+|z|^{2k+1})\exp(-\varepsilon'/2|z|^{\alpha})$;

v$_{24}$) $\left|V_a\dfrac{\partial}{\partial z}f_n(z;a)\right|<(1+|z|^{4k+1})\exp(-\varepsilon'/2|z|^{\alpha})$.

Now we shall formulate three lemmas from which we shall deduce Theorem 1. In the formulations λ is a certain constant larger than 1.

Lemma 1. *Let $n=n_i$, and for $n=n_i$ the conditions (U_{n_i}) are valid for the family $\{f_{n_i}(z;a),\ a\in\mathfrak{B}_{n_i}\}$. There exists a subset $\mathfrak{B}'_{n_i}\subset\mathfrak{B}_{n_i}$ and C^1-diffeomorphism φ_n: $\mathfrak{B}'_{n_i}\to\mathfrak{A}_{n_i}$ such that for the family $\{f_{n_i}(z;\varphi_{n_i}^{-1}(a)),\ a\in\mathfrak{A}_{n_i}\}$ the conditions (V_{n_i}) are valid. Moreover $\|\varphi_{n_i}-\mathrm{Id}\|_{C^1}<\varepsilon^2$ where Id is the identity transformation in the k-dimensional space.*

Lemma 2. *Let for n, $n_i\leqq n<n_{i+1}$ the family $\{f_n(z;a), a\in\mathfrak{A}_n\}$ satisfies the conditions (V_n). Then there exists a subset $\mathfrak{A}'_n\subset\mathfrak{A}_n$ and C^1-diffeomorphism $\psi_n:\mathfrak{A}'_n\to\mathfrak{A}_{n+1}$ such that $d(\psi_n(a'),\psi_n(a''))\geqq\lambda d(a',a'')$ and the family $\{Af_n(z;\psi_n^{-1}(a)),\ a\in\mathfrak{A}_{n+1}\}$ satisfies the conditions (V_{n+1}). For $n=n_{i+1}-1$ the conditions (V_{n+1}) are valid with the functions $e_s^{(i)}$.*

Lemma 3. *Let for $n=n_{i+1}$ the family $\{f_{n_{i+1}}(z;a),\ a\in\mathfrak{A}_{n_{i+1}}\}$ satisfies the conditions $(V_{n_{i+1}})$ with the functions $e_s^{(i)}$. Then there exists a subset $\mathfrak{A}'_{n_{i+1}}\subset\mathfrak{A}_{n_{i+1}}$ and C^1-diffeomorphism $\chi_{n_{i+1}}:\mathfrak{A}'_{n_{i+1}}\to\mathfrak{B}_{n_{i+1}}$ such that for the family $\{f_{n_{i+1}}(z;\chi_{n_{i+1}}^{-1}(a)),$ $a\in\mathfrak{B}_{n_{i+1}}\}$ the conditions $(U_{n_{i+1}})$ are valid and $\|\chi_{n_{i+1}}-\mathrm{Id}\|\leqq\varepsilon^2$.*

Proof of Theorem 1. Let us take the initial family of functions

$$f(z;a)=\varphi(z)(1-\varepsilon G_{2k}+\varepsilon^2\textstyle\sum_{i=0}^{4k}b_iG_{2i}+\sum_{i=0}^{k-1}a_ie_i(z)),$$

where $\varphi(z)\in C_0^\infty$, $\varphi(z)=\varphi(-z)$, $\varphi(z)\equiv 1$ for $|z|<d_{n_0}(\varepsilon)$ and $\varphi(z)\equiv 0$ for $|z|>d_{n_0}(\varepsilon)+1$, coefficients b_i are found from the formulae of the perturbation theory

$$A(1-\varepsilon G_{2k}+\varepsilon^2\textstyle\sum_{i=0}^{4k}b_iG_{2i})-(1-\varepsilon G_{2k}+\varepsilon^2\sum_{i=0}^{4k}b_iG_{2i})=O(\varepsilon^3)$$

where $e_i(z)$ are the eigenfunctions of the operator $L_{f(z;0)}$. It is easy to see that this family satisfies the conditions (U_{n_0}). Now we can apply Lemmas 1, 2, 3 and construct a decreasing sequence of sets $\bar{\mathfrak{B}}_n$, $\bar{\mathfrak{B}}_{n_0}=\mathfrak{A}_{n_0}\supset\bar{\mathfrak{B}}_{n_0+1}=\psi_{n_0+1}^{-1}(\mathfrak{B}_{n_0+1})\supset$ $\bar{\mathfrak{B}}_{n_0+2}=\psi_{n_0+2}^{-1}\psi_{n_0+1}^{-1}(\mathfrak{B}_{n_0+2})$ for which $\bigcap_{n=n_0}^{\infty}\mathfrak{B}_n=\bar{a}\in\mathfrak{A}_{n_0}$.

We shall show that the limit $\lim\limits_{n\to\infty}A^nf(z;\bar{a})=h(z;\bar{a})$ exists uniformly on any finite interval and $Ah=h$. Let $f_{n+1}(z)=A^nf(z;\bar{a})$, $f_0(z)=f(z;\bar{a})$. Lemma 2 can be applied to the function $g_n(z)=f_{n+1}(z)-f_n(z)$ from which it follows that

$$|g_n(z)|<\varepsilon^{5/2}(1-2\varepsilon/3)^n,\qquad z\in D_n,$$

$$|g_n(z)|<\exp(-(\varepsilon/2)|z|^\alpha),\qquad z\notin D_n.$$

Therefore for any fixed l the series $f_0(z)+\sum_{n=1}^{\infty}g_n(z)$ converges uniformly on D_l and for its limit $h(z)\geqq 0$ the following estimate is valid

$$h(z)\leqq\exp(-(\varepsilon'/2)|z|^\alpha),\qquad z\in D_l.$$

From this estimate we have $Ah=h$. Theorem 1 is proved.

§ 5. Proof of Lemma 1

Let be $n=n_i$. Let us denote $h_j(z)=\chi_{D_n}(z)e_j^{(i)}(z)$, $j=0,\ldots,N$, where χ_{D_n} is the indicator of the interval $D_n=[-d_n,d_n]$, $e_j^{(i)}(z)$ is the j-th eigenfunction of the operator $L_{\bar{f}_i}$.

Firstly we shall show that

$$|\int_{-\infty}^{\infty} h_j(z)h_k(z)e^{-\gamma z^2}dz - \delta_k^j| < \varepsilon^{2/3} \tag{5.1}$$

where δ_k^j is the Kronecker symbol. According to the Theorem 3.1

$$\|e_j^{(i)}(z) - G_{2j}(z\sqrt{\gamma})\|_{L^2(R^1; e^{-\gamma z^2})} < \varepsilon^{4/5}$$

and for $|z| > d_0 = 10(c-1)^{-1}\sqrt{\ln\varepsilon^{-1}}$ $|e_j^{(i)}(z)| < |z|^{2j+1}$. Therefore

$$\|h_j(z) - e_j^{(i)}(z)\|^2_{L^2(R^1; e^{-\gamma z^2})} = \int_{|z|>d_n} |e_j^{(i)}(z)|^2 e^{-\gamma z^2}dz$$
$$< \int_{|z|>d_0} |z|^{4j+2} e^{-\gamma z^2}dz < \varepsilon .$$

Consequently,

$$\|h_j(z) - G_{2j}(z\sqrt{\gamma})\|_{L^2(R^1; \exp(-\gamma z^2))} \leqq 2\varepsilon^{4/5} . \tag{5.2}$$

The Hermite polynomials $\{G_{2j}(z\sqrt{\gamma})\}$ are orthogonal in the space $L^2(R^1;\exp(-\gamma z^2))$, therefore inequality (5.1) follows from the last inequality. Then from (5.2) we may readily obtain:

$$|\int_{-\infty}^{\infty} h_j(z)h(z)e^{-\gamma z^2}dz| < 2\varepsilon^{4/5} \tag{5.3}$$

for $j=0, \ldots, N$ and $h(z) \in H_{\bar{f}_i, N}$.

Inequalities (5.1) and (5.3) allow to expand the function $R_n(z; a)$ in functions $\{h_j(z)\}$ for small ε [see the condition (U_{n_i})]

$$R_n(z; a) = \sum_{j=0}^{N} \bar{\delta}_n^{(j)}(a)h_j(z) + \bar{h}_n(z; a) ,$$

where for $a \in \mathfrak{B}_n$ $\operatorname{supp}\bar{h}_n(z; a) \subset D_n$ and $\bar{h}_n(z; a) \in H_{\bar{f}_i, N}$ and obtain for any $j, 0 \leqq j \leqq N$, the following estimates

$$|\bar{\delta}_n^{(j)}(a)| \leqq (1 + \varepsilon^{2/3})\|R_n(z; a)\| , \tag{5.4}$$

$$\|\bar{h}_n(z; a)\| \leqq (1 + \varepsilon^{2/3})\|R_n(z; a)\| , \tag{5.5}$$

$$|V_a\bar{\delta}_n^{(j)}(a)| \leqq (1 + \varepsilon^{2/3})\|V_a R_n(z; a)\| , \tag{5.6}$$

$$\|V_a\bar{h}_n(z; a)\| \leqq (1 + \varepsilon^{2/3})\|V_a R_n(z; a)\| , \tag{5.7}$$

where $\|\cdot\| = \|\cdot\|_{L^2(R^1; \exp(-\gamma z^2))}$.

As a result, we have the expansion of the function $g_n(z; a)$ as follows

$$g_n(z; a) = \sum_{j=0}^{k-1} (a_j + \bar{\delta}_n^{(j)}(a))e_j^{(i)}(z) + (\delta_n(a) + \bar{\delta}_n^{(k)}(a))e_k^{(i)}(z)$$
$$+ \sum_{j=k+1}^{N} \bar{\delta}_n^{(j)}(a)e_j^{(i)}(z) + \bar{h}_n(z; a) . \tag{5.8}$$

The estimates u_{22}) in the condition (U_{n_i}), and the estimates (5.4), (5.6) show that

$$|\bar{\delta}_n^{(j)}(a)| < 2\gamma^{-\frac{1}{2}}\varepsilon^{8/3}(1 - 2\varepsilon/3)^{n_i} , \tag{5.9}$$

$$|V_a\bar{\delta}_n^{(j)}(a)| < 0.9 \cdot 2\gamma^{-\frac{1}{2}}\varepsilon^{5/2}(1 - 2\varepsilon/3)^{n_i} . \tag{5.10}$$

Let us define the mapping $\varphi_n : \mathfrak{B}_n \to R^k$ by the formula:

$$(a_0, \ldots, a_{k-1}) = a \xrightarrow{\varphi_n} b = (a_0 + \bar{\delta}_n^{(0)}(a), \ldots, a_{k-1} + \bar{\delta}_n^{(k-1)}(a)) .$$

Then the estimates (5.9) and (5.10) mean that

$$\|\varphi_n - \mathrm{Id}\|_{C^1} \leqq \varepsilon^{5/2} \tag{5.11}$$

that is, φ_n is C^1-diffeomorphism which is close to the identical one. For the point $x \in \partial\mathfrak{B}_n$, the boundary of the cube $\mathfrak{B}_n$,

$$|\varphi_n(x) - x| < \varepsilon^{\frac{1}{2}}|x| .$$

Let us denote $\mathfrak{A}_n = \{|a_j| < \frac{1}{2}\varepsilon^{5/3}(1-\varepsilon/2)^n\}$. From the last inequality it follows that

$$\mathfrak{A}_n \subset \varphi_n(\mathfrak{B}_n)$$

where $\varphi_n(\mathfrak{B}_n)$ is the image of the cube $\mathfrak{B}_n$ under the mapping φ_n. Let us put $\mathfrak{B}'_n = \varphi_n^{-1}(\mathfrak{A}_n)$. We have proved that the mapping $\varphi_n : \mathfrak{B}'_n \to \mathfrak{A}_n$ satisfies the estimates formulated in the lemma. It should be verified that expansion (5.8) of the function $g_n(z; a)$ satisfies all the requirements of the condition (V_n), provided the variables $a = \varphi_n^{-1}(b)$ are substituted in this expansion. Let us denote

$$\begin{aligned}
&\delta_n^{(k)}(b) = \delta_n(\varphi_n^{-1}(b)) + \bar{\delta}_n^{(k)}(\varphi_n^{-1}(b)) ,\\
&\delta_n^{(j)}(b) = \bar{\delta}_n^{(j)}(\varphi_n^{-1}(b)), j = k+1, \ldots, N ,\\
&h_n(z; b) = \bar{h}_n(\varphi_n^{-1}(b)) ,\\
&t_n(z; b) = 0 .
\end{aligned}$$

We have for $k+1 \leqq j \leqq N$

$$\begin{aligned}
|\delta_n^{(j)}(b)| &= |\bar{\delta}_n^{(j)}(\varphi_n^{-1}(b))| < 2\gamma^{-\frac{1}{2}}\varepsilon^{8/3}(1-2\varepsilon/3)^{n_i} ,\\
|\nabla_b \delta_n^{(j)}(b)| &\leqq |\nabla_{\varphi_n^{-1}(b)} \bar{\delta}_n^{(j)}(\varphi_n^{-1}(b))| \cdot |\nabla_b \varphi_n^{-1}(b)|\\
&\leqq 2 \cdot 0.9\gamma^{-\frac{1}{2}}\varepsilon^{5/2}(1-2\varepsilon/3)^{n_i}(1+\varepsilon^{2/3}) < 2\gamma^{-\frac{1}{2}}\varepsilon^{5/2}(1-2\varepsilon/3)^{n_i} .
\end{aligned}$$

In the same way we verify the remaining parts of the condition (V_n). Lemma 1 is proved.

§ 6. Proof of Lemma 2

We have

$$\begin{aligned}
g_{n+1} &= f_{n+2} - f_{n+1} = Af_{n+1} - Af_n = L_{f_n}(f_{n+1} - f_n)\\
&\quad + A(f_{n+1} - f_n) = L_{\bar{f}_i} g_n + (L_{f_n - \bar{f}_i}) g_n + A g_n\\
&= L_{\bar{f}_i} g_n + t'_n + t''_n .
\end{aligned} \tag{6.1}$$

Firstly we show that for $z \in D_n$

$$|t'_n(z; a)| < (1/4)\varepsilon^{7/2}(1-2\varepsilon/3)^{n_i+1}3^{n-n_i+1} , \tag{6.2}$$

$$|t''_n(z; a)| < (1/4)\varepsilon^{7/2}(1-2\varepsilon/3)^{n_i+1} \cdot 3^{n-n_i+1} , \tag{6.3}$$

$$|\nabla_a t'_n(z; a)|, |\nabla_a t''_n(z; a)| < (1/4)\varepsilon^{10/3}(1-\varepsilon/2)^{n_i+1}3^{n-n_i+1} . \tag{6.4}$$

In order to derive (6.2) let us first establish, that for $z \in D_{n_i}$

$$|f_n(z;a) - \bar{f}_i(z)| < \varepsilon^{7/3}(1-\varepsilon/3)^{n_i}. \tag{6.5}$$

Let us denote $a^{(n)} = a$, $a^{(j)} = \psi_j^{-1}(a^{(j+1)})$ for $j = n-1, \ldots, n_i$, $b^{(n_i)} = \varphi_{n_i}^{-1}(a^{(n_i)})$. From v_{14})

$$|f_n(z;a) - f_{n_i}(z;b^{(n_i)})| \leqq \textstyle\sum_{j=n_i}^{n-1} |g_j(z;a^{(j)})| \leqq 2\varepsilon^{7/3}(1-2\varepsilon/3)^n \tag{6.6}$$

and from u_3) it follows

$$\begin{aligned}|f_{n_i}(z;b^{(n_i)}) - \bar{f}_i(z)| &= |f_{n_i}(z;b^{(n_i)}) - f_{n_i}(z;0)| \\ &\leqq |b^{(n_i)}| \sup_{b \in \mathfrak{B}_{n_i}} |\nabla_b f_{n_i}(z;b)| \leqq \varepsilon^{5/2}(1-\varepsilon/2)^{n_i}\varepsilon^{-1/20} d_{n_i}^{2k+1} \\ &< \varepsilon^{7/3}(1-\varepsilon/3)^{n_i}\varepsilon^{1/10}(1-\varepsilon/2)^{n_i}(1-\varepsilon/3)^{-n_i} d_{n_i}^{2k+1}.\end{aligned} \tag{6.7}$$

Inequality (6.5) will result from (6.6), (6.7), if we show that

$$\varepsilon^{1/10}(1-\varepsilon/6)^{n_i} d_{n_i}^{2k+1} < 1. \tag{6.8}$$

From the form of d_{n_i}, denoting $x = \varepsilon(1-\varepsilon)^{n_i}$, we have

$$\varepsilon^{1/10}(1-\varepsilon/6)^{n_i} d_{n_i}^{2k+1} < (\varepsilon(1-\varepsilon)^{n_i})^{1/10} d_{n_i}^{2k+1} = L x^{1/10} \ln^{k+\frac{1}{2}} x^{-1}$$

where L is limited, and $x \to 0$ for $\varepsilon \to 0$. Thus, (6.8) and therefore (6.5) are proved.

Let us consider now (6.2). Suppose $z \in D_{n+1}$. Then

$$\begin{aligned}|(L_{f_n} - L_{\bar{f}_i})g_n(z;a)| &= |2/\sqrt{\pi} \textstyle\int_{-\infty}^{\infty} e^{-u^2}(f_n(z/\sqrt{c}-u;a) \\ &\quad - \bar{f}_i(z/\sqrt{c}-u))g_n(z/\sqrt{c}+u;a)du| \\ &\leqq (2/\sqrt{\pi})|\textstyle\int_{|u|<0.9(1-c^{-\frac{1}{2}})d_{n+1}}| + (2/\sqrt{\pi})|\textstyle\int_{|u|>0.9(1-c^{-\frac{1}{2}})d_{n+1}}| \\ &= I_1 + I_2.\end{aligned}$$

From u_3) it obviously follows that $f_n \leqq 2$, and, putting $u_n = 0.9(1-c^{-\frac{1}{2}})d_{n+1}$, we have

$$I_2 < L\exp(-u_n^2) < (\varepsilon(1-\varepsilon)^{n+1})^{12}.$$

In case $|u| < u_n$, and $z \in D_{n+1}$, we have

$$\begin{aligned}|z/\sqrt{c} \pm u| &\leqq |z/\sqrt{c}| + |u| \leqq d_{n+1}/\sqrt{c} + u_n = d_{n+1}(1/\sqrt{c} + 0.9(1-1/\sqrt{c})) \\ &= d_{n+1}(0.9 + 0.1/\sqrt{c}) = d_{n_i} d_{n+1}/d_{n_i}(0.9+0.1/\sqrt{c}) < d_{n_i}\end{aligned} \tag{6.9}$$

since

$$\begin{aligned}d_{n+1}/d_{n_i} &= d_{n_{i+1}}/d_{n_i} = ((\ln\varepsilon^{-1} + n_{i+1}\ln(1-\varepsilon)^{-1})/(\ln\varepsilon^{-1} + n_i\ln(1-\varepsilon)^{-1}))^{\frac{1}{2}} \\ &< ((\ln\varepsilon^{-1} + n_i(\omega-1)\ln(1-\varepsilon)^{-1} + n_i\ln(1-\varepsilon)^{-1})/(\ln\varepsilon^{-1} + n_i\ln(1-\varepsilon)^{-1}))^{\frac{1}{2}} \\ &< \sqrt{1+(\omega-1)} = \sqrt{\omega} \underset{\varepsilon\to 0}{\longrightarrow} 1.\end{aligned}$$

From (6.9) it follows that in estimating the value I_1 we may employ the properties of functions $f_n(z;a) - \bar{f}_i(z)$ and $g_n(z;a)$, $z \in D_{n_i}$. Using v_{13}) and (6.5), we obtain

$$\begin{aligned}I_1 &= (2/\sqrt{\pi})|\textstyle\int_{|u|<u_n} e^{-u^2}(f_n(z/\sqrt{c}-u;a) - \bar{f}_i(z/\sqrt{c}-u))g_n(z/\sqrt{c}+u;a)du \\ &\leqq \varepsilon^{7/3}(1-\varepsilon/3)^{n_i}\varepsilon^{7/3}(1-\varepsilon/2)^n 3^{n-n_{i+1}} \leqq (\varepsilon^4/4)(1-2\varepsilon/3)^n 3^{n-n_{i+1}}.\end{aligned}$$

Performing summation of the estimates for I_1 and I_2 we obtain (6.2). The relation (6.3) can be proved analogously.

Let us now prove (6.4). We have

$$\nabla_a t_n'(z;a) = L\nabla_{af_n} g_n + (L_{f_n} - L_{\bar{f}_i})\nabla_a g_n\,.$$

We shall use the inequalities, resulting from (V_n):

$$|\nabla_a f_n(z;a)| \le K_0(1+|z|^{2k+1})\,,$$
$$|\nabla_a g_n(z;a)| \le K_0(1+|z|^{2k+1})\,,$$
$$|\nabla_a g_n(z;a)| \le \varepsilon^{5/2}(1-3\varepsilon/5)^{n_i}3^{n-n_i} \quad \text{for} \quad |z| \le d_n\,.$$

Let us divide the integral, which determines $L_{\Delta_a f_n} g_n$, into the sum of two integrals:

$$L_{\Delta_a f_n} g_n(z;a) = (2/\sqrt{\pi})\left[\int_{-u_n}^{u_n} + \int_{|u|\ge u_n}\right] e^{-u^2}\nabla_a f_n(z/\sqrt{c}-u;a)\,g_n(z/\sqrt{c}+u;a)\,du\,.$$

For the external integral we have the estimate:

$$K_1\int_{|u|>u_n} e^{-u^2}(1+|z/\sqrt{c}-u|)^{\alpha+1}(1+|z/\sqrt{c}+u|)^{\alpha+1}du$$
$$\le K_2\int_{|u|>u_n} e^{-u^2}|u|^{2\alpha+2}du < \varepsilon^{7/2}(1-2\varepsilon/3)^n\,,$$

so far as for $z\in D_n$ and $|u|>u_n$, we shall evidently have $|z/\sqrt{c}\pm u| < K_3 u$. The last inequality follows from the fact that $\exp(-u_n^2) < \varepsilon^4(1-\varepsilon)^n$ due to the definition of u_n. For the internal integral on the base of the condition v_{14}):

$$(2/\sqrt{\pi})\int_{|u|<u_n} e^{-u^2}\nabla_a f_n(z/\sqrt{c}+u;a)\,g_n(z/\sqrt{c}-u;a)\,du$$
$$<(2/\sqrt{\pi})\int_{|u|<u_n} e^{-u^2}du\cdot\varepsilon^{8/3}(1-2\varepsilon/3)^n\cdot u_n^{\alpha+1} \le \varepsilon^{5/2}(1-5\varepsilon/8)^n\,.$$

Summing the estimates of the external and internal integrals, we obtain the estimate for the value $|L_{\Delta_a f_n} g_n|$. Analogously $(L_{f_n} - L_{f_i})\nabla_a g_n$ can be estimated. So, the first inequality in (6.4) is proved. In a similar way we prove the second inequality for $|\nabla_a t_n''|$ in (6.4).

Now we make use of the representation for $g_n(z;a)$ involved in v_1). In the expression

$$L_{\bar{f}_i} g_n(z;a) = \sum_{j=0}^{k-1} a_j L_{\bar{f}_i} e_j^{(i)} + \sum_{j=k}^{N} \delta_n^{(j)}(a) L_{\bar{f}_i} e_j^{(i)} + L_{\bar{f}_i} h_n + L_{\bar{f}_i} t_n$$

we consider each term separately, beginning from the right one. Let us introduce the operator

$$\bar{L}_{\bar{f}_i} g = (2/\sqrt{\pi})\int_{-u_n}^{u_n} e^{-u^2}\bar{f}_i(z/\sqrt{c}-u)\,g(z/\sqrt{c}+u)\,du\,.$$

We shall show that for $z\in D_{n+1}$

$$|(L_{\bar{f}_i} - \bar{L}_{\bar{f}_i})t_n(z;a)| < \varepsilon^4(1-\varepsilon)^{n+1}\,. \tag{6.10}$$

From the conditions V_1)

$$\|\bar{f}_i\|_{C(R^1)} \le 2\,, \qquad \|t_n(z;a)(1+|z|)^{-2N-2}\|_{C(R^1)} < 1\,.$$

Hence

$$|(L_{\bar{f}_i} - \bar{L}_{\bar{f}_i})t_n(z;a)| \le (4/\sqrt{\pi})\int_{|u|>u_n} e^{-u^2}(1+|z/\sqrt{c}-u|)^{2N+2}du$$
$$\le K\int_{|u|>u_n} e^{-u^2}|u|^{2N+2}du < K_1\exp(-u_n^2)u_n^{2N+2}$$
$$<\varepsilon^4(1-\varepsilon)^{n+1}K_1\exp(-0.1u_n^2)u_n^{2N+2}\,,$$

that proves (6.10), since $K_1 \exp(-0.1u_n^2)u_n^{2N+2} \to 0$ at $u_n \to \infty$, whereas due to smallness of ε we may consider all u_n to be sufficiently large.

For $z \in D_{n+1}$, $|u| < u_n$ it is obvious that $z/\sqrt{c} \pm u \in D_n$. Thus,

$$\|\bar{L}_{\bar{f}_i} t_n(z; a)\|_{C(D_{n+1})} \leqq 2,1 \|t_n(z; a)\|_{C(D_n)} . \tag{6.11}$$

Let us put $t_{n+1}(z; a) = t_n'(z; a) + t_n''(z; a) + L_{\bar{f}_i} t_n(z; a)$. Having summed the estimates (6.2), (6.3), (6.10), (6.11), we obtain the resulting estimate for $t_{n+1}(z; a)$:

$$\|t_{n+1}(z; a)\|_{C(D_{n+1})} \leqq \varepsilon^3 (1 - 2\varepsilon/3)^{n_{n_i+1}} 3^{n-n_i+1}$$

The same consideration allows to obtain an estimate for the vector-function $V_a t_{n+1}(z; a)$:

$$\|V_a t_{n+1}(z; a)\|_{C(D_{n+1})} \leqq \varepsilon^{5/2} (1 - 3\varepsilon/5)^{n_{i+1}} 3^{n-n_{i+1}} .$$

Now we turn to the function $\bar{h}_{n+1}(z; a) = \chi_{D_{n+1}}(z) L_{\bar{f}_i} h_n(z; a)$.

According to the assumption of v_{12}) $h_n \in H_{\bar{f}_i, N}$. Therefore from the Theorem 3.1 it follows:

$$\begin{aligned} \|\bar{h}_{n+1}(z; a)\|_{L^2(R^1; \exp(-\gamma z^2))} &\leqq \|L_{\bar{f}_i} h_n(z; a)\|_{L^2(R^1; \exp(-\gamma z^2))} \\ &\leqq v_{N+1} \|h_n(z; a)\|_{L^2(R^1; \exp(-\gamma z^2))} . \end{aligned} \tag{6.12}$$

So far as $h_n(z; a) \in H_{\bar{f}_i, N}$ for any $a \in \mathfrak{A}_n$, then

$$\partial h_n(z; a)/\partial a_j \in H_{\bar{f}_i, N}, \quad j = 0, \ldots, k-1 .$$

So, analogously to (6.12), we shall have

$$\|\partial \bar{h}_{n+1}(z; a)/\partial a_j\|_{L^2(R^1; \exp(-\gamma z^2))} \leqq v_{N+1} \|\partial h_n(z; a)/\partial a_j\|_{L^2(R^1; \exp(-\gamma z^2))} .$$

From the inequalities

$$\|L_{\bar{f}_i} h_n(z; a)\|_{C(D_{n+1})} \leqq 2,1 \|h_n(z; a)\|_{C(D_n)}$$
$$\|L_{\bar{f}_i} h_n(z; a)\|_{C^1(D_{n+1})} \leqq K \|h_n(z; a)\|_{C(D_n)}$$

we get directly the following

$$\|\bar{h}_{n+1}(z; a)\|_{C(D_{n+1})} \leqq 2,1 \|h_n(z; a)\|_{C(D_n)} , \tag{6.13}$$

$$\|\bar{h}_{n+1}(z; a)\|_{C^1(D_{n+1})} \leqq K \|h_n(z; a)\|_{C(D_n)} , \tag{6.14}$$

$$\|V_a \bar{h}_{n+1}(z; a)\|_{C(D_{n+1})} \leqq 2,1 \|V_a h_n(z; a)\|_{C(D_n)} , \tag{6.15}$$

$$\|V_a \bar{h}_{n+1}(z; a)\|_{C^1(D_{n+1})} \leqq K \|V_a h_n(z; a)\|_{C(D_n)} . \tag{6.16}$$

Now we have

$$g_{n+1}(z; a) = \sum_{j=0}^{k-1} \lambda_j a_j e_j(z) + \sum_{j=k}^{N} \lambda_j \delta_n^{(j)}(a) e_j(z) + \bar{h}_{n+1}(z; a) + t_{n+1}(z; a) \tag{6.17}$$

where $\lambda_0, \ldots, \lambda_N$ are the eigenvalues of the operator $L_{\bar{f}_i}$. From the invariance of $H_{\bar{f}_i, N}$ it follows that $L_{\bar{f}_i} h_n \in H_{\bar{f}_i, N}$ but, generally speaking, $h_{n+1} \notin H_{\bar{f}_i, N}$.

Thus, we consider the expansion

$$\bar{h}_{n+1}(z; a) = \sum_{j=0}^{N} \delta_{n+1}^{(j)}(a) \chi_{D_{n+1}}(z) e_j(z) + h_{n+1}(z; a) , \tag{6.18}$$

where $h_{n+1} \in H_{\bar{j}_i, N}$. From the Theorem 3.1

$$\|h_{n+1}(z;a)\|_{L^2(R^1;\exp(-\gamma z^2))} \leqq 1, 1 \|\bar{h}_{n+1}(z;a)\|_{L^2(R^1;\exp(-\gamma z^2))}$$

$$\|\nabla_a h_{n+1}(z;a)\|_{L^2(R^1;\exp(-\gamma z^2))} \leqq 1, 1 \|\nabla_a \bar{h}_{n+1}(z;a)\|_{L^2(R^1;\exp(-\gamma z^2))}$$

$$|\bar{\delta}_{n+1}^{(j)}(a)| \leqq 1, 1 \|\bar{h}_{n+1}(z;a)\|_{L^2(R^1;\exp(-\gamma z^2))}$$

$$|\nabla_a \bar{\delta}_{n+1}^{(j)}(a)| \leqq 1, 1 \|\nabla_a \bar{h}_{n+1}(z;a)\|_{L^2(R^1;\exp(-\gamma z^2))} .$$

Let us introduce the mapping $\psi_n : \mathfrak{A}_n \to R^k$, putting

$$\psi_n(a_0, \ldots, a_{k-1}) = (\lambda_0 a_0 + \bar{\delta}_{n+1}^{(0)}(a), \ldots, \lambda_{k-1} a_{k-1} + \delta_{n+1}^{(k-1)}(a)) .$$

From the last estimates, from (6.13), (6.15) and from the condition $v_{12})$

$$\|\psi_n - \psi^{(0)}\|_C \leqq \varepsilon^{7/3}(1-2\varepsilon/3)^n \tag{6.19}$$

$$\|\psi_n - \psi^{(0)}\|_{C^1} \leqq \varepsilon^{7/3}(1-2\varepsilon/3)^n \tag{6.20}$$

where $\psi^{(0)}(a_0, \ldots, a_{k-1}) = (\lambda_0 a_0, \ldots, \lambda_{k-1} a_{k-1})$. Let us insert expansion (6.18) into (6.17) and denote the vector $\psi(a)$ by a:

$$g_{n+1}(z;a) = \sum_{j=0}^{k-1} a_j e_j(z) + \sum_{j=k}^{N} (\lambda_j \delta_n^{(j)}(\psi^{-1}(a)) + \bar{\delta}_{n+1}^{(j)}(\psi^{-1}(a)) e_j(z)$$
$$+ h_{n+1}(z;\psi^{-1}(a)) + t_{n+1}(z;\psi^{-1}(a)) .$$

From (6.19) it follows that $\psi(\mathfrak{A}_n) \supset \mathfrak{A}_{n+1}$. Let us put

$$\delta_{n+1}^{(j)}(a) = \lambda_j \delta_n^{(j)}(\psi^{-1}(a)) + \bar{\delta}_{n+1}^{(j)}(\psi^{-1}(a)),\ h_{n+1}(z;a)$$
$$= h_{n+1}(z;\psi^{-1}(a)),\ t_{n+1}(z;a) = t_{n+1}(z;\psi^{-1}(a)) .$$

Lemma 2 is proved.

§ 7. Proof of Lemma 3

In this section we assume $n = n_{i+1}$

$$\bar{R}_n(z;a) = \sum_{j=k+1}^{N} \delta_n^{(j)}(a) e_j(z) + h_n(z;a) + t_n(z;a) .$$

We shall estimate firstly $|\bar{R}_n(z;a)|$ for $z \in D_n = [-d_n, d_n]$. From the condition $v_{11})$ and the Theorem 3.1 we obtain the estimate

$$|\delta_n^{(j)}(a) e_j(z)| < v_j^{n-n_i} \varepsilon^{8/3}(1-2\varepsilon/3)^{n_i} d_n^{2j+1} .$$

Hence

$$\sum_{j=k+1}^{N} |\delta_n^{(j)}(a) e_j(z)| \leqq \varepsilon^{8/3}(1-2\varepsilon/3)^{n_i}(1-3\varepsilon/4)^{n-n_i} \sum_{j=k+1}^{N} d_n^{2j+1}(v_j/(1-3\varepsilon/4))^{n-n_i} \tag{7.1}$$

Let us show that for $j > k$

$$(v_j/(1-3\varepsilon/4))^{n-n_i} d_n^{2j+1} < c^{-\frac{1}{8}(j-k)n_0} . \tag{7.2}$$

Indeed:

$$v_j/(1-3\varepsilon/4) \leqq c^{-2(j-k)/3}/(1-3\varepsilon/4) \leqq c^{-5(j-k)/8},\ n - n_i = [(\omega-1)n_i] + n_0 .$$

The left-hand side of Eq. (7.2) does not thus exceed

$$c^{-(5/8)(j-k)([(\omega-1)n_i]+n_0)}d_n^{2j+1} \leqq c^{-\frac{1}{2}(j-k)n_0}c^{-(1/8)(j-k)([(\omega-1)n_i]+n_0)}d_n^{2j+1}.$$

In order to prove the inequality (7.2) it is sufficient to show, that

$$c^{-(1/8)(j-k)([(\omega-1)n_i]+n_0)}d_n^{2N+1}<1.$$

From the definition of the numbers $n_0, n_1, \ldots, \omega, d_n$ we have

$$[(\omega-1)n_i] \geqq (\omega-1)n_i-1 \geqq (\omega-1)n/\omega-1-(\omega-1)n_0/\omega,$$

$$c^{-(1/8)(\omega-1)/\omega}=(5^{\omega-1})^{(\log_5 c)/(8\omega)} \leqq (1-\varepsilon)^{\lambda}, \qquad \lambda=(\log_5 c)/320,$$

$$c^{-n_0/8} \leqq \varepsilon^{\mu}, \qquad \mu=(\log_5 c)/50>\lambda,$$

$$d_n=4/(1-c^{-\frac{1}{4}})\sqrt{\ln\varepsilon^{-1}+n\ln(1-\varepsilon)^{-1}}=\text{const}\sqrt{\ln\varepsilon^{-\lambda}+n\ln(1-\varepsilon)^{-\lambda}}.$$

Consequently, if we denote $x=\varepsilon^{\lambda}(1-\varepsilon)^{n\lambda}$, then

$$c^{-(1/8)(j-k)([(\omega-1)n_i]+n_0)}d_n^{2N+1}<\text{const}\, x\ln^{(2N+1)/2}x \underset{\varepsilon\to 0}{\longrightarrow} 0$$

the fact that should have been shown. Turning to the inequality (7.1) we can see that

$$\begin{aligned}\textstyle\sum_{j=k+1}^{N}|\delta_n^{(j)}(a)e_j(z)| &< \varepsilon^{8/3}(1-2\varepsilon/3)^{n_i}(1-3\varepsilon/4)^{n-n_i}\textstyle\sum_{j=k+1}^{N}c^{-\frac{1}{2}(j-k)n_0}\\ &<\varepsilon^{8/3}(1-2\varepsilon/3)^{n_i}(1-3\varepsilon/4)^{n-n_i}c^{-\frac{1}{2}n_0}/(1-c^{-\frac{1}{2}n_0})\\ &<\varepsilon^{8/3+\log_5 c}(1-2\varepsilon/3)^{n_i}(1-3\varepsilon/4)^{n-n_i},\end{aligned} \tag{7.3}$$

since $c^{-\frac{1}{2}n_0}=5^{-\frac{n_0}{2}\log_5 c}<\varepsilon^{\log_5 c}$ due to the choice of n_0.

Let us estimate now the other terms entering into $\bar{R}_n(z;a)$. From the condition v_{15}) we have $|t_n(z;a)|<\varepsilon^{7/2}(1-2\varepsilon/3)^n$. From v_{12})

$$\begin{aligned}\|h_n(z;a)\|_{L^2(R^1)} &\leqq \exp(\gamma d_n^2)\|h_n(z;a)\|_{L^2(R^1;\exp(-\gamma z^2))}\\ &\leqq e^{\gamma d_n^2}c^{-(2/3)(N-k)(n-n_i)}\varepsilon^{8/3}(1-2\varepsilon/3)^{n_i}=S.\end{aligned}$$

It may be shown now that $S<\varepsilon^3(1-2\varepsilon/3)^n$. The idea of proving consists in the fact that by choosing the number N sufficiently large the increase of $e^{+\gamma d_n^2}$ will be compensated by the decrease of the value $c^{-(2/3)(N-k)(n-n_i)}$. We have

$$\begin{aligned}n-n_i=[(\omega-1)n_i]+n_0 &\geqq (\omega-1)n_i+n_0-1 \geqq (\omega-1/\omega)(n-n_0+1)\\ &+n_0-1 \geqq ((\omega-1)/\omega)n+n_0/\omega-2,\end{aligned}$$

$$c^{-(2/3)(\omega-1)/\omega}=(5^{\omega-1})^{2(\log_5 c)/(3\omega)} \leqq (1-\varepsilon)^{\lambda}, \quad \lambda=(\log_5 c)/(60\omega), \quad c^{-2n_0/3\omega}=\varepsilon^{\mu},$$

$$\mu=(\log_5 c)/(72\omega), \quad e^{\gamma d_n^2}=[(1-\varepsilon)^{-n}\varepsilon^{-1}]^{\nu}, \quad \nu=16(\sqrt{c}+1)/(\sqrt{c}-1).$$

Hence

$$S \leqq \varepsilon^{-\nu+(N-k)\mu+8/3}(1-\varepsilon)^{n(-\nu+(N-k)\lambda)}(1-2\varepsilon/3)^{n_i}c^2.$$

The required inequality for S is obtained, provided $(N-k)\mu$, $(N-k)\lambda \geqq \nu+1$.

Let us estimate now $|h_n(z;a)|$ for $z\in D_n$. The discussion presented below has been already employed in [2]. Suppose $0\leqq z\leqq d_n$ and

$$\Pi(u)=\begin{cases}0, & u\notin[0,-2]\\ 1+\frac{1}{2}u, & u\in[0,-2].\end{cases}$$

From v_{13})

$$|h_n(u;a)|>|h_n(z;a)|\Pi(u-z)$$

and, therefore:

$$|h_n(z;a)|<\|h_n(u;a)\|_{L^2(R^1)}/\|\Pi(u)\|_{L^2(R^1)}<\|h_n(u;a)\|_{L^2(R^1)}<\varepsilon^3(1-2\varepsilon/3)^n. \tag{7.4}$$

Performing the summation of the estimates (7.3), (7.4), we obtain

$$|\bar{R}_n(z;a)|<\varepsilon^{8/3+\frac{1}{2}\log_5 c}(1-2\varepsilon/3)^n.$$

Analogously the following inequality may be proved

$$|\nabla_a\bar{R}_n(z;a)|<\varepsilon^{5/2+\frac{1}{2}\log_5 c}(1-5\varepsilon/8)^n.$$

Let us turn now directly to proving the conditions (U_n). We have

$$g_n(z;a)=\textstyle\sum_{j=0}^{k-1}a_je_j(z)+\delta_n^{(k)}(a)e_k(z)+\bar{R}_n(z;a). \tag{7.5}$$

It should be recalled that here e_j are the eigenfunctions of the operator $L_{\bar{f}_i}$. We verify first that the function $f_{n_{i+1}}(z;0)=f_n(z;0)$ satisfies all the conditions of the Theorem 3.1. Suppose $b^{(j)}=\varphi_j^{-1}(\varphi_{j+1}^{-1}(\dots\varphi_{n-1}^{-1}(0)\dots))$, $j=n_i,\dots,n-1$ where $\varphi_l:\mathfrak{A}_l\to\mathfrak{A}_{l+1}$ are the mappings constructed in Lemma 2. Then from v_{14}) for $z\in D_n$ we have:

$$|f_{j+1}(z;b^{(j+1)})-f_j(z;b^{(j)})|=|g_j(z;b^{(j)})|<\varepsilon^{7/3}(1-2\varepsilon/3)^n3^{j-n}.$$

Whence

$$|f_n(z;0)-f_{n_i}(z;b^{(n_i)})|<2\varepsilon^{7/3}(1-2\varepsilon/3)^n. \tag{7.6}$$

Then, so far as $\|f_{n_i}(z;a^{(0)})-(1-\varepsilon G_{2k}(z;\gamma))\|_{C^1(D_0)}=O(\varepsilon^{4/3})$ the analogous equality is valid for $f_n(z;0)$ also. Thus, in the segment D_0 function f_n satisfies the condition of the theorem. From the conditions (V_n) it also follows that it satisfies the conditions of the Theorem 3.1 outside D_0 too. Consequently, Theorem 3.1 is applicable, and we may introduce the eigenfunctions $e_j(z;\bar{f}_{i+1})$, $\bar{f}_{i+1}=f_{n_{i+1}}(z;0)$, $j=0,1,\dots,N$. From (7.5), v_2) and u_{23})

$$\|f_n(z;0)-f_{n_i}(z;b^{(n_i)})\|_{L^2(R^1;\exp(-\gamma z^2))}<(4/\sqrt{\gamma})\varepsilon^{7/3}(1-2\varepsilon/3)^n$$

$$\|f_{n_i}(z;b^{(n_i)})-f_{n_i}(z;a^{(0)})\|_{L^2(R^1;\exp(-\gamma z^2))}\leqq\sup_a|\nabla_af_{n_i}(z;a)|\cdot|b^{(n_i)}-a^{(0)}|<\varepsilon^3(1-2\varepsilon/3)^n$$

where $a^{(0)}$ is introduced in u_1). Thus,

$$\|f_n(z;0)-f_{n_i}(z;a^{(0)})\|_{L^2(R^1;\exp(-\gamma z^2))}<(5/\sqrt{\gamma})\varepsilon^{7/3}(1-2\varepsilon/3)^n.$$

From this inequality, using the consideration of the Theorem 3.1, it is easy to derive the estimates

$$\|e_j(z;\bar{f}_{i+1})-e_j(z;\bar{f}_i)\|_{L^2(R^1;\exp(-\gamma z^2))}<\varepsilon^{9/4}(1-2\varepsilon/3)^n$$

for $j=0,1,\dots,N$. From here

$$e_j(z;\bar{f}_i)=\sum_{m=0}^{k}(\delta_m^j+c_{jm})e_m(z;\bar{f}_{i+1})+\tilde{R}_n(z;a),$$

where δ_m^j is the Kronecker symbol,

$$|c_{jm}|,\quad \|\tilde{R}_n(z;a)\|_{C(D_n)}<\varepsilon^{11/5}(1-2\varepsilon/3)^n,\quad \tilde{R}_n(z;a)\in H_{\bar{f}_{i+1},k}.$$

Inserting this expansion into the equality (7.3), and performing the substitution of the variables similar to the identical one in the space of the parameters $a=(a_0,\dots,a_{k-1})$ we obtain the condition (U_n) at $n=n_{i+1}$. Lemma 3 is proved.

§ 8. Derivation of Formulas for Indices

In papers [2, 8] there have been obtained results concerning the indices of the asymptotic hierarchical models under the condition $\sqrt{r}<c<r$. As it will be seen in what follows the cases $\sqrt{r}<c<r$ and $c=\sqrt{r}-\varepsilon$ differ essentially. For the sake of simplicity we consider the case $r=2$.

The values of the critical indices we derive by studying the asymptotic behaviour of the recursive relations (1.2) when $n\to\infty$. Function $f_n(z;\beta)$ in (1.2) is defined on the discrete finite lattice of points $M_n=\{c^{n/2}(-1+i/2^{n-1})\}_{i=0}^{2^n}$ with the step $\Delta_n=2(\sqrt{c}/2)^n$, since $\sum_{x\in V_n}u(x)$ is an even number, which does not exceed 2^n in modulus. The summation is carried out in (1.2) so, that $z/\sqrt{c}\pm u\in M_n$.

As in the papers [2] and [8] we obtain the critical indices for a.h.m., their initial distribution $f_{n_0}(z;\beta)$ satisfying some relations of the inequality type for a sufficiently large value of n_0. These inequalities determine the open set Ω which is deliberately non-empty in the space of all a.h.m. In this way we show that the branch g_{c_1} is thermodynamically stable.

Let us suppose c be fixed, and $\sqrt{2}-c>0$ is small. Let $f^{(0)}(z;\beta)\not\equiv\text{const}$ be the solution of Eq. (1.3) constructed in the Theorem 1, and $e_i(z;\beta)$ are the eigenfunctions of the operator $L_{g^{(0)}}$ (see the Theorem 3.1) with the eigen numbers λ_i. The eigenfunctions are considered to be normalized by the condition

$$\|e_i(z;\beta)\|^{-1}_{L^2(R^1;\exp(-\gamma z^2)}=(\gamma^{\frac{1}{2}}/\pi^{\frac{1}{4}})\cdot((2j)!^{\frac{1}{2}}/2^j)\cdot(2\gamma^{\frac{1}{2}})^{2j}/(2j)!,$$

so that $e_i(z;\beta)$ for $z\sim\sqrt{\ln\varepsilon^{-1}}$ has the asymptotics z^{2j}. The set Ω consists of the families of the probability distributions $f_{n_0}(z;\beta)=\exp(-a_0(\beta)z^2)p_{n_0}(z;\beta)$ depending on β, which satisfy the following conditions (the number n_0 is large, it is enough for it to exceed $10^7\varepsilon^{-2}$):

The condition (U). There exists a segment of inverse temperatures $[\beta^-,\beta^+]$ and C^1-function $b(\beta)$ defined on $[\beta^-,\beta^+]$, such that

$$p_{n_0}(z;\beta)=p^{(0)}(z;\beta)+\lambda_1^{n_0}b(\beta)e_1(z;p^{(0)})+R(z;\beta) \tag{8.1}$$

and in this case

u$_1$) $\lambda_1^{n_0} b(\beta^\pm) = \mp\varepsilon^{3/2}$; $\quad |b'(\beta)| > \sqrt{\varepsilon}$ for $\beta \in [\beta^-, \beta^+]$;

u$_2$) $|R(z;\beta)| + |\partial R(z;\beta)/\partial\beta| + |\partial R(z;\beta)/\partial z| < \varepsilon^2$

for $|z| < 4\beta\sqrt{\ln \varepsilon^{-1}}$, $\beta \in [\beta^-, \beta^+]$;

u$_3$) $0 < p_{n_0}(z;\beta) < \exp(-(\beta\varepsilon'/2)|z|^\alpha)$;

$$|\partial p_{n_0}(z;\beta)/\partial z| < |z|^4 \exp(-(\beta\varepsilon'/2)|z|^\alpha);$$

$$|\partial p_{n_0}(z;\beta)/\partial \beta| < |z|^5 \exp(-(\beta\varepsilon'/2)|z|^\alpha);$$

$$|\partial^2 p_{n_0}(z;\beta)/\partial z\partial \beta| < |z|^9 \exp(-(\beta\varepsilon'/2)|z|^\alpha)$$

for $|z| > 4\beta\sqrt{\ln(1/\varepsilon)}$, $\quad \beta \in [\beta^-, \beta^+]$.

Theorem 8.1. *Suppose the value $\sqrt{2} - c$ is sufficiently small, and the condition (U) is fulfilled. Then in the segment $[\beta^-, \beta^+]$ there is one and only one critical point β_{cr}, for which $f_{n_0}(z;\beta_{cr}) \Rightarrow f_c^{(0)}(z;\beta_{cr})$.*

Note. It follows from Theorem 8.1 that the value of the critical index $\eta = 0$, provided the dimension of the model is $d_a = \dfrac{2}{\log_2 2/c}$. ┘

Theorem 8.2. *Suppose the value $\sqrt{2} - c$ is sufficiently small, the condition (U) is fulfilled, and $\beta \in [\beta_+, \beta_{cr})$. Then*

$$(c/2)^n f_n(z(c/2)^{n/2};\beta) \Rightarrow (2\pi\sigma_1(\beta))^{-\frac{1}{2}} \exp(-(\sigma_1(\beta)/2)z^2).$$

For $\beta \to \beta_{cr}$ asymptotically $\sigma_1(\beta) \sim |\beta_{cr} - \beta|^{-\gamma}$, $\gamma = 1 - \log_{\lambda_1}(c\lambda_1/2)$. ┘

Theorem 8.3. *Suppose the value $\sqrt{2} - c$ is sufficiently small, the condition (U) is fulfilled, and $\beta \in [\beta_-, \beta_{cr})$. Then there exists a sequence of the numbers $0 < M_1(\beta) < M_2(\beta) < \ldots$, $\lim_{n\to\infty} M_n(\beta) = M(\beta)$ such, that $(c/2)^{n/2} f_n(z(c/2)^{n/2};\beta) - G_n(z;\beta) \Rightarrow 0$, where*

$$G_n(z;\beta) = \tfrac{1}{2}(2\pi\sigma_2(\beta))^{-\frac{1}{2}}(\exp(-(z - 2^{n/2}M_n(\beta))^2/2\sigma_2(\beta))$$
$$+ \exp(-(z + 2^{n/2}M_n(\beta))^2/2\sigma_2(\beta))).$$

For $\beta \to \beta_{cr}$ the asymptotical formulas

$$M(\beta) \sim |\beta - \beta_{cr}|^\omega, \quad \omega = \tfrac{1}{2}\log_{\lambda_1} c; \quad \sigma_2(\beta) \sim |\beta - \beta_{cr}|^\gamma, \quad \gamma = 1 - \log_{\lambda_1}(c\lambda_1/2)$$

are valid.

Refinement of Theorem 8.3 (calculation of the correlation radius). In the assumptions of the Theorem 8.3 there is a number $N = N(\beta)$ such that for $n < N$ the condition $|(c/2)^n f_n(z(c/2)^n;\beta) - f_0(z;\beta)| < \varepsilon^{2/3}$ is fulfilled, and for $n > N$ the condition $|(c/2)^n f_n(z(c/2)^n;\beta) - G_n(z;\beta)| < \varepsilon^{2/3}$ is fulfilled; the value $\xi = \xi(\beta) = 2^{N(\beta)/d_a}$ is the correlation radius, and for $\beta \to \beta_{cr}$ $\xi(\beta) \sim |\beta - \beta_{cr}|^{-\nu}$, $\nu = \frac{1}{2}\log_{\lambda_1}(2/c)$.

Let us consider the Gibbs distribution in the volume V_n at the external field value H and at the inverse temperature β and put

$$f_n(z;\beta,H) = ((2/\sqrt{c})^n/\Xi_n) \sum_{(\sqrt{c}/2)^n \sum_{x\in V_n}\sigma(x) = z} \exp(-\beta H_n(\sigma) + H \textstyle\sum_{x\in V_n}\sigma(x)). \quad (8.2)$$

Theorem 8.4. *Suppose the value* $\sqrt{2}-c$ *is sufficiently small, the condition* (*U*) *is fulfilled,* $\beta\in[\beta^-,\beta^+]$, *and* $|H|<\varepsilon|\beta^+-\beta^-|$, $H\neq 0$. *Then there is a sequence of numbers* $M_n(\beta,H)\xrightarrow[n\to\infty]{} M(\beta,H)$ *such that*

$$(c/2)^n f_n(z(c/2)^n;\beta)-(2\pi\sigma(\beta,H))^{-\frac{1}{2}}\exp(-(z+2^{n/2}M_n(\beta,H))^2/2\sigma(\beta,H))\Rightarrow 0\,.$$

Theorem 8.5. *In the assumptions of Theorem 8.4 we have:*

a) $M(\beta,H)$ *is the monotonously increasing function of* H;

b) $\lim_{H\to\pm 0} M(\beta,H)=\pm M(\beta)$ *(see Theorem 8.3);*

c) *function* $H=H(\beta,M)$, *which is an inverse one to the function* $M(\beta,H)$, *permits for* β, $M\to 0$ *the expansions;*

c_1) *in the region* $|M|/|\tau|^{\omega}>\ln\varepsilon^{-1}$, $\tau=(\beta-\beta_{cr})/\beta_{cr}$, $\omega=\frac{1}{2}\log_{\lambda_1}c$

$$H=(L_1(\beta_{cr}-\beta)|M|^{\delta_1}+L_2|M|^{\delta_2}+\ldots)\mathrm{sgn}M\,, \tag{8.3}$$

where $\delta_1=1-2\log_c(c\lambda_1/2)$, $\delta_2=3+2\log_c(2/c^2)$, $L_1>0$, $L_2>0$ *are constants,* ... *are the terms of higher order in the expansions; for* $c\to\sqrt{2}$ $L_1\sim 1$, $L_2\sim(\sqrt{2}-c)$;

c_2) *in the region* $|M|/|\tau|^{\omega}<(\ln\varepsilon^{-1})^{-1}$, $\beta<\beta_{cr}$

$$H=L_3(\beta_{cr}-\beta)^{-\gamma}M+\ldots \tag{8.4}$$

where $\gamma=1-\log_{\lambda_1}(c\lambda_1/2)$, L_3 *is a constant; for* $c\to\sqrt{2}$ $L_3\sim 1$.

Note. The presence of two asymptotical expansions in different regions of the equation of state $H(\beta,M)$ in the neighbourhood of the critical point is a very important phenomenon. It shows the type of the expansion $H(\beta,M)$, when the Landau theory cannot be applied.

The Theorem 8.1 is derived in the same way as the proof of the basic theorem (see also [2]), and we shall omit it. The proofs of the remaining theorems also involve essentially the technique of paper [8]. We present two lemmas without proof which elucidate the derivation of Theorem 8.3. These lemmas are proved analogously to the corresponding lemmas in paper [8], and we shall omit it too. Let us denote $N=N(\beta)=\min\{n:\lambda_1^n b'(\beta_{cr})|\cdot|\beta-\beta_{cr}|>(4/5)\beta_{cr}/(2/c-1)\}$, $\varepsilon_0(u)=\varepsilon^{1/3}$ for $\sqrt{\ln(1/\varepsilon)}<|u|\leq\varepsilon^{-2/3}$, $\varepsilon_0(u)=\varepsilon^{1/3}|\varepsilon^{2/3}u|^{-1,5}$ for $\varepsilon^{-2/3}<|u|\leq c^{(n-n_0)/2}\sqrt{\ln\varepsilon^{-1}}$.

Lemma 8.1. *Suppose* $n_0<n<N(\beta)$, $\sqrt{\ln\varepsilon^{-1}}\leq|z^{(0)}|\leq c^{(n-n_0)/2}\sqrt{\ln\varepsilon^{-1}}$. *Then there exist the numbers* $L_n=L_n(\beta,z^{(0)})$, $\mu_n=\mu_n(\beta,z^{(0)})$, $s_n=s_n(\beta,z^{(0)})$ *independent on* z, *such that*

$$f_n(z;\beta)=L_n\exp(-\mu_n(z-s_n)^2)(1+R_n(z))\,, \tag{8.5}$$

where $|R_n(z)|=|R_n(z;z^{(0)},\beta)|<\varepsilon_0(z_0)$ *for* $|z-z^{(0)}|<\sqrt{(1/\mu_n)\ln(z^{(0)}/\varepsilon)}$ *and the recursive relations are fulfilled*

$$\mu_{n+1}(\beta,\sqrt{c}z^{(0)})=(a_0(\beta)+(2/c)(\mu_n(\beta,z^{(0)})-a_0(\beta)))(1+O(\varepsilon_0(z^{(0)})))\,, \tag{8.6}$$

$$s_{n+1}(\beta,\sqrt{c}z^{(0)})=(2/\sqrt{c})(\mu_n(\beta,z^{(0)})/\mu_{n+1}(\beta,\sqrt{c}z^{(0)}))s_n(\beta,z^{(0)})(1+O(\varepsilon_0(z^{(0)})))\,. \tag{8.7}$$

Moreover for $\sqrt{\ln\varepsilon^{-1}}<|z^{(0)}|<2\sqrt{\ln\varepsilon^{-1}}$

$$\mu_n(\beta,z^{(0)})=(a_0(\beta)-L_1\lambda_1^n(\beta-\beta_{cr})/\beta_{cr}+L_2|z^{(0)}|^2)(1+O(|z^{(0)}|^{-1}))\,, \tag{8.8}$$

$$s_n(\beta,z^{(0)})=(L_3|z^{(0)}|^3/\mu_n(\beta,z^{(0)}))(1+O(|z^{(0)}|^{-1}))\,, \tag{8.9}$$

where $L_1, L_2, L_3>0$ *are independent on* β *and* $z^{(0)}$, *and for* $\varepsilon\to 0$ $L_1\sim 1$, $L_2, L_3\sim\varepsilon$, $L_2L_3^{-1}\underset{\varepsilon\to 0}{\longrightarrow}\frac{3}{2}$. *And finally, for* $z^{(0)}=c^{(n-n_0)/2}\sqrt{\ln\varepsilon^{-1}}$ *and* $z>z^{(0)}+\sqrt{(1/\mu_n)\ln(z^{(0)}/\varepsilon)}$ $(1+R_n(z))<1$.

Lemma 8.2. *For* $\beta\in[\beta^-,\beta_{cr}]$ *there is a sequence of the numbers* $z_{N(\beta)+1}(\beta)<z_{N(\beta)+2}(\beta)<\ldots$, $\varepsilon^{-0.49}<z_{N(\beta)+1}<\varepsilon^{-0.51}$ *such that all the statements of the Lemma 8.1 are valid for* $n>N(\beta)$ *for the points* $z^{(0)}$, $|z^{(0)}|>u_n(\beta)=z_n(\beta)-\sqrt{\mu_n^{-1}(\beta,z_n(\beta))\ln z_n(\beta)}$ *and* $t=z_n(\beta)$ *is the solution of the equation* $s_n(\beta,t)=t$. *Besides, for* $|z|<u_n(\beta)$ $f_n(z;\beta)<2f_n(u_n(\beta);\beta)$. *For* $n\to\infty$ *there exists the limit* $c^{-n/2}z_n(\beta)\to M(\beta)$. ⌟

Let us elucidate the derivation of the critical index, connected with magnetization, and the equation of state in the vicinity of the critical point.

Suppose $n>N(\beta)$. Using Lemma 8.2 we may show [8], that

$$\|f_n(z;\beta)-G_n(z;z_n,\mu_n)\|_{C(R^1)}<\varepsilon_0(z_n),$$

where

$$G_n(z;z_n,\mu_n)=L_n[\exp(-\mu_n(z-z_n)^2)+\exp(-\mu_n(z+z_n)^2)],$$

$z_n=z_n(\beta)$ is the solution of the equation $s_n(\beta,t)=t$, $\mu_n=\mu_n(z_n(\beta))$. Therefore, the spontaneous magnetization is determined by the formula $M(\beta)=\lim_{n\to\infty}c^{-n/2}z_n(\beta)$. Denote $M_n(\beta)=z_n(\beta)/c^{n/2}$ and consider such m, that $\sqrt{\ln\varepsilon^{-1}}<c^{m/2}M_n(\beta)<2\sqrt{\ln\varepsilon^{-1}}$. It may be shown that $m<N(\beta)$ [8]. From the asymptotical formulas (8.6), (8.7) it follows that

$$s_n(\beta,z_n(\beta))=(2/\sqrt{c})^{n-m}(\mu_m(\beta,z^{(0)})/\mu_n(\beta,z_n(\beta)))s_m(\beta,z^{(0)})(1+O(\varepsilon^{1/3})),\tag{8.10}$$

$$\mu_n(\beta,z_n(\beta))=a_0(\beta)+(2/c)^{n-m}(\mu_m(\beta,z^{(0)})-a_0(\beta))(1+O(\varepsilon^{1/3})).\tag{8.11}$$

$$z^{(0)}=c^{m/2}M_n(\beta).$$

So far as $\sqrt{\ln\varepsilon^{-1}}<z^{(0)}<2\sqrt{\ln\varepsilon^{-1}}$, we may use the formulas (8.8) and (8.9). The errors in the relations (8.8)–(8.11) can be neglected. Then for $M_n(\beta)$ we derive the equation:

$$s_n=z_n,$$

$$\mu_ns_n=\mu_nz_n,$$

$$(2/\sqrt{c})^{n-m}\mu_ms_m=(a_0+(2/c)^{n-m}(\mu_m-a_0))c^{(n-m)/2}z^{(0)},$$

$$\mu_ms_m=(\mu_m-a_0)z^{(0)},$$

[the term $(\sqrt{c/4})^{n-m}a_0\underset{n\to\infty}{\longrightarrow}0$, which is inessential in deriving the asymptotics, may be omitted]

$$L_3(z^{(0)})^3=(-L_1\lambda_1^m\tau+L_2(z^{(0)})^2)z^{(0)},$$

$$L_3M_n^2=-L_1(\lambda_1/c)^m\tau+L_2M_n^2,\tag{8.12}$$

$$M_n^2=L_4(\lambda_1/c)^m\tau,$$

where $\tau=(\beta-\beta_{cr})/\beta$, $L_4=(L_2-L_3)/L_1>0$, since, as has been formulated in Lemma 8.1, $L_2L_3^{-1}\to\frac{3}{2}$ for $\varepsilon\to 0$. Further $\sqrt{\ln\varepsilon^{-1}}<c^{m/2}M_n<2\sqrt{\ln\varepsilon^{-1}}$, whence it follows that

$$(\lambda_1/c)^m=L_5M_n^{-2\log_c(\lambda_1/c)}, \qquad (\sqrt{\ln\varepsilon^{-1}})^{2\log_c(\lambda_1/c)}<L_5<(2\sqrt{\ln\varepsilon^{-1}})^{2\log_c(\lambda_1/c)}.$$

Thus,

$$\begin{aligned} &M_n^2=L_6M_n^{-2\log_c(\lambda_1/c)}\tau, \qquad L_6=L_4\cdot L_5,\\ &M_n=\tau^{[2(1+\log_c(\lambda_1/c))]^{-1}}. \end{aligned} \tag{8.13}$$

As a result we have found the critical index $\beta=[2(1+\log_c(\lambda_1/c)]^{-1}$. The neglecting of errors in the formulas (8.8)–(8.11) is substantiated as it has been done in the paper [8].

Let us derive now the equation of state. Suppose $f_n(z;\beta,H)$ is the density of the distribution of the random value $(\sqrt{c/2})^n\sum_{x\in V_n}u(x)$ in the Gibbs ensemble at the inverse temperature β, and at the external field H. It may be easily seen that

$$f_n(z;\beta,H)=L_n\exp(\beta H(2/\sqrt{c})^n)f_n(z;\beta,0).$$

For large values of n the function, as may be derived from the Lemmas 8.1, 8.2 [8] is close to the Gaussian density with the average $z_n=z_n(\beta,H)$ satisfying the equation

$$s_n(z_n)+(2c^{-\frac{1}{2}})^n\beta H/(2\mu_n(z_n))=z_n.$$

Let us denote $M_n=M_n(\beta,H)=c^{-n/2}z_n(\beta,H)$. It is clear that $M_n(\beta,H)$ is an odd function of M, thus, we may consider $H>0$.

Let us consider such m that $\sqrt{\ln\varepsilon^{-1}}<z^{(0)}<2\sqrt{\ln\varepsilon^{-1}}$, $z^{(0)}=M_nc^{m/2}$. It is easy to show that $m<N(\beta)$ for $\beta>\beta_{cr}$ [8]. Then analogously to (8.13) we obtain the equation

$$H=(L_7\tau M^{1-2\log_c(c\lambda_1/2)}+L_8M^{3+2\log_c(2/c^2)})\operatorname{sgn}M. \tag{8.14}$$

$\tau=(\beta-\beta_{cr})/\beta_{cr}$, L_7, $L_8>0$ are the constants, which gives the asymptotics of the equation of state in the neighbourhood of the critical point in the region $\beta\geqq\beta_{cr}$, $|M|\geqq M(\beta)$.

At $\beta<\beta_{cr}$ the asymptotics (8.14) holds true, provided $|M|c^{N(\beta)/2}>\sqrt{\ln\varepsilon^{-1}}$. Since

$$N(\beta)=\min\{n\cdot|\lambda_1^n b'(\beta)\tau|\}>(4/5)(2/c-1)^{-1},$$

this condition is equivalent to the following one

$$|M|\,|\tau|^{-\frac{1}{2}\log_{\lambda_1}c}<(\ln\varepsilon^{-1})^{-1}.$$

In fulfilling this condition the number m, determined from the condition $\sqrt{\ln\varepsilon^{-1}}<Mc^{m/2}<2\sqrt{\ln\varepsilon^{-1}}$, is less than $N(\beta)$, and therefore the asymptotics (8.14) takes place.

If the following condition is fulfilled

$$|M|/(|\tau|^{\frac{1}{2}}\log_{\lambda_1}c)<(\ln\varepsilon^{-1})^{-1}$$

then $m \gg N(\beta)$, and

$$f_m(z;\beta)=L_m\exp(-\mu_m z^2)(1+O(\varepsilon/2^{(m-N(\beta))/2}))\,.$$

Thus, in this case the derivation of the equation state is reduced to the case of the Gaussian fixed point studied in [8]. The asymptotics of the equation of state in this case can be given by:

$$H=\text{const.}|\tau|^{-\gamma}M\,,$$

where $\gamma=1-\log_{\lambda_1}(c\lambda_1/2)$ is the critical index calculated in Theorem 2.

Appendix 1
Calculation of the Number

$$a=\int_{-\infty}^{\infty}e^{-\gamma z^2}G_{2k}(\sqrt{\gamma}z)\pi^{-\frac{1}{2}}\int_{-\infty}^{\infty}e^{-u^2}G_{2k}(\sqrt{\gamma/c}z-\sqrt{\gamma}u)G_{2k}(\sqrt{\gamma/c}z+\sqrt{\gamma}u)du dz\,,$$

$$\gamma=1-c^{-1}\,.$$

Let us make a substitution of $z=t\gamma^{-\frac{1}{2}}$:

$$a=\int_{-\infty}^{\infty}e^{-t^2}G_{2k}(t)\Phi(t)dt\,,$$

where

$$\Phi(t)=(\pi\gamma)^{-\frac{1}{2}}\int_{-\infty}^{\infty}e^{-u^2}G_{2k}(t/\sqrt{c}-\sqrt{\gamma}u)G_{2k}(t/\sqrt{c}+\sqrt{\gamma}u)du\,.$$

Using the equality $e^{-t^2}G_{2k}(t)=(\pi^{\frac{1}{4}}2^k(2k)!^{\frac{1}{2}})^{-1}(d^{2k}e^{-t^2}/dt^{2k})$ and integrating by parts, we obtain

$$a=(\pi^{\frac{1}{4}}2^k(2k)!^{\frac{1}{2}})^{-1}\int_{-\infty}^{\infty}e^{-t^2}(d^{2k}\Phi(t)/dt^{2k})dt=(\pi^{\frac{1}{4}}2^k(2k)!^{\frac{1}{2}}(\pi\gamma)^{\frac{1}{2}})^{-1}\int_{-\infty}^{\infty}\int_{-\infty}^{\infty}e^{-t^2-u^2}$$
$$(d^{2k}/dt^{2k})\,[G_{2k}(t/\sqrt{c}-\sqrt{\gamma}u)G_{2k}(t/\sqrt{c}+\sqrt{\gamma}u)]du dt\,.$$

Lemma. $\int_{-\infty}^{\infty}\int_{-\infty}^{\infty}e^{-t^2-u^2}G_j(t/\sqrt{c}+\sqrt{\gamma}u)G_i(t/\sqrt{c}-\sqrt{\gamma}u)dt du=(2/c-1)^i\delta_j^i\sqrt{\pi}$.

Proof. Suppose $i\leq j$. We have $(1/\sqrt{c})^2+(\sqrt{\gamma})^2=c^{-1}+1-c^{-1}=1$. Therefore, the matrix

$$U=\begin{pmatrix}1/\sqrt{c} & -\sqrt{\gamma}\\ \sqrt{\gamma} & 1/\sqrt{c}\end{pmatrix}$$

is orthogonal. Let us make a substitution of the variables in the integral

$$\begin{pmatrix}w\\ v\end{pmatrix}=U\begin{pmatrix}t\\ u\end{pmatrix}.$$

We obtain

$$\int_{-\infty}^{\infty}\int_{-\infty}^{\infty} e^{-w^2-v^2} G_j(w) G_i((-1+2/c)w + 2\sqrt{\gamma/c}v)\,dw\,dv$$
$$= \int_{-\infty}^{\infty}\int_{-\infty}^{\infty} e^{-w^2-v^2} G_j(w)\,[2/c-1)^i G_i(w) + Q(w,v)]\,dw\,dv$$

where the degree of the polynomial $Q(w, v)$ with respect to w does not exceed $i-1<j$, and, therefore,

$$\int_{-\infty}^{\infty} e^{-t^2} G_j(t) Q(t, v)\,dt = 0\,.$$

Since $\int_{-\infty}^{\infty} G_j(t) G_i(t) e^{-t^2} dt = \delta_i^j$, the lemma is proved.

Let us use the known property of the Hermite polynomials:

$$d^i G_j(z)/dz^i = \sqrt{2^i j(j-1)\dots(j-i+1)}\,G_{j-i}(z)\,.$$

In a combination with lemma this gives the following:

$$a = (\pi^{\frac{1}{4}} 2^k (2k)!^{\frac{1}{2}} (\gamma\pi)^{\frac{1}{2}})^{-1} C_{2k}^k 2^k \cdot ((2k)!/k!) c^{-k} \int_{-\infty}^{\infty}\int_{-\infty}^{\infty} e^{-u^2-z^2} G_k(z/\sqrt{c} - u\sqrt{\gamma})$$
$$\cdot G_k(z/\sqrt{c} + u\sqrt{\gamma})\,dz\,du = ([(2k)!]^{\frac{3}{2}}/(k!)^3)\,(2/c^2 - 1/c)^k (\pi^{\frac{1}{4}} (\pi\gamma)^{\frac{1}{2}})^{-1}\,.$$

The number a is calculated.

Appendix 2

One of the authors (Bleher) has investigated the renormalization group transformation for the hierarchical model in the case $d=1$, $r=2$, with the help of the computer. As a result all the critical indices for all the values of the parameter of the hierarchical model were found.

In the case under consideration the renormalization group transformation can be considered as the nonlinear integral mapping:

$$Q: f(z) \to \text{const} \int_{-\infty}^{\infty} e^{-u^2} f(z/\sqrt{c} + u) f(z/\sqrt{c} - u)\,du$$

where $1<c<2$ is a parameter of the hierarchical model. The first aim of the computations was to find all the thermodinamically-stable fixed points (TSFP) of the mapping Q. From the mathematical point of view it means that we seek fixed points for which the linearized mapping $L_f Q$ has explicitly one eigenvalue the modulo of which is bigger than one.

It is one of the results of the computations that for all the values of the parameter c, $1<c<2$, there exists one and only one TSFP of the transformation Q. For $\sqrt{2}<c<2$ this is the evident fixed point $f(x) \equiv \text{const}$. The graphs of the TSFP for various values of the parameter c, $1<c<\sqrt{2}$, are shown on the Figs. 1–5. Probably for $c \to 1$ TSFP degenerates in a discrete measure. The branch of nonconstant TSFP have been considered rigorously before for sufficiently small $\varepsilon = \sqrt{2} - c > 0$, where $\varepsilon = 0$ is the point of the bifurcation of TSFP. The numerical computations show that there are not bifurcations of this branch of TSFP for all ε, $0<\varepsilon<\sqrt{2}-1$.

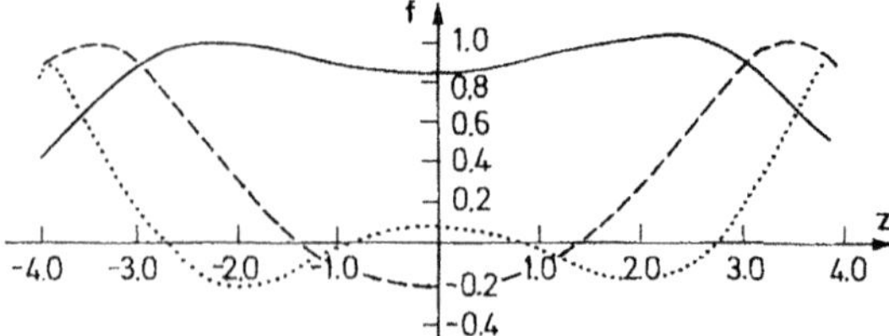

Fig. 1. The graphs of the TSFP (the continuous line), of the first eigenfunction (the interrupted line) and of the second eigenfunction (the dotted line) are plotted for $c = 2^{0.45}$

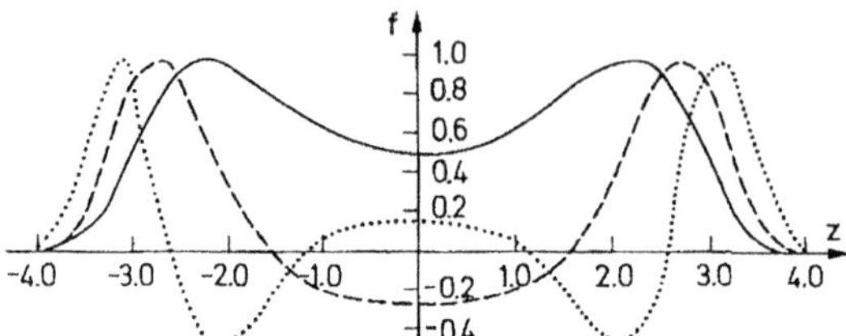

Fig. 2. $c = 2^{1/3}$

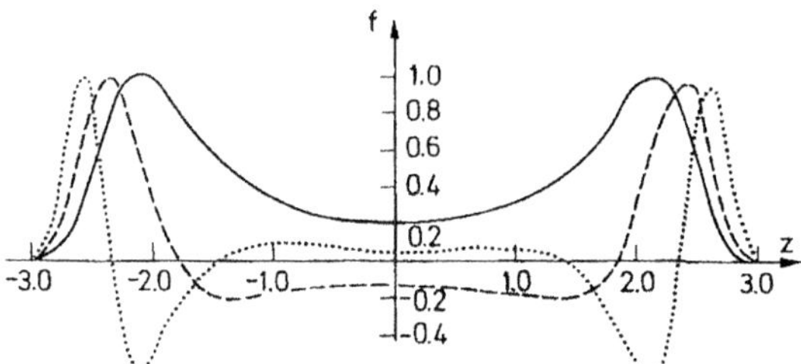

Fig. 3. $c = 2^{0.2}$

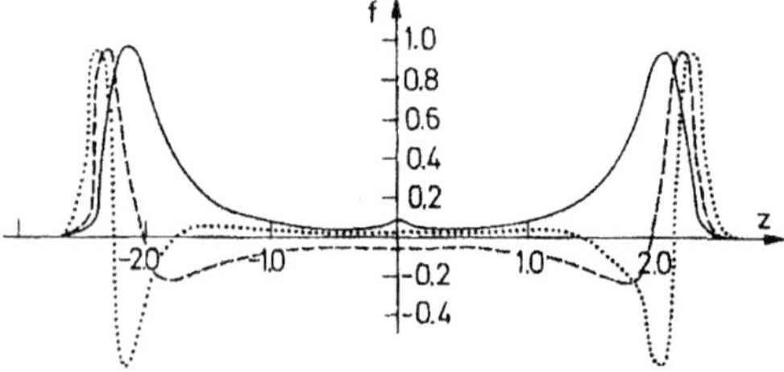

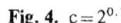

Fig. 4. $c = 2^{0.1}$

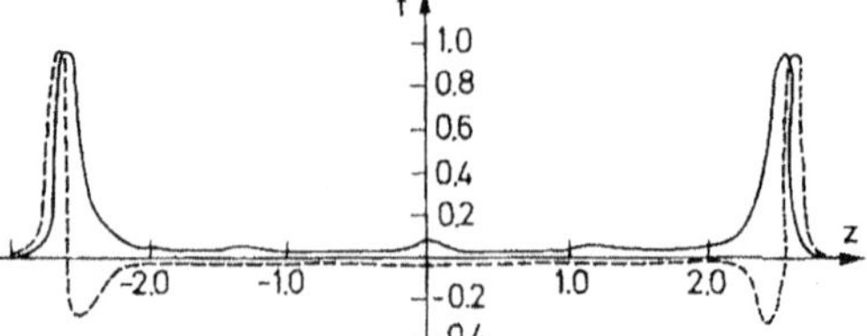

Fig. 5. $c = 2^{0.03}$

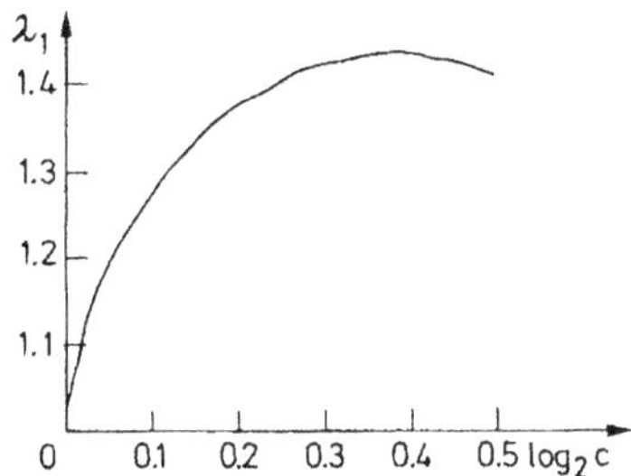

Fig. 6. The dependence of the first eigenvalue on the parameter $\log_2 c$

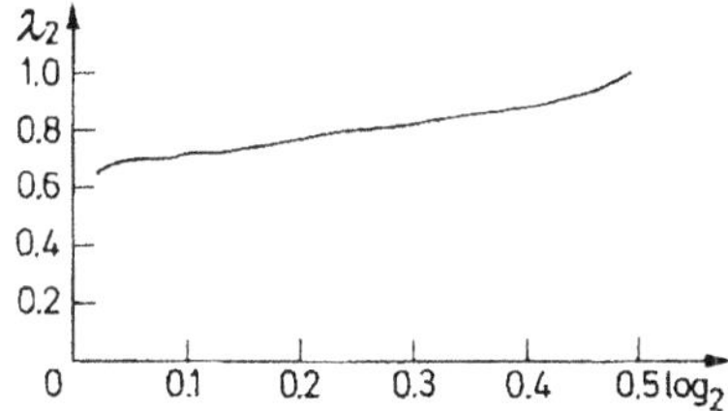

Fig. 7. The dependence of the second eigenvalue on the parameter $\log_2 c$

Our second aim was to compute the spectrum of the linearized mapping $L_f Q$ for TSFP. It is a very interesting problem because of as it was pointed out before all the critical indices of the asymptotically hierarchical models can be expressed via the first eigenvalue $\lambda_1 > 1$ of the operator $L_f Q$. On the Fig. 6 the dependence of λ_1 on the parameter c is plotted. One can see that there is a good agreement of this curve with the theoretical ε-expansions $\lambda_1 = (1 + \varepsilon/3 + O(\varepsilon^2))\sqrt{2}$ for $c = \sqrt{2} - \varepsilon$ and $\lambda_1 = 1 + \sqrt{\varepsilon}$ for $c = 1 + \varepsilon$. The last expansion is taken from the paper by Kosterlitz [13].

Finally on the Fig. 7 it is plotted the dependence of the second eigenvalue of the linearized operator $L_f Q$ on the parameter c. It is evident that $0 < \lambda_2 < 1$ for all the values c, $1 < c < \sqrt{2}$. This points out that the considering branch of TSFP has not any other bifurcation for $1 < c < \sqrt{2}$ indeed.

On the Fig. 1–5 the two first eigenfunctions of the operator $L_f Q$ are also plotted for some values of the parameter c.

References

1. Dyson, F. J.: Existence of a phase transition in a one-dimensional Ising ferromagnet. Commun. math. Phys. **12**, 91 (1969)
2. Bleher, P. M., Sinai, Ja. G.: Investigation of the critical point in models of the type of Dyson's hierarchical models. Commun. math. Phys. **33**, 23 (1973)
3. Wilson, K. G.: Renormalization group and critical phenomena, I, II. Phys. Rev. **B4**, 9, 3174 (1971)
4. Fisher, M. E.: General scaling theory for critical point. Proceedings Nobel Symposium **24**, 16–37 (1973)
5. Jona-Lasinio, G.: The renormalization group: a probabilistic view. Nuovo Cimento **26B**, 99 (1975)
6. Gallavotti, G., Knops, H. J. F.: The hierarchical model and the renormalization group. Inst. Theor. Fys. Univ. Nijmegen, preprint (1974)
7. Wilson, K. G., Kogut, I.: The renormalization group. Phys. Rep. **12**C, 75 (1974)

8. Bleher, P. M.: Investigation of the vicinity of the critical point in models of the type of Dyson's hierarchical models. Inst. Appl. Math. AS USSR, Moscow, preprint N 45 (1973); Tr. Mosc. Mat. Ob., v. 33, to be published (Russian)
9. Stanley, H.: Introduction to phase transitions and critical phenomena. Oxford: Clarendon Press 1971
10. Bleher, P. M.: Critical indices of long-range interaction models (numerical results). Inst. Appl. Math. AS USSR, Moscow, preprint N 3 (1975) (Russian)
11. Bateman, G., Erdelyi, A.: Higher transcendental functions, v. 2. New York, Toronto, London: McGraw-Hill Book Company 1955
12. Arnold, V. I.: Lectures on bifurcations and versal families. Usp. Mat. Nauk **27**, 5, 119 (1972)
13. Kosterlitz, J. M.: Critical properties of the one-dimensional Ising model with long-range interactions. University of Birmingham, preprint

Communicated by G. Gallavotti

(Received May 23, 1975)

Comments

The Gaussian solution of the main equation for hierarchical models becomes unstable in the corresponding region of parameter values. Therefore, generically it cannot appear as a limiting distribution for the normalized spin at β_{cr}. Using some methods from the bifurcation theory we show the existence of non-Gaussian solutions which decay in the simplest case as $\exp\{-x^4\}$. More detailed exposition of the whole theory can be found in the book by P. Collet and J.-P. Eckmann (see [CE]).

Commun. math. Phys. 33, 23—42 (1973)

Investigation of the Critical Point in Models of the Type of Dyson's Hierarchical Models

P. M. Bleher and Ja. G. Sinai

Landau Institute for Theoretical Physics, Academy of Sciences, Moscow, USSR

Received March 27, 1973

Abstract. We consider the classical spin models where the Hamiltonians are small modifications of the Hamiltonians of Dyson's hierarchical models. Under some assumptions we investigate rigorously the neighbourhood of the critical point and find the critical indices. It follows that in the cases under consideration phenomenological Landau's theory of phase transitions is valid.

Introduction

In classical lattice ferromagnets the critical temperature T_{cr} separates the domains with zero and non-zero spontaneous magnetization. The behaviour of different thermodynamical parameters near T_{cr} was considered rigorously for the two-dimensional Ising model using Onsager's exact solution (see [1, 2]), and some other models (see [3]) also using the exact formula for the free energy.

Recently Dyson introduced so-called hierarchical models immitating in many respects the lattice systems with pairwise long-range power interaction. Under several natural assumptions Dyson proved that the spontaneous magnetization in his models is non-zero for sufficiently large β ([4–6]).

We consider in this paper a class of models slightly generalizing hierarchical models and find rigorously under some conditions critical indices for them. Recently some non-rigorous results in this direction were obtained by Baker [7]. Our case corresponds to the "gaussian case" of his paper. Closely related results were presented also in [8] (see also [9]).

Now we want to describe briefly our main results. We impose certain conditions on the distribution of the mean spin in a finite volume of fixed size for an interval of temperatures. Under these conditions we prove the existence of a critical temperature T_{cr} inside this interval and for $T = T_{cr}$ we establish the limit distribution for the mean spin which is a gaussian distribution with a non-usual normalization. This permits us to find the asymptotic of binary correlation functions for $T = T_{cr}$.

For $T > T_{cr}$ we obtain the asymptotic expression for the susceptibility and binary correlation functions. For $T < T_{cr}$ we find an asymptotic

Y.G. Sinai, *Selecta II: Probability Theory, Statistical Mechanics, Mathematical Physics and Mathematical Fluid Dynamics*, DOI 10.1007/978-1-4419-6205-8_10, © Springer Science+Business Media, LLC 2010

expression for the spontaneous magnetization. The critical indices are just as predicted by Landau's theory ([10, 9]).

The case $T=T_{cr}$ was considered by both authors. The cases $T<T_{cr}$, $T>T_{cr}$ were considered by the first author.

1. Hierarchical Models and Asymptotically-Hierarchical Models

We begin with the description of general hierarchical models. Assume to be given a sequence of positive integers $r_n, r_n \geqq 2$, $n=1, \ldots, q_n=\prod_{i=1}^{n} r_i$, and for any integer $r \geqq 2$ some positive quadratic form $Q_r(t_1, \ldots, t_r)$. In general hierarchical models one considers the sequence of volumes $V_n=V_{n,0}$ in which each volume $V_{n,0}$ consists of q_n points. The volume $V_{n,0}$ is decomposed into r_n subvolumes $V_{n-1,i}$, $i=1, \ldots, r_n$, each of which consists of q_{n-1} points; the subvolume $V_{n-1,i}$ is decomposed into r_{n-1} subvolumes $V_{n-2,i}$ each of which consists of q_{n-2} points, and so on. For any $k \geqq 0$ the volume $V_{n,0}$ is decomposed into $r_n \cdot \ldots \cdot r_{n-k}$ subvolumes $V_{n-k-1,i}$ each of which consists of q_{n-k-1} points. The Hamiltonian is defined by a sequence of positive numbers b_p and has the form

$$H_n=-\sum_{p=1}^{n} b_p \sum_{i=0,\ \bigcup_{k=1}^{r_p} V_{p-1,i_k}=V_{p,i}}^{q_n/q_p-1} Q_{r_p}(s_{i_1}, s_{i_2}, \ldots, s_{i_{r_p}}).$$

Here s_{i_l} is the mean spin in the volume V_{p-1,i_l}. We assume that the spin s_x, $x \in V_{n,0}$ takes values ± 1.

Dyson ([4–6]) considered the case $r_n \equiv 2$ and $Q_2(t_1, t_2)=\left(\dfrac{t_1+t_2}{2}\right)^2$. Baker considered the cases $r_n \equiv 2, 4, 8$ ([7]).

We shall consider below slightly more general models which we shall call asymptotically-hierarchical models (a.h.m.). The Hamiltonian in these models is defined by an integer n_0 and can be written for $n>n_0$ in the following form

$$H_n=-\sum_{i=1}^{q_n/q_{n_0}} H_{n_0}(S_{n_0,i})-\sum_{p=n_0+1}^{n} b_p \sum_{i,\ \bigcup_{k=1}^{r_p} V_{p-1,i_k}=V_{p,i}} Q_{r_p}(s_{i_1}, \ldots, s_{i_{r_p}}).$$

Here $S_{n_0,i}$ is the configuration of spins in $V_{n_0,i}$ and H_{n_0} is an arbitrary Hamiltonian in the volume of q_{n_0} points, satisfying the symmetry condition: $H_{n_0}(S_{n_0,i})=H_{n_0}(-S_{n_0,i})$.

In the sequel only the case $r_n \equiv r$ and $b_p=c^p$ will be discussed. Moreover we shall assume that the quadratic form $Q_r(t_1, \ldots, t_r)$ is symmetric in its arguments and has the form

$$Q_r(t_1, \ldots, t_r) = g_r\left(\frac{t_1 + \cdots + t_r}{r}\right)^2 + h_r\left(\frac{t_1^2 + \cdots + t_r^2}{r}\right);$$

we shall assume that $g_r = g$, $h_r = h$, $h + g > 0$, $rg + ch > 0$. The extension of our results to other cases will be done in another paper.

For a.h.m. the limit free energy per particle χ exists and is a convex function of its arguments if $c < r$. An extension of Dyson's arguments shows that spontaneous magnetization for sufficiently large β is non-zero if $c > 1$.

2. Formulation of Results

For a.h.m. let us denote $g_k(s; \beta) = \mathrm{Prob}_k(s_{k,0} = s \mid \beta)$, where $\mathrm{Prob}_k(\cdot \mid \beta)$ means the Gibbs probability distribution with parameter β and zero magnetic field in the volume $V_{k,0}$.

For $k > n_0$ one can write the following recurrence relations:

$$g_k(s; \beta) = \frac{\Xi_{k-1}^r(\beta)}{\Xi_k(\beta)} \cdot \sum_{\frac{s_1 + \cdots + s_r}{r} = s} \exp(\beta c^{k-1} Q(s_1, s_2, \ldots, s_r)) g_{k-1}(s_1; \beta) \ldots g_{k-1}(s_r; \beta) \tag{1}$$

where $\Xi_{k(\beta)}$ is the grand partition function in the volume $V_{k,0}$. For all k the probabilities $g_k(s; \beta)$ are even functions of s.

Let us introduce some normalization for spins $s_{k,0}$ putting $c^{\frac{k}{2}} s_{k,0} = z_k$. It is evident that z_k takes the values $\frac{-r^k + 2m}{r^k} \cdot c^{\frac{k}{2}}$, $0 \leq m \leq r^k$. Therefore for $\Delta_k = \frac{c^{\frac{k}{2}}}{r^k}$ and $f_k(z; \beta) = g_k(zc^{-\frac{k}{2}}; \beta)$ we have instead of (1)

$$f_k(z; \beta) = L_k \sum_{\frac{z_1 + \cdots + z_r}{r} = \frac{z}{\sqrt{c}}} \exp(\beta Q(z_1, \ldots, z_r)) \cdot f_{k-1}(z_1; \beta) \cdot \ldots \cdot f_{k-1}(z_r; \beta) \Delta_{k-1}^{r-1}. \tag{2}$$

Here L_k is a normalization constant.

It is natural to assume that for large k the behaviour of the functions f_k is the same as the behaviour of the iterates of the non-linear integral transformation

$$Q_\beta(f) = \frac{\int_{\sum_{i=1}^r z_i = \frac{r}{\sqrt{c}} z} \exp(\beta Q(z_1, \ldots, z_r)) f(z_1) \cdot \ldots \cdot f(z_r) \prod_i dz_i}{\int \exp(\beta Q(z_1, \ldots, z_r) f(z_1) \cdot \ldots \cdot f(z_r) \prod_i dz_i}.$$

The investigation of such transformations begins with the search for the fixed points or the eigenvectors, i.e. the functions f which satisfy the equality

$$f(z)=\lambda \int_{\sum z_i=\frac{r}{\sqrt{c}}z} \exp(\beta Q(z_1,\ldots,z_r))\, f(z_1)\cdot\ldots\cdot f(z_r)\prod_i dz_i=\tilde{Q}_\beta(f) \qquad (3)$$

for some constant λ. Having such a fixed point it is necessary to investigate its stability properties. Only fixed points with some stability properties can appear as limits of our difference transformations (2).

It is easy to verify that the function $\exp(-a_0(\beta)z^2)$ satisfies the Eq. (3) with $a_0(\beta)=\dfrac{g+h}{r-c}\beta$, i.e. it is an eigenfunction of the transformation $\tilde{Q}_\beta$. Thus the function $\sqrt{\dfrac{a_0}{\pi}}\exp(-a_0(\beta)z^2)=e_\beta(z)$ is a fixed point for the transformation Q_β.

To investigate the stability properties of $e_\beta(z)$ let us consider now the differential of the transformation Q_β near the point $e_\beta(z)$. It is easier to consider the differential of the transformation $\tilde{\tilde{Q}}_\beta$ near the point $e_\beta(z)$. Taking $e(z)=e_\beta(z)+\varepsilon h(z)$ one finds easily

$$\begin{aligned}\mathscr{D}\tilde{Q}_\beta h &= \frac{d}{d\varepsilon}\tilde{Q}_\beta(e_\beta+\varepsilon h)|_{\varepsilon=0}\\ &= \frac{d}{d\varepsilon}\left(\frac{a_0}{\pi}\right)^{\frac{r}{2}} \int_{\sum z_i=\frac{r}{\sqrt{c}}z} \exp(\beta Q(z_1,\ldots,z_r))\,(e_\beta(z_1)+\varepsilon h(z_1))\\ &\quad \cdot\ldots\cdot(e_\beta(z_r)+\varepsilon h(z_r))\prod_i dz_i\\ &= \exp(-a_0 z^2)\, r\left(\frac{\zeta(\beta)}{\pi}\right)^{\frac{1}{2}} \int_{-\infty}^{+\infty} \exp\left(\zeta(\beta)\left(\frac{z}{\sqrt{c}}-z_1\right)^2+a_0(\beta)z_1^2\right)\\ &\quad \cdot h(z_1)\,dz_1\end{aligned}$$

where

$$\zeta(\beta)=\frac{h\cdot\beta-a_0(\beta)\,r}{r-1},$$

which is the well-known integral transformation with Gaussian kernel (see [11]). This operator is a compact selfadjoint operator in the Hilbert space $\mathscr{H}(\gamma_0)=L^2_{\mathrm{ev}}(\mathbb{R}^1;e^{-\gamma_0 z^2})$ of real valued even square-integrable functions with the weight $e^{-\gamma_0 z^2}$, where $\gamma_0=\gamma_0(\beta)=\zeta(\beta)\left(1-\dfrac{1}{c}\right)+2a_0(\beta)$; the spectrum of this integral operator in the space $\mathscr{H}(\gamma_0)$ consists of the infinite sequence of numbers $r, \dfrac{r}{c}, \dfrac{r}{c^2}, \dfrac{r}{c^3}, \ldots$. The corresponding eigenfunctions are $\{e^{-a_0(\beta)z^2}G_{2i}(z;\gamma)\}_{i=0}^{\infty}$,

$$\gamma=\gamma(\beta)=\zeta(\beta)\left(1-\frac{1}{c}\right),$$

where

$$G_{2i}(z;\gamma)=\sqrt[4]{\frac{\gamma}{\pi}}\,\frac{\sqrt{(2i)!}}{2^i}\sum_{j=0}^{i}\frac{(-1)^{i-j}(2\sqrt{\gamma}\,z)^{2j}}{(2j)!\,(i-j)!}$$

is the $2i$-th Hermite polynomial.

The first eigenvalue r is determined by the fact that our transformation is r-linear. The second eigenvalue is always more than one. If $c>\sqrt{r}$ then all other eigenvalues are less than one. We shall show that in an a.h.m. the convergence of $f_k(z;\beta)$ to a gaussian distribution for $\beta=\beta_{cr}$ is in general possible if $c>\sqrt{r}$. Thus we impose the first important condition.

Condition 1. $c>\sqrt{r}$.

Now we are going to discuss our next condition. As the reader will see its exact formulation is not too short. Therefore we want to explain its meaning. One can hope that $f_n(z;\beta)$ for $\beta=\beta_{cr}$ will tend to $e_{\beta_{cr}}(z)$ if the functions $f_{n_0}(z;\beta_{cr})$ are sufficiently close to $e_{\beta_{cr}}(z)$ in an appropriate sense for some sufficiently large $n=n_0$. Here we give the exact formulation of what we mean by "sufficiently close" for our problem. The number $n_0=n_0(c,Q)$ depends only on the constant c and the quadratic form Q. Its exact value is defined by some number (near ten) of inequalities appearing during the proof. Therefore we shall not give its explicit expression here.

Furthermore, if we are given a family of probability distributions $f_{n_0}(z;\beta)$ for some interval of temperatures $\boldsymbol{\beta}=[\beta^-,\beta^+]$ we don't know the value of the critical temperature without solving the whole sequence of recurrence Eqs. (2). However it is possible to formulate conditions which guarantee that the critical temperature will lie inside the interval $\boldsymbol{\beta}$.

Now we proceed to the exact formulation.

Condition 2. Let us choose three numbers $0<\varrho,\ q,\ \xi<1$ depending only on c in such a way that

$$\frac{r}{c^2}<\varrho<1\,;$$

$$\frac{r^{\frac{1}{2}}}{c^{1\,3/4}},\ \frac{r^2}{c^4}<q<\frac{r}{c^2}\,; \tag{4}$$

$$c^{-\frac{1}{3}},\ q^{\frac{1}{4}},\ \frac{qc^2}{r}<\xi<1\,.$$

There exist a number $n_0(c)$, an interval of inverse temperatures $\boldsymbol{\beta}_{n_0}=[\beta_{n_0}^-,\beta_{n_0}^+]$, and a differentiable function $b_{n_0}(\beta)=b(\beta)$, defined on this interval, the number d satisfying the inequalities

$$0 < d,\ r+1 < d_1^2 \quad \text{for} \quad d_1 = \frac{d(\sqrt{c}-1)}{2\sqrt{c}},$$

for which

a$_1$) $b(\beta_{n_0}^-) = -\left(\frac{c}{2}\varrho\right)^{n_0}$, $b(\beta_{n_0}^+) = \left(\frac{c}{2}\varrho\right)^{n_0}$, $|b(\beta)| < \left(\frac{c}{2}\varrho\right)^{n_0}$ for $\beta \in (\beta_{n_0}^-, \beta_{n_0}^+)$.

a$_2$) for each $\beta \in \boldsymbol{\beta}_{n_0}$ the function $f_{n_0}(z;\beta)$ can be represented in the following form

$$f_{n_0}(z;\beta) = L_{n_0}(\beta)\exp\left(-\left(a_0(\beta) + \left(\frac{2}{c}\right)^{n_0} b(\beta)\right)z^2\right)(1 + q_{n_0}(z;\beta))$$

with $L_{n_0}(\beta)$ being a constant factor (with respect to z) and for the "small perturbation" $q_{n_0}(z;\beta)$ satisfying

a$_2'$) for $|z| \leqq \dfrac{d\sqrt{n_0}}{\sqrt{a_0}}$

$$q_{n_0}(z;\beta) = \delta_{n_0}(\beta)\, G_4(z;\gamma) + R_{n_0}(z;\beta)$$

with $\gamma = \gamma(\beta) = \zeta(\beta)\left(1 - \frac{1}{c}\right)$ and $\delta_{n_0}(\beta)$, $R_{n_0}(z;\beta)$ are differentiable functions of $\beta \in \boldsymbol{\beta}_{n_0}$,

$$|R_{n_0}(z;\beta)| \leqq q^{n_0} \leqq -\delta_{n_0}\xi^{n_0},$$

$$-\delta_{n_0} \leqq q^{\frac{n_0}{2}}\xi^{n_0}.$$

a$_2''$) for $|z| > \dfrac{d\sqrt{n_0}}{\sqrt{a_0}}$

$$0 \leqq 1 + q_{n_0}(z;\beta) \leqq \exp(-v_{n_0} z^4),$$

$$0 < v_{n_0} < -0{,}1 \cdot \delta_{n_0}.$$

Remark. The proof of Theorem 1 below shows that the Condition 2 is stable in the sense that it is fulfilled for $n > n_0$ if it is fulfilled for $n = n_0$.

Now we can give the exact formulation of our theorems.

Theorem 1. *If the Conditions 1, 2 hold, then there exists one and only one $\beta_{cr} \in \boldsymbol{\beta}_{n_0}$ for which*

$$\lim_{n\to\infty} \operatorname{Prob}_n\{t_1 < s_{n,0}c^{\frac{n}{2}} < t_2;\beta_{cr}\} = \frac{1}{\sqrt{2\pi\sigma}}\int_{t_1}^{t_2} e^{-\frac{u^2}{2\sigma}}\,du$$

for some positive σ and arbitrary t_1, t_2.

Theorem 1 shows that for $\beta=\beta_{cr}$ the mean spin $s_{n,0}$ has gaussian distribution with non-usual normalization.

Corollary 1. *For $\beta=\beta_{cr}$ the binary correlation function $\langle s_x, s_y\rangle_{n,\beta_{cr}}$, $x, y\in V_{n,0}$ satisfies the inequalities*

$$C_1\cdot c^{-d(x,y)}\leqq\langle s_x, s_y\rangle_{n,\beta_{cr}}\leqq C_2 c^{-d(x,y)}$$

with some constants C_1, C_2, where $d(x, y)$ is the least number k such that $x, y\in V_{k,i}$ for some i.

Theorem 2. *Under the conditions of Theorem 1 let $\beta_{n0}^{+}<\beta<\beta_{cr}$. Then for the mean spin $s_{n,0}$*

$$\lim_{n\to\infty}\operatorname{Prob}_n\{t_1<s_{n,0}r^{\frac{n}{2}}<t_2;\beta\}=\frac{1}{\sqrt{2\pi\sigma_1(\beta)}}\int_{t_1}^{t_2}e^{-\frac{u^2}{2\sigma_1(\beta)}}\,du$$

where $\sigma_1(\beta)\sim\operatorname{const}(\beta_{cr}-\beta)^{-1}$.

Corollary 2. *Under the conditions of Theorem 2 the binary correlation functions $\langle s_x, s_y\rangle_{n,\beta}$, $x, y\in V_{n,0}$ satisfy for all n the inequalities*

$$\frac{C_1'}{(\beta_{cr}-\beta)^2}\left(\frac{r^2}{c}\right)^{-d(x,y)}\leqq\langle s_x, s_y\rangle_{n,\beta}\leqq\frac{C_2'}{(\beta_{cr}-\beta)^2}\left(\frac{r^2}{c}\right)^{-d(x,y)}$$

with some constants C_1', C_2'.

Theorem 3. *Under the conditions of Theorem 1 let $\beta_{cr}<\beta<\beta_{n0}^{-}$. Then there exist positive functions $m(\beta)$, $\sigma_2(\beta)$, $m(\beta)\sim\operatorname{const}(\beta-\beta_{cr})^{\frac{1}{2}}$, $\sigma_2(\beta)\sim\operatorname{const}(\beta-\beta_{cr})$, such that for a sequence $m_n(\beta)$, $\lim_{n\to\infty}m_n(\beta)=m(\beta)$,*

$$\lim_{n\to\infty}\operatorname{Prob}_n\{t_1<(s_{n,0}-m_n(\beta))r^{\frac{n}{2}}<t_2;\beta\}=\frac{1}{2}\,\frac{1}{\sqrt{2\pi\sigma_2(\beta)}}\int_{t_1}^{t_2}e^{-\frac{u^2}{2\sigma_2(\beta)}}\,du\,,$$

$$\lim_{n\to\infty}\operatorname{Prob}_n\{t_1<(s_{n,0}+m_n(\beta))r^{\frac{n}{2}}<t_2;\beta\}=\frac{1}{2}\,\frac{1}{\sqrt{2\pi\sigma_2(\sigma)}}\int_{t_1}^{t_2}e^{-\frac{u^2}{2\sigma_2(\beta)}}\,du\,.$$

3. Thermodynamical Limit of Hierarchical and Asymptotically-Hierarchical Models

Let V be a countable set and r_n be a sequence of integers, $r_n>1$. Following Vershik let us say that a hierarchical $\{r_n\}$-structure is defined on V if there is defined a decreasing sequence $\xi_1\geqq\xi_2\geqq\cdots$ of partitions of V with the following properties:

a$_1$) $\xi_1=\varepsilon$ where ε is as usual the partition of V into separate points;

a$_2$) any element of ξ_i consists of r_i elements of ξ_{i-1};

a$_3$) for any two points $x, y\in V$ there exists a number $d(x, y)$ such that x, y belong to the same element of any partition ξ_i, $i\geqq d(x, y)$.

In the following $d(x, y)$ will be the least number with this property. The number $\prod_{i=1}^{d(x,y)} r_i$ plays the role of a distance between x, y. All spaces with $\{r_n\}$-structure are naturally isomorphic.

If V is the space with $\{r_n\}$-structure then $V^{(k)} = V \mid \xi_k$ is the space with $\{r'_n\}$-structure where $r'_n = r_{n+k}$.

Let us denote by $G = G(V)$ the group of all finite permutations of V leaving each ξ_i invariant, and let $\Omega(V)\,(\Omega_0(V))$ be the space of all real-valued (± 1-valued) functions on V. We can define in the usual way the probability distributions on $\Omega(V)$ and $\Omega_0(V)$, gaussian probability distributions on $\Omega(V)$, the distributions invariant under the group G, and so on.

Now let us return to the definition of Dyson's hierarchical models. One can consider the volumes $V_{n,0}$ as an increasing sequence of subsets of the infinite space V with $\{r_n\}$-structure such that $\bigcup_{n=1}^{\infty} V_{n,0} = V$. The sequence of probability distributions $\mathrm{Prob}_n(\cdot \mid \beta)$ can be considered as a sequence of probability distributions on $\Omega_0(V)$ defined on an increasing sequence of corresponding σ-algebras. Dyson in [4] proved in fact the following theorem.

Theorem 4. *For any $\beta > 0$ the sequence of probability distributions $\mathrm{Prob}_n(\cdot \mid \beta)$ converges in a natural sense to a limit Gibbsian distribution $P(\cdot \mid \beta)$ defined on the σ-algebra of measurable subsets of $\Omega_0(V)$ and invariant under the group G.*

For the asymptotically-hierarchical models the same considerations lead to the following theorem.

Theorem 5. *Under the conditions of Theorem 1 the probability distributions $\mathrm{Prob}_n(\cdot \mid \beta)$ converge in a natural way to a limit Gibbsian distribution $P(\cdot \mid \beta)$. The distribution on $\Omega(V^{(n_0)}) = \Omega(V \mid \xi_{n_0})$ induced by a map*

$$\pi_{n_0} : \Omega_0(V) \to \Omega(V_{n_0}) \quad \textit{where} \quad \pi_{n_0}(f)(x) = \frac{1}{\prod_{i=1}^{n_0} r_i} \sum_{y \in x} f(y)$$

is invariant under the corresponding group G.

Now we shall formulate the results concerning the limit distributions $P(\cdot \mid \beta)$ which are equivalent to the Theorems 1, 2, 3. Let be $s_n = \frac{1}{r^n} \sum_{x \in C_n} s(x)$ where C_n is an arbitrary element of the partition ξ_n entering into the definition of a hierarchical structure on V. We shall consider the distribution of s_n which depends only on n but not on C_n.

Theorem 1′. *Under the conditions of Theorem 1 for $\beta=\beta_{cr}$ and arbitrary fixed t_1, t_2*

$$\lim_{n\to\infty} \mathrm{P}\{t_1 < s_n \cdot c^{\frac{n}{2}} < t_2 ; \beta_{cr}\} = \frac{1}{\sqrt{2\pi\sigma'}} \int_{t_1}^{t_2} e^{-\frac{u^2}{2\sigma'}} du$$

for some positive constant σ'.

Corollary 1′. *There exist constants C_1, C_2 such that for $\beta=\beta_{cr}$*

$$C_1 c^{-d(x,y)} \leqq \langle s_x, s_y \rangle_{\beta_{cr}} \leqq C_2 c^{-d(x,y)} .$$

Theorem 2′. *Under the conditions of Theorems 1 and 2 for arbitrary fixed t_1, t_2*

$$\lim_{n\to\infty} \mathrm{P}\{t_1 < s_n r^n < t_2 ; \beta\} = \frac{1}{\sqrt{2\pi\sigma_1'(\beta)}} \int_{t_1}^{t_2} e^{-\frac{u^2}{2\sigma_1'(\beta)}} du$$

where $\sigma_1'(\beta) \sim \mathrm{const}(\beta_{cr}-\beta)^{-1}$.

Corollary 2′. *Under the conditions of Theorem 2 the binary correlations functions $\langle s_x, s_y \rangle_\beta$, $x, y \in V$ satisfy the inequalities*

$$\frac{C_1'}{\beta_{cr}-\beta} r^{-d(x,y)} \leqq \langle s_x, s_y \rangle_\beta \leqq \frac{C_2'}{\beta_{cr}-\beta} r^{-d(x,y)}$$

with some constants C_1', C_2'.

Theorem 3′. *In the notation of Theorem 3 under the conditions of the Theorem 1 for $\beta_{cr} < \beta < \beta_{n_0}^-$ and arbitrary t_1, t_2*

$$\lim_{n\to\infty} \mathrm{P}\{t_1 < (s_n - m_n(\beta)) r^{\frac{n}{2}} < t_2\} = \frac{1}{2} \frac{1}{\sqrt{2\pi\sigma_2'(\beta)}} \int_{t_1}^{t_2} e^{-\frac{u^2}{2\sigma_2'(\beta)}} du$$

where $\sigma_2'(\beta) \sim \mathrm{const}(\beta-\beta_{cr})^{-1}$.

4. Proof of Theorem 1

We shall consider only the case $r=2$ and $Q(z_1, z_2) = c\left(\frac{z_1+z_2}{2}\right)^2$. The generalization to other cases is straight-forward. We shall construct a sequence of imbedded segments $\boldsymbol{\beta}_n = [\beta_n^-, \beta_n^+]$ and a sequence of differentiable functions $b_n(\beta)$ defined on $\boldsymbol{\beta}_n$ such that for all $\beta \in \boldsymbol{\beta}_n$ the functions $f_n(z;\beta)$ can be represented in the following form

$$f_n(z;\beta) = L_n(\beta) \exp\left[-\left(a_0(\beta) + \left(\frac{2}{c}\right)^n b_n(\beta)\right) z^2\right] (1 + q_n(z;\beta)) \qquad (5)$$

where $L_n(\beta)$, $b_n(\beta)$, $q_n(z;\beta)$ satisfy some estimates listed below. From these estimates it follows that

$$\left(\frac{2}{c}\right)^n b_n(\beta)\xrightarrow[n\to\infty]{}0;\ 0\leqq 1+q_n(z;\beta)\leqq 2 \quad \text{for} \quad \beta\in\boldsymbol{\beta}_n$$

and $-\infty<z<\infty$;

$$q_n(z;\beta)\xrightarrow[n\to\infty]{}0 \quad \text{uniformly for} \quad |z|<\frac{d\sqrt{n}}{\sqrt{a_0}},$$

$\beta\in\boldsymbol{\beta}_n$, d is a constant depending only on c, $L_n(\beta)\to L=\sqrt{\frac{a_0(\beta)}{\pi}}$. Thus we get the assertion of Theorem 1 for $\beta_{cr}=\bigcap_n \boldsymbol{\beta}_n$.

The representation (5) is non-unique because one can change b_n or L_n and include the difference in q_n without changing f_n. The most crucial part of the proof is the special choice of L_n and b_n. This will be described precisely during the proof.

The variable z in (5) varies from $-c^{\frac{n}{2}}$ to $c^{\frac{n}{2}}$ with the step $\Delta_n=2c^{\frac{n}{2}}\cdot 2^{-n}$. The set of values of z will be denoted as M_n.

The substitution of (5) into the formula (2) gives $\Big($one should remember that $r=2$, $Q(z_1,z_2)=c\left(\frac{z_1+z_2}{2}\right)^2\Big)$

$$f_{n+1}(z;\beta)=\tilde{L}_{n+1}\exp(-\tilde{\lambda}_{n+1}z^2)\sqrt{\frac{2\lambda_n}{\pi}}\sum_u e^{-2\lambda_n u^2}\left(1+q_n\left(\frac{z}{\sqrt{c}}-u;\beta\right)\right)$$
$$\cdot\left(1+q_n\left(\frac{z}{\sqrt{c}}+u;\beta\right)\right)\Delta_n, \tag{6}$$

where

$$\tilde{\lambda}_{n+1}=\tilde{\lambda}_{n+1}(\beta)=a_0(\beta)+\left(\frac{2}{c}\right)^{n+1}b_n(\beta),$$

$$\lambda_n=\lambda_n(\beta)=a_0(\beta)+\left(\frac{2}{c}\right)^n b_n(\beta),\ \Delta_n=2c^{\frac{n}{2}}2^{-n}.$$

The summation goes over u for which $\frac{z}{\sqrt{c}}-u\in M_n$, $\frac{z}{\sqrt{c}}+u\in M_n$ and $z\in M_{n+1}$. Let us put

$$1+\tilde{q}_{n+1}(z;\beta)=\sqrt{\frac{2\lambda_n}{\pi}}\sum_u e^{-2\lambda_n u^2}\left(1+q_n\left(\frac{z}{\sqrt{c}}-u;\beta\right)\right)$$
$$\cdot\left(1+q_n\left(\frac{z}{\sqrt{c}}+u;\beta\right)\right)\Delta_n. \tag{7}$$

If we introduce the linear integral operator

$$\mathscr{A}_n q = 2\sqrt{\frac{2\lambda_n}{\pi}} \int_{-\infty}^{+\infty} e^{-2\lambda_n u^2} q\left(\frac{z}{\sqrt{c}} - u\right) du \tag{8}$$

and assume, that the function $q_n(z;\beta)$ is extended to the whole line $\mathbb{R}^1$ as a nice function of $z \in \mathbb{R}^1$ then we can rewrite (7) in the following way

$$\tilde{q}_{n+1} = \mathscr{A}_n q_n + S_n(q_n)$$

where $S_n(q_n)$ will be treated as a remainder term. In fact $S_n(q_n)$ is the sum of two terms. The first one appears from the non-linear part of the formula (7):

$$S_n^{(1)}(q_n) = \sqrt{\frac{2\lambda_n}{\pi}} \sum_u e^{-2\lambda_n u^2} q_n\left(\frac{z}{\sqrt{c}} - u;\beta\right) q_n\left(\frac{z}{\sqrt{c}} + u;\beta\right) \Delta_n . \tag{9}$$

The second term $S_n^{(2)}(q_n)$ has the form

$$S_n^{(2)}(q_n) = (\tilde{\mathscr{A}}_n - \mathscr{A}_n)\,(q_n + \tfrac{1}{2}), \tag{9'}$$

where

$$\tilde{\mathscr{A}}_n q = 2\sqrt{\frac{2\lambda_n}{\pi}} \sum_u e^{-2\lambda_n u^2} q_n\left(\frac{z}{\sqrt{c}} - u;\beta\right) \Delta_n . \tag{10}$$

These properties of $\mathscr{A}_n$ can be easily verified:

1°. $\mathscr{A}_n$ is a self-adjoint compact operator in the Hilbert space $\mathscr{H}(\gamma_n) = L^2_{\mathrm{ev}}(\mathbb{R}^1; e^{-\gamma_n z^2})$ of real-valued even square-integrable functions with the weight $e^{-\gamma_n z^2}$, $\gamma_n = 2\lambda_n\left(1 - \frac{1}{c}\right)$.

2°. $\mathscr{A}_n \mathbb{1} = 2 \cdot \mathbb{1}$.

3°. The spectrum of $\mathscr{A}_n$ in the space $\mathscr{H}(\gamma_n)$ consist of the numbers $\left\{\frac{2}{c^j}\right\}_{j=0}^{\infty}$; the corresponding eigenfunctions are even Hermite polynomials $G_{2j}(z;\gamma_n)$, $j = 0, 1, \ldots$.

4°. $\|\mathscr{A}_n q\|_C \leqq 2\|q\|_C$

$\|\mathscr{A}_n q\|_{C^1} \leqq K\|q\|_C$

where K is some constant.

From 3° it follows that two eigenvalues $2, 2\cdot c^{-1}$ are always greater than 1; the others are smaller than one because of the condition $c > \sqrt{2}$. The main idea is to choose $b_n(\beta)$ and $L_n(\beta)$ in such a way that the projections of $q_n(z;\beta)$ on the expanding subspaces $\mathscr{H}_{\exp}(\gamma_n)$ generated by eigenvectors $G_0(z;\gamma_n)$, $G_2(z;\gamma_n)$ are equal to zero or at least very small. The possibility of doing so for $n = n_0$ follows easily from the Condition 2.

Let us fix $\omega = \omega(c) > 1$ so that $\omega - 1$ is sufficiently small and consider the sequence $n_i = [n_0 \omega^i]$, $i = 0, 1, \ldots$. We shall prove that for $n = n_0, n_1, n_2, \ldots$, the representations

$$f_n(z;\beta) = L_n(\beta)\exp\left(-\left(a_0(\beta) + \left(\frac{2}{c}\right)^n b_n(\beta)\right) z^2\right)(1 + q_n(z;\beta))$$

can be chosen in such a way that

1) $L_n(\beta)$, $b_n(\beta)$ are differentiable functions of $\beta \in \boldsymbol{\beta}_n$,

$$b_n(\beta_n^-) = -\left(\frac{c}{2}\varrho\right)^n, \; b_n(\beta_n^+) = \left(\frac{c}{2}\varrho\right)^n, \; |b_n(\beta)| \leqq \left(\frac{c}{2}\varrho\right)^n, \; \beta \in \boldsymbol{\beta}_n.$$

2) $q_n(z;\beta)$, $z \in M_n$, $|z| < \dfrac{d\sqrt{n}}{\sqrt{a_0}}$, can be represented in the form

$$q_n(z;\beta) = \delta_n G_4(z;\gamma_n) + R_n(z;\beta) \tag{11}$$

where $\delta_n = \delta_n(\beta)$ is a differentiable function of $\beta \in \boldsymbol{\beta}_n$ and

$$K_n^{(1)}\left(\frac{2}{c^2}\right)^{n-n_0} < -\delta_n < K_n^{(2)}\left(\frac{2}{c^2}\right)^{n-n_0}$$

where $K_n^{(i)} = \delta_{n_0}\left(\prod_{j=n_0}^{n}(1+\alpha^j)\right)^{(-1)^i}$, $i = 1, 2$, $0 < \alpha = \alpha(c) < 1$, $R_n(z;\beta)$ is a differentiable function of $\beta \in \boldsymbol{\beta}_n$,

$$|R_n(z;\beta)| < q^n. \tag{12}$$

3) for $|z| > \dfrac{d\sqrt{n}}{\sqrt{a_0}}$

$$0 \leqq 1 + q_n(z;\beta) \leqq \exp(-v_n z^4)$$

with

$$v_n = v_{n_0}\left(\frac{2}{c^2}\right)^{n-n_0}\prod_{j=n_0}^{n}(1+\alpha^j)^{-1}.$$

In order to use the contractive properties of $\mathscr{A}_n$ we shall employ a more detailed representation of $q_n(z;\beta)$ for $n_i < n \leqq n_{i+1}$. Namely, let us write

2′) for $|z| < \dfrac{d\sqrt{n}}{\sqrt{a_0}}$

$$q_n(z;\beta) = \sum_{j=2}^{N} \delta_n^{(j)} G_{2j}(z;\gamma_n) + H_n(z;\beta) + T_n(z;\beta)$$

where $N = N(c)$ is so big that

$$\xi^N < \frac{1}{2}, \left(\frac{2}{c_0^N}\right)^{\omega - 1} e^{d^2} < \frac{q}{2}, \quad c_0 = \xi c > 1 ;$$

$$K_n^{(1)} \left(\frac{2}{c^2}\right)^n < -\delta_n^{(2)} < K_n^{(2)} \left(\frac{2}{c^2}\right)^n ,$$

$$|\delta_n^{(j)}| < 2q^{n_i} \left(\frac{2}{c_0^j}\right)^{n-n_i-1} , \quad j = 3, \ldots, N .$$

The function $H_n(z; \beta)$ is extended to the whole segment

$$D_n = \left[-\frac{d\sqrt{n}}{\sqrt{a_0}}, \frac{d\sqrt{n}}{\sqrt{a_0}}\right]$$

as a C^1-function and $H_n(z; \beta) = 0$ for $z \notin D_n$ and

$$\left\| \sum_{j=2}^{N} \delta_n^{(j)} G_{2j}(z; \gamma_n) + H_n(z; \beta) \right\|_{C^m(D_n)} \leqq K_m 3^{n-n_i} q^{n_i} ,$$

$m = 0, 1$, $K_m = K_m(c)$ is a constant,

$$(H_n, G_{2j})_{\mathscr{H}(\gamma_n)} = \int_{z \in D_n} H_n(z; \beta) G_{2j}(z; \gamma_n) e^{-\gamma_n z^2} dz = 0, \quad j = 0, 1, \ldots, N ,$$

$$\|H_n\|_{\mathscr{H}(\gamma_n)} \leqq 2q^{n_i} \left(\frac{2}{c^{N+1}}\right)^{n-n_i-1} .$$

All the functions $\delta_n^{(j)} = \delta_n^{(j)}(\beta)$, $2 \leqq j \leqq N$, $H_n(z; \beta)$, $T_n(z; \beta)$ are differentiable functions of $\beta \in \boldsymbol{\beta}_n$ and

$$|T_n(z; \beta)| < (\xi q)^{n_{i+1}} 3^{n-n_{i+1}} .$$

It is very important that in 2), 2′) the main role is played by the projection of q_n on the third eigenspace generated by $G_4(z; \gamma_n)$ and that this projection is negative. From this fact it follows that $1 + q_n(z; \beta)$ has the form drawn in Fig. 1:

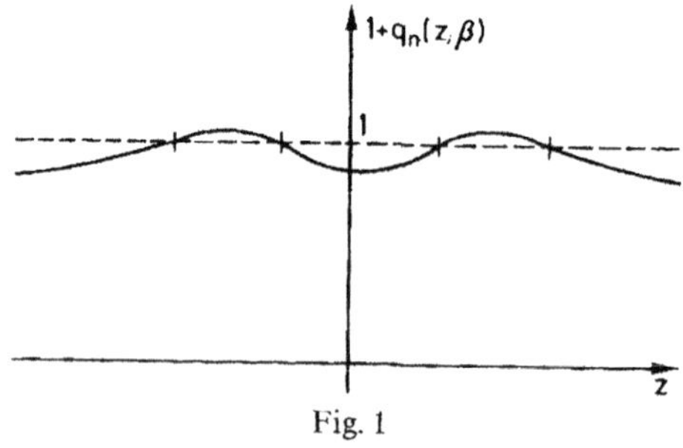

Fig. 1

Let us denote by $(\mathscr{U}_n)$ the properties 1), 2), 3) and by $(\mathscr{V}_n)$ the properties 1), 2′), 3). Now Theorem 1 will follow from the next three lemmas.

Lemma 1. *$(\mathscr{V}_n)$ implies $(\mathscr{U}_n)$ for $n=n_{i+1}$ if $N=N(c)$ is sufficiently large.*

Lemma 2. *For $n_i<n\leqq n_{i+1}$, q_{n+1} can be chosen in such a way that $(\mathscr{V}_n)$ implies $(\mathscr{V}_{n+1})$.*

Lemma 3. *For $n=n_i$, q_{n+1} can be chosen in such a way that $(\mathscr{U}_n)$ implies $(\mathscr{V}_{n+1})$.*

Proof of Lemma 1. Only the implication 2′)$\Rightarrow$2) must be proved. From 2′) one has

$$\int_{z\in D_n} |H_n(z;\beta)|^2\,dz \leqq e^{d^2 n} \int_{z\in D_n} |H_n(z;\beta)|^2 e^{-\gamma_n z^2}\,dz\,; \tag{13}$$

$$\|H_n(z;\beta)\|_{C^1(D_n)} \leqq \left\| \sum_{j=2}^{N} \delta_n^{(j)} G_{2j}(z;\gamma_n) + H_n(z;\beta)\right\|_{C^1(D_n)}$$
$$+ \sum_{j=2}^{N} |\delta_n^{(j)}|\,\|G_{2j}(z;\gamma_n)\|_{C^1(D_n)} \leqq 1\,,$$

because $N=N(c)$ is fixed, $\|G_{2j}(z;\beta)\|_{C^1(D_n)}$ increase as some power of n, $|\delta_n^{(j)}|$ decrease as a geometric progression and n_0 can be assumed to be sufficiently large.

Let us introduce the function $\varphi(z)$ where

$$\varphi(z)=\begin{cases}1+z & -1\leqq z\leqq 0\\ 0 & \text{for other } z\end{cases}$$

From the inequality $\left|\dfrac{d}{dz}H_n(z;\beta)\right|<1$ one has

$$|H_n(z;\beta)| \geqq |H_n(z_0;\beta)\,\varphi(z-z_0)|$$

for any $z_0>0$, $z_0<\dfrac{d\sqrt{n}}{\sqrt{a_0}}$. This gives

$$\|H_n(z;\beta)\|_{L^2} \geqq |H_n(z_0;\beta)|\,\|\varphi(z-z_0)\|_{L^2} = \sqrt{\tfrac{1}{3}}\,|H_n(z_0;\beta)|\,.$$

Now the desired estimate for the $C(D_n)$-norm of $H_n(z;\beta)$ follows from the $\mathscr{H}(\gamma_n)$-norm of $H_n(z;\beta)$. Other terms in $R_n(z;\beta)$ have good C-norms, as can be seen directly from 2′). Thus we have the desired estimate for the C-norm of $R_n(z;\beta)$. Q.E.D.

Proof of Lemma 2. We shall use an important local property of the expression (7) for $1+\tilde{q}_{n+1}(z;\beta)$. Namely, the part of the sum in (7) for u, $|u|>\dfrac{d_1\sqrt{n}}{\sqrt{a_0}}$, with some constant $d_1=d_1(c)$ gives a value which is less then $\varepsilon = 5e^{-d_1^2 n}$ because from 2′) and 3) one has $0 \leqq 1 + q_n(z;\beta) \leqq 2$. Therefore $\tilde{q}_{n+1}(z_0;\beta)$ depends mainly on the values of $q_n(z;\beta)$ for $\left|z - \frac{z_0}{\sqrt{c}}\right| < \frac{d_1\sqrt{n}}{\sqrt{a_0}}$ with ε-error. In particular for

$$z \in D_{n+1} = \left[-\frac{d\sqrt{n+1}}{\sqrt{a_0}}, \frac{d\sqrt{n+1}}{\sqrt{a_0}}\right]$$

$\tilde{q}_{n+1}(z;\beta)$ is defined with ε-error by the values of $q_n(z;\beta)$ for $z \in D_n$ because it follows from the Condition 2 that

$$1 - \frac{1}{\sqrt{c}} > \frac{d_1}{d}.$$

Now we proceed to construct $q_{n+1}(z;\beta)$. First of all let us include the part of the sum in (7) with $|u| \geqq \frac{d_1\sqrt{n}}{\sqrt{a_0}}$ in $T_{n+1}(z;\beta)$. If d_1 is sufficiently large this term will satisfy the estimates for T_{n+1}.

Now we shall estimate the remainder term $S_n(q_n) = S_n^{(1)}(q_n) + S_n^{(2)}(q_n)$ [see (9)]. Due to the quadratic character of $S_n^{(1)}(q_n)$ one has

$$\|S_n^{(1)}(q_n)\|_{C(D_{n+1})} \leqq 1,1 \cdot \|q_n\|^2_{C(D_n)}.$$

This inequality shows that $S_n^{(1)}(q_n)$ can be also included in $T_{n+1}(z;\beta)$ if the estimate of $\|q_n\|^2_{C(D_n)}$ which follows from 2′) is better than the estimate of $T_{n+1}(z;\beta)$ in 2′). The last assertion is true when $\omega - 1$ is sufficiently small.

Let us denote

$$B_n q(z) = 2\sqrt{\frac{2\lambda_n}{\pi}} \int_{u\in D_n} e^{-2\lambda_n u^2} q\left(\frac{z}{\sqrt{c}} - u\right) du,$$

$$\tilde{B}_n q(z) = 2\sqrt{\frac{2\lambda_n}{\pi}} \sum_{u\in D_n} e^{-2\lambda_n u^2} q\left(\frac{z}{\sqrt{c}} - u\right) \Delta n.$$

It is easy to see that

$$\|\tilde{B}_n q\|_C \leqq 2,1\, \|q\|_C.$$

So we can include the term $\tilde{B}_n T_n(z;\beta)$ into $T_{n+1}(z;\beta)$. Next we consider instead of the function $\tilde{B}_n\left(\frac{1}{2} + \sum_{j=2}^{N} \delta_n^{(j)} G_{2j}(z;\gamma_n) + H_n(z;\beta)\right)$ the function $\mathscr{A}_n\left(\frac{1}{2} + \sum_{j=2}^{N} \delta_n^{(j)} G_{2j}(z;\gamma_n) + H_n(z;\beta)\right)$. The error can be estimated using the C^1-estimate of 2′) of the function $\sum_{j=2}^{N} \delta_n^{(j)} G_{2j}(z;\gamma_n) + H_n(z;\beta)$ and the following inequalities:

$$\|(\tilde{B}_n - B_n)\, q\|_C \leqq \|q\|_{C^1}\, \Delta_n$$

(which follow from a simple interpolation formula) and

$$\|(\mathscr{A}_n - \mathscr{B}_n)\, q\|_C = \left\| 2\sqrt{\frac{2\lambda_n}{\pi}} \int\limits_{|u| \geqq \frac{d_1\sqrt{n}}{\sqrt{a_0}}} e^{-2\lambda_n u^2} q\left(\frac{z}{\sqrt{c}} - u\right) du \right\|_C$$
$$\leqq 2\|q\|_C\, e^{-d_1^2 n}.$$

Due to these inequalities we include the error into $T_{n+1}(z;\beta)$. From the property 3° of the operator $\mathscr{A}_n$ it follows that $\mathscr{A}_n(\frac{1}{2}) = 1$ and

$$\mathscr{A}_n\left(\sum_{j=2}^{N} \delta_n^{(j)} G_{2j}(z;\gamma_n) + H_n(z;\beta)\right) = \sum_{j=2}^{N} \delta_n^{(j)} \frac{2}{c^j} G_{2j}(z;\gamma_n)$$
$$+ \mathscr{A}_n H_n(z;\beta)\,, \qquad H_n \perp G_{2j} \quad \text{in} \quad \mathscr{H}(\gamma_n)\,, \qquad j = 0, 1, \ldots, N\,,$$

and consequently

$$\|\mathscr{A}_n H_n\|_{\mathscr{H}(\gamma_n)} \leqq \frac{2}{c^{N+1}} \|H_n\|_{\mathscr{H}(\gamma_n)}\,.$$

It follows from the last estimate that

$$\mathscr{A}_n H_n(z;\beta) = \sum_{j=0}^{N} \tilde{\delta}_{n+1}^{(j)} G_{2j}(z;\gamma_n) + \tilde{H}_{n+1}(z;\beta)$$

where $\tilde{H}_{n+1}(z;\beta)$ is a smooth function of $z \in D_{n+1}$, $\tilde{H}_{n+1}(z;\beta) = 0$ for

$$z \notin D_{n+1},\ (\tilde{H}_{n+1}, G_{2j})_{\mathscr{H}(\gamma_n)} = 0,\ j = 0, 1, \ldots, N,\ |\tilde{\delta}_{n+1}^{(j)}|\,,$$
$$\|\tilde{H}_{n+1}\|_{\mathscr{H}(\gamma_n)} \leqq \frac{2}{c^N} \|H_n\|_{\mathscr{H}(\gamma_n)}\,.$$

Therefore denoting $\tilde{\tilde{\delta}}_{n+1}^{(j)} = \frac{2}{c^j} \delta_n^{(j)} + \tilde{\delta}_{n+1}^{(j)}$ we have

$$1 + \tilde{q}_{n+1}(z;\beta) = 1 + \sum_{j=0}^{1} \tilde{\delta}_{n+1}^{(j)} G_{2j}(z;\gamma_n) + \sum_{j=2}^{N} \tilde{\tilde{\delta}}_{n+1}^{(j)} G_{2j}(z;\gamma_n)$$
$$+ \tilde{H}_{n+1}(z;\beta) + \tilde{T}_{n+1}(z;\beta)\,,$$

where $\tilde{\tilde{\delta}}_{n+1}^{(j)}, \tilde{H}_{n+1}, \tilde{T}_{n+1}$ satisfy 2′). Let us put

$$\gamma_{n+1} = 2\left(a_0(\beta) + \left(\frac{2}{c}\right)^{n+1} b_n(\beta)\right)\left(1 - \frac{1}{c}\right)$$

and rewrite the last equality putting γ_{n+1} instead of γ_n

$$1 + \tilde{q}_{n+1}(z;\beta) = 1 + \sum_{j=0}^{1} \delta_{n+1}^{(j)} G_{2j}(z;\gamma_{n+1}) + \sum_{j=2}^{N} \delta_{n+1}^{(j)} G_{2j}(z;\gamma_{n+1})$$
$$+ H_{n+1}(z;\beta) + T_{n+1}(z;\beta)\,,$$

where $\delta^{(j)}_{n+1}, j=2, \ldots, N, H_{n+1}, T_{n+1}$ also satisfy 2′) because

$$|\gamma_n - \gamma_{n+1}| < \left(\frac{2}{c}\right)^{n+1} |b_n(\beta)| < \frac{2}{c}\varrho^n, \quad |\delta^{(j)}_{n+1}| < |\gamma_n - \gamma_{n+1}|\, |\delta^{(2)}_n| \leqq \left(\frac{2\varrho}{c^2}\right)^n,$$

$$j=0, 1 .$$

For $z \in D_{n+1}$

$$1 + \sum_{j=0}^{1} \delta^{(j)}_{n+1} G_{2j}(z; \gamma_{n+1}) = e^{\sum_{j=0}^{1} \delta^{(j)}_{n+1} G_{2j}(z; \gamma_{n+1})} + T^{(1)}_{n+1}(z; \beta)$$

where $T^{(1)}_{n+1}$ satisfies the estimate of T_{n+1} in 2′). Now we can define q_{n+1} from the expression

$$1 + q_{n+1}(z; \beta) = e^{-\sum_{j=0}^{1} \delta^{(j)}_{n+1} G_{2j}(z; \gamma_{n+1})} (1 + \tilde{q}_{n+1}(z; \beta)) .$$

It follows from the above estimates that with this definition of $q_{n+1}(z; \beta)$ 2′) is true.

The function $\sum_{j=0}^{1} \delta^{(j)}_{n+1} G_{2j}(z; \gamma_{n+1})$ is the even quadratic polynomial of z. Thus we have the representation (5) for $f_{n+1}(z; \beta)$ where the Gaussian multiplier is

$$\tilde{L}_{n+1}(\beta) \exp\left(-\left(a_0(\beta) + \left(\frac{2}{c}\right)^{n+1} b_n(\beta)\right) z^2 + \sum_{j=0}^{1} \delta^{(j)}_{n+1} G_{2j}(z; \gamma_{n+1})\right)$$

$$= L_{n+1}(\beta) \exp\left(-\left(a_0(\beta) + \left(\frac{2}{c}\right)^{n+1} b_{n+1}(\beta)\right) z^2\right).$$

The last expression gives also the definition of the $b_{n+1}(\beta)$. It follows that

$$\left(\frac{2}{c}\right)^{n+1} |b_n(\beta) - b_{n+1}(\beta)| < K\left(\frac{2\varrho}{c^2}\right)^n, \qquad K = K(c, \beta^-_{n_0}, \beta^+_{n_0})$$

is a constant. Therefore

$$\left(\frac{2}{c}\right)^{n+1} b_{n+1}(\beta^+_n) > \left(\frac{2}{c}\right)^{n+1} b_n(\beta^+_n) - K\left(\frac{2\varrho}{c^2}\right)^n = \varrho^n\left(\frac{2}{c} - K\left(\frac{2}{c^2}\right)^n\right) > \varrho^n$$

and in a similar way $\left(\frac{2}{c}\right)^{n+1} b_{n+1}(\beta^-_n) < -\varrho^n$. Furthermore $b_{n+1}(\beta)$ is a differentiable function of β and we can find a segment $\boldsymbol{\beta}_{n+1} = [\beta^-_{n+1}, \beta^+_{n+1}] \subset \boldsymbol{\beta}_n$ satisfying 1).

Next we consider the estimate 3) for $1 + q_{n+1}(z; \beta)$. It can be easily seen from 2) that the estimate 3) is true not only for $|z| > \frac{d\sqrt{n}}{\sqrt{a}}$ but also for $|z|>d_2$ where d_2 is some constant not depending on n. Therefore $\{z: 1+q_{n+1}(z;\beta)>1\} \subset D=[-d_2, d_2]$ and we can write for all z

$$0 \leqq 1+q_n(z;\beta) \leqq e^{-\nu_n z^4}(1+\mu_n \chi_D(z))$$

where $\nu_n=\nu_{n_0}\left(\frac{2}{c^2}\right)^{n-n_0} \prod_{j=n_0}^{n}(1+\alpha^j)^{-1}$, $\mu_n=2|\delta_n^{(2)}|$, and χ_D is the indicator of D. Substitution of the last inequality into (7) gives

$$1+\tilde{q}_{n+1}(z;\beta)<\sqrt{\frac{2\lambda_n}{\pi}} \sum_u e^{-2\lambda_n u^2} e^{-\nu_n\left(\frac{2}{c^2}z^4+\frac{12}{c}z^2u^2+2u^4\right)}$$

$$\cdot\left(1+\mu_n\chi_D\left(\frac{z}{\sqrt{c}}-u\right)\right)\left(1+\mu_n\chi_D\left(\frac{z}{\sqrt{c}}+u\right)\right)=a_1+a_2\mu_n+a_3\mu_n^2 .$$

It is easy to see that for

$$0 \leqq a_1<\exp\left(-\nu_n \frac{2}{c^2} z^4\right)\left(1-\frac{K_0}{n+1}\right)$$

where K_0 is a constant not depending on n,

$$0 \leqq a_2<K_1 \exp\left(-\nu_n \frac{2}{c^2} z^4\right),$$

$$a_3=0 \quad \text{because} \quad \chi_D\left(\frac{z}{\sqrt{c}}-u\right)\chi_D\left(\frac{z}{\sqrt{c}}+u\right)=0 \quad \text{for} \quad |z|>\frac{d_2\sqrt{n+1}}{\sqrt{a_0}}.$$

This gives

$$0 \leqq 1+\tilde{q}_{n+1}(z;\beta) \leqq \exp\left(-\nu_n \frac{2}{c^2} z^4\right)\left(1-\frac{K_0}{n+1}+K_1\mu_n\right)$$

$$\leqq \exp\left(-\frac{2}{c^2}\nu_n z^4\right).$$

Now 3) for $1+q_{n+1}(z;\beta)$ is a simple consequence of the last inequality. Lemma 2 is proved.

Proof of Lemma 3. We rewrite the formula (7) in the following way

$$\tilde{q}_{n+1}(z;\beta)=\tilde{\mathscr{A}}_n q_n(z;\beta)+S_n^{(1)}(q_n)(z;\beta)$$

where

$$\tilde{\mathscr{A}}_n q(z)=2\sqrt{\frac{2\lambda_n}{\pi}} \sum_u e^{-2\lambda_n\left(\frac{z}{\sqrt{c}}-u\right)^2} q(u)\,\Delta_n \tag{14}$$

is the linear part of the transformation (7) and $S_n^{(1)}(q_n)$ defined in (9), is the nonlinear part of this transformation. One can easily verify the following properties of the operator $\tilde{\mathscr{A}}_n$:

1°. The formula (14) is meaningful for all $z \in \mathbb{R}^1$ even if $q_n(z;\beta)$ is defined only for $z \in M_n$ and

$$\|\tilde{\mathscr{A}}_n q\|_{C(\mathbb{R}^1)} < 3\|q\|_{C(M_n)}$$
$$\|\tilde{\mathscr{A}}_n q\|_{C^1(\mathbb{R}^1)} < K\|q\|_{C(M_n)} \qquad K \text{ is a constant}$$

not depending on q and n.

2°. For $|z| < \dfrac{d\sqrt{n+1}}{\sqrt{a_0}}$

$$\tilde{\mathscr{A}}_n G_4(z;\gamma_n) = \mathscr{A}_n G_4(z;\gamma_n) + (\tilde{\mathscr{A}}_n - \mathscr{A}_n)\, G_4(z;\gamma_n) = \frac{2}{c^2} G_4(z;\gamma_n) + T^{(1)}_{n+1}(z;\beta)$$

where $\mathscr{A}_n$ is the integral operator, defined in (8) and $T^{(1)}_{n+1}(z;\beta)$ satisfies the estimate of the $T_{n+1}(z;\beta)$ in 2′).

It follows from these properties of the operator $\tilde{\mathscr{A}}_n$ and from the estimate of $S^{(1)}_n(q_n)(z;\beta)$ during the proof of Lemma 2 that

$$\tilde{q}_{n+1}(z;\beta) = \frac{2}{c^2}\,\delta_n G_4(z;\gamma_n) + \tilde{\mathscr{A}}_n R_n(z;\beta) + T_{n+1}(z;\beta)$$

where the sum $\dfrac{2}{c^2}\,\delta_n G_4(z;\gamma_n) + \tilde{\mathscr{A}}_n R_n(z;\beta)$ satisfies the C^m-estimate, $m=0, 1$, of 2′) because of property 1° of the operator $\tilde{\mathscr{A}}_n$ and $T_{n+1}(z;\beta)$ satisfies the corresponding estimate of 2′) too. Next we decompose the sum $\dfrac{2}{c^2}\,\delta_n G_4(z;\gamma_n) + \tilde{\mathscr{A}}_n R_n(z;\beta)$ (this decomposition is unique):

$$\frac{2}{c^2}\,\delta_n G_4(z;\gamma_n) + \tilde{\mathscr{A}}_n R_n(z;\beta) = \sum_{j=0}^{N} \delta^{(j)}_{n+1} G_{2j}(z;\gamma_{n+1}) + H_{n+1}(z;\beta) \qquad (15)$$

where $H_{n+1}(z;\beta)=0$ for $|z| > \dfrac{d\sqrt{n+1}}{\sqrt{a_0}}$ and

$$(H_{n+1}, G_{2j})_{\mathscr{H}(\gamma_{n+1})} = \int_{z\in D_{n+1}} H_{n+1}(z;\beta)\, G_{2j}(z;\gamma_{n+1})\, e^{-\gamma_{n+1}z^2} dz = 0\,,$$
$$j=0, 1, \dots, N\,.$$

The desired estimates of $\delta^{(j)}_{n+1}$, $j=2, \dots, N$ and $H_{n+1}(z;\beta)$ of 2′) follow from the C-estimate of the $R_n(z;\beta)$ of 2) and from the property 1° of the operator $\tilde{\mathscr{A}}_n$. At last we annul the projections $\delta^{(j)}_{n+1} G_j(z;\gamma_{n+1})$, $j=0, 1$, in the formula (15) on the expanding eigenvectors of the operator $\mathscr{A}_{n+1}$ in the same way as in the proof of the Lemma 2, i.e. changing a little the Gaussian multiplier in the representation (5). One can see that for $q_{n+1}(z;\beta)$ all the estimates of 2′) are true. The proof of the properties 1) and 3) of the function $q_{n+1}(z;\beta)$ is just the same as we used during the proof of Lemma 2. This completes the proof of Lemma 1 and also the proof of Theorem 1.

Proofs of Theorems 2, 3 will be published in another paper.

Acknowledgments. The authors express their gratitude to Professors R. L. Dobrushin, O. Lanford, I. M. Lifshitz, R. A. Minlos for valuable discussions. We thank very much Prof. Lanford for his help in preparing this text for publication.

References

1. Onsager, L.: Phys. Rev. **65**, 117 (1944)
2. Huang, K.: Statistical Mechanics. New York-London: John Wiley and Sons Inc. 1963
3. Lieb, E., Mattis, D.: Mathematical Physics in one dimension. New York: Academic Press 1966
4. Dyson, F.: Commun. math. Phys. **12**, 91 (1969)
5. Dyson, F.: Commun. math. Phys. **12**, 212 (1969)
6. Dyson, F.: Commun. math. Phys. **21**, (4), 269—283 (1971)
7. Baker, Jr. G. A.: Phys. Rev. **B5**, 2622 (1972)
8. Fisher, M., Shang-Keng Ma, Nickel, .: Phys. Rev. Letters **29**, (4) 917 (1972)
9. Wilson, K., Kogut, I.: The Renormalization group and ε-expansion. Institute for Advanced Studies, Princeton, July, 1972
10. Kadanoff, L. *et al.*: Rev. Mod. Phys. **39**, 395 (1967)
11. Bateman, G., Erdelyi, A.: Higher transcendental functions, Vol. II. New York, Toronto, London: McGraw-Hill Book Company 1955

P. M. Bleher
Ja. G. Sinai
Landau Institute
for Theoretical Physics
Academy of Sciences
Vorobiev Chausse, 2
Moscow V-233, USSR

Comments

Hierarchical models of statistical mechanics were introduced by F. Dyson (see [D]) to prove the existence of phase transitions in one-dimensional lattice models with long-range interactions.

Later they become interesting in connection with the theory of critical point and the Renormalization group method. The first results along this line were obtained in the paper by G. Baker (see [B]).

In this paper, we formulate the basic hypothesis for the critical point and derive the equation describing the distribution of the normalized spin at β_{cr}. In this paper, we consider the case when the solutions of this equation are Gaussian.

Following the general approach of the Renormalization group method we find conditions under which the Gaussian solution is stable and proves some limit theorems showing the convergence of distributions of normalized spins to the Gaussian limit. Later P. M. Bleher studied in more detail the behavior of hierarchical models near β_{cr} (see [Bl]).

References

[B] G. Baker, Ising model with a scaling interaction, *Phys. Rev. B5* (1972), Issue 7, 2622–2633.

[Bl] P. M. Bleher, The phenomenon of phase separation in the φ^4-hierarchical model, *Teoret. Mat. Fiz.*, **61** (1984), no. 2, 226–240; translated in *Theor. Math. Phys.*

[CE] P. Collet, J-P. Eckmann, A renormalization group analysis of the hierarchical model in statistical mechanics, *Lecture Notes in Phys.*, **74** (1978), Springer-Verlag, 119 p.

[D] F. Dyson, Existence of a phase transition in a one-dimensional Ising ferromagnet, *Comm. Math. Phys.*, **12** (1969), 91; **12** (1969), 212; **21** (1971), no. 4, 269–283.

[DF] R. L. Dobrushin, J. Fritz, Non-equilibrium dynamics of two-dimensional infinite particle systems with a singular interaction, *Comm. Math. Phys.*, **57** (1977), no. 1, 67–81.

[DG] R. L. Dobrushin, V. M. Gertzik, Gibbs stats in a lattice model with interaction at two stages (Russian), *Funkcional Anal. i Priložen.*, **8** (1974), no. 4, 23–30.

[FS] M. Fisher, W. Selke, Low temperature analysis of the axial next-nearest neighbour Ising model near its multiphase point, *Phil. Trans. Royal Society*, **302** (1981), no. 1.

[GKSZ] A. Gabrielov, V. Keilis-Borok, Ya. G. Sinai, I. Zaliapin, Statistical properties of the cluster dynamics of the systems of statistical mechanics, in *Boltzmann's Legacy* published by *European Math. Society*, 2008, 203–216.

[K] A. Kerimov, Interface sharpness in the Ising model with long-range interaction, *J. Stat. Phys.*, **52** (1988), no. 1–2, 69–98.

[L1] O. Lanford, The classical mechanics of one-dimensional systems of infinitely many particles. An existence theorem, *Comm. Math. Phys.*, **9** (1968), 176–191.

[MPP] C. Marchioro, A. Pellegrinotti, E. Presutti, Existence of time evolution for ν-dimensional statistical mechanics, *Comm. Math. Phys.*, **40** (1975), 175–185.

[PS1] S. A. Pigorov, Ya. G. Sinai, Phase transitions of the first kind for small perturbations of the Ising model, *Funkcional. Anal. i Priložen.*, **8** (1974), no. 1, 25–30; translated in *Funct. Anal. Appl.*, **8** (1974), no. 1, 21–25.

[PS2] S. A. Pigorov, Ya. G. Sinai, Phase diagrams of classical lattice systems, *Teoret. Mat. Fiz.*, **25** (1975), no. 3, 358–369; translated in *Theor. Math. Phys.*, **25** (1975), no. 3, 1185–1192.

[PS3] S. A. Pigorov, Ya. G. Sinai, Phase diagrams of classical lattice systems (continuation), *Teoret. Mat. Fiz.*, **26** (1976), no. 1, 61–76; translated in *Theor. Math. Phys.*, **26** (1976), no. 1, 39–49.

[PS4] S. A. Pigorov, Ya. G. Sinai, Ground states in the two-dimensional boson quantum field theory, *Ann. Phys.*, **109** (1977), no. 2, 393–400.

[Pr] E. Presutti, *Scaling Limits in Statistical Mechanics and Microstructures in Continuum Mechanics*, Springer-Verlag, 2008, 467 p.

[S1] Ya. G. Sinai, Theory of phase transitions, *International Series in Natural Philosophy*, **108**, Pergamon Press, Oxford-Elmsford, 1982, 150 p.

[Z1] M. Zahradnik, An alternate version of Pirogov-Sinai theory, *Comm. Math. Phys.*, **94** (1984), 559–581.

In 1993, Ya. G. Sinai received the Honorary Degree from Warsaw University. On this photo Ya. G. Sinai is shown with the rector of Warsaw University (left) and Professor K. K. Krzyżewsky from Mathematics Department of Warsaw University.

In 2005, Ya. G. Sinai was awared the Honorary Degree from Hebrew University of Jerusalem. This is the view of Israel and Jordan river from the place where the ceremony took place.

Ya. G. Sinai and J. Lebowitz during one of the famous Rutgers Meetings (2001).

Ya. G. Sinai and L. A. Pastur.

Ya. G. Sinai and V. E. Zakharov.

Moscow, 1965.

Lena and Yakov Sinai are crossing a mountain river in Tyan-Shan mountain in Kyrgystan in the 70s.

Part III
Mathematical Physics

THE ONE-DIMENSIONAL SCHRÖDINGER EQUATION WITH A QUASIPERIODIC POTENTIAL

E. I. Dinaburg and Ya. G. Sinai

§1. Statement of the Problem and Formulation of the Results

The spectrum and the eigenfunctions of the Schrödinger equation with a periodic potential were described by Bloch (see, for example, [1], and the mathematical theory can be found in [2]). Each eigenfunction obtained by Bloch is a product of a plane wave exp(i(k, x)) and of a periodic function. The spectrum has a zone structure, i.e., it decomposes into infinitely many sections separated by "forbidden zones." In general there are infinitely many forbidden zones, and in the one-dimensional case they are asymptotically located in the vicinity of the points E_n, where $\pm\sqrt{E_n} = n\omega/2 = n/2T$, T being the period of the potential. The asymptotic properties of the decreasing widths of forbidden zones under various assumptions with regard to the potential u(x) were studied in [3-5].

In this paper we shall study the one-dimensional stationary Schrödinger equation

$$-\frac{d^2\psi}{dx^2} + u(x)\psi = E\psi, \qquad -\infty < x < \infty, \tag{1.1}$$

with a quasiperiodic potential u(x). The latter signifies that u(x) admits a representation

$$u(x) = \sum c_{n_1 n_2 \ldots n_k} e^{i(n_1\omega_1 + \ldots + n_k\omega_k)x},$$

where $\omega_1, \ldots, \omega_k$ is a given finite set of frequencies, and the Fourier coefficients $c_{n_1 \ldots n_k} = c_n$ are such that $\sum_n |c_n|^2 < \infty$.

For the method used below we need the following additional assumptions:

A_1) The condition of noncommensurability of the frequencies $\omega_1, \ldots, \omega_k$: $\left|\sum_{i=1}^{k} n_i\omega_i\right| \geqslant \frac{C_1}{|n|^{k+1}}$, where C_1 is a positive constant, and $|n| = \sum_{i=1}^{k} |n_i|$.

A_2) The analyticity of the potential u(x): $|c_n| \leqslant C_2 e^{-\rho|n|}$, where C_2 and ρ are positive constants.

Without further stipulation, we shall henceforth assume that the conditions A_1 and A_2 are satisfied.

It is easy to show (see, for example, [6]) that condition A_1 holds for almost every point $\omega = (\omega_1, \ldots, \omega_k)$ of k-dimensional space. Condition A_2 signifies that there exists an analytic function $U(z_1, \ldots, z_k)$ inside the strip $\Pi_\rho = \{(z_1, \ldots, z_k) : |\operatorname{Im} z_i| < \rho\}$ that is periodic with a period 2π in each variable z_i, that is real for $|\operatorname{Im} z_i| = 0$, $i = 1, \ldots, k$, and such that $u(x) = U(\omega_1 x, \ldots, \omega_k x)$.

Equation (1.1) can be regarded as a particular case of a one-dimensional Schrödinger equation with a random potential, although the properties of randomness are not very well exhibited here.

Some examples of one-dimensional Schrödinger equations with a quasiperiodic potential have recently been given by Novikov [7].

Moscow State University. Translated from Funktsional'nyi Analiz i Ego Prilozheniya, Vol. 9, No. 4, pp. 8-21, October-December, 1975. Original article submitted May 24, 1974.

Y.G. Sinai, *Selecta II: Probability Theory, Statistical Mechanics, Mathematical Physics and Mathematical Fluid Dynamics*, DOI 10.1007/978-1-4419-6205-8_11, © Springer Science+Business Media, LLC 2010

Let us write $\lambda_n = 1/2(n, \omega)$. The principal result of this paper is the following

THEOREM 1. For any positive ε there exist two positive constants $C'(\varepsilon)$ and $C''(\varepsilon)$ such that in any neighborhood $O_n = \left\{E: |\sqrt{E} - |\lambda_n|| < \frac{C'(\varepsilon)}{|\lambda_n|+1}\right\}$ there exists a unique neighborhood $\tilde{O}_n = \left\{E: |\sqrt{E} - \lambda_n'| \leqslant C''(\varepsilon)\exp\left[-\frac{|n|}{\ln^{1+\varepsilon}|n|}\right]\right\}$ such that if $E \notin \bigcup_{n=(n_1,\ldots,n_k)} \tilde{O}_n = \mathfrak{M}$, then Eq. (1.1) will have two linearly independent solutions $\psi_1(x; E) = \exp(iax)\chi_1(x; E)$, $a = a(E)$, $\psi_2 = \bar{\psi}_1$, where χ_1 is a quasiperiodic function with the vector $\omega = \{\omega_1, \ldots, \omega_k\}$ serving as the vector of independent frequencies, and $|a - \sqrt{E}| \leqslant \frac{C'''}{\sqrt{E}}$ for a constant C'''.

The neighborhoods $\tilde{O}_n$ serve as forbidden zones. Their position is slightly shifted with respect to the points λ_n, and it is rather difficult to specify the value of the center of the neighborhood λ_n'. Let us also note that Theorem 1 is of asymptotic type, since the values of E that satisfy the conditions of the theorem must be fairly large ($E \geq E_0$).

Theorem 1 is proved in §2. In §3 we study the spectral decomposition in terms of the eigenfunctions constructed in Theorem 1. More precisely, let h(x) be a strongly bounded real function with compact support. For any $E \in R^+ \setminus \mathfrak{M}$ let us form the Fourier coefficients

$$\rho_1(E) = \int_{-\infty}^{\infty} h(x)\overline{\psi_1(x; E)}\,dx, \quad \rho_2(E) = \int_{-\infty}^{\infty} h(x)\overline{\psi_2(x; E)}\,dx = \overline{\rho_1(E)}.$$

Let σ be a spectral measure corresponding to the function h, i.e., if L is a linear self-adjoint operator in $\mathcal{L}^2(-\infty, \infty)$, generated by Eq. (1.1), then $(L^r h, h) = \int_0^{\infty} E^r d\sigma(E)$, $r = 0, 1, 2, \ldots$.

THEOREM 2. The measure σ contains a component that is absolutely continuous with respect to a Lebesgue measure whose density on the set $R^+ \setminus \mathfrak{M}$ coincides almost everywhere with $\frac{|\rho_1(E)|^2}{2\pi\sqrt{E}(1+w(E))}$, where w(E) does not depend on h and $w(E) \to 0$ for $E \to \infty$. (The explicit form of w(E) is related to certain details of the proof of Theorem 1 and, therefore, it is not given here.)

The study of the solutions of Eq. (1.1) will be reduced to the study of the reducibility of a system of ordinary differential equations with almost periodic coefficients that depend on a parameter. The parameter values for which it is possible to prove the reducibility of the form needed by us correspond to the permissible values of E. The reducibility will be studied by the Kolmogorov–Arnol'd–Moser method of faster convergence. The application of this method to reducibility problems has been described in detail by Bogolyubov, Mitropol'skii, and Samoilenko in [8]; see also [9] and [10].

The authors express their gratitude to S. P. Novikov for useful discussions of the problems examined here.

§2. Proof of Theorem 1

Setting $\psi = \varphi_1$ and $\psi' = \varphi_2$, let us consider instead of Eq. (1.1) the system of linear equations:*

$$\frac{d\varphi}{dt} = A(y)\varphi, \quad \frac{dy}{dt} = \omega, \tag{2.1}$$

where $\varphi = (\varphi_1, \varphi_2)$, $y = (y_1, \ldots, y_k)$ is a point of the space R^k with coordinates $y_1, \ldots, y_k$, $A(y) = \bar{A} + P(y)$, $\bar{A} = \begin{pmatrix} 0 & 1 \\ -E & 0 \end{pmatrix}$, $P(y) = \begin{pmatrix} 0 & 0 \\ U(y) & 0 \end{pmatrix}$, $U(y) = \sum c_n e^{i(n,y)}$. In (2.1) let us effect a linear change of variables $\varphi = B\zeta_0$ with a constant matrix $B = \begin{pmatrix} 1 & 1 \\ i\sqrt{E} & -i\sqrt{E} \end{pmatrix}$. In the new variables the system (2.1) assumes the form

$$\frac{d\zeta_0}{dt} = iA_0\zeta_0 + P_0(y)\zeta_0, \quad \frac{dy}{dt} = \omega, \tag{2.2}$$

*In going over from Eq. (1.1) to the system (2.1), we are replacing the independent variable x by t. Henceforth, the independent variable in systems of equations of type (2.1) will be t, and the transition from x to t and conversely will be performed without additional explanations.

where

$$A_0 = \begin{pmatrix} \sqrt{E} & 0 \\ 0 & -\sqrt{E} \end{pmatrix}, \quad P_0(y) = \frac{U(y)}{2\sqrt{E}} \begin{pmatrix} -i & -i \\ i & i \end{pmatrix}.$$

We shall assume that for a certain value of E there exists an invertible C^1 change of variables $\zeta_0 = V(y)\,\zeta$, that reduces the system (2.2) to the form

$$\frac{d\zeta}{dt} = iC\zeta, \quad \frac{dy}{dt} = \omega, \tag{2.3}$$

where C is a diagonal matrix with constant real elements, and V(y) a periodic function of the variables $y_1, \ldots, y_k$ with period 2π. Then it is easy to see that with such a value of E the Eq. (1.1) will have solutions of the required form (see Remark 2 below). Thus, for proving Theorem 1 it suffices to study the reducibility of system (2.2) to the form (2.3).

Let us write $\Pi_\rho = \{y : |\operatorname{Im} y| = \max_{1\leqslant i\leqslant k} |\operatorname{Im} y_i| \leqslant \rho\}$ and for any matrix P(y) whose elements $P_{ij}(y)$ are periodic functions with period 2π that are analytic inside the strip Π_ρ and continuous up to its boundary, let us write

$$\|P(y)\|_\rho = \max_i \sum_j \max_{y\in\Pi_\rho} |P_{ij}(y)|, \quad \|P(y)\|_0 = \max_i \sum_j \max_{|\operatorname{Im} y|=0} |P_{ij}(y)|.$$

Let $0 < \varkappa < 1$, $\alpha > 1$ be fixed, and let $r_\alpha = \sum_{s=0}^{\infty} s^{-\alpha}$. The proof of the reducibility of system (2.2) is based on the following inductive assertion.

Inductive Lemma 2.1. Let us assume that for an $s \geq 0$ we are given the system

$$\frac{d\zeta_s}{dt} = iA_s\zeta_s + P_s(y)\zeta_s, \quad \frac{dy}{dt} = \omega, \tag{2.4$_s$}$$

where $A_s = \begin{pmatrix} a_s & 0 \\ 0 & b_s \end{pmatrix}$ is a constant matrix, and $P_s(y)$ is a 2×2 matrix that is periodic with a period 2π in each variable, analytic in the strip Π_{ρ_s}, and continuous up to its boundary, $\|P_s(y)\|_{\rho_s} \leqslant M_s$.

Let us write $l_s = a_s - b_s$, $\delta_s = \delta_0 s^{-\alpha}$, $\delta_0 = \min\left(\frac{\rho_0}{4r_\alpha}, \frac{1}{2r_\alpha}\right)$, $\rho_{s+1} = \rho_s - 2\delta_s$.

We shall assume that for a constant $C_s^{(1)} \geqslant M_s^{\frac{1-\varkappa}{4}}$ we have the inequalities $|(n,\omega) \pm l_s| \geqslant C_s^{(1)}(|n|+1)^{-(k+1)}$ for $|n| \leqslant N_s = -\frac{1}{\delta_s}\ln M_s$.

Then there exists a constant $M = M(\varkappa, \alpha) > 0$ such that for $M_s \leqslant M^{(1+\varkappa)^s}$ there exists a change of variables $\zeta_s = U_s(y)\,\zeta_{s+1}$ that reduces the system (2.4$_s$) to the form (2.4$_{s+1}$), where $a_{s+1} = a_s - id_{s,1}$, $b_{s+1} = b_s - id_{s,2}$, $d_{s,i} = \left(\frac{1}{2\pi}\right)^k \int_0^{2\pi} \ldots \int_0^{2\pi} P_{s,ii}(y)\,dy$, $i = 1, 2$. The functions $P_{s+1}(y)$ and $V_s(y)$ are periodic with a period 2π in each variable, analytic in the strip $\Pi_{\rho_{s+1}}$, and continuous up to its boundary, $\|P_{s+1}(y)\|_{\rho_{s+1}} \leqslant M_{s+1} = M_s^{1+\varkappa}$, $\|V_s(y) - \mathrm{Id}\|_{\rho_{s+1}} \leqslant 0{,}1\, M_s^{\varkappa}$.

Remark 1. In its formulation, Lemma 2.1 is close to the inductive lemma contained in [8], and it primarily differs from it by the fact that the constant $C_s^{(1)}$ is fast tending to zero for $s \to \infty$, and also by the fact that the change of variables has a finite number of nonzero Fourier components which is usual in the case of a singular [6] or in the case of a smooth, but not analytic right-hand side [8, 11].

Proof. The method of proof is similar to the method used in [8], and therefore we shall omit some of the details.

Let us write $P_{s,N_s} = \sum_{|n|\leqslant N_s} P^{(n)} e^{i(n,y)}$, $R_{s,N_s} = P_s - P_{s,N_s}$. We shall seek the change of variables in the form $\zeta_s = V_s(y)\,\zeta_{s+1} = (\mathrm{Id} + W_s(y))\,\zeta_{s+1}$. By effecting this change in (2.4$_s$), we find that if W_s is a solution of the matrix equation

$$\left(\frac{dW_s}{dy}, \omega\right) = i(A_sW_s - W_sA_s) + (P_{s,N_s} - D_s), \tag{2.5}$$

we can reduce the system (2.4_s) to the form (2.4_{s+1}) with $A_{s+1} = A_s - iD_s$, $P_{s+1} = (\mathrm{Id} + W_s)^{-1}[P_{s,N_s}W_s - W_sD_s + R_{s,N_s}(\mathrm{Id} + W_s)]$.

Here D_s is the mean of the diagonal part of the matrix P_s, and $\left(\frac{dW_s}{dy}, \omega\right)$ is a matrix such that $\left(\frac{dW_s}{dy}, \omega\right)_{i,j} = (\mathrm{grad}\, W_{s,ij}, \omega)$.

For solving Eq. (2.5), we shall expand W_s in a Fourier series: $W_s = \sum_n W_s^{(n)} e^{i(n,\nu)}$, and substitute this expansion into (2.5). Then we find that

$$W_{s,kk}^{(n)} = \frac{R_{s,kk}^{(n)}}{i(n,\omega)}, \quad k = 1, 2, \quad 0 < |n| \leqslant N_s,$$

$$W_{s,12}^{(n)} = \frac{P_{s,12}^{(n)}}{i((n,\omega) - l_s)}, \quad W_{s,21}^{(n)} = \frac{P_{s,21}^{(n)}}{i((n,\omega) + l_s)}, \quad 0 \leqslant |n| \leqslant N_s,$$

$$W_{s,ij}^{(n)} = 0, \quad i, j = 1, 2, \quad |n| > N_s,$$

$$W_{s,11}^{(0)} = W_{s,22}^{(0)} = 0.$$

If $C_s^{(1)}$ is sufficiently small, the condition $|(n,\omega)| > C_s^{(1)}(|n| + 1)^{-k-1}$ will be satisfied. Hence, follows that

$$\|W_s\|_{\rho_{s+1}} \leqslant 2\sum_n \frac{M_s}{C_s^{(1)}}(|n| + 1)^{k+1} e^{-2\delta_s |n|}.$$

It is easy to show (see [6] and [8]) that this inequality yields the bound

$$\|W_s\|_{\rho_{s+1}} \leqslant \frac{M_s}{C_s^{(1)}} \frac{C(k)}{\delta_s^{2k+1}} \leqslant M_s^{\frac{3+\varkappa}{4}} C(k) s^{(2k+1)\varkappa} \delta_0^{-(2k+1)},$$

where C(k) is a constant that depends only on k. If $M = M(\varkappa, \alpha)$ is sufficiently small, then $\|W_s\|_{\rho_{s+1}} \leqslant 0,1 M_s^{\varkappa}$.

In the same way we can prove that $\|R_{s,N_s}\|_{\rho_{s+1}} \leqslant \frac{1}{2} M_s^{1+\varkappa}$ for sufficiently small M.

Now let us estimate $\|P_{s+1}\|_{\rho_{s+1}}$. It is evident that

$$\|P_{s+1}\|_{\rho_{s+1}} \leqslant \|(\mathrm{Id} + W_s)^{-1}\|_{\rho_{s+1}}((\|P_s\|_{\rho_{s+1}} + \|R_{s,N_s}\|_{\rho_{s+1}})\|W_s\|_{\rho_{s+1}} + \|W_s\|_{\rho_{s+1}}\|D_s\|_{\rho_{s+1}} + \|R_{s,N_s}\|_{\rho_{s+1}}\|\mathrm{Id} + W_s\|_{\rho_{s+1}}).$$

Hence

$$\|P_{s+1}\|_{\rho_{s+1}} \leqslant \left(1 + \sum_{l>0} 0.1 M_s^{\varkappa l}\right)\left(\left(M_s + \frac{1}{2}M_s^{1+\varkappa}\right)0.1M_s^{\varkappa} + 0{,}1M_s^{1+\varkappa} + \frac{1}{2}M_s^{1+\varkappa}(1 + 0.1M_s^{\varkappa})\right).$$

If M is sufficiently small, then $\|P_{s+1}\|_{\rho_{s+1}} \leqslant M_s^{1+\varkappa}$. This completes the proof of the lemma.

Remark 2. Let us assume that: 1) The matrix A_s is real; 2) the matrix $P_s(y)$ has the form $\begin{pmatrix} \alpha & \beta \\ \bar{\beta} & \bar{\alpha} \end{pmatrix}$; 3) the mean $\overline{\mathrm{tr}\, P_s(y)} = \mathrm{tr}_0\, P_s(y)$ of the trace of the matrix $P_s(y)$ is equal to zero. Then A_{s+1} and P_{s+1} will also have the properties 1-3.

Indeed, from the form of the matrix $P_s(y)$ it follows that $d_{s,11} = \bar{d}_{s,22}$. From property 3 it follows that $d_{s,11} + d_{s,22} = 0$. Hence, the numbers $d_{s,11}$ and $d_{s,22}$ are purely imaginary, and therefore, the matrix A_{s+1} has property 1.

Let us denote the set of matrix functions of the form $\begin{pmatrix} \alpha & \beta \\ \bar{\beta} & \bar{\alpha} \end{pmatrix}$ by G. It is easy to verify that G is a ring with respect to the ordinary operations of matrix addition and multiplication, and if $g \in G$ is invertible then $g^{-1} \in G$. It therefore follows from the expression for W_s that if $P_s \in G$, then $W_s \in G$. Since

$$P_{s+1} = (\mathrm{Id} + W_s)^{-1}(P_{s,N_s}W_s - W_sD_s) + (\mathrm{Id} + W_s)^{-1}R_{s,N_s}(\mathrm{Id} + W_s),$$

it follows that $P_{s+1} \in G$.

Finally, $\mathrm{tr}_0\,[(\mathrm{Id}+W_s)^{-1}R_{s,N_s}(\mathrm{Id}+W_s)] = \mathrm{tr}_0\,R_{s,N_s} = 0$, since the Fourier expansion for the elements of the matrix R_{S,N_S} begins with the Fourier components for which $|n| > N_S$.

Hence

$$\mathrm{tr}_0\,P_{s+1} = \mathrm{tr}_0\,[(\mathrm{Id}+W_s)^{-1}(P_{s,N_s}W_s - W_sD_s)] =$$

$$= \mathrm{tr}_0\,[(\mathrm{Id}+W_s)^{-1}(P_{s,N_s}-D_s)\,W_s] + \mathrm{tr}_0\,[(\mathrm{Id}+W_s)^{-1}(D_sW_s - W_sD_s)].$$

The last term is equal to zero by virtue of the fact that tr AB = tr BA for any A and B. Next,

$$\mathrm{tr}_0\,[(\mathrm{Id}+W_s)^{-1}(P_{s,N_s}-D_s)\,W_s] =$$

$$= \mathrm{tr}_0\,[(\mathrm{Id}+W_s)^{-1}(P_{s,N_s}-D_s)(\mathrm{Id}+W_s)] - \mathrm{tr}_0\,[(\mathrm{Id}+W_s)^{-1}(P_{s,N_s}-D_s)] =$$

$$= -\mathrm{tr}_0\,[(\mathrm{Id}+W_s)^{-1}(P_{s,N_s}-D_s)].$$

By virtue of (2.5) this expression is equal to

$$\mathrm{tr}_0\left[(\mathrm{Id}+W_s)^{-1}\left(\frac{dW}{dy},\,\omega\right)\right] = \mathrm{tr}_0\,\frac{d}{dt}\ln(\mathrm{Id}+W_s) = 0.$$

Returning to the system (2.2), let us note that the matrices A_0 and P_0 satisfy the conditions 1-3. Hence, if the system (2.2) is reducible to the form (2.3) with the aid of successive substitutions described in the inductive lemma, then we find at each step of this process that the elements of the matrix A_S are real and with opposite signs. Hence, the elements of the matrix C are also real and with opposite signs, and the solutions of Eq. (1.1) for an appropriate value of E have the form described in Theorem 1.

COROLLARY. There exists an $E_0 = E_0(\varkappa, \alpha)$ such that for any $E > E_0$ satisfying the "noncommensurability conditions" for all $s \geq 0$

$$|(n,\omega) \pm l_s| > M_s^{\frac{1-\varkappa}{4}}(|n|+1)^{-\varkappa-1}, \quad |n| \leqslant N_s = -\frac{1}{\delta_s}\ln M_s,$$

the system (2.2) will be reducible to (2.3). Here $M_0 = M_0'/\sqrt{E_0}$, where $M_0' = \|U(y)\|_{\rho_0}$, $\rho_0 < \rho$ is a number, and l_S, M_S, and δ_S are defined in Lemma 2.1.

The reduction is realized with the aid of an invertible change of variables $\zeta_0 = V(y)\zeta$, where V(y) is an analytic matrix-valued function in the strip $\Pi_{\rho_0/2}$.

Indeed, for a given value of E the norm $\|P_0(y)\|_{\rho_0}$ in the system (2.2) does not exceed $M_0'/\sqrt{E}$. Hence, by setting $M_0 = M_0'/\sqrt{E_0}$, where E_0 is a sufficiently large number, we can perform at the s-th step a successive change of variables described in the inductive lemma, dropping the values of E that violate the conditions

$$|(n,\omega) \pm l_s| > M^{\frac{1-\varkappa}{4}}(|n|+1)^{-\varkappa-1}, \quad |n| \leqslant N_s. \tag{2.6}$$

For completing the proof of Theorem 1, it remains to study the structure of the set to be dropped.

For any $n = (n_1, \ldots, n_k)$ let us denote by s(n) the least number s such that $|n| \leq N_S$. Let us introduce the set

$$\Delta_n^{(s)} = \{l_0\colon\ |(n,\omega) \pm l_s| \leqslant M_s^{\frac{1-\varkappa}{4}}(|n|+1)^{-\varkappa-1}\}.$$

LEMMA 2.2. For a sufficiently large E_0 and $s \geq s(n)$ we have the inclusion

$$\Delta_n^{(s+1)} \subset \Delta_n^{(s)}\ \ u\ \ l\,(\Delta_n^{s(n)}) < 4\,M_s^{\frac{1-\varkappa}{4}}\,(|n|+1)^{-\varkappa-1}.$$

Proof. Let $\bar{\Delta}_n^{(s)}$ be an $M_s^{\frac{1-\varkappa}{4}}(|n|+1)^{-\varkappa-1}$-neighborhood of the point (n, ω) on the l_S-axis. Let us assume that for a fixed n it is possible to find in any neighborhood of the set $\bar{\Delta}_n^{s+1}$ such values of l_{S+1} that satisfy the conditions (2.6) up to the s-th step inclusive. It follows from the inductive lemma that $|l_{s+1} - l_s| \leqslant 2M_s$. Then we must have

$$2M_s + M_{s+1}^{\frac{1-\varkappa}{4}}(|n|+1)^{-\varkappa-1} \geqslant M_s^{\frac{1-\varkappa}{4}}(|n|+1)^{-\varkappa-1}. \tag{2.7}$$

But the left-hand side is equal to $M_s^{\frac{1-\varkappa}{4}}(|n|+1)^{-k-1}(2(|n|+1)^{k+1}M_s^{\frac{3+\varkappa}{4}}+M_s^{1+\varkappa})$. Since the values of E are sufficiently large, s(n) will be sufficiently large and the expression in parentheses will be smaller than unity. Hence (2.7) cannot hold, which proves the first assertion of the lemma.

It follows from this assertion that we must drop at the s-th step the set of values of E for which $\pm l_s$ has a distance from (n, ω) not greater than $M_{s(n)}^{\frac{1-\varkappa}{4}}(1+|n|)^{-k-1}$.

Now let us show that for a sufficiently large E_0,

$$\frac{1}{2}<\frac{dl_r}{dl_0}<2, \quad r=1,2,\dots. \tag{2.8}$$

We have

$$\frac{dl_r}{dl_0}=\frac{dl_{r-1}}{dl_0}+\frac{d(d_{r-1,11}+d_{r-1,22})}{dl_0}.$$

Hence

$$\frac{dl_{r-1}}{dl_0}-\left|\frac{d(d_{r-1,11}+d_{r-1,22})}{dl_0}\right|\leqslant\frac{dl_r}{dl_0}<\frac{dl_{r-1}}{dl_0}+\left|\frac{d(d_{r-1,11}+d_{r-1,22})}{dl_0}\right|,$$

$$\frac{dl_{r-1}}{dl_0}-2\left\|\frac{dP_{r-1}}{dl_0}\right\|_{\rho_0/2}\leqslant\frac{dl_r}{dl_0}\leqslant\frac{dl_{r-1}}{dl_0}+2\left\|\frac{dP_{r-1}}{dl_0}\right\|_{\rho_0/2}.$$

Let us show that

$$\left\|\frac{dP_r}{dl_0}\right\|_{\rho_0/2}<M_r^{\varkappa}. \tag{2.9}$$

We shall prove (2.9) by induction. It is easy to see that for r = 0 the inequality (2.9) will hold. Let us assume that (2.8) and (2.9) are valid for any r ≤ s − 1. For r = s, we then have

$$\begin{aligned}\left\|\frac{dP_s}{dl_0}\right\|_{\rho_0/2}&=\left\|\frac{d}{dl_0}\{(\mathrm{Id}+W_{s-1})^{-1}(R_{s-1,N_{s-1}}(\mathrm{Id}+W_{s-1})+\right.\\&\quad\left.+P_{s-1,N_{s-1}}W_{s-1}-W_{s-1}D_{s-1})\}\right\|_{\rho_0/2}\leqslant\\&\leqslant\left\|\frac{dW_{s-1}}{dl_0}\right\|_{\rho_0/2}\|(\mathrm{Id}+W_{s-1})^{-1}\|_{\rho_0/2}M_{s-1}^{1+\varkappa}+\|(\mathrm{Id}+W_{s-1})^{-1}\|_{\rho_0/2}\times\\&\quad\times\left(\left\|\frac{dR_{s-1,N_{s-1}}}{dl_0}\right\|_{\rho_0/2}(1+0.1M_{s-1}^{\varkappa})+M_{s-1}^{1+\varkappa}\left\|\frac{dW_{s-1}}{dl_0}\right\|_{\rho_0/2}+\right.\\&\left.+0{,}1\left\|\frac{dP_{s-1,N_{s-1}}}{dl_0}\right\|_{\rho_0/2}M_{s-1}^{\varkappa}+(M_{s-1}+M_{s-1}^{1+\varkappa})\left\|\frac{dW_{s-1}}{dl_0}\right\|_{\rho_0/2}+0{,}1M_{s-1}^{2\varkappa}+M_{s-1}\left\|\frac{dW_{s-1}}{dl_0}\right\|_{\rho_0/2}\right).\end{aligned}$$

If E_0 is sufficiently large, then $\|(\mathrm{Id}+W_{s-1})^{-1}\|_{\rho_0/2}<2$ for $E>E_0$. Let us also note that

$$\left\|\frac{dP_{s-1,N_{s-1}}}{dl_0}\right\|_{\rho_0/2}\leqslant\left\|\frac{dP_{s-1}}{dl_0}\right\|_{\rho_0/2}+\left\|\frac{dR_{s-1,N_{s-1}}}{dl_0}\right\|_{\rho_0/2}.$$

Hence

$$\left\|\frac{dP_s}{dl_0}\right\|_{\rho_0/2}<\left\|\frac{dW_{s-1}}{dl_0}\right\|_{\rho_0/2}(8M_{s-1}^{\varkappa+1}+4M_{s-1})+\left\|\frac{dR_{s-1,N_{s-1}}}{dl_0}\right\|_{\rho_0/2}(2+0{,}4M_{s-1})+0{,}4M_{s-1}^{2\varkappa}.$$

By differentiating the explicit Fourier-series expansions of W_{s-1} and $R_{s-1,N_{s-1}}$ with respect to l_0 and using (2.8) for r = s − 1, it is easy to show that for sufficiently large E,

$$\left\|\frac{dW_{s-1}}{dl_0}\right\|_{\rho_0/2}<M_{s-1}^{\varkappa/2},\quad\left\|\frac{dR_{s-1,N_{s-1}}}{dl_0}\right\|_{\rho_0/2}<M_{s-1}^{2\varkappa}.$$

Therefore $\left\|\frac{dP_s}{dl_0}\right\|_{\rho_0/2}<M_{s-1}^{\varkappa+\varkappa'}=M_s^{\varkappa}$. Hence easily follows the second assertion of the lemma.

It follows from Lemma 2.2 that the set $\Delta_n^{s(n)}$ is a segment whose length on the l_0 axis does not exceed $4M_{s(n)}^{\frac{1-\varkappa}{4}}(|n|+1)^{-k-1}=4M^{(1+\varkappa)^{s(n)}\left(\frac{1-\varkappa}{4}\right)}(|n|+1)^{-k-1}$.

This is precisely the forbidden zone in the assertion of Theorem 1. It is easy to see that $|l_{s-1} - l_s| < 2\frac{\sqrt{E_0}}{\sqrt{E}} M_s$. This shows that this zone lies in a $\frac{C'''}{|(n, \omega)|}$ -neighborhood of the point (n, ω), where C‴ is a constant. By using the expression for N_S, it is easy to see that $(1 + \varkappa)^{s(n)} > C(\varkappa, \alpha) \frac{|n|}{\ln^{\alpha}|n| \cdot |\ln M_0|}$. Hence follows the assertion of the theorem for n such that $|(n, \omega)| > 2\sqrt{E_0}$, $\varepsilon = \alpha - 1$.

By appropriate selection of the constants C′(ε) and C″(ε) it is possible to establish the validity of Theorem 1 for any n.

<u>Remark 3.</u> It follows from Theorem 1 that for any E we have

$$l([E, \infty) \cap (\bigcup_n \tilde{O}_n)) < K_1(\varepsilon) \exp\left(-\frac{K_2(\varepsilon)\sqrt{E}}{\ln^{1+\varepsilon} E}\right),$$

where $K_1(\varepsilon)$ and $K_2(\varepsilon)$ are positive constants.

<u>Remark 4.</u> It follows from Remark 2 that the matrix C in (2.3) has the form $\begin{pmatrix} a & 0 \\ 0 & -a \end{pmatrix}$, where $a > 0$ is a function of E defined on the set of admissible values of E.

Let us show that the function $a(E)$ is uniformly continuous on the set of such values of E.

Let us take a positive δ(s) such that the admissible values of E_1 and E_2 for $|E_1 - E_2| < \delta(s)$ belong to the same connected component of the set obtained from the semiaxis $E > 0$ by dropping all the forbidden zones $\tilde{O}_n$ for which $|n| \leq N_S$. It then follows from Lemma 2.2 that $\frac{1}{2} < \frac{da_s(E)}{da_0(E)} < 2$ for $E_1 < E < E_2$, and therefore $|a_s(E_1) - a_s(E_2)| < 2|\sqrt{E_1} - \sqrt{E_2}|$. On the other hand, $|a(E_1) - a(E_2)| < |a_s(E_1) - a_s(E_2)| + |a_s(E_1) - a(E_1)| + |a_s(E_2) - a(E_2)| < 2|\sqrt{E_1} - \sqrt{E_2}| + 4M_{s+1}$. By taking s sufficiently large and $\delta(\varepsilon) < \delta(s)$ sufficiently small, it is possible to achieve that the last expression will be smaller than any preassigned ε for $|E_1 - E_2| < \delta(\varepsilon)$.

Now let us show that the change of variables $\zeta_0 = (\mathrm{Id} + W)\zeta$, which reduces the system (2.2) to the form (2.3), is a uniformly continuous function of E on the set of admissible values of E.

It is easy to see that

$$\|(\mathrm{Id} + W(E_1)) - (\mathrm{Id} + W(E_2))\| = \left\|\prod_{j=0}^{\infty}(\mathrm{Id} + W_j(E_1)) - \prod_{j=0}^{\infty}(\mathrm{Id} + W_j(E_2))\right\| =$$

$$= \left\|\sum_{k=-1}^{s-1}\prod_{j=0}^{k}(\mathrm{Id} + W_j(E_1))(W_{k+1}(E_1) - W_{k+1}(E_2))\prod_{j=k+2}^{\infty}(\mathrm{Id} + W_j(E_2)) + \right.$$

$$\left. + \prod_{j=0}^{s}(\mathrm{Id} + W_j(E_1))\left(\prod_{j>s}(\mathrm{Id} + W_j(E_1)) - \prod_{j>s}(\mathrm{Id} + W_j(E_2))\right)\right\|.$$

By selecting δ(s), E_1, and E_2 in the same way as above, and by using the previous equation and the bound occurring in the proof of Lemma 2.2,

$$\left\|\frac{dM_s}{dl_0}\right\|_{\rho_0/2} = \left\|\frac{dW_s}{2d\eta_0}\right\|_{\rho_0/2} \leqslant M_s^{\varkappa},$$

it is easy to obtain the desired assertion.

<u>§3. Proof of Theorem 2</u>

Let us take a sufficiently large number E_0 and denote by $\mathfrak{M}_s$ the set of values of $E > E_0$ for which the "noncommensurability conditions" are violated at the s-th step of the reduction process described in the proof of Theorem 1,

$$\mathfrak{M}_s = \{E\colon E > E_0, |(n, \omega) \pm l_s| \leqslant M_s^{\frac{1-\varkappa}{4}}(|n| + 1)^{-k-1}, |n| < N_s\},$$

where $M_0 = \frac{M_0'}{\sqrt{E_0}}$, $\varkappa$, M_s, l_s, and N_S are defined in Lemma 2.1. (The value of E_0 will be evident from the subsequent analysis.) It is obvious that $\mathfrak{M}_s$ is a union of finitely many segments $\Delta_{s,n}$, $|n| \leqslant N_s$, $\Delta_{s,n} =$

$\{E: |(n,\omega) \pm l_s| \leqslant M_s^{\frac{1-\varkappa}{4}} (|n|+1)^{-k-1}\}$. Let us write $\mathfrak{M} = \bigcup_s \mathfrak{M}_s$. We shall say that $\Delta_{s,n}$ are segments of rank s. By virtue of Theorem 1, for sufficiently large E_0 and any $E \in [E_0, \infty) \setminus \mathfrak{M}$, the system (2.2) can be reduced to a system with constant coefficients. Let R_λ be the resolvent of the operator $-d^2/dx^2 + u(x)$ at the point $\lambda = E + ig$, $g > 0$.

LEMMA 3.1. For any $E \in [E_0, \infty) \setminus \mathfrak{M}$ and any strongly bounded finite function h(x) there exists a uniform limit

$$r(E) = \frac{1}{2\pi i} \lim_{g\to 0} [(R_\lambda h, h) - (R_{\bar\lambda} h, h)].$$

The limit function r(E) is uniformly continuous on $[E_0, \infty) \setminus \mathfrak{M}$.

Proof. The resolvent R_λ has a representation (see [2])

$$R_\lambda h = \psi_1(\lambda, x) \int_{-\infty}^{x} \frac{\psi_2(\lambda, y)}{I(\lambda)} h(y)\,dy + \psi_2(\lambda, x) \int_{x}^{\infty} \frac{\psi_1(\lambda, y)}{I(\lambda)} h(y)\,dy,$$

where $\psi_1(\lambda, x)$ and $\psi_2(\lambda, x)$ are solutions of the equation $-\psi'' + u(x)\psi = \lambda\psi$, such that

$$\int_0^\infty |\psi_1(\lambda, y)|^2 dy < \infty, \quad \int_{-\infty}^0 |\psi_2(\lambda, y)|^2 dy < \infty, \quad I(\lambda) = \det \begin{pmatrix} \psi_1 & \psi_2 \\ \psi_1' & \psi_2' \end{pmatrix}.$$

Let us write down the expression for $R_\lambda h$ in our case for $E \in [E_0, \infty) \setminus \mathfrak{M}$ and sufficiently small g. For constructing ψ_1 and ψ_2 let us again examine the reducibility of the system

$$\frac{d\zeta}{dt} = i \begin{pmatrix} \sqrt{E+ig} & 0 \\ 0 & -\sqrt{E+ig} \end{pmatrix} \zeta + \frac{U(y)}{2\sqrt{E+ig}} \begin{pmatrix} -i & -i \\ i & i \end{pmatrix} \zeta, \quad \frac{dy}{dt} = \omega. \tag{3.1}$$

Here $\sqrt{E+ig}$ is a square root of E + ig for which $\operatorname{Re}\sqrt{E+ig} > 0$.

Let us rewrite (3.1) in the form

$$\frac{d\zeta}{dt} = i \begin{pmatrix} \sqrt{E} & 0 \\ 0 & -\sqrt{E} \end{pmatrix} \zeta + \frac{U(y)}{2\sqrt{E}} \begin{pmatrix} -i & -i \\ i & i \end{pmatrix} \zeta +$$
$$+ i \begin{pmatrix} \sqrt{E+ig} - \sqrt{E} & 0 \\ 0 & -(\sqrt{E+ig} - \sqrt{E}) \end{pmatrix} \zeta + \frac{U(y)}{2} \frac{\sqrt{E} - \sqrt{E+ig}}{\sqrt{E(E+ig)}} \begin{pmatrix} -i & -i \\ i & i \end{pmatrix} \zeta,$$
$$\frac{dy}{dt} = \omega.$$

According to the results of §2, the "truncated" system

$$\frac{d\zeta}{dt} = i \begin{pmatrix} \sqrt{E} & 0 \\ 0 & -\sqrt{E} \end{pmatrix} \zeta + \frac{U(y)}{2\sqrt{E}} \begin{pmatrix} -i & -i \\ i & i \end{pmatrix} \zeta, \quad \frac{dy}{dt} = \omega$$

can be reduced for the above values of E to the form

$$\frac{d\xi}{dt} = i \begin{pmatrix} a & 0 \\ 0 & -a \end{pmatrix} \zeta, \quad \frac{dy}{dt} = \omega$$

with the aid of a substitution $\zeta = (\mathrm{Id} + W)\xi$, that is analytic in the strip $\Pi_{\rho_0/2}$ and satisfies the inequality $\|W(y)\|_{\rho_0/2} < 2M_0^{\varkappa}$. Here a is a function of E. Let us write $f(E, g) = i(\sqrt{E+ig} - \sqrt{E}) = -(g/2\sqrt{E})(1 + \varepsilon(g))$, where $|\varepsilon(g)| < |g|/E$ for sufficiently small g. It is easy to see that in the variables ξ the system (3.1) assumes the form

$$\frac{d\xi}{dt} = i \begin{pmatrix} a & 0 \\ 0 & -a \end{pmatrix} \xi + (\mathrm{Id} + W)^{-1} \begin{pmatrix} f(E,g) & 0 \\ 0 & -f(E,g) \end{pmatrix} (\mathrm{Id} + W)\xi +$$
$$+ \frac{if(E,g)}{\sqrt{E+ig}} \frac{U(y)}{2\sqrt{E}} (\mathrm{Id} + W)^{-1} \begin{pmatrix} 1 & 1 \\ -1 & -1 \end{pmatrix} (\mathrm{Id} + W)\xi, \quad \frac{dy}{dt} = \omega$$

or

$$\frac{d\xi}{dt}=\begin{pmatrix} ia+f(E,g) & 0 \\ 0 & -ia-f(E,g)\end{pmatrix}\xi+f(E,g)P_g(y)\xi,\quad \frac{dy}{dt}=\omega, \tag{3.2}$$

where $\|P_g(y)\|_{\rho_0/2}\leqslant 6M_0^{\varkappa}$.

In the system (3.2) the constant matrix has eigenvalues with distinct real parts; hence, it is easy to see that we can apply to it the reducibility theorem of [8] in the case of sufficiently small values of M_0, i.e., for sufficiently large E_0. Hence, with the aid of the substitution $\xi = (\mathrm{Id} + W_g(y))\eta$ we can reduce the system (3.2) to the form

$$\frac{d\eta}{dt}=\begin{pmatrix} ia+f_1(E,g) & 0 \\ 0 & -ia-f_2(E,g)\end{pmatrix}\eta,\quad \frac{dy}{dt}=\omega, \tag{3.3}$$

where $\|W_g\|_{\rho_0/4}\leqslant 12M_0^{\varkappa^2}$, $|f_i(E,g)-f(E,g)|<12M_0^{\varkappa}|f(E,g)|$, $i=1,\ 2$.

By returning to the equation $-\psi'' + u(x)\psi = \lambda\psi$, we find that this equation has two linearly independent solutions $\psi_1(\lambda, x)$ and $\psi_2(\lambda, x)$ for which

$$\begin{pmatrix}\psi_1 & \psi_2\\ \psi_1' & \psi_2'\end{pmatrix}=B_g(\mathrm{Id}+W)(\mathrm{Id}+W_g)\begin{pmatrix} e^{(ia+f_1(E,g))x} & 0\\ 0 & e^{(-ia-f_2(E,g))x}\end{pmatrix},$$

where $B_g=\begin{pmatrix}1 & 1\\ i\sqrt{E+ig} & -i\sqrt{E+ig}\end{pmatrix}$, and $\int\limits_0^\infty|\psi_1(\lambda,y)|^2dy<\infty$, $\int\limits_{-\infty}^0|\psi_2(\lambda,y)|^2dt<\infty$. Hence $\psi_1(\lambda, x)$ and $\psi_2(\lambda, x)$ are the sought solutions.

We shall show that the thus-constructed solutions $\psi_1(\lambda, x)$ and $\psi_2(\lambda, x)$ are uniformly tending, in $E\in[E_0,\infty)\setminus\mathfrak{M}$ and in x belonging to any finite interval, to the solutions of Eq. (1.1) mentioned in Theorem 1. Hence, follows the existence of a uniform limit of $(R_\lambda h, h)$ for $g \to 0$. In the same way we can prove the existence of a uniform limit of $(R_{\bar\lambda} h, h)$ for $g \to 0$. Thus, we have proved the first assertion of the lemma. The second assertion of the lemma is a consequence of Remark 4 of §2.

The existence of a uniform limit of $\psi_1(\lambda, x)$ and $\psi_2(\lambda, x)$ with respect to $E\in[E_0,\infty)\setminus\mathfrak{M}$ and x belonging to any finite interval, and the above-mentioned form of this limit, can be obtained from the following assertion.

LEMMA 3.2. For sufficiently large E_0 we have uniformly on $[E_0,\infty)\setminus\mathfrak{M}$

$$\lim_{g\to 0}\|W_g\|_{\rho_0/4}=0.$$

Proof. It can be assumed that the reduction of the system (3.2) to the form (3.3) has been carried out with the aid of the process described in the inductive Lemma 2.1 in which the parameters δ_0, ρ_0, and M_0 have been replaced by $\delta_0/2$, $\rho_0/2$, and $M_{g,0}=6M_0^{\varkappa}$. The thus-obtained W_g, $f_1(E, g)$, and $f_2(E, g)$ satisfy the above analysis. Then $\mathrm{Id}+W_g=\prod\limits_{s=0}^{\infty}(\mathrm{Id}+W_{g,s})$, where $W_{g,s}$ is defined in the same way as W_s in Lemma 2.1. It is easy to see that the "small denominators"

$$i(n,\omega)\pm l_{g,s}=i(n,\omega)\pm 2ia\pm C(s)g\pm\varepsilon_s(g),$$

where $C(s)>C_1$, $|\varepsilon_s(g)|<C_2g^2$, C_1 and C_2 being positive constants. From these formulas we find that $|i(n,\omega)\pm l_{g,s}|>\frac{C_1}{2}g$ for small g, and hence $\|W_{g,s}\|_{\rho_0/4}<0{,}1M_{g,s}^{\varkappa}$, where $M_{g,s}=M_{g,0}^{(1+\varkappa)^s}$. If $|n| \leq N_S$ and $g \leq M_S^2$, then $\frac{1}{|f(E,g)|}|i(n,\omega)\pm l_{g,s}|>\frac{C_3}{M_s}-C_4$, where C_3 and C_4 are positive constants.

In fact,

$$\frac{1}{|f(E,g)|}|i(n,\omega)\pm l_{g,s}|>\frac{1}{|f(E,g)|}(|(n,\omega)-2a_s|-|2a_s-2a|-$$
$$-C(s)g-|\varepsilon_1(g)|)>\frac{1}{|f(E,g)|}|M_4^{\frac{1-\varkappa}{4}}(|n|+1)^{-k-1}-2M_{s+1}|-C_4>\frac{C_3}{M_s}-C_4.$$

Hence, follows that for $g \leq M_S^2$ we have the inequalities

$$\|W_{g,l}\|_{\rho_0/4}<C_5\sqrt{g}\,M_{g,l}^{\varkappa},\quad l\leqslant s.$$

With the aid of these inequalities it is easy to prove the validity of the lemma.

From Lemmas 3.1 and 3.2 it follows that

$$r(E)=\frac{1}{2\pi\sqrt{E}}\frac{|\rho(E)|^2}{\det(\mathrm{Id}+W)},$$

where $\rho(E)=\int_{-\infty}^{\infty}h(x)\bar{\psi}_1(E,x)\,dx$, $\psi_1(E,x)$ being the solution described in Theorem 1.

Let $\sigma_h = \sigma$ be the spectral measure of the function h. For $\lambda = E + ig$ we have the equation

$$\frac{1}{2\pi i}((R_\lambda h,h)-(R_{\bar\lambda}h,h))=\frac{1}{\pi}\int_{-\infty}^{\infty}\frac{g\,d\sigma(\bar E)}{(\bar E-E)^2+g^2}.$$

LEMMA 3.3.* The measure σ_h is absolutely continuous with respect to Lebesgue's measure on the set $[E_0,\infty)\setminus\mathfrak{M}$.

Proof. Let g be so small that for $E\in[E_0,\infty)\setminus\mathfrak{M}$

$$\frac{1}{\pi}\int_{-\infty}^{\infty}\frac{g\,d\sigma(\bar E)}{(\bar E-E)^2+g^2}<r(E)+1.$$

Let us consider a segment Δ_g of length g that is centered at the point E. Then

$$r(E)+1>\frac{1}{\pi}\int_{-\infty}^{\infty}\frac{g\,d\sigma(\bar E)}{(\bar E-E)^2+g^2}\geqslant\frac{1}{\pi}\int_{\Delta_g}\frac{g\,d\sigma(\bar E)}{(\bar E-E)^2+g^2}>\frac{1}{\pi}\frac{g\sigma(\Delta_g)}{\frac{g^2}{4}+g^2}=\frac{4}{5\pi}\frac{\sigma(\Delta_g)}{g},$$

whence follows that $\sigma(\Delta g) < Kg$, where K is a constant. From this inequality we easily obtain the assertion of the lemma.

It follows from Lemma 3.3 that there exists an integrable function $r_1(E)$ such that for any measurable set $A\in[E_0,\infty)\setminus\mathfrak{M}$ we have the equation $\sigma(A)=\int_A r_1(E)\,dE$.

LEMMA 3.4. For almost every point $E\in[E_0,\infty)\setminus\mathfrak{M}$ we have $r(E) = r_1(E)$.

Proof. Let $\varepsilon > 0$ be fixed, and let E be a condensation point of the set $[E_0,\infty)\setminus\mathfrak{M}$. We shall take a g_1 so small that for any point E_1 of the segment $\Delta_{g_1}=\left[E-\frac{g_1}{2},E+\frac{g_1}{2}\right]$ belonging to $[E_0,\infty)\setminus\mathfrak{M}$, we have $|r(E_1)-r(E)|<\varepsilon$. It is possible to select a $g_2 < g_1$ such that segment $\Delta_{g_2}=\left[E-\frac{g_2}{2},E+\frac{g_2}{2}\right]$ will be entirely contained in $[E_0,\infty)\setminus\bigcup_{p=0}^{s}\mathfrak{M}_p$ for any fixed s. Since the distance between any two segments of rank s is larger than CN_s^{-k-1}, it can be assumed that $g_2 > CN_s^{-k-1}$, where C is a positive constant. Let us estimate the σ-measure $\Delta_{g_2}\cap\mathfrak{M}$. This intersection is contained in $\bigcup_{s_1>s}\bigcup_{r=0}^{N_{s_1}}\Delta_{s_1,r}$. We have

$$l\Big(\bigcup_{r=0}^{N_{s_1}}\Delta_{s_1,r}\Big)<C_1\sqrt{E}\,M_{s_1}^{\frac{1-\varkappa}{4}},\quad l(\Delta_{g_2}\cap\mathfrak{M})<\sum_{s_1>s}C_1\sqrt{E}M_{s_1}^{\frac{1-\varkappa}{4}}<C_2\sqrt{E}M_{s+1}^{\frac{1-\varkappa}{4}},$$

where C_1 and C_2 are positive constants, and l denotes Lebesgue's measure. It follows from this estimate that $\Delta_{g_2}\cap\mathfrak{M}$ is a union of not more than N_{s+1}^{k-1} nonintersecting segments of length smaller than $C_2\sqrt{E}\cdot M_{s+1}^{\frac{1-\varkappa}{4}}$, N_{s+2}^{k-1} intersecting segments of length smaller than $C_2\sqrt{E}M_{s+2}^{\frac{1-\varkappa}{4}}$, etc. For any such segment there exists, at a distance not exceeding its length, a point belonging to the complement $\Delta_{g_2}\cap\mathfrak{M}$. Let $\tilde\Delta$ be one such segment of length $\tilde g$ and let E' be a point belonging to $\Delta_{g_2}\setminus(\Delta_{g_2}\cap\mathfrak{M})$, whose distance from one of the ends of this segment is smaller than $\tilde g$. Let us consider a segment Δ one of whose ends coincides with E', whereas the other is the end of the segment $\tilde\Delta$ most remote from E'. Then the length of Δ will not exceed $2\tilde g$. For sufficiently small $\tilde g$ we have

*A similar assertion can be found in F. Atkinson's book Discrete and Continous Boundary Problems, Academic Press, New York–London (1964).

$$r(E') + \varepsilon + 1 > r(E') + 1 > \frac{1}{\pi} \int_{-\infty}^{\infty} \frac{\tilde{g}\, d\sigma(E)}{(E - E')^2 + \tilde{g}^2} \geqslant \frac{1}{\pi} \int_{\Delta} \frac{\tilde{g}\, d\sigma(E)}{(E - E')^2 + \tilde{g}^2} \geqslant \frac{\sigma(\Delta)}{5\tilde{g}} \geqslant \frac{\sigma(\tilde{\Delta})}{5\tilde{g}}.$$

Hence, $\sigma(\tilde{\Delta}) < K\tilde{g}$, where K is a constant. Hence follows that $\sigma(\Delta_{g_s} \cap \mathfrak{M}) < KC_2 \sqrt{E} \sum_{s_1 > s} N_{s_1}^{k-1} M_{s_1}^{\frac{1-\varkappa}{4}}$. But $N_s = -\frac{1}{\delta_s} \ln M_s$. For sufficiently large E we therefore have $\sigma(\Delta_{g_s} \cap \mathfrak{M}) < M_{s+1}^{\frac{1-\varkappa}{8}}$.

By setting $g_3 = CN_s^{-k}$, we obtain

$$\frac{1}{\pi} \int_{-\infty}^{\infty} \frac{g_3\, d\sigma(E)}{(E - E)^2 + g_3^2} < \frac{1}{\pi} \int_{\Delta_{g_s}} \frac{g_3 (r_1(E) + \varepsilon)\, dE}{(E - E)^2 + g_3^2} + \frac{1}{\pi} \int_{R^1 \setminus \Delta_{g_s}} \frac{g_3\, d\sigma(E)}{(E - E)^2 + g_3^2} + \frac{\sigma(\Delta_{g_s} \cap \mathfrak{M})}{g_3}.$$

Hence follows that for sufficiently large s,

$$\frac{1}{\pi} \int_{-\infty}^{\infty} \frac{g_3\, d\sigma(E)}{(E - E)^2 + g_3^2} < r_1(E) + \varepsilon_1,$$

where $\varepsilon_1 \to 0$ when $s \to \infty$. In the same way we can obtain the lower bound. This completes the proof of the lemma.

From Lemmas 3.1-3.4 follows Theorem 2.

LITERATURE CITED

1. F. Bloch, "Über die Quantenmechanik der Elektronen in Kristallgittern," Z. Physik, 52, 555-560 (1928).
2. E. C. Titchmarsh, Eigenfunction Expansions Associated with Second-Order Differential Equations, Oxford University Press (1962).
3. A. M. Dykhne, "Quasiclassical particle in one-dimensional periodic potential," Zh. Eksperim. i Teor. Fiz., 40, No. 5, 1423-1426 (1961).
4. S. G. Simonyan, "Asymptotic properties of the width of gaps in the spectrum of the Sturm–Liouville operator with a periodic potential," Differents. Uravnen., 6, No. 7, 1265-1272 (1970).
5. V. F. Lazutkin and T. F. Pankratova, "Asymptotic properties of the width of gaps in the spectrum of the Sturm–Liouville operator with a periodic potential," Dokl. Akad. Nauk SSSR, 215, No. 5, 1048-1051 (1974).
6. V. I. Arnol'd, "Small denominators and the problem of stability of motion in classical and celestial mechanics," Usp. Matem. Nauk, 18, No. 6, 91-192 (1963).
7. S. P. Novikov, "The periodic problem for the Kortweg–DeVries equation, I," Funktsional. Analiz i Ego Prilozhen., 8, No. 3, 54-66 (1974).
8. N. N. Bogolyubov, Yu. A. Mitropol'skii, and A. M. Samoilenko, Method of Faster Convergence in Nonlinear Mechanics [in Russian], Naukova Dumka, Kiev (1969).
9. É. G. Belaga, "Reducibility of systems of ordinary differential equations in the neighborhood of almost periodic motion," Dokl. Akad. Nauk SSSR, 143, No. 2, 255-258 (1962).
10. J. Moser, "Perturbation theory for almost periodic solutions for nonlinear differential equations," Intern. Symposium Nonlinear Differential Equations and Nonlinear Mechanics, Academic Press, New York (1963).
11. J. Moser, "A new technique for the construction of solutions of nonlinear differential equations," Proc. National Academy of Sci. USA, 47 (1961).

Comments

This paper was based on a new idea according to which Bloch eigenfunctions of the one-dimensional Schrödinger operator with quasi-periodic potential are connected with the reducibility problem for the corresponding two-dimensional linear system of differential equations on the two-dimensional torus. This latter problem can be solved with the help of KAM-theory. On the spectral axis there appear some intervals where the reducibility problem cannot be solved. They correspond to the so-called "forbidden zones."

J. Moser studied spectral properties of the Schrödinger operator with quasi-periodic potential (see, e.g., his paper [Mo]). Final results were obtained by H. Eliasson (see [E]). He showed that for sufficiently small quasi-periodic potentials the whole spectrum consists of Bloch eigenfunctions.

In [S1], the Bloch eigenfunctions near the left boundary of the spectrum were constructed for different Schrödinger operators with quasi-periodic potentials.

Commun. Math. Phys. 170, 375–403 (1995)

Communications in Mathematical Physics

Distribution of Energy Levels of Quantum Free Particle on the Liouville Surface and Trace Formulae

Pavel M. Bleher[1], **Denis V. Kosygin**[2], **Yakov G. Sinai**[2]

[1] Department of Mathematical Sciences, Indiana University – Purdue University at Indianapolis, 402 N. Blackford Street, Indianapolis, IN 46202, USA

[2] Princeton University, Department of Mathematics, Princeton, NJ 08544, USA

Received: 6 June 1994/in revised form: 2 November 1994

Abstract: We consider the Weyl asymptotic formula

$$\#\{E_n \leqq R^2\} = \frac{\operatorname{Area} Q}{4\pi} R^2 + n(R)\,,$$

for eigenvalues of the Laplace–Beltrami operator on a two-dimensional torus Q with a Liouville metric which is in a sense the most general case of an integrable metric. We prove that if the surface Q is non-degenerate then the remainder term $n(R)$ has the form $n(R) = R^{1/2}\theta(R)$, where $\theta(R)$ is an almost periodic function of the Besicovitch class B^1, and the Fourier amplitudes and the Fourier frequencies of $\theta(R)$ can be expressed via lengths of closed geodesics on Q and other simple geometric characteristics of these geodesics. We prove then that if the surface Q is generic then the limit distribution of $\theta(R)$ has a density $p(t)$, which is an entire function of t possessing an asymptotics on a real line, $\log p(t) \sim -C_{\pm}t^4$ as $t \to \pm\infty$. An explicit expression for the Fourier transform of $p(t)$ via Fourier amplitudes of $\theta(R)$ is also given. We obtain the analogue of the Guillemin–Duistermaat trace formula for the Liouville surfaces and discuss its accuracy.

1. Introduction

The question about the relation of a quantum system to its classical limit has been discussed since the moment of appearance of quantum mechanics. Recently this question became popular again both in physics and mathematics due to the theory of quantum chaos. It turns out that statistical properties of quantum energy levels depend strongly on the ergodic properties of the underlying classical system. As an important and instructive example one may think of the Laplace–Beltrami operator on a Riemannian manifold and of the geodesic flow in the manifold as its classical counterpart. This case has been widely discussed in many physical and mathematical papers (see, e.g., [1–11, 13–17, 20, 21]).

Let X be a smooth compact Riemannian manifold and let $-\Delta$ be its Laplace–Beltrami operator. The eigen-states of a free quantum particle moving on X are the eigenfunctions of the Laplace–Beltrami operator, i.e., the solutions of the equation

Y.G. Sinai, *Selecta II: Probability Theory, Statistical Mechanics, Mathematical Physics and Mathematical Fluid Dynamics*, DOI 10.1007/978-1-4419-6205-8_12, © Springer Science+Business Media, LLC 2010

$$-\Delta f = Ef .$$

The positive square root of this operator, $\sqrt{-\Delta}$ is a selfadjoint elliptic pseudo-differential operator of order one. Its spectrum is discrete and each eigenvalue has finite multiplicity. If we enumerate them in non-decreasing order then we have

$$0 = R_0 < R_1 \leqq R_2 \leqq \cdots \leqq R_k \leqq \cdots \to \infty \text{ as } k \to \infty .$$

Note that the corresponding eigenvalue E_k of the Laplace–Beltrami operator is equal to R_k^2. One of the possible approaches to the study of statistics of the spectrum is to consider the trace of the operator $\psi(-\Delta)$ for various functions ψ (see [1, 2, 10, 11]). Properties of this trace, for some particular function ψ, are closely related to the properties of the Green function of the wave equation

$$u_t = i\sqrt{-\Delta}\, u$$

on X which allows to use this relation reciprocally (see [9, 10, 13, 14]).

The behavior of the sum $\sum_{k=1}^{\infty} \psi(R_k)$ depends on the geometry of X and in particular on the structure of the set of closed geodesics on X. Before formulating corresponding results let us introduce some definitions. Denote by M the unit cotangent bundle over X and denote by ξ the Hamilton vector field on M whose generating function is the symbol of $\sqrt{-\Delta}$. We recall that the Hamilton flow generated by ξ is the geodesic flow acting on M whose trajectories, when projected onto X, are geodesics. Let γ be a periodic trajectory of period T. The map $\exp(T\xi)$ maps M onto M and it is the identity map on γ. Thus for $x \in \gamma$, the differential $d(\exp T\xi)_x$ is a linear map on T_x onto inself such that the tangent space to γ is its one-dimensional eigenspace. By P_γ we denote the induced linear transformation of the orthogonal complement to the tangent space to γ at the point $x \in \gamma$. P_γ is the linearization of the Poincaré map defined on a small submanifold of codimension 1 transversal to γ. Two linear Poincaré maps taken at two distinct points x and x' of γ are conjugate. A periodic trajectory γ is called nondegenerate if it is isolated and $I - P_\gamma$ is invertible. A periodic trajectory $\{\gamma(t),\ 0 \leqq t \leqq T\}$, $\gamma(T)$, is called primitive if it does not contain periodic trajectories of smaller period, i.e., $\gamma(0) \neq \gamma(T')$ for $0 < T' < T$. Each periodic trajectory is obviously an iterate of a primitive periodic trajectory. Given a periodic trajectory we denote by T_γ its period and by T_γ^* the period of the primitive trajectory of which it is the iterate.

If all periodic trajectories of ξ are nondegenerate then as shown in [9], the difference of the distributions

$$\left(1 + \sum_{\lambda^2 \in \operatorname{spec}(-\Delta)} \exp(i\lambda t)\right) - \left(\sum_{\text{all periodic trajectories}} i^{\operatorname{ind}_\gamma} T_\gamma^* |I - P_\gamma|^{-1/2} \delta(t - T_\gamma)\right)$$

is a locally L^1-summable function of t on the interval $(0, \infty)$. Here $\operatorname{ind}_\gamma$ is the Maslov index of the trajectory γ. (Note that $\operatorname{ind}_\gamma$ is an integer mod 4).

When X is a surface of constant negative curvature the exact expression of this function is given by the Selberg trace formula ([16, 18]).

Surfaces with completely integrable geodesic flow are in a sense opposite to the last mentioned case. Namely, the phase space M is foliated onto n-dimensional invariant tori with quasi-periodic motion defined by ξ. When the ratio of frequencies of this motion on a torus is rational then this torus is filled up with periodic trajectories of the same period. It is not surprising that the exact expressions of the type of Selberg trace formula are not true any longer. We remark that general methods described in [9, 10, 13, 14] allows one to obtain some trace-type formulae even in the case when closed trajectories fill submanifolds of nontrivial codimension. In this paper we take a different approach.

In the present paper we continue to study spectral properties of the Laplace–Beltrami operator on Liouville surfaces started in [15] and [6]. The geodesic flow on a Liouville surface is completely integrable and it is believed that this is the most general class of surfaces for which this is the case (cf. [22]). It is to be noted also that contrary to surfaces of revolution a generic Liouville surface has no symmetries.

Let (Q, dq^2) be a two-dimensional compact closed Riemannian manifold homeomorphic to a torus. We may think of Q as being represented by unit square with coordinates q_1, q_2, $0 \leqq q_1, q_2 \leqq 1$. A surface Q is called Liouville if its metric dq^2 has the form

$$dq^2 = [U_1(q_1) - U_2(q_2)](dq_1^2 + dq_2^2)\,, \tag{1.1}$$

where $U_1(q_1), U_2(q_2)$ are periodic functions of period 1 satisfying the inequality

$$U_1(q_1) - U_2(q_2) > 0 \text{ for all } q = (q_1, q_2), 0 \leqq q_1, q_2 \leqq 1\,. \tag{1.2}$$

The Laplace–Beltrami operator $-\Delta$ defined by this metric is given by the formula

$$-\Delta f = -[U_1(q_1) - U_2(q_2)]^{-1}\left(\frac{\partial^2 f}{\partial q_1^2} + \frac{\partial^2 f}{\partial q_2^2}\right)\,. \tag{1.3}$$

This is a non-negative selfadjoint operator in the Hilbert space $L^2(Q, \mu(dq))$ having a discrete spectrum. Here

$$\mu(dq) = (U_1(q_1) - U_2(q_2))dq_1 dq_2\,.$$

The first eigenvalue of $-\Delta$ is simple and corresponds to the constant eigenfunction. If we enumerate the eigenvalues in a non-decreasing order then we have

$$0 = E_0 < E_1 \leqq E_2 \leqq \cdots \leqq E_k \leqq \cdots \to \infty \text{ as } k \to \infty\,.$$

Denote by $N(x)$ the number of eigenvalues E_k which are not greater than x,

$$N(x) = \#\{k : E_k \leqq x\}\,. \tag{1.4}$$

Then due to the Weyl asymptotic formula (see [14])

$$N(x) = \frac{\text{Area}\, Q}{4\pi} x + n(x)\,, \tag{1.5}$$

where $\frac{n(x)}{x} \to 0$ as $x \to \infty$. The behavior of the remainder term $n(x)$ as $x \to \infty$ for Liouville surfaces was studied in [15], and it was shown that under a certain condition of non-degeneracy on the Liouville surface, the function $\theta(R) = R^{-1/2} n(R^2)$ is an almost periodic function of the Besicovitch class B^1. In addition, if the surface (Q, dq^2) is generic then the probability distribution corresponding to the function $\theta(R)$ possesses nice analytical properties. In particular, it has a density which is an entire function of its argument decaying on the real axis at least as $\exp(-|t|^{(16/9)-\varepsilon})$ when $t \to \pm\infty$.

Surfaces of revolution are Liouville surfaces but they do not satisfy the non-degeneracy conditions assumed in [15]. This case was considered separately in [3–7] (see especially [6]). It was proven that under similar conditions of non-degeneracy the function $\theta(R)$ is an almost periodic function of the Besicovitch class B^2, and the set of the Fourier frequencies of $\theta(R)$ coincides with the geodesic spectrum of Q, i.e., with the set of the lengths of all closed geodesics on Q. The Fourier amplitudes of $\theta(R)$ have a nice geometric interpretation as well. In addition, for generic surfaces of revolution, which are characterized by linear independence of the lengths of all primitive geodesics on Q, the density of the distribution corresponding to $\theta(R)$ is an entire function which decays on the real axis roughly as $\exp(-\lambda t^4)$ when $|t| \to \infty$.

It is to be noted that almost periodic functions appear naturally in various problems of number theory (see e.g. [4, 7, 12, 15] and references therein). In particular, one interesting application of the theory of almost periodic functions to the problem of the distribution of primes in arithmetic progressions was described in [19]. As was shown in [15], the spectrum of the Laplace–Beltrami operator on a Liouville surface has a nice geometric interpretation in terms of points of two-dimensional lattices. Then the study of the function $N(x)$ is reduced to the counting lattice points in a planar domain which is essentially a problem in number theory.

In this paper we extend the results of [6] to the case treated in [15] and we obtain a refined estimate on the rate of decay of the distribution density corresponding to the function $\theta(R)$. This proves a conjecture stated in [15]. In addition, we prove a trace formula for the Liouville surfaces. The structure of this paper is the following. All necessary properties of the geodesic flow on a Liouville surface are described in Sect. 2 and the main results are formulated in Sect. 3. Section 4 contains a reduction of the main results to corresponding problems in number theory. The main steps of the proofs are exposed in Sect. 5 with some technical lemmas proven in Sect. 6. In Sect. 7 we prove some properties of the kernel of fractional integration of order 3/2 which were used in the other parts of the paper.

2. Properties of the Geodesic Flow of Liouville Surface

Consider a torus Q with the Liouville metric (1.1). Denote by $Q^r, 0 \leqq r \leqq \infty$, the space of pairs of C^r-functions (U_1, U_2) on a circle satisfying (1.2) and the following condition:

A1. For each function U_1, U_2 there are exactly two points where its derivative vanishes. One of these points is its maximum, and the other one is minimum. Both of these points are non-degenerate (i.e., the second derivative does not vanish at these points).

Q^r is a topological space with the C^r-topology. One may think of points in the space Q^r as of corresponding Liouville surfaces, defined by means of the functions U_1, U_2 and the formula (1.1). The structure of the geodesic flow on a surface Q with the metric from the space Q^r was described in detail in [15] and for proofs we refer the reader to that paper.

The geodesic flow is embedded into the Hamilton flow acting on the cotangent bundle T^*Q and defined by the Hamiltonian

$$H(p,q) = \frac{1}{2(U_1(q_1) - U_2(q_2))}(p_1^2 + p_2^2), \tag{2.1}$$

where

$$p_i = (U_1(q_1) - U_2(q_2))dq_i, \ i = 1,2 \ .$$

This Hamilton flow possesses the first integral different from the energy integral,

$$S(p,q) = \frac{1}{2(U_1(q_1) - U_2(q_2))}(U_2(q_2)p_1^2 + U_1(q_1)p_2^2) \ , \tag{2.2}$$

and thus it is completely integrable. The function $S(p,q)$ (as well as H) is a homogeneous function of order 2 of the momentum $p = (p_1, p_2)$. We associate with $S(p,q)$ the quantum operator

$$\sigma = [U_1(q_1) - U_2(q_2)]^{-1}\left(U_2(q_2)\frac{\partial^2}{\partial q_1^2} + U_1(q_1)\frac{\partial^2}{\partial q_2^2}\right) .$$

A direct computation shows that Δ and σ commute.

Introduce the constants $c_1, c_2, c_3, c_4, m_1, M_1, m_2, M_2$ as

$$c_1 = \max_{0 \leq x \leq 1} U_1(x) = U_1(M_1) \ , \quad c_2 = \min_{0 \leq x \leq 1} U_1(x) = U_1(m_1) \ ,$$

$$c_3 = \max_{0 \leq x \leq 1} U_2(x) = U_1(M_2) \ , \quad c_4 = \min_{0 \leq x \leq 1} U_2(x) = U_1(m_2) \ ,$$

For $E > 0$ and $c \in [c_4, c_1]$, define

$$M_{E,c} = \{(p,q) \in T^*Q : \ H(p,q) = E, \ S(p,q) = cE\} \ .$$

Due to the complete integrability, every set $M_{E,c}$ consists of two two-dimensional tori when $c \in (c_4, c_3) \cup (c_2, c_1)$, and of four two-dimensional tori when $c_3 < c < c_2$. The phase trajectories of the Hamiltonian system are windings over these tori, and the projection $\pi : (p,q) \longrightarrow q$ maps them onto geodesics on the configuration torus Q.

In the case $c \in (c_4, c_3) \cup (c_2, c_1), \pi$ projects each component of $M_{E,c}$ onto either a horizontal or vertical strip, and the projection of any trajectory on $M_{E,c}$ is a geodesic which rotates along the strip and oscillates between its two boundaries. Two components of $M_{E,c}$ correspond to two possible directions of the rotation along the strip. If $c \in (c_3, c_2)$, π projects each component of $M_{E,c}$ in a one-to-one way onto the whole configurational torus Q, and every trajectory on $M_{E,c}$ is projected onto a geodesic, which is a curvilinear winding over Q. Four components of $M_{E,c}$ correspond in this case to four possible combinations of the direction of rotation along q_1 and q_2.

The values $c = c_1, c_2, c_3, c_4$ are exceptional: for $c = c_1, c_4$ the tori degenerate into two circles while for $c = c_2, c_3$ they degenerate into four cylinders corresponding to the direct product of the separatrix and the circle, with different directions of motion along separatrix and the circle (for more detailed description see [15]). In what follows we will call freely the components of $M_{E,c}$ when $c = c_1, c_2, c_3, c_4$ invariant tori as well.

The Hamiltonian flow can be described with the help of action-angle variables $I_1, I_2, \varphi_1, \varphi_2$. Let ζ_1, ζ_2 be the circles on Q defined by the equations $q_1 = M_1, q_2 = m_2$, respectively, with the orientation induced by the coordinate axes. For each invariant torus corresponding to a nonexceptional value of the parameter c a connected component of its intersection with $\pi^{-1}(\zeta_i)$ is a cycle α_i homeomorphic to circle ($i = 1,2$). The canonical form $pdq = p_1dq_1 + p_2dq_2$ determines uniquely an orientation of the cycle α_i. Namely, the tangent vector to α_i is positively oriented if the value of pdq on it is positive. The action variables I_1, I_2 related to the cycles α_1, α_2 are given by the formulae:

$$I_i = \begin{cases} -\int_{\alpha_i} pdq, & \text{if } \pi : \alpha_i \to \zeta_i \text{ is bijective and } \pi(\alpha_i) \text{ and } \zeta_i \text{ have different orientations ;} \\ \int_{\alpha_i} pdq, & \text{otherwise .} \end{cases}$$

By φ_1, φ_2 we denote the angle variables canonically conjugate to I_1, I_2.

As follows from (2.1) and (2.2), $(p,q) \in M_{E,c}$ implies

$$p_1^2 = 2E(U_1(q_1) - c), \quad p_2^2 = 2E(c - U_2(q_2)), \tag{2.3}$$

hence the action variables are

$$I_1(E,c) = \oint_{\pi(\alpha_1)} p_1 dq_1 = \begin{cases} 2(2E)^{1/2} \int_{q_1 : U_1(q_1) - c \geqq 0} (U_1(q_1) - c)^{1/2} dq_1, \\ \text{when } c_2 < c \leqq c_1 ; \\ \pm(2E)^{1/2} \int_0^1 (U_1(q_1) - c)^{1/2} dq_1, \\ \text{when } c_4 \leqq c < c_2 ; \end{cases} \tag{2.4}$$

and

$$I_2(E,c) = \oint_{\pi(\alpha_2)} p_2 dq_2 = \begin{cases} \pm(2E)^{1/2} \int_0^1 (c - U_2(q_2))^{1/2} dq_2, \\ \text{when } c_3 < c \leqq c_1 ; \\ 2(2E)^{1/2} \int_{q_2 : c - U_2(q_2) \geqq 0} (c - U_2(q_2))^{1/2} dq_2, \\ \text{when } c_4 \leqq c < c_3 ; \end{cases} \tag{2.5}$$

According to the definition, in formulae (2.4) and (2.5) we put the plus sign if the motion goes in the positive direction (with respect to a given coordinate) and the minus sign otherwise. More precisely, the formulae (2.4), (2.5) enable us to define I_1, I_2 as functions on the phase space T^*Q which are constant on each invariant torus, and the value of which on the given torus is defined by (2.4), (2.5). It is noteworthy that the mapping which maps an invariant torus to the corresponding pair (I_1, I_2) is one-to-one. Note also, that the actions $I_1(E,c), I_2(E,c)$ are multivalued functions of E and c, namely, the values of I_1 and I_2 on the invariant tori corresponding to the same pair (E,c) are equal in absolute value but may differ in sign.

The frequencies of the motion along an invariant torus defined by a pair (I_1, I_2) are given by the formulae

$$\omega_1(E,c) = \frac{\partial H}{\partial I_1}, \quad \omega_2(E,c) = \frac{\partial H}{\partial I_2}. \tag{2.6}$$

The frequencies $\omega_1(E,c), \omega_2(E,c)$ like the actions I_1, I_2 are multivalued functions of E and c. Namely, frequencies of motion along invariant tori corresponding to the same pair (E,c) are equal in absolute value but may differ in sign. We define the functions ω_1, ω_2 at $c = c_1, c_2, c_3, c_4$ by continuity. The functions $I_i(E, c), \omega_i(E, c)$ are homogeneous functions of E of order 1/2, hence

$$I_i(E, c) = (2E)^{1/2} f_i(c), \quad \omega_i(E, c) = (2E)^{1/2} \xi_i(c)\,, \tag{2.7}$$

where $f_i(c) = I_i(1/2, c)$, $\xi_i(c) = \omega_i(1/2, c)$. Recall that an invariant torus is called non-degenerate if on this torus,

$$\det\left(\frac{\partial^2 H}{\partial I_k \partial I_l}\right) = \det\left(\frac{\partial \omega_k}{\partial I_l}\right) \neq 0\,.$$

If the invariant torus is non-degenarate then in its small neighborhood the frequencies ω_j can serve as local coordinates, instead of the action variables. A straightforward computation gives

$$\begin{aligned} \xi_1 &= \frac{f_2'}{\Delta_1}, \quad \xi_2 = -\frac{f_1'}{\Delta_1}\,; \\ \xi_1' &= -f_2\frac{\Delta_2}{\Delta_1^2}, \quad \xi_2' = f_1\frac{\Delta_2}{\Delta_1^2}\,; \end{aligned} \tag{2.8}$$

where

$$\Delta_1 = f_1 f_2' - f_2 f_1', \quad \Delta_2 = f_1' f_2'' - f_2' f_1''\,. \tag{2.9}$$

Hence

$$\xi_1 f_1' + \xi_2 f_2' = 0\,, \tag{2.10}$$

$$\xi_1 f_1 + \xi_2 f_2 = 1\,, \tag{2.11}$$

and

$$\xi_1 \xi_2' - \xi_2 \xi_1' = \frac{\Delta_2}{\Delta_1^2}\,. \tag{2.12}$$

This implies

$$\det\left(\frac{\partial^2 H}{\partial I_k \partial I_l}\right)(E, c) = \frac{\xi_1 \xi_2' - \xi_2 \xi_1'}{\Delta_1} = \frac{\Delta_2}{\Delta_1^3}\,. \tag{2.13}$$

Observe that $\det\left(\frac{\partial^2 H}{\partial I_k \partial I_l}\right)(E, c)$ does not depend on E. In the sequel we shall need the following functions:

$$b(c) = \frac{f_1' f_2'' - f_2' f_1''}{(f_1'^2 + f_2'^2)^{3/2}}\,, \tag{2.14}$$

$$a(c) = (\xi_1^2 + \xi_2^2)^{3/2} b(c) = \det\left(\frac{\partial^2 H}{\partial I_k \partial I_l}\right)(E, c) \tag{2.15}$$

and

$$F_1(c) = \begin{cases} \frac{1}{2} f_1(c) & \text{when } c_2 \leqq c \leqq c_1\,; \\ f_1(c) & \text{when } c_4 \leqq c \leqq c_2\,; \end{cases} \tag{2.16}$$

$$F_2(c) = \begin{cases} f_2(c) & \text{when } c_3 \leqq c \leqq c_1\,; \\ \frac{1}{2} f_2(c) & \text{when } c_4 \leqq c \leqq c_3\,; \end{cases} \tag{2.17}$$

The functions $a(c), b(c)$ like f_1, f_2, ξ_1, ξ_2 are multivated functions of c and their values on different branches at the same point c are equal in absolute value but may differ in sign.

Denote by W the set of all invariant tori with $E = 1/2$. As was said above, every $\omega \in W$ is characterized by the value of the parameter $c_4 \leqq c \leqq c_1$ and by the choice of one or two signs, which give the direction of rotation on ω along the axes q_1, q_2. It is to be noted that the tori $\omega \in W$ fill up the unit cotangent bundle over Q. Denote by W_+ the set of $\omega \in W$ with positive direction of rotation.

Let $\mathcal{G}$ be the set of all nonzero oriented closed (in general, multiple) geodesics on Q. For $g \in \mathcal{G}$ denote by $\omega(g) \in W$ the invariant torus on which $\pi^{-1}g$ lies and by $c(g)$ the value of the parameter c on $\omega(g)$. Let $n_1(g), n_2(g) \in \mathbb{Z}$ be the rotation numbers of $\pi^{-1}g$ along the axes q_1, q_2, respectively. If π projects $\omega(g)$ onto the whole Q (which occurs when $c_3 \leqq c(g) \leqq c_2$) then we assign to $n_1(g), n_2(g)$ the signs of the direction of rotation of g along the axes q_1, q_2, respectively. If π projects $\omega(g)$ onto a strip along q_1 (which occurs when $c_4 \leqq c(g) < c_3$) then we assign to $n_1(g)$ the sign of the direction of rotation of g along q_1 and the sign $+$ to $n_2(g)$. Observe that in this case $n_2(g)$ describes the number of oscillations of g along the axis q_2 so it is natural that $n_2(g) \geqq 0$. Similarly, if π projects $\omega(g)$ onto a band along q_2 (which occurs when $c_2 < c(g) \leqq c_1$) we assign to $n_2(g)$ the sign of the direction of rotation of g along q_2 and the sign $+$ to $n_1(g)$. We denote by $\xi_1(g), \xi_2(g), f_1(g), f_2(g)$, respectively, the frequencies and the actions of the motion along $\omega(g)$.

If $g \in \mathcal{G}$ then the Hamiltonian flow is periodic on the invariant torus $\omega(g)$ and

$$\xi_1/\xi_2 = n_1(g)/n_2(g) \in \mathbb{Q} \cup \{\pm\infty\} .$$

We will call two closed geodesics equivalent if they correspond to the same torus $\omega(g)$, and we will denote by G the set if geodesics $g \in \mathcal{G}$ factorized by this relation of equivalence. Remark that all equivalent closed geodesics have the same values of $n_1(g), n_2(g)$ and of the length

$$|g| = \frac{n_1}{\xi_1} = \frac{n_2}{\xi_2} = n_1 f_1 + n_2 f_2 , \tag{2.18}$$

so $n_1(g), n_2(g)$ and $|g|$ can be viewed as functions on G.

3. Formulation of the Main Results

We begin with some definitions. We shall call a pair of real numbers α_1, α_2 diophantine if there exist $\tau > 1$ and $C > 0$ such that for any nonzero pair of integers k_1, k_2,

$$|k_1\alpha_1 + k_2\alpha_2| \geqq \frac{C}{|k|(\log|k|)^\tau}, \quad |k| = (k_1^2 + k_2^2)^{1/2} . \tag{3.1}$$

A number α is called diophantine if the pair 1, α is diophantine. The complement to the set of diophantine pairs of numbers (as well as to the set of diophantine numbers) has zero Lebesgue measure. Real numbers $1 = \lambda_0, \lambda_1, \lambda_2, \dots$ are called linearly independent over $\mathbb{Z}$ if

$$k_0 + k_1\lambda_1 + \cdots + k_n\lambda_n = 0$$

with $k_i \in \mathbb{Z}$ implies $k_0 = \cdots = k_n = 0$.

A function $f(t)$ on the positive half-axis $\{t > 0\}$ is called an almost periodic function of the Besicovitch class B^p if for any positive ε there exists a trigonometric polynomial

$$P_\varepsilon(t) = \sum_{n=1}^{N_\varepsilon} a_{n,\varepsilon} \exp(i\lambda_{n,\varepsilon} t)$$

such that

$$\limsup_{T\to\infty} T^{-1} \int_0^T |f(t) - P_\varepsilon(t)|^p dt \leqq \varepsilon .$$

In addition, a trigonometric series

$$\sum_{n=1}^{\infty} a_n \exp(i\lambda_n t)$$

is called the Fourier series of $f(t)$ with respect to B^p if

$$\lim_{N\to\infty} \limsup_{T\to\infty} T^{-1} \int_0^T \left| f(t) - \sum_{n=1}^{N} a_n \exp(i\lambda_n t) \right|^p dt = 0 .$$

The Fourier series of an almost periodic function is always well-defined and unique (see [17]).

Theorem 3.1. *Let a Liouville surface $Q \in \mathcal{Q}^r$, $r \geqq 5$, satisfy the following conditions:*

(i) *The function*

$$a(c) = \det d^2 H(E, c), \quad d^2 H = \left(\frac{\partial^2 H}{\partial I_k \partial I_l} \right) ,$$

has only finitely many zeroes and all these zeroes are zeroes of the first order.

(ii) *If c is a zero of $a(c)$ then the pairs $(\xi_1(c), \xi_2(c))$ and $(f_1(c), f_2(c))$ are diophantine.*

(iii) *The pairs $(f_1(c_2), f_2(c_2))$ and $(f_1(c_3), f_2(c_3))$ are diophantine,*

(iv) *The pairs $(\omega_1(c_1), \omega_2(c_1))$ and $(\omega_2(c_4), \omega_2(c_4))$ are diophantine.*

Then the function $N(x) = \#\{k : E_k \leqq x\}$ has the representation

$$N(x) = \frac{\text{Area } Q}{4\pi} x + x^{1/4} \theta(x^{1/2}) , \tag{3.2}$$

where $\theta(R)$ is an almost-periodic function of the Besicovitch class B^1. The Fourier series of $\theta(R)$ with respect to B^1 is

$$\theta(R) = (2\pi^3)^{-1/2} \sum_{g\in G} |g|^{-3/2} \varkappa(g)^{-1/2} \sin\left(|g|R - \frac{\pi}{2} \text{ind}\, g - \frac{\pi}{4}\, \sigma(g) \right) . \tag{3.3}$$

Here ind g *means the Maslov index of any geodesic from the class $g \in G$ and*

$$\varkappa(g) = |\det d^2 H|, \quad \sigma(g) = \text{sign } \det d^2 H , \tag{3.4}$$

where the value of $\det d^2 H$ *is taken on the torus $\omega(g)$.*

It is noteworthy that the fulfillment of assumptions in Theorem 3.1 does not depend on the choice of the branches of the functions $a, f_1, f_2, \omega_1, \omega_2$. The almost-periodic function $\theta(R)$ was also considered by Berry and Tabor in [2]. As concerns the properties of the function $\theta(R)$, it is plausible that this function belongs actually to the Besicovitch class B^2 (and not only to B^1 as stated), and Theorem 4.3 below strongly supports this conjecture. Also the B^2 property of $\theta(R)$ is established for a class of surfaces of revolution, see [6]. Still in Theorem 3.1 we have a problem with proving the B^2 almost periodicity, when we estimate the difference between the function $\theta(R)$ and the function $\theta_0(R)$ of Theorem 4.3. Namely, to estimate this difference we use semiclassical quantization of the energy levels, and the usual Bohr-Sommerfeld quantization formula turns out to be insufficient for our purposes, because it does not work near unstable periodic orbits. So we derive a more sophisticated semiclassical quantization formula which works near unstable periodic orbits, and a difficult problem is to obtain a good uniform estimate of the error term in this formula, see [15] and [6a].

Denote by G_0 the subset of G consisting of all closed geodesics (up to the equivalence) with non-negative relatively prime rotation numbers $n_1(g), n_2(g)$. In other words, G_0 is the set of all primitive closed geodesics on Q, up to the equivalence and to the choice of orientation.

Theorem 3.2. *Let Q be a Liouville surface which satisfies the assumptions of Theorem* 3.1. *Assume in addition that the numbers $\{|g|, g \in G_0\}$ are linearly independent over $\mathbb{Z}$. Then the limit distribution of the function $\theta(R)$ is absolutely continuous with respect to the Lebesgue measure, and the density $p(t)$ of this distribution is an entire function of t which has the following asymptotics on the real axis*:

$$\lim_{t\to\pm\infty}(-t^{-4}\log p(t)) = C_\pm > 0\ , \tag{3.5}$$

where $C_\pm$ are two (in general different) constants which can be expressed explicitly via geometrical characteristics of the geodesic flow (for an exact formula see (4.25) *below).*

Before formulating the exact statement concerning the trace formula on the Liouville surface we give a heuristic derivation of it. Let $\varphi(t)$ be a test function from the Schwartz space, i.e., $\varphi(t)$ decays faster than polynomially as $|t| \to \infty$, together with all its derivatives. Put

$$F(x) = \sum_{R_k\in\operatorname{spec}\sqrt{-\Delta}} \varphi(x-R_k), \quad M(R) = N(R^2), \quad m(R) = n(R^2)\ .$$

Then integrating by parts we have

$$\begin{aligned} F(x) &= \sum_{R_k\in\operatorname{spec}\sqrt{-\Delta}} \varphi(x-R_k) = \int_{-0}^{\infty} \varphi(x-y)\,dM(y) = \int_{-0}^{\infty} \varphi'(x-y)M(y)\,dy \\ &= \frac{\operatorname{Area} Q}{4\pi}\int_0^\infty \varphi'(x-y)\,y^2 dy + \int_{-0}^{\infty}\varphi'(x-y)\,m(y)\,dy \\ &= \frac{\operatorname{Area} Q}{2\pi}\int_0^\infty \varphi(x-y)\,y\,dy + \int_{-0}^{\infty}\varphi'(x-y)\,y^{1/2}\theta(y)\,dy\ . \end{aligned}$$

Replacing $\theta(y)$ by its Fourier series (3.3) and neglecting the error term coming from this replacement (since the Fourier series of a function from B^1 represents the function only asymptotically at infinity) we obtain

$$F(x) = \frac{\text{Area}\, Q}{2\pi}\int_0^\infty \varphi(x-y)y\,dy + (2\pi^3)^{-1/2}\sum_{g\in G}|g|^{-3/2}\varkappa(g)^{-1/2} \times \int_{-0}^\infty \varphi'(x-y)y^{1/2}\sin\left(|g|y - \frac{\pi}{2}\text{ind}\ g - \frac{\pi}{4}\sigma(g)\right)dy\,. \quad (3.6)$$

Now,

$$\int_{-0}^\infty \varphi'(x-y)y^{1/2}\sin\left(|g|y - \frac{\pi}{2}\text{ind}\ g - \frac{\pi}{4}\sigma(g)\right)dy$$
$$= |g|\int_{-0}^\infty \varphi(x-y)y^{1/2}\cos\left(|g|y - \frac{\pi}{2}\text{ind}\ g - \frac{\pi}{4}\sigma(g)\right)dy + O(x^{-1/2}), \quad x\to\infty\,,$$

so neglecting the error term $O(x^{-1/2})$ in the last formula we arrive at

$$F(x) = \frac{\text{Area}\, Q}{2\pi}\int_0^\infty \varphi(x-y)y\,dy + (2\pi^3)^{-1/2}\sum_{g\in G}|g|^{-1/2}\varkappa(g)^{-1/2} \times \int_{-0}^\infty \varphi(x-y)y^{1/2}\cos\left(|g|y - \frac{\pi}{2}\text{ind}g - \frac{\pi}{4}\sigma(g)\right)dy\,. \quad (3.7)$$

In the language of the theory of distributions this is equivalent to

$$\sum_{R_k\in\text{spec}\sqrt{-\Delta}} \delta(x-R_k) = \frac{\text{Area}\ Q}{2\pi}x_+ + x_+^{1/2}(2\pi^3)^{-1/2}\sum_{g\in G}|g|^{-1/2}\varkappa(g)^{-1/2} \times \cos\left(|g|y - \frac{\pi}{2}\text{ind}\ g - \frac{\pi}{4}\sigma(g)\right)\,,$$

where

$$x_+^\lambda = \begin{cases} x^\lambda & \text{for } x > 0\,, \\ 0 & \text{for } x \leqq 0\,. \end{cases}$$

Passing to the Fourier transform we obtain that

$$\sum_{R_k\in\text{spec}\sqrt{-\Delta}} e^{iR_k\xi} = -\frac{\text{Area}\ Q}{2\pi}(\xi + i0)^{-2} + (2\pi^3)^{-1/2}\sum_{g\in G}\sum_{\pm}|g|^{-1/2}\varkappa(g)^{-1/2} \times e^{\pm\frac{\pi i}{4}(2\,\text{ind}\ g + \sigma(g))}v(\xi \mp |g|)\,, \quad (3.8)$$

where $v(\xi) = \frac{\sqrt{\pi}}{2}e^{\frac{3\pi i}{4}}(\xi + i0)^{-3/2}$ is the Fourier transform of $x_+^{1/2}$. The formula (3.8) is certainly only approximate since in its derivation we neglected different error terms. The sense of the formula (3.8) is that it describes the principal singularities of the tempered distribution

$$\chi(\xi) = \text{tr}\ e^{i\xi\sqrt{-\Delta}} = \sum_{R_k\in\text{spec}\sqrt{-\Delta}} e^{iR_k\xi}\,.$$

In contrast with the case when all closed geodesics are isolated and $\chi(\xi)$ has singularities of the homogeneity order -1 at the points of the geodesic spectrum of the riemannian manifold (see [9,10]), in our case when closed geodesics form the families filling up the invariant tori, $\chi(\xi)$ has stronger singularities at the points of the geodesic spectrum, of the homogeneity order $-3/2$.

Theorem 3.3. *Let Q be a Liouville surface which satisfies the assumptions of Theorem* 3.1. *Consider the tempered distributions*

$$d(x) = \sum_{R_k \in \operatorname{spec}\sqrt{-\Delta}} \delta(x - R_k) - \frac{\operatorname{Area} Q}{2\pi} x + -x_+^{1/2}(2\pi^3)^{-1/2} \sum_{g \in G} |g|^{-3/2} \varkappa(g)^{-1/2}$$
$$\times \cos\left(|g|x - \frac{\pi}{2} \operatorname{ind} g - \frac{\pi}{4}\sigma(g))\right), \tag{3.9}$$

and

$$\widehat{d}(\xi) = \sum_{R_k \in \operatorname{spec}\sqrt{-\Delta}} e^{iR_k\xi} + \frac{\operatorname{Area} Q}{2\pi}(\xi + i0)^{-2} - (2\pi^3)^{-1/2} \sum_{g \in G} \sum_{\pm} |g|^{-1/2} \varkappa(g)^{-1/2}$$
$$\times e^{\pm \frac{\pi i}{4}(2 \operatorname{ind} g + \sigma(g))} v(\xi \mp |g|). \tag{3.10}$$

If a C^3-function $\varphi(t)$ satisfies the inequalities

$$\left|\frac{d^k \varphi}{dt^k}\right| \leqq \operatorname{const}(k)(1 + |t|^{2+k})^{-1}, \quad k = 0, 1, 2, 3, \tag{3.11}$$

then

$$T^{-1} \int_0^T t^{-1/2} |(d * \varphi)|(t) dt \to 0 \quad as \quad T \to \infty, \tag{3.12}$$

and

$$T^{-1} \int_0^T t^{-1/2} |\langle \widehat{d}, \exp(-it\xi)\varphi \rangle| dt \to 0 \quad as \quad T \to \infty. \tag{3.13}$$

It is worthwhile to note that the general trace formula discussed in [9, 10] and the trace formula of Theorem 3.3 in a sense complement each other. Namely, to the local L^1 property of the function $\widehat{d}(\xi)$ in the general trace formula of [9, 10], the formula (3.13) adds some control of this function at infinity.

4. Proofs

As was shown in [15], the study of the spectrum of the Laplace–Beltrami operator on a Liouville surface can be reduced to the study of distribution of lattice points with respect to some particular domains dilated with some factor and then shifted. For the convenience of the reader we describe below this reduction proven in [15], and then we derive the desired theorems from the corresponding results for the lattice-point problem. It should be noted that the latter is of interest in itself.

Here we give a reduction of Theorem 3.1 to Theorems 2.1(a) and 11.1(i) of [15]. According to [15], the study of the function $N(x)$ can be reduced to the following lattice-point problem. Consider the curve Γ on a plane given parametrically as

$$x = (x_1(c), x_2(c)), \quad c_4 \leqq c \leqq c_1,$$

where

$$\begin{aligned} x_1(c) = F_1(c) &= \int_{q_1:\, U_1(q_1)-c \geqq 0} (U_1(q_1) - c)^{1/2} dq_1 \,, \\ x_2(c) = F_2(c) &= \int_{q_2:\, c-U_2(q_2) \geqq 0} (c - U_2(q_2))^{1/2} dq_2 \,. \end{aligned} \tag{4.1}$$

Observe that $x_1'(c) < 0$ and $x_2'(c) > 0$. In addition, $x_1(c_1) = x_2(c_4) = 0$. This implies that Γ is a star-like curve in the first quadrant so it can be written in the polar coordinates ρ, α as a graph of an single-valued function,

$$\rho = G(\alpha), \quad 0 \leqq \alpha \leqq \pi/2 \,.$$

Define the angles $0 = \alpha_4 < \alpha_3 < \alpha_2 < \alpha_1 = \pi/2$ as solution of the equations

$$\tan \alpha_i = \frac{F_2(c_i)}{F_1(c_i)}, \quad i = 1, 2, 3, 4,$$

and partition the first quadrant into three sectors A_1, A_2, A_3 with

$$A_i = \{(\rho, \alpha) : \alpha_{i+1} \leqq \alpha \leqq \alpha_i\} \,.$$

Theorem 4.1. ([15], Theorem 6.3). *Let $Q \in Q^r, r \geqq 3$ be a Liouville surface. Then*
(a) *$G(\alpha) \in C^1\left((0, \frac{1}{2}\pi)\right)$ and the tangent line to Γ at the point with the angle coordinate α_i is parallel to the x_i-axis ($i = 1, 2$).*
(b) *$G(\alpha) \in C^{r+1}\left((0, \alpha_1) \cup (\alpha_1, \alpha_2) \cup (\alpha_2, \frac{1}{2}\pi)\right)$.*
(c) *For all l ($0 \leqq l \leqq r + 1$) there exist finite limits*

$$\lim_{\alpha \to +0} \frac{d^l G}{d\alpha^l}(\alpha) = G^{(l)}(+0) \,, \tag{4.2}$$

$$\lim_{\alpha \to \frac{1}{2}\pi - 0} \frac{d^l G}{d\alpha^l}(\alpha) = G^{(l)}(\frac{1}{2}\pi - 0) \,. \tag{4.3}$$

In addition, $G^{(0)}(+0) = G(0) > 0$, $G^{(0)}(\frac{1}{2}\pi - 0) = G(\frac{1}{2}\pi) > 0$ and $G^{(1)}(+0) \neq 0$, $G^{(1)}(\frac{1}{2}\pi - 0) \neq 0$.
(d) *In the vicinity of the critical angles α_1, α_2 derivatives of G have the following asymptotics:*

$$\frac{d^l G}{d\alpha^l}(\alpha) \sim \frac{\text{const}(i, l)}{(\alpha - \alpha_i)^{l-1} (\log|\alpha - \alpha_i|^{-1})^2}, \quad \text{as} \quad \alpha \to \alpha_i, \; i = 1, 2, \; 1 \leqq l \leqq r - 1 \,. \tag{4.4}$$

(e) *The following inequalities hold for $0 \leqq \alpha \leqq \frac{1}{2}\pi$:*

$$G(\alpha) \geqq \text{const} > 0 \,,$$

$$\left| \frac{dG}{d\alpha}(\alpha) \right| \leqq \text{Const} \,.$$

Denote by D_1, D_2, D_3 the finite sectorial domains cut off by the curve Γ from the sectors A_1, A_2, A_3, respectively. Let RD_i be the image of D_i under the dilation with the factor R with respect to the origin. Consider the lattices

$$L_1 = \{(\pi(m_1 + (1/2)), 2\pi m_2),\ (m_1, m_2) \in \mathbb{Z}^2\} ,$$
$$L_2 = \{(2\pi m_1, 2\pi m_2), (m_1, m_2) \in \mathbb{Z}^2\} ,$$
$$L_3 = \{(2\pi m_1, \pi(m_2 + (1/2))), (m_1, m_2) \in \mathbb{Z}^2\} ,$$

and define

$$N_i(R) = \#\{L_i \cap RD_i\}, \quad i = 1,2,3 ,$$
$$N_0(R) = 2N_1(R) + 4N_2(R) + 2N_3(R) .$$

Theorem 4.2. ([15]), Theorem 6.2). *If a Liouville surface $Q \in \mathcal{Q}^r, r \geqq 5$, then the function*

$$\theta_1(R) = R^{-1/2}(N(R^2) - N_0(R))$$

is an almost periodic function equivalent to 0 *in the Besicovitch class B^1, i.e.,*

$$\lim_{T\to\infty} \frac{1}{T}\int_0^T |\theta_1(R)| dR = 0 . \tag{4.5}$$

Denote $\Gamma_i = \Gamma \cap A_i$. Let

$$D_4 = \{(x_1, x_2) \in \mathbb{R}^2 \ :\ (\pi x_1, 2\pi|x_2|) \in D_1\} ,$$
$$D_5 = \{(x_1, x_2) \in \mathbb{R}^2 \ :\ (2\pi|x_1|, 2\pi|x_2|) \in D_2\} ,$$
$$D_6 = \{(x_1, x_2) \in \mathbb{R}^2 \ :\ (2\pi|x_1|, \pi x_2) \in D_3\} , \tag{4.6}$$

and for a point $a \in \mathbb{R}^2$ and a domain $D \subset \mathbb{R}^2$, put

$$N(R, a, D) = \#\{(a + \mathbb{Z}^2) \cap RD\} ,$$
$$n(R, a, D) = N(R, a, D) - R^2 \operatorname{Area} D .$$

Obviously,

$$N(R, a_4, D_4) = 2N_1(R), \quad N(R, a_5, D_5) = 4N_2(R), \quad N(R, a_6, D_6) = 2N_3(R) ,$$

with

$$a_4 = ((1/2), 0), \quad a_5 = (0, 0), \quad a_6 = (0, 1/2)) ,$$

hence

$$N_0(R) = N(R, a_4, D_4) + N(R, a_5, D_5) + N(R, a_6, D_6)$$

and by (4.5),

$$\lim_{T\to\infty} \frac{1}{T}\int_0^T R^{-1/2} |N(R^2) - N(R, a_4, D_4) - N(R, a_5, D_5) - N(R, a_6, D_6)| dR = 0 . \tag{4.7}$$

By the Weyl law (see (1.5)),

$$N(R^2) = \frac{\operatorname{Area} Q}{4\pi} R^2 + n(R^2)$$

with $n(R^2) = o(R^2)$, and it is easy to see that

$$N(R, a_i, D_i) = (\text{Area}\, D_i)R^2 + n(R, a_i, D_i), \quad i = 4, 5, 6\ ,$$

with $n(R, a_i, D_i) = o(R^2)$. Hence (4.7) implies

$$\frac{\text{Area}\, Q}{4\pi} = \text{Area}\, D_4 + \text{Area}\, D_5 + \text{Area}\, D_6$$

and

$$\lim_{T\to\infty} \frac{1}{T}\int_0^T R^{-1/2}|n(R^2) - n(R, a_4, D_4) - n(R, a_5, D_5) - n(R, a_6, D_6)|dR = 0\ . \qquad (4.8)$$

This equation shows that we have reduced Theorem 3.1 to a problem of number theory about the behaviour of the number of lattice points inside a dilated domain.

Let D_0 be a star-like domain given in the polar coordinates by the inequalities

$$0 \leqq \rho \leqq G_0(\beta)\ ,\ \beta_1 \leqq \beta \leqq \beta_2\ , \qquad (4.9)$$

and denote by Γ_0 a part of its boundary given by equation:

$$\rho = G_0(\beta)\ ,\ \beta_1 \leqq \beta \leqq \beta_2\ . \qquad (4.10)$$

Assume that:

(A1) *$\Gamma_0 \in C^1([\beta_1, \beta_2]) \cap PC^5((\beta_1, \beta_2))$ where PC^r denotes the space of piecewise C^r-functions, and for $\beta_1 \leqq \beta \leqq \beta_2$,*

$$0 < c \leqq G_0(\beta) \leqq C, \quad |G'(\beta)| \leqq C\ .$$

(A2) *The curvature $\varkappa_0(\beta)$ of Γ_0 has only finitely many zeroes and each zero has the first order. More precisely, if $\varkappa_0(\beta_0) = 0$ then*

$$\varkappa_0(\beta) = (\beta - \beta_0)\varkappa_1(\beta) \quad \text{with} \quad \varkappa_1(\beta_0) \neq 0\ .$$

(A3) *There are only finitely many points on Γ_0 with infinite curvature. If β^* is such a point then in the vicinity of β^* the function $\varkappa_0$ and its derivatives have the following asymptotics:*

$$\frac{d^k\varkappa_0(\beta)}{d\beta^k} \sim \frac{C(\beta^*, k)}{(\beta - \beta^*)^{1+k}(\log|\beta - \beta^*|^{-1})^2}\ ,\ k = 0, 1, 2, 3\ .$$

Here $C(\beta^*, k)$ are nonvanishing real numbers depending on the critical point β^*.

Denote by V the set of all points of Γ_0 with infinite or zero curvature and include also in V the endpoints of Γ_0 if it is not closed.

(A4) *If $V \neq \emptyset$ then for each point from V with finite curvature the directions of its radius-vector and the tangent line to Γ_0 at this point are diophantine (i.e. the tangents of the slope angles are diophantine). For points where the curvature is infinite the directions of the radius vectors of these points are diophantine and the directions of the tangent lines at these points are either diophantine or rational.*

For a given curve Γ_0 define the function $S_0(\eta)$, which maps a vector $\eta \in \mathbb{R}^2 \setminus \{0\}$ onto the set of all points x on Γ_0 for which the vector of outer normal $n(x)$ to D_0 has the same direction as η, i.e.,

$$S_0(\eta) = \left\{ x \in \Gamma_0 : n(x) = \frac{\eta}{|\eta|} \right\} . \tag{4.11}$$

For some η the set $S_0(\eta)$ can be empty. Note that the assumptions A1, A2 imply that the curve Γ_0 consists of finitely many concave and convex arcs. This in turn implies $S_0(\eta)$ to be a finite set whose cardinality is uniformly bounded from above.

Define

$$E = \{ n = (n_1, n_2) \in \mathbb{N} : n_1, n_2 \text{ are coprime} \} \cup \{(1,0),(0,1)\} , \tag{4.12}$$

and

$$\Lambda = \cup_{\eta \in E} \{ \langle x, \eta \rangle | x \in S_0(\eta) \} . \tag{4.13}$$

A(5) *The elements of* Λ *are linearly independent over* $\mathbb{Z}$.

Theorem 4.3. ([15]), Theorem 11.1 (i), Lemmas 13.1–13.8). *If the domain* D_0 *satisfies the assumptions* A1–A4 *then for any* $a \in \mathbb{R}^2$ *the function* $\theta_0(R) = R^{-1/2} n(D_0, a, R)$ *is an almost periodic function of the Besicovitch class* B^2. *The Fourier series of* θ_0 *with respect to* B^2 *has the form*:

$$\theta_0(R) = \pi^{-1} \sum_{m \in \mathbb{Z}^2 \setminus 0} \sum_{x \in S_0(m)} |m|^{-3/2} |\kappa_0(x)|^{-1/2} \times \sin\left(2\pi R \langle x, m \rangle - 2\pi \langle a, m \rangle - \frac{1}{4} \pi \operatorname{sign} \varkappa_0(x) \right) . \tag{4.14}$$

In addition, for any T *and* K *sufficiently large and* $\beta > 0$

$$T^{-1} \int_0^T |\theta_0(R) - P_K(R)|^2 dR \leqq \text{const}(\beta) \left(T^{-\frac{1}{2}+\beta} + K^{-\frac{1}{6}+\beta} + K^{\frac{5}{12}} T^{-12} \right) , \tag{4.15}$$

where

$$P_K(R) = \pi^{-1} \sum_{\substack{m \in \mathbb{Z}^2 \setminus 0 \\ |m| \leqq K \\ x \in S_0(m)}} |m|^{-3/2} |\varkappa_0(x)|^{-1/2} \times \sin\left(2\pi R \langle x, m \rangle - 2\pi \langle a, m \rangle - \frac{1}{4} \pi \operatorname{sign} \varkappa_0(x) \right) . \tag{4.16}$$

Theorem 4.4. ([15], Theorem 11.1 (*ii*)). *If in addition to the assumptions of Theorem* 4.3 *the condition* A5 *is also fulfilled, then the limit distribution* $\mu(dy)$ *of* $\theta_0(R)$ *does not depend on* $a \in \mathbb{R}^2$ *and has a density* $p(y)$ *with respect to the Lebesgue measure. The density* $p(y)$ *is an entire function of its argument and satisfies the following inequalities on the real line*: $\forall \varepsilon > 0 \exists y(\varepsilon) > 0$ *such that*

$$p(y) \leqq \exp\left(-|y|^{\frac{16}{9} - \varepsilon} \right) \quad \text{when} \quad |y| \geqq y(\varepsilon) . \tag{4.17}$$

For a finite collection of domains $B_1, B_2, \ldots B_l$, consider the linear combination

$$n_s(R) = \sum s_i \, n(R, a_i, B_i) ,$$

where s_i, $i = 1, \ldots, l$, are arbitrary real numbers. Denote by Λ_i the set defined by (4.13) with $E = B_i$.

Theorem 4.5. ([15]), Corollary 11.2). *If the assumptions A1–A4 are satisfied for each domain B_i then the function $R^{-1/2}n_s(R)$ belongs to the Besicovitch class B^2 of almost periodic functions. In addition, if the assumption A5 holds for the set $\Lambda = \cup \Lambda_i$ then the density $p(y)$ of the limit distribution of $R^{-1/2}n_s(R)$ posesses all the properties formulated in Theorem 4.4.*

In the proof of Theorem 3.2 we use an improvement of the estimate (4.17) which will be proven in Sect. 5 below. For an exact formulation of the improved estimate we need some definitions.

Define

$$\varphi(\lambda) = \int_0^1 \exp(i\lambda B(t))\,dt\,, \tag{4.18}$$

where

$$B(t) = \sum_{m=1}^{\infty} m^{-3/2} \sin(2\pi m t - (\pi/4))\,, \tag{4.19}$$

and

$$A(\lambda) = (4/3)\int_0^{\infty} x^{-7/3} \log\varphi(\lambda x)\,dx\,. \tag{4.20}$$

It follows from (4.18) that $\varphi(iy)$ is a positive real-analytic function of $y \in \mathbb{R}$ with $\varphi(0) = 1$ and $\varphi'(0) = 0$. In addition, $\varphi(iy)$ has exponential asymptotics as $t \to \pm\infty$. This implies that the RHS of (4.20) is well-defined for λ pure imaginary and

$$A(iv) = \begin{cases} A_+ v^{4/3} & \text{if} \quad v > 0\,, \\ A_-(-v)^{4/3} & \text{if} \quad v < 0\,, \end{cases} \tag{4.21}$$

with $A_\pm = A(\pm i)$. We will show in Sect. 6 below that there exists $\delta > 0$ such that the RHS of (4.20) is well-defined for all λ in the sector

$$S_\delta = \{\lambda \in \mathbb{C}\backslash\{0\} : |\text{Re}\,\lambda| \leqq \delta|\text{Im}\lambda|\}\,, \tag{4.22}$$

and

$$A(ie^{i\theta}v) = A_\pm(\pm e^{i\theta}v)^{4/3}$$

for all $|\theta| \leqq \delta$ and $\pm v > 0$.

Define

$$\Phi(\lambda) = \prod_{m \in E} \prod_{x \in S_0(m)} \varphi(\pi^{-1}\rho(m,x)\lambda)\,, \tag{4.23}$$

where $E, S_0(m)$ and $\varphi(\lambda)$ are defined in the formulas (4.12), (4.11) and (4.18), respectively,

$$\rho(m,x) = |m|^{-3/2} \text{sign}\,(\varkappa(x))|\varkappa(x)|^{-1/2}$$

and $\varkappa(x)$ is the curvature at the point x.

Theorem 4.6. *Let all the conditions of Theorem 4.5 be satisfied. Then the characteristic function $\Phi_s(\lambda)$ of the limit distribution of $R^{-1/2}n_s(R)$ is equal to the product*

$$\Phi_s(\lambda) = \Phi^{(1)}(s_1\lambda)\cdots\Phi^{(l)}(s_l\lambda)\,,$$

where the function $\Phi^{(i)}(\lambda)(1 \leqq i \leqq l)$ is defined by (4.23) with the help of the domain B_i, and the density $p(t)$ of this limit distribution is an entire function of t such that for real t,

$$-t^{-4}\log p(t) \to C_{\pm} > 0 \quad as \quad t \to \infty \tag{4.24}$$

with

$$C_{\pm} = 8\pi^4 \left(\sum_{j=1}^{l} \int_{\partial B_j} A(\pm i s_j \operatorname{sgn} \varkappa_j(x)) |\varkappa_j(x)|^{1/3} d l_j(x) \right)^{-3} . \tag{4.25}$$

Here dl stands for the differential of the length of arc along the curve.

Proof of Theorem 3.1. Theorems 4.1, 4.2, 4.3, and 4.5 expressed in terms of the geodesic flow on Liouville surface give the desired result immediately.

Let

$$\begin{aligned} \Gamma_4 &= \{(x_1,x_2) \in \mathbb{R}^2 : (\pi x_1, 2\pi|x_2|) \in \Gamma_1\} , \\ \Gamma_5 &= \{(x_1,x_2) \in \mathbb{R}^2 : (2\pi|x_1|, 2\pi|x_2|) \in \Gamma_2\} , \\ \Gamma_6 &= \{(x_1,x_2) \in \mathbb{R}^2 : (2\pi|x_1|, \pi x_2) \in \Gamma_3\} , \end{aligned} \tag{4.26}$$

so that Γ_i is the curvilinear part of the boundary of D_i (cf. (4.6)). Denote by $S_i(\eta)$, $i = 4,5,6$, the function $S_0(\eta)$ constructed as described above with the help of the curve Γ_i instead of Γ_0 (see the paragraph before (4.12)). According to Theorems 4.2, 4.5 the Fourier series of $\theta(R)$ is

$$\begin{aligned} \theta(R) &= \pi^{-1} \sum_{i=4,5,6} \sum_{\substack{m \in \mathbb{Z}^2 \setminus \{0\} \\ x \in S_i(m)}} |m|^{-3/2} |\varkappa_i(x)|^{-1/2} \\ &\times \sin(2\pi\langle m,x\rangle R - 2\pi\langle m,a_i\rangle - (\pi/4) \operatorname{sign} \varkappa_i(x)) , \end{aligned} \tag{4.27}$$

where $|m| = (m_1^2 + m_2^2)^{1/2}$, $\langle m,x\rangle = m_1x_1 + m_2x_2$ and $\varkappa_i(x)$ is the curvature of Γ_i at the point $x \in \Gamma_i$.

We can establish a one-to-one correspondence between the set of invariant tori $w \in W$ filling up the unit cotangent bundle and the set

$$\Gamma_* = \Gamma_4 \cup \Gamma_5 \cup \Gamma_6 .$$

Namely, let an invariant torus $w \in W$ correspond to the point $x(w) = (2\pi)^{-1}(I_1, I_2)$, where I_1, I_2 are the values of the action variables on w (see (2.4), (2.5)). Comparing (2.4), (2.5) with (4.1), (4.26) we see that $x(w) \in \Gamma_*$. Observe that $E = 1/2$ on any $w \in W$ hence by (2.7) $I_i = f_i$. Therefore, (2.7), (2.8) imply that the vector (ω_1, ω_2) is collinear to the vector of the outer normal to Γ_* at the point $x(w)$. This allows us to establish a one-to-one correspondence between the set G of families of closed geodesics on Q and summands in (4.27). Namely, every $g \in G$ is characterized by an invariant torus $w(g) \in W$ where it lives and by the rotation numbers $n_1(g), n_2(g)$. So we can define the map

$$g \to (m = (n_1(g), n_2(g)), x = x(w(g))) ,$$

which is the one-to-one correspondence between G and the summands in (4.27).

Since the equation of Γ_* is $x(c) = (2\pi)^{-1}(f_1(c), f_2(c))$ the curvature $\varkappa_i(x)$ is

$$\varkappa_i(x) = 2\pi(f_1' f_2'' - f_2' f_1'')(f_1'^2 + f_2'^2)^{-3/2} .$$

By (2.18),

$$|m|^2 = n_1^2 + n_2^2 = |g|^2(\omega_1^2 + \omega_2^2) = |g|^2(\xi_1^2 + \xi_2^2) ,$$

hence

$$|m|^3\varkappa_i(x) = |g|^3(\xi_1^2 + \xi_2^2)^{3/2} 2\pi(f_1' f_2'' - f_2' f_1'')(f_1'^2 + f_2'^2)^{-3/2} ,$$

and by (2.15),

$$|m|^3 k_i(x) = 2\pi|g|^3 \det(d^2H)((1/2), c) . \tag{4.28}$$

Also, (2.18) implies

$$2\pi\langle m, x\rangle = n_1 f_1 + n_2 f_2 = |g| . \tag{4.29}$$

By (4.28),

$$\operatorname{sgn}\, (d^2H)((1/2), c) = \operatorname{sgn}\, \varkappa_i(x) ,$$

and the number $4\langle m, a_i\rangle$ mod 4 is the Maslov index of $g \in G$. Thus the formula (3.3) follows from (4.27).

In addition, the formulae (2.3)–(2.5), (2.8), (2.14), (2.15) and Theorem 4.1 show that the conditions of Theorem 3.1 coincide with the conditions of Theorem 4.3 formulated in terms of the geodesic flow on Q which completes the proof of Theorem 3.1.

Proof of Theorem 3.2. Invariant tori $w \in W_+$ of the unit cotangent bundle with non-negative frequencies correspond to the points of

$$\Gamma_0 = \Gamma_* \cap \{x = (x_1, x_2) \in \mathbb{R}^2 \,:\, x_1, x_2 \geqq 0\} .$$

We define the multiplicity function $h(x)$ on Γ^0 as

$$h(x) = \begin{cases} 2 & \text{if } \; x \in \Gamma_4 , \\ 4 & \text{if } \; x \in \Gamma_5 , \\ 2 & \text{if } \; x \in \Gamma_6 . \end{cases}$$

Consider the set

$$E = \{n = (n_1, n_2) \in \mathbb{N} \,:\, n_1, n_2 \text{ are relatively prime}\} \cup \{(1,0), (0,1)\} .$$

Then

$$\Omega = \{|g|, g \in G_0\} = \bigcup_{m \in E} \{2\pi\langle m, x\rangle, x \in S_0(m)\} .$$

Theorem 3.2 follows directly from Theorems 4.5, 4.6 in view of the formula (4.28).

Proof of Theorem 3.3. To prove (3.12) let us first observe that if $\varphi(t)$ satisfies (3.11) then the series

$$\sum_{g \in G} |g|^{-3/2}\varkappa(g)^{-1/2} \int_0^\infty \varphi'(x-y) y^{1/2} \sin\Big(|g|y - \frac{\pi}{2}\operatorname{ind} g - \frac{\pi}{4}\sigma(g)\Big)\, dy$$

$$= \sum_{g \in G} |g|^{-5/2}\varkappa(g)^{-1/2} \int_0^\infty [\varphi'(x-y) y^{1/2}]' \cos\Big(|g|y - \frac{\pi}{2}\operatorname{ind} g - \frac{\pi}{4}\sigma(g)\Big)\, dy$$

is absolutely convergent because the sum over g is basically a two-dimensional lattice sum and $\sum_g |g|^{-5/2} < \infty$. This implies that if we define the tempered distribution $d_K(x)$ as

$$d_K(x) = \sum_{R_k \in \operatorname{spec}\sqrt{-\Delta}} \delta(x - R_k) - \frac{\operatorname{Area} Q}{2\pi} x_+ - (2\pi^3)^{-1/2} \sum_{g \in G:\ |g| \leq K} |g|^{-3/2} \times \varkappa(g)^{-1/2} \left[x_+^{1/2} \sin\left(|g|\, x - \frac{\pi}{2} \operatorname{ind} - \frac{\pi}{4} \sigma(g) \right) \right]',$$

then

$$\lim_{K \to \infty} \lim_{T \to \infty} \frac{1}{T} \int_0^T t^{-1/2} |(d - d_K) * \varphi(t)|\, dt = 0 . \tag{4.30}$$

On the other hand,

$$d_K * \varphi(t) = \varepsilon_K * \varphi'(t) , \tag{4.31}$$

where

$$\varepsilon_K(t) = N(t^2) - \frac{\operatorname{Area} Q}{4\pi} t_+^2 - (2\pi^3)^{-1/2} \sum_{g \in G:\ |g| \leq K} |g|^{-3/2} \varkappa(g)^{-1/2} \times t_+^{1/2} \sin\left(|g| x - \frac{\pi}{2} \operatorname{ind} - \frac{\pi}{4} \sigma(g) \right) ,$$

and by Theorem 3.1

$$\lim_{K \to \infty} \lim_{T \to \infty} \frac{1}{T} \int_0^T t^{-1/2} |\varepsilon_K(t)|\, dt = 0 ,$$

hence

$$\lim_{K \to \infty} \lim_{T \to \infty} \frac{1}{T} \int_0^T t^{-1/2} |\varepsilon_K * \varphi'(t)|\, dt = 0. \tag{4.32}$$

From (4.30)–(4.32) we obtain (3.12). Since $\widehat{d}$ is the Fourier transform of d, (3.13) follows from (3.12). Theorem 3.3 is proved.

5. Proof of Theorem 4.6

Proof of Theorem 4.6. We consider a particular case when $l = 1, s_1 = 1$ and $D = D_1$ is convex. The proof for the case of several convex and concave arcs goes *mutatis mutandis.*

Using the inverse Fourier transform we obtain

$$p(t) = (2\pi)^{-1} \int_{-\infty}^{\infty} \Phi(\lambda) e^{-it\lambda} d\lambda . \tag{5.1}$$

Lemma 5.1. (see [15], Lemma 15.1). *If the conditions of Theorem 4.4 are satisfied then the function $\Phi(\lambda)$ is an entire function of $\lambda = \lambda_1 + i\lambda_2$ satisfying the estimation*

$$|\Phi(\lambda)| \leqq \exp\left(-C_1 |\lambda_1|^2 |\lambda|^{-2/3} + C_2 |\lambda_2| |\lambda|^{3/4} \right) \quad \text{as } |\lambda| \geqq 1 ,$$

where C_1, C_2 are some positive constants.

By Cauchy's theorem and with the help of Lemma 5.1 we can shift the axis of integration in (5.1),

$$p(t) = (2\pi)^{-1} \int_{-\infty}^{\infty} \Phi(\mu + iv)\exp(-it(\mu + iv))\,d\mu\,. \tag{5.2}$$

We evaluate the asymptotics of $p(t)$ as $t \to \pm\infty$ with the help of the saddle-point method. To this end we put

$$\Psi(\lambda) = \log \Phi(\lambda) = \sum_{m \in E} \psi(\pi^{-1}\rho(m)\lambda)\,, \tag{5.3}$$

where $\psi(\lambda) = \log \varphi(\lambda)$ and $\rho(m) = \rho(m, S_0(m))$ (recall that we assume that $S_0(m)$ consists of one point). Due to Lemma 7.5 formulated in Sect. 7 below, there exists a $\delta > 0$ so that for λ such that $|\lambda_1| \leqq \delta|\lambda_2|$, the function $\psi(\lambda)$ is well defined and analytic. Let us choose $\delta > 0$ so that all the statements of Sect. 7 are valid and rewrite (5.2) as

$$\begin{aligned} p(t) &= (2\pi)^{-1} \int_{|\mu| \leqq \delta|v|} \exp(\Psi(\mu + iv) - it(\mu + iv))\,d\mu \\ &\quad + (2\pi)^{-1} \int_{|\mu| \geqq \delta|v|} \Phi(\mu + iv)\exp(-it(\mu + iv))\,d\mu \\ &= I_1(\delta) + I_2(\delta)\,. \end{aligned} \tag{5.4}$$

Lemma 5.2. *If* $\lambda \in S_\delta = \{z \in \mathbb{C} \setminus \{0\} : |\mathrm{Re}\, z| \leqq \delta|\mathrm{Im}\, z|\}$, *then*

$$\Psi(\lambda) = (\varkappa/2)A(\pi^{-1}\lambda)(1 + o(1)) \quad as \quad |\lambda| \to \infty\,, \tag{5.5}$$

uniformly in arg λ. *Here*

$$\varkappa = \varkappa(\Gamma) = \int_\Gamma |\varkappa(x)|^{1/3} dl(x)$$

is the affine length of Γ, the curvilinear part of ∂D.

Corollary 5.3. *If* $\lambda \in S_\delta$ *then*

$$\left.\begin{aligned} \Psi'(\lambda) &= (4/3)\ (\pi\lambda)^{-1}\varkappa\ A(\pi^{-1}\lambda)\ (1 + o(1))\,, \\ \Psi''(\lambda) &= (4/9)\ (\pi\lambda)^{-2}\varkappa\ A(\pi^{-1}\lambda)\ (1 + o(1))\,, \end{aligned}\right\} \quad as \quad |\lambda| \to \infty\,, \tag{5.6}$$

uniformly in arg λ.

Corollary 5.3 is an obvious consequence of the Cauchy integration formula and Lemma 5.2, and we leave it without proof.

Let us apply the saddle-point method to the integral $I_1(\delta)$. If δ is chosen small enough then Lemma 7.5 below shows that the function $\Psi(\lambda) - it\lambda$ has in S_δ for large $|t|$ a unique critical point $\lambda_c(t) = \mu_c(t) + iv_c(t)$ with $\mu_c(t) = 0$. Taking $v = v_c(t)$ in (5.2) we obtain, with the help of the saddle-point method, the following lemma:

Lemma 5.4.

$$\log I_1(\delta) = -8\pi^4(\varkappa A_\pm)^{-3}t^4(1 + o(1))\,.$$

The following lemma completes the proof of Theorem 4.6:

Lemma 5.5.

$$I_2(\delta) = o(I_1(\delta)), \quad as \quad t \to \pm\infty .$$

6. Proof of Technical Lemmas

Proof of Lemma 5.2. Introduce the set of measures μ_R, $R > 0$, in the first quadrant,

$$\mu_R = (\pi^2/6)R^{-2} \sum_{m \in E} \delta(x - R^{-2}m) . \tag{6.1}$$

The family μ_R weakly converges to the Lebesgue measure as $R \to \infty$ so that for any continuous function g with compact support,

$$\lim_{R \to \infty} R^{-2} \sum_{m \in E} g(R^{-1}m) = (6/\pi^2) \int_0^\infty \int_0^\infty g(x, y)\, dxdy . \tag{6.2}$$

Consider a sequence of C^∞ functions $\zeta_n(\lambda)$ equal 1 inside the disk $|\lambda| \leqq n$ and having the support in the disk $|\lambda| \leqq 2n$. Applying (6.2) to the sums

$$\Psi_n(\lambda) = \sum_{m \in E} \psi(\pi^{-1}\rho(m)\lambda)\, \zeta_n(\lambda)$$

and letting $n \to \infty$ for $\lambda \in S_\delta$, we get uniformly in arg λ,

$$\sum_{m \in E} \Psi(\pi^{-1}\rho(m)\lambda) = \pi^{-4/3}|\lambda|^{4/3} \int_0^\infty \int_0^\infty \Psi(\rho(m)\exp(i \arg\lambda))\, dm\, (1 + o(1)) . \tag{6.3}$$

Recall that the geometrical meaning of the vector m is that it is a normal vector to Γ. Since Γ is a convex curve it can be smoothly parametrized by the angle β between the direction of the normal vector to Γ and polar axis. Then we can split the variables of integration in (6.3):

$$\begin{aligned} &|\lambda|^{4/3} \int_0^\infty \int_0^\infty \Psi(\rho(m)\exp(i \arg\lambda))\, dm \\ &\quad = |\lambda|^{4/3} \int_0^\infty \int_\Gamma \Psi(\exp(i \arg\lambda) r^{-3/2} \varkappa(\beta)^{-1/2})\, r d\beta dr \\ &\quad = \frac{2}{3}|\lambda|^{4/3} \int_0^\infty \int_\Gamma \Psi(\exp(i \arg\lambda)\, s \varkappa(\beta)^{-1/2})\, s^{-7/3} d\beta ds \\ &\quad = \frac{2}{3} \int_0^\infty \int_\Gamma \Psi(\lambda t) t^{-7/3} \varkappa(\beta)^{2/3})\, d\beta dt . \end{aligned} \tag{6.4}$$

We changed in (6.4) the variables by putting $r^{-3/2} = s$ and $s = \varkappa(\beta)^{1/2}|\lambda|t$. Thus substituting the integration with respect to the arclength parametrization for the integration with respect to β finally we get

$$\begin{aligned} \sum_{m \in E} \Psi(\pi^{-1}\rho(m)\lambda) &= (2/3)\varkappa \int_0^\infty \Psi(\pi^{-1}\lambda y) y^{-7/3} dy (1 + o(1)) \\ &= (\varkappa/2) A(\pi^{-1}\lambda)(1 + o(1)) . \end{aligned} \tag{6.5}$$

Lemma 5.2 is proved.

Proof of Lemma 5.4. Corollary 5.3 shows that $v_c(t)$ is the unique solution of the equation

$$\Psi'(iv_c(t)) - it = 0\,. \tag{6.5}$$

Hence using (4.21) we obtain

$$v_c(t) = -\left(\frac{2}{3}\varkappa A_\pm\right)^{-3}\pi^4 t^3(1+o(1)), \quad t\to\pm\infty\,. \tag{6.6}$$

Thus by (4.20),

$$\Psi(iv_c(t)) + t\,v_c(t) = -(\pi^4/4)((2/3)\varkappa A_\pm)^{-3}t^4(1+o(1)) \quad t\to\pm\infty\,. \tag{6.7}$$

In addition, by (5.6)

$$\Psi''(iv_c(t)) = C_\pm t^{-6}(1+o(1)), \quad t\to\pm\infty\,, \tag{6.8}$$

with some $C_\pm > 0$. From (6.7), (6.8) with the help of the saddle-point method we get that as $t\to\pm\infty$,

$$\begin{aligned}\log I_1(\delta) &= \log\Big[(2\pi)^{-1}\exp\Big(\Psi(iv_c(t)) + t\,v_c(t) - \frac{1}{4}\pi\,\mathrm{sign}\,\Psi''(iv_c(t))\Big)\\ &\quad\times|\Psi''(iv_c(t))|^{-1/2}\Big](1+o(1)) = -(\pi^4/4)((2/3)\varkappa A_\pm)^{-3}t^4(1+o(1))\,.\end{aligned} \tag{6.9}$$

Lemma 5.4 is proved.

Proof of Lemma 5.5. Fix a large number N. Then from Lemmas 7.2, 7.4 and formulae (6.7), (6.8) we have for large $|t|$,

$$\begin{aligned}\frac{|I_2(\delta)|}{|I_1(\delta)|} &\leqq ct^4\int\limits_{|\mu|>\delta|v_c(t)|}\left|\frac{\Phi(\mu+iv_c(t))}{\Phi(iv_c(t))}\right|d\mu\\ &\leqq ct^4\int\limits_{|\mu|>\delta|v_c(t)|}\prod_{m\in E}\prod_{x\in S(m)}\left|\frac{\varphi(\pi^{-1}\rho(m)(\mu+iv_c(t)))}{\varphi(\pi^{-1}\rho(m)(iv_c(t)))}\right|d\mu\,.\end{aligned} \tag{6.10}$$

If $|\lambda|\geqq N$, $|\mathrm{Re}\,\lambda|\geqq\delta|\mathrm{Im}\,\lambda|$ and N is large enough then due to Corollary 7.5,

$$\left|\frac{\varphi(\lambda)}{\varphi(\mathrm{Re}\,\lambda)}\right|\leqq\left(\frac{|\mathrm{Re}\,\lambda|}{|\lambda|}\right)^{1/2}(1+(\delta/2))\,.$$

In addition, if $|t|$ is large then the number of $m\in E$ for which $S(m)\neq\emptyset$ and $|\pi^{-1}\rho(m)(\mu+iv_c(t))|\geqq 1$ is bounded from below by $ct^2N^{-2/3}$. Hence

$$\left|\frac{\Phi(\mu+iv_c(t))}{\Phi(iv_c(t))}\right|\leqq\left[(1+(\delta/2))\left(1+\left|\frac{\mu}{v_c(t)}\right|\right)^{-1}\right]^{cN^{-2/3}t^2}$$

and

$$\begin{aligned}\frac{|I_2(\delta)|}{|I_1(\delta)|} &\leqq c_0 t^4 \int_{\delta|v_c(t)|}^{\infty} \left[(1+(\delta/2))\left(1+\left|\frac{\mu}{v_c(t)}\right|\right)^{-1}\right]^{cN^{-2/3}t^2} d\mu \\ &\leqq c_1 t^7 (1-(\delta/2))^{cN^{-2/3}t^2} \to 0\end{aligned}$$

as $|t| \to \infty$. Lemma 5.5 is proved.

7. Properties of Auxiliary Functions

Define the functions

$$\begin{aligned} b(x) &= \sum_{m=1}^{\infty} m^{-3/2} \exp(2\pi i m x), \\ B(x) &= \sum_{m=1}^{\infty} m^{-3/2} \sin(2\pi m x - (\pi/4)), \end{aligned} \tag{7.1}$$

which are continuous periodic functions of period 1. In addition, $b(x)$ can be extended to the upper complex half-plane as a bounded periodic analytic function of period 1. Remark that

$$B(x) = \operatorname{Im}\left(\frac{1-i}{\sqrt{2}}\, b(x)\right). \tag{7.2}$$

Lemma 7.1. *The function $B(x)$ is a continuous periodic function of period 1 with*

$$\int_0^1 B(x)\, dx = 0, \tag{7.3}$$

and moreover $B(x)$ is a real analytic and strictly concave function on $(0,1)$, so that

$$B''(x) < 0, \quad \forall\, 0 < x < 1.$$

Also the function $B(x)$ has a unique maximum point $x_{\max}$ on the interval $(0,1)$ and

$$B(x_{\max}) > 0. \tag{7.4}$$

In addition, $B(x)$ is minimal at the point $x_{\min} = 0$ and

$$B(0) < 0. \tag{7.5}$$

Near 0 the function $B(x)$ has the form

$$B(x) = 2\pi\, (2x)_+^{1/2} + b_1(x), \tag{7.6}$$

where

$$y_+^{1/2} = \begin{cases} y^{1/2} & \text{if} \quad y > 0, \\ 0 & \text{if} \quad y \leq 0, \end{cases}$$

and $b_1(x)$ is analytic near 0 with $b_1'(0) < 0$.

It is to be noted that by (7.6), $B(x)$ is real-analytic from the left of 0 with $B'(-0) < 0$ and it has a square-root singularity from the right of 0.

Proof of Lemma 7.1. For $z = x + iy$ with $y > 0$ we have from (7.1) that

$$b''(z) = -4\pi^2 \sum_{m=1}^{\infty} m^{1/2} \exp(2\pi i m z) = -4\pi^2 \sum_{m=-\infty}^{\infty} m_+^{1/2} \exp(2\pi i m z) . \tag{7.7}$$

Define for $z \in \mathbb{C}$ with $\operatorname{Im} z > 0$,

$$\beta(z) = \int_{-\infty}^{\infty} m_+^{1/2} \exp(2\pi i m z) dm = \int_0^{\infty} m^{1/2} \exp(2\pi i m z) dm = \frac{i-1}{8\pi} z^{-3/2} . \tag{7.8}$$

Then by the Poisson summation formula

$$\sum_{m=1}^{\infty} m^{1/2} \exp(2\pi i m z) = \sum_{m=-\infty}^{\infty} \beta(m+z), \quad \operatorname{Im} z > 0 , \tag{7.9}$$

hence

$$b''(z) = \frac{1-i}{2}\pi \sum_{m=-\infty}^{\infty} (z+m)^{-3/2}, \quad \operatorname{Im} z > 0 . \tag{7.10}$$

This formula defines $b''(z)$ as an analytic function in the strip $\{z = x + iy : 0 < x < 1\}$. Now from (7.2) we obtain

$$B''(x) = \operatorname{Im}\left(\frac{1-i}{\sqrt{2}} b''(x)\right) = -\frac{\pi}{\sqrt{2}} \sum_{m=0}^{\infty} (m+x)^{-3/2}, \quad 0 < x < 1 . \tag{7.11}$$

This proves that $B''(x)$ (and hence $B(x)$) is a real analytic function on (0, 1) and $B''(x) < 0$ so that $B(x)$ is concave.

Equation (7.3) follows directly from (7.1), and this implies (7.4) and (7.5). Finally, (7.6) follows from (7.11). Lemma 7.1 is proved.

Introduce the constants $B_\pm$ and $D_\pm$ as follows:

$$\begin{aligned} B_+ &= B(x_{\max}), \quad B_- = -B(0) ; \\ D_+ &= -B''(x_{\max}), \quad D_- = -B'(-0) , \end{aligned} \tag{7.12}$$

and consider the function

$$\varphi(\lambda) = \int_0^1 \exp(i\lambda B(x))\, dx . \tag{7.13}$$

Lemma 7.2. *$\varphi(\lambda)$ is an entire function for which*

$$\varphi(0) = 1, \quad \varphi'(0) = 0 ; \tag{7.14}$$

$$\varphi(\lambda) > 0, \ \operatorname{Re}\varphi'(\lambda) = 0, \ \varphi''(\lambda) < 0 \ \textit{ if } \ \operatorname{Re}\lambda = 0 ; \tag{7.15}$$

$$|\varphi(\mu + i\nu)| \le \varphi(i\nu) \quad \forall\, (\mu + i\nu) \in \mathbb{C} . \tag{7.16}$$

Proof. Equation (7.13) implies that $\varphi(\lambda)$ is an entire function and

$$\varphi'(\lambda) = i\int_0^1 B(x)\exp(i\lambda B(x))\,dx\,;$$

$$\varphi''(\lambda) = -\int_0^1 B^2(x)\exp(i\lambda B(x))\,dx\,, \tag{7.17}$$

so Lemma 7.2 follows.

Corollary 7.3. *The function $\varphi(it)$ is a strictly convex positive function of $t \in \mathbb{R}$ which achieves it minimum at $t = 0$.*

Lemma 7.4. *The following asymptotics holds as $|\lambda| \to \infty$: if $\lambda = \mu + i\nu$ with $\nu \geqq 0$, then*

$$\varphi(\lambda) = i(D_-\lambda)^{-1}e^{-i\lambda B_-}(1 + O(|\lambda|^{-1} + |\lambda|^{1/2}e^{-c\nu}))\,, \tag{7.18}$$

where $c > 0$ is independent of λ; if $\nu \leqq 0$ then

$$\varphi(\lambda) = (2\pi)^{1/2}(i\lambda D_+)^{-1/2}e^{i\lambda B_+}(1 + O(|\lambda|^{-1}))\,. \tag{7.19}$$

Corollary 7.5. *As $|\lambda| = |\mu + i\nu| \to \infty$,*

$$\left|\frac{\varphi(\mu + i\nu)}{\varphi(i\nu)}\right| = O\left(\left|\frac{\mu + i\nu}{\mu}\right|^{-1/2}\right)\,. \tag{7.20}$$

Proof of Lemma 7.4. Let $\chi(x)$ be a C^∞ function which is equal to 1 in ε-neighborhood of the point $x_{\max}$ and which is equal to 0 outside of 2ε-neighborhood of $x_{\max}$. Then $\varphi(\lambda) = \varphi_1(\lambda) + \varphi_2(\lambda) + \varphi_3(\lambda)$, where

$$\begin{aligned}
\varphi_1(\lambda) &= \int_0^1 \chi(x)e^{i\lambda B(x)}dx\,,\\
\varphi_2(\lambda) &= \int_0^{x_{\max}} [1 - \chi(x)]e^{i\lambda B(x)}dx\,,\\
\varphi_3(\lambda) &= \int_{x_{\max}}^1 [1 - \chi(x)]e^{i\lambda B(x)}dx\,.
\end{aligned} \tag{7.21}$$

Let us first evaluate $\varphi_1(\lambda)$.

Assume $\nu \geqq 0$. Let $y = y(x)$ be a smooth increasing function near $x_{\max}$ such that

$$y^2 = B(x_{\max}) - B(x) = B_+ - B(x)\,.$$

Then

$$\varphi_1(\lambda) = e^{i\lambda B_+}\int_{-\infty}^{\infty} \chi(y)e^{-i\lambda y^2}x'(y)\,dy\,.$$

We can write

$$x'(y) = t + ys(y), \qquad t = (|B''(x_{\max})|/2)^{-1/2} = (D_+/2)^{-1/2}$$

with smooth $s(y)$, hence

$$\varphi_1(\lambda) = e^{i\lambda B_+}[t\varphi_4(\lambda) + \varphi_5(\lambda)]\,, \tag{7.22}$$

where

$$\varphi_4(\lambda) = \int_{-\infty}^{\infty} \chi(y) e^{-i\lambda y^2} dy ,$$

$$\varphi_5(\lambda) = \int_{-\infty}^{\infty} \chi(y) e^{-i\lambda y^2} ys(y)\, dy .$$

Now we can write $\varphi_4(\lambda)$ as follows: $\varphi_4(\lambda) = \varphi_6(\lambda) + \varphi_7(\lambda)$ with

$$\varphi_6(\lambda) = \int_{-\infty}^{\infty} \chi(y) e^{-i\mu y^2} dy ,$$

$$\varphi_7(\lambda) = \int_{-\infty}^{\infty} \chi(y) e^{-i\mu y^2} (e^{\nu y^2} - 1)\, dy , \tag{7.23}$$

and

$$\varphi_6(\lambda) = O\left((1 + |\mu|)^{-1/2}\right) , \tag{7.24}$$

$$\varphi_7(\lambda) = O\left((1 + |\mu|)^{-1} e^{\varepsilon_0 \nu}\right) , \tag{7.25}$$

where

$$\varepsilon_0 = 2 \max_{|x - x_{\max}| \leqq 2\varepsilon} |y(x)|^2 .$$

Indeed, (7.24) is obvious when $|\mu| \leqq 1$ and when $|\mu| > 1$,

$$\int_{-\infty}^{\infty} \chi(y) e^{-i\mu y^2} dy = \int_{-\infty}^{\infty} e^{-i\mu y^2} dy + \int_{-\infty}^{\infty} [1 - \chi(y)] e^{-i\mu y^2} dy = c_0 |\mu|^{-1/2} + O(|\mu|^{-1}) ,$$

which implies (7.24). Also (7.25) is obvious when $|\mu| \leqq 1$, and when $|\mu| > 1$ we integrate by parts in (7.23) which gives

$$\varphi_7(\lambda) = \int_{-\infty}^{\infty} e^{-i\mu y^2} \left(\chi(y) \frac{e^{\nu y^2} - 1}{2i\mu y} \right)' dy = O(|\mu|^{-1} e^{\varepsilon_0 \nu}) .$$

Thus (7.24) and (7.25) are proved. Since also

$$\varphi_5(\lambda) = (2i\lambda)^{-1} \int_{-\infty}^{\infty} e^{-i\lambda y^2} [\chi(y) s(y)]' dy = O(|\lambda|^{-1} e^{\varepsilon_0 \nu}) ,$$

we obtain from (6.22)–(6.25) that

$$\varphi_1(\lambda) = O((1 + |\mu|)^{-1/2} e^{(-B_+ + \varepsilon_0)\nu}) .$$

We may assume ε_0 is small, say, $\varepsilon_0 < B_+/2$. Then the last relation implies

$$\varphi_1(\lambda) = O\left(|\lambda|^{-1/2} e^{B_- \nu}\right) . \tag{7.26}$$

Let us estimate now $\varphi_2(\lambda)$.

By Lemma 7.1 $B(x)$ is a smooth increasing function on the interval $0 < x < x_{\max} - \varepsilon$ with $B'(x) > 0$ and $B'(x)$ has a square-root singularity at $x = 0$ (see (7.6)).

Therefore $B_0(y) = B(y^2)$ is a smooth increasing function with $B_0'(y) > 0$ on $[0, y_0]$, where $y_0 = (x_{\max} - \varepsilon)^2$. Making the change of variable $x = y^2$ in (7.21) we have

$$\varphi_2(\lambda) = \int_0^{y_0} [1 - \chi(y^2)] e^{i\lambda B_0(y)} 2y dy .$$

Since $B_0'(y) > 0$ we can integrate by parts twice in the last formula, integrating the exponent and differentiating the rest. The main contribution comes from the boundary term at $y = 0$ and this gives

$$\varphi_2(\lambda) = O(|\lambda|^{-2} e^{B_-\nu}) . \tag{7.27}$$

Similarly, the function $B(x)$ is a smooth decreasing function on $[x_{\max} + \varepsilon, 1]$ with $B'(x) < 0$. So we can differentiate by parts in the third formula in (7.21), integrating the exponent and differentiating the rest. The main contribution comes from the boundary term at $x = 1$ and this gives

$$\varphi_3(\lambda) = i(D_-\lambda)^{-1} e^{-i\lambda B_-}(1 + O(|\lambda|^{-1})) . \tag{7.28}$$

From (7.26)–(7.28), (7.19) follows. Equation (7.20) is proved similarly. Lemma 7.4 is proved.

References

1. Berry, M.V., Mount, K.E.: Semiclassical approximations in wave mechanics. Rep. Progr. Phys. **35**, 315–397 (1972)
2. Berry, M.V., Tabor, M.: Closed orbits and the regular bound spectrum. Proc. R. Soc. Lond. **A349**, 101–123 (1976)
3. Bleher, P.M.: Quasiclassical expansion and the problem of quantum chaos. Lect. Notes in Math **1469**, 60–89 (1991)
4. Bleher, P.M.: On the distribution of the number of lattice points inside a family of convex ovals. Duke Math. J. **67**, 461–481 (1993)
5. Bleher, P.M.: Distribution of the error in the Weyl asymptotics for the Laplace operator on a two-dimensional torus and related lattice problems. Duke Math. J. **70**, 655–682 (1993)
6. Bleher, P.M.: Distribution of energy levels of a quantum free particle on a surface of revolution. Duke Math. J. **74**, 1–49 (1994)

6a. Bleher, P.M.: Semiclassical quantization rules near separatrices. Commun. Math. Phys. **165**, 621–640 (1994)

7. Bleher, P.M., Cheng, Zh., Dyson, F.J., Lebowitz, J.L.: Distribution of the error term for the number of lattice points inside a shifted circle. Commun. Math. Phys. **154**, 433–469 (1993)
8. Colin de Verdiére, Y.: Spectre du laplacien et longueurs des geodesiques periodiques. Comp. Math. **27**, 159–184 (1973)
9. Duistermaat, J.J., Guillemin, V.: Spectrum of elliptic operators and periodic bicharacteristics. Invent. Math. **29**, 39–79 (1975)
10. Guillemin, V.: Lectures on spectral theory of elliptic operators. Duke Math. J. **44**, 485–517 (1977)
11. Gutzwiller, G.: Chaos in classical and quantum mechanics. New York: Springer-Verlag, 1990
12. Heath-Brown, D.R.: The distribution and the moments of the error term in the Dirichlet divisor problem. Acta Arithm. **60**, 389–415 (1992)
13. Hörmander, L.: Fourier integral operators I. Acta Math. **127**, 79–183 (1971)
14. Hörmander, L.: The spectral function of an elliptic operator. Acta Math. **121**, 193–218 (1968)
15. Kosygin, D.V., Minasov, A.A., Sinai, Ya.G.: Statistical properties of the Laplace–Beltrami operator on Liouville surfaces. Uspekhi Mat. Nauk **48**, no. 4, 3–130 (1993)
16. Lax, P., Phillips, R.: Scattering theory for automorphic functions. Princeton, NJ: Princeton University Press, 1976

17. Levitan, B.M., Zhikov, V.V.: Almost periodic functions and differential equations. Cambridge: Cambridge Univ. Press, 1968
18. McKean, H.B.: Selberg's trace formula as applied to a compact Riemann surface. Comm. Pure Appl. Math. **25**, 225–246 (1972)
19. Rubinstein, M., Sarnak, P.: Chebyshev bias (to appear)
20. Sarnak, P.: Arithmetic quantum chaos. Schur Lectures, Tel Aviv, 1992
21. Sinai, Ya.G.: Mathematical problems in the theory of quantum chaos. Lect. Notes in Math. **1469**, 41–59 (1991)
22. Nguen Tien Zung, Polyakova L.S., Selivanova, E.N.: Topological classification of integrable geodesic flows with additional quadratic or linear in momenta integral on two-dimensional oriented riemannian manifolds. Funct. Analysis and Appl. **27**, 42–56 (1993)

Communicated by A. Jaffe

Comments

In our joint paper with D. Kosygin and A. Minasov (see [KMS]), we constructed semiclassical approximations of eigenfunctions of Laplacians on Liouville surfaces. In this paper, using these approximations we derive trace formula for Liouville surfaces analogous to the famous Selberg formula for manifolds of negative curvature.

MATHEMATICAL PROBLEMS IN THE THEORY OF QUANTUM CHAOS

Ya.G. Sinai

Landau Institute for Theoretical Physics
Academy of Sciences
Moscow, USSR

§1 Introduction

Theory of quantum chaos has both experimental and theoretical aspects. It contains in its title the word "chaos", not because of some erratic temporal behaviour of a quantum system, but because of its analysis of the role of classical chaos to properties of quantum systems. Theoretical works on quantum chaos are split into two parts. In one part people start with a classical chaotic dynamic system and try to quantize it. In contrast to this, in the other part everything starts with a quantum system and the problem is to try to find properties which are determined by the properties of the corresponding classical limit. It is natural that in the last case the methods are based upon the theory of quasi-classical approximation, and, in particular, on the precision of this approximation.

The main researchers in quantum chaos are M. Berry, I. Percival, M. Gutzwiller, J. Ford, A. Voros, G. Casati, B. Chirikov, G. Zaslavski, F. Izraelev, D. Shepeljanski, G. Berman. Since I am a beginner in this field I include only the names of those whose works have influenced my studies in this topic. One of the first works in quantum chaos was the paper by A. Einstein (1917). The main works in the theory of quasi-classical approximations include the well-known papers by J. Keller, R. Balyan and C. Bloch, V.P. Maslov. I was very glad to meet Professor U. Smilansky of the Weizmann Institute during my stay in Tel Aviv, and to have several very interesting discussions with him concerning the works done by him and his colleagues on quantum chaos.

Mostly, we shall deal in these lectures with two-dimensional quantum systems. The corresponding models describe quantum particles on two-dimensional surfaces. Mathematically this means that we shall consider two-dimensional closed compact Riemannian surfaces Q equipped with some Riemannian metrics ds^2. The C^2-smoothness of the metrics is sufficient for our goals.

It has just been brought to our attention by the author that he recently submitted this paper also to the American Institute of Physics for publication and it has just appeared in "Chaos/XAOC" 1990, edited by D.K. Campbell. We thank the AIP for permission to go ahead with this publication.

Y.G. Sinai, *Selecta II: Probability Theory, Statistical Mechanics, Mathematical Physics and Mathematical Fluid Dynamics*, DOI 10.1007/978-1-4419-6205-8_13,

The metrics ds^2 generates the Laplace operator Δ and the stationary states of the quantum particle are described by eigen-functions of Δ, i.e. by the solutions of the equation

$$-\Delta\psi_k = E_k\psi_k \ .$$

Here E_k are eigen-values of $-\Delta$ which are labelled in the increasing order:

$$0 < E_1 \le E_2 \le \ldots \le E_k \le \ldots \ .$$

Generically only strict inequalities hold.

We introduce the function $N(x)$ equal to the number of E_k which are less than x. It is the famous Weyl theorem which says that asymptotically

$$N(x) = \frac{\mathrm{Area}(Q)}{4\pi} x + o(x) \ .$$

The detailed study of the asymptotic expansion of $o(x)$ is sometimes a very complex mathematical problem.

One of the main ideas of quantum chaos is that even for regular surfaces the function $N(x)$ behaves on small distances as a random function. There are at least two possible ways of describing this randomness.

1. Fix an interval (α,β), $0<\alpha<\beta<\infty$, and introduce $\mu_x\big((a,b)\big) = \frac{1}{N(x)}\#\{E_k < x,\ E_k - E_{k-1} \in (\alpha,\beta)\}$. Then for every x we have a probability measure on R^1. The Weyl asymptotics easily implies that the family of measures μ_x is weakly compact. The weak limits of μ_x describe the limiting distributions of spacings between the nearest eigen-values.
2. Take $c>0$ and for any integer $k \ge 0$ introduce the set $A_k(x) = \{E \le x \mid (E, E+c) \text{ has } k \text{ eigen-values}\}$. Then the probabilities $\pi_x(k) = \frac{\ell(A_k(x))}{x}$ where ℓ is the Lebesgue measure and their limiting points characterize the clustering properties of E_k.

Many qualitative arguments and numerical studies show that weak limits of μ_x, and $\pi_x = \{\pi_x(k)\}$ exist as $x \to \infty$. We consider these limiting distributions as describing statistical properties of $N(x)$ on scales of order of unity or the local randomness of $N(x)$. Mathematically it is a difficult problem to show the existence of weak limits because we have no convenient functional representations for the introduced probabilities. Moreover, for some special cases it might even happen that the limits really are not unique and there are several limit points.

The main statement in the theory of quantum chaos which was expressed explicitly by I. Percival, M. Berry, G. Zaslavsky, B. Chirikov and maybe by some other researchers is that the form and the properties of these limiting distributions are determined completely by the ergodic properties of the corresponding geodesic flow acting on the unit tangent bundle over Q generated by the metrics ds^2.

The paper by M. Berry and Tabor [1] contains very convincing arguments in favor of a very striking statement which says that if the geodesic flow is completely integrable than $w - \lim_{x\to\infty} \mu_x = \mu$ is an exponential distribution, i.e. $\mu(\alpha,\beta) = \int_\alpha^\beta \rho e^{-\rho u} du$ where ρ can be found from the Weyl asymptotics, and $w - \lim_{x\to\infty} \pi_x = \pi$ is Poisson, i.e. $\pi(\kappa) = e^{c\rho} \frac{(c\rho)^\kappa}{\kappa!}$. The density of the exponential distribution $\rho e^{-\rho u}$ remains positive as $u \to 0$. This property is interpreted in the physical literature as the absence of repulsion of levels because the probability of having two close eigen-values is approximately the same as the probability of having two neighboring eigen-values at any distance of order of unity.

It is widely believed that if the geodesic flow is ergodic, mixing, K-flow, etc., then the density of the limiting distribution μ behaves as Const u^γ as $u \to 0$ which is interpreted as a repulsion of levels. This type of behavior is known for the distribution of spacing between levels in various ensembles of matrices of large dimension which were studied by Wigner, Dyson, Mehta and others. Apparently it is the most interesting part of the theory of quantum chaos but unfortunately at the moment I have no definite results here and therefore shall not discuss it any more.

§2 Surfaces of Revolution, Their Laplace Operators and Quasi-classical Expressions of Eigen-values

In this and the next sections we shall discuss the above-formulated statement by Berry and Tabor. The simplest example of an integrable Hamiltonian system with two-degrees of freedom is the geodesic flow on a two-dimensional surface of revolution. Its integrability follows from the extra symmetry of the surface and the existence of the corresponding Clairot integral. We shall deal with the surfaces having the topology of the torus. Such surfaces are naturally described by continuous positive periodic functions f of some period h, $f(0) = f(h)$. The properties of smoothness of f will be clear from the context. The whole surface Q is the result of the rotation of the graph of f along the r-axis and the subsequent gluing of boundary circles (see Fig. 1).

For some technical reasons which we will discuss later we shall assume that f has only two non-degenerate critical points, being strictly monotone outside them. The natural coordinates on Q are r, φ, where φ is the angle counted from some fixed direction. The Riemannian metrics

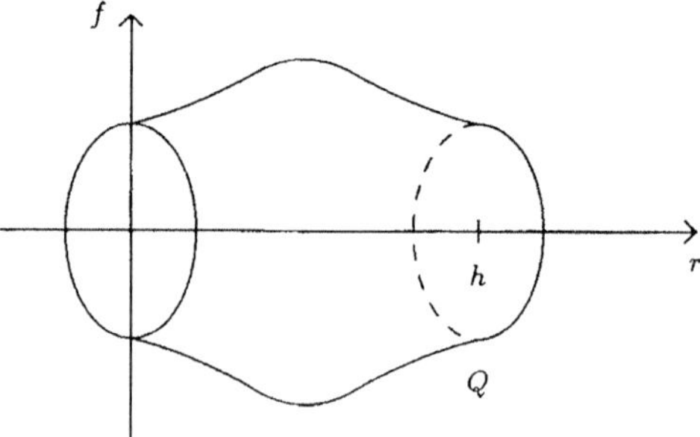

Fig. 1

in these coordinates has the form

$$ds^2 = f^2(r)d\varphi + \left(1 + \left(f'(r)\right)^2\right)dr^2 \ .$$

Now we can write down the explicit expression for the Laplace operator

$$\Delta\psi = \frac{1}{f\sqrt{1+(f')^2}}\frac{\partial}{\partial r}\left(\frac{f}{\sqrt{1+(f')^2}}\frac{\partial}{\partial r}\psi\right) + \frac{1}{f^2}\frac{\partial^2\psi}{\partial\varphi^2} \ .$$

Look for the eigen-functions in the form $\psi_{n,m} = e^{2\pi i n\varphi} \cdot v_{n,m}(r)$. This representations follows from the symmetry of the metrics which is preserved on the quantum level as some symmetry of the Laplacian. For $v_{n,m}$ we get the equation

$$\frac{1}{f\sqrt{1+(f')^2}}\frac{d}{dr}\frac{f}{\sqrt{1+(f')^2}}\frac{d}{dr}v_{n,m} - \frac{4\pi^2n^2}{f^2}v_{n,m} = -E_{n,m}v_{n,m}$$

where $E_{n,m}$ is the corresponding eigen-value. Introduce the new variable s through the formula

$$\frac{dr}{ds} = \frac{f}{\sqrt{1+(f')^2}} \ ,$$

or $s = \int_0^r \frac{1}{f(t)}\sqrt{1+(f't)^2}\,dt$. Then f becomes the periodic function of period $h_1 = \int_0^h \frac{1}{f(t)}\sqrt{1+\left(f'(t)\right)^2}\,dt$ and the equation for $v_{n,m}$ takes its final form

$$\frac{d^2}{ds^2}v_{n,m} + (E_{n,m}f^2 - 4\pi^2n^2)v_{n,m} = 0 \ .$$

We shall study large eigen-values $E_{n,m}$. Take a parameter L which will later tend to infinity. Assume that n is of order L. To be more precise, fix two positive numbers a_1 and a_2 and consider $a_1L \leq n < a_2L$. The eigen-value $E_{n,m}$ will be of order L^2. In these circumstances we can use the quasi-classical approximation for the eigen-functions $v_{n,m}$ and write them in the form $v_{n,m} = \exp\{iL(\sigma^{(0)}_{n,m} + \frac{1}{iL}\sigma^{(1)}_{n,m} + \dots)\}$ where the dots mean terms of a smaller order. For $\sigma^{(0)}_{n,m}$ we have the equation

$$\left((\sigma^{(0)}_{n,m})'\right)^2 = (\varepsilon_{n,m} f^2 - 4\pi^2 \nu_n^2)$$

where we put $\varepsilon_{n,m} = E_{n,m} L^{-2}$, $\nu_n = nL^{-1}$, or

$$\sigma^{(0)}_{n,m} = \pm \int \sqrt{\varepsilon_{n,m} f^2 - 4\pi^2 \nu_n^2}\, dt \ . \tag{1}$$

The interpretation of the square root in (1) is well-known (see, e.g. [2]). Fix ν_n and consider for any ε the domain $U(\varepsilon, \nu_n)$ where $U(\varepsilon, \nu) = \{x \mid \varepsilon f^2 - 4\pi^2 \nu_n^2 > 0\}$. Due to our assumptions about f the domain $U(\varepsilon, \nu_n)$ consists of a single interval (see Fig. 2). It is the domain of a possible motion of the corresponding classical particle. The quasi-classical expressions give strongly oscillating

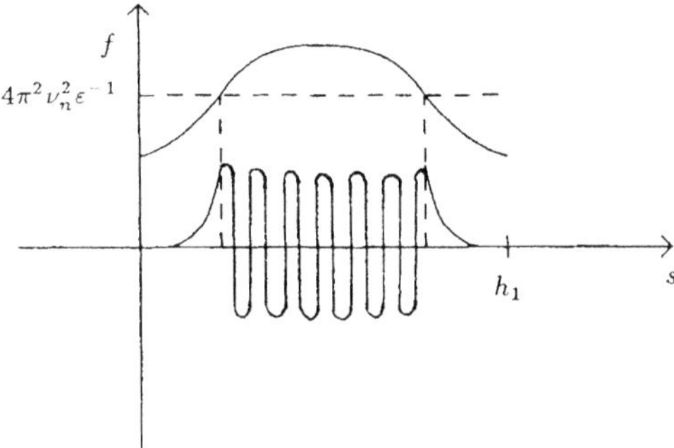

Fig. 2

expressions on $U(\varepsilon, \nu_n)$ which decay exponentially outside $U(\varepsilon, \nu_n)$. If f could have several critical points then $U(\varepsilon, \nu_n)$ might be more complicated. For some values of ε where the number of components changes, the structure of $U(\varepsilon, \nu_n)$ has singularities. In the vicinity of these values, quantum resonances take place and the formulas of the quasi-classical approximation become more complicated. We shall not discuss this here.

For the next term $\sigma^{(1)}_{n,m}$ we have the equation

$$(\sigma^{(1)}_{n,m})' = -\frac{1}{2}\left(\frac{(\sigma^{(0)})''}{(\sigma^{(0)})'}\right)$$

which gives $\sigma^{(1)}_{n,m} = -\frac{1}{2}\ln(\sigma^{(0)}_{n,m})' = -\frac{1}{4}\ln(\varepsilon^2 f^2 - 4\pi^2\nu^2)$. Now we can write down the usual quasi-classical expression for the eigen-functions

$$v_{n,m} = \frac{C_1}{\sqrt{\varepsilon f^2 - 4\pi^2\nu^2}} e^{iL\int\sqrt{\varepsilon f^2 - 4\pi^2\nu^2}\,dt} + \frac{C_2}{\sqrt{\varepsilon f^2 - 4\pi^2\nu^2}} e^{-iL\int\sqrt{\varepsilon^2 f^2 - 4\pi^2\nu^2}\,dt} .$$

The integration starts at the left end-point of $U(\varepsilon,\nu_n)$ or at the left solution of the equation $\varepsilon f^2 = 4\pi^2\nu^2$. It is worth mentioning that these formulas work successfully far from the boundary of $U(\varepsilon,\nu_n)$. Now we write down the famous quantization Bohr-Sommerfeld rules which give quasi-classical expressions for the eigen-values

$$\int_{s_1}^{s_2} \sqrt{(E_{n,m} f^2(s) - 4\pi^2 n^2)}\,ds = \pi\left(m + \tfrac{1}{2}\right) . \tag{2}$$

Here $s_1 = s_1(E_{m,n})$, $s_2 = s_2(E_{m,n})$ are the end-points of $U(\varepsilon_{m,n},\nu_n)$. Equation (2) should be understood in the following way. Fix n, m of order L and start to increase E. Since the left-hand part of (2) also increases with E we can always find $E_{n,m}$ which solves (2). We shall call the solutions of (2) quasi-classical eigen-values (qce).

Discuss briefly the problem of the precision of the quasi-classical approximation. Since sometimes the eigen-values can be untypically close to each other the approximation cannot be uniformly small. But apparently it is small at least in the average. Let us formulate a quasi-theorem, i.e. a statement which is undoubtedly true but for which I do not have a complete proof at the moment.

Quasi-Theorem. *For generic surfaces of revolution Q take qce $E_{n,m}$ which are less than x and fix any $\varepsilon > 0$. Then if $N^{(1)}(x)$ is the number of such qce $E_{n,m}$ that* $\operatorname{dist}\left(E_{n,m}, \operatorname{spec}(-\Delta)\right) \le \varepsilon$ *then* $\frac{N^{(1)}(x)}{N(x)} \to 1$ *as* $x \to \infty$.

Apparently a stronger statement is valid which describes in a more precise way $\operatorname{dist}\left(E_{n,m}, \operatorname{spec}(-\Delta)\right)$. Anyway further we shall deal only with qce. Also we neglect the term $\frac{1}{2}\pi$ in the right-hand part of (2). Later this error can be easily taken into account. Consider the functional equation

$$\int_{s_1(r)}^{s_2(r)} \sqrt{\left(r^2 f^2(s) - 4\pi^2 \sin^2\alpha\right)}\,ds = \pi\cos\alpha , \tag{3}$$

which determines r as an implicit function of α in the domain $\cos\alpha > 0$. Here $s_1(r), s_2(r)$ are the roots of the equation $r^2 f^2(s) = 4\pi^2\sin^2\alpha$. If $r^2 f^2(s) \ge 4\pi^2\sin^2\alpha$ everywhere then

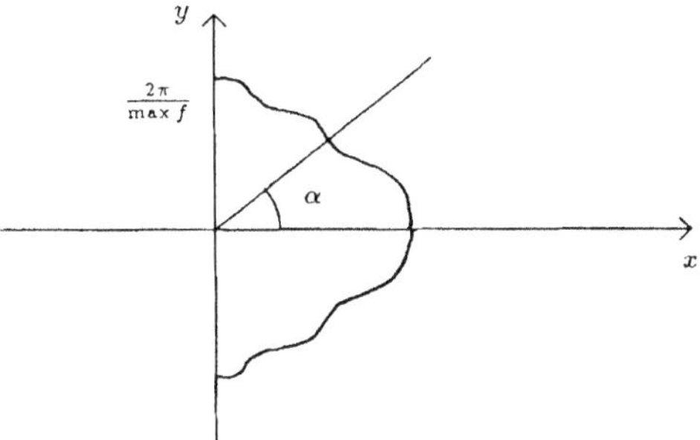

Fig. 3

$s_1(r) = 0$, $s_2(r) = h_1$. A typical form of the function $r = F(\alpha)$, is presented in Fig. 3. It is important also that it is symmetric with respect to the x-axis.

For any $(m, n) \in \mathbb{Z}^2$ put $\sin\alpha = \frac{n}{\sqrt{m^2+n^2}}$, $\cos\alpha = \frac{m}{\sqrt{m^2+n^2}}$, $\alpha = \alpha(m, n)$. Take two positive numbers E, c, and consider the inequality

$$E \leq E_{n,m} \leq E + c\ .$$

Further, $E_{n,m}$ will be of order L^2. Since we want to consider terms of order L it is more convenient to rewrite these inequalities in another form

$$\sqrt{E} \leq \sqrt{E_{n,m}} \leq \sqrt{E+c} = \sqrt{E} + \frac{c}{2\sqrt{E}} + \cdots \tag{4}$$

The dots mean as before terms of smaller order. We shall neglect them as well and as before the error can easily be taken into account. Since $E_{m,n}$ are the solutions of

$$\int_{s_1}^{s_2} \sqrt{E_{m,n} f^2(s) - 4\pi^2 n^2}\, ds = \pi m\ , \tag{2'}$$

i.e. $E_{m,n} = F\big(\alpha(m,n)\big)\sqrt{m^2+n^2}$, we have

$$\sqrt{E} \leq \sqrt{m^2+n^2}\, F\big(\alpha(m,n)\big) \leq \sqrt{E} + \frac{c}{2\sqrt{E}}$$

or

$$\frac{\sqrt{E}}{F\big(\alpha(m,n)\big)} \leq \sqrt{m^2+n^2} \leq \left(\sqrt{E} + \frac{c}{2\sqrt{E}}\right) \cdot \frac{1}{F\big(\alpha(m,n)\big)}\ . \tag{5}$$

We shall see that (5) have a beautiful geometric interpretation. Put $G_1(\alpha) = 1/F(\alpha)$ and consider for any $R \geq 1$ the curve γ_R whose equation in the polar coordinates takes the form

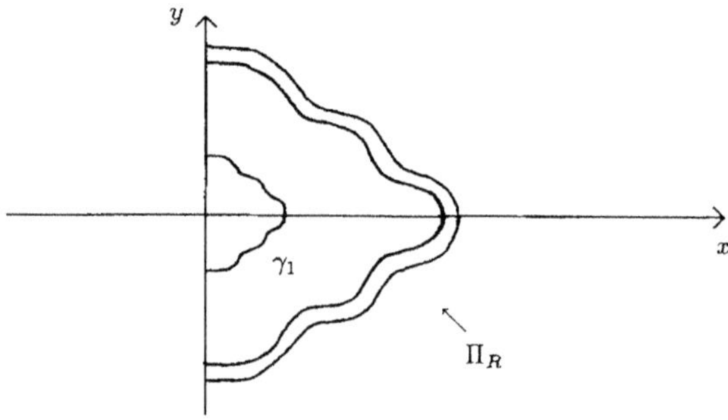

Fig. 4

$r = RG_1(\alpha)$. Denote by Π_R the closed curvilinear strip bounded by the curves $\gamma_{R+\frac{c}{2R}}$ and γ_R (see Fig. 4). Then the number of solutions of (5) is equal exactly to the number of points of the lattice $\mathbb{Z}^2$ belonging to the strip $\Pi_{\sqrt{E}}$. It follows from the symmetry of γ_1 that this number is always even.

Now we can formulate differently the problem of distribution of spacings between qce. Take $L \to \infty$ and fix two positive numbers a_1, a_2, $a_1 < a_2$. Consider the solutions of (2′) such that $a_1^2 L^2 \leq E_{n,m} \leq a_2^2 L^2$. Fix also a number $c > 0$ and put $\xi(E, G_1)$ equal to the number of $E_{n,m} \in [E, E+c]$. The previous discussion shows it is equal to the number $\eta(\sqrt{E}, G_1)$ of points of the lattice $\mathbb{Z}^2$ belonging to $\Pi_{\sqrt{E}}$. Denote $\mathcal{P}_\kappa(L; G_1)$ the Lebesgue measure of such $R = \sqrt{E}$, $a_1 L \leq R \leq a_2 L$, for which $\eta(R, G_1) = 2k$ and $p_\kappa(L, G_1) = \frac{1}{L(a_2 - a_1)} \mathcal{P}_\kappa(L; G_1)$.

Individual Poisson distribution. $\lim\limits_{L\to\infty} p_\kappa(L, G_1) = e^{-\lambda} \frac{\lambda^\kappa}{\kappa!}$ for some parameter $\lambda = \lambda(G_1)$. It is easy to see that in our notations $\lambda(G_1) = \frac{c}{2} \int G_1^2(\alpha) d\alpha$.

Averaged Poisson distribution. Assume that a probability distribution Prob on the Borel σ-algebra of the space of smooth curves γ_1 is given. Then

$$\lim_{L\to\infty} \int p_\kappa(L, G_1) d\,\mathrm{Prob}(G_1) = \int e^{-\lambda(G_1)} \frac{\left(\lambda(G_1)\right)^\kappa}{\kappa!} d\,\mathrm{Prob}(G_1)\ .$$

In the next section we discuss the conditions under which we can prove the Averaged Poisson distribution and some version of the individual Poisson distribution.

Problems.

1. To prove the quasi-theorem.
2. To study the case of functions f having several critical points.
3. To generalize the described construction to the case of Liouville integrable metrics.

4. To study the quantum analogy of the KAM-theorem in the following form: for small generic perturbations of metrics of revolution a big fraction of eigen-values can be obtained with the help of the usual Bohr-Sommerfeld quantization rules.

§3 Poisson Distribution

Let us recall how the Poisson distribution appears in the traditional probability theory. One considers N independent random variables $\xi_1, \ldots, \xi_N$ taking the values 0 and 1 and $p_N = \text{Prob}\{\xi_\kappa = 1\} \sim \frac{\rho}{N}$ as $N \to \infty$. For the characteristic function $u_N(\lambda) = Ee^{i\lambda\zeta_N}$, $\zeta_N = \xi_1 + \ldots + \xi_N$ we have the expression $u_N(\lambda) = \left(1 + \frac{\rho}{N}(e^{i\lambda} - 1) + o(N)\right)^N$ which tends to $\exp\left\{\rho(e^{i\lambda} - 1)\right\}$ which is the characteristic function of the Poisson distribution. A direct way to extend this derivation to dependent random variables is to write $u_N(\lambda)$ in the form

$$u_N(\lambda) = Ee^{i\lambda\zeta_N} = E\prod_{\kappa=1}^{N}\left(1 + (e^{i\lambda\xi_\kappa} - 1)\right) =$$

$$= \sum_{r=0}^{N}(e^{i\lambda} - 1)^r \sum_{\{k_1,\ldots,k_r\}} \text{Prob}\{\xi_{k_1} = \ldots = \xi_{k_r} = 1\}\ .$$

and impose some conditions which permit replacing $\sum_{\{k_1,\ldots,k_r\}} \text{Prob}\{\xi_1 = \ldots = \xi_{k_r} = 1\}$ by $C_N^r\left(\frac{\rho}{N}\right)^r +$ some corrections. Following this direction we must assume that $\text{Prob}\{\xi_{k_1} = \ldots = \xi_{k_r} = 1\}$ behave as $\left(\frac{\rho}{N}\right)^r$. We shall see that in our problems it is not the case and the reasons for the Poisson distribution are in a sense quite different.

Since $G_1(\alpha)$ are even functions of α we shall consider only the angles $0 \le \alpha \le \frac{\pi}{2}$. Starting with this moment we consider a probability distribution P on the space $\mathcal{G}$ of positive functions $G_1(\alpha)$, $o \le \alpha \le \frac{\pi}{2}$. We list below the conditions which we have to assume about P.

1°. There exist positive constants C_1, C_2, C_3 such that with P – probability 1

$$C_1 \le G_1(\alpha) \le C_2\ , \qquad \left|G_1'(\alpha)\right| \le C_3\ .$$

2°. For any r and for any $0 < \alpha_1 < \alpha_2 < \ldots < \alpha_r < \frac{\pi}{2}$ there exists a density $\pi_r(y_r, \ldots, y_1 \mid \alpha_r, \alpha_{r-1}, \ldots, \alpha_1)$ of the joint probability distribution of the random variables $G_1(\alpha_1), G_1(\alpha_2), \ldots, G_1(\alpha_r)$ which is the C^1-function of its variables.

3°. The conditional density $\pi_{r+1}(y_{r+1} \mid G_1(\alpha_r) = y_r, \ldots, G_1(\alpha_1) = y_1; \alpha_r, \ldots, \alpha_1)$ satisfies the inequality

$$\pi_{r+1}\left(y_{r+1} \mid G_1(\alpha_r) = y_r, \ldots, G_1(\alpha_1) = y; \alpha_r, \ldots, \alpha_1\right) \le \frac{C_4}{\ell}$$

for a constant C_4, where ℓ is the interval of possible values of y_{r+1}.

Condition 1° can apparently be weakened. Condition 2° is essential. It shows that G_1 depends on infinitely many random parameters. It also means that we cannot consider the case of G_1 being polynomials of a bounded degree. Condition 3° means that G_1' is more or less uniformly distributed and G_1 has now preferable directions. It would be logically more consistent to consider a probability distribution on the space of functions f and to derive the needed properties of the distribution of G_1 but in my opinion the transition from the probability distribution on f to the probability distribution on G_1 is a technical problem.

Return now to our situation. For any positive function G_1 and any R we consider the strip Π_R bounded by the curves γ_R and $\gamma_{R+\frac{c}{2R}}$ where γ_R is described by the equation $r = RG_1(\alpha)$ in the polar coordinates. The value of the parameter R belongs to the interval $(a_1 L, a_2 L)$ where a_1, a_2 are two fixed numbers and $L \to \infty$. We consider the random variable $\eta(R, G_1) = \#(\Pi_R \cap \mathbb{Z}^2)$ which is a random variable defined on the direct product $(a_1 L, a_2 L) \times \mathcal{G}$. The measure on $[a_1 L, a_2 L]$ is the normalized Lebesgue measure. Denote $p_L(n) = \operatorname{Prob}\{\eta(R, G_1) = n\}$.

Take $(m,n) \in \mathbb{Z}^2$ and introduce the interval $\mathcal{D}_{m,n}$ of such values of R for which $(m,n) \in \Pi_R$. It is easy to calculate that up to terms of a higher order of smallness the center of $\mathcal{D}_{m,n}$ is $d_{m,n} = \frac{\sqrt{m^2+n^2}}{G_1(\alpha(m,m))}$ and the length $\ell(\mathcal{D}_{m,n}) = \ell_{m,n} = \frac{c}{2\sqrt{m^2+n^2}} G(\alpha(m,n))$. Thus on the interval $(a_1 L, a_2 L)$ we have intervals $\mathcal{D}_{m,n}$ whose length is of order L^{-1} and the total number of such intervals is of order L^2. Thus a mean number of intervals covering a point is finite. It is easy to see that $\eta(R, G_1)$ equals exactly to the number of intervals $\mathcal{D}_{m,n}$ covering R. Introduce the random variable $\chi_{(m,n)}(R, G_1)$ which is equal to 1 if $R \in \mathcal{D}_{m,n}$ and 0 otherwise. Thus $\eta(R, G_1) = \sum_{(m,n)\in\mathbb{Z}^2} \chi_{(m,n)}(R, G_1)$ and we are in a situation typical for the Poisson limit theorem since $E\chi_{(m,n)}(R, G_1) = \mathcal{O}(L^{-2})$ and the number of summands in the last sum is $\mathcal{O}(L^2)$. However the product $\chi_{(m_1,n_1)}(R, G_1) \cdot \ldots \cdot \chi_{(m_r,n_r)}(R, G_1)$ takes the value 1 on the interval $\bigcap_{j=1}^{r} \mathcal{D}_{(m_j,n_j)}$ and is 0 otherwise. Typically the length $\ell(\bigcap_{j=1}^{r} \mathcal{D}_{(m_j,n_j)})$ is either 0 or of order L^{-1} and thus the direct generalization of the usual proof of the Poisson limit theorem does not work.

We are now going to discuss the proof of the averaged Poisson distribution. The precise formulation of the theorem is the following.

Theorem 1. *Under the conditions 1-3 there exists a weak limit of the distributions of the random variables $\eta(R, G_1)$ as $L \to \infty$ which is the mixture of the Poisson distributions with the parameter $\frac{c}{2}\int_0^{\pi/2} G_1^2(\alpha)d\alpha$.*

The proof is based upon the following simple lemma. Let ξ be a random variable taking non-negative integer values. For any $k \geq 0$ introduce the new random variable $\xi^{(k)}$ which is equal to zero if $\xi < k$ and $\xi^{(k)} = C_\xi^k$ otherwise.

Lemma 1. *Random variable ξ has the Poisson distribution with parameter ρ iff $E\xi^{(k)} = \frac{1}{k!}\rho^k$ for any k. Random variable ξ has the distribution which is a mixture of Poisson distributions iff $E\xi^{(k)} = \frac{1}{k!}\int \rho^\kappa dS(\rho)$ for any k where S is the probability distribution for ρ.*

In view of the lemma, our goal is to show that for any k

$$\lim_{L\to\infty} \sum p_L(n) \cdot C_n^k = \frac{c^\kappa}{2^\kappa k!} E\Big(\int G_1^2(\alpha) d\alpha \Big)^k =$$
$$= \frac{c^k}{2^k k!} \int \dots \int \pi_\kappa(y_1, \dots, y_\kappa \mid \alpha_1, \dots, \alpha_\kappa) y_1^2 \dots y_\kappa^2 \prod_{j=1}^k dy_j \, d\alpha_j \,. \tag{6}$$

In order to prove (6) we shall study the statistics of mutual intersections of different intervals $\mathcal{D}_{m,n}$. Fix an interval B on the R-axis.

Definition 1. The intervals $\mathcal{D}_{m',n'}$ and $\mathcal{D}_{m'',n''}$ have B-intersection if $d_{m',n'} - d_{m'',n''} \in B$.

We shall consider intervals B of the form $B = B^{(s)} = \left[\frac{s}{L}\delta, \frac{s+1}{L}\delta\right]$ where δ is a small constant. Take now k intervals $\Delta^{(j_1)} < \Delta^{(j_2)} < \dots < \Delta^{(j_\kappa)}$, $\Delta^{(j_t)} \subset \left[0, \frac{\pi}{2}\right]$, $1 \leq t \leq k$, and $(k-1)$ intervals $B_2, \dots, B_k$, each $B_j = B^{(s_j)}$. Let $N_{(m_1,n_1)} = N_{(m_1,n_1)}(\Delta^{(j_1)}, \dots, \Delta^{(j_k)}; B_2, \dots, B_\kappa \mid G_1)$ be the number of k-tuples $(m_1, n_1), \dots, (m_k, n_k)$ such that

1) $\alpha(m_t, n_t) \in \Delta^{j_t)}$, $t = 1, \dots, k$;

2) $\mathcal{D}_{(m_t,n_t)}$ and $\mathcal{D}_{(m_1,n_1)}$ have B_t-intersection, $t = 2, \dots, k$.

Lemma 2.

$$EN_{(m_1,n_1)} = \int \dots \int dy_1 dz_2 \dots dz_k \cdot z_2^2 \cdot \dots \cdot z_\kappa^2 \left(\frac{\sqrt{m_1^2 + n_1^2}}{y_1 L} \right)^{k-1}$$
$$\cdot \pi_k\big(y_1, z_2, \dots, z_k \mid \alpha(m_2, n_1), \overline{\alpha}^{(2)}, \dots, \overline{\alpha}^{(k)}\big) \cdot \prod_{s=2}^k \ell(\Delta^{(j_s)}) \delta^{k-1} (1 + \gamma_1) \,.$$

Here γ_1 is a remainder term which tends to zero as both $\delta \to 0$ and $\ell(\Delta^{(j_s)}) \to \mathcal{O}$ provided that all distances $\operatorname{dist}(\Delta^{(j_{s_1})}, \Delta^{(j_{s_2})}) \geq \beta$ where β is an arbitrary constant.

The proof of the lemma consists of direct calculations.

Definition 2. A set of $(k-1)$ intervals $\{B_2,\ldots,B_k\}$ is essential if for any possible positions of the intervals $\mathcal{D}_{(m_j,n_j)}$, $j=2,\ldots,k$, having B_j-intersection with $\mathcal{D}_{(m_1,n_1)}$ the length $\ell(\bigcap_{j=2}^{k}\mathcal{D}_{(m_j,n_j)})>0$.

The variance of this length is not more than const $k\delta\cdot L^{-1}$ while $\ell(\mathcal{D}_{(m_j,n_j)})=\mathcal{O}(L^{-1})$. Denote

$$\ell(B_2,\ldots,B_k)=\frac{\max\ell\left(\bigcap_{j=1}^{k}\mathcal{D}_{(m_j,n_j)}\right)+\min\ell\left(\bigcap_{j=1}^{k}\mathcal{D}_{(m_j,n_j)}\right)}{2}$$

where max and min are taken over all possible positions of $\mathcal{D}_{(m_j,n_j)}$ having B_j-intersection with $\mathcal{D}_{(m_1,n_1)}$, $j=2,\ldots,k$. The notation $\sum^{(e)}$ means further the summation over all essential $(k-1)$-tuples $\{B_2,\ldots,B_k\}$.

Lemma 3. $\sum^{(e)}\ell(B_2,\ldots,B_k)\cdot\delta^{k-1}=L^{-1}\left(\left(\frac{cL}{2\sqrt{m_1^2+n_1^2}}y_1\right)^k+\mathcal{O}(\delta)\right)$.

Now we describe the last step in the proof of Theorem 1. Let $\chi_{(m_1,n_1),\ldots,(m_k,n_k)}(G_1;B_2,\ldots,B_k)$ be equal to 1 if $\mathcal{D}_{(m_j,n_j)}$ and $\mathcal{D}_{(m_1,n_1)}$ have B_j-intersection, $2\le j\le k$. Let $A_s(L;G_1)$ be the set of $R\in[a_1L,a_2L]$ for which $\eta(R,G_1)=s$. For such R one can find $(m_1,n_1),\ldots,(m_s,n_s)\in\mathbb{Z}^2$ such that $R\in\bigcap_{j=1}^{s}\mathcal{D}_{(m_j,n_j)}$. The expectation (see Lemma 1)

$$E_L^{(k)}=E\eta^{(k)}(R,G_1)=\frac{1}{L(a_2-a_1)}\sum_{s\ge k}C_s^kE\ell(A_s(G_1;L)).$$

The crucial step is based upon the possibility of rewriting $E_L^{(k)}$ in a different way:

$$\begin{aligned}E_L^{(k)}&=\frac{1}{L(a_2-a_1)}\sum_{(m_1,n_1),\ldots,(m_k,n_k)}E\ell(\mathcal{D}_{(m_1,n_1)}\cap\ldots\cap\mathcal{D}_{(m_k,n_k)})=\\&=\frac{1}{L(a_2-a_1)}\sum_{(m_1,n_1)}\sum_{B_2,\ldots,B_k}^{(e)}\sum_{(m_2,n_2),\ldots,(m_k,n_k)}E\big[\ell(\mathcal{D}_{(m_1,n_1)}\cap\ldots\cap\mathcal{D}_{(m_k,n_k)})\cdot\\&\quad\cdot\chi_{(m_1,n_1),\ldots,(m_k,n_k)}(G_1;B_2,\ldots,B_k)\big].\end{aligned}\tag{7}$$

It is a technical part of the proof to show that one can restrict oneself by such $(m_1,n_1),\ldots,(m_k,n_k)$ that $\alpha(m_j,n_j)-\alpha(m_{j-1},n_{j-1})\ge\beta$ for some fixed β and after tending $L\to\infty$ let $\beta\to0$ (see [3]). If we replace in the last expression $\ell(\mathcal{D}_{(m_1,n_1)}\cap\ldots\cap\mathcal{D}_{(m_k,n_k)})$ by $\ell(B_2,\ldots,B_k)$, the error has an absolute value not more than const $k\delta\cdot L^{-1}$. Take a partition of $[0,\frac{\pi}{2}]$ onto small intervals $\Delta^{(j)}$.

The main term in (7) takes the form

$$\begin{aligned}\widetilde{E}_L^{(k)}&=\frac{1}{L(a_2-a_1)}\sum_{m^{(1)}}\sum_{\Delta^{(j_1)}<\ldots<\Delta^{(j_k)}}^{(\beta)}\sum_{B_2,\ldots,B_k}^{(e)}\cdot\\&\quad\cdot\ell(B_2,\ldots,B_k)\cdot EN_{(m_1,n_1)}(\Delta^{(j_1)},\ldots,\Delta^{(j_k)};B_2,\ldots,B_k)\end{aligned}$$

where $\sum^{(\beta)}$ means that $\text{dist}(\Delta^{(j_{s_1})}, \Delta^{(j_{s_2})}) \geq \beta$. Using Lemmas 2 and 3 one easily derives that

$$\widetilde{E}_L^{(k)} = \frac{1}{n!}\int z_1^2 \ldots z_k^2 \pi_k(z_1, \ldots, z_\kappa \mid \alpha_1, \ldots, \alpha_\kappa)\Pi dz_j d\alpha_j + \gamma_2$$

where γ_2 tends to zero as $\beta \to 0$.

This gives theorem 1. Concerning the Poisson distribution for individual functions G_1 we have a weaker statement.

Theorem 2. *Under the conditions 1°-3° there exists a subsequence $\{L_j\}$, $L_j \to \infty$ as $j \to \infty$, and a set $\mathcal{G}_0 \subset \mathcal{G}$, $P(\mathcal{G}_0) = 1$ such that for any $G_1 \in \mathcal{G}_0$ the weak limit of distributions of $\eta(R, G_1)$, $a_1 L_j \leq R \leq a_2 L_j$ tends to the Poisson distribution with parameter $\frac{c}{2}\int G_1^2(\alpha)d\alpha$.*

Detailed proofs of Theorems 1 and 2 can be found in the paper [3].

§4 Quantum Kicked Rotator Model

One of the most popular models in quantum chaos is the model of the so-called quantum kicked rotator. It was introduced in the paper by Casati, Chirikov and Izraelev, Ford [4] as a quantum analog of the so-called Chirikov's or standard map. We shall describe now a slightly more general model and we claim that in a sense it is more natural. Write down the time-dependent Schrödinger equation

$$i\frac{\partial \psi(t,x)}{\partial t} = a\frac{\partial^2 \psi(t,x)}{\partial x^2} + ib\frac{\partial \psi(t,x)}{\partial x} + V(t,x)\psi(t,x)\ . \tag{8}$$

Here V is periodic in t and x, the period in time and x are taken to be equal to one. The function $\psi(t,x)$ is periodic in x with the same period 1. Most often people consider the case of $V(t,x) = \kappa \cos 2\pi x \sum_n \delta(t-n)$ which corresponds to a periodic sequence of kicks. The natural way to study the solutions of (8) is to consider the monodromy operator W, where $W\psi(t,x) = \psi(t+1,x)$. It is a unitary operator acting in $\mathcal{L}^2(S^1)$ and for the case of the kicks $W = U_{a,b} \cdot W_{1,\kappa}$ where W_κ is the multiplication to $\exp\{i\kappa \cos 2\pi x\}$ while $U_{a,b} = e^{i(a\frac{d^2}{dx^2} + ib\frac{d}{dx})}$. After the Fourier transform $W_{1,\kappa}$ becomes a Toeplitz operator whose matrix elements decay very quickly outside the diagonal while $U_{a,b}$ is a diagonal operator with the matrix elements $\exp\left\{i(-4\pi^2 an^2 - 2\pi bn)\right\}\delta_{nm}$. It is convenient to change slightly the notations and consider a slightly more general case

$$W_x = U_{\alpha,x_1,x_2} W_{1,\kappa} \tag{9}$$

where $x = (x_1, x_2)$, $U_{\alpha,x_1,x_2} = \left\| \exp\left\{2\pi i\left(\alpha\frac{n(n-1)}{2} + nx_1 + x_2\right)\right\}\delta_{nm}\right\|$. Recall now that if $T_\alpha(x_1, x_2) = (x_1+\alpha, x_2+x_1)$ is the skew rotation of the two-dimensional torus then $T_\alpha^n(x, y) = \left(x_1 + n\alpha, x_2 + nx_1 + \frac{n(n-1)}{2}\alpha\right)$ and we see a close connection of (9) with the skew rotation.

We shall introduce now some general definitions which cover the case of (9) and the case of one-dimensional discrete Schrödinger operators with random potentials and which lead us to the notion of a random system. Start with a measure-preserving transformation T defined on a probability space $(M, \mathcal{M}, \mu)$, Assume that for each $x \in M$ we are given either a self-adjoint or a unitary operator B_x acting on the space ℓ_2 of sequences $f = \{f_n\}$. Denote by S the shift acting on ℓ_2, i.e. $(Sf)_n = f_{n+1}$.

Definition 3. The family $\{B_x\}$, $x \in M$ is connected with the dynamical system $(M, \mathcal{M}, \mu, T)$ if

$$SB_x = B_{Tx}S \ . \tag{10}$$

Let us give three examples which show the usefulness of the introduced notion.

1. The one-dimensional discrete Anderson model. Let $(M, \mathcal{M}, \mu)$ be the direct product of measure spaces, i.e. $x = \{x_n\}$ is a sequence of identically distributed random variables. Put

$$(B_x f)_n = -(f_{n+1} + f_{n-1}) + x_n \cdot f_n \ .$$

Definition 3 holds if T is the shift acting on M.

Thus M is infinite-dimensional. We have a rare situation where the infinite-dimensional case is simpler than a finite-dimensional one.

2. The one-dimensional discrete Schrödinger operator with a quasi-periodic potential. Here

$$(B_x f)_n = -(f_{n+1} + f_{n-1}) + V(n\omega + \alpha)f_n$$

and $V(\alpha)$ is a periodic function of period 1. Take $M = S^1$ and T to be a rotation to the angle ω. Again Definition 3 is applied.

3. In the two previous examples B_x were self-adjoint operators. Now take $M = \mathrm{Tor}^2$ with the Lebesgue measure and $T = T_2$ being the skew rotation. Then the unitary operator W_x satisfies (10).

One of the first notions in the theory of random systems is the notion of the limiting density of states. It is well-known in the theory of Schrödinger operators with random potentials. For operators (9) it was constructed only recently by my student I. Koshovetz. The main idea was the following. Replace $W_{1,\kappa}$ by an operator which acts on the space of periodic sequences of the period N discretizing the Fourier transform. Let the result be denoted by $W_{1,\kappa}^{(N)}$. Then restrict also the diagonal operator U_{α,x_1,x_2} to the N-dimensional subspace and continue it periodically. If $U_{\alpha,x_1,x_2}^{(N)}$ is the corresponding finite-dimensional operator then the product $W_x^{(N)} = U_{\alpha,x_1,x_2}^{(N)} \cdot W_{1,\kappa}^{(N)}$ is the N-dimensional unitary operator and we may consider its spectrum. After that we can define the limiting density of state as the weak limit of the appearing finite-dimensional distributions. Koshovetz proves the existence of this limit as $N \to \infty$. I would be interesting to give a construction of the limiting density of states in the most general situation. The assumptions should include some assumptions concerning the dependence of B_x on x and properties of the approximation of T by periodic transformations.

Now we shall deal with the unitary operators W_x in (9) for $\kappa = 0$. It follows from the definitions that U_{α,x_1,x_2} is the diagonal operator with the diagonal matrix elements $\exp\left\{2\pi i\left(\frac{n(n-1)}{2}\alpha + nx_1 + x_2\right)\right\} = \lambda_n(\alpha, x_1, x_2)$. Following the general strategy we shall consider $-N \leq n \leq N$. Then we can introduce the characteristics of clustering of $\lambda_n(\alpha, x_1 x_2)$ in the same manner as in §1. In particular, take a constant c and introduce the set $A_k^{(N)}(\alpha, x_1, x_2)$ of $\lambda \in S^1$ such that $\Delta_\lambda = \left(\lambda, \lambda + \frac{c}{N}\right)$ contains k numbers $\lambda_n(\alpha, x_1, x_2)$. We are interested in the behaviour of $\pi_k(N) = \int \ell(A_k^N(\alpha, x_1, x_2))d\alpha\, dx_1\, dx_2$. Let $f(x) = \exp\{2\pi i x_2\}$, where $x = (x_1, x_2) \in \mathrm{Tor}^2$. Then $\lambda_n(\alpha, x_1 x_2) = f(T_\alpha^n x)$. Thus $\lambda \in A_k^{(N)}(\alpha, x_1, x_2)$ iff the inclusion $T_\alpha^n x \in \Delta_\lambda$ holds for k values of n among $-N \leq n \leq N$. Denote $p_k(N)$ the probability of this event assuming that we have the Lebesgue measure on the three-dimensional torus. By symmetry it does not depend on λ and $\pi_k(N) = p_k(N)$. Therefore we can consider $\lambda = 0$.

Now we shall show that the problem of study of the probability $p_k(N)$ is of the same nature as the problem which we discussed in §2,§3. Indeed, take the three-parameter family of functions $G(t; \alpha, x_1, x_2) = \frac{t(t-1)}{2}\alpha + tx_1 + x_2$. We are interested in the number of such n, $-N \leq n \leq N$, that

$\lambda_n(\alpha, x_1, x_2) \in \left(0, \frac{c}{N}\right)$ or, in other words, in the number of such n that $m < \frac{\alpha n(n-1)}{2} + nx_1 + x_2 \leq m + \frac{c}{N}$ for some integer m. For every (α, x_1, x_2) consider the strip $\Pi_N(\alpha, x_1, x_2)$ bounded by the curves $G(t; \alpha, x_1 x_2)$ and $G(t; \alpha, x_1, x_2) + \frac{c}{N}$, $|t| \leq -N$. Again this number is equal to the number of points of the two-dimensional lattice $\mathbb{Z}^2$ belonging to $\Pi_N(\alpha, x_1, x_2)$. But now the strips $\Pi_N(\alpha, x_1, x_2)$ depend only on three independent random parameters α, x_1, x_2. This makes the analysis of the situation much harder.

It turns out that our problem is ultimately connected with some problem in number theory.

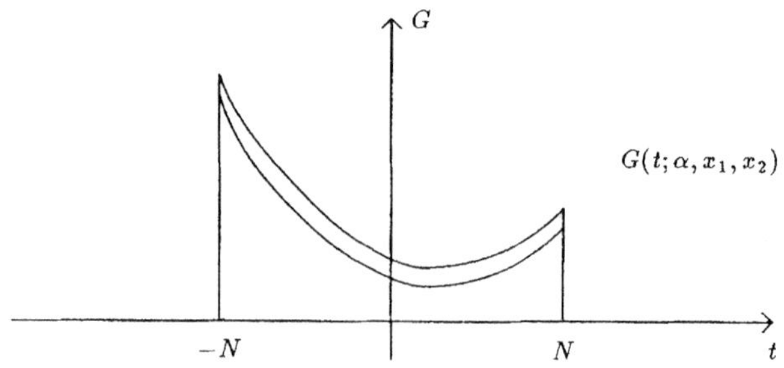

Fig. 5

Consider the so-called double trigonometric sum of Weyl

$$S_N(a) = \frac{1}{N} \sum_{m=-N}^{N} \sum_{n=-N}^{N} \exp\left\{2\pi i m(a,n)\right\} ,$$

$a = (a_0, a_1, a_2)$, $(a,n) = a_0 n^2 + a_1 n + a_2$. It is easy to see that

$$\int S_N^2(a) da \leq \text{const} .$$

Therefore the sequence of probability distributions of $S_N(a)$ is weakly compact and one may pose the problem of finding the limiting probability distribution of $S_N(a)$ as $N \to \infty$. The summation over m can be done explicitly:

$$S_N(a) = \frac{1}{N} \sum_{n=-N}^{N} \frac{\exp\left\{2\pi i(N+1)(a,n)\right\} - \exp\left\{-2\pi i N(a,n)\right\}}{\exp\left\{2\pi i(a,n)\right\} - 1} =$$

$$= \frac{1}{N} \sum_{n=-N}^{N} \frac{\sin 2\pi N(a,n)}{tg\pi(a,N)} + \frac{1}{N} \sum_{n=-N}^{N} \cos 2\pi N(a,n) .$$

The last sum converges to zero at least in probability and we shall not consider it any more. The first sum can be rewritten as follows

$$S_N^{(1)}(a) = \frac{1}{N} \sum_{n=-N}^{N} \frac{\sin 2\pi N(a,n)}{tg\pi(a,n)} = \frac{1}{N} \sum_{n=-N}^{N} f_N(T_\alpha^n(x))$$

where $\alpha = 2a_0$, $x = (x_1, x_2)$ and $x_1 = a_1 + a_0$, $x_2 = a_2$, $f_N(x) = \frac{\sin 2\pi N x_2}{tg\pi x_2}$. Introduce

$$S_N^{(2)}(a) = \frac{1}{N} \sum_{\substack{n \mid |n| \leq N \\ \text{and } |(a,n)| \geq \frac{R}{N}}} f_N(T_\alpha^n x) .$$

Theorem 3. *$\int |S_N^{(2)}(a)|da \le \varepsilon_R$ for all sufficiently large N where $\varepsilon_R \to 0$ as $R \to \infty$.*

Proof: Introduce the horizontal strip Π_0 of the width $2RN^{-1}$, $\Pi_0 = \{x \in \mathrm{Tor}^2 \mid |x_2| \le RN^{-1}\}$ and the strips Π_k, $\Pi_k = \{x \in \mathrm{Tor}^2 \mid x_2 \in \Delta_k\}$ where, 1) $\Delta_{-k} = -\Delta_k$; 2) $\ell(\Delta_k) = \ell(\Delta_{-k}) = \frac{k}{N}$ except, perhaps, the last strip, $\Delta_1 < \Delta_2 < \ldots < \Delta_{k_0}$, $k_0 = \mathcal{O}(\sqrt{N})$. If θ_k is the left (right) boundary of Δ_k for $k > 0$ ($k < 0$) then

$$\theta_k = \frac{1}{N}\left(R + \frac{k(k-1)}{2}\right), \qquad k > 0,$$

$$\theta_k = -\frac{1}{N}\left(R + \frac{k(k-1)}{2}\right), \qquad k < 0.$$

Thus

$$S_N^{(2)}(a) = \sum_{k|1\le|k|\le k_0} \frac{1}{N} A_k, \qquad A_k = \sum_{n|T_\alpha^n(x)\in\Pi_k} f_N(T_\alpha^n x).$$

Denote also $h_N(x) = \sin 2\pi N x_2$. Now we can write

$$A_k = \frac{1}{tg\pi\theta_k} \sum_{n|T_\alpha^n x\in\Pi_k} h_N(T_\alpha^n x) + \sum_{n|T_\alpha^n x\in\Pi_k} h_N(T_\alpha^n x)\left(\frac{1}{tg\pi(a,n)} - \frac{1}{tg\pi\theta_k}\right) = A_k^{(1)} + A_k^{(2)}.$$

First we estimate $A_k^{(2)}$. We have

$$|ctg\pi(a,n) - ctg\pi\theta_k| \le \frac{|(a,n)-\theta_k|}{\sin^2\pi\theta_k} \le \frac{\ell(\Delta_k)}{\sin^2\pi\theta_k}.$$

Therefore

$$\left|\sum_k \frac{1}{N} A_k^{(2)}\right| \le \frac{1}{N}\sum_k \frac{|\ell(\Delta_k)|}{\sin^2\pi\theta_k}\nu_\kappa(x)$$

where $\nu_\kappa(x) = \#\{n \mid T_\alpha^n x \in \Pi_k,\ |n| \le N\}$. Thus $\int \nu_\kappa(x) d\alpha\, dx_1 dx_2 = (2N+1)$ mes $\Pi_k = \frac{2N+1}{N}k$ and

$$\int \left|\sum_k \frac{1}{N} A_k^{(2)}\right| d\alpha\, dx_1\, dx_2 \le \text{const} \sum \frac{k^2}{(R+k^2)^2} = \varepsilon_R^{(1)}.$$

Now come back to $A_k^{(1)} = \frac{1}{tg\pi\theta_k} \sum_{n|T_\alpha^n x\in\Pi_k} h_N(T_\alpha^n x)$. Decompose Π_k onto k^2 equal strips

$$\Pi_{k,j,\ell} = \left\{\frac{R}{N} + \frac{k(k-1)}{2N} + \frac{j}{N} + \frac{\ell}{kN} \le x_2 < \frac{R}{N} + \frac{k(k-1)}{2} + \frac{j}{N} + \frac{\ell+1}{kN}\right\},$$

$0 \le j < k$, $0 \le \ell < k$, put $\Pi_{k,\ell}^{(1)} = \bigcup_{j=0}^{k-1} \Pi_{k,j,\ell}$ and write

$$A_k^{(1)} = \frac{1}{tg\pi\theta_k}\sum_{\ell=0}^{k-1} \sin 2\pi\frac{\ell}{k}\cdot\nu_{k,\ell}^{(1)}(x) + \frac{1}{tg\pi\theta_k}\sum_{\ell=0}^{k-1}\sum_{n|T_\alpha^n x\in\Pi_{k,\ell}^{(1)}}\left[h_N(T_\alpha^n x) - \sin 2\pi\frac{\ell}{k}\right] = A_k^{(3)} + A_k^{(4)}.$$

Here $\nu_{k,\ell}^{(1)}(x) = \#\{n \mid T_\alpha^n x \in \Pi_{k,\ell}^{(1)},\ |n| \le N\}$. The sum $A_k^{(4)}$ is estimated in the same manner as above and we shall not reproduce the corresponding arguments in detail.

Consider $A_k^{(3)}$. Put

$$\chi_{\kappa,\ell}(x) = \begin{cases} 1 & x \in \Pi_{\kappa,\ell}^{(1)} \\ 0 & \text{otherwise} . \end{cases}$$

Then

$$\nu_{\kappa,\ell}^{(1)}(x) = \sum_{|n|\le N} \chi_{\kappa,\ell}(T_\alpha^n x) = \sum_{|n|\le N} \left[\chi_{\kappa,\ell}(T_\alpha^n x) - \mathrm{mes}(\Pi_{k,\ell}^{(1)})\right] + (2N+1)\mathrm{mes}\,\Pi_{k,\ell}^{(1)} .$$

This yields

$$A_k^{(3)} = \frac{1}{tg\pi\theta_\kappa} \sum_{\ell=0}^{k-1} \sin 2\pi\frac{\ell}{k} \cdot \mathrm{mes}(\Pi_{k,\ell}^{(1)} + \frac{1}{tg\pi\theta_\kappa} \sum_{\ell=0}^{k-1} \sin 2\pi \frac{\ell}{k} \sum_{|n|\le N} \left[\chi_{k,\ell}(T_\alpha^n x) - \mathrm{mes}(\Pi_{k,\ell}^{(1)})\right] .$$

The measure of $\Pi_{k,\ell}^{(1)} = N^{-1}$ does not depend on ℓ. Therefore the first sum is equal to zero. Concerning the second sum we remark that it follows from the properties of the skew rotation T_α that the functions $\left[\chi_{\kappa,\ell}(T_\alpha^n x) - \mathrm{mes}(\Pi_{k,\ell}^{(1)})\right]$ are mutually orthogonal for different n. Therefore

$$\begin{aligned} \|A_k^{(3)}\|_2^2 &= \int |A_k^{(3)}|^2 d\alpha\, dx_1 dx_2 = \\ &= \frac{1}{tg^2\pi\theta_k} \sum_{\ell=0}^{k-1} \left(\sin 2\pi \frac{\ell}{k}\right)^2 \cdot \int \sum_{\substack{|n_1|\le N \\ |n_2|\le N}} \left[\chi_{\kappa,\ell}(T_\alpha^{n_1} x) - \mathrm{mes}(\Pi_{k,\ell}^{(1)})\right] \cdot \\ &\cdot \left[\chi_{k,\ell}(T_\alpha^{n_2} x) - \mathrm{mes}(\Pi_{k,\ell}^{(1)})\right] d\alpha\, dx_1\, dx_2 \le \frac{\mathrm{const}\, k}{tg^2\pi\theta_k} \end{aligned}$$

and $\|A_k^{(3)}\|_2 \le \frac{\mathrm{const}\sqrt{k}}{|tg\pi\theta_k|}$. Finally it gives

$$\left\| \sum_k \frac{1}{N} A_k^{(3)} \right\|_2 \le \mathrm{const} \sum \frac{\sqrt{k}}{R+k^2} = \varepsilon_R^{(2)} .$$

which implies the statement of the theorem.

We shall use this theorem for the description of the limiting probability distributions for double trigonometric sums. Put $X_2 = x_2 N$ and fix R. Take such n_j that $x_2^{(j)} = \frac{n_j(n_j-1)}{2}\alpha + n_j x_1 + x_2 \in \left[-\frac{R}{N}, \frac{R}{N}\right](\mathrm{mod}\, 1)$. Then

$$S_N(a) = 2\sum_j \frac{\sin 2\pi X_j}{2\pi X_j} + \gamma_R$$

where γ_R tends to zero in probability.

Hypothesis. The probability distribution of the random field $\{X_j\}$ converges to the Poisson field.

If this hypothesis is true then the limiting distribution for the double trigonometric sums is the distribution of the random variable $\zeta = 2\sum \frac{\sin 2\pi X_j}{2\pi X_j}$ where $X = \{X_j\}$ is a random realization of the Poisson field. The hypothesis also implies the exponential distribution of spacings between $\lambda_{\alpha,x_1,x_2}(n)$. I had a text with the proof of this hypothesis but P. Major found a gap in my calculations. I am sure that the results is true and have some ideas of how to correct the proof but at the moment it is better to consider the statement as an open question.

Final Remark. V.F. Lazutkin has recently informed me about his results concerning the quantum analog of KAM-Theory. They will be published in his forthcoming book. Previous results can be found in [5].

I thank the R. and B. Sackler foundation and Professor S. Abarbanel for the invitation to present these lectures. I also thank V. Milman for his efforts and constant help in arranging this visit.

References

1. M.V. Berry and M. Tabor. Proc. Roy. Soc. A 349 (1976), 101-123.
2. L.D. Landau, E.M. Lifschitz. Quantum Mechanics.
3. Ya.G. Sinai, Poisson Distribution in a Geometrical Problem. Advances in Soviet Mathematics, Publications of AMS (in press).
4. G. Casati, B.V. Chirikov, J. Ford, F.M. Izraelev, in: Stochastic Behaviour in Classical and Quantum Hamiltonian Systems, eds., G. Casati, J. Ford. Lecture Notes in Physics, 93, Springer-Verlag, p. 334-352.
5. V.F. Lazutkin, Convex Billiard and Eigenfunctions of the Laplace Operator, Publ. Leningrad University, Leningrad, 1981, 196pp. (in Russian).

Comments

I attended many talks by physicists on problems of quantum chaos and in particular on the problem of distribution of spacings between the nearest eigenvalues. The hypothesis on the Poisson distribution of the spacings in the case of integrable systems was formulated for the first time by M. Berry and M. Tabor (see [BT1]) and was especially popular.

The text of this paper explains in more detail the BT-hypothesis in the case of surfaces of revolution.

Theoretical and Mathematical Physics, Vol. 121, No. 1, 1999

DYNAMICS OF A HEAVY PARTICLE SURROUNDED BY A FINITE NUMBER OF LIGHT PARTICLES

Ya. G. Sinai[1]

We consider a dynamic system consisting of a heavy particle of mass M and surrounding light particles moving independently with elastic reflections in the segment $[0,1]$. At a given total energy value independent of M, we find the dynamics of the heavy particle with respect to the slow time $t/\sqrt{M}$.

In connection with the problem of the dynamics of an adiabatic piston, E. Lieb [1] proposed the following model. Let the segment $[0,1]$ and a heavy particle of mass M with the initial velocity $V(0)=0$ at the point $x=1/2$ be given. Of course, instead of $1/2$, we could take any other point of the interval $(0,1)$. We also assume that $r^{(-)}$ particles of masses $m_i^{(-)}$ with the initial coordinates $y_i^{(-)}(0)<1/2$ and the initial velocities $v_i^{(-)}(0)>0$, $1\le i\le r^{(-)}$, are on the left of the heavy particle and $r^{(+)}$ similar particles of masses $m_j^{(+)}$ with the initial coordinates $y_j^{(+)}(0)>1/2$ and the initial velocities $v_j^{(+)}(0)<0$, $1\le j\le r^{(+)}$, are on the right of the heavy particle. Each light particle moves uniformly and independently of the other particles and is elastically reflected from the heavy particle and the corresponding boundary $x=0$ or 1. In what follows, we consider $\left(m_i^{(-)},y_i^{(-)},v_i^{(-)}\right)$, $1\le i\le r^{(-)}$, and $\left(m_j^{(+)},y_j^{(+)},v_j^{(+)}\right)$, $1\le j\le r^{(+)}$, to be fixed whereas $M\to\infty$. Let $V(t)$ denote the velocity of the heavy particle. Then the energy conservation law is

$$H=\frac{1}{2}MV^2(t)+\frac{1}{2}\sum_{i=1}^{r^{(-)}}m_i^{(-)}\left(v_i^{(-)}(t)\right)^2+\frac{1}{2}\sum_{j=1}^{r^{(+)}}m_j^{(+)}\left(v_j^{(+)}(t)\right)^2=$$
$$=\frac{1}{2}\sum_{i=1}^{r^{(-)}}m_i^{(-)}\left(v_i^{(-)}(0)\right)^2+\frac{1}{2}\sum_{j=1}^{r^{(+)}}m_j^{(+)}\left(v_j^{(+)}(0)\right)^2.$$

It follows that $|V(t)|\le\sqrt{2H/M}$, i.e., the heavy particle can move with a velocity never exceeding $O\left(1/\sqrt{M}\right)$. We write its coordinate as $X(t)=x\left(t/\sqrt{M}\right)$. It is natural to assume that as a function of the slow time $\tau=t/\sqrt{M}$, $x(\tau)$ has a limit as $M\to\infty$. The main purpose of this paper is to find the limit function $x(\tau)$. We note that within any fixed small time interval on the τ axis, many collisions of light particles with the heavy particle occur. Therefore, the problem of finding $x(\tau)$ can be considered as an averaging theory problem. Our analysis is based on the elastic collision formulas. If v,m and V,M are the velocity-mass pairs of the colliding particles before the collision, then their velocities v' and V' after the collision are

$$v'=-\left(1-\frac{2m}{M+m}\right)v+\frac{2M}{M+m}V,\qquad V'=\left(1-\frac{2m}{M+m}\right)V+\frac{2m}{M+m}v. \tag{1}$$

[1] Department of Mathematics, Princeton University, Princeton, NJ, USA; Landau Institute of Theoretical Physics, RAS, Chernogolovka, Moscow Oblast, Russia.

Translated from Teoreticheskaya i Matematicheskaya Fizika, Vol. 121, No. 1, pp. 110–116, October, 1999. Original article submitted June 30, 1999.

0040-5779/99/1211-1351$22.00

Y.G. Sinai, *Selecta II: Probability Theory, Statistical Mechanics, Mathematical Physics and Mathematical Fluid Dynamics*, DOI 10.1007/978-1-4419-6205-8_14,

We take two instants $t = \tau\sqrt{M}$ and $t_1 = t+\delta\sqrt{M} = (\tau+\delta)\sqrt{M}$. In this time interval, the displacement of the heavy particle is of the order of δ, and the number of collisions of each light particle with the heavy one grows as $\sqrt{M}$. We assume that all collisions of the heavy particle are enumerated in the order they occur. Let the numbers in the run of the heavy particle collisions within the segment $[t, t_1]$ vary from k_0+1 to k_1. It follows from Eq. (1) that

$$V_{k+1} = \left(1 - \frac{2m_k}{M+m_k}\right) V_k + \frac{2m_k}{M+m_k} v_k, \tag{2}$$

where m_k takes one of the values $m_i^{(-)}$ or $m_j^{(+)}$. We set

$$\varepsilon_i^{(-)} = \frac{2m_i^{(-)}}{M+m_i^{(-)}}, \qquad \varepsilon_j^{(+)} = \frac{2m_j^{(+)}}{M+m_j^{(+)}}.$$

It follows from Eq. (2) that

$$\begin{aligned} V_{k_1} - \prod_{\substack{1\le i\le r^{(-)}\\ 1\le j\le r^{(+)}}} \left(1-\varepsilon_i^{(-)}\right)^{n_i^{(-)}} \left(1-\varepsilon_j^{(+)}\right)^{n_j^{(+)}} V_{k_0} = \\ = \sum_{i=1}^{r^{(-)}} \varepsilon_i^{(-)} \sum_{k\in\mathcal{N}_i^{(-)}} v_k \prod_{\substack{1\le i'\le r^{(-)}\\ 1\le j'\le r^{(+)}}} \left(1-\varepsilon_{i'}^{(-)}\right)^{n_{i'}^{(-)}(k)} \left(1-\varepsilon_{j'}^{(+)}\right)^{n_{j'}^{(+)}(k)} + \\ + \sum_{j=1}^{r^{(+)}} \varepsilon_j^{(+)} \sum_{k\in\mathcal{N}_j^{(+)}} v_k \prod_{\substack{1\le i''\le r^{(-)}\\ 1\le j''\le r^{(+)}}} \left(1-\varepsilon_{i''}^{(-)}\right)^{n_{i''}^{(-)}(k)} \left(1-\varepsilon_{j''}^{(+)}\right)^{n_{j''}^{(+)}(k)}. \end{aligned} \tag{3}$$

Here, $n_i^{(-)}$ and $n_j^{(+)}$ are the total numbers of collisions of the ith particle, $1 \le i \le r^{(-)}$, and the jth one, $1 \le j \le r^{(+)}$, within the time interval under consideration; $n_i^{(-)}(k)$ and $n_j^{(+)}(k)$ are the similar numbers of collisions in the run with the numbers from k to k_1; and $\mathcal{N}_i^{(-)}$ and $\mathcal{N}_j^{(+)}$ are the sets of the numbers corresponding to these collisions. We must emphasize that in the general situation, we have $v_k > 0$ for the collisions with the left particles and $v_k < 0$ for the collisions with the right ones.

Our further reasoning is as follows. If $x(t/\sqrt{M})$ has a limit as $M \to \infty$, then we have

$$\frac{d}{dt} x\left(\frac{t}{\sqrt{M}}\right) = \frac{1}{\sqrt{M}} x'\left(\frac{t}{\sqrt{M}}\right).$$

We write the heavy particle velocity as

$$V(t) = \frac{1}{\sqrt{M}} W\left(\frac{t}{\sqrt{M}}\right).$$

We now derive the differential equation for $W(\tau)$, first passing to the limit $M \to \infty$ and then to the limit

$\delta \to 0$. Applying this method to Eq. (3), we obtain

$$W(\tau+\delta)-W(\tau)=\sum_{i=1}^{r^{(-)}}\frac{2m_i^{(-)}\sqrt{M}}{M+m_i^{(-)}}\sum_{k\in\mathcal{N}_i^{(-)}}v_k+\sum_{j=1}^{r^{(+)}}\frac{2m_j^{(+)}\sqrt{M}}{M+m_j^{(+)}}\sum_{k\in\mathcal{N}_j^{(+)}}v_k+$$

$$+\left[\prod_{i=1}^{r^{(-)}}\left(1-\varepsilon_i^{(-)}\right)^{n_i^{(-)}}\left(1-\varepsilon_j^{(+)}\right)^{n_j^{(+)}}-1\right]W(\tau)+$$

$$+\sqrt{M}\sum_{i=1}^{r^{(-)}}\varepsilon_i\sum_{k\in\mathcal{N}_i^{(-)}}v_k\left[\prod_{i=1}^{r^{(-)}}\prod_{j=1}^{r^{(+)}}\left(1-\varepsilon_i^{(-)}\right)^{n_i^{(-)}(k)}\left(1-\varepsilon_j^{(+)}\right)^{n_j^{(+)}(k)}-1\right]+$$

$$+\sqrt{M}\sum_{j=1}^{r^{(+)}}\varepsilon_j\sum_{k\in\mathcal{N}_j^{(+)}}v_k\left[\prod_{i=1}^{r^{(-)}}\prod_{j=1}^{r^{(+)}}\left(1-\varepsilon_i^{(-)}\right)^{n_i^{(-)}(k)}\left(1-\varepsilon_j^{(+)}\right)^{n_j^{(+)}(k)}-1\right].$$

Each expression in the square brackets is of the order of $O\left(1/\sqrt{M}\right)$; $\varepsilon_i^{(-)}$ and $\varepsilon_j^{(+)}$ are of the order of $O(1/M)$. Therefore, we have

$$W(\tau+\delta)-W(\tau)=\sum_{i=1}^{r^{(-)}}\frac{2m_i^{(-)}\sqrt{M}}{M+m_i^{(-)}}\sum_{k\in\mathcal{N}_i^{(-)}}v_k+\sum_{j=1}^{r^{(+)}}\frac{2m_j^{(+)}\sqrt{M}}{M+m_j^{(+)}}\sum_{k\in\mathcal{N}_j^{(+)}}v_k+O\left(\frac{1}{\sqrt{M}}\right). \quad (4)$$

From the formulas for the velocities of the light particles (see below), it follows that within the time interval on the t axis from $\tau\sqrt{M}$ to $(\tau+\delta)\sqrt{M}$, the light particle velocities at the instants preceding the collisions with the heavy particle vary only slightly. Relying on this fact, we can write

$$\sum_{k\in\mathcal{N}_i^{(-)}}v_k=v_i^{(-)}\left(\tau\sqrt{M}\right)n_i^{(-)}(1+o(1)),\qquad \sum_{k\in\mathcal{N}_j^{(+)}}v_k=v_j^{(+)}\left(\tau\sqrt{M}\right)n_j^{(+)}(1+o(1)).$$

Here and in what follows, $o(1)$ is a quantity tending to zero uniformly in M as $\delta\to 0$. The respective time interval between two successive collisions of a left or a right particle with the heavy particle is

$$\frac{2x(\tau)}{v_i^{(-)}\left(\tau\sqrt{M}\right)}(1+o(1))\qquad\text{or}\qquad\frac{-2(1-x(\tau))}{v_j^{(+)}\left(\tau\sqrt{M}\right)}(1+o(1)).$$

Here, $v_i^{(-)}\left(\tau\sqrt{M}\right)$ and $v_j^{(+)}\left(\tau\sqrt{M}\right)$ are the velocities of the corresponding particles at the instants preceding their first collision in the run, and $v_i^{(-)}\left(\tau\sqrt{M}\right)>0$ and $v_j^{(+)}\left(\tau\sqrt{M}\right)<0$. Therefore, we have

$$n_i^{(-)}=\frac{\delta\sqrt{M}v_i^{(-)}\left(\tau\sqrt{M}\right)}{2x(\tau)}(1+o(1)),\qquad n_j^{(+)}=-\frac{\delta\sqrt{M}v_j^{(+)}\left(\tau\sqrt{M}\right)}{2(1-x(\tau))}(1+o(1)). \quad (5)$$

Substituting these expressions in Eq. (4) and passing to the limit $M\to\infty$ and then to the limit $\delta\to 0$, we obtain a remarkably simple differential equation for the limit function $x(\tau)$:

$$W'(\tau)=x''(\tau)=\sum_{i=1}^{r^{(-)}}\frac{m_i^{(-)}\left(V_i^{(-)}(\tau)\right)^2}{x(\tau)}-\sum_{j=1}^{r^{(+)}}\frac{m_j^{(+)}\left(V_j^{(+)}(\tau)\right)^2}{1-x(\tau)}, \quad (6)$$

where $V_i^{(-)}(\tau) = v_i^{(-)}\left(\tau\sqrt{M}\right)$ and $V_j^{(+)}(\tau) = v_j^{(+)}\left(\tau\sqrt{M}\right)$.

We now turn to deriving equations for the light particle velocities. We consider only one of the left particles, the first particle; the others are considered similarly. Let $k \in \mathcal{N}_1^{(-)}$, and let the velocity v_k be of the order of unity. Then it is positive, i.e., $v_k > 0$. After the collision with the heavy particle, the velocity becomes negative. The light particle flies to the wall $x = 0$ and is reflected from it, after which its velocity again becomes positive. We introduce the ordinal index s for the instants $k \in \mathcal{N}_1^{(-)}$. In accordance with Eq. (1), we can write

$$v_{s+1}^{(-)} = \left(1 - \varepsilon_1^{(-)}\right) v_s^{(-)} - \frac{2M}{M + m_1^{(-)}} V_s. \tag{7}$$

We emphasize that here V_s is the velocity of the heavy particle at the instant preceding the $(s+1)$th collision with the left particle, the index s varies from 0 to $n_i^{(-)} - 1$. From Eq. (7), we obtain

$$v_{n_1^{(-)}}^{(-)} - \left(1 - \varepsilon_1^{(-)}\right)^{n_1^{(-)}} v_1^{(-)}\left(\tau\sqrt{M}\right) = -\frac{2M}{M + m_1^{(-)}} \sum_{s=0}^{n_1^{(-)}-1} V_s \left(1 - \varepsilon_1^{(-)}\right)^{n_1^{(-)}-s-1}. \tag{8}$$

The difference between the coefficient $\left(1 - \varepsilon_1^{(-)}\right)^{n_1^{(-)}-s-1}$ and unity is of an order not exceeding $O\left(1/\sqrt{M}\right)$, each velocity $V_s = O\left(1/\sqrt{M}\right)$, and the number of summands $n_1^{(-)} = O\left(\sqrt{M}\right)$. Therefore, the right-hand side of Eq. (8) is of an order not exceeding $O(\delta)$. Isolating the main terms, we rewrite Eq. (8) as

$$v_1^{(-)}\left((\tau + \delta)\sqrt{M}\right) - v_1^{(-)}\left(\tau\sqrt{M}\right) = -2 \sum_{s=0}^{n_1^{(-)}-1} V_s + O\left(\frac{1}{\sqrt{M}}\right). \tag{9}$$

We now write

$$V_s = \sum_{t=1}^{s} (V_t - V_{t-1}) + V_0$$

and substitute this expansion in Eq. (9),

$$v_1^{(-)}\left((\tau + \delta)\sqrt{M}\right) - v_1^{(-)}\left(\tau\sqrt{M}\right) = -2n_1^{(-)} V_0 + \sum_{s=0}^{n_1^{(-)}} \sum_{t=1}^{s} (V_t - V_{t-1}). \tag{10}$$

The value of $n_1^{(-)}$ is calculated previously:

$$n_1^{(-)} = \frac{\delta\sqrt{M} v_1^{(-)}\left(\tau\sqrt{M}\right)}{2x(\tau)}(1 + o(1)).$$

We show that the double sum in Eq. (10) is of the order of δ^2 and can therefore be neglected in the limit $\delta \to 0$. It follows from Eq. (2) that

$$V_s - V_{s-1} = \sum_{i=2}^{r^{(-)}}{}' \varepsilon_i^{(-)} v_i^{(-)}\left(\tau\sqrt{M}\right)(1 + o(1)) + \sum_{j=1}^{r^{(+)}}{}' \varepsilon_j^{(+)} v_j^{(+)}\left(\tau\sqrt{M}\right)(1 + o(1)) +$$

$$+ \sum_{i=2}^{r^{(-)}}{}' \varepsilon_i^{(-)} n_i^{(-)}(s) v_i^{(-)}\left(\tau\sqrt{M}\right)(1 + o(1)) + \sum_{j=1}^{r^{(+)}}{}' \varepsilon_j^{(+)} n_j^{(+)}(s) v_j^{(+)}\left(\tau\sqrt{M}\right)(1 + o(1)),$$

where $n_i^{(-)}(s)$, $i \geq 2$, and $n_j^{(+)}(s)$, $j \geq 1$, are the total numbers of collisions of the heavy particle with the corresponding light particles between the $(s-1)$th and sth collisions with the first left particle and the sums $\sum'$ are taken over those light particles that collide with the heavy particle within the time interval under consideration. Therefore, we have

$$\left|\sum_{s=1}^{n^{(-)}}\sum_{t=1}^{s}(V_t - V_{t-1})\right| \leq \left|\sum_{i=1}^{r^{(-)}}\varepsilon_i^{(-)}v_i^{(-)}\left(\tau\sqrt{M}\right)\frac{\left(n_1^{(-)}\right)^2}{2}(1+o(1))\right| +$$

$$+\left|\sum_{i=2}^{r^{(-)}}\varepsilon_i^{(-)}v_i^{(-)}\left(\tau\sqrt{M}\right)\sum_{s=1}^{n_1^{(-)}}\bar{n}_i^{(-)}(s)(1+o(1))\right| +$$

$$+\left|\sum_{j=1}^{r^{(+)}}\varepsilon_j^{(+)}v_j^{(+)}\left(\tau\sqrt{M}\right)\sum_{s=1}^{n_1^{(-)}}\bar{n}_j^{(+)}(s)\right|,$$

where $\bar{n}_i^{(-)}(s)$ and $\bar{n}_j^{(+)}(s)$ are the numbers of the collisions of the heavy particle with the corresponding particles within the time interval before the sth collision of the first left particle. Considerations identical to the previous ones show that

$$\bar{n}_i^{(-)}(s) \leq \frac{v_i^{(-)}\left(\tau\sqrt{M}\right)s}{v_1^{(-)}\left(\tau\sqrt{M}\right)}(1+o(1)),$$

$$\bar{n}_j^{(+)}(s) \leq -\frac{x(\tau)v_j^{(+)}\left(\tau\sqrt{M}\right)s}{v_1^{(-)}\left(\tau\sqrt{M}\right)(1-x(\tau))}(1+o(1)),$$

and we therefore have

$$\sum_{s=1}^{n_1^{(-)}}\bar{n}_i^{(-)}(s) = \frac{x(\tau)v_i^{(-)}\left(\tau\sqrt{M}\right)}{2v_1^{(-)}\left(\tau\sqrt{M}\right)}\left(n_1^{(-)}\right)^2(1+o(1)),$$

$$\sum_{s=1}^{n_1^{(-)}}\bar{n}_j^{(+)}(s) = \frac{x(\tau)\left|v_j^{(+)}\left(\tau\sqrt{M}\right)\right|}{v_1^{(-)}\left(\tau\sqrt{M}\right)(1-x(\tau))}\left(n_1^{(-)}\right)^2(1+o(1)).$$

We see that the last expressions are of the order of $O(\delta^2)$ and therefore do not contribute to the differential equation. It therefore follows from Eq. (5) that

$$v_1^{(-)}\left((\tau+\delta)\sqrt{M}\right) - v_1^{(-)}\left(\tau\sqrt{M}\right) = -\frac{v^{(-)}\left(\tau\sqrt{M}\right)\delta}{x(\tau)}x'(\tau) + O\left(\delta^2\right). \tag{11}$$

We return to the previous notation: $V_i^{(-)}(\tau) = v_i^{(-)}\left(\tau\sqrt{M}\right)$, $1 \leq i \leq r^{(-)}$, $V_j^{(+)}(\tau) = v_j^{(+)}\left(\tau\sqrt{M}\right)$, $1 \leq j \leq r^{(+)}$. From Eq. (11), we obtain

$$\frac{dV_1^{(-)}(\tau)}{d\tau} = -V_1^{(-)}(\tau)\frac{x'(\tau)}{x(\tau)}. \tag{12}$$

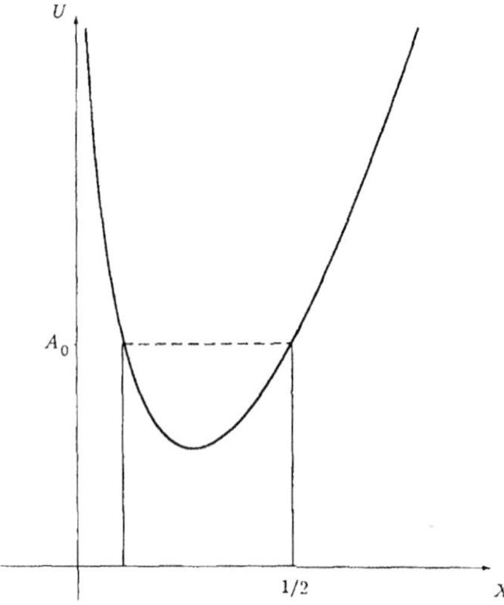

Fig. 1

We can also write similar equations for the other left particles. For the right particles, the equations are

$$\frac{dV_j^{(+)}(\tau)}{d\tau} = V_j^{(+)}(\tau)\frac{x'(\tau)}{1 - x(\tau)}. \tag{13}$$

It follows from Eqs. (12) and (13) that

$$V_i^{(-)}(\tau) = \frac{C_i^{(-)}}{x(\tau)}, \qquad V_j^{(+)}(\tau) = \frac{C_j^{(+)}}{1 - x(\tau)},$$

where the constants $C_i^{(-)}$, $1 \leq i \leq r^{(-)}$, and $C_j^{(+)}$, $1 \leq j \leq r^{(+)}$, can be found from the initial conditions. Substituting these expressions in Eq. (6), we obtain the final equation for $x(\tau)$:

$$x''(\tau) = \frac{\sum_{i=1}^{r^{(-)}} m_i^{(-)} \left(C_i^{(-)}\right)^2}{x^3(\tau)} - \frac{\sum_{j=1}^{r^{(+)}} m_j^{(+)} \left(C_j^{(+)}\right)^2}{(1 - x(\tau))^3}. \tag{14}$$

Equation (14) is the equation of motion for a one-dimensional particle in the field with the potential

$$U(x) = \frac{1}{2}\left[\frac{1}{x^2}\sum_{i=1}^{r^{(-)}} m_i^{(-)} \left(C_i^{(-)}\right)^2 + \frac{1}{(1-x)^2}\sum_{j=1}^{r^{(+)}} m_j^{(+)} \left(C_j^{(+)}\right)^2\right].$$

Therefore, for the solutions of Eq. (14), "the energy conservation law"

$$\frac{1}{2}(x')^2 + U(x) = A_0$$

holds, where the constant A_0 can also be found from the initial conditions:

$$A_0 = 2\left[\sum_{i=1}^{r^{(-)}} m_i^{(-)} \left(C_i^{(-)}\right)^2 + \sum_{j=1}^{r^{(+)}} m_j^{(+)} \left(C_j^{(+)}\right)^2\right].$$

The form of the function U is shown in Fig. 1. The constant $A_0 = U(x(0)) = U(1/2)$. The heavy particle oscillates inside the segment defined by the inequality $U(x) \leq A_0$. One of the endpoints of this segment is the point $x = 1/2$.

Acknowledgments. The author is grateful to A. I. Neischtadt for the useful discussions of this work and for the indication for the relation to the theory of adiabatic invariants, to B. Voronov for various comments on the text, and to E. Lieb, J. Lebowitz, and Ya. Pyasetski for many useful discussions of the adiabatic piston problem.

This work was supported by the Russian Foundation for Basic Research (Grant No. 99-01-00314) and the NSF (Grant No. DMS-9706794).

REFERENCES

1. E. Lieb, *Physica A*, **263**, 491–499 (1999).

Comments

In his talk during the International Conference on Statistical Physics, E. Lieb cited a famous problem on the dynamics of the adiabatic piston (see [L1]).

In the case considered in this paper, the limit is described as a one-dimensional Hamiltonian system with potential $\frac{C_1}{x^2}+\frac{C_2}{(1-x)^2}$, $0 \leq x \leq 1$, where C_1, C_2 are constants.

It is a finite-dimensional version of the adiabatic piston problem. Some results related to the adiabatic piston problem can be found in [LPS, PS, CLS1, CLS2].

Journal of Statistical Physics, Vol. 116, Nos. 1/4, August 2004 (© 2004)

Adiabatic Piston as a Dynamical System[1]

A. I. Neishtadt[2] and Y. G. Sinai[3, 4]

Received August 18, 2003; accepted October 15, 2003

We consider systems of finitely many interacting particles in a cube with a separating wall having a big mass M (adiabatic piston). Assuming that the particles reflect elastically from the ball and the initial velocity of the piston is zero we prove that as M tends to infinity the dynamics of the piston converges to periodic oscillations.

KEY WORDS: Adiabatic piston; adiabatic invariant; averaging method.

Recently E. Lieb attracted the attention of many people to the problem of dynamics of adiabatic piston (see ref. 7). It became so popular that J. Lebowitz even coined the words "notorious piston."

Thermodynamical aspects of the dynamics of the adiabatic piston were considered in many papers. As few examples we can mention[(3, 4, 8)] where the equations of motion of the piston in the thermodynamical limit were derived and analyzed.

In this paper we study the finite-dimensional version of the piston problem assuming that the mass M of the piston tends to infinity and the number of gas particles stays fixed. More precisely, the piston is the separating wall inside a volume $V \subset R^d$, where d is the dimension, $V = \{(x_1,..., x_d) \mid a_i \leqslant x_i \leqslant b_i, 1 \leqslant i \leqslant d\}$ and a_i, b_i are fixed. V is cut by the piston Π_X onto two parts $V^{(l)}, V^{(r)}$

$$V^{(l)} = \{(x_1,..., x_d) \mid x_1 \leqslant X\}, \qquad V^{(r)} = \{(x_1,..., x_d) \mid x_1 \geqslant X\}.$$

[1] Dedicated to E. Lieb on the occasion of his 70th birthday.

[2] Space Research Institute, Russian Academy of Sciences.

[3] Mathematics Department, Princeton University; e-mail: sinai@math.princeton.edu

[4] Landau Institute of Theoretical Physics, Russian Academy of Sciences.

Y.G. Sinai, *Selecta II: Probability Theory, Statistical Mechanics, Mathematical Physics and Mathematical Fluid Dynamics*, DOI 10.1007/978-1-4419-6205-8_15, © Springer Science+Business Media, LLC 2010

The piston can move along the x_1-axis and changes its velocity under the action of elastic collisions with the "gas" particles inside $V^{(l)}$ and $V^{(r)}$ so that X becomes a function of time. The number and the masses of gas particles inside each part are fixed while the mass M tends to infinity. If initially the velocity $v(0)$ of the piston is zero then the total energy of the system does not depend on M and the absolute value of the piston velocity $v(t)$ at time t is $O(\frac{1}{\sqrt{M}})$. Therefore it is natural to introduce "slow time" $\tau = \frac{t}{\sqrt{M}}$ and study the limiting form of dynamics of the piston in the rescaled time. The main result of this paper states that this limiting dynamics is periodic and its form depends on the Hamiltonians of the left and right particles. For the case of non-interacting particles ("ideal gas") this statement was proven in ref. 9. It turns out that the oscillatory regime of the piston is quite universal and in some sense resembles Carnot cycles.

The Hamiltonian of the whole system "left gas" + "right gas" + piston can be written in the form

$$H = \frac{1}{2}\frac{P^2}{M} + H^{(l)} + H^{(r)}, \tag{1}$$

where $P = Mv$ is the momentum of the piston, $H^{(l)}$, $H^{(r)}$ are Hamiltonians of the left and right particles. To define completely the dynamics we should include "boundary" conditions by assuming elastic collisions of gas particles with the boundary of V and with the moving piston. Formally this can be done by adding the potential equal to zero inside $V^{(l)}, V^{(r)}$ and infinity on the boundary.

As was mentioned above we consider the slow time $\tau = \frac{t}{\sqrt{M}}$. Also we need the limiting dynamics which formally corresponds to $M = \infty$. For this dynamics the piston stays fixed and gas particles undergo elastic collisions with the piston during which the x_1-component of the velocity changes its sign but the others remain unchanged.

The microcanonical distribution $\frac{d\sigma_\alpha}{|\mathrm{grad}\, H^{(\alpha)}|}$, $\alpha = l$ or r, is an invariant measure for the α-part concentrated on the manifold $\Sigma^{(\alpha)}$ of constant energy $H^{(\alpha)} = h$ provided that the piston is fixed. We shall use phase averages:

$$\langle f \rangle_\alpha = \frac{1}{\displaystyle\int_{\Sigma^{(\alpha)}} \frac{d\sigma_\alpha}{\mathrm{grad}\, H^{(\alpha)}}} \int_{\Sigma^{(\alpha)}} \frac{f(x)\, d\sigma_\alpha}{|\mathrm{grad}\, H^{(\alpha)}|}.$$

Basic Assumption. For almost every values of the energies $h^{(l)}$ of $H^{(l)}$ and $h^{(r)}$ of $H^{(r)}$ and almost every position of the piston X the dynamics on each $\Sigma^{(\alpha)}$ is ergodic wrt the microcanonical distribution.

Under this assumption we can use averaging method. Denote by $I^{(\alpha)} = I^{(\alpha)}(h, X)$, $\alpha = l, r$ the phase volume of the set $\{H^{(\alpha)} < h\}$ provided that the position of the piston is X. It is easy to derive the formulas (see ref. 6)

$$\frac{\partial I^{(\alpha)}}{\partial h} = \int_{\Sigma^{(\alpha)}} \frac{1}{|\text{grad } H^{(\alpha)}|} \, d\sigma_\alpha, \tag{2}$$

$$\frac{\partial I^{(\alpha)}}{\partial X} = -\int_{\Sigma^{(\alpha)}} \frac{\partial H^{(\alpha)}}{\partial X} \frac{d\sigma_\alpha}{|\text{grad } H^{(\alpha)}|}. \tag{3}$$

Strictly speaking the last formula has a well defined meaning when the hard-core potential at the piston is replaced by a "soft potential." The limiting form as the soft potential tends to the hard-core is a δ-function integral over a submanifold of $\Sigma^{(\alpha)}$ where the coordinate of one of the particles coincides with the coordinate of the piston. We do not discuss this in more detail and only remark that from (2) and (3)

$$\frac{\partial I^{(\alpha)}}{\partial X} = -\frac{\partial I^{(\alpha)}}{\partial h} \left\langle \frac{\partial H^{(\alpha)}}{\partial X} \right\rangle_\alpha. \tag{4}$$

Denote $\bar{P} = P/\sqrt{M}$, $\varepsilon = \frac{1}{\sqrt{M}}$. Then

$$H = \tfrac{1}{2} \bar{P}^2 + H^{(l)} + H^{(r)}$$

and we can write down the equations of motion

$$\dot{X} = \varepsilon \bar{P}, \tag{5$'$}$$

$$\dot{\bar{P}} = -\varepsilon \left(\frac{\partial H^{(l)}}{\partial X} + \frac{\partial H^{(r)}}{\partial X} \right), \tag{5$''$}$$

$$\frac{dH^{(l)}}{dt} = \varepsilon \frac{\partial H^{(l)}}{\partial X} \bar{P}, \tag{5$'''$}$$

$$\frac{dH^{(r)}}{dt} = \varepsilon \frac{\partial H^{(r)}}{\partial X} \bar{P}. \tag{5$''''$}$$

In the last two equations $H^{(l)}$, $H^{(r)}$ are considered as functions of phase coordinates of gas particles and of the position of the piston. Since gas particles interact with the piston $H^{(\alpha)}$ are not the first integrals of the exact dynamics.

The limiting dynamics corresponding to $M=\infty$ is ergodic. Therefore, using the averaging method, we can replace Eqs. (5′)–(5″″) by the averaged equations of motion:

$$\dot{X}=\varepsilon\bar{P}, \tag{6′}$$

$$\dot{\bar{P}}=-\varepsilon\left(\left\langle\frac{\partial H^{(l)}}{\partial X}\right\rangle_l+\left\langle\frac{\partial H^{(r)}}{\partial X}\right\rangle_r\right), \tag{6″}$$

$$\frac{dH^{(l)}}{dt}=\varepsilon\left\langle\frac{\partial H^{(l)}}{\partial X}\right\rangle_l\bar{P}, \tag{6‴}$$

$$\frac{dH^{(r)}}{dt}=\varepsilon\left\langle\frac{\partial H^{(r)}}{\partial X}\right\rangle_r\bar{P}. \tag{6″″}$$

The right-hand sides of these equations depend on X, $\bar{P}$ and the values of $H^{(l)}$, $H^{(r)}$. Thus in slow time we get a closed system of equations for these variables. The theorem by Anosov (see ref. 1) says that on time intervals $O(\frac{1}{\varepsilon})=O(\sqrt{M})$ the measure of the set of initial phase points where solutions of the exact equations are arbitrarily close to solutions of the averaged equations tends to 1 as $M\to\infty$.

Remark. Anosov theorem was proved for system with smooth Hamiltonians. It is possible to check that it remains valid in our case as well.

Return back to the variables $I^{(\alpha)}$. Differentiation of these variables along solutions of the averaged systems gives (see (4))

$$\dot{I}^{(\alpha)}=\varepsilon\left(\frac{\partial I^{(\alpha)}}{\partial H^{(\alpha)}}\left\langle\frac{\partial H^{(\alpha)}}{\partial X}\right\rangle_\alpha+\frac{\partial I^{(\alpha)}}{\partial X}\right)=0.$$

Therefore $I^{(\alpha)}(H^{(\alpha)},X)$ are the first integrals of the averaged system. By this reason they are approximately conserved for majority of initial conditions by the exact system on time intervals $O(\sqrt{M})$ and are called sometimes almost adiabatic invariants (see ref. 2).

Let us consider again the functions $I^{(\alpha)}(H^{(\alpha)},X)$. Assume that they are non-degenerate and we can invert them to write $H^{(\alpha)}=F^{(\alpha)}(X,I^{(\alpha)})$, $\alpha=l,r$. Equation (4) implies that $\partial F^{(\alpha)}/\partial X=\langle\partial H^{(\alpha)}/\partial X\rangle_\alpha$.

Introduce the effective Hamiltonian for the piston

$$\mathscr{H}(\bar{P},X,I^{(l)},I^{(r)})=\tfrac{1}{2}\bar{P}^2+W(X,I^{(l)},I^{(r)}),$$

where $W(X,I^{(l)},I^{(r)})=F^{(l)}(X,I^{(l)})+F^{(r)}(X,I^{(r)})$.

As was said above, $I^{(l)}$ and $I^{(r)}$ are almost adiabatic invariants and $\mathcal{H}$ gives the Hamiltonian of the limiting dynamics of the piston.

Consider several examples.

1. Assume that the gases consist of hard balls or disks in the two-dimensional case of the same radius (see ref. 5). Then $I^{(\alpha)}(H, X) = C^{(\alpha)} H^{\frac{d}{2} n^{(\alpha)}} Q^{(\alpha)}(X)$, where $Q^{(\alpha)}(X)$ is the partition function corresponding to the system of $n^{(\alpha)}$ particles, $C^{(\alpha)}$ is an absolute constant depending on $n^{(\alpha)}$. Thus

$$H = ((Q^{(\alpha)}(X)\, C^{(\alpha)})^{-1}\, I^{(\alpha)})^{\frac{2}{dn^{(\alpha)}}}$$

and the potential W takes the form

$$W = ((C^{(l)})^{-1}\, I^{(l)})^{\frac{2}{dn^{(l)}}}\, (Q^{(l)}(X))^{-\frac{2}{dn^{(l)}}} + ((C^{(r)})^{-1}\, I^{(r)})^{\frac{2}{dn^{(r)}}}\, (Q^{(r)}(X))^{-\frac{2}{dn^{(r)}}}.$$

If we write formally $Q^{(\alpha)}(X) = e^{n^{(\alpha)} f^{(\alpha)}(x)}$, where $f^{(\alpha)}(x)$ is proportional to the free energy, then

$$(Q^{(\alpha)}(X))^{\frac{2}{dn^{(\alpha)}}} = e^{\frac{2}{d} f^{(\alpha)}(X)}.$$

2. In the cases of the ideal gases the dynamics of left and right particles are non-ergodic on the manifold of constant energy. However, if we fix all additional first integrals we get the effective potential $W = \frac{C_l}{X^2} + \frac{C_r}{(L-X)^2}$, $a_1 = 0$, $L = b_1$.

ACKNOWLEDGMENTS

The first author thanks RFFI, Grants 03-01-00158, NSh-136.2003.1 and "Integration" program, grant B0053 for the financial support. The second author thanks NSF, Grant DMS–0070698 and RFFI, Grant 99-01-00314 for the financial support.

REFERENCES

1. D. V. Anosov, Averaging in systems of ordinary differential equations with rapidly oscillating solutions, *Izv. Akad. Nauk SSSR, Ser. Mat.* **24**:721–742 (1960).
2. V. I. Arnold, *Geometrical Methods in the Theory of Ordinary Differential Equations* (Springer–Verlag, 1983), p. 334.
3. N. I. Chernov, J. L. Lebowitz, and Y. G. Sinai, Dynamics of a massive piston in an ideal gas, *Russian Math. Surveys* **57**:1045–1125 (2002).
4. C. Gruber, Thermodynamics of systems with internal adiabatic constraints: Time evolution of the adiabatic piston, *European J. Phys.* **20**:259–266 (1999).
5. *Hard Ball Systems and Lorentz Gas*, Encyclopaedia Math. Sci., Vol. 101 (Springer-Verlag, 2000).

6. T. Kasuga, On the adiabatic theorem for the Hamiltonian system of differential equations in classical mechanics. I, II, III, *Proc. Japan Acad.* **37**:366–382 (1961).
7. E. Lieb, Some problems in statistical mechanics that I would like to see solved, *Phys. A* **263**:491–499 (1999).
8. J. L. Lebowitz, J. Piasecki, and Y. G. Sinai, Scaling dynamics of a massive piston in an ideal gas, in *Hard Ball Systems and Lorentz Gas*, Encyclopaedia Math. Sci., Vol. 101 (Springer-Verlag, 2000), pp. 217–227.
9. Y. G. Sinai, Dynamics of a massive particle surrounded by a finite number of light particles, *Theor. Math. Phys.* **125**:1351–1357 (1999).

Comments

In this paper, we discuss the problem of dynamics of the adiabatic piston from the point of view of the averaging method and explain the result obtained in the previous paper.

ADVANCES IN SOVIET MATHEMATICS
Volume 3, 1991

Poisson Distribution in a Geometric Problem

YA. G. SINAI

§1. Introduction

The problem which we address in this work arose from the analysis of some physical papers in the theory of quantum chaos [1]. We feel it is also very interesting in itself independently of its origin. We intend to discuss its connections with quantum chaos elsewhere.

Let γ_1 be a star-like curve containing the origin O inside the domain it bounds. In polar coordinates it is described by a continuous strictly positive function $\rho = f(\varphi)$, where φ is the angle variable, $\varphi \in [0, 2\pi)$ with the natural identification of the end-points. Denote by γ_R the curve given by the equation $r = Rf(\varphi)$. Fix a constant $c > 0$ and introduce the strip Π_R contained between the curves $\gamma_{R+c/2R}$ and $\gamma_{R-c/2R}$. It is easy to see that the area of Π_R tends to $c\int f^2(\varphi)d\varphi$ as $R \to \infty$.

The main object of our study is $\zeta(R, \gamma_1)$—the number of points of the lattice $\mathbb{Z}^2$ belonging to Π_R. Further we assume that both R and γ_1 are random. Thus $\zeta(R, \gamma_1)$ becomes a random variable. Let us formulate the assumptions concerning the joint probability distribution of R and γ_1.

I) R and γ_1 are independent.

II) R is uniformly distributed on the interval (a_1L, a_2L). Here $0 < a_1 < a_2 < \infty$ are two fixed numbers and L is the main parameter (which will tend to infinity). Our main goal is to study the limiting probability distribution of $\zeta(R, \gamma_1)$ as $L \to \infty$.

III) Denote by P the corresponding probability distribution on the space of positive continuous functions $f(\varphi)$ on the circle S^1, i.e. $f(0) = f(2\pi)$. There exist two positive constants $0 < f_1 < f_2 < \infty$ such that with P-probability 1

$$f_1 \leqslant f(\varphi) \leqslant f_2.$$

IV) For any $\alpha > 0$ and any set of values $\varphi_1, \dots, \varphi_s$, $|\varphi_{j_1} - \varphi_{j_2}| \geqslant \alpha > 0$, $1 \leqslant j_1, j_2 \leqslant s$, the joint probability distribution of the random variables $f(\varphi_1), \dots, f(\varphi_s)$ is given by a density $\pi_s(y_1, \dots, y_s|\varphi_1, \dots, \varphi_s)$. This

1991 *Mathematics Subject Classification*. Primary 60D05, 60E99.

Y.G. Sinai, *Selecta II: Probability Theory, Statistical Mechanics, Mathematical Physics and Mathematical Fluid Dynamics*,
DOI 10.1007/978-1-4419-6205-8_16, © Springer Science+Business Media, LLC 2010

function belongs to the space $C^1(M_\alpha)$, where M_α is the domain in $2s$-dimensional space defined by the inequalities

$$f_1 \leqslant y_j \leqslant f_2, \quad 1 \leqslant j \leqslant s; \qquad \operatorname{dist}(\varphi_{j_1}, \varphi_{j_2}) \geqslant \alpha, \quad 1 \leqslant j_1, j_2 \leqslant s.$$

Certainly the bounds of the derivatives may depend on α.

Moreover, for any $\overline{\varphi}_1, \dots, \overline{\varphi}_l$, $l \leqslant 5$, fix the values $f(\overline{\varphi}_1), \dots, f(\overline{\varphi}_l)$ and consider the conditional distribution of $f(\varphi_1), \dots, f(\varphi_k)$ provided that

$$\operatorname{dist}(\varphi_{j_1}, \varphi_{j_2}) \geqslant \alpha, \quad 1 \leqslant j_1, j_2 \leqslant k,$$

and

$$\operatorname{dist}(\varphi_{j_1}, \overline{\varphi}_{j_2}) \geqslant \alpha, \quad 1 \leqslant j_1 \leqslant k, 1 \leqslant j_2 \leqslant l.$$

Then this distribution is given by the density

$$\pi_k(y_1, \dots, y_k | \varphi_1, \dots, \varphi_k; \quad \overline{\varphi}_1, \dots, \overline{\varphi}_l)$$

which has the same properties as π_k in IV) with respect to all variables $y_1, \dots, y_k, \varphi_1, \dots, \varphi_k$. The corresponding bounds of the first derivatives depend only on α and not on $\overline{\varphi}_1, \dots, \overline{\varphi}_l$.

V) Take two values $\varphi', \varphi'', \varphi' > \varphi''$, the corresponding random variables $f(\varphi')$, $f(\varphi'')$, and the conditional distribution $P(\cdot|f(\varphi''))$ of $f(\varphi')$ provided that $f(\varphi'')$ is fixed. Then it has the density $\pi_1(y_1|f(\varphi''))$. There exist two constants $f_1' > 0$, $f_2' < 0$ such that π_1 is equal to zero outside the interval $f_2'(\varphi' - \varphi'') < y_1 - f(\varphi'') < f_1'(\varphi' - \varphi'')$. Inside this interval it satisfies the inequality

$$\pi_1(y|f(\varphi'')) \leqslant \frac{\text{const}}{\varphi' - \varphi''}.$$

We shall also need a sixth property of the distribution P which has a more technical character and will be formulated at the end of §3, where it will be used.

Discussion. The assumptions I), II) mean that we choose a random curve γ_1 with distribution P and then independently a value R uniformly distributed on the interval $[a_1 L, a_2 L]$. The assumption III) is taken for simplicity. It can be essentially weakened. The assumption V) is very important. If $\zeta(R, \gamma_1) = k$, then one can find $m^{(1)}, \dots, m^{(k)}$ such that $m^{(j)} \in \Pi_R \cap \mathbb{Z}^2$, $1 \leqslant j \leqslant k$. The assumption V) implies that it is highly improbable that the angles corresponding to $m^{(j)}$ are close to each other. Thus it is analogous to the property of "ordinarity" in the theory of Poisson fields. Also it shows that $f(\varphi)$ satisfies the Lipschitz condition with a uniformly bounded constant equal to $\max(f_1, |f_2|)$.

We shall denote by $\mathscr{P}_L$ the joint probability distribution of R and γ_1. The expectations with respect to $\mathscr{P}_L$ and P are denoted by $\mathsf{E}_{\mathscr{P}_L}$, E_P.

An example. Consider the Wiener bridge, i.e. the probability distribution P_0 on $C^0(S^1)$ which is the conditional measure of the Wiener measure under the condition that $\omega(0) = \omega(2\pi) = 0$. Take the induced conditional measure under the condition $\int_0^{2\pi}\omega(\varphi)d\varphi = 0$ and its restriction to the subset of functions $\omega(\varphi)$ such that $f_1' < \omega(\varphi) < f_2'$. Put $f(\varphi) = f_0 + \int_0^{\varphi}\omega(u)du$, where $f_0 > (f_2' - f_1')2\pi$. Then the distribution P that arises satisfies all the required assumptions.

It is easy to show (see also §§2 and 3) that

$$\mathsf{E}_{\mathscr{P}_L}\zeta(R, \gamma_1) \sim c\int_0^{2\pi}\mathsf{E}_P f^2(\varphi)d\varphi$$

as $L \to \infty$. This asymptotic expression shows that the family $\mathscr{F}_L$ of the induced probability distributions of the random variables $\zeta(R, \gamma_1)$ is weakly compact. Therefore there exists at least one weak limit of these distributions as $L \to \infty$. The first result of this paper is the following theorem.

THEOREM 1. *Let $L \to \infty$. Then the probability distributions $\mathscr{F}_L$ converge weakly to the mixture of Poisson distributions with the random parameter $c\int f^2(\varphi)d\varphi$ whose distribution is determined by P.*

In §4 we prove

THEOREM 2. *There exists a sequence L_j, $L_j \to \infty$ such that for P-almost every γ_1 the distributions $\mathscr{F}_{L_j,\gamma}$ of $\zeta(R, \gamma_1)$ as a function of $R \in [a_1 L, a_2 L]$ (γ_1 is fixed) converge to the Poisson distribution with the parameter $c\int f^2(\varphi)d\varphi$.*

In §4 there are other statements concerning the convergence for typical γ_1.

§2. General remarks and preliminary constructions

Let $\xi = \xi(\omega)$ be an arbitrary random variable on a probability space $\Omega = \{\omega\}$ taking nonnegative integer values. For any $k \geqslant 0$ introduce the new random variable $\xi^{(k)}$, where $\xi^{(k)}(\omega) = 0$ if $\xi(\omega) < k$ and $\xi^{(k)}(\omega) = \binom{\xi(\omega)}{k}$ if $\xi(\omega) \geqslant k$.

LEMMA 1. *The random variable ξ has the distribution which is the mixture of Poisson distributions with random parameter λ taking values in a bounded interval and having the probability distribution Q iff*

$$h_k = \mathsf{E}\xi^{(k)} = \frac{1}{k!}\int \lambda^k dQ(\lambda). \tag{1}$$

PROOF. Denote $p_n = \mathrm{Prob}\{\xi = n\}$. Then for any $\{h_k\}$

$$\sum_{k\geqslant 0} h_k z^k = \sum_{k=0}^{\infty} z^k \sum_{n\geqslant k} p_n\binom{n}{k} = \sum_{n=0}^{\infty} p_n \sum_{k=0}^{n} z^k\binom{n}{k} = \sum_{n=0}^{\infty} p_n(1+z)^n.$$

This shows that knowing h_k we can construct the generating function and the whole distribution p_n. Assume that the h_k satisfy (1). Then

$$\sum_{k\geqslant 0} h_k z^k = \sum_{k\geqslant 0} \frac{z^k}{k!}\int \lambda^k dQ(\lambda) = \int e^{\lambda z} dQ(\lambda)$$
$$= \int \sum_{n\geqslant 0} e^{-\lambda} \frac{\lambda^n (1+z)^n}{k!} dQ(\lambda)$$

which gives the statement of the lemma. In the opposite direction the statement follows from direct calculations. q.e.d.

It was already mentioned that the family of probability distributions $\mathcal{F}_L$ is weakly compact. Take any sequence $L_j \to \infty$ for which the probabilities $p_{L_j}(n) = \mathcal{P}_{L_j}\{\zeta(R, \gamma_1) = n\}$ converge to limits $p(n)$. We intend to show that for any k

$$\sum_{n\geqslant k} \binom{n}{k} p(n) = \frac{c^k}{k!} E_P \left(\int f^2(\varphi)\, d\varphi\right)^k$$
$$= \frac{c^k}{k!} \int \cdots \int y_1^2 \cdots y_k^2 \pi_k(y_1, \dots, y_k | \varphi_1, \dots, \varphi_k) \prod_{j=1}^k dy_j d\varphi_k. \quad (2)$$

Then the statement of Theorem 1 follows from Lemma 1. It is enough to show that given $\varepsilon > 0$ for all sufficiently large K the sum $\sum_{k\leqslant n\leqslant K} \binom{n}{k} p(n)$ differs from the right-hand term in (2) by not more than ε.

Assume that $K > K(\varepsilon)$ is fixed. Then for all sufficiently large j the difference satisfies

$$\left| \sum_{k\leqslant n\leqslant K} \binom{n}{k} p(n) - \sum_{k\leqslant n\leqslant K} \binom{n}{k} p_{L_j}(n) \right| \leqslant \frac{\varepsilon}{4},$$

where $p_L(n) = \mathcal{P}_L\{\zeta(R, \gamma_1) = n\} = \mathcal{F}_L(n)$.

Taking any curve γ_1 and any point $m = (m_1, m_2) \in \mathbb{Z}^2$, consider the interval $\mathcal{D}_m$ on the R-axis whose center is $d_m = |m|/f(\varphi(m))$ and whose length $l(\mathcal{D}_m) = l_m = (c/|m|) f(\varphi(m))$. Here $\varphi(m)$ is the angle corresponding to m, $|m| = \sqrt{m_1^2 + m_2^2}$, and $l(\cdot)$ is the length of the interval in the brackets. It is easy to see that $m \in \Pi_R$ iff $r \in \mathcal{D}_m$. Thus $\zeta(R, \gamma_1)$ is equal to the number of $\mathcal{D}_m$ covering R.

Denote $H_s = H_s(\gamma_1)$ the set of those R where $\zeta(R, \gamma_1) = s$. For $R \in H_s(\gamma_1)$ one can find $m^{(1)}, \dots, m^{(s)} \in \mathbb{Z}^2$ such that $R \in \bigcap_{j=1}^s \mathcal{D}_{m^{(j)}}$. The expectation is

$$\mathsf{E}_l^{(k)} = \mathsf{E}_{\mathcal{P}_L} \xi^{(k)}(R, \gamma_1) = \frac{1}{L(a_2 - a_1)} \sum_{s\geqslant k} \binom{s}{k} \mathsf{E}_p l(H_s \cap [a_1 L, a_2 L]).$$

Our analysis is based upon the possibility of rewriting $\mathsf{E}_L^{(k)}$ in a different way:

$$\mathsf{E}_L^{(k)} = \frac{1}{L(a_2 - a_1)} \sum_{\{m^{(1)}, \dots, m^{(k)}\}} \mathsf{E}_P l(\mathscr{D}_{m^{(1)}} \cap \dots \cap \mathscr{D}_{m^{(k)}}).$$

The last sum is taken over unordered k-tuples $\{m^{(1)}, \dots, m^{(k)}\}$. Therefore we may assume that $\varphi(m^{(1)}) < \varphi(m^{(2)}) < \dots < \varphi(m^{(k)})$.

Take $\alpha = 2\pi/M$, where M is an integer, and decompose the circle S^1 into M equal parts $\Delta^{(j)}$. Let $H_s(\alpha) \subset H_s$ be the subset of those R that if $R \in \bigcap_{t=1}^{s} \mathscr{D}_{m^{(t)}}$ and $\varphi(m^{(t)}) \in \Delta^{(j_t)}$ then $\operatorname{dist}(\Delta^{(j_{t_1})}, \Delta^{(j_{t_2})}) \geqslant \alpha$. Put also $p_L(s; \alpha) = \mathscr{P}_L(H_s(\alpha)) = (1/L(a_2 - a_1))\mathsf{E}_P l(H_s(\alpha) \cap [a_1 L, a_2 L])$. We shall show that

I) For any K one can find $\alpha_1 = \alpha_1(K, \varepsilon)$ such that for all $0 < \alpha < \alpha_1$ and all L

$$\left| \sum_{k \leqslant n \leqslant K} \binom{n}{k} p_L(n) - \sum_{k \leqslant n \leqslant K} \binom{n}{k} p_L(n; \alpha) \right| \leqslant \frac{\varepsilon}{4}.$$

II) One can find $K_1 = K_1(\varepsilon)$ and $\alpha_2 = \alpha_2(\varepsilon)$ such that for all L, $0 < \alpha < \alpha_2$ and all $K > K_1(\varepsilon)$

$$\left| \frac{1}{L(a_2 - a_1)} \sum_{\{m^{(1)}, \dots, m^{(k)}\}}^{(\alpha)} \mathsf{E}_P l(\mathscr{D}_{m^{(1)}} \cap \dots \cap \mathscr{D}_{m^{(k)}}) - \sum_{k \leqslant n \leqslant K} \binom{n}{k} p_L(n; \alpha) \right| \leqslant \frac{\varepsilon}{4}.$$

Here $\Sigma^{(\alpha)}$ is the summation over such $m^{(1)}, \dots, m^{(k)}$ that if $\varphi(m^{(t)}) \in \Delta^{(j_t)}$, $1 \leqslant t \leqslant k$, then $\operatorname{dist}(\Delta^{(j_{t_1})}, \Delta^{(j_{t_2})}) \geqslant \alpha$.

III) There exists an $\alpha_3 = \alpha_3(\varepsilon)$ such that for all $0 < \alpha < \alpha_3$ and all sufficiently large L

$$\left| \frac{1}{L(a_2 - a_1)} \sum_{\{m^{(1)}, \dots, m^{(k)}\}}^{(\alpha)} \mathsf{E}_P l(\mathscr{D}_{m^{(1)}} \cap \dots \cap \mathscr{D}_{m^{(k)}}) \right.$$
$$\left. - \frac{c^k}{k!} \int y_1^2 \cdots y_k^2 \pi_k(y_1, \dots, y_k | \varphi_1, \dots, \varphi_k) \prod_{j=1}^{k} dy_j d\varphi_j \right| \leqslant \frac{\varepsilon}{4}.$$

Assuming that I), II), III) are proved, let us derive the statement of the theorem. Then we shall prove I) and II). The statement III) will be proved in the next section.

We may assume that $K \geqslant K_1(\varepsilon)$. Take α so small that I) holds, II) holds for K, and III) holds. Using these statements in this order we can write

$$\left|\sum_{k\leqslant n\leqslant K}\binom{n}{k}p(n)-\frac{c^k}{k!}\int y_1^2\cdots y_k^2\pi_k(y_1,\ldots,y_k|\varphi_1,\ldots,\varphi_k)\prod_{j=1}^{k}dy_j d\varphi_j\right|$$

$$\leqslant\left|\sum_{k\leqslant n\leqslant K}\binom{n}{k}p(n)-\sum_{k\leqslant n\leqslant K}\binom{n}{k}p_{L_j}(n)\right|$$

$$+\left|\sum_{k\leqslant n\leqslant K}\binom{n}{k}p_{L_j}(n)-\sum_{k\leqslant n\leqslant K}\binom{n}{k}p_{L_j}(n;\alpha)\right|$$

$$+\left|\sum_{k\leqslant n\leqslant K}\binom{n}{k}p_{L_j}(n;\alpha)-\frac{1}{L(a_2-a_1)}\sum_{\{m^{(1)},\ldots,m^{(k)}\}}^{(\alpha)}\mathsf{E}_p l(\mathscr{D}_{m^{(1)}}\cap\ldots\cap\mathscr{D}_{m^{(k)}})\right|$$

$$+\left|\frac{1}{L(a_2-a_1)}\sum_{\{m^{(1)},\ldots,m^{(k)}\}}^{(\alpha)}\mathsf{E}_p l(\mathscr{D}_{m^{(1)}}\cap\cdots\cap\mathscr{D}_{m^{(k)}})\right.$$

$$\left.-\frac{c^k}{k!}\int\prod_{j=1}^{k}y_j^2\pi_R(y_1,\ldots,y_k|\varphi_1,\ldots,\varphi_k)\prod_{j=1}^{k}dy_jd\varphi_j\right|\leqslant\varepsilon,$$

q.e.d.

Now we shall prove I). It is sufficient to show that

$$I=\frac{1}{L(a_2-a_1)}\sum_{\overline{m},\overline{\overline{m}}}{}' E_p(l(\mathscr{D}_{\overline{m}}\cap\mathscr{D}_{\overline{\overline{m}}}))\leqslant\gamma_1(\alpha), \tag{3}$$

where Σ' is taken over such pairs $\overline{m},\overline{\overline{m}}$ that $|\varphi(\overline{m})-\varphi(\overline{\overline{m}})|\leqslant 2\alpha$ and $\gamma_1(\alpha)\to 0$ as $\alpha\to 0$. Indeed, this will imply that

$$\frac{1}{L(a_2-a_1)}\mathsf{E}_L l(H_s)-\frac{1}{L(a_2-a_1)}\mathsf{E}_L l(H_s(\alpha))$$

can be made arbitrarily small is α is sufficiently small.

Rewrite (3) differently:

$$\begin{aligned}I&=\frac{1}{L(a_2-a_1)}\sum_{\overline{m}}\sum_{\overline{\overline{m}}}{}'\mathsf{E}_p(l(\mathscr{D}_{\overline{m}}\cap\mathscr{D}_{\overline{\overline{m}}}))\\&\leqslant\frac{\text{const}}{L^2}\sum_{\overline{m}}\sum_{\overline{\overline{m}}}{}' P\{l(\mathscr{D}_{\overline{m}}\cap\mathscr{D}_{\overline{\overline{m}}})>0\}.\end{aligned} \tag{4}$$

Further

$$\begin{aligned}&P\{l(\mathscr{D}_{\overline{m}}\cap\mathscr{D}_{\overline{\overline{m}}})>0\}\\&\quad=\int_{f_1}^{f_2}\pi_1(\overline{y}|\varphi(\overline{m}))\,dy\int_{A(\overline{y},\overline{\overline{m}})}\pi_1(\overline{\overline{y}};\varphi(\overline{\overline{m}})|\overline{y};\varphi(\overline{m}))d\overline{\overline{y}},\end{aligned}$$

where $A(\overline{y}, \overline{m})$ is the interval of values $\overline{\overline{y}}$ for which $l(\mathscr{D}_{\overline{m}} \cap \mathscr{D}_{\overline{\overline{m}}}) > 0$. It is easy to see that the length of this interval is not greater than $\text{const}\, L^{-2}$. The property V) of the distribution P gives

$$\pi_1(\overline{\overline{y}}; \varphi(\overline{\overline{m}})|\overline{y}; \varphi(\overline{m})) \leqslant \frac{\text{const}}{|\varphi(\overline{m}) - \varphi(\overline{\overline{m}})|}.$$

Thus

$$\int_{A(\overline{y}, \overline{m})} \pi_1(\overline{\overline{y}}; \varphi(\overline{\overline{m}})|\overline{y}; \varphi(\overline{m}))d\overline{\overline{y}} \leqslant \frac{\text{const}}{L^2|\varphi(\overline{\overline{m}}) - \varphi(\overline{m})|}.$$

Decompose the interval $[-2\alpha, 2\alpha]$ into intervals U_j whose length is contained between $1/L$ and $2/L$. Let $V_j = \{\overline{\overline{m}}|\varphi(\overline{\overline{m}}) - \varphi(\overline{m}) \in U_j\}$. The index j changes within the interval $-\text{const}\, L\alpha \leqslant j \leqslant \text{const}\, L\alpha$. The number of points $\overline{\overline{m}}$ belonging to V_j is not greater than $\text{const}\, j$. Therefore

$$\begin{aligned}\sum_{\overline{\overline{m}}} \frac{1}{L^2|\varphi(\overline{\overline{m}}) - \varphi(\overline{m})|} &\leqslant \frac{\text{const}}{L^2} \sum_{|j| \leqslant \text{const}\, L\alpha} \sum_{\overline{\overline{m}} \in V_j} \frac{1}{|\varphi(\overline{\overline{m}}) - \varphi(\overline{m})|} \\ &\leqslant \frac{\text{const}}{L^2} \sum_{|j| \leqslant \text{const}\, L\alpha} \frac{\text{const}\, L(|j| + 1)}{|j| + 1} \leqslant \text{const}\, \alpha.\end{aligned}$$

The substitution of the last inequality into (4) gives the desired result.

Now we shall prove II) assuming that III) is already proved. The proof of III) does not depend on II). Introduce the event $C_{m^{(1)}, \dots, m^{(k)}}$ consisting of such γ_1 that $\bigcap_{j=1}^{k} \mathscr{D}_{m^{(j)}} \neq \varnothing$ and $C = \bigcup' C_{\overline{m}, \overline{\overline{m}}}$, where $\bigcup'$ is the union over such pairs $\overline{m}, \overline{\overline{m}}$ that if $\varphi(\overline{m}) \in \Delta^{(j_1)}$, $\varphi(\overline{\overline{m}}) \in \Delta^{(j_2)}$, then $\text{dist}(\Delta^{(j_1)}, \Delta^{(j_2)}) \leqslant \alpha$. We can write

$$\begin{aligned}\sum_{n \geqslant k} \binom{n}{k} p_L(n; \alpha) &= \frac{1}{L(a_2 - a_1)} \sum_{\{m^{(1)}, \dots, m^{(k)}\}} \mathsf{E}_P l \left(\bigcap_{j=1}^{k} (\mathscr{D}_{m^{(j)}} \setminus C) \right) \\ &= \frac{1}{L(a_2 - a_1)} \sum_{\{m^{(1)}, \dots, m^{(k)}\}}^{(\alpha)} \mathsf{E}_P l \left(\bigcap_{j=1}^{k} (\mathscr{D}_{m^{(j)}} \setminus C) \right)\end{aligned}$$

and

$$\sum_{n \geqslant k} \binom{n}{k} p_L(n; , \alpha) \leqslant \frac{1}{L(a_2 - a_1)} \sum_{\{m^{(1)}, \dots, m^{(k)}\}}^{(\alpha)} \mathsf{E}_P l \left(\bigcap_{j=1}^{k} \mathscr{D}_{m^{(j)}} \right). \tag{5}$$

On the other hand

$$\begin{aligned}\sum_{n \geqslant k} \binom{n}{k} p_L(n; \alpha) \geqslant{}& \frac{1}{L(a_2 - a_1)} \sum_{\{m^{(1)}, \dots, m^{(k)}\}}^{(\alpha)} \mathsf{E}_P l \left(\bigcap_{j=1}^{k} \mathscr{D}_{m^{(j)}} \right) \\ &- \frac{1}{L(a_2 - a_1)} \sum_{\{m^{(1)}, \dots, m^{(k)}, \overline{m}, \overline{\overline{m}}\}}^{(\alpha)} \mathsf{E}_P l \left(\bigcap_{j=1}^{k} \mathscr{D}_{m^{(j)}} \cap \mathscr{D}_{\overline{m}} \cap \mathscr{D}_{\overline{\overline{m}}} \right).\end{aligned} \tag{6}$$

Here $\overline{m}, \overline{\overline{m}}$ are such that if

$$\varphi(\overline{m}) \in \Delta^{(s_1)}, \quad \varphi(\overline{\overline{m}}) \in \Delta^{(s_2)},$$

then $\operatorname{dist}(\Delta^{(s_1)}, \Delta^{(s_2)}) \leqslant \alpha$. The second sum (6) will be estimated in §3. We shall show that it is not greater than $\gamma_2(\alpha)$, where $\gamma_2(\alpha) \to 0$ as $\alpha \to 0$.

Take $k_1 = 2k$. Then III) implies that for all sufficiently large L

$$\frac{1}{L(a_2 - a_1)} \sum\nolimits^{(\alpha)}_{\{m^{(1)}, \dots, m^{(k)}\}} \mathsf{E}_P l(\mathscr{D}_{m^{(1)}} \cap \dots \cap \mathscr{D}_{m^{(k_1)}}) \leqslant \text{const},$$

where const depends on k. For any $n \geqslant 4k_1$

$$\binom{k}{n} = \binom{k_1}{n} \frac{\binom{n}{k}}{\binom{n}{k_1}} = \frac{k_1!}{k!} \frac{n(n-1)\cdots(n-k)}{n(n-1)\cdots(n-k_1)} \binom{n}{k_1} \leqslant \frac{k_1!}{k!} \frac{2^k}{n^k} \binom{n}{k_1}.$$

Thus for $K_1 \geqslant 4k_1$

$$\sum_{n \geqslant N} \binom{n}{k} p_L(n;\alpha) \leqslant \frac{k_1!}{k!} \frac{2^k}{K_1^k} \sum_{n \geqslant k_1} \binom{n}{k_1} p_L(n;\alpha) \leqslant \frac{k_1!}{k!} 2^k \text{const} \frac{1}{K_1^k} \leqslant \frac{\varepsilon}{8}.$$

This gives the result in view of the promised estimate of (6).

§3. The main estimates

It was mentioned in §2 that $\zeta(R, \gamma_1)$ is equal to the number of intervals $\mathscr{D}_m$ covering R. As in §2, take $\alpha = 2\pi/M$ and the corresponding intervals $\Delta^{(j)}$. Also let B be an interval on the R-axis.

DEFINITION 1. We say that the intervals $\mathscr{D}_{m'}$ and $\mathscr{D}_{m''}$ have B-intersection if $d_{m'} - d_{m''} \in B$.

We shall consider later the intervals B having the form

$$B = B^{(s)} = [(s/L)\delta, ((s+1)/L)\delta],$$

where $\delta > 0$ is a constant. However in the beginning it is more convenient to assume that B is arbitrary, $B \subset [-\text{const}/L, \text{const}/L]$, $l(B) = (1/L)\delta(B)$.

Take k intervals $\Delta^{(j_1)} < \Delta^{(j_2)} < \dots < \Delta^{(j_k)}$ such that

$$\operatorname{dist}(\Delta^{(j_s)}, \Delta^{(j_t)}) \geqslant \beta = (2\pi/M)q,$$

where q is an integer, $j_s \neq j_t$, and $(k-1)$ intervals $B_2, \dots, B_k$, $B_j = (b_j', b_j'')$, $l(B_j) = (1/L)\delta_j$. The main object of our study is

$$N = N(\Delta^{(j_1)}, \dots, \Delta^{(j_k)}, \; B_2, \dots, B_k)$$

equal to the number of k-tuples $(m^{(1)}, \dots, m^{(k)})$ such that

1) $m^{(s)} \in O_s \cap \mathbb{Z}^2$, $1 \leqslant s \leqslant k$, where O_s is the plane domain bounded by the rays $\varphi_1 = (2\pi/M)j_s$, $\varphi_2 = (2\pi/M)(j_s + 1)$ and by the arcs of the circles of radii $a_1 f_1 L$ and $a_2 f_2 L$ respectively;

2) $\mathscr{D}_{m^{(j)}}$ and $\mathscr{D}_{m^{(1)}}$ have B_j-intersection, $j = 2, \dots, k$. N depends certainly on γ_1 and thus it is a random variable. We shall find an asymptotic expression for the expectation $\mathsf{E}_P(N)$ as $L \to \infty$.

Let $\mathscr{D}_{m^{(1)}, \dots, m^{(k)}}$ be the event consisting of γ_1 such that $\mathscr{D}_{m^{(j)}}$ and $\mathscr{D}_{m^{(1)}}$ have B_j-intersection, $j = 2, \dots, k$. If $\chi_{m^{(1)}, \dots, m^{(k)}}(\gamma_1)$ is the indicator of $\mathscr{D}_{m^{(1)}, \dots, m^{(k)}}$, then

$$N = \sum_{\substack{m^{(j)} \in O_j \cap \mathbb{Z}^2, \\ 1 \leqslant j \leqslant k}} \chi_{m^{(1)}, \dots, m^{(k)}}(\gamma_1), \qquad \mathsf{E}_P N = \sum_{\substack{m^{(j)} \in O_j \cap \mathbb{Z}^2 \\ 1 \leqslant j \leqslant k}} \mathsf{E}_P \chi_{m^{(1)}, \dots, m^{(k)}}(\gamma_1).$$

Further

$$\mathsf{E}_P \chi_{m^{(1)}, \dots, m^{(k)}} = \int_{f_1}^{f_2} dy_1 \int \cdots \int \pi_k(y_1, y_2, \dots, y_k | \varphi(m^{(1)}, \dots, \varphi(m^{(k)})) \prod_{j=1}^{k} dy_j.$$

The domain of integration of y_j is the interval Y_j having (up to higher order terms) the form

$$\frac{|m^{(j)}|}{|m^{(1)}|} y_1 \left(1 - \frac{y_1}{|m^{(1)}|} b_j''\right) \leqslant y_j \leqslant \frac{|m^{(j)}|}{|m^{(1)}|} y_1 \left(1 - \frac{y_1}{|m^{(1)}|} b_j'\right).$$

With the same precision

$$\mathsf{E}_P \chi_{m^{(1)}, \dots, m^{(k)}} = \int_{f_1}^{f_2} dy_1 \pi_k \left(y_1, y_1 \frac{|m^{(2)}|}{|m^{(1)}|}, \dots, y_1 \frac{|m^{(k)}|}{|m^{(1)}|} \middle| \varphi(m^{(1)}), \dots, \varphi(m^{(k)})\right) \times \left\{ \prod_{j=2}^{k} \left[y_1^2 \frac{|m^{(j)}|}{|m^{(1)}|} (b_j'' - b_j') \right] \left(1 + O\left(\frac{1}{L}\right)\right) \right\}.$$

Certainly $O(1/L)$ depends also on α.

Now fix $m^{(1)}$ and perform the summation over $m^{(2)}, \dots, m^{(k)}$. We shall do this in the following way. Let us carry out a contraction of the plane with coefficient $y_1/|m^{(1)}|$. Then we get the lattice $(y_1/|m^{(1)}|) \cdot \mathbb{Z}^2$. The function $\pi_k(y_1, z_2, \dots, z_k | \varphi_1, \dots, \varphi_k)$ depends continuously on its variables provided that $|\varphi_{s_1} - \overline{\varphi}_{s_1}| \leqslant \alpha$, $1 \leqslant s_1, s_2 \leqslant k$, where $\overline{\varphi}_{j_s}$ is the center of $\Delta^{(j_s)}$. If α is small as compared to β, then

$$\mathsf{E} \chi_{m^{(1)}, \dots, m^{(k)}} = \int_{f_1}^{f_2} dy_1 \pi_k \left(y_1, y_1 \frac{|m^{(2)}|}{|m^{(1)}|}, \dots, y_1 \frac{|m^{(k)}|}{|m^{(1)}|} \middle| \overline{\varphi}^{(j_1)}, \dots, \overline{\varphi}^{(j_k)}\right) \times \left\{ \prod_{j=2}^{k} \left[y_1^2 \frac{|m^{(j)}|}{|m^{(1)}|^2} (b_j'' - b_j') \right] + \gamma_3(\alpha) \overline{L}^{-k+1} \prod_{j=2}^{k} (b_j'' - b_j') \right\},$$

where $\gamma_3(\alpha)$ and all other $\gamma_j(\alpha)$, $j \geqslant 3$, tend to zero as $\alpha \to 0$ uniformly in $m^{(1)}, \dots, m^{(k)}$. Thus

$$\begin{aligned}
&\sum\nolimits^{(\beta)}_{m^{(2)},\dots,m^{(k)}} \mathsf{E}\chi_{m^{(1)},\dots,m^{(k)}} \\
&= \int_{f_1}^{f_2} dy_1 \sum_{m^{(2)},\dots,m^{(k)}} \pi_k\left(y_1, y_1\frac{|m^{(2)}|}{|m^{(1)}|}, \dots, y_1\frac{|m^{(k)}|}{|m^{(1)}|}\Big|\overline{\varphi}^{(j_1)}, \dots, \overline{\varphi}^{(j_k)}\right) \\
&\quad\times\left\{\prod_{j=2}^{k} y_1\frac{|m^{(j)}|}{|m^{(1)}|}\left(\frac{y_1^2}{|m^{(1)}|^2}\right)^{k-1}\left(\frac{|m^{(1)}|}{y_1}\right)^{k-1}\prod_{s=2}^{k-1}(b_s'' - b_s')|\Delta^{(j_s)}|\right. \\
&\qquad\left. + \gamma_4(\alpha)\prod_{s=2}^{k}\delta(B_s)l(\Delta^{(j_s)})\right\} \\
&= V_{m^{(1)}}(\Delta^{(j_1)}, \dots, \Delta^{(j_k)}; B^{(s_2)}, \dots, B^{(s_k)}) = V. \qquad (7)
\end{aligned}$$

Further

$$\begin{aligned}
V = \int_{f_1}^{f_2} dy_1 \int\cdots\int \pi_k(y_1, z_2, \dots, z_k|\overline{\varphi}^{(j_1)}, \dots, \overline{\varphi}^{(j_k)})z_2^2\cdots z_k^2 \\
\times\left[\frac{|m^{(1)}|}{y_1 L}\right]^{k-1}\prod_{s=2}^{k}\delta_s l(\Delta^{(j_s)}) + \gamma_4(\alpha)\prod_{s=2}^{k}\delta(B_s)l(\Delta^{(j_s)}). \quad (8)
\end{aligned}$$

Since $|m^{(1)}| = O(L)$, the total sum takes value of order of $\prod_{s=2}^{k}\delta_s l(\Delta^{(j_s)})$.

Remark that the differences between the lengths of intervals $\mathscr{D}_{m^{(j)}}$, $j = 1, \dots, k$, having a nonempty intersection are $o(1/L)$. Therefore we may assume that

$$l(\mathscr{D}_{m^{(j)}}) = l(\mathscr{D}_{m^{(1)}}), \quad j = 2, \dots, k.$$

Also $l_0 = l(\mathscr{D}_{m^{(1)}})$ is defined as soon as we know the values of $m^{(1)}$ and $y_1 = f(\varphi(m^{(1)}))$.

Now we recall that the intervals B which we consider have in fact the form $B^{(s)} = [\delta s/L, \delta(s+1)/L]$.

DEFINITION 2. A set of $(k-1)$ intervals $\{B^{(s_2)}, \dots, B^{(s_k)}\}$ is *essential* if for any position of intervals $\mathscr{D}_{m^{(j)}}$, having $B^{(s_j)}$-intersection with $\mathscr{D}_{m^{(1)}}$, $2 \leqslant j \leqslant k$, the length of the intersection $\mathscr{D}_{m^{(1)}} \cap \cdots \cap \mathscr{D}_{m^{(k)}}$ is positive.

The variation of this length for different positions of all $\mathscr{D}_{m^{(j)}}$, $2 \leqslant j \leqslant k$, is not greater than $\operatorname{const} k\delta L^{-1}$, while $l(\mathscr{D}_{m^{(j)}})$ is $O(1/L)$. Denote

$$l(B^{(s_2)}, \dots, B^{s_k)}) = \frac{\max l(\cap_{j=1}^{k}\mathscr{D}_{m^{(j)}}) + \min l(\cap_{j=1}^{k}\mathscr{D}_{m^{(j)}})}{2},$$

where max and min are taken over all possible positions of $\mathscr{D}_{m^{(j)}}$ having $B^{(s_j)}$-intersection with $\mathscr{D}_{m^{(1)}}$, $j = 2, \dots, k$. The notation $\Sigma^{(e)}$ further indicates summation over all essential $(k-1)$-tuples $\{B^{(s_2)}, \dots, B^{(s_k)}\}$.

LEMMA 2.

$$\sum^{(e)} l(B^{(s_2)}, \dots, B^{(s_k)})\delta^{k-1} = L^{-1}\left(\left(\frac{cL}{|m^{(1)}|}y_1\right)^k + O(\delta)\right).$$

PROOF. Let us expand the real line with coefficient L. Then the interval $\mathscr{D}_{m^{(1)}}$ is transformed into the interval $\widetilde{\mathscr{D}}_{m^{(1)}}$ of length

$$\lambda = (cL/|m^{(1)}|)f(\varphi(m^{(1)})),$$

while the intervals $B^{(s_j)}$ are transformed into the intervals $\tilde{B}^{(s_j)}$ of the length δ. The sum in the formulation of the lemma becomes the Riemannian integral sum for the integral

$$I_k = \int_{-\lambda}^{\lambda}\cdots\int_{-\lambda}^{\lambda} du_2\cdots du_k \lambda(u_2, \dots, u_k).$$

The function $\lambda(u_1, \dots, u_k)$ is defined in the following way. Having the values $u_2, \dots, u_k$, construct the intervals $\tilde{\mathscr{D}}_j$ whose centers are at u_j and lengths are equal to λ. Then

$$\lambda(u_2, \dots, u_k) = l\left(\bigcap_{j=1}^{k}\widetilde{\mathscr{D}}_j\right)$$

is the interval of length λ with center at 0. It follows easily from this description that

$$\sum^{(e)} l(B^{(s_2)}, \dots, B^{(s_k)})\delta^{k-1} = L_{-1}(I_k + O(\delta)).$$

We shall show that $I_k = \lambda^k$. Take $u_1 \in \tilde{\mathscr{D}}_1$ and integrate over values $u_2, \dots, u_k$ such that all $\tilde{\mathscr{D}}_j$ cover u_1 and then integrate the result over u_1. It is easy to see that the first integral is equal to λ^{k-1}, while the last integration gives the extra factor λ. This gives the value of I_k and the final result

$$\sum^{(e)} l(B^{(s_2)}, \dots, B^{(s_k)})\delta^{k-1} = L^{-1}\left(\left(\frac{cL}{|m^{(1)}|}y_1\right)^k + O(\delta)\right). \quad \text{q.e.d.}$$

Now we shall estimate the difference between

$$S_k = \frac{1}{L(a_2 - a_1)}\sum^{(\beta)}_{\{m^{(1)}, \dots, m^{(k)}\}} \mathsf{E}_P l(\mathscr{D}_{m^{(1)}} \cap \cdots \cap \mathscr{D}_{m^{(k)}})$$

and the integral

$$\frac{c^k}{k!}\int y_1^2\cdots y_k^2 \pi_k(y_1, \dots, y_k|\varphi_1, \dots, \varphi_k)\prod_{j=1}^{k} dy_j d\varphi_j.$$

Rewrite S_k in the following way:

$$S_k = \frac{1}{L(a_2 - a_1)} \sum_{m^{(1)}} \sum_{B^{(s_2)}, \dots, B^{(s_k)}}^{(e)} \sum_{\{m^{(1)}, \dots, m^{(k)}\}} E_P l(\mathscr{D}_{m^{(1)}} \cap \cdots \cap \mathscr{D}_{m^{(k)}}).$$

The inner sum is taken over $m^{(1)}, \dots, m^{(k)}$ such that $\mathscr{D}_{m^{(j)}}$ and $\mathscr{D}_{m^{(1)}}$ have $B^{(s_j)}$-intersection. Replace

$$l(\mathscr{D}_{m^{(1)}} \cap \cdots \cap \mathscr{D}_{m^{(k)}}) \quad \text{by} \quad l(B^{(s_2)}, \dots, B^{(s_k)}).$$

The error which we make has absolute value not greater than $\text{const}\,\delta L^{-1}$, where const may depend on k and does not depend on $m^{(1)}, \dots, m^{(k)}$. From the next estimations it will follow easily that the expectation with respect to $\mathscr{P}_L$ of this error is not greater than $\text{const}\,\delta$. We shall choose δ so small that $\text{const}\,\delta \leqslant \varepsilon/12$. Thus we must consider the expression

$$\begin{aligned} S_k^{(1)} = {} & \frac{1}{L(a_2 - a_1)} \\ & \times \sum_{m^{(1)}} \sum_{\Delta^{(j_1)} < \cdots < \Delta^{(j_k)}}^{(\beta)} \sum_{B^{(s_2)}, \dots, B^{(s_k)}}^{(e)} l(B^{(s_2)}, \dots, B^{(s_k)}) \\ & \qquad \times V_{m^{(1)}}(\Delta^{(j_1)}, \dots, \Delta^{(j_k)}; B^{(s_2)}, \dots, B^{(s_k)}). \end{aligned} \tag{9}$$

Here $\sum^{\beta}_{\Delta^{(j_1)} < \cdots < \Delta^{(j_k)}}$ means summation over the $\Delta^{(j_s)}$ such that the distances between them are greater than β (see also (7), (8)). Choose β so small that the summation of the above-mentioned error gives not more than $\varepsilon/12$. Using Lemma 2, we see that the main term in (9) is equal to

$$\begin{aligned} S_k^{(2)} = {} & \frac{c^k}{L(a_2 - a_1)} \sum_{\Delta^{(i_1)} < \cdots < \Delta^{(i_k)}}^{(\beta)} \prod_{s=2}^{k} l(\Delta^{(j_s)}) \\ & \times \int \cdots \int y_1 \pi_k(y_1, z_2, \dots, z_k), \overline{\varphi}^{(j_1)}, \dots, \overline{\varphi}^{(j_k)}) \\ & \qquad \times z_2^2 \cdots z_k^2 \sum_{m^{(1)}} \frac{1}{|m^{(1)}|}. \end{aligned} \tag{10}$$

Assuming that y_1 is fixed, let us make the summation over $m^{(1)}$. Remark that

$$La_1 \leqslant \frac{|m^{(1)}|}{y_1} \leqslant La_2, \quad \text{or} \quad La_1 y_1 \leqslant |m^{(1)}| \leqslant La_2 y_1.$$

Therefore

$$\sum_{\substack{\varphi(m^{(1)}) \in \Delta^{(j_1)} \\ La_1 y_1 \leqslant |m^{(1)}| \leqslant La_2 y_1}} \frac{1}{|m^{(1)}|} = (a_2 - a_1) L y_1 |\Delta^{(i_1)}| \left(1 + O\left(\frac{1}{\sqrt{L}}\right)\right).$$

Putting this expression into (10), we finally get

$$S_k^{(3)} = c^k \sum_{\Delta^{(i_1)}<\cdots<\Delta^{(i_k)}}^{(\beta)} \prod_{s=1}^{k} l(\Delta^{(j_s)})$$
$$\times \int \cdots \int z_1^2 \cdots z_k^2 \pi_k(z_1, \ldots, z_k | \overline{\varphi}^{(j_1)}, \ldots, \overline{\varphi}^{(j_k)}) dz_1 \cdots dz_k .$$

This sum differs from the integral

$$\frac{c^k}{k!} \int \cdots \int z_1^2 \cdots z_k^2 \pi_k(z_1, \ldots, z_k | \overline{\varphi}^{(j_1)}, \ldots, \overline{\varphi}^{(j_k)}) dz_1 \cdots dz_k$$

by a number tending to zero as $\beta \to 0$. This gives III). In order to complete the proof of theorem 1 we have to estimate the second sum in (6), i.e. the expression

$$\frac{1}{L(a_2 - a_1)} \sum_{\{m^{(1)}, \ldots, m^{(k)}, \overline{m}, \overline{\overline{m}}\}}^{(\alpha)} \mathsf{E}_P l \left(\bigcap_{j=1}^{k} \mathscr{D}_{m^{(j)}} \cap \mathscr{D}_{\overline{m}} \cap \mathscr{D}_{\overline{\overline{m}}} \right). \tag{11}$$

Here the summation is taken over $m^{(1)}, \ldots, m^{(k)}$, $\overline{m}, \overline{\overline{m}}$ such that if

$$\varphi(m^{(s)}) \in \Delta^{(j_s)}, \quad 1 \leqslant s \leqslant k, \quad \varphi(\overline{m}) \in \Delta^{(\bar{j})}, \quad \varphi(\overline{\overline{m}}) \in \Delta^{(\dot{j})},$$

then

$$\operatorname{dist}(\Delta^{(j_{s_1})}, \Delta^{(j_{s_2})} \geqslant \alpha, \quad \operatorname{dist}(\Delta^{(\bar{j})}, \quad \Delta^{(\dot{j})}) \leqslant \alpha.$$

Decompose the sets $m^{(j)}$, $1 \leqslant j \leqslant k$, and $\overline{m}, \overline{\overline{m}}$ into two parts. The first one consists of $\tilde{m}^{(s)}$, $1 \leqslant s \leqslant k_1 \leqslant k$, $k - k_1 \leqslant 5$, while the second one consists of $\overline{m}^{(l)}$, $2 \leqslant l \leqslant 4$. The second part includes $\overline{m}, \overline{\overline{m}}$, and if $\varphi(\overline{m}^{(l)}) \in \Delta^{(j_l')}$, $j_1' < j_2' < \cdots < j_{k_1-k}'$, then

$$\operatorname{dist}\left(\Delta^{(j_s)}, \bigcup_{s' \neq s} \Delta^{(j_{s'})} \cup \bigcup_{l} \Delta^{(j_l')}\right) \geqslant \alpha.$$

The arguments for all k_1 are the same. Firstly we fix $\{\overline{m}^{(l)}\}$, $2 \leqslant l \leqslant k_1 - k \leqslant 5$ and make the summation over all $\tilde{m}^{(s)}$. This estimation is done in the same way as the estimation of the sum over $m^{(2)}, \ldots, m^{(k)}$ in this section and gives a number whose absolute value is not greater than const. Instead of the densities π_k, one should use the conditional densities π_k under the conditions that $f(\varphi(\overline{m}^{(l)}))$, $2 \leqslant l \leqslant k_1 - k$, are fixed and under the assumption IV) concerning P (see §1).

The rest of the proof is the estimate of the expression

$$Q = \frac{1}{L^2} \sum_{\{\overline{m}^{(l)}\}} P\left\{ \bigcap_{l} \mathscr{D}_{\overline{m}^{(l)}} \neq \varnothing \right\}.$$

We rewrite it as follows:

$$Q = \frac{1}{L^2}\sum_{\overline{m}^{(1)}}\int \pi_1(y_1|\varphi(\overline{m}^{(1)}))\sum_{\overline{m}^{(2)}}\int dP(y_2|y_1\,,\,\varphi(\overline{m}^{(1)}))$$

$$\times\sum_{\overline{m}^{(k_1-k)}}\int dP(y_{k_1-k}|\varphi(\overline{m}^{(k_1-k)}),y_1,\varphi(\overline{m}^{(1)}),y_2,\varphi(\overline{m}^{(2)}),\dots,y_{k_1-k},\varphi(\overline{m}^{(k_1-k)})).$$

Here

$$P(\cdot|\varphi(\overline{m}^{(j)})\,,\,y_1\,,\,\varphi(\overline{m}^{(1)})\,,\,y_2\,,\,\varphi(\overline{m}^{(2)})\,,\,\dots\,,\,y_{j-1}\,,\,\varphi(\overline{m}^{(j-1)}))$$

is the conditional distribution of $f(\varphi(\overline{m}^{(j)}))$ provided that

$$y_s = f(\varphi(\overline{m}^{(s)}))\,,\qquad s<j\,,$$

are fixed. Now we shall formulate the sixth assumption concerning the distribution P (see §1).

VI) Let $\overline{m}^{(1)}\,,\,\overline{m}^{(2)}\,,\,\dots\,,\,\overline{m}^{(j-1)}\,,\,\overline{m}^{(j)}$ be given such that $\varphi(\overline{m}^{(j)})$ lies between $\varphi(\overline{m}^{(j-1)})$ and $\varphi(\overline{m}^{(1)})$ and

$$\operatorname{dist}(\varphi(\overline{m}^{(j)})\,,\,\varphi(\overline{m}^{(j-n)}))\leqslant\alpha\,,\qquad \operatorname{dist}(\varphi(\overline{m}^{(j)})\,,\,\varphi(\overline{m}^{(1)}))\geqslant\alpha.$$

Denote by $P\{\cdot|f(\varphi(\overline{m}^{(1)}))\,,\,\dots\,,\,f(\varphi(\overline{m}^{(j-1)}))\}$ the conditional distribution of $f(\varphi(\overline{m}^{(j)}))$ provided that all $f(\varphi(\overline{m}^{(1)}))\,,\,\dots\,,\,f(\varphi(\overline{m}^{(j-1)}))$ are given. Then

$$\sum_{\overline{m}^{(j)}} P\left\{\mathscr{D}_{\overline{m}^{(j)}}\cap\left(\bigcup_{s=1}^{j-1}\mathscr{D}_{\overline{m}^{(s)}}\right)\neq\varnothing\middle| f(\varphi(\overline{m}^{1}))\,,\,\dots\,,\,f(\varphi(\overline{m}^{(j-1)}))\right\}\leqslant\gamma_5(\alpha)\,,$$

where $\gamma_5(\alpha)\to 0$ as $\alpha\to 0$.

Under the assumption VI) the required estimate of (11) immediately follows. This completes the proof of Theorem 1.

§4. Limiting behavior for typical γ_1

Recall that $H_s(\gamma_1)\subset[a_1L\,,\,a_2L]$ is the set of those R where $\zeta(R\,,\,\gamma_1)=s$. Denote

$$\zeta_L^{(k)}(\gamma_1)=\frac{1}{L(a_2-a_1)}\sum_{s\geqslant k}\binom{s}{k}l(H_s(\alpha)).$$

Here l as before means the length. In this section we shall prove that

$$\zeta_L^{(k)}(\gamma_1)\to\frac{c^k}{k!}\left(\int f^2(\varphi)d\varphi\right)$$

in P-probability as $L\to\infty$. Let us show that this implies Theorem 2.

Find positive $\varepsilon_j\to 0\,,\,\delta_j\,,\,\sum_j\delta_j<\infty$, and a sequence $L_j\to 0$ such that

$$P\left\{\left|\zeta_{L_j}^{(k)}(\gamma_1)-\frac{c^k}{k!}\left(\int f^2(\varphi)d\varphi\right)^k\right|\geqslant\varepsilon_j\right\}\leqslant\delta_j.$$

In view of the Borel-Cantelli lemma, this means that for P-almost every γ_1 and any k

$$\lim_{j\to\infty} \zeta_{L_j}^{(k)}(\gamma_1) = \frac{c^k}{k!}\left(\int f^2(\varphi)d\varphi\right)^k .$$

It follows from the results of §2 that for P-almost every γ_1 the distribution of the random variable $\zeta(R, \gamma_1)$ as a function of R converges to the Poisson distribution with the parameter $c\int f^2(\varphi)d\varphi$.

As in §3, we take $\alpha = 2\pi/M$ and introduce the random variable

$$\zeta_\alpha(\gamma_1) = \frac{1}{L(a_2 - a_1)} l(H_s(\alpha)).$$

We shall show that

$$\left| \mathsf{E}_P \left(\zeta_\alpha^{(k)}(\gamma_1) - \frac{c^k}{k!}\left(\int f^2(\varphi)d\varphi\right)^k\right)^2\right| \leqslant \gamma_6(\alpha).$$

Obviously, this gives the required result. In fact, in §3 we used the formula

$$\begin{aligned}\xi_L^{(k)}(\gamma_1) = &\sum_{B^{(s_2)},\dots,B^{(s_k)}}^{(e)} l(B^{(s_2)}, \dots, B^{(s_k)}) \\ &\times \sum_{\Delta^{(j_1)},\dots,\Delta^{(j_k)}}^{(\alpha)} N(\Delta^{(j_1)}, \dots, \Delta^{(j_k)}; B^{(s_2)}, \dots, B^{(s_k)}) + \gamma_7(\alpha, \gamma_1), \qquad (12)\end{aligned}$$

where $\gamma_7(\alpha, \gamma_1)$ tends to zero in P-probability as $\alpha \to 0$.

We must estimate the terms $\mathsf{E}_P(\zeta_L^{(k)}(\gamma_1))^2$, $\mathsf{E}_P(\zeta_L^{(k)}(\gamma_1)(\int f^2(\varphi)d\varphi)^k)$, or $\mathsf{E}_P(\int f^2(\varphi)d\varphi)^{2k}$. The most difficult part is the estimation of $\mathsf{E}_P(\zeta_\alpha^{(k)}(\gamma_1))^2$ or, using (12), the estimation of

$$\mathsf{E}_P = \mathsf{E}_P N(\Delta^{(j_1')}, \dots, \Delta^{(j_k')}; B_1^{(s_2')}, \dots, B_2^{(s_k')}) N(\Delta^{(j_1'')}, \dots, \Delta^{(j_k'')}; B_2^{(s_2'')}, \dots, B_2^{(s_k'')})$$

for two different sets $(\Delta^{(j_1')}, \dots, \Delta^{(j_k')}; B^{(s_2')}, \dots, B^{(s_k')})$ and $(\Delta^{(j_1'')}, \dots, \Delta^{(j_k'')}; B_2^{(s_2'')}, \dots, B_2^{(s_k'')})$. We shall discuss only this estimate, the others are done in a similar way.

If $\Delta^{(j_1')}, \dots, \Delta^{(j_k')}, \Delta^{(j_1'')}, \dots, \Delta^{(j_k'')}$ are such that their pairwise distances are greater than α, then the expression for the E_P is obtained in the same way as in §3 and its substitution into $\mathsf{E}_P(\xi^{(k)}(\gamma_1))^2$ gives the main term. We must show that the rest is a remainder term.

Using (13), we come to the estimate of the expectations of the form

$$\mathsf{E}_P \chi_{m^{(1)},\dots,m^{(k)}} \chi_{n^{(1)},\dots,n^{(k)}},$$

$m^{(s)} \in \Delta^{(j_s')}$, $n^{(s)} \in \Delta^{(j_s'')}$, where for at least one pair $\Delta^{(j_s')}$, $\Delta^{(j_t'')}$ we have $\operatorname{dist}(\Delta^{(j_s')}, \Delta^{(j_t'')}) \leqslant \alpha$. Write all Δ in increasing order: $\Delta^{(j_1)}, \dots, \Delta^{(j_{2k})}$. Now we shall make the summation just as at the end of the previous section, fixing $m^{(1)} \in \Delta^{(j_1)}, \dots, m^{(r)} \in \Delta^{(j_r)}$ and summing over $\Delta^{(j_{r+1})}$. When the distance from $\Delta^{(j_{r+1})}$ to the other $\Delta^{(j_s)}$, $s \leqslant r$ is bigger than α, we get a const independent of α. If this distance is less than α, we get $\gamma(\alpha)$, where $\gamma(\alpha) \to 0$ as $\alpha \to 0$. This gives the desired result.

§5. A concluding remark

We do not know any concrete curve for which the statement of Theorem 2 is true. In the simplest cases of algebraic curves the problem is reduced to difficult questions of number theory concerning the estimations of multiple trigonometric sums.

References

1. M.V. Berry and M. Tabor, *Level clustering in the regular spectrum*, Proc. Roy. London Soc. Ser. A **356** (1977), 375–394.

Translated by THE AUTHOR

Comments

The problem which was considered in this paper appeared in the theory of quantum chaos (see [S2]). As is well-known, the theory of quantum chaos studies statistical properties of eigenfunctions and eigenvalues of quantum systems which are limits of classical systems. M. Berry and M. Tabor (see [BT2]) formulated a general hypothesis according to which the spacings between eigenvalues of systems which are integrable in the classical limit satisfy the Poisson limit theorem. In [S2], it was explained that in the two-dimensional case the problem can be reduced to a number-theoretic problem of the distribution of the number of integer points in thin strips. In general, this problem remains open. I considered the case of "random strips" and proved some results concerning the Poissonian behavior. An improved version of the corresponding results can be found in [M] and [Mi].

Many results about the behavior of the number of integer points in thin strips were considered in the works by Bleher, Cheng, Dyson, and Lebowitz (see, e.g., [Bl1, BCDL]).

Let me also mention our joint paper with A. Mazel (see [MS]) where we proved the Poissonian behavior for some number of moments of considered lattice points.

References

[BT1] M. Berry, M. Tabor, Closed orbits and the regular bound spectrum, *Proc. Royal Society*, London, Ser. A. **349** (1976), 101–123.

[BT2] M. Berry, M. Tabor, Level clustering in the regular spectrum, *Proc. Royal Society*, London, Ser. A. **356** (1977), 375–394.

[Bl1] P. Bleher, Trace formula for quantum integrable systems, the lattice-point problem and small divisors, *IMA Vol. Math. Appl.*, **109** (1999), Springer, I.

[BCDL] P. Bleher, Z. Cheng, F. Dyson, J. Lebowitz, Distribution of the error term for the number of lattice points inside a shifted circle, *Comm. Math. Phys.*, **154** (1993), no. 3, 433–469.

[CLS1] N. Chernov, J. Lebowitz, Ya. G. Sinai, Dynamics of a massive piston immersed in an ideal gas, *Russian Math. Surv.*, **57** (2002), no. 6, 1045–1125.

[CLS2] N. Chernov, J. Lebowitz, Ya. G. Sinai, Scaling dynamics of a massive piston in a cube filled with ideal gas: exact results, *J. Stat. Phys.*, **109** (2002), no. 3–4, 529–548.

[E] L. H. Eliasson, Linear quasi-periodic systems-reducibility and almost reducibility, *XIVth International Congress on Mathematical Physics*, (2005), World Sci. Publ., 195–205.

[KMS] D. Kosygin, A. Minasov, Ya. G. Sinai, Statistical properties of the Laplace-Beltrami operator on Liouville surfaces, *Russian Math. Surv.*, **48** (1993), no. 4, 3–130.

[LPS] J. Lebowitz, J. Piasecki, Ya. G. Sinai, Dynamics of an adiabatic piston, *Doklady*, **375** (2000), no. 6, 734–736.

[L1] E. Lieb, Some problems in statistical mechanics that I would like to see solved, *Phys. A*, **263** (1999), 491–499.

[M] P. Major, Poisson law for the number of lattice points in a random strip with finite area, *Prob. Theory Related Fields*, **92**, 1992, pp. 423–464.

[MS] A. E. Mazel, Ya. G. Sinai, A limiting distribution connected with fractional parts of linear forms, in: *Ideas and Methods in Mathematical Analysis Stochastics, and Applications*, Cambridge University Press, Oslo, 1988, pp. 220–229.

[Mi] Nariyuki Minami, On the Poisson limit theorems of Sinai and Major, *Comm. Math. Phys.*, **213** (2000), no. 1, 203–247.

[Mo] J. Moser, An example of a Schrödinger equation with almost periodic potential and nowhere dense spectrum, *Comment. Math. Helv.*, **56** (1981), no. 2, 198–224.

[PS] J. Piasecki, Ya. G. Sinai, A model of non-equilibrium statistical mechanics, *Nato Sci. Ser. E Appl. Sci.***371**, Kluwer Acad. Publ, Dordrecht, 2000.

[1] Ya. G. Sinai, The structure of the spectrum of a difference Schrödinger operator, *Funct. Anal. Appl.*, **19** (1985), no. 1, 42–48.

[S2] Ya. G. Sinai, Mathematical problems in the theory of quantum chaos, *Lecture Notes in Math.*, **1469** (1991), 41–59, Springer Verlag.

Part IV
Mathematical Fluid Dynamics

Invariant Measures for Burgers Equation with Stochastic Forcing

Weinan E; K. Khanin; A. Mazel; Ya. Sinai

The Annals of Mathematics, 2nd Ser., Vol. 151, No. 3. (May, 2000), pp. 877-960.

Y.G. Sinai, *Selecta II: Probability Theory, Statistical Mechanics, Mathematical Physics and Mathematical Fluid Dynamics*, DOI 10.1007/978-1-4419-6205-8_17, © Springer Science+Business Media, LLC 2010

Annals of Mathematics, **151** (2000), 877–960

Invariant measures for Burgers equation with stochastic forcing

By WEINAN E, K. KHANIN, A. MAZEL, and YA. G. SINAI

1. Introduction

In this paper we study the following Burgers equation

$$\frac{\partial u}{\partial t} + \frac{\partial}{\partial x}\left(\frac{u^2}{2}\right) = \varepsilon\frac{\partial^2 u}{\partial x^2} + f(x,t) \tag{1.1}$$

where $f(x,t) = \frac{\partial F}{\partial x}(x,t)$ is a random forcing function, which is periodic in x with period 1, and with white noise in t. The general form for the potentials of such forces is given by:

$$F(x,t) = \sum_{k=1}^{\infty} F_k(x)\dot{B}_k(t) \tag{1.2}$$

where the $\{B_k(t),\, t \in (-\infty,\infty)\}$'s are independent standard Wiener processes defined on a probability space $(\Omega, \mathcal{F}, P)$ and the F_k's are periodic with period 1. We will assume for some $r \geq 3$,

$$f_k(x) = F_k'(x) \in C^r(\mathbb{S}^1),\ ||f_k||_{C^r} \leq \frac{C}{k^2}\,. \tag{1.3}$$

Here $\mathbb{S}^1$ denotes the unit circle, and C, a generic constant. Without loss of generality, we can assume that for all k, $\int_0^1 F_k(x)dx = 0$. We will denote the elements in the probability space Ω by $\omega = (\dot{B}_1(\cdot), \dot{B}_2(\cdot), \ldots)$. Except in Section 8 where we study the convergence as $\varepsilon \to 0$, we will restrict our attention to the case when $\varepsilon = 0$:

$$\frac{\partial u}{\partial t} + \frac{\partial}{\partial x}\left(\frac{u^2}{2}\right) = \frac{\partial F}{\partial x}(x,t)\,. \tag{1.4}$$

Besides establishing existence and uniqueness of an invariant measure for the Markov process corresponding to (1.4), we will also give a detailed description of the structure and regularity properties for the solutions that live on the support of this measure.

The randomly forced Burgers equation (1.1) is a prototype for a very wide range of problems in nonequilibrium statistical physics where strong nonlinear effects are present. It arises in studies of various one-dimensional systems such as vortex lines in superconductors [BFGLV], charge density waves [F], directed polymers [KS], etc. (1.1) with its high-dimensional analog is the differentiated version of the well-known KPZ equation describing, among other things, kinetic roughening of growing surfaces [KS]. Most recently, (1.1) has received a great deal of interest as the testing ground for field-theoretic techniques in hydrodynamics [CY], [Pol], [GM], [BFKL], [GK2]. In fact, we expect that the randomly forced Burgers equation will play no lesser a role in the understanding of nonlinear non-equilibrium phenomena than that of the Burgers equation in the understanding of nonlinear waves.

Before proceeding further let us give an indication why an invariant measure is expected for (1.1) even when $\varepsilon = 0$. Since energy is continuously supplied to the system, a dissipation mechanism has to be present to maintain an invariant distribution. In the case when $\varepsilon > 0$, the viscous term provides the necessary energy dissipation, and the existence of an invariant measure has already been established in [S1], [S2]. When $\varepsilon = 0$, it is well-known that discontinuities are generally present in solutions of (1.4) in the form of shock waves [La]. These weak solutions are limits of solutions of (1.1) as $\varepsilon \to 0$, and satisfy an additional entropy condition: $u(x+,t) \le u(x-,t)$, for all (x,t). It turns out that this entropy condition enforces sufficient energy dissipation (in the shocks) for maintaining an invariant measure. We will always restrict our attention to weak solutions of (1.4) that satisfy the entropy condition.

The starting point of our analysis is the following variational characterization of solutions of (1.4) satisfying the entropy condition [Li]:

For any Lipschitz continuous curve ξ: $[t_1, t_2] \to \mathbb{S}^1$, define its action

$$(1.5) \qquad \mathcal{A}_{t_1,t_2}(\xi) = \int_{t_1}^{t_2} \left\{ \frac{1}{2} \dot{\xi}(s)^2 ds + \sum_k F_k(\xi(s)) dB_k(s) \right\} .$$

Then for $t > \tau$, solutions of (1.4) satisfy

$$(1.6) \qquad u(x,t) = \frac{\partial}{\partial x} \inf_{\xi(t)=x} \left\{ \mathcal{A}_{\tau,t}(\xi) + \int_0^{\xi(\tau)} u(z,\tau) dz \right\}$$

where the infimum is taken over all Lipschitz continuous curves on $[\tau, t]$ satisfying $\xi(t) = x$.

Here and below, we avoid in the notation explicit indication of the dependence on realization of the random force when there is no danger of confusion. Otherwise we indicate such dependence by a super- or subscript ω. In addition, we will denote by θ^τ the shift operator on Ω with increment τ: $\theta^\tau \omega(t) = \omega(t + \tau)$, and by $S^\tau_\omega w$ the solution of (1.1) at time $t = \tau$ when the realization of the force is ω and the initial datum at time $t = 0$ is w. We will denote by D the Skorohod space on $\mathbb{S}^1$ (see [B], [Pa]) consisting of functions having only discontinuities of the first kind; i.e., both left and right limits exist at each point, but they may not be equal.

It is easy to see that the dynamics of (1.4) conserves the quantity $\int_0^1 u(x,t)dx$. Therefore to look for unique invariant measure, we must restrict attention to the subspace

$$D_c = \{u \in D, \int_0^1 u(x)dx = c\}.$$

In this paper we will restrict most of our attention to the case when $c = 0$ but it is relatively easy to see that all of our results continue to hold for the case when $c \neq 0$. We will come back to this point at the end of this section. At the end of Section 3, we will outline the necessary changes for the case when $c \neq 0$.

Our basic strategy for the construction of an invariant measure is to show that the following "*one force, one solution*" principle holds for (1.4): For almost all ω, there exists a unique solution of (1.4), u^ω, defined on the time interval $(-\infty, +\infty)$. In other words, the random attractor consists of a single trajectory almost surely. Furthermore, if we denote the mapping between ω and u^ω by Φ:

$$u^\omega = \Phi(\omega), \tag{1.7}$$

then Φ is invariant in the sense that

$$\Phi(\theta^\tau \omega) = S_\omega^\tau \Phi(\omega). \tag{1.8}$$

It is easy to see that if such a map exists, then the distribution of $\Phi_0 : \Omega \to D$:

$$\Phi_0(\omega)(x) = u^\omega(x, 0),$$

is an invariant measure for (1.4). Moreover, this invariant measure is necessarily unique.

This approach of constructing the invariant measure has the advantage that many statistical properties of the forces, such as ergodicity and mixing, carry over automatically to the invariant measure. More importantly, it facilitates the study of solutions supported by the invariant measure, i.e. the associated stationary Markov process. This study will be carried out in the second half of the present paper.

The construction of u^ω will be accomplished in Section 3. The variational principle (1.6) allows us to restrict our attention to $t = 0$.

Our construction of Φ relies heavily on the notion of one-sided minimizer. A curve ξ $(-\infty, 0] \to \mathbb{S}^1$ is called a one-sided minimizer if it minimizes the action (1.5) with respect to all compact perturbations. More precisely, we introduce:

Definition 1.1. A piecewise C^1-curve $\{\xi(t),\ t \leq 0\}$ is a *one-sided minimizer* if for any Lipschitz continuous $\tilde{\xi}$ defined on $(-\infty, 0]$ such that $\tilde{\xi}(0) = \xi(0)$ and $\tilde{\xi} = \xi$ on $(-\infty, \tau]$ for some $\tau < 0$,

$$\mathcal{A}_{s,0}(\xi) \leq \mathcal{A}_{s,0}(\tilde{\xi})$$

for all $s \leq \tau$.

It is important to emphasize that the curves are viewed on the cylinder $\mathbb{R}^1 \times \mathbb{S}^1$. Similarly, we define one-sided minimizers on $(-\infty, t]$, for $t \in \mathbb{R}^1$.

The interest of this notion lies in the fact that we are considering an infinite interval. It is closely related to the notion of geodesics of type A introduced and studied by Morse [Mo] and Hedlund [H] and the notion of global minimal orbits in Aubry-Mather theory [A], [M]. In the geometric context, it has been studied by Bangert (see [Ba]) as geodesic rays. A somewhat surprising result is that, in the random case, one-sided minimizers are almost unique. More precisely, we have:

THEOREM 1.1. *With probability* 1, *except for a countable set of x values, there exists a unique one-sided minimizer ξ, such that $\xi(0) = x$.*

This theorem states that one-sided minimizers are intrinsic objects to (x, ω). It allows us to construct $\Phi_0(\omega)$ by patching together all one-sided minimizers:

$$\Phi_0\{\omega(\tau),\ \tau < 0\}(x) = u^\omega(x, 0) = \dot{\xi}(0) \tag{1.9}$$

where ξ is the unique one-sided minimizer such that $\xi(0) = x$. In (1.9) we emphasized the fact that Φ_0 depends only on the realization of ω in the past $\tau < 0$. Now (1.9) defines $u^\omega(\cdot, 0)$ except on a countable subset of $\mathbb{S}^1$. Similarly we construct $u^\omega(\cdot, t)$ for other values of $t \in \mathbb{R}^1$. It is easy to verify that this construction is self-consistent and satisfies the invariance condition (1.8), as a consequence of the variational principle (1.6).

The existence part of Theorem 1.1 is proved by studying limits of minimizers on finite intervals $[-k, 0]$ as $k \to +\infty$. The uniqueness part of Theorem 1.1 is proved by studying the intersection properties of one-sided minimizers. The absence of two intersections of two different minimizers is a general fact in calculus of variations. However, we will prove the absence of even one intersection which is a consequence of randomness.

We are now ready to define formally the invariant measure. There are two alternative approaches. Either we can define the invariant measure on the product space $(\Omega \times D_0, \mathcal{F} \times \mathcal{D})$ with a skew-product structure, or we can define it as an invariant distribution of the Markov process on $(D_0, \mathcal{D})$ defined by (1.4), where $\mathcal{D}$ is the σ-algebra generated by Borel sets on D_0. The skew-product structure is best suited for the exploration of the "one force, one solution" principle.

Definition 1.2. A measure $\mu(du, d\omega)$ on $(\Omega \times D_0, \mathcal{F} \times \mathcal{D})$ is called an *invariant measure* if it is preserved under the skew-product transformation $F^t\colon \Omega \times D_0 \to \Omega \times D_0$,

$$F^t(\omega, u_0) = (\theta^t \omega, S^t_\omega u_0), \tag{1.10}$$

and if its projection to Ω is equal to P.

Alternatively we may consider a homogeneous Markov process on D_0 with the transition probability

$$P_t(u, A) = \int_\Omega \chi_A(u, \omega) P(d\omega) \tag{1.11}$$

where $u \in D_0$, $A \in \mathcal{D}$, and

$$\chi_A(u, \omega) = \begin{cases} 1 & \text{if } S^t_\omega u \in A \\ 0 & \text{otherwise}. \end{cases} \tag{1.12}$$

Definition 1.3. An *invariant measure* $\kappa(du)$ *of the Markov process* (1.11) is a measure on $(D_0, \mathcal{D})$ satisfying

$$\kappa(A) = \int_{D_0} P_t(u, A) \kappa(du) \tag{1.13}$$

for any Borel set $A \in \mathcal{D}$ and any $t > 0$.

Let $\delta^\omega(du)$ be the atomic measure on $(D_0, \mathcal{D})$ concentrated at $\Phi_0(\omega) = u^\omega(\cdot, 0)$, and let $\mu(du, d\omega) = \delta^\omega(du) P(d\omega)$; we then have:

THEOREM 1.2. *If μ is an invariant measure for the skew-product transformation F^t, it is the unique invariant measure on $(\Omega \times D_0, \mathcal{F} \times \mathcal{D})$ with the given projection $P(d\omega)$ on $(\Omega, \mathcal{F})$.*

THEOREM 1.3. *For the Markov process* (1.11), $\kappa(du) = \int_\Omega \mu(du, d\omega)$ *is the unique invariant measure.*

The uniqueness result is closely related to the uniqueness of one-sided minimizers and reflects the lack of memory in the dynamics of (1.4): Consider solutions of (1.4) with initial data $u(x, -T) = u_0(x)$. Then for almost all $\omega \in \Omega$ and any $t \in \mathbb{R}^1$, $\lim_{T \to +\infty} u(\cdot, t)$ exists and does not depend on u_0. The key step in the proof of uniqueness is to prove a strengthened version of this statement.

In the second half of this paper, we study in detail the properties of solutions supported by the invariant measure. The central object is the two-sided minimizer which is defined similarly to the one-sided minimizer but for the interval $(-\infty, +\infty) = \mathbb{R}^1$. Under very weak nondegeneracy conditions, we prove that almost surely, the two-sided minimizer exists and is unique. In Section 6, we show that the two-sided minimizer is a hyperbolic trajectory of the dynamical system corresponding to the characteristics of (1.4):

$$\frac{dx}{dt} = u, \quad \frac{du}{dt} = \frac{\partial F}{\partial x}(x, t).$$

We can therefore consider the stable and unstable manifolds of the two-sided minimizer using Pesin theory [Pes]. As a consequence, we show:

THEOREM 1.4. *With probability* 1, *the graph of* $\Phi_0(\omega)$ *is a subset of the unstable manifold* (*at* $t = 0$) *of the two-sided minimizer.*

We use this statement to show that, almost surely, $u^\varepsilon(\cdot, 0)$ is piecewise smooth and has a finite number of discontinuities. This is done in Section 7.

Dual to the two-sided minimizer is an object called the main shock which is a continuous shock curve $x^\omega\colon \mathbb{R}^1 \to \mathbb{S}^1$ defined on the whole line $-\infty < t < \infty$. The main shock is also unique. Roughly speaking, the main shock plays the role of an attractor for the one-sided minimizers while the two-sided minimizer plays the role of a repeller.

Finally in Section 8, we show that as $\varepsilon \to 0$, the invariant measures of (1.1) constructed in [S1], [S2] converge to the invariant measure of (1.4).

The results of this paper have been used to analyze the asymptotic behavior of tail probabilities for the gradients and increments of u (see [EKMS]). It also provides the starting point for the work in [EV] on statistical theory of the solutions. These results are of direct interest to physicists since they can be compared with predictions based on field-theoretic methods (see [Pol], [GM], [GK2], [CY]).

Our theory is closely related to the Aubry-Mather theory [A], [M] which is concerned with special invariant sets of twist maps obtained from minimizing the action

$$\tfrac{1}{2}\sum_i (x_i - x_{i-1} - \gamma)^2 + \lambda \sum_i V(x_i) \tag{1.14}$$

where γ is a parameter and V is a periodic function. The continuous version of (1.14) is

$$\int \{\tfrac{1}{2}(\dot{\xi}(t) - a)^2 + F(\xi(t), t)\} dt \tag{1.15}$$

where F is a periodic function in x and t [Mo]. The main result of the Aubry-Mather theory is the existence of invariant sets with arbitrary rotation number, with a suitable a. Such invariant sets are made from the velocities of the two-sided minimizers defined earlier. It can be proved that such an invariant set lies on the graph of the periodic solutions of (1.4) [E], [JKM], [So]. In this connection, the results of this paper apply to the random version of (1.15):

$$\int \{\tfrac{1}{2}(\dot{\xi}(t) - a)^2 dt + \sum_k F_k(\xi(t))dB_k(t)\}\,. \tag{1.16}$$

Although only $a = 0$ is considered in this paper, extension to arbitrary a is straightforward and the results are basically the same for different values of a. This is because, over a large interval of duration T, the contribution of the kinetic energy is of order T, and the contribution from the potential is typically of order $\sqrt{T}$ for the random case but of order T for the periodic case. This gives rise to subtle balances between kinetic and potential energies in the latter case. Consequently the conclusions for the random case become much simpler. While in the deterministic case, there are usually many different two-sided minimizers in the invariant set and they are not necessarily hyperbolic, there is only one two-sided minimizer in the random case and it is always hyperbolic.

The value of a is closely related to the value of c discussed earlier. In the setting of Aubry-Mather theory, a is the average speed of the global minimizers and is related to c through the Legendre transform of the homogenized Hamiltonian. In the random case, $a = c$ for the reason given in the last paragraph.

2. The variational principle

Let us first define in the probabilistic context the notion of weak solutions of (1.4) with (deterministic) initial data $u(x, t_0) = u_0(x)$. We will always assume $u_0 \in L^\infty(\mathbb{S}^1)$.

Definition 2.1. Let u_ω be a random field parametrized by $(x,t) \in \mathbb{S}^1 \times [t_0, +\infty)$ such that for almost all $\omega \in \Omega$, $u_\omega(\,\cdot\,, t) \in D$ for all $t \in (t_0, \infty)$. Then u_ω is a *weak solution* of (1.4) if:

(i) For all $t > t_0$, $u_\omega(\,\cdot\,, t)$ is measurable with respect to the σ-algebra $\mathcal{F}_{t_0}^t$ generated by all $\dot{B}_k(s)$, $t_0 \le s \le t$.

(ii) $u_\omega \in L^1_{\text{loc}}(\mathbb{S}^1 \times [t_0, \infty))$ almost surely.

(iii) With probability 1, the following holds for all $\varphi \in C^2(\mathbb{S}^1 \times \mathbb{R}^1)$ with compact support:

$$\int_0^1 u_0(x)\varphi(x,t_0)dx + \int_{t_0}^\infty \int_0^1 \frac{\partial \varphi}{\partial t} u_\omega(x,t)\,dx\,dt + \frac{1}{2}\int_{t_0}^\infty \int_0^1 \frac{\partial \varphi}{\partial x} u_\omega^2(x,t)dx\,dt$$
$$= -\int_0^1 \sum_k \Big\{ F_k(x) \int_{t_0}^\infty \frac{\partial^2 \varphi}{\partial x \partial t}(x,t)(B_k(t) - B_k(t_0))dt \Big\} dx\,.$$

Also, u_ω is an entropy-weak solution if, for almost all $\omega \in \Omega$,

$$u_\omega(x+,t) \le u_\omega(x-,t)$$

for all $(x,t) \in \mathbb{S}^1 \times (t_0,\infty)$.

Our analysis is based on a variational principle characterizing entropy weak solutions of (1.4). To formulate this variational principle, we redefine the action in order to avoid using stochastic integrals. Given $\omega \in \Omega$, for any Lipschitz continuous curve ξ: $[t_1,t_2] \to \mathbb{S}^1$, define

(2.1)
$$\begin{aligned} \mathcal{A}_{t_1,t_2}(\xi) = \int_{t_1}^{t_2} \left\{ \tfrac{1}{2}\dot\xi(s)^2 - \sum_k f_k(\xi(s))\dot\xi(s)(B_k(s) - B_k(t_1)) \right\} ds \\ + \sum_k F_k(\xi(t_2))(B_k(t_2) - B_k(t_1)). \end{aligned}$$

(2.1) can be formally obtained from (1.5) with an integration by parts. It has the advantage that the integral in (2.1) can be understood in the Lebesgue sense instead of the Ito sense, for example.

LEMMA 2.1. *Let $u_0(x) \in D$. For almost all $\omega \in \Omega$, there exists a unique weak solution of* (1.4) *satisfying the entropy condition, such that $u(x,t_0) = u_0(x)$. For $t \ge t_0$, this solution is given by*:

$$u(x,t) = \frac{\partial}{\partial x} \inf_{\xi(t)=x} \left\{ \mathcal{A}_{t_0,t}(\xi) + \int_0^{\xi(t_0)} u_0(z)dz \right\} \tag{2.2}$$

and $u(\,\cdot\,,t) \in D$.

This type of result was obtained for the first time in [Ho], [La] and [Ol] for scalar conservation laws. The generalization to multi-dimensional Hamilton-Jacobi equations is given in [Li]. Extension to the random case is straightforward, but requires some additional arguments which we present in Appendix A.

Any action minimizer γ satisfies the following Euler-Lagrange equation:

$$\dot\gamma(s) = v(s), \quad dv(s) = \sum_{k=1}^{\infty} f_k(\gamma(s))dB_k(s). \tag{2.3}$$

Under the assumptions in (1.3), the stochastic differential equation (2.3) has a unique solution starting at any point x. It is nothing but the equation of characteristics for (1.4). Therefore the variational principle (2.2) can be viewed as the generalization of the method of characteristics to weak solutions. In general, characteristics intersect each other forward in time, resulting in the formation of shocks. Given initial data at time t_0: $u(x,t_0) = u_0(x)$, to find the solution at (x,t), consider all characteristics γ that arrive at x at time t

and choose among them the ones that minimize $\mathcal{A}_{t_0,t}(\gamma) + \int_0^{\gamma(t_0)} u_0(z)dz$. If such a minimizing characteristic is unique, say $\gamma(\cdot)$, then $u(x,t) = \dot{\gamma}(t)$. In the case when there are several such minimizing characteristics, $\{\gamma_\alpha(\cdot)\}$, the solution $u(\cdot,t)$ has a jump discontinuity at x, with $u(x-,t) = \sup_\alpha \dot{\gamma}_\alpha(t)$ and $u(x+,t) = \inf_\alpha \dot{\gamma}_\alpha(t)$.

This characterization is closely related to the notion of backward characteristics developed systematically by Dafermos (see [D]).

Our task of finding the invariant measure for (1.4) is different from what is usually asked about (1.4). Instead of solving (1.4) with given initial data, we look for a special distribution of the initial data that has the invariance property. Translated into the language of the variational principle, we will look for special minimizers or characteristics.

3. One-sided minimizers

A fundamental object needed for the construction of invariant measures for (1.4) is the one-sided minimizer. These are curves that minimize the action (2.1) over the semi-infinite interval $(-\infty, t]$.

In the following we will study the existence and intersection properties of one-sided minimizers. Before doing this, we formulate some basic facts concerning the effect on the action as a result of reconnecting and smoothing of curves.

Fact 1. Let ξ_1, ξ_2 be two C^1-curves on $[t_1, t_2]$ with values in $\mathbb{S}^1$. Then one can find a reconnection of the two curves, ξ_r, such that $\xi_r(t_1) = \xi_1(t_1), \xi_r(t_2) = \xi_2(t_2)$ and

$$(3.1) \qquad |\mathcal{A}^\omega_{t_1,t_2}(\xi_1) - \mathcal{A}^\omega_{t_1,t_2}(\xi_r)|, \quad |\mathcal{A}^\omega_{t_1,t_2}(\xi_2) - \mathcal{A}^\omega_{t_1,t_2}(\xi_r)|$$
$$\leq C\{\omega(\tau), \tau \in [t_1,t_2]\} \|\xi_1(t) - \xi_2(t)\|_{C^1} (1 + |t_2 - t_1|) \Big(1 + \max_{t\in[t_1,t_2]} (|\dot{\xi}_1(t)|,\ |\dot{\xi}_2(t)|)\Big).$$

Here and in the following we will use norms such as $\|\cdot\|_{C^1}$ for functions that take values on $\mathbb{S}^1$. These will always be understood as the norms of a particular represention of the functions on $\mathbb{R}^1$. The choice of the representation will either be immaterial or obvious from the context.

Fact 2. If ξ is a curve containing corners, i.e. jump discontinuities of $\dot{\xi}$, smoothing out a corner in a sufficiently small neighborhood strictly decreases the action.

Both facts are classical and are more or less obvious.

The following lemma provides a bound on the velocities of minimizers over a large enough time interval.

LEMMA 3.1. *For almost all* $\omega \in \Omega$ *and any* $t \in (-\infty, \infty)$ *there exist random constants* $T(\omega, t)$ *and* $C(\omega, t)$ *such that if* γ *minimizes* $\mathcal{A}^{\omega}_{t_1,t}(\cdot)$ *and* $t_1 < t - T(\omega, t)$, *then*

$$|\dot{\gamma}(t)| \le C(\omega, t). \tag{3.2}$$

Proof. Denote

$$C_1(\omega, t) = \tfrac{1}{4} + \max_{t-1\le s\le t} \sum_{k=1}^{\infty} \|F_k(x)\|_{C^2} |B_k(s) - B_k(t)| \tag{3.3}$$

and set $C(\omega, t) = 20C_1(\omega, t)$, $T(\omega, t) = (4C_1(\omega, t))^{-1}$. Clearly $T(\omega, t) < 1$. If $|\dot{\gamma}(t)| \le 16C_1$ then (3.2) is true with $C = 16C_1$.

If $|\dot{\gamma}(t)| > 16C_1$, we first show that the velocity $\dot{\gamma}(s)$ cannot be too large inside the interval $[t - T, t]$. Denote

$$v_0 = |\dot{\gamma}(t)| \qquad \text{and} \qquad v = \max_{t-T\le s\le t} |\dot{\gamma}(s)|\,. \tag{3.4}$$

Integrating by parts from (2.3), one gets for $s \in [t - T, t]$

$$\begin{aligned} |\dot{\gamma}(s)| &= \left|\dot{\gamma}(t) - \int_s^t \sum_{k=1}^{\infty} f_k(\gamma(r))dB_k(r)\right| \\ &\le v_0 + \left|\sum_{k=1}^{\infty} f_k(\gamma(s))(B_k(s) - B_k(t))\right| \\ &\quad + \left|\int_s^t \dot{\gamma}(r) \sum_{k=1}^{\infty} f'_k(\gamma(r))(B_k(r) - B_k(t))dr\right| \\ &\le v_0 + C_1 + C_1 vT \\ &= v_0 + C_1 + \frac{1}{4} v\,. \end{aligned} \tag{3.5}$$

Hence

$$v \le v_0 + C_1 + \frac{v}{4}, \tag{3.6}$$

implying

$$v \le \frac{4}{3}(v_0 + C_1) \le \frac{3}{2} v_0 \tag{3.7}$$

since $v_0 > 16C_1$.

Next we check that $|\dot\gamma(s)|$ remains of order v_0, i.e. sufficiently large, for $s \in [t-T, t]$. As before, we have

$$
\begin{aligned}
(3.8) \qquad |\dot\gamma(s) - \dot\gamma(t)| &= \left| \int_s^t \sum_{k=1}^{\infty} f_k(\gamma(r)) dB_k(r) \right| \\
&\le C_1 + C_1 v T \\
&\le C_1 + \frac{3}{8} v_0 \\
&\le \frac{1}{2} v_0 \,.
\end{aligned}
$$

The last step is to show that (3.8) contradicts the minimization property of $\gamma(s)$ if $v_0 > 20C_1$. Consider a straight line $\gamma_1(s)$ joining $\gamma(t)$ and $\gamma(t-T)$. Clearly $|\gamma(t) - \gamma(t-T)| \le 1$ since $\gamma(t), \gamma(t-T) \in \mathbb{S}^1$. Then

(3.9)

$$
\mathcal{A}^{\omega}_{t-T,t}(\gamma_1) \le \frac{1}{2}\left(\frac{\gamma(t) - \gamma(t-T)}{T}\right)^2 T + C_1 + C_1 \left|\frac{\gamma(t) - \gamma(t-T)}{T}\right| T \le \frac{1}{2T} + 2C_1
$$

while

$$
(3.10) \qquad \mathcal{A}^{\omega}_{t-T,t}(\gamma) \ge \frac{1}{2}\left(\frac{v_0}{2}\right)^2 T - C_1 - \frac{3}{2} v_0 C_1 T \,.
$$

It is easy to see that $\frac{1}{2}\frac{v_0^2}{4} T - \frac{3}{2} v_0 C_1 T > \frac{1}{2T} + 3C_1$ for $v_0 > 20C_1$; i.e.,

$$
(3.11) \qquad \mathcal{A}^{\omega}_{t-T,t}(\gamma_1) < \mathcal{A}^{\omega}_{t-T,t}(\gamma) \,.
$$

This contradicts the minimization property of γ. Hence $v_0 \le 20C_1$. □

Now we are ready to prove the existence of one-sided minimizers that arrive at any given point $x \in \mathbb{S}^1$.

THEOREM 3.1. *With probablity* 1, *the following holds. For any* $(x,t) \in \mathbb{S}^1 \times \mathbb{R}^1$, *there exists at least one one-sided minimizer* $\gamma \in C^1(-\infty, t]$, *such that* $\gamma(t) = x$.

Proof. Given $\omega \in \Omega$, fix $(x,t) \in \mathbb{S}^1 \times \mathbb{R}^1$. Consider a family of minimizers $\{\gamma_\tau\}$ for $\tau < t - T(\omega, t)$, where γ_τ minimizes $\mathcal{A}^{\omega}_{\tau,t}(\xi)$ subject to the constraint that $\xi(t) = x$, $\xi(\tau) \in \mathbb{S}^1$. From Lemma 3.1, we know that $\{\dot\gamma_\tau(t)\}$ is uniformly bounded in τ. Therefore, there exists a subsequence $\{\tau_j\}$, $\tau_j \to -\infty$, and $v \in \mathbb{R}^1$, such that

$$
\lim_{\tau_j \to -\infty} \dot\gamma_{\tau_j}(t) = v \,.
$$

Furthermore, if we define γ to be a solution of (2.3) on $(-\infty, t]$ such that $\gamma(t) = x$, $\dot\gamma(t) = v$, then γ_{τ_j} converges to γ uniformly, together with their derivatives, on compact subsets of $(-\infty, t]$. We will show that γ is a one-sided minimizer.

Assume that there exists a compact perturbation $\gamma_1 \in C^1(-\infty, t]$, of γ such that $\gamma_1(t) = x$, support $(\gamma_1 - \gamma) \subset [t_2, t_3]$, and

$$\mathcal{A}^{\omega}_{t_2,t_3}(\gamma) - \mathcal{A}^{\omega}_{t_2,t_3}(\gamma_1) = \varepsilon > 0 .$$

Let j be sufficiently large such that $\tau_j \leq t_2$ and

$$|\mathcal{A}^{\omega}_{t_2,t}(\gamma) - \mathcal{A}^{\omega}_{t_2,t}(\gamma_{\tau_j})| \leq \frac{\varepsilon}{3} \tag{3.12}$$

and

$$\|\gamma(s) - \gamma_{\tau_j}(s)\|_{C^1[t_2-1,t_2]} \leq \delta \tag{3.13}$$

(δ will be chosen later). Define a new curve γ_2 by

$$\gamma_2(s) = \begin{cases} \gamma_{\tau_j}(s), & \text{for } s \in [\tau_j, t_2 - 1]; \\ \gamma_r(s), & \text{for } s \in [t_2 - 1, t_2]; \\ \gamma_1(s), & \text{for } s \in [t_2, t], \end{cases} \tag{3.14}$$

where γ_r is the reconnecting curve described in Fact 1. We have

$$\begin{aligned} \mathcal{A}^{\omega}_{\tau_j,t}(\gamma_{\tau_j}) - \mathcal{A}^{\omega}_{\tau_j,t}(\gamma_2) &= \mathcal{A}^{\omega}_{t_2,t}(\gamma_{\tau_j}) - \mathcal{A}^{\omega}_{t_2,t}(\gamma) \\ &\quad + \mathcal{A}^{\omega}_{t_2,t}(\gamma) - \mathcal{A}^{\omega}_{t_2,t}(\gamma_1) \\ &\quad + \mathcal{A}^{\omega}_{t_2-1,t_2}(\gamma_{\tau_j}) - \mathcal{A}^{\omega}_{t_2-1,t_2}(\gamma_2) \\ &\geq -\frac{\varepsilon}{3} + \varepsilon - C\delta \\ &\geq \frac{\varepsilon}{3}, \end{aligned} \tag{3.15}$$

if δ is small enough. Here the constant C depends only on ω and γ_1. This contradicts the minimization property of $\gamma_{\tau_j}(s)$. □

Now we study the intersection properties of one-sided minimizers. We use $C^1_x(-\infty, t]$ to denote the set of C^1 curves γ on $(-\infty, t]$ such that $\gamma(t) = x$. We start with a general fact for minimizers (see [A], [M]).

LEMMA 3.2. *Two different one-sided minimizers $\gamma_1 \in C^1(-\infty, t_1]$ and $\gamma_2 \in C^1(-\infty, t_2]$ cannot intersect each other more than once.*

In other words, if two one-sided minimizers intersect more than once, they must coincide on their common interval of definition.

Proof. Suppose that γ_1 and γ_2 intersect each other twice at times t_3 and t_4, with $t_4 > t_3$. Assume without loss of generality

$$\mathcal{A}^{\omega}_{t_3,t_4}(\gamma_1) \leq \mathcal{A}^{\omega}_{t_3,t_4}(\gamma_2) . \tag{3.16}$$

Then for the curve

$$\gamma_3(s) = \begin{cases} \gamma_2(s), & \text{for } s \in (-\infty, t_3] \cup [t_4, t_2]; \\ \gamma_1(s), & \text{for } s \in [t_3, t_4], \end{cases} \tag{3.17}$$

one has

$$\mathcal{A}^{\omega}_{t_3,t_4}(\gamma_3) \le \mathcal{A}^{\omega}_{t_3,t_4}(\gamma_2)\,, \tag{3.18}$$

where γ_3 has two corners at t_3 and t_4. Smoothing out these corners, we end up with a curve $\gamma^* \in C^1(-\infty, t_2]$ for which

$$\mathcal{A}^{\omega}_{t_3-\tau,t_2}(\gamma^*) - \mathcal{A}^{\omega}_{t_3-\tau,t_2}(\gamma_2) < 0 \tag{3.19}$$

for some $\tau > 0$. This contradicts the assumption that $\gamma_2(s)$ is a one-sided minimizer. □

Exploiting the random origin of the force f, we can prove a result which is much stronger than Lemma 3.2.

THEOREM 3.2. *The following holds for almost all ω. Let γ_1, γ_2 be two distinct one-sided minimizers on the intervals $(-\infty, t_1]$ and $(-\infty, t_2]$, respectively. Assume that they intersect at the point (x,t). Then $t_1 = t_2 = t$, and $\gamma_1(t_1) = \gamma_2(t_2) = x$.*

In other words, two one-sided minimizers do not intersect except for the following situation: they both come to the point (x,t), having no intersections before and they both are terminated at that point as minimizers. Of course they can be continued beyond time t as the solution of SDE (2.3) but they are no longer one-sided minimizers.

The proof of Theorem 3.2 resembles that of Lemma 3.2 with an additional observation that, because of the randomness of f, two minimizers always have an "effective intersection at $t = -\infty$." The precise formulation of this statement is given by:

LEMMA 3.3. *With probability 1, for any $\varepsilon > 0$ and any two one-sided minimizers $\gamma_1 \in C^1(-\infty, t_1]$ and $\gamma_2 \in C^1(-\infty, t_2]$, there exist a constant $T = T(\varepsilon)$ and an infinite sequence $t_n(\omega, \varepsilon) \to -\infty$ such that*

$$\begin{aligned} &|\mathcal{A}^{\omega}_{t_n-T,t_n}(\gamma_1) - \mathcal{A}^{\omega}_{t_n-T,t_n}(\gamma_{1,2})|, \quad |\mathcal{A}^{\omega}_{t_n-T,t_n}(\gamma_2) - \mathcal{A}^{\omega}_{t_n-T,t_n}(\gamma_{1,2})|, \\ &|\mathcal{A}^{\omega}_{t_n-T,t_n}(\gamma_1) - \mathcal{A}^{\omega}_{t_n-T,t_n}(\gamma_{2,1})|, \quad |\mathcal{A}^{\omega}_{t_n-T,t_n}(\gamma_2) - \mathcal{A}^{\omega}_{t_n-T,t_n}(\gamma_{2,1})| < \varepsilon\,, \end{aligned} \tag{3.20}$$

where $\gamma_{1,2}$ is the reconnecting curve defined in Fact 1 with

$$\gamma_{1,2}(t_n - T) = \gamma_1(t_n - T), \gamma_{1,2}(t_n) = \gamma_2(t_n),$$

and $\gamma_{2,1}$ is the reconnecting curve satisfying

$$\gamma_{2,1}(t_n - T) = \gamma_2(t_n - T), \gamma_{2,1}(t_n) = \gamma_1(t_n).$$

Proof. Fix T sufficiently large. With probability 1, there exists a sequence $t_n(\omega, \varepsilon) \to -\infty$ such that

$$\max_{s \in \bigcup_n [t_n - T, t_n]} \sum_{k=1}^{\infty} \|F_k(x)\|_{C^2} |B_k(s) - B_k(t_n)| \leq C_1 = \frac{1}{4T}. \tag{3.21}$$

Repeating the proof of Lemma 3.1, one can check that for any n

$$\max_{t_n - T \leq s \leq t_n} (|\dot{\gamma}_1(s)|, |\dot{\gamma}_2(s)|) \leq \frac{4}{3}(20C_1 + C_1) = \frac{7}{T}. \tag{3.22}$$

Using (3.22), we can choose $\gamma_{1,2}$, $\gamma_{2,1}$ such that

$$\max_{t_n - T \leq s \leq t_n} (|\dot{\gamma}_{1,2}(s)|, |\dot{\gamma}_{2,1}(s)|) \leq \frac{7}{T} + \frac{1}{T} = \frac{8}{T}. \tag{3.23}$$

We then have

(3.24)

$$\begin{aligned}
&|\mathcal{A}^{\omega}_{t_n - T, t_n}(\gamma_1) - \mathcal{A}^{\omega}_{t_n - T, t_n}(\gamma_{1,2})| \\
&\quad \leq \left| \sum_{k=1}^{\infty} (F_k(\gamma_1(t_n)) - F_k(\gamma_{1,2}(t_n)))(B_k(t_n) - B_k(t_n - T)) \right| \\
&\qquad + \int_{t_n - T}^{t_n} \left| \left(\frac{1}{2}\dot{\gamma}_1(t)^2 - \frac{1}{2}\dot{\gamma}_{1,2}(t)^2 \right) \right. \\
&\qquad - \sum_{k=1}^{\infty} (B_k(t) - B_k(t_n - T)) \Big(f_k(\gamma_1(t))(\dot{\gamma}_1(t) - \dot{\gamma}_{1,2}(t)) \\
&\qquad \left. + (f_k(\gamma_1(t)) - f_k(\gamma_{1,2}(t)))\dot{\gamma}_{1,2}(t) \Big) \right| dt \\
&\quad \leq \frac{1}{4T} + T\left(\frac{1}{2}\left(\frac{7}{T}\right)^2 + \frac{1}{2}\left(\frac{8}{T}\right)^2 + C_1\left(\frac{7}{T} + \frac{8}{T}\right) + C_1 \frac{8}{T} \right) \\
&\quad = \frac{125}{2T} \\
&\quad \leq \varepsilon,
\end{aligned}$$

if $T \geq \frac{125}{2\varepsilon}$. Similarly, one proves other inequalities in (3.20). □

Proof of Theorem 3.2. We will use τ to denote a sufficiently large negative number. Suppose that γ_1 and γ_2 intersect each other at time $t < \max(t_1, t_2)$ and for definiteness let $t_1 > t$. Then the curve

$$\gamma_3(s) = \begin{cases} \gamma_2(s), & \text{for } s \in (-\infty, t]; \\ \gamma_1(s), & \text{for } s \in [t, t_1] \end{cases} \tag{3.25}$$

has a corner at time t. This corner can be smoothed out according to Fact 2, and the resulting curve $\gamma^* \in C^1(-\infty, t_1]$ satisfies

$$\mathcal{A}^{\omega}_{\tau, t_1}(\gamma_3) - \mathcal{A}^{\omega}_{\tau, t_1}(\gamma^*) = \delta > 0. \tag{3.26}$$

Set $\varepsilon = \delta/4$. Choose sufficiently negative $t_n(\omega, \varepsilon)$ defined in Lemma 3.3 such that $\gamma^*(s) = \gamma_2(s)$ for $s \in (-\infty, t_n]$.

Assume that

$$\mathcal{A}^{\omega}_{t_n,t}(\gamma_2) - \mathcal{A}^{\omega}_{t_n,t}(\gamma_1) > 2\varepsilon\,. \tag{3.27}$$

Then in view of Lemma 3.3

$$\gamma_4(s) = \begin{cases} \gamma_2(s), & \text{for } s \in (-\infty, t_n - T]; \\ \gamma_{2,1}(s), & \text{for } s \in [t_n - T, t_n]; \\ \gamma_1(s), & \text{for } s \in [t_n, t], \end{cases} \tag{3.28}$$

is a local perturbation of $\gamma_2 \in C^1(-\infty, t]$ with

$$\begin{aligned} \mathcal{A}^{\omega}_{\tau,t}(\gamma_2) - \mathcal{A}^{\omega}_{\tau,t}(\gamma_4) &= \mathcal{A}^{\omega}_{t_n-T,t_n}(\gamma_2) - \mathcal{A}^{\omega}_{t_n-T,t_n}(\gamma_{2,1}) \\ &\quad + \mathcal{A}^{\omega}_{t_n,t}(\gamma_2) - \mathcal{A}^{\omega}_{t_n,t}(\gamma_1) \\ &> -\varepsilon + 2\varepsilon \\ &= \varepsilon\,. \end{aligned} \tag{3.29}$$

This contradicts the assumption that γ_2 is a one-sided minimizer. Thus

$$\mathcal{A}^{\omega}_{t_n,t}(\gamma_1) - \mathcal{A}^{\omega}_{t_n,t}(\gamma_2) \geq -2\varepsilon\,, \tag{3.30}$$

and

$$\gamma_5(s) = \begin{cases} \gamma_1(s), & \text{for } s \in (-\infty, t_n - T]; \\ \gamma_{1,2}(s), & \text{for } s \in [t_n - T, t_n]; \\ \gamma^*(s), & \text{for } s \in [t_n, t_1] \end{cases} \tag{3.31}$$

is a local perturbation of $\gamma_1 \in C^1(-\infty, t_1]$ with

$$\begin{aligned} \mathcal{A}^{\omega}_{\tau,t_1}(\gamma_1) - \mathcal{A}^{\omega}_{\tau,t_1}(\gamma_5) &= \mathcal{A}^{\omega}_{t_n-T,t_n}(\gamma_1) - \mathcal{A}^{\omega}_{t_n-T,t_n}(\gamma_{1,2}) \\ &\quad + \mathcal{A}^{\omega}_{t_n,t}(\gamma_1) - \mathcal{A}^{\omega}_{t_n,t}(\gamma_2) \\ &\quad + \mathcal{A}^{\omega}_{\tau,t_1}(\gamma_3) - \mathcal{A}^{\omega}_{\tau,t_1}(\gamma^*) \\ &\geq -\varepsilon - 2\varepsilon + \delta \\ &= \varepsilon > 0\,. \end{aligned} \tag{3.32}$$

This contradicts the assumption that γ_1 is a one-sided minimizer and proves the theorem. □

Theorem 3.2 implies the following remarkable properties of one-sided minimizers. Given ω and t, denote by $J(\omega, t)$ the set of points $x \in \mathbb{S}^1$ with more than one one-sided minimizer coming to (x, t).

LEMMA 3.4. *The following holds with probability* 1. *For any t, the set $J(\omega, t)$ is at most countable.*

Proof. Any $x \in J(\omega, t)$ corresponds to a segment $[\gamma_-(t-1), \gamma_+(t-1)]$, where γ_- and γ_+ are two different one-sided minimizers coming to (x,t) and $\gamma_+(s) > \gamma_-(s)$, for $s < t$. In view of Theorem 3.2, these segments are mutually disjoint. This implies the lemma. □

LEMMA 3.5. *Given ω and t, consider a sequence of one-sided minimizers $\gamma_n(s)$ defined on $(-\infty, t]$ such that $\gamma_n(t) \to x$ and $\dot\gamma_n(t) \to v$ as $n \to \infty$. Let γ be the solution of the* SDE (2.3) *on $(-\infty, t]$ with the initial data $\gamma(t) = x$ and $\dot\gamma(t) = v$. Then γ is a one-sided minimizer.*

Proof. Suppose that $\gamma^* \in C^1(-\infty, t]$ coincides with γ outside an interval $[t_1, t_2] \subset (-\infty, t]$ and $\mathcal{A}^\omega_{t_1,t_2}(\gamma) - \mathcal{A}^\omega_{t_1,t_2}(\gamma^*) = \varepsilon > 0$. It is clear that by taking sufficiently large n one can make $\|\gamma(s) - \gamma_n(s)\|_{C^1[t_1-1,t]}$ arbitrarily small. Let γ_1 be the reconnecting curve on $[t_1 - 1, t_1]$ between $\gamma_n(t_1 - 1)$ and $\gamma(t_1)$, and for some $\delta > 0$, let γ_2 be the reconnecting curve on $[t-\delta, t]$ between $\gamma(t-\delta)$ and $\gamma_n(t)$. Then the curve

$$\gamma^{**}(s) = \begin{cases} \gamma_n(s), & \text{for } s \in (-\infty, t_1 - 1]; \\ \gamma_1(s), & \text{for } s \in [t_1 - 1, t_1]; \\ \gamma^*(s), & \text{for } s \in [t_1, t_2]; \\ \gamma(s), & \text{for } s \in [t_2, t-\delta]; \\ \gamma_2(s), & \text{for } s \in [t-\delta, t] \end{cases} \tag{3.33}$$

satisfies $\mathcal{A}^\omega_{-\infty,t}(\gamma_n) - \mathcal{A}^\omega_{-\infty,t}(\gamma^{**}) > 0$ if δ and $\|\gamma(s) - \gamma_n(s)\|_{C^1[t_1-1,t]}$ are small enough. This contradicts the assumption that γ_n is a one-sided minimizer since γ^{**} is a local perturbation of γ_n. Note that (3.33) cannot be used if $t_2 = t$. In this case in the segment $[t-\delta, t]$ one can directly reconnect γ_n and γ^* and it is not hard to check that for δ small enough, $|\mathcal{A}^\omega_{t-\delta,t}(\gamma^*) - \mathcal{A}^\omega_{t-\delta,t}(\gamma^{**})|$ can be made arbitrarily small. □

LEMMA 3.6. *With probability one, the following holds. Fix an arbitrary sequence $t_n \to -\infty$ and a sequence of functions $\{v_n\}$, $v_n \in D_0$, $\int_0^1 v_n(z)dz = 0$. Consider* (1.4) *on the time interval $[t_n, t]$ with the initial condition $u(x, t_n) = v_n(x)$. Take any $x \in \mathbb{S}^1$ and a sequence of characteristics $\gamma_n \in C^1[t_n, t]$, $\gamma_n(t) = x$ minimizing $\mathcal{A}^\omega_{t_n,t}(\xi) + \int_0^{\xi(t_n)} v_n(z)dz$. Suppose that v is a limiting point of the set $\{\dot\gamma_n(t)\}$. Then the solution γ of* SDE (2.3) *with initial data $\gamma(t) = x$ and $\dot\gamma(t) = v$ is a one-sided minimizer on $(-\infty, t]$.*

Proof. The proof of this lemma is the same as the final part of the proof of Theorem 3.1. □

Next we study the measurability issues. Fix a time t and consider all integer times $-n \le t$. Introduce

$$A^{\omega}_{-n,t}(x) = \min_{\substack{\xi \in C^1[-n,t] \\ \xi(t)=x}} \mathcal{A}^{\omega}_{-n,t}(\xi)\,. \tag{3.34}$$

LEMMA 3.7. *The following statement holds with probability* 1. *Suppose that $\gamma \in C^1_x(-\infty, t]$ is a one-sided minimizer. Then for any $\varepsilon > 0$ there exist an infinite number of integer times $-n \le t$ such that*

$$|\mathcal{A}^{\omega}_{-n,t}(\gamma) - A^{\omega}_{-n,t}(x)| \le \varepsilon\,. \tag{3.35}$$

Conversely, if a curve $\xi \in C^1_x(-\infty, t]$ has the property that for any $\varepsilon > 0$ there exist an infinite number of integer times $-n \le t$ such that

$$|\mathcal{A}^{\omega}_{-n,t}(\xi) - A^{\omega}_{-n,t}(x)| \le \varepsilon\,, \tag{3.36}$$

then ξ is a one-sided minimizer.

Proof. Suppose that for some $\varepsilon > 0$ and n_0

$$|\mathcal{A}^{\omega}_{-n,t}(\gamma) - A^{\omega}_{-n,t}(x)| > \varepsilon \tag{3.37}$$

for all $-n \le -n_0$. Consider the curves $\xi_{-n} \in C^1_x[-n, t]$ such that $\mathcal{A}^{\omega}_{-n,t}(\xi_{-n}) = A^{\omega}_{-n,t}(x)$. Then, according to Lemma 3.3, there exist an interval $[-n_1, -n_2] \subset (-\infty, -n_0]$ and a reconnecting curve γ_r with $\gamma_r(-n_1) = \gamma(-n_1), \gamma_r(-n_2) = \xi_{-n_1}(-n_2)$, such that

$$|\mathcal{A}^{\omega}_{-n_1,-n_2}(\gamma_{-n_1}) - \mathcal{A}^{\omega}_{-n_1,-n_2}(\gamma_r)| \le \frac{\varepsilon}{2}\,. \tag{3.38}$$

Then

$$\gamma_1(s) = \begin{cases} \gamma(s), & \text{for } s \in (-\infty, -n_1]; \\ \gamma_r(s), & \text{for } s \in [-n_1, -n_2]; \\ \xi_{-n_1}(s), & \text{for } s \in [-n_2, t] \end{cases} \tag{3.39}$$

is a local perturbation of γ which lowers the action by at least $\varepsilon/2$. This contradicts the assumption that γ is a one-sided minimizer.

Note that formally Lemma 3.3 cannot be applied here since ξ_{-n_1} is not a one-sided minimizer but Lemma 3.1 remains valid for all ξ_{-n} with sufficiently negative $-n$. Thus the same argument as in the proof of Lemma 3.3 proves (3.38).

To prove the second statement, observe that if ξ_1 is a local perturbation of ξ lowering $\mathcal{A}^{\omega}_{-n,t}(\xi)$ by some $\varepsilon > 0$, then $\mathcal{A}^{\omega}_{-n,t}(\xi) \ge A^{\omega}_{-n,t}(x) + \varepsilon$ for all sufficiently negative $-n$. This contradicts (3.36). $\square$

Now we are ready to define the main object of this paper. We will denote by $\{\gamma_{x,t,\alpha}(s)\}$ the family of all one-sided minimizers coming to (x,t) indexing them by α.

Definition 3.1.

$$u_+^\omega(x,t) = \inf_\alpha \dot{\gamma}_{x,t,\alpha}(t)\,, \tag{3.40}$$

$$u_-^\omega(x,t) = \sup_\alpha \dot{\gamma}_{x,t,\alpha}(t)\,. \tag{3.41}$$

It is clear that $u_+^\omega(x,t) = u_-^\omega(x,t)$ for $x \notin J(\omega,t)$.

LEMMA 3.8. *With probability* 1, *for every* $x \in \mathbb{S}^1$

$$\lim_{y\uparrow x} u_+^\omega(y,t) = u_-^\omega(x,t)\,, \tag{3.42}$$

$$\lim_{y\downarrow x} u_+^\omega(y,t) = u_+^\omega(x,t)\,, \tag{3.43}$$

and hence $u_+^\omega(\cdot,t) \in D$ *for fixed* t.

Proof. We will prove (3.42). The proof of (3.43) is similar. It was shown in Lemma 3.1 that $|u_+^\omega(y,t)| \le C(\omega,t)$. Suppose that there exists a sequence $y_n \uparrow x$ such that $u_+^\omega(y_n,t) \to v \ne u_-^\omega(x,t)$. Then, according to Lemma 3.5, the solution γ of SDE (2.3) with the initial data $\gamma(t) = x$ and $\dot{\gamma}(t) = v$ is a one-sided minimizer. Theorem 3.2 implies that $\dot{\gamma}(t) > u_-^\omega(x,t)$ which contradicts the definition of $u_-^\omega(x,t)$. □

It follows immediately from the construction that on any finite time interval $[t_1,t_2]$, u_+^ω is a weak solution of (1.4) with initial data $u_0(x) = u_+^\omega(x,t_1)$. Moreover, the following statement holds:

LEMMA 3.9. *Given* t, *the mapping* $u_+^\omega(\,\cdot\,,t)\colon \Omega \mapsto D$ *is measurable.*

Proof. Without loss of generality, let $t = 0$. Since $\mathcal{D}$ is generated by cylinder sets of the type $A(x_1,\dots,x_n)$ with x_i from a dense subset of $\mathbb{S}^1$, it is enough to show that $u_+^\omega(x,0)\colon \Omega \to \mathbb{R}^1$ are measurable for a dense set of x values. For any positive integer n, denote by $u_{-n,+}^\omega$ the right continuous weak solution of (1.4) on the time interval $[-n,0]$ with the initial data $u_{-n,+}^\omega(x,-n) \equiv 0$. For any $x \in \mathbb{S}^1$ and $v \in \mathbb{R}^1$ denote by $\xi_{x,v}^\omega(s)$, $s \in [-n,0]$ the backward solution of (2.3) with the initial data $\xi_{x,v}^\omega(0) = x$ and $\dot{\xi}_{x,v}^\omega(0) = v$. The action $A_{-n,0}^\omega(x,v) = A_{-n,0}^\omega(\xi_{x,v}^\omega)$ is a continuous function on $\Omega \times \mathbb{S}^1 \times \mathbb{R}^1$. Hence the set $M = \{(\omega,x,v)\colon\ A_{-n,0}^\omega(x,v) = A_{-n,0}^\omega(x)\}$ is closed. Let $M_{\omega,x} = \{v \in \mathbb{R}^1\colon (\omega,x,v) \in M\}$. We conclude that $u_{-n,+}^\omega(x,0) = \max_v M_{\omega,x}$ is a measurable function on $\Omega \times \mathbb{S}^1$ and $u_{-n,+}^\omega(\,\cdot\,,0)$ is a measurable mapping $\Omega \mapsto D$.

As in the proof of Theorem 3.1, it is easy to check that for

$$v_+^\omega(x,0) = \limsup_n u_{-n,+}^\omega(x,0) \tag{3.44}$$

the corresponding curve $\gamma_x^\omega = \xi_{x,v_+^\omega(x,0)}^\omega$ is a one-sided minimizer.

For positive integers k and m, introduce measurable subsets of $\Omega \times \mathbb{S}^1 \times \mathbb{R}^1$:

(3.45)
$$M_n(k,m) = \Big\{(\omega,x,v) \in \Omega \times \mathbb{S}^1 \times \mathbb{R}^1\colon |v - u_{-n,+}^\omega(x,0)| \geq \frac{1}{k},$$
$$\mathcal{A}_{-n,0}^\omega(x,v) - A_{-n,0}^\omega(x) \leq \frac{1}{m}\Big\}.$$

Let Π be the projection of the measurable set

$$\cup_k \cap_m \left(\cap_l \cup_{n=l}^{\infty} M_n(k,m)\right) \tag{3.46}$$

on $\Omega \times \mathbb{S}^1$. The points (ω, x) in Π are characterized by the following property. There exists a backward solution $\xi_{x,v}^\omega$ different from the one-sided minimizer γ_x^ω such that at infinitely many negative integer times $-n$, the action $\mathcal{A}_{-n,0}^\omega(\xi_{x,v}^\omega)$ is arbitrarily close to its minimal value $A_{-n,0}^\omega(x)$. In view of Lemma 3.7, Π is precisely the set (ω, x) having at least two one-sided minimizers coming to $(x, 0)$; i.e.,

$$\Pi = J = \{\omega, x) \in \Omega \times \mathbb{S}^1\colon u_+^\omega(x,0) \neq u_-^\omega(x,0)\}\,. \tag{3.47}$$

Consider the sections $J_x = \{\omega \in \Omega\colon u_+^\omega(x,0) \neq u_-^\omega(x,0)\}$ of J and define $I = \{x \in \mathbb{S}^1\colon P(J_x) > 0\}$. Using the measurability of J, Lemma 3.4 and Fubini's theorem, we conclude that the Lebesgue measure of I is 0.

Fix $x \in \mathbb{S}^1 \setminus I$. Then for almost all ω one has

$$u_+^\omega(x,0) = \lim_{n\to\infty} u_{n,+}^\omega(x,0)\,. \tag{3.48}$$

The functions $u_{n,+}^\omega(x,0)\colon \Omega \mapsto \mathbb{R}^1$ are measurable. Hence $u_+^\omega(x,0)\colon \Omega \mapsto \mathbb{R}^1$ is also measurable. This proves the lemma, since $\mathbb{S}^1 \setminus I$ is dense in $\mathbb{S}^1$. □

Before ending this section, we formulate some corollaries of Lemmas 3.1 and 3.6 that will be useful later.

LEMMA 3.10. *The following estimate holds:*

$$\|u^\omega(\,\cdot\,,t)\|_{L^\infty(\mathbb{S}^1)} \leq \bar{C}(\{\omega(s), s \in [t-1,t)\})\,.$$

The stationary random variable $\bar{C}(\{\omega(s),\ s \in [t-1,t]\})$ *has finite moments of all orders.*

This estimate says that $|u^\omega(x,t)|$ can be bounded by a quantity that depends only on $\{\omega(s), s \in [t-1,t)\}$.

LEMMA 3.11. *Let $\{v_n\}$ be a sequence of functions in D_0. Let $\{t_n\}$ be a sequence such that $t_n \to -\infty$, and let u_n be the solution of (1.4) with initial condition $u_n(x,t_n) = v_n(x)$. Then for almost all ω, $\lim_{t_n \to -\infty} u_n(\,\cdot\,, 0) = u_+^\omega(\cdot, 0)$ almost everywhere.*

This result follows directly from Lemma 3.6.

So far we restricted our attention to the case when $\int_0^1 u(x,t)dx = 0$. This is reflected in the variational principle used in (2.1). In the case when $\int_0^1 u(x,t)dx = c \neq 0$, the relevant action is replaced by

$$\mathcal{A}_{t_1,t_2}(\xi) = \int_{t_1}^{t_2} \left\{ \tfrac{1}{2}(\dot{\xi}(s) - c)^2 - \sum_k f_k(\xi(s))\dot{\xi}(s)(B_k(s) - B_k(t_1)) \right\} ds$$
$$+ \sum_k F_k(\xi(t_2))(B_k(t_2) - B_k(t_1)).$$

In this more general case, we can also define one-sided minimizers in an analogous way, and all the results obtained so far hold in the general case. Instead of giving all the details, let us just comment on the most important aspect when $c \neq 0$. When $c = 0$, the one-sided minimizers stay roughly inside one period so that their asymptotic speed (which is the analog of rotation number in Aubry-Mather theory) is zero when lifted to the universal cover:

$$\alpha(c=0) = \lim_{t \to -\infty} \frac{\xi(t)}{t} = 0.$$

In the general case,

$$\alpha(c) = \lim_{t \to -\infty} \frac{\xi(t)}{t} = c. \tag{3.49}$$

Roughly speaking, this is because, if a curve has different asymptotic speed, the cost to the action grows linearly in time, whereas the savings from the random potential can at most grow as $O(\sqrt{t})$. For example, in Lemma 3.3 and Theorem 3.2, we showed that there are large time intervals on which the one-sided minimizers are almost parallel. These results are still true except that the minimizers are parallel with an average slope $1/c$ in the $x-t$ plane. Similarly, the reconnection used in Lemma 3.1 to prove a bound for the velocity of minimizers will also have to be done with curves that have an average slope of $1/c$.

(3.49) is in sharp contrast with the case of periodic (in x and t) potential studied in Aubry-Mather theory. There the function α is usually a very complicated function of c with Cantor-like structures.

4. Construction and uniqueness of the invariant measure

In this section we prove Theorems 1.2 and 1.3 stated in Section 1. Let $\delta^\omega(du)$ be the atomic measure on $(D_0, \mathcal{D})$ concentrated at $u_+^\omega(\cdot, 0)$.

THEOREM 4.1. *The measure*

$$\mu(d\omega, du) = \delta^\omega(du)P(d\omega) \tag{4.1}$$

is an invariant measure for the skew-product transformation F^t (see Definition 1.2), and the measure $\kappa(du) = \int_\Omega \mu(d\omega, du)$ is a stationary distribution for the Markov process corresponding to (1.4).

Proof. The second statement of the theorem follows trivially from the first one since the $\delta^\omega(du)$'s are measurable with respect to the σ-algebra $\mathcal{F}_{-\infty}^0$. It is a general fact that any measure which satisfies the measurability property and which is invariant under F^t generates a stationary distribution for the Markov process (1.4). The first statement is an immediate consequence of the construction of $u_+^\omega(x,t)$. Indeed, $u_+^{\theta^t\omega}(x,0) = u_+^\omega(x,t)$ by construction. Hence $S_\omega^t \delta^\omega = \delta^{\theta^t\omega}$. This is exactly the condition for the invariance of μ. □

The "one force, one solution" principle not only gives the existence of an invariant measure, it also implies uniqueness.

THEOREM 4.2. *The measure $\mu(d\omega, du)$ is the unique invariant measure on $(\Omega \times D_0, \mathcal{F} \times \mathcal{D})$ with given projection $P(d\omega)$ on $(\Omega, \mathcal{F})$. The measure $\kappa(du) = \int_\Omega \mu(d\omega, du)$ is the unique stationary distribution for the Markov process (1.4).*

Proof. Assume λ is another invariant measure on $\Omega \times D_0$. Write λ as

$$\lambda(d\omega, du) = \int_\Omega \lambda_\omega(du)P(d\omega). \tag{4.2}$$

For $t < 0$, let H_t^ω be the operator which maps the solution of (1.4) at time t to the solution at time 0 when the realization of the force is ω. By definition, the invariance of λ implies that

$$(H_t^\omega)^* \lambda_{\theta^t\omega}(du) = \lambda_\omega(du) \tag{4.3}$$

for $t < 0$, where $(H_t^\omega)^*$ is the push forward action on the spaces of measures. This means that there exists a subset B of D_0 of full measure with respect to $\lambda_\omega(du)$, such that for every $u \in B$ and $n \in \mathbb{N}$, there exists a v_n such that

$$H_{-n}^\omega v_n = u.$$

From Lemma 3.6, if a solution of (1.4) can be extended backward to arbitrary negative times, that solution must coincide with u_+^ω at $t = 0$, for all $x \in \mathbb{S}^1 \setminus J(\omega, 0)$. In particular, we have

$$u(x) = u_+^\omega(x, 0)$$

for $x \in \mathbb{S}^1 \setminus J(\omega, 0)$. Hence $\lambda_\omega(du) = \delta_{u_+^\omega}(du)$ and $\lambda = \mu$.

To prove the second statement suppose that $\nu(du) \neq \kappa(du)$ is another stationary distribution for the Markov process (1.4). Let

$$A = \{u\colon (u(x_1), \dots, u(x_k)) \in C \subset \mathbb{R}^k\}$$

be an arbitrary cylindrical set based on the points $x_1, \dots, x_k \in I^c$. By definition

$$\nu(A) = \int_{D_0} P_n(u_{-n}, A)\nu(du_{-n}) \tag{4.4}$$
$$= \int_{D_0}\left(\int_{\Omega} \chi_A(u_{-n}, \omega)P(d\omega)\right)\nu(du_{-n})$$
$$= \int_{\Omega}\left(\int_{D_0} \chi_A(u_{-n}, \omega)\nu(du_{-n})\right)P(d\omega).$$

Denote by $\chi_A(\omega)$ the indicator function of the event that $u_+^\omega \in A$. Then for $\eta_n(\omega) = \int_{D_0} \chi_A(u_{-n}, \omega)\nu(du_{-n})$ one has

$$\lim_{n\to\infty} \eta_n(\omega) = \chi_A(\omega) \tag{4.5}$$

implying $\nu(A) = \int_\Omega \chi_A(\omega)P(d\omega) = \kappa(A)$.

Indeed, in view of the uniqueness of one-sided minimizers, coming to each of the points $(x_1, 0), \dots, (x_k, 0)$, one observes that

$$\lim_{n\to\infty} \eta_n(\omega) = \begin{cases} 1 & \text{for } (A \setminus \partial A) \ni (u_+^\omega(x_1,0), \dots, u_+^\omega(x_k,0)), \\ 0 & \text{for } (A^c \setminus \partial A) \ni (u_+^\omega(x_1,0), \dots, u_+^\omega(x_k,0)). \end{cases} \tag{4.6}$$

Thus to obtain (4.3) one need only show that $P\{\omega\colon (u_+^\omega(x_1,0), \dots, u_+^\omega(x_k,0)) \in \partial A\} = 0$. Clearly it is enough to check that:

LEMMA 4.1. *For any $v \in \mathbb{R}^1$ and for any $x \in X$,*

$$P\{\omega\colon u_+^\omega(x,0) = v\} = 0. \tag{4.7}$$

Proof. For fixed x and v the backward solution (for $t \leq 0$) of the SDE (2.3) with the initial data $\gamma(0) = x$ and $\dot\gamma(0) = v$ is a random process with variance $\operatorname{Var}\dot\gamma(t)$ that grows like $|t|$ as $t \to -\infty$. Thus with probability 1 this solution cannot be a one-sided minimizer. This completes the proof of Theorem 4.2. □

5. The two-sided minimizer and the main shock

Sections 5–7 will be devoted to the study of the structure of the solution u^ω. In this section, we will define the two basic objects needed for this study, the two-sided minimizer and the main shock.

Let (x, t) be a point of shock, i.e., $x \in J(\omega, t)$. Denote by $\Delta_{x,t}(t_1)$, $t_1 < t$, an open interval at time t_1 generated by the shock at (x, t):

$$\Delta_{x,t}(t_1) = (\gamma^-_{x,t}(t_1),\ \gamma^+_{x,t}(t_1)) \tag{5.1}$$

where we use $\gamma^-_{x,t}$ and $\gamma^+_{x,t}$ to denote respectively the left-most and right-most one-sided minimizers starting at (x,t). Roughly speaking, $\Delta_{x,t}(t_1)$ is the set of points at time t_1 that merge into the shock at x before time t, i.e. the one-sided minimizer that passes through $(y,t_1), y \in \Delta_{x,t}(t_1)$, intersects the past history of the shock at (x,t) before time t.

LEMMA 5.1. *For almost all ω the following statements hold for any fixed t_1 and t such that $t_1 < t$.*

(a) *For arbitrary $x_1 \in \mathbb{S}^1$, either there exists a unique one-sided minimizer at time t which passes through (x_1,t_1) or there exists a unique shock at (x,t) for some $x \in \mathbb{S}^1$, such that $(x_1,t_1) \in \Delta_{x,t}(t_1)$. In the second case we will say that (x_1,t_1) is covered by the shock at (x,t). In particular, if (x_1,t_1) is a point of shock, i.e., $x_1 \in J(\omega,t_1)$, then there exists a unique shock (x,t) which covers (x_1,t_1); i.e., $(x_1,t_1) \in \Delta_{x,t}(t_1)$.*

(b) *Let $(x_1,t_1) \in J(\omega,t_1)$ be a point of shock. Denote by $x(t)$, $t \geq t_1$, the position of the shock at time t which covers (x_1,t_1). Then $x(t)$, $t \geq t_1$, is a Lipschitz continuous function.*

Proof. Denote by $A(t_1,t)$ the set of points $x_1 \in \mathbb{S}^1$ that correspond to the first situation; i.e., there exists a one-sided minimizer at time t which passes through (x_1,t_1). Obviously this minimizer is unique, and $A(t_1,t)$ is closed. Hence $B(t_1,t) = \mathbb{S}^1 \setminus A(t_1,t)$ is open and consists of nonintersecting open intervals. Let (x_1',x_1'') be one of these intervals. Both (x_1',t_1) and (x_1'',t_1) are reached by one-sided minimizers which start at (x',t) and (x'',t). It is easy to see that $x' = x''$. Otherwise a minimizer which starts from some point between x' and x'' will reach a point inside (x_1',x_1''). It follows that (x',t) is a point of shock, and $(x_1',x_1'') \subset \Delta_{x',t}(t_1)$. Obviously a point of shock cannot be reached by a one-sided minimizer that extends to time t. Thus if $t_1 < t$, any point of shock at (x_1,t_1) must be covered by a shock at (x,t) for some $x \in \mathbb{S}^1$. Clearly such a covering shock (x,t) is unique. This completes the proof of (a).

(b) basically follows from the fact that the velocities of minimizers are bounded. It is enough to show that $x(\cdot)$ is Lipschitz continuous at $t = t_1$. It follows from Lemma 3.1 that there exists a constant $C_1(t_1,\omega)$ such that for all one-sided minimizers γ and $t \in [t_1,t_1+1]$, $|\dot\gamma(t)| \leq C_1(t_1,\omega)$. Therefore for any shock at $(x(t),t)$, $t \in [t_1,t_1+1]$, $|x(t) - \gamma^-_{x(t),t}(t_1)| \leq C_1(t_1,\omega)(t-t_1)$, $|x(t) - \gamma^+_{x(t),t}(t_1)| \leq C_1(t_1,\omega)(t-t_1)$. Since $x_1 \in (\gamma^-_{x(t),t}(t_1),\ \gamma^+_{x(t),t}(t_1))$, we also have $|x_1 - x(t)| \leq C_1(t_1,\omega)(t-t_1)$. This estimate implies that $x(t)$ is Lipschitz continuous. □

Remark. It may happen for some points (x_1, t_1) that they are covered by shocks and at the same time there exist one-sided minimizers passing through them. This is possible only if there are more than two minimizers starting at covering shock. This situation occurs when two shocks merge.

In this and following sections we study the detailed structure and regularity properties of solutions supported by the invariant measure. For this purpose, we need certain nondegeneracy conditions on the forcing.

Nondegeneracy condition. If x^* is the local maximum of some F_k, we will denote by $I(x^*)$ a closed interval on $\mathbb{S}^1$ containing x^* which is contained in the basin of attraction of x^* for the potential F_k. In other words, $f_k < 0$ on $I(x^*)$ to the right of x^*, and $f_k > 0$ on $I(x^*)$ to the left of x^*. Assume that

(A1) There exists a finite set of points $X^* = \{x^*\}$, $x^* \in \mathbb{S}^1$, each of which is a local maximum of some F_k with the following property: for any $(x_1, x_2) \in \mathbb{S}^1 \times \mathbb{S}^1$ there exists an $x^* \in X^*$, such that $x_1, x_2 \in I(x^*)$.

Below we will always assume that (A1) holds. Obviously (A1) fails if there is only one term in the sum of $F(x,t)$. Nevertheless (A1) is fulfilled in generic situations. In particular, it is easy to see that three such intervals $I(x^*)$ suffice. However, by refining the argument in Appendix D, we can show that the basic collision lemma below also holds when the potential contains two shifted cosine functions, for example. We will come back to this point at the end of this section.

Consider two points x_1^0 and x_2^0 at time $t = 0$. We say that x_1^0 and x_2^0 merge before $t = \tau > 0$ if there exists a shock at (y, τ), $y \in \mathbb{S}^1$, which covers both x_1^0 and x_2^0; i.e., $x_1^0, x_2^0 \in \Delta_{y,\tau}(0)$. The following lemma is of fundamental importance to what follows.

LEMMA 5.2 (basic collision lemma). *For any $\tau > 0$, there exists a positive number $p_0(\tau)$ with the following property. Let $u(\,\cdot\,, 0) \in D_0$ and x_1^0, x_2^0 be two positions at $t = 0$ which are measurable with respect to $\mathcal{F}_{-\infty}^0$. Then the conditional probability under $\mathcal{F}_{-\infty}^0$ that x_1^0 merges with x_2^0 before $t = \tau$ is no less than $p_0(\tau)$.*

The proof of this lemma is given in Appendix D.

LEMMA 5.3. *The set of ω's for which $u^\omega(\,\cdot\,, t_0)$ is continuous for some $t_0 \in \mathbb{R}^1$ has probability zero.*

Proof. It follows from Lemma 5.1 that if $u^\omega(\,\cdot\,, t_0)$ is continuous, then $u^\omega(\,\cdot\,, t)$ is continuous for all $t \le t_0$. Denote by $C(t)$ the set of ω such that $u^\omega(\,\cdot\,, s)$ is continuous for $s \le t$. Then $\theta^s C(t) = C(t+s) \subseteq C(t)$ for all $s \ge 0$. Using ergodicity we conclude that either $P(C(t)) = 1$ for all t, or $P(C(t)) = 0$

for all t. Assume that $P(C(t)) \equiv 1$. It follows from Lemma 5.2 that there is a positive conditional probability $p(\{\omega(t'), t' < t\})$ under $\mathcal{F}^t_{-\infty}$ that $u(\,\cdot\,, t+1)$ has at least one shock. Hence we have

$$P(C(t+1)) = \int_{C(t)} \left(1 - p\{\omega(t'), t' \le t\}\right) dP < P(C(t))\,. \tag{5.2}$$

Therefore $P(C(t+1)) < 1$. Hence $P(C(t)) \equiv 0$ for all t. □

It follows from Lemma 5.3 that for almost all ω and arbitrary t_0 there exists at least one shock at t_0. Since the number of shocks is at most countable and the sum of their sizes is bounded, i.e.,

$$\sum_{x \in J(\omega, t_0)} \left(u^\omega_-(x, t_0) - u^\omega_+(x, t_0)\right) \le \operatorname{Var} u^\omega_+(x, t_0) < +\infty,$$

we can numerate all the shocks in a measurable way. The first shock is the largest one, the second is next in size and so on. If two or more shocks have the same size then we numerate them according to their order of occurrence on the semi-interval $[0, 1)$. Denote by $\xi_i(t_0) = (x_i(t_0), t_0)$ the position of the i-th shock. Obviously $\xi_i(t_0)$ is a measurable function with respect to $\mathcal{F}^{t_0}_{-\infty}$.

We will use $\ell(I)$ to denote the length of the interval I.

LEMMA 5.4. *There exist positive constants $C_1, C_2, K_1, K_2 > 0$ such that for all j, and $t > t_0$,*

$$\begin{aligned} P\{\omega\colon \ell_j(t) = \ell\left(\Delta_{x_j(t), t}(t_0)\right) \le 1 - K_1 \exp(-C_1(t - t_0))\} \\ \le K_2 \exp(-C_2(t - t_0)). \end{aligned} \tag{5.3}$$

Proof. Fix any $j \in \mathbb{N}$. The position of the j-th shock at time t will be denoted by $x(t)$. The following estimates are independent of j.

Consider a sequence of times $t_i = t_0 + i$, $i = 0, 1, 2, \dots$. For each i let $I_i = \mathbb{S}^1 \setminus \Delta_{x(t_i), t_i}(t_0)$ and z_i be the mid-point of I_i. Denote by y_i a point on $\mathbb{S}^1$ at time t_i which corresponds to z_i at $t = t_0$; i.e., either (y_i, t_i) is a point of shock which covers (z_i, t_0), or there is a unique one-sided minimizer at (y_i, t_i) which passes through (z_i, t_0). Clearly, y_i is measurable with respect to $\mathcal{F}^{t_i}_{-\infty}$. Denote by η_i a random variable which takes the value 1 if y_i is covered by $\Delta_{x(t_{i+1}), t_{i+1}}(t_i)$ and 0 otherwise. Obviously $\eta_i = 1$ if and only if y_i merges with $x(t_i)$ before t_{i+1}.

Notice that if $\eta_i = 1$ then the length of the complement to the interval $\Delta_{x(t_{i+1}), t_{i+1}}(t_0)$ is no more than half of the length of the complement to

$\Delta_{x(t_i),t_i}(t_0)$. For any fixed positive integer K, with $n = t - t_0$,

(5.4)

$$P\{\omega: \ell(I_n) \geq 2^{-K}\} \leq P\Big(\sum_{i=0}^{n-1} \eta_i \leq K\Big)$$
$$\leq \sum_{m=0}^{K} \sum_{0<i_1<i_2<\ldots<i_{n-m}\leq n-1} P\{\eta_{i_1} = \eta_{i_2} = \ldots = \eta_{i_{n-m}} = 0\}$$
$$\leq \sum_{m=0}^{K} C_m^n (1 - p_0(1))^{n-m}$$

where we used Lemma 5.2 and the Markov property to conclude that

$$P(\eta_{i_1} = \eta_{i_2} = \ldots = \eta_{i_s} = 0) \leq (1 - p_0(1))^s, \tag{5.5}$$

for all $0 \leq i_1 < i_2 < \ldots < i_s$.

It follows from (5.4) that for $K \leq \frac{n}{2}$

$$P\{\omega: \ell(I_n) \geq 2^{-K}\} \leq (K+1) C_n^K (1 - p_0(1))^{n-K}. \tag{5.6}$$

Let $K = [\alpha n]$ and choose α so small that

$$q_1 = \left(\frac{1}{\alpha}\right)^{\alpha} \left(\frac{1}{1-\alpha}\right)^{1-\alpha} (1 - p_0(1))^{1-\alpha} < 1\,. \tag{5.7}$$

Then,

$$P\{\omega: \ell(I_n) \geq e^{-[\alpha n]\ln 2}\} \leq M\sqrt{n}\, q_1^n \tag{5.8}$$

where M is an absolute constant. It follows that

$$P\{\omega: \ell\left(\Delta_{x(t),t}(t_0)\right) \leq 1 - e^{-[\alpha n]\ln 2}\} \leq M\sqrt{t - t_0}\, q_1^{t-t_0}. \tag{5.9}$$

Take any q such that $q_1 < q < 1$ and let $C_1 = \alpha \ln 2$, $K_1 = 4$, $C_2 = -\ln q$. Then (5.3) follows from (5.9) for large enough K_2. □

Using the Borel-Cantelli lemma, one gets from Lemma 5.4 the following:

LEMMA 5.5. *For almost all* ω

$$\ell\left(\Delta_{x_j(t),t}(t_0)\right) \to 1 \qquad as \qquad t \to \infty. \tag{5.10}$$

Moreover for any shock at $\xi_j(t_0)$ *there exists a random constant* $T_j(\omega, t_0)$ *such that for all* $t > T_j(\omega, t_0)$

$$\ell\left(\Delta_{x_j(t),t}(t_0)\right) \geq 1 - 2K_1 \exp(-C_1(t - t_0))\,. \tag{5.11}$$

Remark. Since the intervals $\Delta_{x_j(t),t}(t_0)$ do not intersect each other, Lemma 5.5 implies that the shocks $\xi_{j_1}(t_0), \xi_{j_2}(t_0)$ merge with each other after time $T = \max(T_{j_1}, T_{j_2})$.

Let us define now an object which will play a very important role in the remaining part of this paper.

Definition 5.1. A C^1 curve γ: $(-\infty,+\infty) \to \mathbb{S}^1$ is called a *two-sided minimizer* if for any C^1 compact perturbation $\gamma+\delta\gamma$: $(-\infty,+\infty) \to \mathbb{S}^1$ of γ

$$\mathcal{A}_{-s,s}(\gamma+\delta\gamma) \geq \mathcal{A}_{-s,s}(\gamma)$$

for all sufficiently large s.

In other words, a curve γ is called a two-sided minimizer if and only if for arbitrary $t_0 \in \mathbb{R}^1$, its restriction on $(-\infty, t_0]$ is a one-sided minimizer.

THEOREM 5.1. *With probability* 1 *there exists a unique two-sided minimizer.*

Proof. Existence of the two-sided minimizer follows from a compactness argument. Consider a sequence of curves $\gamma^{(n)}$: $[-n,n] \to \mathbb{S}^1$ which minimize $\mathcal{A}_{-n,n}(\gamma)$ in the class of C^1 curves. It follows from Lemma 3.3 that $|\dot\gamma^{(n)}(0)| \leq C(\omega,0)$. Hence the sequence of points $(x_0^{(n)}, v_0^{(n)}) = (\gamma^{(n)}(0), \dot\gamma^{(n)}(0))$ belongs to a compact set $\mathbb{S}^1 \times [-C(\omega,0), C(\omega,0)]$. Then there exists at least one limiting point (x_0, v_0). A standard argument as in the proof of Theorem 3.1 shows that the solution of the Euler-Lagrange equation (2.3) with initial conditions $x(0)=x_0$, $v(0)=v_0$ defines a two-sided minimizer.

For uniqueness notice that points on a two-sided minimizer γ do not belong to the intervals $\Delta_{x_j(t),t}(t_0)$ for any j. Since $\ell\left(\Delta_{x_j(t),t}(t_0)\right) \to 1$ as $t \to \infty$, the two-sided minimizer is unique. $\square$

Denoting the two-sided minimizer by y_ω, we now construct another important object, the main shock. For arbitrary $t_0 \in \mathbb{R}^1$ consider a sequence of non-intersecting intervals $\Delta_{x_j,t_0}(t)$, $t \leq t_0$ corresponding to shocks $\xi_j(t_0) = (x_j, t_0)$ at time t_0, $x_j \in J(\omega, t_0)$. Notice that here we consider intervals $\Delta_{x_j,t_0}(t)$ for $t \leq t_0$. It turns out that for almost all ω there exists a unique shock at $(z(t_0), t_0)$ for which $\ell(t) = \ell(\Delta_{z(t_0),t_0}(t)) \to 1$ as $t \to -\infty$.

THEOREM 5.2. *For almost all ω the following statements hold.*

(a) *For any $t_0 \in \mathbb{R}^1$ there exists a unique shock at $(z(t_0), t_0)$ such that*

$$\ell(t) = \ell\left(\Delta_{z(t_0),t_0}(t)\right) \to 1 \quad \text{as} \quad t \to -\infty. \tag{5.12}$$

Moreover, for any $\delta > 0$ there exists a random constant $T_{\delta,t_0}(\omega)$ such that for all $t < T_{\delta,t_0}(\omega)$

$$\ell(t) \geq 1 - \exp(-(C_1 - \delta)(t_0 - t)). \tag{5.13}$$

The position of the shock $(z(t_0), t_0)$ is measurable with respect to the σ-algebra $\mathcal{F}_{-\infty}^{t_0}$.

(b) *For all other shocks* $\xi_j(t_0) = (x_j(t_0), t_0)$

$$\ell_j(t) = \ell\left(\Delta_{x_j(t_0),t_0}(t)\right) \to 0 \qquad as \qquad t \to -\infty.$$

(c) $\{(z(t), t),\ t \in \mathbb{R}^1\}$ *is a Lipschitz continuous curve.*

Proof. Consider a sequence of times $\bar{t}_i = t_0 - i$. It follows from Lemma 5.4 that with probability greater than $1 - K_2 \exp(-C_2 i)$ there exists a shock at some point $(x_0(i), t_0)$ such that

$$\ell\left(\Delta_{x_0(i),t_0}(\bar{t}_i)\right) \geq 1 - K_1 \exp(-C_1 i)\,. \tag{5.14}$$

By the Borel-Cantelli lemma, there exists $N_1(\omega)$ such that for all $i > N_1(\omega)$, (5.14) holds for some shock at $(x_0(i), t_0)$. We will show that for i large enough $x_0(i)$ does not depend on i. Suppose $x_0(i+1) \neq x_0(i)$ for some $i > N_1$. Then,

$$\ell\left(\Delta_{x_0(i+1),t_0}(\bar{t}_i)\right) \leq K_1 \exp(-C_1 i)$$

and

$$\ell\left(\Delta_{x_0(i+1),t_0}(\bar{t}_{i+1})\right) \geq 1 - K_1 \exp(-C_1(i+1)).$$

Denote by a_i, b_i, a_{i+1}, b_{i+1} the end points of $\Delta_{x_0(i+1),t_0}(\bar{t}_i)$ and $\Delta_{x_0(i+1),t_0}(\bar{t}_{i+1})$, respectively, and by $v(a_i)$, $v(b_i)$, $v(a_{i+1})$, $v(b_{i+1})$ the velocities of the corresponding one-sided minimizers. It follows from Lemma B.8 that

$$|v(a_i) - v(b_i)| \leq L_i |a_i - b_i|\,,$$

where $L_i = L_0(\theta^{t_0 - i}\omega)$. Thus,

(5.15)

$$\mathcal{D}_i = \mathrm{dist}((a_i, v(a_i)), (b_i, v(b_i))) \leq \sqrt{1 + L_i^2}\, K_1 \exp(-C_1 i),$$

$$\mathcal{D}_{i+1} = \mathrm{dist}((a_{i+1}, v(a_{i+1})), (b_{i+1}, v(b_{i+1}))) \geq 1 - K_1 \exp(-C_1(i+1))\,.$$

On the other hand, we have $\mathcal{D}_{i+1} \leq \exp(d_i)\mathcal{D}_i$, where $d_i = d_1(\theta^{t_0 - i}\omega)$ and $d_1(\omega)$ is as defined in Lemma B.5. It follows that

$$\exp(d_i) \geq \frac{\mathcal{D}_{i+1}}{\mathcal{D}_i} \geq \frac{1}{2\sqrt{1 + L_i^2}\, K_1} \exp(C_1 i). \tag{5.16}$$

Since L_i and d_i have finite expectations, it follows that for any $\varepsilon > 0$ there exists $N_\varepsilon(\omega)$ such that

$$|d_i| \leq \varepsilon i, \ |L_i| \leq \varepsilon i, \quad \text{for all} \quad i > N_\varepsilon(\omega) \tag{5.17}$$

(see Lemma 6.2). Take $\varepsilon < C_1$. Then for $i > \bar{N}_\varepsilon(\omega) > N_\varepsilon(\omega)$, (5.16) and (5.17) contradict each other. Hence, for $i > \bar{N}_\varepsilon(\omega)$, $x_0(i+1) = x_0(i)$. Define

$z(t_0) = x_0(i)$, $i > \bar{N}_\varepsilon(\omega)$. Obviously, $\ell(\Delta_{z(t_0),t_0}(\bar{t}_i)) \geq 1 - K_1 \exp(-C_1 i)$, $i > \bar{N}_\varepsilon(\omega)$. It follows from the estimates (5.15)–(5.17) that for any $\delta > 0$ there exists a random constant $T_{\delta,t_0}(\omega)$ such that for all $t < T_{\delta,t_0}(\omega)$, (5.13) holds. A shock that satisfies (5.12) is obviously unique. Clearly $z(t_0)$ is measurable with respect to $\mathcal{F}^{t_0}_{-\infty}$. Notice that for any $t > t_0$, the shock which covers $(z(t_0), t_0)$ also satisfies (5.12), (5.13). Hence for almost all ω such a shock exists for all $t_0 \in \mathbb{R}^1$. (a) is now proven. (b) follows from (a), since

$$0 \leq \sum_{x_j \in J(\omega,t_0)} \ell\left(\Delta_{x_j(t_0),t_0}(t)\right) \leq 1 \quad \text{and} \quad \ell\left(\Delta_{z(t_0),t_0}(t)\right) \to 1 \quad \text{as} \quad t \to -\infty .$$

Since the shock at $\left(z(\tilde{t}_0), \tilde{t}_0\right)$ covers $(z(t_0), t_0)$ for all $\tilde{t}_0 \geq t_0$, we get (c). □

Definition 5.2. The shock $(z(t), t)$ constructed in Theorem 5.2 is called the *main shock* at time t.

As remarked in the introduction, the two-sided minimizer and the main shock play dual roles. The former acts as a repeller, the latter acts as an attractor. Indeed, it follows from Theorem 5.2 that for any two one-sided minimizers γ_1, γ_2, $\operatorname{dist}(\gamma_1(t), \gamma_2(t)) \to 0$ as $t \to -\infty$. One can say that all one-sided minimizers approach the two-sided minimizer as $t \to -\infty$.

LEMMA 5.6. (a) *For any two minimizers* γ_1, γ_2

$$\operatorname{dist}(\gamma_1(t), \gamma_2(t)) \to 0 \quad as \quad t \to -\infty . \tag{5.18}$$

Moreover, for any $\delta > 0$ *there exists a random constant* $T^{1,2}_\delta(\omega)$ *such that for all* $t < T^{1,2}_\delta(\omega)$

$$\operatorname{dist}(\gamma_1(t), \gamma_2(t)) \leq \exp(-(C_1 - \delta)t) .$$

If γ_1 *starts at time* t_1, *and* γ_2 *starts at* t_2, *then* $T^{1,2}_\delta(\omega) = T_{\delta,t^{1,2}}(\omega)$, *where constant* $T_{\delta,t}(\omega)$ *is defined as in Theorem* 5.2 *and* $t^{1,2} = \min(t_1, t_2)$.

For minimizers starting at the same time convergence in (5.18) *is uniform.*

(b) *Any shock at a given time t will be eventually absorbed by the main shock.*

This is obvious.

Another way to characterize the curve of the main shock is to say that it is the only shock curve defined for all $t \in \mathbb{R}^1$.

LEMMA 5.7. *For almost all* ω *there exists a unique shock curve*

$$x^\omega : (-\infty, +\infty) \to \mathbb{S}^1$$

such that $u^\omega_+(x^\omega(t), t) < u^\omega_-(x^\omega(t), t)$ *and* $x^\omega(t)$ *is measurable with respect to* $\mathcal{F}^t_{-\infty}$. *This curve is the curve of the main shock.*

Proof. The existence follows from the existence of the main shock. Suppose now that there exists another measurable shock curve $x^{\omega}(\cdot)$ defined on $\mathbb{R}^1$. Fix arbitrary $t_0 \in \mathbb{R}^1$. It follows from Lemma 5.2 that with probability 1 the curves $z(t)$ and $x^{\omega}(t)$ merge before $t = t_0$. Since t_0 is arbitrary, $x^{\omega}(t)$ coincides with $z(t)$ with probability 1. □

Remark. Later we will prove a stronger result: the curve of the main shock is the only shock curve which is defined for all sufficiently negative times.

We end this section with some discussion on the assumption (A1). Obviously, a necessary condition for the main result of this section to hold, namely the existence of a unique main shock and two-sided minimizer, is that the minimum period of all the F_k's is equal to 1. However, this condition is not sufficient, as we show now.

THEOREM 5.3. *If* $F(x,t) = \cos 2\pi x\, dB(t)$, *then with probability* 1, *there are at least two main shocks, i.e., shock curves defined for all negative times. There are also at least two two-sided minimizers on* $\mathbb{S}^1 \times \mathbb{R}^1$.

We will give an outline of the proof showing that with probability one, $x = 0$ and $x = 1/2$ are points of shock for any time t. The main point is:

LEMMA 5.8. *With probability* 1, $\xi\colon (-\infty, t] \to \mathbb{R}^1$, $\xi(s) \equiv 0$, *is not a one-sided minimizer.*

This follows from the observation that with probability 1, there are large intervals on which $B(t) - B(s) > 0$. On such intervals, one-sided minimizers are close to $x = \frac{1}{2}$. Hence $\xi \equiv 0$ is not minimizing.

As a consequence of symmetry, if $\xi\colon (-\infty, t] \to \mathbb{R}^1$ is a one-sided minimizer such that $\xi(t) = 0$, then $-\xi$ is also a one-sided minimizer. Therefore, with probability 1, $x = 0$ is a point of shock for all $t \in \mathbb{R}^1$. The same argument applies to $x = \frac{1}{2}$. So there are at least two main shocks. The rest of the statement in Theorem 5.3 follows from the same argument as in the proof of Theorems 5.1 and 5.2.

It is easy to check that (A1) holds for

$$F(x,t) = \cos 2\pi(x + x_1)dB_1(t) + \cos 2\pi(x + x_2)dB_2(t) + \cos 2\pi(x + x_3)dB_3(t) \tag{5.19}$$

where x_1, x_2, x_3 are fixed constants such that their differences are not integral multiples of $\frac{1}{2}$. By refining the argument in Appendix D, one can actually show that Lemma 5.2 also holds if there are only two terms in (5.19). On the other hand, without shifting phases, (A1) does not hold if all of the F_k's are of the form $\cos 2\pi kx$. It fails when $x_1^0 = 0$, $x_2^0 = \frac{1}{2}$. However, the following argument shows that Lemma 5.2 still holds if:

The set $\{F_k\}$ contains either

$\{\sin 2\pi x, \cos 2\pi lx$, for some $l \neq 0\}$, or $\{\cos 2\pi x, \sin 2\pi lx$, for some $l \neq 0\}$.

We will illustrate how this claim can be proved when $\{F_k\}$ contains $\{\cos 2\pi x, \sin 4\pi x\}$. The only situation we have to reconsider is when x_1^0 is close to a critical point of $F_1(x) = \cos 2\pi x$, and x_2^0 is close to a critical point of $F_2(x) = \sin 4\pi x$. Without loss of generality, let us assume that x_1^0 is close to 0, and x_2^0 is close to $\frac{1}{8}$. Heuristically we can first use F_1 to move x_2^0 to a small neighborhood of $\frac{1}{2}$. If in this process x_1^0 has moved out of the neighborhood of 0, then we can use Lemma 5.2 with F_1. If not, we use F_2 to move x_1^0 to a small neighborhood of $-\frac{1}{8}$. The forces dB_2 can be chosen such that x_2^0 will stay inside $\left(\frac{1}{8}, \frac{5}{8}\right)$. Now both x_1^0 and x_2^0 are inside the region where F_1' is bounded away from 0, so we can apply the proof of Lemma 5.2 to x_1^0 and x_2^0 with the potential F_1. We will omit the detailed proof of these statements since they follow closely the proof of Lemma 5.2.

6. Hyperbolicity and unstable manifolds of the two-sided minimizer

In this section we prove that the two-sided minimizer $y_\omega(\cdot)$ constructed in Section 5 is a hyperbolic trajectory of the dynamical system (2.3), and we establish the existence of its stable and unstable manifolds. The main technical difficulty is associated with the fact that the hyperbolicity is nonuniform, as in many other dynamical systems with noises. This is overcome by using Pesin's theory (see [Pes], [En]). Note in passing that

$$y_\omega(t+s) = y_{\theta^s\omega}(t).$$

Denote by G_t^ω the stochastic flow generated by the solutions of (2.3). Let $J_s^t(\omega)$ be the Jacobi map, i.e., the tangent map that maps the tangent plane $T(y_\omega(s), u^\omega(y_\omega(s),\, s))$ onto the tangent plane $T(y_\omega(t), u^\omega(y_\omega(t), t))$. This is well-defined since $y_\omega(t)$ is a point of continuity of $u^\omega(\,\cdot\,, t)$ for all t. Moreover the Jacobi map has determinant 1 since the dynamical system (2.3) preserves the Lebesgue measure. Obviously, we have

$$J_0^{t_2}(\omega) = J_0^{t_2-t_1}(\theta^{t_1}\omega) J_0^{t_1}(\omega)$$

for all t_1, t_2. In the terminology of ergodic theory, $\{J_0^t(\omega)\}$ is a cocycle (see [O]).

LEMMA 6.1. *Define* $\log^+ x = \max(\log x, 0)$, *for* $x > 0$. *Then*

$$\sup_{-1\le t\le 1} \log^+ \|J_0^t(\omega)\| \in L^1(dP)\,.$$

This result, together with some other technical estimates, is proved in Appendix B (Lemma B.5).

As a consequence of the multiplicative ergodic theorem [O], [En], we conclude that with probability 1:

(A) either

$$\lim_{t\to\pm\infty}\frac{1}{t}\ln\|J_{t_1}^{t+t_1}(\omega)e\| = 0$$

for all $e \in T_{t_1} = T(y_\omega(t_1), u^\omega(y_\omega(t_1), t_1))$;

(B) or there exist a constant $\lambda > 0$ and a measurable normalized basis $\{e_t^u(\omega), e_t^s(\omega)\}$ of $T_t = T(y_\omega(t), u^\omega(y_\omega(t), t))$, such that

$$J_{t_1}^t(\omega)e_{t_1}^u(\omega) = a^u(t, t_1;\omega)e_t^u(\omega),\ \ J_{t_1}^t(\omega)e_{t_1}^s(\omega) = a^s(t, t_1;\omega)e_t^s(\omega)\,,$$

where the functions $a^u(t, t_1;\omega)$ and $a^s(t, t_1;\omega)$ are also cocycles satisfying

$$a^u(t+s, 0;\omega) = a^u(s, 0;\theta^t\omega)a^u(t, 0;\omega)\,,$$
$$a^s(t+s, 0;\omega) = a^s(s, 0;\theta^t\omega)a^s(t, 0;\omega)\,.$$

Furthermore,

$$\lim_{t\to\infty}\frac{\ln a^u(t, t_1;\omega)}{t-t_1} = \lambda,\ \lim_{t\to\infty}\frac{\ln a^s(t, t_1;\omega)}{t-t_1} = -\lambda\,.$$

If (B) holds, the cocycle $\{J_s^t(\omega)\}$ is said to be hyperbolic and the basis $\{e_t^u(\omega), e_t^s(\omega)\}$ is called the Oseledetz basis.

THEOREM 6.1. *With probability* 1, *the cocycle* $\{J_s^t(\omega)\}$ *is hyperbolic.*

We will prove Theorem 6.1 later. It is useful to recall the following simple result:

LEMMA 6.2. *Let* $\{\eta_i\}$ *be a sequence of identically distributed random variables such that* $E|\eta_i| < +\infty$. *Then for any* $\varepsilon > 0$, *there exists a random variable* $N_\varepsilon > 0$, *such that for all* i, $|i| \geq N_\varepsilon$,

$$|\eta_i| \leq \varepsilon|i|\,.$$

This is a simple consequence of the Chebyshev inequality and the Borel-Cantelli lemma. Lemma 6.2 is equivalent to the statement that

$$\lim_{i\to\infty}\frac{\eta_i}{i} = 0$$

with probability 1. However, we will use it in the form of Lemma 6.2.

Let $x(\cdot)$ be an arbitrary one-sided minimizer defined on $(-\infty, 0]$. Fix a positive integer k and consider a sequence of times $t_i = -ki$.

Denote $(y_i, u_i) = (y_\omega(t_i), u^\omega(y_\omega(t_i), t_i))$, $(x_i, v_i) = (x(t_i), u^\omega(x(t_i), t_i))$, $J_i = J_{t_i}^{t_{i-1}}(\omega)$, $\ell_i = \text{dist}(x_i, y_i)$, $\rho_i = \text{dist}((x_i, v_i), (y_i, u_i))$.

LEMMA 6.3. *For any $\varepsilon > 0$ there exists a random constant $\ell_{\varepsilon,k}(\omega)$ such that with probability* 1

$$\rho_i \le (1+\varepsilon)\|J_{i+1}\|\rho_{i+1}, \quad i \ge 0, \tag{6.1}$$

provided that $\ell_0 \le \ell_{\varepsilon,k}(\omega)$.

Proof. Let $L_0(i) = L_0(\theta^{t_i}\omega)$, $d(i) = d_k(\theta^{t_i}\omega)$, $\bar{d}(i) = \bar{d}_k(\theta^{t_i}\omega)$, where L_0 is as defined in Lemma B.8, and d_k, $\bar{d}_k$ are as defined in (B.29–30) and Lemma B.5. It follows from Theorem 5.2 that for any $\delta > 0$ there exists a random constant $N_\delta(\omega)$ such that for all $i > N_\delta(\omega)$

$$\ell_i \le \exp((C_1 - \delta)t_i). \tag{6.2}$$

Since $|v_i - u_i| \le L_0(i)\ell_i$, we have

$$\rho_i \le \sqrt{L_0^2(i) + 1}\ \ell_i \le \sqrt{L_0^2(i) + 1}\ \exp((C_1 - \delta)t_i). \tag{6.3}$$

Let $\Delta_i = \{(x, v) = \alpha(x_i, v_i) + (1 - \alpha)(y_i, u_i),\ 0 \le \alpha \le 1\}$ be the interval connecting (x_i, v_i) and (y_i, u_i). Clearly $\Delta_i \in B_k(\theta^{t_i}\omega)$. It follows from the definition of $d(i)$ that for any $(x, v) \in \Delta_i$

$$\|G_t^{\theta^{t_i}\omega}(x, v) - G_t^{\theta^{t_i}\omega}(y_i, u_i)\| \le \exp(d(i))\rho_i, \ -k \le t \le 0. \tag{6.4}$$

Since $d(i)$, $\bar{d}(i)$, $L_0(i)$ have finite expectations, for any $\nu > 0$, there exists $N_\nu(\omega)$ such that

$$|d(i)|,\ |\bar{d}(i)|,\ |L_0(i)| \le \nu i \quad \text{for} \quad i > N_\nu(\omega). \tag{6.5}$$

Hence, for $i > \max(N_\delta(\omega), N_\nu(\omega))$

$$\|G_t^{\theta^{t_i}\omega}(x, v) - G_t^{\theta^{t_i}\omega}(y_i, u_i)\| \le \sqrt{\nu^2 i^2 + 1}\ \exp(\nu i + (C_1 - \delta)t_i), \ -k \le t \le 0. \tag{6.6}$$

Take $\nu < (C_1 - \delta)$. Then (6.6) implies that there exists

$$N_{\delta,\nu}(\omega) > \max(N_\delta(\omega), N_\nu(\omega))$$

such that $\Delta_i \subset O_k(\theta^{t_i}\omega)$ for all $i > N_{\delta,\nu}(\omega)$. Clearly, if ρ_0 is small enough, then $\Delta_i \subset O_k(\theta^{t_i}\omega)$ for $i \le N_{\delta,\nu}(\omega)$. Since the two-sided minimizer corresponds to a point of continuity of u_+^ω, we have $\rho_0 \to 0$ as $\ell_0 \to 0$. Thus, there exists $\overline{\ell_0}(\omega) > 0$ such that $\Delta_i \in O_k(\theta^{t_i}\omega)$ for all i, provided that $\ell_0 \le \overline{\ell_0}(\omega)$. Denote now $D(i) = D_{k,2}(\omega)$, $\bar{D}(i) = \bar{D}_{k,2}(\omega)$, where $D_{T,r}(\omega)$, $\bar{D}_{T,r}(\omega)$ are defined as in (B.22). Since $\Delta_i \in O_k(\theta^{t_i}\omega)$, we have for all $i \ge 0$

$$\rho_i \le \|J_{i+1}\|\rho_{i+1} + \frac{1}{2}\exp(\bar{D}(i))\rho_{i+1}^2 = \|J_{i+1}\|\rho_{i+1}\left(1 + \frac{\exp(\bar{D}(i))}{2\|J_{i+1}\|}\rho_{i+1}\right).$$

Since $\frac{1}{\|J_{i+1}\|} \le \|J_{i+1}^{-1}\| \le \exp(D(i))$,

$$\rho_i \le \|J_{i+1}\|\rho_{i+1}\left(1 + \frac{1}{2}\exp(D(i) + \bar{D}(i))\rho_{i+1}\right). \tag{6.7}$$

Again, since $D(i)$ and $\bar{D}(i)$ have finite expectations, for any $\nu > 0$, there exists $\overline{N_\nu}(\omega)$ such that

$$|D(i)|,\ |\bar{D}(i)| \le \nu i \quad \text{for} \quad i > \overline{N_\nu}(\omega). \tag{6.8}$$

Thus, by (6.3), (6.5), (6.8), for $i > \max(N_\delta(\omega), N_\nu(\omega), \overline{N_\nu}(\omega))$:

$$\frac{1}{2}\exp(D(i) + \bar{D}(i))\rho_{i+1} \le \frac{1}{2}\sqrt{\nu^2 i^2 + 1}\ \exp(2\nu i + (C_1 - \delta)t_i). \tag{6.9}$$

Take $\nu < \frac{C_1-\delta}{2}$. It follows from (6.9) that there exists $N(\omega)$ such that for $i > N(\omega)$

$$\frac{1}{2}\exp(D(i) + \bar{D}(i))\rho_{i+1} \le \varepsilon. \tag{6.10}$$

This implies (6.1) for $i > N(\omega)$. Now, in order to get (6.1) for all i, take $\rho_0(\omega)$ so small that for $\rho_0 \le \rho_0(\omega)$

$$\frac{1}{2}\exp(D(i) + \bar{D}(i))\rho_{i+1} \le \varepsilon, \quad 0 \le i \le N(\omega). \tag{6.11}$$

As above, we can choose $\ell_{\varepsilon,k}(\omega) < \overline{\ell_0}(\omega)$ so small that $\ell_0 < \ell_{\varepsilon,k}(\omega)$ implies $\rho_0 \le \rho_0(\omega)$. (6.1) obviously follows from (6.10), (6.11). □

Proof of Theorem 6.1. Assume that (A) holds. It follows from the subadditive ergodic theorem that

$$\lim_{n\to\infty} \frac{\int \ln \|J_0^n(\omega)\| P(d\omega)}{n} = 0. \tag{6.12}$$

Then, for any $\varepsilon > 0$ there exists $k \in \mathbb{N}$ such that $A_k = \frac{1}{k}\int \ln \|J_0^k(\omega)\| P(d\omega) < \varepsilon$. By the ergodic theorem, with probability 1,

$$\frac{1}{kn}\sum_{i=1}^{n} \ln \|J_0^k(\theta^{t_i}\omega)\| \underset{n\to\infty}{\longrightarrow} \frac{1}{k}\int \ln \|J_0^k(\omega)\| P(d\omega) = A_k. \tag{6.13}$$

Hence there exists a random constant $n_\varepsilon(\omega)$ such that, with probability 1,

$$\frac{1}{kn}\sum_{i=1}^{n} \ln \|J_0^k(\theta^{t_i}\omega)\| \le A_k + \varepsilon \le 2\varepsilon \tag{6.14}$$

for all $n > n_\varepsilon(\omega)$.

Consider now a one-sided minimizer $x(\cdot)$ at time $t_0 = 0$ such that $\ell_0 = |x(0) - y_\omega(0)| \le \ell_{\varepsilon,k}(\omega)$, where $\ell_{\varepsilon,k}(\omega)$ is defined as in Lemma 6.3. Then, by Lemma 6.3 for all $n > 0$:

$$\ell_0 \le \rho_0 \le (1+\varepsilon)^n \prod_{i=1}^{n} \|J_{t_i}^{t_{i-1}}(\omega)\| \rho_n \,. \tag{6.15}$$

Thus for $n > n_\varepsilon(\omega)$

(6.16)

$$\rho_n \ge \frac{\rho_0}{(1+\varepsilon)^n \prod_{i=1}^{n} \|J_{t_i}^{t_{i-1}}(\omega)\|} = \frac{\rho_0}{(1+\varepsilon)^n \exp\left(\sum_{i=1}^{n} \ln \|J_0^k(\theta^{t_i}\omega)\|\right)} \ge \rho_0 e^{-\varepsilon n} \exp(-2\varepsilon k n)\,.$$

On the other hand, it follows from Theorem 5.2 that for large enough n

(6.17)

$$\rho_n \le \sqrt{L_0^2(n)+1}\; \ell_n \le \sqrt{L_0^2(n)+1}\; \exp(-(C_1-\delta)kn) \le \sqrt{\nu^2 n^2+1}\; \exp(-(C_1-\delta)kn)\,.$$

Here, as in the proof of Lemma 6.3, we used again that $|L_0(n)| \le \nu n$ for n large enough. Take ε so small that $3\varepsilon < C_1 - \delta$. Then, (6.16) and (6.17) are contradictory to each other. □

Remark. It follows from the proof of Theorem 6.1 that $\lambda \ge C_1 - \delta$. Since δ is arbitrarily small, $\lambda \ge C_1$.

Next we construct stable and unstable manifolds of the two-sided minimizer. We will denote by Γ_ω the trajectory in the phase space of the two-sided minimizer $\Gamma_\omega = \{(y_\omega(t), u^\omega(y_\omega(t),t)),\, t \in \mathbb{R}^1\}$, and let $(x(t;x_0,u_0), u(t;x_0,u_0))$ be the solution of the SDE (2.3) with initial data $x(0) = x_0$, $u(0) = u_0$. We will concentrate on $t = 0$ but the same holds for any other t.

Definition 6.1. A *local stable manifold* of Γ_ω at $t = 0$ is the set

$$W^s_{\delta,\varepsilon} = \{(x_0,u_0),\ \operatorname{dist}((x(t;x_0,u_0),u(t;x_0,u_0)), (y_\omega(t), u^\omega(y_\omega(t),t))) \le \delta e^{-(\lambda-\varepsilon)t}\}$$

for some $\varepsilon > 0$, $\delta > 0$ and all $t > 0$. A *local unstable manifold* of Γ_ω at $t = 0$ is the set

$$W^u_{\delta,\varepsilon} = \{(x_0,u_0),\ \operatorname{dist}((x(t;x_0,u_0),u(t;x_0,u_0)), (y_\omega(t), u^\omega(y_\omega(t),t))) \le \delta e^{-(\lambda-\varepsilon)|t|}\}$$

for some $\varepsilon > 0$, $\delta > 0$ and all $t < 0$.

Pesin [Pes] gave general conditions under which such local stable and unstable manifolds exist for smooth maps of compact manifolds. It is easy to check that his results can be extended directly to the current situation of stochastic flows. Below we will formulate Pesin's theorem and later verify that its conditions are satisfied for our problem.

Denote by S_i the Poincaré map at $t = i$ associated with the SDE (2.3). In other words, S_i maps (x_i, u_i) at $t = i$ to the solution of (2.3), (x_{i+1}, u_{i+1}) at $t = i+1$. Similarly we denote by S_i^n, S_i^{-n} the maps that map the solution of (2.3) at $t = i$ to the solution at $t = i+n,\ t = i-n$ respectively.

Define λ_i^u, λ_i^s by the relations

$$J_i^{i+1}(\omega)e_i^u = e^{\lambda_i^u} e_{i+1}^u,\ J_i^{i+1}(\omega)e_i^s = e^{-\lambda_i^s} e_{i+1}^s$$

where $\{e_i^s, e_i^u\}$ constitutes the Oseledetz basis.

PESIN'S THEOREM. *Assume that there exist constants $\lambda, \mu > 0$, and $\varepsilon_0 \in (0,1)$, and for $\varepsilon \in (0, \varepsilon_0)$, one can find a positive random variable $C(\varepsilon, \omega)$ such that for $i \in \mathbb{Z}$*

(I)
$$\|DS_i^n e_i^s\| \le C(\varepsilon,\omega) e^{-(\lambda-\varepsilon)n} e^{\varepsilon|i|}$$
$$\|DS_i^{-n} e_i^u\| \le C(\varepsilon,\omega) e^{-(\mu-\varepsilon)n} e^{\varepsilon|i|},$$

(II)
$$|\sin\langle e_i^s, e_i^u\rangle| \ge \frac{1}{C(\varepsilon,\omega)} e^{-\varepsilon|i|}.$$

(III) *Let* $r_i = \frac{1}{C(\varepsilon,\omega)} e^{-\varepsilon|i|}$, *and*

$$\mathcal{B}_i(\omega) = \{(x,u), \|(x,u) - (y_\omega(i), u^\omega(y_\omega(i), i))\| \le r_i\}.$$

Then for some $r \ge 2$,

$$\sup_{(x,u)\in\mathcal{B}_i} \max_{1\le j\le r} \left(\|D^j S_i(x,u)\|,\ \|D^j S_i^{-1}(x,u)\|\right) \le C(\varepsilon,\omega) e^{\varepsilon|i|}.$$

Under these assumptions, there exist positive $\varepsilon_1(\lambda, \mu, \varepsilon_0)$ and $\delta(\varepsilon)$, defined for $0 < \varepsilon < \varepsilon_1$, and C^{r-1} curves $W_{\delta,\varepsilon}^s$, $W_{\delta,\varepsilon}^u$ in the phase space of the dynamical system (2.3), such that

(i) *$W_{\delta,\varepsilon}^s$ and $W_{\delta,\varepsilon}^u$ are respectively the stable and unstable manifolds of Γ_ω at $t = 0$. Moreover, they are C^{r-1} graphs on the interval $[-\delta_1(\varepsilon), \delta_1(\varepsilon)]$ for some $\delta_1(\varepsilon) > 0$.*

(ii) $W_{\delta,\varepsilon}^s \cap W_{\delta,\varepsilon}^u = (y_\omega(0), u^\omega(y_\omega(0), 0))$.

(iii) *The tangent vectors to $W_{\delta,\varepsilon}^s$ and $W_{\delta,\varepsilon}^u$ at $(y_\omega(0), u^\omega(y_\omega(0), 0))$ are respectively e_0^s and e_0^u.*

(iv) *If* $(x,u) \in \mathcal{B}_0$, *and* $n \geq 0$,

$$\operatorname{dist}\left(S_0^{-n}(x,u), (y_\omega(-n), u^\omega(y_\omega(-n), -n))\right) \leq \bar{\delta} e^{-\chi n}$$

for some constants $\chi > 0$ *and* $\bar{\delta} > 0$, *then* $(x,u) \in W^u_{\delta,\varepsilon}$.

Our task is reduced to checking the assumptions (I), (II) and (III) in Pesin's Theorem.

To begin with, let us observe that (II) follows from the next argument (see [R]). Since

$$\lim_{t\to\infty} \frac{\ln a^s(0,t)}{t} = \lambda, \quad \lim_{t\to\infty} \frac{\ln a^u(0,t)}{t} = -\lambda$$

and the area of the parallelogram generated by e_t^s and e_t^u is independent of t, we have

$$\lim_{t\to\infty} \frac{\ln|\sin\langle e_t^s, e_t^u\rangle|}{t} = 0\,.$$

To see that (III) holds, define

$$d_i(\omega) = \sup_{-1\leq t\leq 0} \sup_{(x,u)\in\mathcal{B}_0(\theta^i\omega)} \|DG_t^{\theta^i\omega}(x,u)\|\,,$$

$$\bar{d}_i(\omega) = \sup_{0\leq t\leq 1} \sup_{(x,u)\in\mathcal{B}_0(\theta^i\omega)} \|DG_t^{\theta^i\omega}(x,u)\|\,,$$

where G_t^ω is the stochastic flow defined earlier.

LEMMA 6.4. *For any* $\varepsilon > 0$, *there exist random constants* $C_1(\varepsilon,\omega)$, $C_2(\varepsilon,\omega)$ *such that, with probability* 1,

$$d_i(\omega) \leq C_1(\varepsilon,\omega)e^{\varepsilon|i|}, \quad \bar{d}_i(\omega) \leq C_2(\varepsilon,\omega)e^{\varepsilon|i|}$$

for $i \in \mathbb{Z}$.

Proof. Assume $C(\varepsilon,\omega) > 1$. Then it follows from Lemma B.5 that

$$\int \log^+ d_0(\omega)dP < +\infty, \quad \int \log^+ \bar{d}_0(\omega)dP < +\infty\,.$$

Now Lemma 6.4 follows directly from Lemma 6.2. □

Let $C(\varepsilon,\omega) > \max(C_1(\varepsilon,\omega), C_2(\varepsilon,\omega))$, and

$$d_{i,r}(\omega) = \sup_{(x,v)\in\mathcal{B}_i} \max_{1\leq j\leq r} \|D^j S_i(x,v)\|\,,$$

$$\bar{d}_{i,r}(\omega) = \sup_{(x,v)\in\mathcal{B}_i} \max_{1\leq j\leq r} \|D^j (S_{i-1})^{-1}(x,v)\|\,.$$

Statement (III) follows from the next result.

LEMMA 6.5. *For any* $\varepsilon > 0$, *there exist random constants* $C_3(\varepsilon,\omega)$ *and* $C_4(\varepsilon,\omega)$ *such that*

$$d_{i,r}(\omega) \le C_3(\varepsilon,\omega)e^{\varepsilon|i|}, \quad \bar{d}_{i,r}(\omega) \le C_4(\varepsilon,\omega)e^{\varepsilon|i|}$$

for $i \in \mathbb{Z}$.

Proof. Let $(x_0, v_0) \in \mathcal{B}_i(\omega)$. Consider the solution of (2.3), $(x(t), v(t))$, such that $x(i) = x_0$, $v(i) = v_0$, for $t \in [i, i+1]$,

$$|v(t) - u^\omega(y_\omega(i+t), i+t)| \le d_i(\omega)\|(x_0, v_0) - (y_\omega(i), u^\omega(y_\omega(i), i))\| \le 1$$

from Lemma 6.4. Therefore $\mathcal{B}_i(\omega) \subset O_1(\theta^i\omega)$ where $O_1(\omega)$ is defined as in Appendix B. Lemma 6.5 now follows directly from Lemma B.4 and Lemma 6.2. The second estimate can be proved in the same way. □

Finally, we prove statement (I).

LEMMA 6.6. *For any* $\varepsilon > 0$, *one can find random constants* $C_5(\varepsilon,\omega)$ *and* $C_6(\varepsilon,\omega)$ *such that*

$$\|DS_i^n e_i^s\| \le C_5(\varepsilon,\omega)e^{-(\lambda-\varepsilon)n}e^{\varepsilon|i|}, \qquad n \ge 1,$$
$$\|DS_i^{-n} e_i^u\| \le C_6(\varepsilon,\omega)e^{-(\lambda-\varepsilon)|n|}e^{\varepsilon|i|}, \qquad n \le -1.$$

Proof. We will prove the first statement. The second one can be proved in the same way.

From the ergodic theorem,

$$\lim_{n\to+\infty} \frac{1}{n}\sum_{j=0}^{n-1} \lambda_j^s = -\lambda < 0.$$

Thus for any $\varepsilon \in (0,\lambda)$, there exists a constant $C_7(\varepsilon,\omega) \ge 0$ such that

$$e^{\sum_{j=0}^{n-1}\lambda_j^s} \le C_7(\varepsilon,\omega)e^{-n(\lambda-\varepsilon)}.$$

For any $\delta \in (0,1)$, define $K(\delta)$ by:

$$K(\delta) = \inf\{K: \quad P(C(\omega) \le K) \ge \delta\}.$$

Denote

$$\delta_1 = P(C(\omega) \le K(\delta)) \ge \delta, \ m_1(m,\omega) = \max\{i: \ 1 \le i \le m, \ C(\theta^i\omega) \le K(\delta)\}.$$

Notice that $m_1(m,\omega)$ is defined for large enough m. By the ergodic theorem,

$$\lim_{m\to\infty} \frac{\#\{i: \ 1 \le i \le m, \ C(\theta^i\omega) \le K(\delta)\}}{m} = \delta_1$$

where # denotes cardinality. Thus for any $\delta_2 \in (0, \delta_1)$, there exists a random constant $M(\delta_2, \omega)$ such that for all $m > M(\delta_2, \omega)$,

$$\frac{\#\{i:\ 1 \le i \le m,\ C(\theta^i\omega) \le K(\delta)\}}{m} \ge \delta_1 - \delta_2 .$$

Hence

$$m_1 \ge (\delta_1 - \delta_2)m .$$

Consequently for $m > M(\delta_2, \omega)$,

$$\begin{aligned}
\|DS^n_m e^s_m\| &= e^{\sum_{j=m}^{m+n-1} \lambda^s_j} = e^{\sum_{j=m_1}^{m+n-1} \lambda^s_j} e^{-\sum_{j=m_1}^{m-1} \lambda^s_j} \\
&\le C(\theta^{m_1}\omega) e^{-(m+n-m_1)(\lambda-\varepsilon)} e^{-\sum_{j=m_1}^{m-1} \lambda^s_j} \\
&\le K(\delta) e^{-n(\lambda-\varepsilon)} e^{-(m-m_1)(\lambda-\varepsilon)} e^{\sum_{j=m_1}^{m-1}(-\lambda^s_j)} .
\end{aligned}$$

We also have, with $\delta_3 = \delta_1 - \delta_2$

$$\begin{aligned}
\sum_{j=m_1}^{m-1} (-\lambda^s_j) &\le \max_{\delta_3 m \le k \le m-1} \sum_{k}^{m-1} (-\lambda^s_j) \\
&\le \max_{\delta_3 m \le k \le m-1} \sum_{j=k+1}^{m} \log^+ \|D(S_j^{-1})(y_\omega(j), u^\omega(y_\omega(j), j))\| \\
&\le \max_{\delta_3 m \le k \le m-1} \sum_{j=k+1}^{m} d_1(\theta^j\omega)
\end{aligned}$$

where $d_1(\omega)$ is defined as in (B.29). Using Lemma B.5 and standard probabilistic estimates, one can show that for appropriate $\delta_3 < 1$ there exists a constant $M_1(\omega)$ such that for all $m > M_1(\omega)$,

$$\sum_{j=m_1}^{m-1} (-\lambda^s_j) \le \max_{\delta_3 m \le k \le m-1} \sum_{j=k+1}^{m} d_1(\theta^j\omega) \le \varepsilon m .$$

In fact, it is enough to have δ_3 so close to 1 that $(1 - \delta_3)\mathbb{E}d_1(\omega) < \frac{\varepsilon}{2}$, where $\mathbb{E}d_1(\omega) = \int d_1(\omega)P(d\omega)$. Hence we choose $\delta = 1 - \frac{\varepsilon}{8\mathbb{E}d_1(\omega)}, \delta_2 = \frac{\varepsilon}{8\mathbb{E}d_1(\omega)}$. Then for $m > \max(M_1(\omega), M(\delta_2, \omega))$,

$$\|DS^n_m e^s_m\| \le K(\delta) e^{-n(\lambda-\varepsilon)} e^{\varepsilon m} . \qquad \square$$

This completes the verification of the assumptions in Pesin's theorem, and establishes the existence of local stable and unstable manifolds $W^s_{\delta,\varepsilon}$, $W^u_{\delta,\varepsilon}$.

One can also define, in a standard way, global stable and unstable manifolds W^s, W^u:

$$W^s = \bigcup_{i=1}^{\infty} S_i^{-i} W_{\delta,\varepsilon}^s(y_{\theta^i\omega}(0), u^\omega(y_{\theta^i\omega}(0),0)),$$

$$W^u = \bigcup_{i=1}^{\infty} S_{-i}^{i} W_{\delta,\varepsilon}^u(y_{\theta^{-i}\omega}(0), u^\omega(y_{\theta^{-i}\omega}(0),0)).$$

Obviously, W^s, W^u are also C^{r-1} curves which coincide with $W_{\delta,\varepsilon}^s$, $W_{\delta,\varepsilon}^u$ in some neighborhood of $(y_\omega(0), u^\omega(y_\omega(0),0))$.

The following theorem is a consequence of the properties of unstable manifolds.

THEOREM 6.2. a. *The graph* $\{(x, u_+^\omega(x,0)),\ x \in \mathbb{S}^1\}$ *is a subset of the global unstable manifold* W^u.

b. *There exists a (random) neighborhood of* $(y_\omega(0), u^\omega(y_\omega(0),0))$ *such that* $W_{\delta,\varepsilon}^u$ *consists of one-sided minimizers in this neighborhood, i.e., the solutions of* (2.3) *with initial data on* $W_{\delta,\varepsilon}^u$ *in this neighborhood give rise to one-sided minimizers.*

Proof. a. As was shown earlier, any one-sided minimizer (x,u) converges exponentially fast to $(y_\omega(t), u^\omega(y_\omega(t),t))$ as $t \to -\infty$. It follows that $S_0^{-i}(x,u) \in \mathcal{B}_0(\theta^{-i}\omega)$ for some $i > 0$. Hence $S_0^{-i}(x,u)$ lies on the local unstable manifold $W_{\delta,\varepsilon}^u(y_{\theta^{-i}\omega}(0), u^\omega(y_{\theta^{-i}\omega}(0),0))$, as a consequence of Pesin's theorem (iv), and (x,u) lies on the global unstable manifold.

b. The local unstable manifold $W_{\delta,\varepsilon}^u$ is a C^{r-1} curve with the tangent vector e_0^u at $(y_\omega(0),\ u^\omega(y_\omega(0),0))$. Let $M_\alpha = \{(x,u)\colon x \in (y_\omega(0) - \alpha, y_\omega(0) + \alpha),\ (x,u)$ corresponds to a one-sided miminizer$\}$. Since $y_\omega(0)$ is a point of (Lipschitz) continuity of $u_+^\omega(x)$ it follows that there exists $\alpha_0(\omega)$ such that for all $\alpha < \alpha_0(\omega)$, $M_\alpha \subset W_{\delta,\varepsilon}^u$. Hence e_0^u is not a vertical vector; i.e., $e_0 \neq (0,1)$. Therefore there exists a neighborhood O_ω of $(y_\omega(0), u^\omega(y_\omega(0),0))$ such that in this neighborhood $W_{\delta,\varepsilon}^u$ is a graph of a C^{r-1} function; i.e.,

$$W_{\delta,\varepsilon}^u \cap O_\omega = \{(x,u)\colon x \in (y_\omega(0) - \alpha_1(\omega), y_\omega(0) + \alpha_2(\omega)),\ u = \bar{u}(x)\},$$

where $\alpha_1(\omega)$, $\alpha_2(\omega) > 0$, and $\bar{u}(x)$ is a C^{r-1} function. Now choose α so small that $M_\alpha \subset W_{\delta,\varepsilon}^u \cap O_\omega$. It follows that for $x \in (y_\omega(0) - \alpha, y_\omega(0) + \alpha)$, $\bar{u}(x) = u_+^\omega(x,0)$, which proves b. $\square$

COROLLARY 6.3. *There exists* $\alpha(\omega) > 0$ *such that there are no shocks inside the interval* $(y_\omega(0) - \alpha, y_\omega(0) + \alpha)$. *Moreover,*

$$u_+^\omega \in C^{r-1}(y_\omega(0) - \alpha, y_\omega(0) + \alpha).$$

7. Regularity of solutions

In this section, we give a complete description of the solution u_+^ω in terms of the unstable manifold W^u, and prove that the number of shocks is finite for almost all ω. We also prove a stronger version of Lemma 5.8, namely, that all shocks except the main shock have finite prehistory. We will start with the latter statement.

THEOREM 7.1. *Fix arbitrary $t_0 \in \mathbb{R}^1$. With probability 1 there exists a random constant $T_{t_0}(\omega)$ such that all shocks at time t_0, except the main shock, are generated after the time $t_0 - T_{t_0}(\omega)$. In other words, all shocks at time $t_0 - T_{t_0(\omega)}$ merge with the main shock before $t = t_0$.*

Proof. Fix any $t \in \mathbb{R}^1$. It follows from Theorem 6.2 that there exists $\varepsilon_1(t,\omega) > 0$ such that the velocities of all one-sided minimizers in $(y_\omega(t) - \varepsilon_1,\, y_\omega(t) + \varepsilon_1)$ lie on a C^{r-1} curve. Hence there are no shocks in the ε_1-neighborhood of $y_\omega(t)$. Notice that the random constant $\varepsilon_1(t,\omega)$ has stationary distribution. We can choose an $\alpha > 0$ so small that $P(\omega\colon \varepsilon_1(t,\omega) > \alpha) > 0$. Then there exists an infinite sequence $t_i \to -\infty$, such that $\varepsilon_1(t_i,\omega) > \alpha, i \in \mathbb{N}$. Since minimizers at $t = t_0$ converge uniformly to the two-sided minimizer as $t \to -\infty$, there exists $I_{t_0}(\omega)$ such that for $i \geq I_{t_0}(\omega)$, all minimizers starting at $t = t_0$ pass through the ε_1-neighborhood of $y_\omega(t_i)$. Now let $T_{t_0}(\omega) = t_0 - t_{I_{t_0}(\omega)}$. We conclude that the complement (on $\mathbb{S}^1$) of the ε_1-neighborhood of $y_\omega(t_{I_{t_0}(\omega)})$ will merge into the main shock before time t_0. Since the ε_1-neighborhood of $y_\omega(t_{I_{t_0}(\omega)})$ contains no shocks, this completes the proof of Theorem 7.1. □

Let s be the signed arc-length parameter for the unstable manifold W^u of the two-sided minimizer at $t = 0$:

$$W^u = \{(x(s), u(s)),\ x(s) \in \mathbb{S}^1,\ u(s) \in \mathbb{R}^1\}, \tag{7.1}$$

with $s = 0$ at $(y_\omega(0), u^\omega(y_\omega(0), 0))$. From the proof of Theorem 6.2, $\frac{dx}{ds}(0) \neq 0$. We will fix orientation of the parameter s by the assumption $\frac{dx}{ds}(0) > 0$. Let $\widetilde{\Gamma}_0$ be the lifting of W^u to the universal cover

$$\widetilde{\Gamma}_0 = \{(\tilde{x}(s), u(s)),\ \tilde{x}(s) \in \mathbb{R}^1,\ u(s) \in \mathbb{R}^1\}. \tag{7.2}$$

Also, denote by $(x_s(t,\omega), v_s(t,\omega))$ the solution of (2.3) with initial data $x_s(0,\omega) = x(s)$, $v_s(0,\omega) = u(s)$. Since for all s, the solutions $(x_s(t,\omega), v_s(t,\omega))$ converge exponentially fast to $(y_\omega(t), \dot{y}_\omega(t))$ as $t \to -\infty$, we can define the function

$$A(s) = \int_{-\infty}^{0} \left\{ \frac{1}{2}(v_s^2(t,\omega) - \dot{y}_\omega^2(t)) + (F(x_s(t,\omega),t) - F(y_\omega(t),t)) \right\} dt. \tag{7.3}$$

Since W^u is almost surely a C^{r-1} manifold, $A(s)$ is almost surely a C^{r-1} function. Let

$$\bar{A}(x) = \min_{s:\, x(s)=x} A(s)\,. \tag{7.4}$$

In the following, we will enumerate the shocks as in Section 5, except we number the main shock (which is not necessarily the strongest shock at a given time) as the zeroth shock.

Our next theorem describes the following picture. When we view the unstable manifold W^u as a curve on the cylinder $\{x \in \mathbb{S}^1,\ u \in \mathbb{R}^1\}$, the two-sided minimizer divides W^u into left and right pieces. It turns out that all shocks correspond to double folds of W^u, i.e., graphs of a multi-valued function. A single-valued function is obtained by introducing jump discontinuities which are vertical cuts on the double fold. These are the shocks in the solution. The end points of the cut define two points on W^u with the same x-coordinate (namely the position of the shock) and the same value of the action A in (7.3). If x denotes the position of the shock, then the end points of the cut are $(x, u_+^\omega(x,0))$ and $(x, u_-^\omega(x,0))$. Except for the main shock, the one-sided minimizers starting from $(x, u_+^\omega(x,0))$ and $(x, u_-^\omega(x,0))$ approach the two-sided minimizer as $t \to -\infty$ from the same side. However, for the main shock, they approach the two-sided minimizer from different sides. We formulate this as:

THEOREM 7.2. *Fix arbitrary* $t_0 \in \mathbb{R}^1$.

I. *Let* $(x(s), u(s)) \in W^u$. *Now* $(x(s), u(s))$ *gives rise to a one-sided minimizer if and only if* $A(s) = \bar{A}(x(s))$. *With probability* 1, $\bar{A}(x)$ *is defined for all* $x \in \mathbb{S}^1$; *i.e., the minimum in* (7.4) *is attained. Moreover* $\bar{A}$ *is a continuous function on* $\mathbb{S}^1$.

II. *Let* x_i *be the position of the* i^{th} *shock,* $i \geq 1$ *(not the main shock!). Then there exists an interval* $\ell_i = [\underline{s}_i, \overline{s}_i]$ *such that*

$$\underline{s}_i = \min\{s\colon x(s) = x_i,\ A(s) = \bar{A}(x_i)\}\,,$$
$$\overline{s}_i = \max\{s\colon x(s) = x_i,\ A(s) = \bar{A}(x_i)\}\,.$$

Also, ℓ_i *lies either to the left or to the right of the two-sided minimizer, i.e.* $s = 0 \notin \ell_i$, $i \geq 1$ *and* $\ell_i \cap \ell_j = \emptyset$, $i \neq j$. *If* $\tilde{x}$ *is the* x*-coordinate of points on the unstable manifold lifted to the universal cover, then* $\tilde{x}(\underline{s}_i) = \tilde{x}(\overline{s}_i)$.

III. *The main shock corresponds to the only point* $z(t_0,\omega) \in \mathbb{S}^1$ *such that there exist* $s^{(1)} < 0$, $s^{(2)} > 0$ *for which* $A(s^{(1)}) = A(s^{(2)}) = \bar{A}(z(t_0,\omega))$, $x(s^{(1)}) = x(s^{(2)}) = z(t_0,\omega)$. *Denote*

$$\underline{S} = \max\{s < 0\colon x(s) = z(t_0,\omega),\ A(s) = \bar{A}(z(t_0,\omega))\}$$
$$\overline{S} = \min\{s > 0\colon x(s) = z(t_0,\omega),\ A(s) = \bar{A}(z(t_0,\omega))\}\,.$$

Then, $\tilde{x}(\overline{S}) - \tilde{x}(\underline{S}) = 1$.

IV. *Let* $\Delta^\omega = [\underline{S}, \overline{S}) \setminus \bigcup_{i\geq 1} [\underline{s}_i, \overline{s}_i)$. *Then for almost all* ω *the graph of* $u_+^\omega(x,0)$, $x \in \mathbb{S}^1$, *coincides with* $\{(x(s), u(s)),\ s \in \Delta^\omega\}$.

Proof. I. Clearly minimizers correspond to minima of A when the x coordinate is fixed. Since with probability 1, minimizers exist for all $x \in \mathbb{S}^1$, $\bar{A}(x)$ attains its minimum. Since the set of minimizers is closed, $\bar{A}(x)$ is continuous.

II. For every shock (except the main shock), denote by $\underline{s}_i$, $\overline{s}_i$ the values of the parameter s corresponding to the left-most and right-most minimizers. Since both the left-most and right-most minimizers approach the two-sided minimizer from the same side, the interval $l_i = [\underline{s}_i, \overline{s}_i]$ does not contain $s = 0$. Since minimizers do not intersect, the intervals ℓ_i do not intersect. It follows from Theorem 7.1 that all shocks, except the main shock, have finite past history. Notice that, at the moment of creation of a shock, $\underline{s}_i = \overline{s}_i$. Hence, $\tilde{x}(\underline{s}_i) = \tilde{x}(\overline{s}_i)$. Since $\tilde{x}(\underline{s}_i) - \tilde{x}(\overline{s}_i)$ is a continuous function of time between merges and it takes only integer values, $\tilde{x}(\underline{s}_i) - \tilde{x}(\overline{s}_i) \equiv 0$ for all shocks except the main shock.

III. The main shock is the only shock for which the two extreme one-sided minimizers approach the two-sided minimizer from different sides. Thus $[\underline{S}, \overline{S}]$ has nonempty interior. Clearly the intervals ℓ_i constructed above belong to $[\underline{S}, \overline{S}]$. As a consequence of periodicity, we have $\tilde{x}(\overline{S}) = 1 + \tilde{x}(\underline{S})$.

IV. IV follows easily from I–III. □

We next prove that for fixed time t_0 the number of shocks is finite. Consider time $t_0 - 1$. Although the position of the two-sided minimizer at time $t_0 - 1$ is not measurable with respect to $\mathcal{F}_{-\infty}^{t_0-1}$, the position of the main shock and the unstable manifold are measurable with respect to $\mathcal{F}_{-\infty}^{t_0-1}$. Consider the unstable manifold $W^u(t_0)$ at time t_0 as the image of $W^u(t_0 - 1)$ under the time-1 stochastic flow $G_1 = G_1^{\theta^{t_0-1}\omega}$:

(7.5)

$$W^u(t_0) = \{(x(s), v(s)) = G_1(y(s), w(s)),\ (y(s), w(s)) \in W^u(t_0 - 1), s \in \mathbb{R}^1\}.$$

Let E be the event that there exists $s_0 \in \mathbb{R}^1$ such that $\frac{dx}{ds}(s_0) = \frac{d^2x}{ds^2}(s_0) = 0$.

LEMMA 7.3.

$$P\{E|\mathcal{F}_{-\infty}^{t_0-1}\} = 0$$

for almost all conditions.

Proof. Consider an arbitrary interval $[s_1, s_2], s_1, s_2 \in \mathbb{R}^1$. Denote by E_1 the event that there exists $s_0 \in [s_1, s_2]$ such that $\frac{dx}{ds}(s_0) = \frac{d^2x}{ds^2}(s_0) = 0$. It is enough to show that $P\{E_1|\mathcal{F}_{-\infty}^{t_0-1}\} = 0$ for all s_1, s_2. Let $G_1 = (G_1^{(1)}, G_1^{(2)})$. We have

$$\frac{dx}{ds} = \frac{\partial G_1^{(1)}}{\partial y}\frac{dy}{ds} + \frac{\partial G_1^{(1)}}{\partial w}\frac{dw}{ds}, \quad \frac{dv}{ds} = \frac{\partial G_1^{(2)}}{\partial y}\frac{dy}{ds} + \frac{\partial G_1^{(2)}}{\partial w}\frac{dw}{ds} \tag{7.6}$$

$$\frac{d^2x}{ds^2} = \frac{\partial^2 G_1^{(1)}}{\partial y^2}\left(\frac{dy}{ds}\right)^2 + 2\frac{\partial^2 G_1^{(1)}}{\partial y \partial w}\frac{dy}{ds}\frac{dw}{ds} + \frac{\partial^2 G_1^{(1)}}{\partial w^2}\left(\frac{dw}{ds}\right)^2 + \frac{\partial G_1^{(1)}}{\partial y}\frac{d^2y}{ds^2} + \frac{\partial G_1^{(1)}}{\partial w}\frac{d^2w}{ds^2}.$$

Denote by

$$A(\omega) = \max_{s_1 \le s \le s_2}\left\{\left|\frac{\partial^k G_1^{(1,2)}}{\partial y^i \partial w^j}(y(s), w(s))\right|, \left|\frac{d}{ds}\left(\frac{\partial^k G_1^{(1,2)}}{\partial y^i \partial w^j}(y(s), w(s))\right)\right|, \; 1 \le k \le 2,\ i+j=k,\ i \ge 0,\ j \ge 0\right\}, \tag{7.7}$$

$$B(\omega) = \max_{s_1 \le s \le s_2}\left\{1, \left|\frac{d^2y}{ds^2}\right|, \left|\frac{d^2w}{ds^2}\right|\right\}.$$

Notice that $B(\omega)$ is measurable with respect to $\mathcal{F}_{-\infty}^{t_0-1}$. Take a small $\varepsilon > 0$ and divide the interval $[s_1, s_2]$ into subintervals of length ε. Denote by $s(j)$, $1 \le j \le \left[\frac{s_2-s_1}{\varepsilon}\right] + 2$, the end-points of the elements of the partition. Assume that there exists $s_0 \in [s(j), s(j+1)]$, such that $\frac{dx}{ds}(s_0) = \frac{d^2x}{ds^2}(s_0) = 0$. Then

$$\left|\frac{dx}{ds}(s(j))\right| = \left|\frac{dx}{ds}(s_0) + \frac{d^2x}{ds^2}(\xi)(s(j) - s_0)\right| = \left|\frac{d^2x}{ds^2}(\xi)(s(j) - s_0)\right|, \tag{7.8}$$

where $\xi \in (s(j), s_0)$. Denote by

$$V(\varepsilon, \omega) = \max_{s_1 \le s', s'' \le s_2, |s'-s''| \le \varepsilon}\left\{\left|\frac{dy}{ds}(s') - \frac{dy}{ds}(s'')\right|, \left|\frac{dw}{ds}(s') - \frac{dw}{ds}(s'')\right|, \left|\frac{d^2y}{ds^2}(s') - \frac{d^2y}{ds^2}(s'')\right|, \left|\frac{d^2w}{ds^2}(s') - \frac{d^2w}{ds^2}(s'')\right|\right\}. \tag{7.9}$$

Obviously $V(\varepsilon, \omega)$ is measurable with respect to $\mathcal{F}_{-\infty}^{t_0-1}$ and $V(\varepsilon, \omega) \to 0$ as $\varepsilon \to 0$. Using $\left(\frac{dy}{ds}\right)^2 + \left(\frac{dw}{ds}\right)^2 = 1$ and (7.6) it is easy to show that for all $s_1 \le s', s'' \le s_2$, $|s' - s''| \le \varepsilon$,

$$\left|\frac{d^2x}{ds^2}(s') - \frac{d^2x}{ds^2}(s'')\right| \le A(\omega)(10V(\varepsilon, \omega) + 2\varepsilon B(\omega) + 4\varepsilon). \tag{7.10}$$

Notice that $|\xi - s_0| \le \varepsilon$. Hence, using (7.8), (7.9), we have

$$\left|\frac{dx}{ds}(s(j))\right| \le A(\omega)(10V(\varepsilon,\omega) + 2\varepsilon B(\omega) + 4\varepsilon)\varepsilon\,. \tag{7.11}$$

Fix arbitrary $\delta > 0$. We will show that the conditional probability that s_0 exists is less than δ for almost all conditions. Clearly there exists a random constant $K(\omega) > 0$ which is measurable with respect to $\mathcal{F}_{-\infty}^{t_0-1}$ such that

$$P(A(\omega) > K(\omega)|\mathcal{F}_{-\infty}^{t_0-1}) < \frac{\delta}{2} \tag{7.12}$$

for almost all conditions. If $A(\omega) \le K(\omega)$ then

$$\left|\frac{dx}{ds}(s(j))\right| \le T(\varepsilon,\omega) = K(\omega)(10V(\varepsilon,\omega) + 2\varepsilon B(\omega) + 4\varepsilon)\varepsilon, \tag{7.13}$$

where $T(\varepsilon,\omega)$ is measurable with respect to $\mathcal{F}_{-\infty}^{t_0-1}$.

Fix $\tilde{s} \in [s_1, s_2]$ and consider the random process $x_s(t) = \frac{\partial x}{\partial s}(t,\tilde{s})$, $v_s(t) = \frac{\partial v}{\partial s}(t,\tilde{s})$, where $(x(t,s), v(t,s)) = G_t^{\theta^{t_0-1}\omega}(y(s), w(s))$. Clearly $(x_s(t), v_s(t))$ satisfies the stochastic differential equations

$$\begin{aligned} \dot{x}_s(t) &= v_s(t),\ x_s(0) = \frac{dy}{ds}(\tilde{s})\,, \\ \dot{v}_s(t) &= \sum_k f_k'(x(t,\tilde{s}))x_s(t)\dot{B}_k(t),\ v_s(0) = \frac{dw}{ds}(\tilde{s})\,. \end{aligned} \tag{7.14}$$

It follows from Lemma B.9 that the joint probability distribution for $\left(\frac{\partial x}{\partial s}(1,\tilde{s}) = x_s(1), \frac{\partial v}{\partial s}(1,\tilde{s}) = v_s(1)\right)$ has density $p(x_s, v_s)$ which is uniformly bounded inside any compact set for all $\tilde{s} \in [s_1, s_2]$. If $A(\omega) \le K(\omega)$ then, as follows from (7.6),

$$\max_{s_1 \le s \le s_2} \max\left(\left|\frac{dx}{ds}\right|, \left|\frac{dv}{ds}\right|\right) \le 2K(\omega)\,. \tag{7.15}$$

Denote by

$$\begin{aligned} O(\omega) &= \{(x_s, v_s) \in \mathbb{R}^2 : x_s^2 + v_s^2 \le 8K^2(\omega)\}, \\ \Pi_\varepsilon(\omega) &= \{(x_s, v_s) \in \mathbb{R}^2 : |x_s| \le T(\varepsilon,\omega)\}\,. \end{aligned}$$

Let $R(\omega) = \max_{s_1 \le \tilde{s} \le s_2} \sup_{(x_s,v_s)\in O(\omega)} p(x_s, v_s)$. Then, for any $\tilde{s} \in [s_1, s_2]$,

(7.16)
$$P\left(\left(\frac{\partial x}{\partial s}(1,\tilde{s}), \frac{\partial v}{\partial s}(1,\tilde{s})\right) \in O(\omega) \cap \Pi_\varepsilon(\omega)|\mathcal{F}_{-\infty}^{t_0-1}\right) \le R(\omega)(4K(\omega))2T(\varepsilon,\omega)\,.$$

Clearly, $R(\omega)$ is measurable with respect to $\mathcal{F}_{-\infty}^{t_0-1}$. Using (7.12), (7.16), we have

$$P(E_1|\mathcal{F}_{-\infty}^{t_0-1}) \le \frac{\delta}{2} + 8K(\omega)R(\omega)T(\varepsilon,\omega)\left(\left[\frac{s_2 - s_1}{\varepsilon}\right] + 1\right). \tag{7.17}$$

Choose ε so small that

$$8K^2(\omega)R(\omega)\varepsilon\left(\left[\frac{s_2-s_1}{\varepsilon}\right]+1\right)(10V(\varepsilon,\omega)+2\varepsilon B(\omega)+4\varepsilon)\le\frac{\delta}{2}.$$

Then, $P(E_1|\mathcal{F}_{-\infty}^{t_0-1})\le\delta$ for almost all conditions. □

Lemma 7.3 easily implies the following theorem.

THEOREM 7.4. *Fix $t_0\in\mathbb{R}^1$. With probability* 1, *the number of shocks at time t_0 is finite.*

Proof. As above, consider the unstable manifold $W^u(t_0)$ parametrized by the arc-length parameter of the unstable manifold $W^u(t_0-1)$. Denote by $\underline{S}',\overline{S}'$ the values of the parameter corresponding to the main shock. For every shock at time t_0 (except the main shock) there exists an interval $\ell=[s',s'']\subset[\underline{S}',\overline{S}']$, such that $x(s')=x(s'')$. Thus there exists a point $\hat{s}\in(s',s'')$ for which $\frac{dx}{ds}(\hat{s})=0$. Notice that the intervals ℓ corresponding to different shocks do not intersect. If there are infinitely many shocks, then there exists an infinite sequence of $\hat{s}_i$'s in $[\underline{S}',\overline{S}']$, such that $\frac{dx}{ds}(\hat{s}_i)=0$. Let s_0 be an accumulation point for the sequence $\{\hat{s}_i\}$. Obviously, $\frac{dx}{ds}(s_0)=\frac{d^2x}{ds^2}(s_0)=0$. It follows from Lemma 7.3 that the conditional probability for the existence of such an s_0 is equal to zero. This immediately implies the theorem. □

8. The zero viscosity limit

In this section, we study the limit as $\varepsilon\to 0$ of the invariant measures for the viscous equation

$$\frac{\partial u}{\partial t}+\frac{\partial}{\partial x}\left(\frac{u^2}{2}\right)=\frac{\varepsilon}{2}\frac{\partial^2 u}{\partial x^2}+\frac{\partial F}{\partial x}. \tag{8.1}$$

Under the same assumptions on F, it was proved in [S2] that for $\varepsilon>0$, there exists a unique invariant measure κ_ε defined on the σ-algebra of Borel sets of D_0. Furthermore, as in the inviscid case studied in this paper, κ_ε can be constructed as the probability distribution of an invariant functional

$$u_\varepsilon^\omega(\cdot,0)=\Phi_0^\varepsilon(\omega)(\cdot) \tag{8.2}$$

such that u_ε^ω is a solution of (8.1) when the realization of the forces is given by ω. The main result of this section is the following:

THEOREM 8.1. *With probability* 1,

$$u_\varepsilon^\omega(x,0)\to u^\omega(x,0),$$

for almost all $x \in \mathbb{S}^1$, as $\varepsilon \to 0$. More precisely, let $x \in I(\omega) = \{y \in [0,1]$: there exists a unique one-sided minimizer passing through $(y,0)\}$. Then

$$u_\varepsilon^\omega(x,0) \to u^\omega(x,0) \qquad as \qquad \varepsilon \to 0\,. \tag{8.3}$$

As a simple corollary, we have:

THEOREM 8.2. *κ_ε converges weakly to κ as $\varepsilon \to 0$.*

Our proof of Theorem 8.1 relies on the Hopf-Cole transformation: $u_\varepsilon^\omega = -\varepsilon(\log\varphi)_x$ where φ satisfies the stochastic heat equation

$$\frac{\partial \varphi}{\partial t} = \frac{\varepsilon}{2}\frac{\partial^2 \varphi}{\partial x^2} - \frac{1}{\varepsilon}\varphi \circ F\,. \tag{8.4}$$

As explained in Appendix C, the product $F \circ \varphi$ should be understood in the Stratonovich sense. The solution of (8.4) has the Feynman-Kac representation

$$\varphi(x,t) = \mathbb{E}\left\{e^{-\frac{1}{\varepsilon}\int_s^t F(x+\beta(\tau),\tau)d\tau}\varphi(x+\beta(s),s)\right\} \tag{8.5}$$

for $s < t$, where $\mathbb{E}$ denotes expectation with respect to the Wiener measure with variance ε, $\beta(t) = 0$, and

$$\begin{aligned}\int_s^t F(x+\beta(\tau),\tau)d\tau &= \sum_k F_k(x+\beta(t))B_k(t) - \sum_k F_k(x+\beta(s))B_k(s) \\ &\quad - \sum_k \int_s^t f_k(x+\beta(\tau))B_k(\tau)d\beta(\tau)\,.\end{aligned} \tag{8.6}$$

The integrals in (8.6) are understood in the Ito sense.

For $x, y \in \mathbb{R}^1$, $\tau_1 > \tau_2$, define

$$\begin{aligned}K_\varepsilon(x,\tau_1,y,\tau_2) &= e^{\frac{1}{\varepsilon}\left(\sum\limits_k F_k(x)B_k(\tau_1) - \sum\limits_k F_k(y)B_k(\tau_2)\right)} \\ &\quad \times \int e^{\frac{1}{\varepsilon}\sum\limits_k \int_{\tau_2}^{\tau_1} f_k(\beta(s))B_k(s)d\beta(s)}\, dW_{(y,\tau_2)}^{(x,\tau_1)}(\beta)\end{aligned}$$

where $dW_{(y,\tau_2)}^{(x,\tau_1)}(\beta)$ is the probability measure defined by the Brownian bridge: $\beta(\tau_1) = x$, $\beta(\tau_2) = y$, with variance ε. Using (8.5), for $s < t$, we can write the solution of (8.1) as

$$u^\varepsilon(x,t) = -\varepsilon \frac{\int_0^1 \frac{\partial}{\partial x} M(x,t,y,s)e^{-\frac{1}{\varepsilon}h^\varepsilon(y,s)}dy}{\int_0^1 M(x,t,y,s)e^{-\frac{1}{\varepsilon}h^\varepsilon(y,s)}dy} \tag{8.7}$$

where $h^\varepsilon(y,s) = \int_0^y u^\varepsilon(z,s)dz$, and

$$M(x,\tau_1,y,\tau_2) = \sum_{m=-\infty}^{\infty} K_\varepsilon(x,\tau_1,y+m,\tau_2)$$

for $x, y \in [0,1]$. M is the transfer matrix for Brownian motion on the circle $\mathbb{S}^1$.

Define also

$$A(x,\tau_1,y,\tau_2) = \inf_{\xi(\tau_1)=x,\xi(\tau_2)=y} \mathcal{A}_{\tau_2,\tau_1}(\xi)\,.$$

LEMMA 8.1. *For almost every $\omega \in \Omega$, there exists a $\tau = \tau(\omega) > 0$, and $C_0 = C_0(\omega) > 0$, such that*

$$\frac{1}{g_m} \le K_\varepsilon(x,\tau,z+m,0)e^{\frac{1}{\varepsilon}A(x,\tau,z+m,0)} \le g_m \tag{8.8}$$

for all $x,z \in [0,1]$, $m \in \mathbb{N}$, where

$$g_m = \begin{cases} C_1(\omega), & \text{if } (|m|+1)\,\|F\|_\tau \le C_0(\omega) \\ C_2(\omega)e^{\frac{8(|m|+1)^2}{\varepsilon}\|F\|_\tau^2\tau}, & \text{if } (|m|+1)\,\|F\|_\tau > C_0(\omega)\,. \end{cases} \tag{8.9}$$

$C_1(\omega)$ and $C_2(\omega)$ are constants depending only on ω and $\|F\|_\tau$ is defined as in Appendix B.

Proof. We will assume $m \ge 0$. Let γ: $[0,\tau] \to R^1$ be a minimizer such that $\gamma(0) = z+m$, $\gamma(\tau) = x$, and

$$A(x,\tau,z+m,0) = \mathcal{A}_{0,\tau}(\gamma)\,.$$

Now, γ satisfies the Euler-Lagrange equation:

$$\int_0^\tau \dot\gamma(s)d\eta(s) - \int_0^\tau \sum_k B_k(s)\Big(f_k'(\gamma(s))\dot\gamma(s)ds + f_k(\gamma(s))d\eta(s)\Big) = 0 \tag{8.10}$$

for test functions η on $[0,\tau]$. Performing a change of variable $\beta = \gamma + \sqrt{\varepsilon}\eta$ in the functional integral in K_ε, we obtain, using (8.10) and the Cameron-Martin-Girsanov formula:

$$K_\varepsilon(x,\tau,z+m,0) = e^{-\frac{1}{\varepsilon}A(x,\tau,z+m,0)}\mathbb{E}_\eta e^{\frac{1}{\varepsilon}H} \tag{8.11}$$

where the exponent H is given by

$$\begin{aligned} H &= H_1 + \sqrt{\varepsilon}H_2\,,\\ H_1 &= \int_0^\tau \sum_k \left\{f_k\left(\gamma+\sqrt{\varepsilon}\eta\right) - f_k(\gamma) - f_k'(\gamma)\sqrt{\varepsilon}\eta\right\} B_k(s)\dot\gamma(s)ds\,,\\ H_2 &= \int_0^\tau \sum_k B_k(s)\left\{f_k\left(\gamma+\sqrt{\varepsilon}\eta\right) - f_k(\gamma)\right\} d\eta(s)\,. \end{aligned} \tag{8.12}$$

In (8.11), $\mathbb{E}_\eta$ denotes expectation with respect to the standard Brownian bridge $\eta(0) = 0$, $\eta(\tau) = 0$. □

We now estimate H. A simple Taylor expansion to second order gives:

$$|H_1| \le \varepsilon\|F\|_\tau \max_{0\le s\le\tau} |\dot\gamma(s)| \int_0^\tau \eta^2(s)ds\,. \tag{8.13}$$

Using Lemma B.1, we get for $\tau = \tau(\omega)$,

$$(8.14)\quad |H_1| \le \varepsilon \|F\|_\tau \frac{C(\omega)(|m|+1)}{\tau} \int_0^\tau \eta^2(s)ds \le C(\omega)\varepsilon\|F\|_\tau(|m|+1)(\eta^*)^2,$$

where $\eta^* = \max_{0\le s\le \tau} \eta(s)$. For H_2, we use the mean value theorem to write

$$(8.15)\qquad \frac{1}{\sqrt{\varepsilon}} H_2 = \int_0^\tau \sum_k B_k(s) f_k'\left(\gamma + \sqrt{\varepsilon}\theta_k \eta\right)\eta(s)d\eta(s) = H_{21} + H_{22}$$

where $\theta_k \in [0,1]$, $H_{21} = \frac{1}{\sqrt{\varepsilon}} H_2 - H_{22}$ with

$$(8.16)\qquad H_{22} = \frac{\alpha}{2}\int_0^\tau \Big(\sum_k B_k(s) f_k'\left(\gamma + \sqrt{\varepsilon}\theta_k\eta\right)\Big)^2 \eta^2(s)ds \le \frac{\alpha}{2}\|F\|_\tau^2 \tau(\eta^*)^2 .$$

We will choose the value of α later ($\alpha = 3$ will suffice). Note that (8.14) and (8.16) can be combined to give:

$$\frac{1}{\varepsilon}|H_1| + |H_{22}| \le C(\omega)\|F\|_\tau(|m|+1)(\eta^*)^2 .$$

The constant $C(\omega)$ is changed to a different value in the last step. Now we have

(8.17)

$$\mathbb{E}_\eta e^{\frac{1}{\varepsilon}H} = \mathbb{E}_\eta e^{\frac{1}{\varepsilon}H_1 + H_{21} + H_{22}} \le \left(\mathbb{E}_\eta e^{C(\omega)\|F\|_\tau(|m|+1)(\eta^*)^2}\right)^{1/2} \left(\mathbb{E}_\eta e^{2H_{21}}\right)^{1/2} .$$

Using the fact that for $a > 0$

$$(8.18)\qquad P\{\eta^* > a\} \le \frac{C}{\sqrt{\pi\tau}} \int_a^{+\infty} e^{-\frac{\lambda^2}{2\tau}} d\lambda$$

we have

$$(8.19)\qquad \mathbb{E}_\eta e^{C(\omega)\|F\|_\tau(|m|+1)(\eta^*)^2} \le \text{Constant}$$

if

$$(8.20)\qquad C(\omega)\|F\|_\tau(|m|+1) < \frac{1}{2\tau} .$$

For the second factor in (8.17), we use the inequality (see [McK])

$$(8.21)\qquad P\{H_{21} > \beta\} < e^{-\alpha\beta} .$$

Then

$$(8.22)\qquad \mathbb{E}_\eta e^{2H_{21}} \le \sum_{N=0}^{\infty} e^{2(N+1)} P\{H_{21} > N\} \le \sum_{N=1}^{\infty} e^{2(N+1)-\alpha N} < +\infty$$

if we choose $\alpha > 2$.

When m violates (8.20), we estimate H_1 using Lemma B.1:

(8.23)
$$H_1 = \sqrt{\varepsilon}\int_0^\tau \sum_k B_k(s)\Big(f'_k\left(\gamma+\sqrt{\varepsilon}\theta_k\eta\right) - f'_k(\gamma)\Big)\eta(s)\dot\gamma(s)ds\,,$$

(8.24)
$$|H_1| \le 2\sqrt{\varepsilon}\,\|F\|_\tau \frac{(|m|+1)}{\tau}\int_0^\tau |\eta(s)|ds \le 2\sqrt{\varepsilon}\,\|F\|_\tau(|m|+1)\eta^*\,.$$

Hence

$$\mathbb{E}_\eta\left(e^{\frac{1}{\varepsilon}H_1}\right)^2 \le C(\omega)\int_0^{+\infty} e^{\frac{4(|m|+1)}{\sqrt{\varepsilon}}\|F\|_\tau\sqrt{\tau}\lambda-\frac{\lambda^2}{2}}d\lambda \tag{8.25}$$
$$\le C(\omega)e^{\frac{16(|m|+1)^2}{\varepsilon}\|F\|_\tau^2\tau}\,.$$

As before we have then

$$\mathbb{E}_\eta e^{\frac{1}{\varepsilon}H} \le C(\omega)e^{\frac{8(|m|+1)^2}{\varepsilon}\|F\|_\tau^2\tau}\,.$$

These give the upper bounds.

Similarly we can prove the same bounds for $(\mathbb{E}_\eta e^{\frac{1}{\varepsilon}H})^{-1}$. This completes the proof of Lemma 8.1. □

It is easy to see that for fixed τ, z and m, $A(x,\tau,z+m,0)$ is differentiable at x, if and only if there exists a unique minimizer γ such that $A(x,\tau,z+m,0) = \mathcal{A}_{0,\tau}(\gamma)$ and $\gamma(\tau)=x$, $\gamma(0)=z+m$. In this case we have

$$\dot\gamma(\tau) = \frac{\partial}{\partial x}A(x,\tau,z+m,0)\,. \tag{8.26}$$

When the minimizer is not unique, $A(x,\tau,z+m,0)$ has both left and right derivatives. Moreover

$$D_x^+A(x,\tau,z+m,0) = \dot\gamma_+(\tau)\,,$$
$$D_x^-A(x,\tau,z+m,0) = \dot\gamma_-(\tau)$$

where γ_+ and γ_- are the right-most and left-most minimizers. In either case, let us define

$$v(x,z+m,\tau) = D_x^-A(x,\tau,z+m,0)\,.$$

LEMMA 8.2. *The following inequality holds:*

(8.27)
$$\left|\frac{1}{K_\varepsilon(x,\tau,z+m,0)}\Big(\varepsilon\frac{\partial K_\varepsilon}{\partial x}(x,\tau,z+m,0)+v(x,z+m,\tau)K_\varepsilon(x,\tau,z+m,0)\Big)\right|$$
$$\le \sqrt{\varepsilon}\,\|F\|_\tau g_m$$

where g_m is as defined in Lemma 8.1.

Proof. For simplicity, we will write

$$K_m = K_\varepsilon(x,\tau;z+m,0), \quad v_m = v(x,z+m,\tau).$$

A straightforward computation gives

(8.28)
$$\varepsilon\frac{\partial K_m}{\partial x} = e^{-\frac{1}{\varepsilon}\sum_k F_k(x)B_k(\tau)}\int G(\beta)e^{\frac{1}{\varepsilon}\sum_k\int_0^\tau f_k(\beta(s))B_k(s)d\beta(s)}dW_{(z+m,0)}^{(x,\tau)}(\beta)$$

where

$$G(\beta) = -\sum_k f_k(x)B_k(\tau) - \frac{x-(z+m)}{\tau} + \sum_k\int_0^\tau B_k(s)\Big[f_k'(\beta)\frac{s}{\tau}d\beta + f_k(\beta)\frac{1}{\tau}ds\Big], \tag{8.29}$$

and

$$\varepsilon\frac{\partial K_m}{\partial x} + v_m K_m = e^{-\frac{1}{\varepsilon}\sum_k F_k(x)B_k(\tau)} \times \int (G(\beta)-G(\gamma_-))e^{\frac{1}{\varepsilon}\sum_k\int_0^\tau f_k(\beta(s))B_k(s)d\beta(s)}dW_{(z+m,0)}^{(x,\tau)}(\beta).$$

Performing a change of variables $\beta = \gamma_- + \sqrt{\varepsilon}\eta$, we get

$$\varepsilon\frac{\partial K_m}{\partial x} + v_m K_m = e^{-\frac{1}{\varepsilon}A(x,\tau,z+m,0)}\mathbb{E}_\eta\Big((G(\gamma_- + \sqrt{\varepsilon}\eta) - G(\gamma_-))e^{\frac{1}{\varepsilon}H}\Big) \tag{8.30}$$

where H is defined as before. Write $G(\gamma_- + \sqrt{\varepsilon}\eta) - G(\gamma_-)$ as

$$\begin{aligned}
&G(\gamma_- + \sqrt{\varepsilon}\eta) - G(\gamma_-)\\
&\quad = \frac{\sqrt{\varepsilon}}{\tau}\sum_k\Big[\int_0^\tau B_k(\tau)f_k''(\gamma_- + \theta_k\sqrt{\varepsilon}\eta)s\eta(s)\dot{\gamma}_-(s)ds\\
&\qquad + \int_0^\tau B_k(s)f_k'(\gamma_- + \sqrt{\varepsilon}\eta)sd\eta(s) + \int_0^\tau B_k(\tau)f_k'(\gamma_- + \theta_k\sqrt{\varepsilon}\eta)\eta(s)ds\Big].
\end{aligned}$$

We can then follow the steps in the proof of Lemma 8.1 to establish (8.27). □

Remark. The estimates in Lemmas 8.1 and 8.2 are proved for the time interval $[0,\tau]$. We see easily that they hold for arbitrary intervals of the type $[t,t+\tau]$ and $[t-\tau,t]$ by choosing suitable τ which in general depend on t. For t in a compact set, we can choose τ to be independent of t such that (8.8) holds.

Our next lemma gives uniform estimates of u^ε.

LEMMA 8.3. *There exist positive constants* $\varepsilon_0(\omega,t)$, $C(\omega,t)$, *such that*

$$|u^{\varepsilon}_{\omega}(x,t)| \le C(\omega,t) \tag{8.31}$$

for $x \in [0,1]$, $0 < \varepsilon \le \varepsilon_0(\omega,t)$. *Furthermore,* $\varepsilon_0(\,\cdot\,,t)$ *and* $C(\,\cdot\,,t)$ *are stationary random processes in* t.

Proof. The basic idea is to use the fact that for ε small, the functional integral is concentrated near minimizers whose velocities are estimated in Lemma B.1. We will prove (8.31) for $t=0$ by working on the time interval $[-\tau,0]$ where τ is as defined in Lemma 8.1. It will be clear that the proof works with little change for arbitrary t. □

Let $N = \frac{C_0(\omega)}{\|F\|_\tau}$, where $C_0(\omega)$ is as defined in Lemma 8.1. Denote $A^*(x,z,\tau) = \inf_{m\in\mathbb{N}} A(x,0,z+m,-\tau)$ for $x,z\in[0,1]$. It is easy to see that for $\tau \ll 1$,

$$A(x,0,z+m,-\tau) - A^*(x,z,\tau) \ge \frac{1}{3}\frac{(|m|+1)^2}{\tau} - C(\omega) \tag{8.32}$$

for $|m| > N$.

Again for simplicity of notation, we will denote $K_m = K_\varepsilon(x,0,y+m,-\tau)$, $\mu(dy) = e^{-\frac{1}{\varepsilon}h^\varepsilon(y,0)}dy$. Using Lemma 8.2 and (8.7), we have

$$|u^\varepsilon(x,0)| \le I_1 + \sqrt{\varepsilon}\|F\|_\tau I_2$$

where

$$I_1 = \frac{\sum_m \int_0^1 v_m K_m \mu(dy)}{\sum_m \int_0^1 K_m \mu(dy)}, \tag{8.33}$$

$$I_2 = \frac{\sum_m \int_0^1 g_m K_m \mu(dy)}{\sum_m \int_0^1 K_m \mu(dy)}.$$

For $|m| > N$, we can use (8.32) and Lemma 8.1 to get

$$\begin{aligned} K_m &\le e^{-\frac{1}{\varepsilon}A(x,0,y+m,-\tau)} g_m \\ &\le C(\omega) e^{-\frac{1}{\varepsilon}A^*(x,y,\tau)} e^{-\frac{1}{3\varepsilon}\frac{(|m|+1)^2}{\tau} + \frac{8(|m|+1)^2}{\varepsilon}\|F\|_\tau^2 \tau} \\ &\le C(\omega)^{-\frac{1}{\varepsilon}A^*(x,y,\tau)} e^{-\frac{1}{4\varepsilon}\frac{(|m|+1)^2}{\tau}} \end{aligned}$$

if τ is small enough. Hence we get, using the fact that $|v_m| \le \frac{C(\omega)(|m|+1)}{\tau}$,

$$\begin{aligned}|I_1| &\le \frac{\sum\limits_{|m|<N} \int_0^1 |v_m| K_m \mu(dy)}{\sum\limits_m \int_0^1 K_m \mu(dy)} + \frac{\sum\limits_{|m|>N} \int_0^1 |v_m| K_m \mu(dy)}{\sum\limits_m \int_0^1 K_m \mu(dy)} \\ &\le C(\omega) N \cdot \frac{1}{\tau} + C(\omega) \frac{\left(\sum\limits_{|m|>N} \frac{C(\omega)(|m|+1)}{\tau} e^{-\frac{1}{4\epsilon}\frac{(|m|+1)^2}{\tau}}\right) \int_0^1 e^{-\frac{1}{\epsilon}A^*(x,y,\tau)} \mu(dy)}{\int_0^1 e^{-\frac{1}{\epsilon}A^*(x,y,\tau)} \mu(dy)} \\ &\le C(\omega).\end{aligned}$$

In the last step we used the fact that τ depends only on ω. Similarly

$$\begin{aligned}|I_2| &\le C(\omega) + \frac{\sum\limits_{|m|>N} \int_0^1 g_m K_m \mu(dy)}{\sum\limits_m \int_0^1 K_m \mu(dy)} \\ &\le C(\omega)\end{aligned}$$

where we used

$$K_m g_m \le C(\omega) e^{-\frac{1}{\epsilon}A^*(x,y,\tau)} e^{-\frac{1}{4\epsilon}\frac{(|m|+1)^2}{\tau}}$$

for small enough τ. This completes the proof of Lemma 8.3. □

Define for $C > 0$

$$Q_C = \Big\{ h \in \mathrm{Lip}[0,1], \quad \text{such that} \\ h(y) = \int_0^y u(z)dz, \ |u| \le C, \ \int_0^1 u(z)dz = 0 \Big\}. \tag{8.34}$$

Take $x \in I(\omega)$. Denote the unique minimizer that passes through $(x, 0)$ by ξ^*. For $h \in Q_C$, $T < 0$, define the modified action as

$$\mathcal{A}^h_{T,0}(\xi) = \mathcal{A}_{T,0}(\xi) + h(\xi(T)) \tag{8.35}$$

and denote by ξ^{**}_h the minimizer of $\mathcal{A}^h_{T,0}$. Obviously ξ^{**}_h in general depends on h and T.

LEMMA 8.4. *Fix a constant $C > 0$. For any $\delta > 0$, there exists $T^* = T^*(\delta) < 0$, such that*

$$|\dot{\xi}^{**}_h(0) - \dot{\xi}^*(0)| < \frac{\delta}{2} \tag{8.36}$$

for $T < T^*$ and all $h \in Q_C$ (T^* in general depends on C).

Proof. Assume to the contrary that there exists a sequence $T_j \to -\infty$, $h_j \in Q_C$, such that

$$|\dot{\xi}^{**}_{h_j}(0) - \dot{\xi}^*(0)| \ge \frac{\delta}{2}.$$

Then from Lemma 3.3, the $\{\dot{\xi}^{**}_{h_j}(0)\}$'s are uniformly bounded and we can choose a subsequence, still denoted by $\{\xi^{**}_{h_j}\}$, such that $\xi^{**}_{h_j}$ converges (uniformly on compact sets of $(-\infty, 0]$ and $\dot{\xi}^{**}_{h_j}(0) \to \dot{\tilde{\xi}}(0)$ to a limiting path $\tilde{\xi}$ defined on $(-\infty, 0]$. From Lemma 3.6, $\tilde{\xi}$ is also a one-sided minimizer. Since $\tilde{\xi}(0) = x$, and

$$|\dot{\tilde{\xi}}(0) - \dot{\xi}^*(0)| \geq \frac{\delta}{2},$$

this violates the assumption that there exists a unique one-sided minimizer passing through $(x, 0)$. □

LEMMA 8.5. *Fix a constant $C > 0$. There exists a function $\alpha(\cdot)$ defined on $(0, +\infty)$, $\alpha > 0$, with the following properties: For any $\delta > 0$, one can find a $T^* = T^*(\delta) < 0$, such that for any path ξ defined on $[T^*, 0]$, with $\xi(0) = x$, the inequality $|\dot{\xi}(0) - \dot{\xi}^*(0)| > \delta$ implies*

$$|\mathcal{A}^h_{T^*,0}(\xi) - \mathcal{A}^h_{T^*,0}(\xi^{**}_h)| > \alpha(\delta) \tag{8.37}$$

for all $h \in Q_C$.

Proof. Assume to the contrary that there exist a $\delta > 0$, and a sequence $\{T_j\}$, $T_j \to -\infty$, $h_j \in Q_C$, and ξ_j defined on $[-T_j, 0]$, such that $|\dot{\xi}_j(0) - \dot{\xi}^*(0)| > \delta$, and

$$\left|\mathcal{A}^{h_j}_{T^*_j,0}(\xi_j) - \mathcal{A}^{h_j}_{T^*_j,0}(\xi^{**}_{h_j})\right| < \frac{1}{j}. \tag{8.38}$$

From the estimates proved in Section 3, $\{\dot{\xi}_j(0)\}$ are uniformly bounded. Therefore we can choose a subsequence, still denoted by $\{\xi_j\}$, such that ξ_j converges (uniformly on compact sets of $(-\infty, 0]$ and $\dot{\xi}_j(0) \to \dot{\tilde{\xi}}(0)$) to $\tilde{\xi}$ defined on $(-\infty, 0]$. From (8.38), $\tilde{\xi}$ is also a one-sided minimizer. Since $\tilde{\xi}(0) = x$, $|\dot{\tilde{\xi}}(0) - \dot{\xi}^*(0)| > \delta$, we arrive at a contradiction to the assumption that there exists a unique one-sided minimizer passing through $(x, 0)$. □

Now we are ready to prove Theorem 8.1.

Proof of Theorem 8.1. Fix an $x \in I(\omega)$. Denote by ξ^* the unique one-sided minimizer passing through $(x, 0)$. Take $\delta > 0$. From Lemmas 8.4 and 8.5 we can find a $T^* < 0$, such that (8.36) and (8.37) hold.

Let n be a sufficiently large integer (depending only on ω and T^*), such that the estimates in Lemma 8.1 hold on the intervals $[(k+1)s, ks]$ where $s = \frac{T^*}{n}$, $k = 0, 1, \ldots, , n-1$. Using Lemma 8.2, we have $\tau = -s$, for,

$$u^\varepsilon(x, 0) = -\frac{\varepsilon \int_0^1 \frac{\partial}{\partial x} M(x, 0, y, T^*)\mu(dy)}{\int_0^1 M(x, 0, y, T^*)\mu(dy)}$$

where

$$\mu(dy) = -\frac{1}{\varepsilon}h(y),\ h(y) = \int_0^y u^\varepsilon(z, T^*)dz\,.$$

Hence

$$\begin{aligned}
u^\varepsilon(x,0) &= \frac{-\varepsilon \int_0^1 \int_0^1 \frac{\partial}{\partial x} M(x,0,z_1,s)M(z_1,s,y,T^*)dz_1\mu(dy)}{\int_0^1 \int_0^1 M(x,0,z_1,s)M(z_1,s,y,T^*)dz_1\mu(dy)} \\
&= \frac{\int_0^1 \int_0^1 \Big(\sum_m v(x,0,z_1+m,s)K_\varepsilon(x,0,z_1+m,s)\Big)M(z_1,s,y,T^*)dz_1\mu(dy)}{\int_0^1 \int_0^1 \Big(\sum_m K_\varepsilon(x,0,z_1+m,s)\Big)M(z_1,s,y,T^*)dz_1\mu(dy)} \\
&\quad + O(1)\sqrt{\varepsilon}\|F\|_\tau \frac{\int_0^1 \int_0^1 \Big(\sum_m g_m K_\varepsilon(x,0,z+m,s)\Big)M(z_1,s,y,T^*)dz_1\mu(dy)}{\int_0^1 \int_0^1 \Big(\sum_m K_\varepsilon(x,0,z+m,s)\Big)M(z_1,s,y,T^*)dz_1\mu(dy)} \\
&= I_3 + O(1)\sqrt{\varepsilon}\|F\|_\tau I_4
\end{aligned}$$

where $O(1)$ denotes a uniformly bounded quantity. As in the proof of Lemma 8.3, we can show:

$$|I_4| \le C(\omega)N = \frac{C(\omega)C_0(\omega)}{\|F\|_\tau}\,.$$

For $z_k, z_{k+1} \in [0,1]$ we have, using Lemma 8.1,

$$\begin{aligned}
&M(z_k, ks,\ z_{k+1}, (k+1)s) \\
&\quad = \sum_{|m|<N} K_\varepsilon(z_k, ks, z_{k+1}+m, (k+1)s) \\
&\qquad + \sum_{|m|>N} K_\varepsilon(z_k, ks, z_{k+1}+m, (k+1)s) \\
&\quad \le e^{-\frac{1}{\varepsilon}A^*(z_k,ks,z_{k+1},(k+1)s)}\Big(C(\omega)N + \sum_{|m|>N} e^{-\frac{(|m|+1)^2}{3\varepsilon}+\frac{8(|m|+1)^2}{\varepsilon}\|F\|_\tau^2\tau}\Big) \\
&\quad \le C(\omega)Ne^{-\frac{1}{\varepsilon}A^*(z_k,ks,z_{k+1},(k+1)s)}\,.
\end{aligned}$$

Letting $x = z_0$, $y = z_n$, we obtain for fixed $\{z_0, z_1, z_2, \dots, z_n\}$,

$$\prod_{k=1}^{n-1} M(z_k, ks, z_{k+1}, (k+1)s) \le (C(\omega)N)^n e^{-\frac{1}{\varepsilon}\sum\limits_{k=0}^{n-1} A^*(z_k,ks,z_{k+1},(k+1)s)}$$

Denote by $\int' dz_1 \sum_m'$ and $\int'' dz_1 \sum_m''$ summation and integration over the sets of (z_1, m) such that $|v_m - \dot\xi^*(0)| > \delta$ and $|v_m - \dot\xi^*(0)| < \delta$ respectively, where $v_m = v(x, 0, z_1+m, s)$. From Lemma 8.4, the second sum and integral $\int'' dz_1 \sum_m''$

cover the set $|v_m - \dot{\xi}_h^{**}(0)| < \frac{\delta}{2}$. We have

$$
\begin{aligned}
I_5 &= \int' dz_1 \sum_m{}' K_\varepsilon(x,0,z_1+m,s) M(z_1,s,y,T^*)\mu(dy) \\
&= \int' dz_1 \sum_m{}' K_\varepsilon(x,0,z_1+m,s) \\
&\quad \times \int_{[0,1]^{n-1}} \prod_{k=1}^{n-1} M(z_k,ks,z_{k+1},(k+1)s) dz_2 \dots dz_n \mu(dy) \\
&\le (C(\omega)N)^n \int' dz_1 \sum_m{}' \\
&\quad \times \int_{[0,1]^{n-1}} e^{-\frac{1}{\varepsilon}(A(x,0,z_1+m,s)+\sum_{k=1}^{n-1} A^*(z_k,ks,z_{k+1},(k+1)s)+h(z_n))} dz_2 \dots dz_n \\
&\le (C(\omega)N)^n \int' dz_1 \sum_m{}' \int_{[0,1]^{n-1}} e^{-\frac{1}{\varepsilon}\sum_{k=0}^{n-1} A^*(z_k,ks,z_{k+1},(k+1)s)+h(z_n)} dz_2 dz_3 \dots dz_n \\
&\le (C(\omega)N)^n e^{-\frac{1}{\varepsilon}(\mathcal{A}^h_{T^*,0}(\xi_h^{**})+\alpha(\delta))} .
\end{aligned}
$$

In the last step, we used (8.39). On the other hand, there exists a $\delta_2 > 0$, such that if $|z - \xi_h^{**}(s)| < \delta_2$, then

$$
|v(x,0,z,s) - \dot{\xi}_h^{**}(0)| < \frac{\delta}{2} .
$$

Choose a $\delta_1 > 0$, such that $\delta_1 < \delta_2$, $\delta_1 < \frac{\alpha(\delta)}{2nC_3(\omega)}$, with $C_3(\omega)$ to be defined later. Using Lemma 8.4 we get

$$
\begin{aligned}
I_6 &= \int'' dz_1 \sum_m{}'' K_\varepsilon(x,0,z_1+m,s) M(z_1,s,y,T^*)\mu(dy) \\
&\ge \int_{|z_k - \xi_h^{**}(ks)| < \delta_1} \prod_{k=1}^{n} K_\varepsilon(z_{k-1},(k-1)s,z_k,ks) e^{-\frac{1}{\varepsilon}h(z_n)} dz_1 \dots dz_n \\
&\ge \frac{1}{C(\omega)^n} \int_{|z_k - \xi_h^*(ks)| < \delta_1} e^{-\frac{1}{\varepsilon}\left(\sum_{k=1}^{n} A^*(z_{k-1},(k-1)s,z_k,ks)+h(z_n)\right)} dz_1 \dots dz_n .
\end{aligned}
$$

It is easy to see that if $|z_k - \xi_h^*(ks)| < \delta_1$, $|z_{k-1} - \xi_h^*((k-1)s)| < \delta_1$, then

$$
|A^*(z_{k-1},(k-1)s,z_k,ks) - \mathcal{A}_{ks,(k-1)s}(\xi_h^{**})| \le C_3(\omega)\delta_1 ,
$$

and if $|z_n - \xi_h^{**}(T^*)| < \delta_1$, then

$$
|h(z_n) - h(\xi_h^{**}(T^*))| \le C_3(\omega)\delta_1 ,
$$

with $C_3(\omega)$ defined by the above estimates. Hence we have

$$I_6 \geq \frac{1}{C(\omega)^n}\delta_1^n e^{-\frac{1}{\varepsilon}(\mathcal{A}^h_{T^*,0}(\xi_h^{**})+nC_3(\omega)\delta_1)}.$$

Therefore

$$\left|\frac{I_5}{I_6}\right| \leq C(\omega)^{2n}\cdot N^n\delta_1^{-n}e^{-\frac{1}{2\varepsilon}\alpha(\delta)}.$$

Similarly, if we define

$$I_7 = \int' dz_1 \sum_m{}' v(x,0,z_1+m,s)K_\varepsilon(x,0,z_1+m,s)M(z_1,s,y,T^*)\mu(dy),$$

$$I_8 = \int'' dz_1 \sum_m{}'' v(x,0,z_1+m,s)K_\varepsilon(x,0,z_1+m,s)M(z_1,s,y,T^*)\mu(dy),$$

then we can also get

$$\left|\frac{I_7}{I_6}\right| \leq C(\omega)^{2n}N^n\delta_1^{-n}e^{-\frac{1}{2\varepsilon}\alpha(\delta)}.$$

Finally we obtain

$$\begin{aligned}
|u^\varepsilon(x,0)-\dot{\xi}^*(0)| &\leq \left|\frac{I_7+I_8-\xi^*(0)(I_5+I_6)}{I_5+I_6}\right| + \sqrt{\varepsilon}C(\omega)\|F\|_\tau \\
&\leq \left|\frac{I_8-\xi^*(0)I_6}{I_6}\right| + \frac{|I_7|}{I_6} + |\dot{\xi}(0)|\frac{I_5}{I_6} + \sqrt{\varepsilon}\|F\|_\tau C(\omega) \\
&\leq \delta + C(\omega)^{2n}N^n\delta_1^{-n}e^{-\frac{1}{2\varepsilon}\alpha(\delta)} + \sqrt{\varepsilon}\|F\|_\tau C(\omega) \\
&\leq \delta+\delta = 2\delta
\end{aligned}$$

if we choose ε sufficiently small. This completes the proof of Theorem 8.1. □

Appendix A. Proof of Lemma 2.1 for the random case

In this appendix we comment on the proof of Lemma 2.1 for the random case. Let $F^\delta(x,t) = \sum_{k=1}^\infty F_k(x)\dot{B}_k^\delta(t)$, where B_k^δ is the standard mollification of B_k. Denote by $u^\delta(x,t)$ the unique entropy solution of

$$\frac{\partial u}{\partial t} + \frac{\partial}{\partial x}\left(\frac{u^2}{2}\right) = \frac{\partial F^\delta}{\partial x} \tag{A.1}$$

with the initial data $u(x,t_0) = u_0(x)$. We will assume that $\|u_0\|_{L^\infty} \leq$ Const, $\int_0^1 u_0(z)dz = 0$. From classical results [Li] we know that $u^\delta(x,t)$ is given by

$$u^\delta(x,t) = \frac{\partial}{\partial x}\inf_{\xi:\,\xi(t)=x}\left\{\mathcal{A}^\delta_{t_0,t}(\xi) + \int_0^{\xi(t_0)} u_0(z)dz\right\}, \tag{A.2}$$

where

(A.3)

$$\begin{aligned}\mathcal{A}^{\delta}_{t_0,t}(\xi) &= \int_{t_0}^{t} \frac{1}{2}\dot{\xi}(s)^2 ds + \int_{t_0}^{t} \sum_{k=1}^{\infty} F_k(\xi(s)) dB^{\delta}_k(s) \\ &= \sum_{k=1}^{\infty} F_k(\xi(t))(B^{\delta}_k(t) - B^{\delta}_k(t_0)) \\ &\quad + \int_{t_0}^{t} \left(\frac{1}{2}\dot{\xi}(s)^2 - \sum_{k=1}^{\infty} f_k(\xi(s))\dot{\xi}(s)(B^{\delta}_k(s) - B^{\delta}_k(t_0)) \right) ds\,.\end{aligned}$$

It is easy to see that the boundary terms at the right hand side of (A.3) resulted from integration by parts do not affect the variational formula (A.2) and can be neglected. Denote

$$U(x,t) = \inf_{\xi:\,\xi(t)=x} \left\{ \mathcal{A}_{t_0,t}(\xi) + \int_0^{\xi(t_0)} u_0(z)dz \right\} \tag{A.4}$$

and

$$U^{\delta}(x,t) = \inf_{\xi:\,\xi(t)=x} \left\{ \mathcal{A}^{\delta}_{t_0,t}(\xi) + \int_0^{\xi(t_0)} u_0(z)dz \right\}. \tag{A.5}$$

It is clear that $U(x,t)$ is well-defined; i.e., the variational problem in (A.4) does have a solution. We will show that $u^{\delta}(x,t) \to u(x,t) = \frac{\partial U}{\partial x}(x,t)$ in $L^1_{\mathrm{loc}}(\mathbb{S}^1 \times [t_0,\infty))$ as $\delta \to 0$ and consequently $u(x,t)$ is a weak entropy solution of (1.1). This follows from:

LEMMA A.1. *For almost all ω, there exist c_1, c_2, c_3, depending only on ω, t and t_0, such that*

$$\|U^{\delta}(\,\cdot\,,t) - U(\,\cdot\,,t)\|_{L^\infty(\mathbb{S}^1)} \le c_1(\omega,t,t_0)\delta^{1/3}, \tag{A.6}$$

$$\|u^{\delta}(\,\cdot\,,t)\|_{L^\infty(\mathbb{S}^1)} \le c_2(\omega,t,t_0)\,, \tag{A.7}$$

and

$$\|u^{\delta}(\,\cdot\,,t)\|_{\mathrm{BV}(\mathbb{S}^1)} \le c_3(\omega,t,t_0) \tag{A.8}$$

where $\mathrm{BV}(\mathbb{S}^1)$ is the space of functions on $\mathbb{S}^1$ with bounded variation.

Proof. For any $\xi \in C^1[t_0,t]$ we have

$$|\mathcal{A}^{\delta}_{t_0,t}(\xi) - \mathcal{A}_{t_0,t}(\xi)| \le \sum_k \int_{t_0}^{t} |f_k(\xi(s))|\; |\dot{\xi}(s)|\; |B^{\delta}_k(s) - B_k(s)| ds\,.$$

For almost every ω, $\{B_k(\,\cdot\,)\}$ is $C^{\frac{1}{3}}$ for all k. Hence

$$|B_k^\delta(s) - B_k(s)| \le C(\omega)\delta^{1/3}\,.$$

This gives

$$|\mathcal{A}_{t_0,t}^\delta(\xi) - \mathcal{A}_{t_0,t}(\xi)| \le \max_{t_0\le s\le t} |\dot\xi(s)|C(\omega)\delta^{1/3}(t-t_0)\,. \tag{A.9}$$

Denote by ξ_δ^* and ξ^* the minimizers in (A.5) and (A.4) respectively. We have, using Lemma B.1,

$$\begin{aligned}
U^\delta(x,t) - U(x,t) &= \mathcal{A}_{t_0,t}^\delta(\xi_\delta^*) + \int_0^{\xi_\delta^*(t_0)} u_0(z)dz - \Big(\mathcal{A}_{t_0,t}(\xi^*) + \int_0^{\xi^*(t_0)} u_0(z)dz\Big)\\
&\le \mathcal{A}_{t_0,t}^\delta(\xi^*) + \int_0^{\xi^*(t_0)} u_0(z)dz - \Big(\mathcal{A}_{t_0,t}(\xi^*) + \int_0^{\xi^*(t_0)} u_0(z)dz\Big)\\
&\le \max_{t_0\le s\le t} |\xi^*(s)|C(\omega)\delta^{1/3}|t-t_0|\\
&\le C(\omega,t,t_0)\delta^{1/3}\,.
\end{aligned}$$

Similarly,

$$U(x,t) - U^\delta(x,t) \le C(\omega,t,t_0)\delta^{1/3}\,.$$

To prove (A.7) we use the theory of backward characteristics (see [D]). If (x,t) is a point of continuity of $u^\delta(\,\cdot\,,t)$, then there exists a unique backward characteristic γ coming to (x,t) and, for $s\in[t_0,t]$,

$$\begin{aligned}
u^\delta(\gamma(s),s) &= u_0(\gamma(t_0),t_0) + \sum_{k=1}^\infty\Big(F_k(\gamma(s))B_k^\delta(s) - F_k(\gamma(t_0))B_k^\delta(t_0)\Big)\\
&\quad - \int_{t_0}^s \sum_{k=1}^\infty f_k(\gamma(r))u^\delta(\gamma(r),r)\dot B_k^\delta(r)dr\,.
\end{aligned} \tag{A.10}$$

Hence

$$|u^\delta(\gamma(s),s)| \le |u_0(\gamma(t_0),t_0)| + c(\omega) + c(\omega)\int_{t_0}^s |u^\delta(\gamma(r),r)|dr \tag{A.11}$$

and $|u^\delta(\gamma(t),t)| \le c_2(\omega,t-t_0)$. Since the points of continuity form a set of full measure in $\mathbb{S}^1$, we have (A.7).

Now consider two points of continuity for $u^\delta(\,\cdot\,,t)$, x_1 and x_2, and let $\gamma_1(s)$ and $\gamma_2(s)$ be the characteristics coming to (x_1,t) and (x_2,t) respectively. For

$i = 1, 2$ denote $u_i(s) = u^\delta(\gamma_i(s), s)$. Then

(A.12)
$$\begin{aligned}\frac{d}{ds}\left(\frac{u_1 - u_2}{\gamma_1 - \gamma_2}\right) &= -\left(\frac{u_1 - u_2}{\gamma_1 - \gamma_2}\right)^2 + \frac{1}{\gamma_1 - \gamma_2}\frac{d}{ds}(u_1 - u_2) \\ &= -\left(\frac{u_1 - u_2}{\gamma_1 - \gamma_2}\right)^2 + \sum_{k=1}^{\infty}\left(\int_0^1 f_k(\gamma_1 + r(\gamma_2 - \gamma_1))dr\right)dB_k^\delta(r)\end{aligned}$$

by the mean value theorem and (2.3). This implies $\frac{u_1 - u_2}{\gamma_1 - \gamma_2} \le c(\omega)$ since it solves an equation of the form $\dot{y} = -y^2 + C$. This, together with (A.7), gives (A.8). □

LEMMA A.2. *For almost every ω, the sequence u^δ converges in $L^1_{\text{loc}}(\mathbb{S}^1 \times [t_0, \infty))$ to a limit u as $\delta \to 0$. Moreover $u(x,t) = \frac{\partial}{\partial x}U(x,t)$ and u is an entropy-weak solution of* (1.4).

Proof. Integrating (A.1) on $\mathbb{S}^1 \times [t, t+\tau]$, we get

$$\begin{aligned}\int_0^1 dx|u^\delta(x, t+\tau) - u^\delta(x,t)| &\le \frac{1}{2}\int_t^{t+\tau} ds\|(u^\delta)^2\|_{\mathrm{BV}(\mathbb{S}^1)} \\ &\quad + \sum_k |B_k(t+\tau) - B_k(t)| \int_0^1 |f_k(x)|dx \\ &\le C_1(\omega)\tau + C_2(\omega)\tau^{1/3}.\end{aligned}$$

In the last step, we used Lemma A.1 and Hölder continuity of the Wiener process. Hence u^δ is uniformly continuous in t, viewed as a function of t in $L^1(\mathbb{S}^1)$. Therefore there exists a subsequence, still denoted by u^δ, and $u \in L^\infty_{\text{loc}}([t_0, \infty), \mathrm{BV}(\mathbb{S}^1)) \cap C([t_0, \infty), L^1(\mathbb{S}^1))$ such that

$$u^\delta \to u \quad \text{in} \quad L^1_{\text{loc}}(\mathbb{S}^1 \times [t_0, \infty)),$$

as $\delta \to 0$. From (A.6), we have

$$u = \frac{\partial U}{\partial x}.$$

From (A.8), the convergence also takes place in $L^p_{\text{loc}}(\mathbb{S}^1 \times [t_0, \infty))$ for $p < +\infty$. Hence u is an entropy weak solution of (1.4). □

Finally, observe that the solution operator for the mollified problem is order-preserving; i.e., $u_1^\delta(\,\cdot\,, t_0) \le u_2^\delta(\,\cdot\,, t_0)$ implies $u_1^\delta(\,\cdot\, t) \le u_2^\delta(\,\cdot\,, t)$ for $t \ge t_0$. Therefore the limiting solution, as $\delta \to 0$, is also order-preserving. Together with the conservation properties, we see that the solution operator is contractive in $L^1(\mathbb{S}^1)$ by the Crandall-Tartar lemma [CM]

(A.13) $$\|u_1(\,\cdot\,, t) - u_2(\,\cdot\,, t)\|_{L^1(\mathbb{S}^1)} \le \|u_1(\,\cdot\,, t_0) - u_2(\,\cdot\,, t_0)\|_{L^1(\mathbb{S}^1)}.$$

This implies uniqueness of order-preserving weak solutions. In particular, since the solutions obtained in the zero-viscosity limit of (1.1) is also order-preserving, as a consequence of the comparison principle, we conclude that $u = \frac{\partial}{\partial x} U$ is the viscosity limit.

Appendix B. Some technical estimates

Denote by $(x(t; x_0, v_0), v(t; x_0, v_0))$ the solution of (2.3) with initial data $x(0; x_0, v_0) = x_0$, $v(0; x_0, v_0) = v_0$. Sometimes we will also use the abbreviation $(x(t), v(t))$. Consider the stochastic flow G_t^ω defined by

$$G_t^\omega(x_0, v_0) = (x(t; x_0, v_0), v(t; x_0, v_0)) \,. \tag{B.1}$$

Since $f_k \in C^r$ the stochastic flow G_t^ω is C^r smooth with probability 1. For $\tau > 0$ and $\omega \in \Omega$, define Γ_τ to be the set of τ-minimizers,

$$\Gamma_\tau = \{\gamma \in C^1[-\tau, 0];\ \gamma(0), \gamma(-\tau) \in \mathbb{S}^1, \mathcal{A}_{-\tau,0}(\gamma) = \min_{\substack{\xi(0)=\gamma(0)\\ \xi(-\tau)=\gamma(-\tau)}} \mathcal{A}_{-\tau,0}(\xi)\} \,. \tag{B.2}$$

We shall also consider the case when endpoints belong to the universal cover $\mathbb{R}^1$, rather than $\mathbb{S}^1$. Denote

$$\begin{aligned}\Gamma_{\tau,m} =& \{\gamma \in C^1[-\tau, 0];\ 0 \le \gamma(0) \le 1, m \le \gamma(-\tau) \le (m+1),\\ &\mathcal{A}_{-\tau,0}(\gamma) = \min_{\substack{\xi(0)=\gamma(0)\\ \xi(-\tau)=\gamma(-\tau)}} \mathcal{A}_{-\tau,0}(\xi)\}.\end{aligned} \tag{B.3}$$

Of course $\Gamma_\tau, \Gamma_{\tau,m}$ depend on ω. Let

$$\begin{aligned} V_\tau(\omega) &= \sup_{\gamma \in \Gamma_\tau} |\dot\gamma(0)| \,, \\ V_{\tau,m}(\omega) &= \sup_{\gamma \in \Gamma_{\tau,m}} |\dot\gamma(0)| \,, \\ \bar V_\tau(\omega) &= \sup_{\gamma \in \Gamma_\tau} \max_{-\tau \le s \le 0} |\dot\gamma(s)| \,, \\ \bar V_{\tau,m}(\omega) &= \sup_{\gamma \in \Gamma_{\tau,m}} \max_{-\tau \le s \le 0} |\dot\gamma(s)| \,. \end{aligned}$$

In Lemma 3.3 it was shown that $V_\tau(\omega) \le C(\omega)$ for $\tau \ge T(\omega)$. We consider now the case of small τ.

LEMMA B.1. *There exists a constant $\tau(\omega)$ such that for $0 < \tau < \tau(\omega)$*

$$V_\tau(\omega) \le \bar V_\tau(\omega) \le \frac{2}{\tau} \,. \tag{B.4}$$

Furthermore, for any $m \in \mathbb{Z}^1$

$$V_{\tau,m}(\omega) \leq \bar{V}_{\tau,m}(\omega) \leq \frac{2(|m|+1)}{\tau}. \tag{B.5}$$

Proof. The proof is similar to the proof of Lemma 3.3. Let

$$\|F\|_\tau = \max_{-\tau\leq s\leq 0} \sum_k \|F_k(x)\|_{C^3(\mathbb{S}^1)} |B_k(s) - B_k(0)|. \tag{B.6}$$

For arbitrary solutions of (2.3), $(x(t), v(t))$, $t \in [-\tau, 0]$, if we denote $\gamma(t) = x(t)$, $v_0 = |\dot{\gamma}(0)|$, $v = \max_{-\tau\leq t\leq 0} |\dot{\gamma}(t)|$, then we have

$$v \leq v_0 + \|F\|_\tau + \|F\|_\tau \tau v \tag{B.7}$$

or

$$v \leq \frac{v_0 + \|F\|_\tau}{1 - \|F\|_\tau \tau} \tag{B.8}$$

provided that $\|F\|_\tau \tau < 1$. Now

$$|\dot{\gamma}(t) - \dot{\gamma}(0)| \leq \|F\|_\tau + \|F\|_\tau v\tau \leq \frac{\|F\|_\tau (1 + v_0\tau)}{1 - \|F\|_\tau \tau}. \tag{B.9}$$

Assume that $\tau \leq 1$ is small enough so that $\|F\|_\tau \leq \epsilon$ (to be chosen later). Then we have

$$v \leq \frac{v_0 + \epsilon}{1 - \epsilon}, \tag{B.10}$$

$$|\dot{\gamma}(t) - \dot{\gamma}(0)| \leq \frac{\epsilon(v_0 + 1)}{1 - \epsilon}. \tag{B.11}$$

Thus, provided that $v_0 \geq 1$, we have

$$|\dot{\gamma}(t)| \geq \frac{1 - 3\epsilon}{1 - \epsilon} v_0 \tag{B.12}$$

for $t \in [-\tau, 0]$. From (B.12), we get

$$\begin{aligned} A_{-\tau,0}(\gamma) &\geq \frac{1}{2}\left(\frac{1-3\epsilon}{1-\epsilon}\right)^2 v_0^2 \tau - \|F\|_\tau \tau v - \|F\|_\tau \\ &\geq \frac{1}{2}\left(\frac{1-3\epsilon}{1-\epsilon}\right)^2 v_0^2 \tau - \frac{\epsilon}{1-\epsilon}(v_0 + 1). \end{aligned}$$

Let γ_1 be the straight line such that $\gamma_1(0) = \gamma(0)$, $\gamma_1(-\tau) = \gamma(-\tau)$. Then

$$A_{-\tau,0}(\gamma_1) \leq \frac{l^2}{2\tau} + l\|F\|_\tau + \|F\|_\tau \leq \frac{l^2}{2\tau} + (l+1)\epsilon,$$

where $l = |\gamma_1(0) - \gamma_1(-\tau)|$. Since $A_{-\tau,0}(\gamma_1) \geq A_{-\tau,0}(\gamma)$, one can easily show that $v_0 \leq \frac{3l}{2\tau}$ if $l \geq 1$ and $\epsilon \leq \frac{1}{40}$. If $l < 1$, then

$$A_{-\tau,0}(\gamma_1) \leq \frac{1}{2\tau} + 2\|F\|_\tau \leq \frac{1}{2\tau} + 2\epsilon,$$

which together with $A_{-\tau,0}(\gamma_1) \geq A_{-\tau,0}(\gamma)$ gives $v_0 \leq \frac{3}{2\tau}$ if $\epsilon \leq \frac{1}{40}$ and $\tau \leq 1$. It follows that $V_\tau(\omega) \leq \frac{3}{2\tau}$, $\bar{V}_\tau(\omega) \leq \frac{2}{\tau}$ and $V_{\tau,m}(\omega) \leq \frac{3(|m|+1)}{2}$, $\bar{V}_{\tau,m}(\omega) \leq \frac{2(|m|+1)}{\tau}$.

In summary $\tau(\omega) \leq 1$ can be chosen such that $\|F\|_{\tau(\omega)} \leq \frac{1}{40}$. □

LEMMA B.2. *For any $K > 1$, there exists $\bar{\tau}(\omega) > 0$, such that for all $0 < \tau \leq \bar{\tau}(\omega)$*

$$\|D^i G_t^\omega(x, v)\| \leq K \tag{B.13}$$

for $1 \leq i \leq r$, $|v| \leq V_\tau(\omega) + 1$, $x \in \mathbb{S}^1$, $t \in [-\tau, 0]$.

Proof. We will prove Lemma B.2 for $i = 1$. For $2 \leq i \leq r$ the proof is similar.

Consider the Jacobi matrix

$$DG_t^\omega = \begin{pmatrix} J_{11}(t), & J_{12}(t) \\ J_{21}(t), & J_{22}(t) \end{pmatrix}$$

where

$$J_{11}(t) = \frac{\partial x(t)}{\partial x_0}, \quad J_{12}(t) = \frac{\partial x(t)}{\partial v_0}, \quad J_{21}(t) = \frac{\partial v(t)}{\partial x_0}, \quad J_{22}(t) = \frac{\partial v(t)}{\partial v_0}.$$

Obviously (J_{11}, J_{12}) and (J_{21}, J_{22}) satisfy

$$\begin{cases} \dot{J}_{11}(t) = J_{21}(t) \\ \dot{J}_{21}(t) = \sum_k f_k(x(t)) J_{11}(t) \dot{B}_k(t), \end{cases} \tag{B.14}$$

$J_{11}(0) = 1$, $J_{21}(0) = 0$; and

$$\begin{cases} \dot{J}_{12}(t) = J_{22}(t) \\ \dot{J}_{22}(t) = \sum_k f_k(x(t)) J_{12}(t) \dot{B}_k(t), \end{cases} \tag{B.15}$$

$J_{12}(0) = 0$, $J_{22}(0) = 1$.

Consider first (B.14). Let $J(\tau) = \max_{-\tau \leq s \leq 0} |J_{21}(s)|$. Then

$$|J_{11}(s)| \leq 1 + J(\tau)\tau \tag{B.16}$$

for $s \in [-\tau, 0]$.

(B.17)

$$\begin{aligned} J_{21}(s) &= \int_s^0 \sum_k f_k(x(t)) J_{11}(t) dB_k(t) \\ &= -\sum_k f_k(x(s)) J_{11}(s) B_k(s) - \int_s^0 \sum_k f_k'(x(t)) v(t) J_{11}(t) B_k(t) dt \\ &\quad - \int_s^0 \sum_k f_k(x(t)) J_{21}(t) B_k(t) dt. \end{aligned}$$

Using (B.8), (B.4) we have

$$\text{(B.18)} \qquad \max_{-\tau\leq s\leq 0}|v(s)| \leq \frac{v_0 + \|F\|_\tau}{1-\|F\|_\tau \tau} \leq \frac{3}{\tau}$$

if τ is sufficiently small. Therefore

$$\text{(B.19)} \qquad J(\tau) \leq \|F\|_\tau\Big(1+J(\tau)\tau\Big) + 3\|F\|_\tau\Big(1+J(\tau)\tau\Big) + \|F\|_\tau \tau J(\tau).$$

It follows from (B.19) that

$$J(\tau) \leq \frac{4\|F\|_\tau}{1-5\|F\|_\tau\tau} \to 0$$

and

$$|J_{11}(s) - 1| \leq J(\tau)\tau \to 0$$

as $\tau \to 0$. Hence we have

$$|J_{11}(s)|, \; |J_{21}(s)| \leq K$$

for $s \in [-\tau, 0]$, if τ is sufficiently small. In the same way, we can prove

$$|J_{12}(s)|, \; |J_{22}(s)| \leq K$$

for $s \in [-\tau, 0]$ if τ is sufficiently small. This completes the proof of Lemma B.2. □

Denote $B_\tau = \{(x,v),\, x\in \mathbb{S}^1,\, |v| \leq V_\tau(\omega)+1\}$, and $B_\tau(t) = G_t^\omega B_\tau$, for $t\in[-\tau,0]$. Then, similar to Lemma B.2, we have:

LEMMA B.3. *For any $K > 1$, there exists $\tilde{\tau}(\omega) > 0$, such that for all $0 < \tau \leq \tilde{\tau}(\omega)$*

$$\text{(B.20)} \qquad \|D^i(G_t^\omega)^{-1}(x,v)\| \leq K$$

for $1 \leq i \leq r$, $(x,v)\in B_\tau(t)$, $t \in [-\tau, 0]$.

Proof. Lemma B.3 follows immediately from Lemma B.2 together with the estimate

$$\text{(B.21)} \qquad \|DG_t^\omega(x,v) - I\| \leq \max\left(\frac{5\|F\|_\tau}{1-5\|F\|_\tau\tau}, \frac{\tau}{1-5\|F\|_\tau\tau}\right)$$

for $t\in[-\bar{\tau}(\omega), 0]$, $(x,v)\in B_\tau$. (B.21) can be proved in the same way as (B.19). □

We will denote by $\tau_{r,K}(\omega)$ the maximum value of $\bar{\tau}(\omega)$ such that both (B.13) and (B.20) hold for all $\tau \leq \bar{\tau}(\omega)$.

Let $O_T(\omega) = \{(x_0, v_0),\ |v(t; x_0, v_0)| \le \sup_{\gamma\in\Gamma_\tau} |v_\gamma(t)| + 1, \text{ for } t \in [-T, 0]\}$ and $O_T(t,\omega) = G_t^\omega O_T(\omega)$. Define

$$D_{T,r}(\omega) = \sup_{-T\le t\le 0} \sup_{(x,v)\in O_T(\omega)} \max_{1\le i\le r} \log^+ \|D^i G_t^\omega(x,v)\|, \tag{B.22}$$

$$\bar{D}_{T,r}(\omega) = \sup_{-T\le t\le 0} \sup_{(x,v)\in O_T(t,\omega)} \max_{1\le i\le r} \log^+ \|D^i (G_t^\omega)^{-1}(x,v)\|.$$

LEMMA B.4. *For any positive integer m,*

$$\int D_{T,r}(\omega)^m P(d\omega) < \infty, \quad \int \bar{D}_{T,r}(\omega)^m P(d\omega) < \infty. \tag{B.23}$$

Proof. The proof is similar to the proofs of similar statements in [Bax], [K2]. Define a sequence of stopping times τ_n:

$$\tau_0 = 0,\ \tau_1 = \tau_{r,K}(\omega),\ \tau_{i+1} = \tau_{r,K}(\theta^{t_i}\omega),$$

where t_i is defined by $t_0 = 0$, $t_1 = -\tau_1$, $t_i = -\sum_{j=0}^{i} \tau_j$. Choose n such that $|t_n(\omega)| \le T < |t_{n+1}(\omega)|$. Then

$$G_{-T}^\omega(x,v) = f_{n+1} \circ f_n \circ \ldots \circ f_1(x,v) \tag{B.24}$$

where $f_1 = G_{-\tau_1}^\omega$, $f_2 = G_{-\tau_2}^{\theta^{t_1}\omega}, \ldots, f_i = G_{-\tau_i}^{\theta^{t_{i-1}}\omega}$ for $1 \le i \le n$, and $f_{n+1} = G_{t_n(\omega)-T}^{\theta^{t_n}\omega}$. Notice that for all $(x,v) \in O_T(\omega)$,

$$\sup_{1\le i\le r} \|D^i f_j((f_{j-1} \circ \ldots \circ f_1)(x,v))\| \le K, \tag{B.25}$$

for $1 \le j \le n+1$, since $f_{j-1} \circ \ldots \circ f_1(x,v) \in B_{\tau_j}$. We now use the following fact. Consider $f \circ g(x,v)$. Assume that

$$\sup_{1\le i\le r} \|D^i g(x,v)\| \le M_1 \text{ and } \sup_{1\le i\le r} \|D^i f(g(x,v))\| \le M_2.$$

Then there exists a constant C_r depending only on r such that

$$\sup_{1\le i\le r} \|D^i (f \circ g)(x,v)\| \le C_r M_2 M_1^r.$$

Using (B.24) and (B.25), we obtain

$$\sup_{(x,v)\in O_T(\omega)} \|D^i G_{-T}^\omega(x,v)\| \le K(C_r K^r)^n \tag{B.26}$$

provided that $|t_n(\omega)| \le T < |t_{n+1}(\omega)|$.

Since for any $t \in [-T, 0]$, we can write

$$G_t^\omega(x,v) = \bar{f}_l \circ f_{l-1} \circ \ldots \circ f_1(x,v) \tag{B.27}$$

for some l, $1 \le l \le n+1$ and (B.25) holds, we get

$$\sup_{-T\le t\le 0,\,(x,v)\in O_T(\omega)} \|D^i G_t^\omega(x,v)\| \le K(C_r K^r)^n . \tag{B.28}$$

We now have

$$\int D_r(\omega)^m dP \le \sum_{n=0}^{\infty} \log^m(K(C_r K^r)^n) P\{|t_n(\omega)| \le T < |t_{n+1}(\omega)|\} .$$

Let $q = P\{\tau_{r,K}(\omega) \le T\}$. Obviously $q < 1$ since there exists a set of ω's with positive probability such that $\tau_{r,K}(\omega) > 1$. Using the strong Markov property we have

$$\begin{aligned} P\{|t_n(\omega)| \le T < |t_{n+1}(\omega)|\} &\le P(\tau_j \le T,\, 1 \le j \le n) \\ &= \prod_{j=1}^{n} P(\tau_j \le T) = q^n . \end{aligned}$$

Thus,

$$\int D_r(\omega)^m dP \le \sum_{n=0}^{\infty} (\log K + n \log(C_r K^r))^m q^n < +\infty .$$

The other estimate can be proved in the same way. □

A stronger estimate holds for the first derivative DG_t^ω. For $T > 0$, define

$$d_T(\omega) = \sup_{-T\le t\le 0}\ \sup_{(x,v)\in B_T} \log^+ \|DG_t^\omega(x,v)\| , \tag{B.29}$$

$$\bar{d}_T(\omega) = \sup_{0\le t\le T}\ \sup_{(x,v)\in B_T} \log^+ \|DG_t^\omega(x,v)\| . \tag{B.30}$$

LEMMA B.5. *Let m be a positive integer, then*

$$\int_\Omega (d_T(\omega))^m dP < +\infty, \qquad \int_\Omega (\bar{d}_T(\omega))^m dP < +\infty . \tag{B.31}$$

Remark. Lemma B.5 is stronger than Lemma B.4 for $r=1$ since $B_T \supset O_T$.

Proof. Let $x(t) = x(t; x_0, v_0)$, $v(t) = v(t; x_0, v_0)$. Now,

$$\begin{aligned} v(t) &= v_0 - \int_t^0 \sum_k f_k(x(s)) dB_k(t) \\ &= v_0 - \sum_k f_k(x(t)) B_k(t) + \int_t^0 \sum_k f_k'(x(s)) B_k(s) v(s) ds . \end{aligned}$$

Let $M^\omega(t) = \max_{t\le s\le 0} |v(s)|$; then for $t \in [-T, 0]$

$$M^\omega(t) \le |v_0| + \|F\|_T + \int_t^0 \|F\|_T M^\omega(s) ds .$$

This implies, for $t \in [-T, 0]$,

$$M^{\omega}(t) \le (|v_0| + \|F\|_T) e^{\|F\|_T |t|}.$$

Consider next (B.14) and (B.15). Let $J^{\omega}(t) = \max_{t \le s \le 0} |J^{\omega}_{21}(s)|$, $\Delta_{t,\tau,k} = \sup_{t-\tau \le s \le t} |B_k(s) - B_k(t)|$, for $\tau > 0$, and $\|f_k\| = \sup_{0 \le x \le 1} |f_k(x)|$. Since

$$J^{\omega}_{21}(s) = J^{\omega}_{21}(t) - \int_s^t \sum_k f_k(x(u)) J^{\omega}_{11}(u) dB_k(u),$$

we have (without loss of generality, we can assume $B_k(t) = 0$)

$$\begin{aligned} J^{\omega}(t-\tau) &\le J^{\omega}(t) + \sup_{t-\tau \le s \le t} \Big| \sum_k f_k(x(s)) B_k(s) J^{\omega}_{11}(s) \Big| \\ &\quad + \int_{t-\tau}^t \Big| \sum_k f_k'(x(u)) B_k(u) v(u) J^{\omega}_{11}(u) \Big| du \\ &\quad + \int_{t-\tau}^t \Big| \sum_k f_k(x(u)) B_k(u) J^{\omega}_{21}(u) \Big| du \\ &\le J^{\omega}(t) + \Big(\sum_k \|f_k\| \Delta_{t,\tau,k} \Big) (1 + J^{\omega}(t-\tau) T) \\ &\quad + (1 + J^{\omega}(t-\tau)T) M^{\omega}(T) 2 \|F\|_T \tau + J^{\omega}(t-\tau) 2 \|F\|_T \tau. \end{aligned}$$

Choosing τ small enough so that

$$\sum_k \|f_k\| \Delta_{t,\tau,k} \le \frac{1}{6T}, \ 2TM^{\omega}(T)\|F\|_T \tau < \frac{1}{6}, \ 2\|F\|_T \tau \le \frac{1}{6}, \tag{B.32}$$

we get

$$J^{\omega}(t-\tau) \le 2\left(J^{\omega}(t) + \frac{1}{3T} \right). \tag{B.33}$$

Define a sequence of stopping times $\bar{\tau}_i$, $i \ge 1$, by

$$\bar{\tau}_1 = \inf\Big\{ \tau \colon \sum_k \|f_k\| \Delta_{0,\tau,k} = \frac{1}{6T} \Big\},$$

$$\bar{\tau}_{i+1} = \inf\Big\{ \tau \colon \sum_k \|f_k\| \Delta_{t_i,\tau,k} = \frac{1}{6T} \Big\},$$

where $t_i = -\sum_{j=1}^{i} \bar{\tau}_j$. Assume that $|t_{k-1}| \le T < |t_k|$. We can divide $[t_{i+1}, t_i]$, $0 \le i \le k-1$, into subintervals such that (B.32) holds on each subinterval. The total number of these subintervals can be estimated from above by

$$R(k,T) = k + 12T\|F\|_T (1 + TM^{\omega}(T)).$$

From (B.33) we get

$$J^\omega(T) \le 2^{R(k,T)}\Big(J^\omega(0) + \frac{2}{3T}\Big) = \frac{2}{3T}\, 2^{R(k,T)}. \tag{B.34}$$

We now show that

$$\int R(k,T)^m dP < +\infty. \tag{B.35}$$

Denote $q = P(\bar\tau_1 \le T) < 1$. Using the strong Markov property, we have

$$P\{|t_{k-1}| \le T \le |t_k|\} \le q^{k-1}. \tag{B.36}$$

Hence

$$\int k^m P(d\omega) \le \sum_{k=1}^{\infty} k^m P\{|t_{k-1}| \le T \le |t_k|\} \le \sum_{k=1}^{\infty} k^m q^{k-1} < \infty. \tag{B.37}$$

On the other hand, since $M^\omega(T) \le (|v_0| + \|F\|_T)e^{\|F\|_T T}$,

$$\begin{aligned}\int (M^\omega(T)\|F\|_T)^m P(d\omega) &\le \int (|v_0| + \|F\|_T)^{2m} e^{m\|F\|_T T} P(d\omega)\\ &\le \int (V_T(\omega) + 1 + \|F\|_T)^{2m} e^{m\|F\|_T T} P(d\omega).\end{aligned}$$

Recall that $\|F\|_T = \sum_k \|F_k\|_{C^3} \max_{-T\le t\le 0} |B_k(t)|$. There exist constants $A, B > 0$, such that

$$P(\|F\|_T \ge x) \le Ae^{-Bx^2}. \tag{B.38}$$

Therefore for any positive integers l and m

$$\int \|F\|_T^l e^{m\|F\|_T T} P(d\omega) < \infty. \tag{B.39}$$

We also have from Lemma 3.3 and (B.38) that

$$\int V_T(\omega)^l e^{m\|F\|_T T} P(d\omega) < +\infty.$$

Hence we obtain

$$\int (M^\omega(T)\|F\|_T)^m P(d\omega) < +\infty.$$

An estimate for $J^\omega_{11}(t)$ follows from (B.16). Similar estimates can also be proved for $J^\omega_{12}(t)$ and $J^\omega_{22}(t)$. Together we obtain the first inequality in (B.31). The second inequality can be proved in the same way. □

Consider two minimizers γ_1 and γ_2 on $(-\infty, 0]$, $\gamma_1(0) = y$, $\gamma_2(0) = x$. Denote $v_1(\tau) = \dot\gamma_1(\tau)$, $v_2(\tau) = \dot\gamma_2(\tau)$.

LEMMA B.6. *Assume* $y - x > 0$, $v_1(0) - v_2(0) = \tilde{L}(y-x)$ *and* $\tilde{L} > 4\|F\|_1(6 + 21\|F\|_1)$. *Then*

$$v_1(t) - v_2(t) \geq 0 \tag{B.40}$$

for $t \in [-\tau_0, 0]$, *where* $\tau_0 = \min\left(1, \frac{1}{2\|F\|_1}\right)$ *and*

$$v_1(t) - v_2(t) > \frac{\tilde{L}}{4}(y-x) \tag{B.41}$$

for $t \in [-\tau_1, 0]$, *where* $\tau_1 = \min\left(1, \frac{1}{4\|F\|_1}\right)$.

Proof. We first prove (B.40). We shall consider γ_1, γ_2 as curves on the universal cover. Suppose that for some $-\tau_0 < t \leq 0$, $v_1(t) - v_2(t) = 0$. Denote by

$$t_1 = \max\{-\tau_0 \leq t \leq 0 \colon v_1(t) - v_2(t) = 0\},$$
$$t_2 = \min\{t \colon -t_1 \leq t \leq 0, v_1(t) - v_2(t) = \tilde{L}(y-x)\}.$$

Clearly $0 \leq v_1(t) - v_2(t) \leq \tilde{L}(y-x)$, $t_1 \leq t \leq t_2$. Also, since minimizers do not intersect, $0 \leq \gamma_1(t) - \gamma_2(t) \leq \gamma_1(0) - \gamma_2(0) = y - x$. Now,

$$v_1(t_1) = v_1(t_2) + \sum_k f_k(\gamma_1(t_1))B_k(t_1) - \sum_k f_k(\gamma_2(t_2))B_k(t_2) + \int_{t_1}^{t_2} \sum_k f_k'(\gamma_1(s))v_1(s)B_k(s)ds\,,$$

$$v_2(t_1) = v_2(t_2) + \sum_k f_k(\gamma_2(t_1))B_k(t_1) - \sum_k f_k(\gamma_2(t_2))B_k(t_2) + \int_{t_1}^{t_2} \sum_k f_k'(\gamma_2(s))v_2(s)B_k(s)ds\,.$$

Thus, $0 = v_1(t_1) - v_2(t_1) = v_1(t_2) - v_2(t_2) + \Delta v$, where

$$\Delta v = \sum_k (f_k(\gamma_1(t_1)) - f_k(\gamma_2(t_1)))B_k(t_1) - \sum_k (f_k(\gamma_1(t_2)) - f_k(\gamma_2(t_2)))B_k(t_2)$$
$$+ \int_{t_1}^{t_2} \sum_k (f_k'(\gamma_1(s)) - f_k'(\gamma_2(s)))v_1(s)B_k(s)ds$$
$$+ \int_{t_1}^{t_2} \sum_k f_k'(\gamma_2(s))(v_1(s) - v_2(s))B_k(s)ds\,.$$

Let $C_1 = \frac{1}{4} + \|F\|_1$, $C = 20C_1$. It follows from Lemma 3.3 and (B.8) that $|v_1(0)|$, $|v_2(0)| \le C$ and for all $-\tau_0 \le s \le 0$:

$$|v_1(s)|,\ |v_2(s)| \le 2(C + \|F\|_1) \le 10 + 42\|F\|_1 .$$

Thus,

$$\begin{aligned}|\Delta v| &\le 2\|F\|_1(y-x) + (t_2 - t_1)\|F\|_1(10 + 42\|F\|_1)(y-x) \\ &\quad + (t_2 - t_1)\|F\|_1 \cdot \tilde{L}(y-x) \le \left((12 + 42\|F\|_1) + \tau_0 \tilde{L}\right)\|F\|_1(y-x) .\end{aligned}$$

Since $\tau_0 \le \frac{1}{2\|F_1\|}$, $(12 + 42\|F\|_1)\|F\|_1 < \frac{\tilde{L}}{2}$ we have from the estimate above:

$$|\Delta v| < \left(\frac{\tilde{L}}{2} + \frac{\tilde{L}}{2}\right)(y-x) < \tilde{L}(y-x) ,$$

which contradicts the fact that $|\Delta v| = \tilde{L}(y-x)$.

Next we prove (B.41). Suppose $-\tau_1 \le t \le 0$. Then $v_1(t) - v_2(t) \ge 0$. Suppose for some $-\tau_1 \le t \le 0$: $v_1(t) - v_2(t) = \frac{\tilde{L}}{4}(y-x)$. Denote

$$\begin{aligned} t_3 &= \max\{-\tau_1 \le t \le 0 : v_1(t) - v_2(t) = \frac{\tilde{L}}{4}(y-x)\} , \\ t_4 &= \min\{-t_3 \le t \le 0 : v_1(t) - v_2(t) = \tilde{L}(y-x)\} . \end{aligned}$$

Clearly, $\frac{\tilde{L}}{4}(y-x) \le v_1(t) - v_2(t) \le \tilde{L}(y-x)$, $0 \le \gamma_1(t) - \gamma_2(t) \le y - x$, $t_3 \le t \le t_4$. Using the same estimates as above, we have

$$\frac{\tilde{L}}{4}(y-x) = v_1(t_3) - v_2(t_3) = v_1(t_4) - v_2(t_4) + \overline{\Delta v} ,$$

where

$$\begin{aligned}|\overline{\Delta v}| &\le 2\|F\|_1(y-x) + (t_4 - t_3)\|F\|_1(10 + 42\|F\|_1)(y-x) \\ &\quad + (t_4 - t_3)\|F\|_1 \tilde{L}(y-x) \\ &\le ((12 + 42\|F\|_1) + \tau_1\tilde{L})\|F\|_1(y-x) < \left(\frac{\tilde{L}}{2} + \frac{\tilde{L}}{4}\right)(y-x) = \frac{3\tilde{L}}{4}(y-x) .\end{aligned}$$

On the other hand, $v_1(t_4) - v_2(t_4) = \tilde{L}(y-x)$, and $\overline{\Delta v} = -\frac{3}{4}\tilde{L}(y-x)$, which contradicts the estimate above. □

LEMMA B.7. *Let* $y - x > 0$, $v_1(0) - v_2(0) = \tilde{L}(y-x)$. *Then, with P-probability* 1, $\tilde{L} \le \max(4\|F\|_1(6 + 21\|F\|_1), 4)$.

Proof. Suppose $\tilde{L} > 4\|F\|_1(6 + 21\|F\|_1)$. Then, for all $-\tau_1 \leq t \leq 0$, $v_1(t) - v_2(t) > \frac{\tilde{L}}{4}(y - x)$. Thus the two minimizers would intersect before the time $\tau_* = -\frac{(y-x)}{\frac{\tilde{L}}{4}(y-x)} = -\frac{4}{\tilde{L}}$, where $-\tau_1 \leq \tau_* \leq 0$ since

$$\tilde{L} > \max\left(4\|F\|_1(6 + 21\|F\|_1), 4\right).$$

This contradiction proves Lemma B.7. □

Denote

$$\|F\|_{-1,1} = \max_{-1 \leq s \leq 1} \sum_k \|F_k(x)\|_{C^3} |B_k(s) - B_k(0)|.$$

Let $L_0 = 4 + 24\|F\|_{-1,1} + 84\|F\|_{-1,1}^2$. Obviously,

$$L_0 > \max\left(4\|F\|_1(6 + 21\|F\|_1), 4\right).$$

Consider two minimizers at time $t = 1$: $\gamma_1(\tau), \gamma_2(\tau)$, $-\infty \leq \tau \leq 1$. Denote $y = \gamma_1(0)$, $v(y) = \gamma_1(0)$, $x = \gamma_2(0)$, $v(x) = \dot{\gamma}_2(0)$.

LEMMA B.8. *With P-probability* 1,

$$|v(y) - v(x)| \leq L_0 |y - x|.$$

Proof. Suppose $y - x > 0$. Then, it follows from Lemma B.7 that: $v(y) - v(x) \leq L_0(y - x)$. Similarly we can prove an estimate from the other side. □

LEMMA B.9. *Consider the process*

$$\begin{aligned}
dx &= v dt\,, \\
dv &= \sum_k f_k(x(t)) dB_k(t)\,, \\
da &= b dt\,, \\
db &= a \sum_k f_k'(x(t)) dB_k(t)\,,
\end{aligned}$$

and let $a(0)$, $b(0)$ satisfy $a(0)^2 + b(0)^2 = 1$. Assume that there exists a constant $\alpha_0 > 0$, such that

$$\sum_k f_k'(x)^2 \geq \alpha_0$$

for all $x \in [0, 1]$. Then the joint probability distribution of $(a(1), b(1))$ has density $\bar{p}(a, b)$ which is uniformly bounded (with respect to $(a(0), b(0))$) on any compact domain.

Proof. We will give only an outline for the proof. The generator $\mathfrak{L}$ for the diffusion process can be written as

$$\mathfrak{L} = \mathfrak{L}_{a,b} + \mathfrak{L}'$$

where

$$\mathcal{L}_{a,b} = b\frac{\partial}{\partial a} + \frac{1}{2}a^2\left(\sum_k f_k'(x)^2\right)\frac{\partial^2}{\partial b^2},$$

$$\mathcal{L}' = v\frac{\partial}{\partial x} + a\sum_k f_k'(x)f_k(x)\frac{\partial^2}{\partial b\partial v} + \frac{1}{2}\sum_k f_k(x)^2\frac{\partial^2}{\partial v^2}.$$

The operator $\mathcal{L}_{a,b}$ is hypoelliptic on $R^2 \setminus \{(0,0)\}$ for each fixed $x \in [0,1]$ (see [IK]). Therefore for each fixed $x \in [0,1]$, the solution of

$$\partial_t p^x = \mathcal{L}^*_{a,b}p^x,$$
$$p^x(a,b,0) = \delta(a-a(0), b-b(0))$$

is smooth for $t > 0$, except at $(a,b) = (0,0)$ [IK]. Since the delta function $p^x(\cdot, 0)$ is concentrated on the unit circle, we have that for $0 \le t \le 1$, p^x is uniformly bounded (with respect to x and $(a(0), b(0))$) on the circle

$$0 \le p^x(a,b,t) \le C^*$$

if $a^2 + b^2 = \frac{1}{4}$, and $0 \le t \le 1$. Using the maximum principle for the operator $\mathcal{L}_{a,b}$ on the domain $\{(a,b), a^2+b^2 \le \frac{1}{4}\} \times [0,1]$, we conclude that

$$p^x(a,b,t) \le C^*$$

if $a^2+b^2 \le \frac{1}{4}$, and $t \le 1$. Since C^* is independent of (x,v) and $(a(0), b(0))$, and since p^x is smooth away from the origin, we obtain the desired result. □

Appendix C. Hopf-Cole transformation and the Feynman-Kac formula

The Hopf-Cole transform and the Feynman-Kac formula are standard tools used in the analysis of (1.1). In the random case, some care has to be taken because of the appearance of stochastic integrals [S2].

Consider the stochastic PDE

$$d\psi = \frac{\varepsilon}{2}\frac{\partial^2\psi}{\partial x^2}dt + \Big(-\frac{1}{\varepsilon}\sum_k F_k(x)dB_k(t) + c(x)dt\Big)\psi \tag{C.1}$$

(the function $c(x)$ to be defined later). In the following, stochastic integrals will be understood in the Ito sense.

Let $v = -\varepsilon \ln \psi$. Using the Ito formula, we have

$$dv = -\frac{\varepsilon^2}{2}\frac{1}{\psi}\frac{\partial^2\psi}{\partial x^2}dt + \sum_k F_k(x)dB_k(t) + c(x)dt + \frac{1}{2\varepsilon}a(x)dt \tag{C.2}$$

where

$$a(x)dt = \mathbb{E}\left(\sum F_k(x)dB_k(t)\right)^2 = \left(\sum_k F_k^2(x)\right)dt\,.$$

Choosing $c(x) = -\frac{1}{2\varepsilon}\,a(x)$, we get

$$dv = -\frac{\varepsilon^2}{2}\frac{1}{\psi}\frac{\partial^2\psi}{\partial x^2}\,dt + \sum_k F_k(x)dB_k(t)\,. \tag{C.3}$$

Let $u = -v_x$. It is straightforward to verify that u satisfies

$$du + \left(u\,\frac{\partial u}{\partial x} - \frac{\varepsilon}{2}\frac{\partial^2 u}{\partial x^2}\right)dt = \sum_k f_k(x)dB_k(t)\,. \tag{C.4}$$

The Feynman-Kac formula for (C.1) takes the form

$$\psi(x,t) = \mathbb{E}_\beta\psi\left(x + \sqrt{\varepsilon}\,\beta(t_0), t_0\right) e^{-\frac{1}{\varepsilon}\int_{t_0}^{t}\sum_k F_k\left(x+\sqrt{\varepsilon}\,\beta(s)\right)dB_k(s)}\,, \tag{C.5}$$

where $\mathbb{E}_\beta$ denotes expectation with respect to the Wiener process on $[t_0, t]$ such that $\beta(t) = 0$. It is easy to verify that the extra terms that occur in the Ito formula for the exponential function in (C.5) are accounted for by the last term $c(x)\psi dt$ in (C.1).

(C.1) can also be rewritten as

$$d\psi = \frac{\varepsilon}{2}\frac{\partial^2\psi}{\partial x^2}\,dt - \frac{1}{\varepsilon}\,\psi \circ \sum_k F_k(x)dB_k(t) \tag{C.6}$$

where "$\circ$" denotes product in the Stratonovich sense.

Appendix D. The basic collision lemma

This appendix is devoted to the proof and discussion of Lemma 5.2. We will use the notion of the backward Lagrangian map. It will be convenient to work with $\mathbb{R}^1$ instead of $\mathbb{S}^1$. Fix $t, s \in \mathbb{R}^1$, $t > s$, and $x \in \mathbb{R}^1$. Let ξ_+, ξ_- be the maximal and minimal backward characteristics (see [D]) such that $\xi_+(t) = \xi_-(t) = x$. We define $Y_{s,t}^+(x) = \xi_+(s)$, $Y_{s,t}^-(x) = \xi_-(s)$.

We will study the case of

$$F(x,t) = -\frac{1}{2\pi}\cos(2\pi x)dB(t)\,,$$
$$f(x,t) = \sin(2\pi x)dB(t)\,.$$

It will be clear that the general situation follows from the same argument. From Lemma B.1, we can assume, without loss of generality, that $\|u(\cdot\,,0)\|_{L^\infty} \leq C$ for some random constant C. Otherwise we change the initial time from $t = 0$ to some positive number, say $\frac{1}{16}$. It follows from Lemma B.1, that $\left\|u\left(\cdot\,,\frac{1}{16}\right)\right\|_{L^\infty} \leq C_1$ for some random constant C_1 depending on the forces on

$\left[0, \frac{1}{16}\right]$. In addition, we will consider a particular case when $x_1^0 = \frac{1}{8}$, $x_2^0 = \frac{7}{8}$. It is easy to see from the proof that the argument works in the general case as well. We will use the notation $O(\delta)$ to denote quantities that are bounded in absolute value by $A\delta$, where A is an absolute constant.

The basic strategy is to construct forces that are large on $[0, t_1]$ and small on $[t_1, 1]$ for some t_1 in order to set up approximately the following picture: At $t = t_1$, u is very positive for $x \in [0, \frac{1}{2}]$ and very negative for $x \in [\frac{1}{2}, 1]$. If the forcing is small on $[t_1, 1]$, a shock must form which will absorb a sufficient amount of mass, if we imagine that there is a uniform distribution of masses on $[0, 1]$ at $t = t_1$. In order to make this intuitive picture rigorous, we must carefully control the value of u when the forcing is small.

On B and t_1, assume

$$B(0) = 0, \max_{0 \le s \le t_1} |B(s)| \le 2B(t_1),\ 4\pi t_1 B(t_1) < \delta_0, B(t_1) > \bar{C} \tag{D.1}$$

with $\bar{C}, \delta_0$ as chosen below. We will show that if B satisfies (D.1), then x_1^0 and x_2^0 merge before $t = 1$. Therefore the probability of merging is no less than the probability of the Brownian paths satisfying (D.1) which is positive.

Fix $x \in [0, 1]$. Let ξ be a genuine backward characteristic emanating from x at $t = t_1$; $\xi(t_1) = x$. Denote $y = \xi(0)$; then

$$u(x, t_1) = u(y, 0) + \sin(2\pi x)B(t_1) - \int_0^{t_1} 2\pi \cos(2\pi\xi(s))\dot{\xi}(s)B(s)ds\,. \tag{D.2}$$

Hence $|u|_\infty = \max\limits_{\substack{0 \le x \le 1 \\ 0 \le t \le t_1}} |u(x, t)|$ satisfies

$$|u|_\infty \le C + B(t_1) + 2\pi|u|_\infty \int_0^{t_1} |B(s)|ds\,.$$

Therefore

$$|u|_\infty \le M = \frac{C + B(t_1)}{1 - \delta_0}\,.$$

We now estimate $u(\cdot\,, t_1)$. The idea is that on the set where the force is bounded away from zero, u is either very negative or very positive, reflected by the term involving δ_2 below. We will bound u on the complement of this set. For $x \in \left[\frac{1}{16}, \frac{1}{2} - \varepsilon\right]$, $0 < \varepsilon \ll 1$, ε to be fixed later, we have

$$\begin{aligned} u(x, t_1) &\ge -C + \delta_1 B(t_1) - 2\pi M \int_0^{t_1} |B(s)|ds \\ &\ge -C + \delta_1 B(t_1) - \delta_0 M \end{aligned}$$

with $\delta_1 = \sin\left(2\pi\left(\frac{1}{2} - \varepsilon\right)\right) = \sin(2\pi\varepsilon)$. For $x \in \left[\frac{1}{16}, \frac{7}{16}\right]$, with $\delta_2 = \sin\frac{\pi}{8}$, this can be improved to

$$u(x, t_1) \ge -C + \delta_2 B(t_1) - \delta_0 M\,.$$

The size of ε is chosen such that there is a finite gap between δ_2 and δ_1. For $x \in \left[\frac{1}{2} - \varepsilon, \frac{1}{2}\right]$, we have

$$|u(x, t_1)| \le C + \delta_1 B(t_1) + \delta_0 M\,.$$

Similarly on $\left[\frac{1}{2}, 1\right]$ we have the estimates

$$|u(x, t_1)| \le C + \delta_1 B(t_1) + \delta_0 M\,,$$

for $x \in \left[\frac{1}{2}, \frac{1}{2} + \varepsilon\right]$;

$$u(x, t_1) \le C - \delta_2 B(t_1) + \delta_0 M$$

for $x \in \left[\frac{9}{16}, \frac{15}{16}\right]$; and

$$u(x, t_1) \le C - \delta_1 B(t_1) + \delta_0 M$$

for $x \in \left[\frac{1}{2} + \varepsilon, \frac{15}{16}\right]$.

Next on $[t_1, 1]$ we will choose $B(t)$ to be so small that

$$\max_{t_1 \le s \le 1} |B(s) - B(t_1)| \le \delta\,. \tag{D.3}$$

The value of δ will be chosen later. We first prove the following approximate monotonicity lemma.

LEMMA D.1. *Let x^* be a point of shock at $t = 1$, $y_1 = Y^-_{t_1,1}(x^*)$, $y_2 = Y^+_{t_1,1}(x^*)$ and $y \in (y_1, y_2)$. Then*

$$\int_{y_1}^{y} (z + tu(z, t_1))dz - x^*(y - y_1) \ge -C\delta |u|_\infty\,,$$

$$x^*(y_2 - y) - \int_{y}^{y^2} (z + tu(z, t_1))dz \ge -C\delta |u|_\infty\,,$$

where $t = 1 - t_1$, $|u|_\infty = ||u(\cdot, t_1)||_{L^\infty}$.

Remark. In the absence of forces, the correct statement is

$$\int_{y_1}^{y} (z + tu(z, t_1))dz - x^*(y - y_1) \ge 0\,.$$

These statements were used in [ERS] as the basis for an alternative formulation of the variational principle. In the presence of force, similar statements appear to be invalid due to the presence of conjugate points. However, when the force is small, the error is also small, as claimed in Lemma D.1.

Proof of Lemma D.1. Define y^* by:

$$y^* + tu(y^*, t_1) = x^*\,. \tag{D.4}$$

Denote by ξ_+, ξ_- the maximal and minimal backward characteristics such that $\xi_+(1) = \xi_-(1) = x^*$. Then

$$y_1 + \int_{t_1}^{1} u(\xi_-(s), s)ds = x^*,$$

$$y_2 + \int_{t_1}^{1} u(\xi_+(s), s)ds = x^*.$$

Furthermore, for $s \in [t_1, 1]$, we have

$$\text{(D.5)} \qquad u(\xi_+(s), s) = u(y_2, t_1) + \sin(2\pi\xi_+(s))(B(s) - B(t_1)) - \int_{t_1}^{s} 2\pi\cos(2\pi\xi_+(s))\dot{\xi}_+(s)(B(s) - B(t_1))ds.$$

Hence

$$\text{(D.6)} \qquad |u(\xi_+(s), s) - u(y_2, t_1)| \le O(\delta)|u|_\infty$$

and

$$\text{(D.7)} \qquad y_2 + tu(y_2, t_1) = x^* + O(\delta)|u|_\infty.$$

Similarly

$$\text{(D.8)} \qquad y_1 + tu(y_1, t_1) = x^* + O(\delta)|u|_\infty$$

and

$$\text{(D.9)} \qquad |u(\xi_-(s), s) - u(y_1, t_1)| \le O(\delta)|u|_\infty.$$

From the action minimizing property of ξ_-, we get, by comparing the action of ξ_- and $\xi(s) = y^* + (s - t_1)u(y^*, t_1)$,

$$\begin{aligned}\int_{t_1}^{1} \frac{u^2(\xi_-(s), s)}{2}ds &- \frac{1}{2\pi}\cos(2\pi x^*)(B(1) - B(t_1)) \\ &+ \int_{t_1}^{1} \dot{\xi}_-(s)\sin(2\pi\xi_-(s))(B(s) - B(t_1))ds \\ \le t\frac{u^2(y^*, t_1)}{2} &- \frac{1}{2\pi}\cos(2\pi x^*)(B(1) - B(t_1)) \\ &+ \int_{t_1}^{1} u(y^*, t_1)\sin 2\pi(y^* + (s - t_1)u(y^*, t_1))(B(s) - B(t_1))ds \\ &+ \int_{y_1}^{y^*} u(y, t_1)dy.\end{aligned}$$

This gives

$$t\frac{u^2(y_1, t_1)}{2} \le t\frac{u^2(y^*, t_1)}{2} + \int_{y_1}^{y^*} u(y, t_1)dy + O(\delta)|u|_\infty.$$

Finally, we get, using (D.4), (D.7) and (D.8):

$$\int_{y_1}^{y^*} u(y, t_1)dy \geq \frac{t}{2}(u^2(y_1, t_1) - u^2(y^*, t_1)) + O(\delta)|u|_\infty ;$$

$$\begin{aligned}
&\int_{y_1}^{y^*} (z + tu(z, t_1))dz - (y^* - y_1)x^* \\
&\qquad = \frac{(y^*)^2}{2} - \frac{y_1^2}{2} + t\int_{y_1}^{y^*} u(z, t_1)dz - (y^* - y_1)x^* \\
&\qquad \geq \frac{(y^*)^2}{2} - \frac{y_1^2}{2} + \frac{t^2}{2}(u^2(y_1, t_1) - u^2(y^*, t_1)) - (y^* - y_1)x^* + O(\delta)|u|_\infty \\
&\qquad = O(\delta)|u|_\infty .
\end{aligned}$$

□

Proof of Lemma 5.2. Let z_1^0 and z_2^0 be the Eulerian positions of x_1^0 and x_2^0 respectively at time t_1 following the forward characteristics defined by u. They are well-defined if the forward characteristics are continued properly by shocks [D]. Moreover, $|x_1^0 - z_1^0| \leq O(\delta_0)$, $|x_2^0 - z_2^0| \leq O(\delta_0)$, since t_1 is small. Assume to the contrary that z_1^0 and z_2^0 do not merge until time 1. Then there exist x_1 and x_2, such that $x_1 < x_2$, and $z_1^0 \in [Y_{t_1,1}^-(x_1), Y_{t_1,1}^+(x_1)]$, $z_2^0 \in [Y_{t_1,1}^-(x_2), Y_{t_1,1}^+(x_2)]$. Let $\alpha_1 = Y_{t_1,1}^-(x_1)$, $\alpha_2 = Y_{t_1,1}^+(x_1)$, $\beta_1 = Y_{t_1,1}^-(x_2)$, $\beta_2 = Y_{t_1,1}^+(x_2)$. We have $\alpha_1 < \alpha_2 < \beta_1 < \beta_2$. Using the estimates obtained earlier on $u(\cdot, t_1)$, we have the following:

If $\alpha_2 < \frac{3}{8}$, then

$$\begin{aligned}
x_1 &= \alpha_2 + (1 - t_1)u(\alpha_2, t_1) + O(\delta)|u|_\infty \\
&\geq \alpha_2 + (1 - t_1)(-C + \delta_2 B(t_1) - \delta_0 M) + O(\delta)|u|_\infty .
\end{aligned} \tag{D.10}$$

Similarly, if $\beta_1 > \frac{5}{8}$, then

$$\begin{aligned}
x_2 &= \beta_1 + (1 - t_1)u(\beta_1, t_1) + O(\delta)|u|_\infty \\
&\leq \beta_1 + (1 - t_1)(C - \delta_2 B(t_1) + \delta_0 M) + O(\delta)|u|_\infty .
\end{aligned} \tag{D.11}$$

□

To deal with the case when either $\beta_1 < \frac{5}{8}$, or $\alpha_2 > \frac{3}{8}$, we introduce the parametrized measure $dQ_s(\cdot)$ which is the pullback of Lebesgue measure by the backward Lagrangian map from $t = t_1$ to $t = s$: $Q_s[x_1, x_2) = Y_{t_1,s}^+(x_2) - Y_{t_1,s}^-(x_1)$.

If we define $\rho = dQ_s$, $u(x, t) = \frac{1}{2}(u(x+, t) + u(x-, t))$, then it is easy to see that (ρ, u) satisfies

$$\rho_t + (\rho u)_x = 0 \tag{D.12}$$

in the distributional sense.

Let ξ_1, ξ_2 be two genuine backward characteristics defined on $[t_1, 1]$, such that $\xi_1 < \xi_2$. Multiplying the above equation by x and integrating over the region: $t_1 \leq s \leq 1$, $\xi_1(s) \leq x \leq \xi_2(s)$, we get

$$\int\limits_{\xi_1(1)}^{\xi_2(1)} x dQ_1(x) - \int\limits_{\xi_1(t_1)}^{\xi_2(t_1)} x dx = \int\limits_{t_1}^{1} ds \int\limits_{\xi_1(s)}^{\xi_2(s)} u(x,s) dQ_s(x) . \tag{D.13}$$

LEMMA D.2. *For* $s \in [t_1, 1]$,

$$\int\limits_{\xi_1(s)}^{\xi_2(s)} u(x,s) dQ_s(x) = \int\limits_{\xi_1(t_1)}^{\xi_2(t_1)} u(y,t_1) dy + O(\delta)|u|_\infty .$$

Proof. First we assume that x is a point of continuity of $u(\cdot\,,s)$. Then from the arguments presented earlier, we have

$$u(x,s) = u(y,t_1) + O(\delta)|u|_\infty \tag{D.14}$$

where $y = Y^+_{t_1,s}(x) = Y^-_{t_1,s}(x)$.

If x is a point of discontinuity of $u(\cdot\ ,s)$, let $y_1 = Y^-_{t_1,s}(x)$, $y_2 = Y^+_{t_1,s}(x)$. We then have

$$\begin{aligned} u(x,s) &= \frac{1}{2}(u(x-,s) + u(x+,s)) \\ &= \frac{1}{2}(u(y_1,t_1) + u(y_2,t_1)) + O(\delta)|u|_\infty . \end{aligned}$$

On the other hand, similar to the proof of Lemma D.1, we also have

$$\begin{aligned} y_1 + (s-t_1)u(y_1,t_1) &= y_2 + (s-t_1)u(y_2,t_1) + O(\delta)|u|_\infty \frac{u^2(y_1,t_1)}{2}(s-t_1) \\ &= \frac{u^2(y_2,t_1)}{2}(s-t_1) + \int_{y_1}^{y_2} u(y,t_1)dy + O(\delta)|u|_\infty . \end{aligned}$$

Hence

$$\begin{aligned} \int_{y_1}^{y_2} u(y,t_1)dy &= \frac{s-t_1}{2}(u^2(y_1,t_1) - u^2(y_2,t_1)) + O(\delta)|u|_\infty \\ &= \frac{s-t_1}{2}(u(y_1,t_1) - u(y_2,t_1))(u(y_1,t_1) + u(y_2,t_1)) + O(\delta)|u|_\infty \\ &= (y_2 - y_1)\frac{u(y_1,t_1) + u(y_2,t_2)}{2} + O(\delta)|u|_\infty . \end{aligned}$$

This can be written as

$$u(x,s)Q_s(\{x\}) = \int_{y_1}^{y_2} u(y,t_1)dy + O(\delta)|u|_\infty. \tag{D.15}$$

Now Lemma D.2 follows from (D.14) and (D.15) when we use a standard approximation argument. □

We now continue with the proof of Lemma 5.2. Using Lemma D.2, we can rewrite (D.13) as

$$\int_{\xi_1(1)}^{\xi_2(1)} x dP_1(x) - \int_{\xi_1(t_1)}^{\xi_2(t_1)} x dx = (1-t_1)\int_{\xi_1(t_1)}^{\xi_2(t_1)} u(y,t_1)dy + O(\delta)|u|_\infty .$$

Applying this to $Y^{\pm}_{t_1,1}(x_1)$, $Y^{\pm}_{t_1,1}(x_2)$, we get

$$x_1 - \frac{1}{\alpha_2-\alpha_1}\int_{\alpha_1}^{\alpha_2}\{x+(1-t_1)u(x,t_1)\}dx = \frac{O(\delta)|u|_\infty}{\alpha_2-\alpha_1},$$

$$x_2 - \frac{1}{\beta_2-\beta_1}\int_{\beta_1}^{\beta_2}\{x+(1-t_1)u(x,t_1)\}dx = \frac{O(\delta)|u|_\infty}{\beta_2-\beta_1}.$$

Assume that $\alpha_2 > \frac{3}{8}$. Then $\alpha_2 - z_1^0 > \frac{1}{4} - O(\delta_0)$. Integrating both sides of the equation $u_t + \left(\frac{u^2}{2}\right)_x = -F_x$ over the region: $0 \le t \le t_1$, $\xi_1(t) \le x \le \xi_2(t)$, where $\xi_1(t) = Y^{-}_{t,t_1}(x_1^0)$, $\xi_2(t) = Y^{+}_{t,t_1}(\alpha_2)$, we get

(D.16)

$$\begin{aligned}\int_{x_1^0}^{\alpha_2} u(x,t_1)dx - \int_{\xi_1(0)}^{\xi_2(0)} u(x,0)dx &= \frac{1}{2}\int_0^{t_1}[u(\xi_1(t),t)^2 - u(\xi_2(t),t)^2]dt \\ &- \frac{1}{2\pi}B(t_1)\Big(\cos(2\pi\xi_2(t_1)) - \cos(2\pi\xi_1(t_1))\Big) \\ &+ \int_0^{t_1} B(t)\Big(\dot\xi_2(t)\sin(2\pi\xi_2(t)) - \dot\xi_1(t)\sin(2\pi\xi_1(t))\Big)dt.\end{aligned}$$

This implies that

$$\begin{aligned}\int_{x_1^0}^{\alpha_2} u(x,t_1)dx \ge & -\frac{1}{2\pi}B(t_1)\Big(\cos(2\pi\alpha_2) - \cos(2\pi x_1^0)\Big) \\ & - C - t_1|u|_\infty^2 - |u|_\infty\int_0^{t_1}|B(s)|ds.\end{aligned}$$

Hence, using Lemma D.1, we obtain

$$
\begin{aligned}
\text{(D.17)} \qquad x_1 &\geq \frac{1}{\alpha_2 - x_1^0} \int_{x_1^0}^{\alpha_2} (z + (1 - t_1)u(z, t_1))dz + O(\delta)|u|_\infty \\
&\geq -\frac{2}{2\pi}(1 - t_1)B(t_1)\Big(\cos(2\pi\alpha_2) - \cos(2\pi x_1^0)\Big) \\
&\quad - C - 4t_1|u|_\infty^2 - 4\delta_0|u|_\infty + O(\delta)|u|_\infty \,.
\end{aligned}
$$

Similarly, if $\beta_1 < \frac{5}{8}$,

$$
\begin{aligned}
\text{(D.18)} \qquad x_2 &\leq \frac{1}{x_2^0 - \beta_1} \int_{\beta_1}^{x_2^0} \Big(z + (1 - t_1)u(z, t_1)\Big)dz + O(\delta)|u|_\infty \\
&\leq \frac{2}{2\pi}(1 - t_1)B(t_1)(\cos 2\pi x_2^0 - \cos 2\pi\beta_1) \\
&\quad + C + 4t_1|u|_\infty^2 + 4\delta_0|u|_\infty + O(\delta)|u|_\infty \,.
\end{aligned}
$$

If $\beta_1 \geq \frac{3}{8}$, we have $\cos(2\pi x_2^0) - \cos(2\pi\beta_1) \geq 0$, and if $\alpha_2 \leq \frac{5}{8}$, we have $\cos(2\pi\alpha_2) - \cos(2\pi x_1^0) \leq 0$. Otherwise, we can use (D.10) and (D.11). In any case, we always have, for some positive constant C^*,

$$
\begin{aligned}
\text{(D.19)} \qquad x_1 - x_2 &\geq C^* B(t_1) - C_0(1 + t_1|u|_\infty^2 + \delta_0|u|_\infty + \delta|u|_\infty) \\
&\geq C^* B(t_1) - C_0 - C_0(4\delta_0 + \delta)|u|_\infty \\
&\geq \left[C^* - \frac{2C_0(4\delta_0 + \delta)}{1 - \delta_0}\right] B(t_1) - C_1 \,.
\end{aligned}
$$

The constants C^*, C_0, C_1 do not depend on δ_0, δ, $B(t_1)$.

If we choose δ_0, δ, such that

$$
\text{(D.20)} \qquad C^* - \frac{2C_0(4\delta_0 + \delta)}{1 - \delta_0} > 0
$$

we can then choose $\bar{C}$, such that

$$
x_1 - x_2 > 0 \,,
$$

contradicting the assumption that $x_1 \leq x_2$. This completes the proof of Lemma 5.2.

We now estimate the location of the shock x^* where x_1^0 and x_2^0 have merged at $t = 1$, assuming that the forces are chosen as in the proof of Lemma 5.2. Let $y_1 = Y_{t_1,1}^-(x^*)$, $y_2 = Y_{t_1,1}^+(x^*)$. Now,

$$
x^* = \frac{1}{y_2 - y_1} \int_{y_1}^{y_2} (y + (1 - t_1)u(y, t_1))dy + O(\delta)|u|_\infty \,.
$$

Also,

(D.21)

$$\left|\frac{1}{y_2-y_1}\int_{y_1}^{y_2} y\,dy-\frac{1}{2}\right|<\frac{1}{2}(1-x_2^0+x_1^0),$$

$$\left|\int_{y_1}^{y_2} u(y,t_1)dy\right|\le(1-x_2^0+x_1^0)|u(\cdot\ .0)|_\infty+t_1|u|_\infty^2$$

$$+\frac{1}{2\pi}|(\cos 2\pi y_2-\cos 2\pi y_1)B(t_1)|+\delta_0|u|_\infty,$$

where we used an analog of (D.16). The factors $1-x_2^0+x_1^0$, $\frac{1}{2}(1-x_2^0+x_1^0)$ can be made arbitrarily small by choosing x_1^0 close to 0, and x_2^0 close to 1.

Notice that in (D.21) the coefficient in front of $B(t_1)$ is approximately equal to $|y_1+1-y_2|\sin 2\pi\bar{y}$ for some $\bar{y}\in(y_2,y_1+1)$, whereas C^* in (D.19) is bounded from below by $\min(\sin 2\pi x_1^0,\sin 2\pi x_2^0)$. Therefore by choosing x_1^0 close to 0, x_2^0 close to 1, and $B(t_1)$ such that (D.20) holds but $|\cos 2\pi y_2 - \cos 2\pi y_1|B(t_1)$ is small, we can make x^* arbitrarily close to $\frac{1}{2}$. We have arrived at:

LEMMA D.3. *Assume that* $F(x,t)=-\frac{1}{2\pi}\cos(2\pi x)dB(t)$. *Fix any* ε_1, $\varepsilon_2>0$. *Then the following event has positive probability* $p_0(\varepsilon_1,\varepsilon_2)$. *There exists* $x^*\in\left[\frac{1}{2}-\varepsilon_1,\frac{1}{2}+\varepsilon_1\right]$, *such that* $[\varepsilon_2,1-\varepsilon_2]\subset[Y_{0,1}^-(x^*),Y_{0,1}^+(x^*)]$. *In other words, the interval* $[\varepsilon_2,1-\varepsilon_2]$ *is mapped to a point* $x^*\in\left[\frac{1}{2}-\varepsilon_1,\frac{1}{2}+\varepsilon_1\right]$ *by the forward Lagrangian map.*

To prove this, we just have to take an $\varepsilon_3<\varepsilon_2$, and $x_1^0=\varepsilon_3$, $x_2^0=1-\varepsilon_3$ and use the argument outlined above. We omit the details.

Acknowledgement. We have discussed the results of this paper with many people. We are especially grateful to P. Baxendale, U. Frisch, Yu. Kifer, J. Mather, J. Mattingly, J. Moser, A. Polyakov, T. Spencer, S. R. S. Varadhan and V. Yakhot for their suggestions and comments. The work of Weinan E is partially supported by an NSF Presidential Faculty Fellowship. The work of Khanin and Mazel is supported in part by RFFI of Russia under grant 96-01-00377, with further support for Khanin from Leverhulme Trust Research Grant RF&G/4/9700445. The work of Sinai is supported in part by NSF grant DMS-9706794.

Courant Institute, New York University, New York, NY
Current address: Princeton University, Princeton, NJ
E-mail address: weinan@math.princeton.edu

Isaac Newton Institute, University of Cambridge, England, UK
BRIMS, Hewlett-Packard Laboratories, Bristol, UK
Herrot-Watt University, Edinburgh, Scotland
Landau Institute of Theoretical Physics, Moscow, Russia
E-mail address: K.Khanin@newton.cam.ac.uk

Rutgers University, Piscataway, NJ
E-mail address: mazel@math.rutgers.edu

Princeton University, Princeton, NJ
Landau Institute of Theoretical Physics, Moscow, Russia
E-mail address: sinai@math.princeton.edu

References

[A] S. Aubry, The twist map, the extended Frenkel-Kontorova model and the devil's staircase, *Physica D* **7** (1983), 240–258.

[BFKL] E. Balkovsky, G. Falkovich, I. Kolokolov, and V. Lebedev, Intermittency of Burgers' turbulence, *Phys. Rev. Lett.* **78** (1997), 1452–1455.

[Ba] V. Bangert, Geodesic rays, Busemann functions and monotone twist maps, *Cal. Var. Partial Diff. Equ.* **2** (1994), 49–63.

[Bax] P. Baxendale, Brownian motions in the diffeomorphism group. I, *Compositio Math.* **53** (1984), 19–50.

[B] P. Billingsley, *Convergence of Probability Measures*, John Wiley & Sons, Inc., New York, 1968.

[BFGLV] G. Blatter, M. V. Feigelman, V. B. Geshkenbein, A. I. Larkin, and V. M. Vinokur, Vortices in high-temperature superconductors, *Rev. Modern Phys.* **66** (1994), 1125–1388.

[BMP] J. P. Bouchaud, M. Mézard, and G. Parisi, *Phys. Rev. E* **52** (1995), 3656–3674.

[CT] M. G. Crandall and A. Majda, Monotone difference approximations for scalar conservation laws, *Math. Comp.* **34** (1980), 1–21.

[CY] A. Chekhlov and V. Yakhot, Kolmogorov Turbulence in a random-force-driven Burgers equation, *Phys. Rev. E* **51** (1995), R2739–R2749.

[D] C. Dafermos, Generalized characteristics and the structure of solutions of hyperbolic conservation laws, *Indiana Math. J.* **26** (1977), 1097–1119.

[E] Weinan E, Aubry-Mather theory and periodic solutions of forced Burgers equations *Comm. Pure Appl. Math.* **52** (1999), 811–828.

[EKMS] Weinan E, K. Khanin, A. Mazel, and Ya. Sinai, Probability distribution functions for the random-forced Burgers equations, *Phys. Rev. Lett.* **78** (1997), 1904–1907.

[ERS] Weinan E, Yu. Rykov, and Ya. Sinai, Generalized variational principles, global weak solutions and behavior with random initial data for systems of conservation laws arising in adhesion particle dynamics, *Comm. Math. Phys.* **177** (1996), 349–380.

[EV] Weinan E and E. Vanden Eijnden, Asymptotic theory for the probability density functions in Burgers turbulence, *Phys. Rev. Lett.* **83** (1999), 2572–2575.

[F] M. V. Feigelman, One-dimensional periodic structures in a weak random potential, *Sov. Phys. JETP* **52** (1980), 555–561.

[FW] M. I. Freidlin and A. D. Wentzell, *Random Perturbations of Dynamical Systems, Fundamental Principles of Math. Sci.* **260**, Springer-Verlag, New York, 1984.

[GK1] T. Gotoh and R. H. Kraichnan, Statistics of decaying Burgers turbulence, *Phys. Fluids A* **5** (1993), 445–457.

[GK2] T. Gotoh and R. H. Kraichnan, Burgers turbulence with large scale forcing, *Phys. Fluids A* **10** (1998), 2859–2866.

[GM] V. Gurarie and A. Migdal, Instantons in Burgers equations, *Phys. Rev. E* **54** (1996), 4908.

[H] G. A. Hedlund, Geodesics on two-dimensional Riemann manifold with periodic coefficients, *Ann. of Math.* **33** (1932), 719–739.

[Ho] E. Hopf, On the right weak solution of the Cauchy problem for a quasilinear equation of first order, *J. Math. Mech.* **19** (1969), 483–487.

[IK] K. Ichihara and H. Kunita, A classification of the second order degenerate elliptic operators and its probabilistic characterization, *Z. Wahrscheinlichkeitstheorie und Verw. Gebiete* **30** (1974), 235–254.

[JKM] H. R. Jauslin, H. O. Kreiss, and J. Moser, On the forced Burgers equation with periodic boundary conditions, 133–153, in *Differential Equations: La Pietra* 1996, *Proc. Sympos. Pure Math.* **65** A.M.S., Providence, RI, 1999.

[K1] Y. Kifer, The Burgers equation with a random force and a general model for directed polymers in random environments, *Probab. Theory Related Fields* **108** (1997), 29–65.

[K2] ———, A note on integrability of C^r-norms of stochastic flows and applications, *Lecture Notes in Math.*, *Stochastic Mechanics and Stochastic Processes* (Swansea, 1986) **1325** Springer-Verlag, New York, 1988, 125–131.

[KS] J. Krug and H. Spohn, Kinetic roughening of growing surfaces, in *Solids Far from Equilibrium* (G. C. Godreche, ed.), Cambridge University Press, England, 1992.

[La] P. D. Lax, Hyperbolic systems of conservation laws. II, *Comm. Pure Appl. Math.* **10** (1957), 537–566.

[LN] C. Licea and C. M. Newman, Geodesics in two-dimensional first-passage percolation, *Ann. of Probab.* **24** (1996), 399–410.

[Li] P. L. Lions, *Generalized Solutions of Hamilton-Jacobi Equations*, *Research Notes in Math.* **69** Pitman Advanced Publishing Program, Boston, 1982.

[M] J. N. Mather, Existence of quasiperiodic orbits for twist homeomorphisms of the annulus, *Topology* **21** (1982), 457–467.

[McK] H. P. McKean, *Stochastic Integrals*, Academic Press, New York, 1969.

[Mo] M. Morse, A fundamental class of geodesics on any closed surface of genus greater than one, *Trans. A.M.S.* **26** (1924), 25–60.

[N] C. M. Newman, A surface view of first-passage percolation, *Proc. ICM*-94 (Zürich, 1994), 1017–1025, Birkhäuser, Boston, 1995.

[O] V. I. Oseledec, A multiplicative ergodic theorem, Lyapunov characteristic numbers for dynamical systems, *Trans. Moscow Math. Soc.* **19** (1968), 197–231.

[Ol] O. A. Oleĭnik, Discontinuous solutions of non-linear differential equations, *Uspekhi Mat. Nauk* **12** (1957), 3–73.

[Pa] K. R. Parthasarathy, *Probability Measures on Metric Spaces*, Academic Press, New York, 1967.

[Pes] Ya. B. Pesin, Characteristic Lyapunov exponents and smooth ergodic theory, *Russian Math. Surveys* **32** (1977), 55–114.

[Pol] A. Polyakov, Turbulence without pressure, *Phys. Rev. E* **52** (1995), 6183–6188.

[R] D. Ruelle, Ergodic theory of differentiable dynamical systems, *Publ. IHES* **50** (1979), 27–58.

[S1] Ya. Sinai, Two results concerning asymptotic behavior of solutions of the Burgers equation with force, *J. Statist. Phys.* **64** (1991), 1–12.

[S2] ———, Burgers system driven by a periodic stochastic flow, in *Itô's Stochastic Calculus and Probability Theory*, 347–353 Springer-Verlag, New York, 1996.

[S3] YA. SINAI (ED.), *Encyclopedia of Mathematical Sciences* **2**, Springer-Verlag, New York, 1989.

[So] A. N. SOBOLEVSKI, Aubry-Mather theory and idempotent eigenfunctions of Bellman operator, *Commun. Contemp. Math.* **1** (1999), 517–533.

[V] S. R. S. VARADHAN, *Large Deviations and Applications*, CBMS-NSF *Regional Conference Series in Applied Mathematics* **46** SIAM, Philadelphia, 1984.

[Ya] I. G. YAKUSHKIN, Description of turbulence in the Burgers model, *Radiophys. and Quantum Electronics* **24** (1981), 41–48.

[Y] L. S. YOUNG, Ergodic theory of differentiable dynamical systems, in *Real and Complex Dynamical Systems* (Branner and Hjorth, eds.), 293–336, Kluwer Acad. Publ., Dordrecht, 1995.

(Received May 1, 1998)

Comments

Burgers equation with stochastic forcing is a popular model in nonequilibrium stochastic dynamics and nonlinear dynamics.

In the case of zero viscosity solutions have shock waves. Therefore, stationary measures are probability measures in the space of discontinuous functions. In our paper [EKMS], we describe some results of the analysis of statistical properties of shock waves.

Communications in Contemporary Mathematics, Vol. 1, No. 4 (1999) 497–516

AN ELEMENTARY PROOF OF THE EXISTENCE AND UNIQUENESS THEOREM FOR THE NAVIER–STOKES EQUATIONS

J. C. MATTINGLY

Department of Mathematics, Stanford University, Stanford CA 94305, USA

YA. G. SINAI

Department of Mathematics, Princeton University, Princeton NJ 08544, USA

Received 19 November 1998
Revised 10 March 1999

1. Introduction

The purpose of this paper is to show that some results concerning solutions of the Navier–Stokes systems can be proven by purely elementary methods. In two-dimensions with periodic boundary conditions, the Navier–Stokes system has the form

$$
\begin{aligned}
\frac{\partial u_1}{\partial t} + u_1 \frac{\partial u_1}{\partial x_1} + u_2 \frac{\partial u_1}{\partial x_2} &= \nu \Delta u_1 - \frac{\partial p}{\partial x_1} + f_1(x_1, x_2, t)\,, \\
\frac{\partial u_2}{\partial t} + u_1 \frac{\partial u_2}{\partial x_1} + u_2 \frac{\partial u_2}{\partial x_2} &= \nu \Delta u_2 - \frac{\partial p}{\partial x_2} + f_2(x_1, x_2, t)\,, \\
\frac{\partial u_1}{\partial x_1} + \frac{\partial u_2}{\partial x_2} &= 0\,.
\end{aligned}
\tag{1.1}
$$

Here ν is the viscosity, p is the pressure, and f_1, f_2 are the components of an external forcing which may be time-dependent. As our setting is periodic, the functions u_1, u_2, ∇p, f_1, and f_2 are all periodic in x. For simplicity, we take the period to be one.

The first existence and uniqueness theorems for weak solutions of (1.1) were proven by Leray [10] in whole plane $\mathbb{R}^2$. Later these results were extended by E. Hopf (see [7]). In 1962, Ladyzenskaya proved existence and uniqueness results for strong solutions for general two-dimensional domains [9]. V. Yudovich, C. Foias, R. Teman, P. Constantin and others developed strong methods which provided deep insights into the dynamics described by (1.1) (see [1, 11, 12, 13]).

Y.G. Sinai, *Selecta II: Probability Theory, Statistical Mechanics, Mathematical Physics and Mathematical Fluid Dynamics*, DOI 10.1007/978-1-4419-6205-8_1, © Springer Science+Business Media, LLC 2010

The purpose of this paper is to present elementary proofs of three theorems. These theorems imply the existence and uniqueness of smooth solutions of (1.1) and shed some additional light on the dissipative character of the dynamics. These results are essentially the same as those of [5], however our point of view and proofs are different. We will also discuss what our techniques can give in the three-dimensional setting.

In two-dimensions, it is useful to consider the vorticity

$$\omega(x_1, x_2, t) = \frac{\partial u_1(x_1, x_2, t)}{\partial x_2} - \frac{\partial u_2(x_1, x_2, t)}{\partial x_1}.$$

The equation governing ω has the form (see [2, 3])

$$\frac{\partial \omega}{\partial t} + u_1 \frac{\partial \omega}{\partial x_1} + u_2 \frac{\partial \omega}{\partial x_2} = \nu \Delta \omega + g(x_1, x_2, t), \tag{1.2}$$

where $g(x_1, x_2, t) = \frac{\partial f_1(x_1, x_2, t)}{\partial x_2} - \frac{\partial f_2(x_1, x_2, t)}{\partial x_1}$. We will need $g(x_1, x_2)$ to posses a modicum of spatial smoothness; this will be made precise shortly.

In our two-dimensional setting, the systems (1.1) and (1.2) are equivalent. Expanding ω in Fourier series where $\omega(x_1, x_2, t) = \sum_{k \in \mathbb{Z}^2} \omega_k(t) e^{2\pi i (x,k)}$ with $x = (x_1, x_2)$, we can write a coupled ODE-system for the modes $\omega_k(t)$ (see [3]),

$$\frac{d\omega_k}{dt} + 2\pi i \sum_{l_1 + l_2 = k} \omega_{l_1} \omega_{l_2} \frac{(k, l_2^{\perp})}{(l_2, l_2)} = -4\pi^2 \nu |k|^2 \omega_k + g_k(t), \tag{1.3}$$

where $k \in \mathbb{Z}^2$, $|k| = \sqrt{k_1^2 + k_2^2}$, $l^{\perp} = (l^{(1)}, l^{(2)})^{\perp} = (-l^{(2)}, l^{(1)})$, and $g_k(t)$ are the spatial Fourier modes of the function $g(x, t)$. Since ω is real, we know $\omega_{-k} = \bar{\omega}_k$. Furthermore, we always assume that $\omega_0 = 0$. The system (1.3) is the Galerkin system corresponding to (1.2). A finite dimensional approximation of this Galerkin system can be associated to any finite subset $\mathcal{Z}$ of $\mathbb{Z}^2$ by setting $\omega_k(t) = 0$ for all k outside of $\mathcal{Z}$. In the following, we will implicitly assume that $\mathcal{Z}$ is centrally-symmetric, that is if $k \in \mathcal{Z}$ then $-k \in \mathcal{Z}$.

In fact, we will study a slightly more general version of (1.2) where the Laplacian is replaced by the operator $|\nabla|^{\alpha}$ with $\alpha > 1$. This leads to a version of (1.3) which we index by the choice of α and by the finite index set $\mathcal{Z}$, $\mathcal{Z} \subset \mathbb{Z}^2$, indicating which modes are included in the Galerkin approximation. In short, we consider the finite dimensional ODE system

$$\frac{d\omega_k}{dt} + 2\pi i \sum_{\substack{l_1 + l_2 = k \\ l_1, l_2 \in \mathcal{Z}}} \omega_{l_1} \omega_{l_2} \frac{(k, l_2^{\perp})}{(l_2, l_2)} = -4\pi^2 \nu |k|^{\alpha} \omega_k + g_k. \tag{$1.3^{\alpha}_{\mathcal{Z}}$}$$

We now state the assumptions on the coefficients $g_k(t)$ to be used at various times during our discussion.

Assumption 1.1. *The forcing* $f(x,t) = (f_1(x,t), f_2(x,t))$ *is such that* $g^* = \sup_{t\in[0,\infty)} |g(\cdot,t)|_{L^2} < \infty$.

Assumption 1.2. *For some* r, *there exists a constant* $\mathcal{G}(r) > 0$ *such that*

$$\sup_{t\in[0,\infty)} |g_k(t)| \le \frac{\mathcal{G}(r)}{|k|^{r-\alpha+\epsilon}}$$

for some $\epsilon > 0$ *and all* $k \in \mathbb{Z}^2\backslash 0$. *The constant* α *is the same as in* $(1.3^{\alpha}_{\mathcal{Z}})$.

Assumption 1.3. *For some* r *and* $\gamma > 0$, *there exists a constant* $\mathcal{G}(r,\gamma) > 0$ *such that*

$$\sup_{t\in[0,\infty)} |g_k(t)| \le \frac{\mathcal{G}(r,\gamma)}{|k|^{r-\alpha+\epsilon}} e^{-\gamma|k|^{1+\delta}}$$

for some $\delta > 0$, $\epsilon > 0$, *and all* $k \in \mathbb{Z}^2\backslash 0$. *Again, the constant* α *is the same as in* $(1.3^{\alpha}_{\mathcal{Z}})$.

Observe that Assumption 1.3 implies Assumption 1.2. Critical to our discussion is that for $(1.3^{\alpha}_{\mathcal{Z}})$ we have the so-called enstrophy estimate. Namely, if $\mathcal{E}(0) = \int \omega^2(x_1,x_2,0)dx_1dx_2 = \sum_k |\omega_k(0)|^2 < \infty$ then one can find $\mathcal{E}^*$ depending only on $\mathcal{E}(0)$, ν, $\sup_{t\in[0,\infty)} |g(\cdot,t)|_{L^2}$, and α such that $\mathcal{E}(t) = \int \omega^2(x_1,x_2,t)dx_1dx_2 \le \mathcal{E}^*$ for all solutions to $(1.3^{\alpha}_{\mathcal{Z}})$. It is important to note that $\mathcal{E}^*$ is independent of the set $\mathcal{Z}$ which defines the Galerkin approximation. This enstrophy estimate holds if the forcing satisfies Assumption 1.1 (see e.g. [1, 3, 11]).

Now we are ready to formulate our theorems.

Theorem 1.1. *Assume the forcing satisfies Assumptions* 1.1 *and* 1.2 *for some* $r > 1$ *and* $\mathcal{G}(r) > 0$. *If for some* $\mathcal{D}_1 < \infty$

$$|\omega_k(0)| \le \frac{\mathcal{D}_1}{|k|^r}$$

then one can find a $\mathcal{D}_1' < \infty$, *depending only on* $\mathcal{D}_1$, ν, g^*, *and* $\mathcal{G}$, *such that any solution to* $(1.3^{\alpha}_{\mathcal{Z}})$ *with these initial conditions satisfies*

$$|\omega_k(t)| \le \frac{\mathcal{D}_1'}{|k|^r}$$

for all $t > 0$. *In particular,* $\mathcal{D}_1'$ *is independent of the set* $\mathcal{Z}$ *defining the Galerkin approximation.*

An existence and uniqueness theorem for (1.3) follows from Theorem 1.1 by now standard considerations (see [1, 3, 11]). We briefly recall the general line of the argument. By the Sobolev embedding theorem, the Galerkin approximations are trapped in a compact subset of L^2 of the 2-torus. This guarantees the existence of a limit point which can be shown to satisfy (1.3). Using the the regularity inherited from the Galerkin approximations, one then shows that there is a unique solution to (1.3). Gallavotti [6] contains a similar proof of a similar statement.

Theorem 1.2. *Assume that Assumption* 1.3 *holds for some* $r > 1$, $\gamma > 0$, *and* $\mathcal{G}(r, \gamma) > 0$. *If the initial conditions satisfy*

$$|\omega_k(0)| \le \frac{\mathcal{D}_2}{|k|^r} e^{-\gamma_2 |k|}$$

for some $\mathcal{D}_2 < \infty$ *and* $\gamma_2 > 0$, *then one can find a* $\mathcal{D}_2' < \infty$ *and a* $\gamma_2' > 0$, *depending only on* $\mathcal{D}_2, \gamma, r, \nu, g^*, \mathcal{G}$, *such that any solution to* $(1.3^{\alpha}_{\mathcal{Z}})$ *starting from these initial conditions satisfies*

$$|\omega_k(t)| \le \frac{\mathcal{D}_2'}{|k|^r} e^{-\gamma_2' |k|}$$

for all $t > 0$. *In particular, the constants* $\mathcal{D}_2'$ *and* γ' *are independent of the set* $\mathcal{Z}$ *defining the Galerkin approximation.*

Theorem 1.2 shows that Eq. (1.2) preserves the class of real analytic functions on the 2-torus.

Theorem 1.3. *Assume that Assumption* 1.3 *holds for some* $r > 1$, $\gamma > 0$, *and* $\mathcal{G}(r, \gamma) > 0$. *If the initial conditions satisfy*

$$|\omega_k(0)| \le \frac{\mathcal{D}_3}{|k|^r},$$

then for any $t_0 > 0$, *one can find a* $\mathcal{D}_3' > 0$ *and a* $\gamma_3' > 0$ *such that any solution to* $(1.3^{\alpha}_{\mathcal{Z}})$ *with these initial conditions satisfies*

$$|\omega_k(t_0)| \le \frac{\mathcal{D}_3'}{|k|^r} e^{-\gamma_3' |k|}.$$

As before, the constant $\mathcal{D}_3'$ *is independent of the set* $\mathcal{Z}$ *defining the Galerkin approximation.*

Theorem 1.3 shows that if the initial conditions $\omega(x, 0)$ for (1.2) are smooth enough then, the solution $\omega(x, t_0)$ is real analytic for arbitrarily small time t_0. Then according to Theorem 1.2, it remains with in this class for all $t > t_0$. Statements close to these were proven in the works by C. Foias and R. Temam [5], C. Doering and E. Titi [4] and H. Kreiss [8]. Theorem 1.1 is proven in Sec. 2 and Theorems 1.2 and 1.3 are proven in Sec. 3.

The proofs of all of the theorems in this paper share a common structure. We consider the system of coupled ODEs for the Fourier coefficients. Then we construct a subset Ω of the phase space (the set of possible configurations of the Fourier modes) so that all points in Ω possess the desired decay properties. In addition, Ω is constructed so that it contains the initial data in its interior. Then we endeavor to show that the dynamics never cause the sequence of Fourier modes to leave the subset Ω. How this is done can be understood geometrically. It amounts to showing that the vector field on the boundary of Ω points into the interior of Ω. If this is true, then the solution can never escape Ω.

2. Proof of Theorem 1.1

Fixing an arbitrary Galerkin approximation corresponding to the modes in some finite subset $\mathcal{Z}$ of $\mathbb{Z}^2$, we write the real version of (1.3). As we already mentioned, we assume $\omega_0 = 0$ and, because the velocity is real, we also have $\omega_{-k} = \bar{\omega}_k$. Setting $\omega_k = \omega_k^{(1)} + i\omega_k^{(2)}$, we separate the equations for $\omega_k^{(1)}$ and $\omega_k^{(2)}$ obtaining

$$\begin{aligned} \frac{d\omega_k^{(1)}}{dt} &= 2\pi \sum_{\substack{l_1+l_2=k \\ l_1,l_2\in\mathcal{Z}}} [\omega_{l_1}^{(1)}\omega_{l_2}^{(2)} + \omega_{l_1}^{(2)}\omega_{l_2}^{(1)}]\frac{(k, l_2^{\perp})}{(l_2, l_2)} - 4\pi\nu|k|^{\alpha}\omega_k^{(1)} + g_k^{(1)}, \\ \frac{d\omega_k^{(2)}}{dt} &= -2\pi \sum_{\substack{l_1+l_2=k \\ l_1,l_2\in\mathcal{Z}}} [\omega_{l_1}^{(1)}\omega_{l_2}^{(1)} + \omega_{l_1}^{(2)}\omega_{l_2}^{(2)}]\frac{(k, l_2^{\perp})}{(l_2, l_2)} - 4\pi\nu|k|^{\alpha}\omega_k^{(2)} + g_k^{(2)}, \end{aligned} \tag{2.1}$$

where $g_k = g_k^{(1)} + ig_k^{(2)}$.

It follows from the enstrophy estimate that $\sum_k[(w_k^{(1)}(t))^2 + (w_k^{(2)}(t))^2] \le \mathcal{E}^*$ and thus $|w_k^{(1)}(t)| \le \sqrt{\mathcal{E}^*}$ and $|w_k^{(2)}(t)| \le \sqrt{\mathcal{E}^*}$ for all $k \in \mathbb{Z}^2$ and $t > 0$. Hence, for any $K_0 > 0$, we can find a $\mathcal{D}_1' = \mathcal{D}_1'(K_0)$ such that for any $t \ge 0 |w_k^{(1)}(t)|$, $|w_k^{(2)}(t)| < \frac{\mathcal{D}_1'}{|k|^r}$ for all $k \in \mathbb{Z}^2$ with $|k| \le K_0$. We also require $\mathcal{D}_1'$ to be greater than $\mathcal{G}$ so later estimates will arrange themselves nicely. Recall that $\mathcal{G}(r)$ was the constant from Assumption 1.2. Since $\mathcal{G}$ is given and only K_0 is ours to vary, we will suppress the dependence of $\mathcal{D}_1'$ on $\mathcal{G}$.

Now consider the subset

$$\Omega_1(K_0) = \left\{(\omega_k^{(1)}, \omega_k^{(2)})_{k\in\mathbb{Z}^2} : |\omega_k^{(j)}| \le \frac{\mathcal{D}_1'(K_0)}{|k|^r} \text{ for all } j \in \{1,2\},\, k \in \mathbb{Z}^2\backslash 0\right\}$$

of $(\mathbb{R}^2)^{\mathbb{Z}^2}$. Its boundary is the subset

$$\partial\Omega_1(K_0) = \left\{(\omega_k^{(1)}, \omega_k^{(2)})_{k\in\mathbb{Z}^2} : \begin{array}{l} |\omega_k^{(j)}| \le \frac{\mathcal{D}_1'}{|k|^r} \text{ for all } j \in \{1,2\}, k \in \mathbb{Z}^2\backslash 0 \\ \text{and equality holds for some } \bar{k} \text{ and } \bar{j}\,. \end{array}\right\}.$$

We shall also need the subset of this boundary

$$\overline{\partial\Omega_1}(K_0) = \left\{(\omega_k^{(1)}, \omega_k^{(2)})_{k\in\mathbb{Z}^2} : \begin{array}{l} |\omega_k^{(j)}| \le \frac{\mathcal{D}_1'}{|k|^r} \text{ for all } j \in \{1,2\}, k \in \mathbb{Z}^2\backslash 0 \\ \text{and equality for some } \bar{k} \text{ and } \bar{j} \text{ with } |\bar{k}| > K_0\,. \end{array}\right\}.$$

Showing that the trajectories of our system remain inside of Ω_1 is equivalent to the statement of the theorem. Recall that using the enstrophy estimate, we picked a $\mathcal{D}_1'(K_0)$ such that if $|k| \le K_0$ then $|w_k^{(1)}(t)|$ and $|w_k^{(2)}(t)|$ were bounded by $\frac{\mathcal{D}_1'}{|k|^r}$ for all $t \in [0, \infty)$. Thus, the only remaining way for a trajectory to leave $\Omega_1(K_0)$, is through the section of the boundary $\overline{\partial\Omega_1}(K^0)$ introduced above. Our basic idea is to show that if K_0 is greater than a specific K_{crit}, then the vector field on $\overline{\partial\Omega_1}(K^0)$ points inward. In other words, the dynamics of (2.1) can never move the system configuration through $\partial\Omega_1(K_0)$. In still different words, Ω_1 is a trapping region. Since the initial data begins in Ω_1, proving this picture would prove the theorem.

To show that the vector field points inward, fix a point on $\overline{\partial\Omega_1}(K_0)$. For definiteness, consider the case when $\omega_{\bar{k}}^{(1)} = \frac{\mathcal{D}_1'}{|\bar{k}|^r}$ for some $\bar{k}$ with $|\bar{k}| > K_0$, $|\omega_{k'}^{(1)}| \le \frac{\mathcal{D}_1'}{|k'|^r}$ for all $k' \in \mathbb{Z}\backslash 0$ with $k' \neq \bar{k}$, and $|\omega_k^{(2)}| \le \frac{\mathcal{D}_1'}{|k|^r}$ for all $k \in \mathbb{Z}^2\backslash 0$. The other cases, namely where $\omega_{\bar{k}}^{(1)} = -\frac{\mathcal{D}_1'}{|\bar{k}|^2}$ or $\omega_{\bar{k}}^{(2)} = \pm\frac{\mathcal{D}_1'}{|\bar{k}|^2}$, are handled in the same manner. We have to show that,

$$2\pi\left|\sum_{l_1+l_2=\bar{k}}[\omega_{l_1}^{(1)}\omega_{l_2}^{(1)} + \omega_{l_1}^{(2)}\omega_{l_2}^{(2)}]\frac{(\bar{k}, l_2^{\perp})}{(l_2, l_2)}\right| + |g_{\bar{k}}^{(2)}| < 4\pi\nu|\bar{k}|^{\alpha}|\omega_{\bar{k}}^{(2)}|. \tag{2.2}$$

We shall see that the restriction that $|\bar{k}| \ge K_0 > K_{crit}$ does not depend on $\mathcal{D}_1$ only on $\mathcal{E}$.

Consider the following three sums which together bound the first absolute value on the left-hand side of (2.2):

$$\Sigma_1 = \sum_{\substack{l_1+l_2=\bar{k}\\ |l_2|\le|\frac{\bar{k}}{2}|}} |\omega_{l_1}^{(1)}\omega_{l_2}^{(1)} + \omega_{l_1}^{(2)}\omega_{l_2}^{(2)}|\left|\frac{(\bar{k}, l_2^{\perp})}{(l_2, l_2)}\right|,$$

$$\Sigma_2 = \sum_{\substack{l_1+l_2=\bar{k}\\ |\frac{\bar{k}}{2}|<|l_2|\le 2|\bar{k}|}} |\omega_{l_1}^{(1)}\omega_{l_2}^{(1)} + \omega_{l_1}^{(2)}\omega_{l_2}^{(2)}|\left|\frac{(\bar{k}, l_2^{\perp})}{(l_2, l_2)}\right|,$$

$$\Sigma_3 = \sum_{\substack{l_1+l_2=\bar{k}\\ |l_2|>2|\bar{k}|}} |\omega_{l_1}^{(1)}\omega_{l_2}^{(1)} + \omega_{l_1}^{(2)}\omega_{l_2}^{(2)}|\left|\frac{(\bar{k}, l_2^{\perp})}{(l_2, l_2)}\right|.$$

We treat each sum separately. For Σ_1, using the Cauchy–Schwartz inequality and the inequalities $|\frac{(\bar{k}, l_2^{\perp})}{(l_2, l_2)}| \le \frac{|\bar{k}|}{|l_2|}$, $|l_1| \ge \frac{|\bar{k}|}{2}$, and $|\omega_{l_1}^{(1)}| \le 2^r\mathcal{D}_1'\frac{1}{|\bar{k}^r}$, $|\omega_{l_1}^{(2)}| \le 2^r\mathcal{D}_1'\frac{1}{|\bar{k}|^r}$ produces

$$|\Sigma_1| \le 2^r\frac{\mathcal{D}_1'}{|\bar{k}|^r}|\bar{k}|\sum_{|l_2|\ge|\frac{\bar{k}}{2}|}[|\omega_{l_2}^{(1)}| + |\omega_{l_2}^{(2)}|]\frac{1}{|l_2|}$$

$$\le 2^r\frac{\mathcal{D}_1'|\bar{k}|}{|\bar{k}|^r}\left(\sqrt{\sum|\omega_{l_2}^{(1)}|^2} + \sqrt{\sum|\omega_{l_2}^{(2)}|^2}\right)\sqrt{\sum_{|l_2|\le|\frac{\bar{k}}{2}|}\frac{1}{|l_2|^2}}$$

$$\le 2^{r+1}(\mathrm{const})(\sqrt{\mathcal{E}^*})|\bar{k}|\left(\sqrt{\ln|\bar{k}|}\right)\left(\frac{\mathcal{D}_1'}{|\bar{k}|^r}\right). \tag{2.3}$$

The (const) in the final line is from the inequality

$$\sum_{|l_2|\le|\frac{\bar{k}}{2}|} \frac{1}{|l_2|^2} \le (\text{const})^2 \ln|\bar{k}| \,.$$

To estimate Σ_2, we use the inequalities $|\frac{(k,l_2^\perp)}{(l_2,l_2)}| \le 2$, $|\omega_{l_2}^{(1)}| \le 2^r \frac{\mathcal{D}_1'}{|\bar{k}|^r}$, and $|\omega_{l_2}^{(2)}| \le 2^r \frac{\mathcal{D}_1'}{|\bar{k}|^r}$ obtaining

$$\begin{aligned}
|\Sigma_2| &\le 2^{r+1}\frac{\mathcal{D}_1'}{|\bar{k}|^r} \sum_{|l_1|\le 3|\bar{k}|} [|\omega_{l_1}^{(1)}| + |\omega_{l_1}^{(2)}|] \\
&\le 2^{r+1}\frac{\mathcal{D}_1'}{|\bar{k}|^r}\left[\sqrt{\sum_{|l_1|\le 3|\bar{k}|} |\omega_{l_1}^{(1)}|^2} + \sqrt{\sum_{|l_1|\le 3|\bar{k}|} |\omega_{l_1}^{(2)}|^2}\right](6|\bar{k}|+1) \\
&\le 2^{r+2}\mathcal{E}(6|\bar{k}|+1)\frac{\mathcal{D}_1'}{|\bar{k}|^r} \,.
\end{aligned} \tag{2.4}$$

The factor $(6|\bar{k}|+1)$ arises as an estimate of the square root of the number of lattice points $l_1 \in \mathbb{Z}^2$ for which $|l_1| \le 3|\bar{k}|$.

In estimating Σ_3, we use $|\frac{(\bar{k},l_2^\perp)}{(l_2,l_2)}| \le |\frac{\bar{k}}{l_2}|$ producing

$$\begin{aligned}
|\Sigma_3| &\le |\bar{k}| \sum_{\substack{l_1+l_2=\bar{k} \\ |l_2|\ge 2|\bar{k}|}} [|\omega_{l_1}^{(1)}||\omega_{l_2}^{(2)}| + |\omega_{l_1}^{(2)}||\omega_{l_2}^{(1)}|]\frac{1}{|l_2|} \\
&\le |\bar{k}| \left[\left(\sum_{|l_1|\ge\bar{k}} (\omega_{l_1}^{(1)})^2\right)^{\frac{1}{2}} \left(\sum_{|l_2|\ge 2\bar{k}} \frac{(\omega_{l_2}^{(2)})^2}{|l_2|^2}\right)^{\frac{1}{2}}\right. \\
&\qquad \left. + \left(\sum_{|l_1|\ge\bar{k}} (\omega_{l_1}^{(2)})^2\right)^{\frac{1}{2}} \left(\sum_{|l_2|\ge 2\bar{k}} \frac{|\omega_{l_2}^{(1)}|^2}{|l_2|}\right)^{\frac{1}{2}}\right] \\
&\le 2\sqrt{\mathcal{E}^*}|\bar{k}|\mathcal{D}_1' \left(\sum_{|l_2|\ge 2|\bar{k}|} \frac{1}{|l_2|^{2(r+1)}}\right)^{\frac{1}{2}} \\
&\le 2\sqrt{\mathcal{E}^*}(\text{const})\ |\bar{k}|\frac{\mathcal{D}_1'}{|\bar{k}|^r} \,,
\end{aligned} \tag{2.5}$$

where (const) is defined by the inequality

$$\sum_{|l_2|\ge 2|\bar{k}|} \frac{1}{|l_2|^{2(r+1)}} \le (\text{const})^2 \frac{1}{|\bar{k}|^{2r}} \,.$$

Adding (2.3), (2.4), and (2.5) together, we obtain the needed bound on the right hand side of (2.2):

$$\begin{aligned}
2\pi & \sum_{l_1+l+2=\bar{k}} |\omega_{l_1}^{(1)}||\omega_{l_2}^{(2)}| + |\omega_{l_1}^{(2)}||\omega_{l_2}^{(1)}| \\
& \leq [2^{r+1}(\text{const})\ \sqrt{\mathcal{E}^*}|\bar{k}|\sqrt{\ln|\bar{k}|} + 2^{r+2}\mathcal{E}^*(6|\bar{k}|+1) \\
& \quad +2\sqrt{\mathcal{E}^*}(\text{const})\ |k|]\frac{\mathcal{D}_1'}{|\bar{k}|^r} \\
& \leq 2^{r+2}\mathcal{E}^*\,\overline{(\text{const})}|\bar{k}|\sqrt{\ln|\bar{k}|}\frac{\mathcal{D}_1'}{|\bar{k}|^r}, \qquad (2.6)
\end{aligned}$$

where $\overline{(\text{const})}$ is a new constant.

By Assumption 1.2 and our requirement that the $\mathcal{D}_1'$ be greater than $\mathcal{G}$ (the constant from Assumption 1.2), we know that $|g_k| \leq \frac{\mathcal{D}_1'}{|k|^{r-\alpha+\epsilon}}$. Thus, inequality (2.2) will be satisfied if

$$\left[2^{r+2}\mathcal{E}^*\,\overline{(\text{const})}\frac{|\bar{k}|\sqrt{\ln|\bar{k}|}}{|\bar{k}|^{\alpha}} + \frac{1}{|\bar{k}|^{\epsilon}}\right]\frac{\mathcal{D}_1'}{|\bar{k}|^{r-\alpha}} \leq 4\pi\nu\frac{\mathcal{D}_1'}{|\bar{k}|^{r-\alpha}}. \qquad (2.7)$$

From this we see that for all $\alpha > 1$, there exists K_{crit} so that if $|\bar{k}| \geq K_{crit}$ then (2.7) holds. Also notice that K_{crit} is independent of our choice of $\mathcal{D}_1'$ except for the condition that $\mathcal{D}_1' > \mathcal{G}$. Thus we can find K_{crit} first and then fix K_0 which determines $\mathcal{D}_1'$.

3. Proofs of Theorems 1.2 and 1.3

We begin by stating the central estimate on which both theorems rely. It requires estimates similar in spirit to the previous theorem and will be proven at the end of the section. We present a d-dimensional version of the lemma because it will be useful in the discussions of the 3-dimensional setting in the next section.

Lemma 3.1. *Let $\{a_k\}$ and $\{b_k\}$ be two sequences with $k \in \mathbb{Z}^d$. If for some $r > d-1$ and some $\mathcal{C} > 0$*

$$|a_k| \leq \frac{\mathcal{C}}{|k|^r}, \quad |b_k| \leq \frac{\mathcal{C}}{|k|^r},$$

then for all $k \in \mathbb{Z}^d$

$$\sum_{\substack{l_1+l_2=k \\ l_1,l_2\in\mathbb{Z}^d}} |a_{l_1}|\,|b_{l_2}|\frac{|k|}{|l_2|} \leq (const)\left(2^r|k| + 2^{r+1}(6|k|+1)^{\frac{d}{2}} + \frac{1}{2}|k|^{d-1-r}\right)\frac{\mathcal{C}^2}{|k|^r},$$

where the constant depends only on r and not on k.

We now turn to the proof of Theorem 1.2.

Proof of Theorem 1.2. If $|\omega_k^{(1)}(0)| \le \frac{\mathcal{D}_2}{|k|^r} e^{-\gamma_2|k|}$, $|\omega_k^{(2)}(0)| \le \frac{\mathcal{D}_2}{|k|^r} e^{-\gamma_2|k|}$, then surely $|\omega_k^{(1)}|, |\omega_k^{(2)}| \le \frac{\mathcal{D}_2}{|k|^r}$. Therefore by Theorem 1.1, one can find a constant $\bar{\mathcal{D}}_2$ such that $|\omega_k^{(1)}(t)|, |\omega_k^{(2)}(t)| \le \frac{\bar{\mathcal{D}}_2}{|k|^r}$ for all $k \in \mathbb{Z}^2\backslash\{0\}$. Let us set $\mathcal{D}'_2 = \max(2\bar{\mathcal{D}}_2, \mathcal{G})$, where $\mathcal{G}$ is the constant from Assumption 1.3. The numerical factor 2 is somewhat arbitrary. We could chose any factor greater than 1; we take 2 for simplicity.

Choose $K_0 \ge 0$ and consider the set

$$\Omega_2(K_0) = \left\{ (\omega_k^{(1)}, \omega_k^{(2)})_{k\in\mathbb{Z}^2\backslash\{0\}} : |\omega_k^{(1)}| \le \frac{\mathcal{D}'_2}{|k|^r} e^{-\gamma'_2|k|}, |\omega_k^{(2)}| \le \frac{\mathcal{D}'_2}{|k|^r} e^{-\gamma'_2|k|} \right\}.$$

The value of $\gamma'_2 = \gamma'_2(K_0)$ is chosen in such a way that the inequalities $|\omega_k^{(i)}(t)| \le \frac{\bar{\mathcal{D}}_2}{|k|^r}$ given by Theorem 1.1 imply that $|\omega_k^{(i)}(t)| \le \frac{\mathcal{D}'_2}{|k|^r} e^{-\gamma'_2|k|}$ for all k, $|k| \le K_0$, and that for $|k| \ge K_0$, $e^{-\gamma'_2|k|} \ge e^{-\gamma|k|^{1+\delta}}$. Here γ and δ are the constants from Assumption 1.3.

As in the proof of Theorem 1.1, we shall show that for sufficiently large K_0 the vector field corresponding to (2.2) is directed inside $\Omega_2(K_0)$ along the part of the boundary $\partial\Omega_2(K_0)$ where $|\omega_k^{(i)}| \le \frac{\mathcal{D}'_2}{|k|^r} e^{-\gamma'_2|k|}$ for all $k \in \mathbb{Z}^2\backslash\{0\}$ with $|k| \ge K_0$ and for at least one of these, say $\bar{k}$, we have equality. It will be shown that our restriction from below on K_0, needed to ensure the vector field points inward, will not depend on γ'_2. This will yield the stated result.

As in Theorem 1.1, consider for definiteness the case where $\omega_{\bar{k}}^{(1)} = \frac{\mathcal{D}'_2}{|\bar{k}|^2} e^{-\gamma'_2|\bar{k}|}$. The other cases are handled in the same manner. As before, we have to show that the vector field points inward. This would be assured if

$$2\pi \left| \sum_{l_1+l_2=\bar{k}} [\omega_{l_1}^{(1)}\omega_{l_2}^{(2)} + \omega_{l_2}^{(2)}\omega_{l_2}^{(1)}] \frac{(\bar{k}, l_2^{\perp})}{(l_2, l_2)} \right| + |g_{\bar{k}}| < 4\pi^2 |\bar{k}|^\alpha \frac{\mathcal{D}'_2}{|\bar{k}|^r} e^{-\gamma'_2|\bar{k}|}. \tag{3.1}$$

This time we do not use the enstrophy estimate as previously. Instead, we use the estimates $|\omega_{l_1}^{(1)}| \le \frac{\mathcal{D}'_2}{|l_1|^r} e^{-\gamma'_2|l_1|}$ and $|\omega_{l_1}^{(2)}| \le \frac{\mathcal{D}'_2}{|l_2|^r} e^{-\gamma'_2[|l_2|}$.

Let us put $v_k^{(j)} = e^{\gamma'_2|k|}\omega_k^{(j)}$, $j = 1, 2$, $k \in \mathbb{Z}^2\backslash 0$. In terms of v, (3.1) becomes

$$2\pi \left| \sum_{l_1+l_2=\bar{k}} [v_{l_1}^{(1)}v_{l_2}^{(2)} + v_{l_2}^{(2)}v_{l_2}^{(1)}] \frac{e^{-\gamma'_2|l_1|-\gamma'_2|l_2|}}{e^{-\gamma'_2|\bar{k}|}} \frac{(\bar{k}, l_2^{\perp})}{(l_2, l_2)} \right| + |g_{\bar{k}}| e^{\gamma'_2|\bar{k}|} < 4\pi^2 |\bar{k}|^\alpha \frac{\mathcal{D}'_2}{|\bar{k}|^r}. \tag{3.2}$$

First notice that $\frac{e^{-\gamma'_2|l_1|}e^{-\gamma'_2|l_2|}}{e^{-\gamma'_2|\bar{k}|}} \le 1$ so it may be neglected. Second notice that for $v_k^{(j)}$, we have the estimate $|v_k^{(j)}| \le \frac{\mathcal{D}'_2}{|k|^r}$ for $k \in \mathbb{Z}^2\backslash 0$. Lastly, we know that

$|\frac{(\bar{k},l_2^\perp)}{(l_2,l_2)}| \le \frac{|k|}{|l_2|}$. These estimates allow us to apply Lemma 3.1, producing

$$2\pi\left|\sum[v_{l_1}^{(1)}v_{l_2}^{(2)}+v_{l_2}^{(2)}v_{l_2}^{(1)}]\frac{e^{-\gamma_2'|l_1|-\gamma_2'|l_2|}}{e^{-\gamma_2'|\bar{k}|}}\frac{(\bar{k},l_2^\perp)}{(l_2,l_2)}\right|$$

$$\le 2\pi \text{ const } (2^{r+1}|\bar{k}|+2^{r+2}(6|\bar{k}|+1)+|\bar{k}|^{1-r})\mathcal{D}_2'\frac{\mathcal{D}_2'}{|\bar{k}|^r}. \tag{3.3}$$

From this estimate, we see that if

$$2\pi(\text{const})\ (2^{r+1}|\bar{k}|+2^{r+2}(6|\bar{k}|+1)+2|\bar{k}|^{1-r})\mathcal{D}_2'$$

$$+\frac{\mathcal{G}}{\mathcal{D}_2'}\frac{e^{-\gamma|\bar{k}|^{1+\delta}}}{e^{-\gamma_2'|\bar{k}|}}|\bar{k}|^{\alpha-\epsilon} < 4\pi^2\nu|\bar{k}|^\alpha, \tag{3.4}$$

then we have established (3.2), which was our goal. Notice that we chose $\mathcal{D}_2' \ge \mathcal{G}$ and γ_2' such that $\frac{e^{-\gamma|\bar{k}|^{1+\delta}}}{e^{-\gamma_2'|\bar{k}|}} \le 1$ for all k with $|k| \ge K_0$. Since $\alpha > 1$ by picking K_0 large enough, we can force (3.4) to hold. This is the criteria which sets the level of K_{crit}. The proof of Theorem 1.2 is concluded. □

We now present the proof of Theorem 1.3. Its structure is very similar to the previous proof and also employs Lemma 3.1 but uses a slightly different change of variable.

Proof of Theorem 1.3. Let $\mathcal{D}_1'$ be the constant given by Theorem 1.1, that is such that $|\omega_k(t)| \le \frac{\mathcal{D}_1'}{|k|^2}$ for all $k \in \mathbb{Z}^2\backslash 0$ and all t. Let us put $v_k^{(j)} = \omega_k^{(j)}e^{\gamma_3 t|k|}, j=1,2$ where the constant γ_3 will be determined later. The evolution of the $v_k^{(1)}(t)$ are described by the following ODEs

$$\frac{dv_k^{(1)}(t)}{dt} = \gamma_3|k|v_k^{(1)}(t) - 4\pi^2\nu|k|^\alpha v_k^{(1)}(t) + g_k^{(1)}e^{\gamma_3 t|k|}$$

$$-2\pi\sum_{l_1+l_2=k}[v_{l_1}^{(1)}(t)v_{l_2}^{(2)}(t)+v_{l_1}^{(2)}(t)v_{l_2}^{(1)}(t)]\frac{e^{-\gamma_3 t|l_1|}e^{-\gamma_3 t|l_2|}}{e^{-\gamma_3 t|k|}}\frac{(k,l_2^\perp)}{(l_2,l_2)}. \tag{3.5}$$

The analogous equations describe the evolution of the $v_k^{(2)}(t)$.

The methods of the previous section can be applied to this coupled system. We fix a time $t_0 > 0$ and an arbitrary positive constant γ_0. For $t=0$, we have the inequalities

$$|v_k^{(1)}(0)| \le \frac{\mathcal{D}_3}{|k|^r}, \quad |v_k^{(2)}(0)| \le \frac{\mathcal{D}_3}{|k|^r}$$

for all k. In light of the definition of $v_k(t)$, Theorem 1.3 would be proven if we show that

$$|v_k^{(1)}(t_0)| \leq \frac{\mathcal{D}_3'}{|k|^r}, \quad |v_k^{(2)}(t_0)| \leq \frac{\mathcal{D}_3'}{|k|^r} \tag{3.6}$$

for some appropriate $\mathcal{D}_3'$.

As in the proof of Theorem 1.3, we put $\mathcal{D}_3' = \max(2\mathcal{D}_1', \mathcal{G})$ where $\mathcal{G}$ is again the constant from Assumption 1.3. For any fixed K_0, we can find a γ_3 so that the following three conditions hold. First, the inequalities $|\omega_k^{(j)}(t)| \leq \frac{\mathcal{D}_1'}{|k|^r}$, imply $|v_k^{(j)}(t)| \leq \frac{\mathcal{D}_3'}{|k|^r}$ for $j = 1, 2$, $t \in [0, t_0]$, and $|k| \leq K_0$. Second, so $e^{\gamma_3 t|k|} \leq e^{\gamma|k|^{1+\delta}}$ for $k \in \mathbb{Z}^2$ with $|k| \geq K_0$ and $t \in [0, t_0]$. In this condition the constants γ and δ are again from Assumption 1.3. Third, we can always assume that $\gamma_3 \leq \gamma_0$. (This last assumption is to simplify the exposition and is not really needed as γ_3 decreases as we increase K_0.)

Now consider the set

$$\Omega_3(K_0) = \left\{ (v_k^{(1)}, v_k^{(2)})_{k \in \mathbb{Z}^2 \setminus 0} \text{ with } |v_k^{(j)}| \leq \frac{\mathcal{D}_3'}{|k|^r} \text{ for } j = 1, 2 \text{ and } |k| > K_0 \right\}.$$

Again we will show that if K_0 is greater than some K_{crit}, the vector field along the boundary of $\Omega_3(K_0)$ points inward. The calculation parallels that in Theorem 1.2. For definiteness, we assume that $v_{\bar{k}}^{(1)}(t) = \frac{\mathcal{D}_3'}{|\bar{k}|^r}$ for some $\bar{k}$ with $|\bar{k}| > K_0$ and that the inequality bounds which define Ω_3 hold for all other k. The other cases proceed analogously.

We wish to show that the vector field points inward. Since $\gamma_3 \leq \gamma_0$, from (3.5), we see that it is sufficient to show that for $t \in [0, t_0]$

$$\begin{aligned}(4\pi^2\nu|\bar{k}|^\alpha &- \gamma_0|\bar{k}|)v_{\bar{k}}^{(1)} \\ &> 2\pi \left| \sum_{l_1+l_2=\bar{k}} [v_{l_1}^{(1)}(t)v_{l_2}^{(2)}(t) + v_{l_1}^{(2)}(t)v_{l_2}^{(1)}(t)] \frac{(k, l_2^\perp)}{(l_2, l_2)} \right| + |g_{\bar{k}}^{(1)}| e^{\gamma_3 t|\bar{k}|}. \end{aligned} \tag{3.7}$$

Here, as before, we have neglected the factor $\frac{e^{-\gamma_3 t|l_1|} e^{-\gamma_3 t|l_2|}}{e^{-\gamma_3 t|k|}}$ as it is always less than 1. After applying the inequalities $\mathcal{G} \leq \mathcal{D}_3'$, $e^{\gamma_3 t|k|} \leq e^{\gamma|k|^{1+\delta}}$ and Lemma 3.1, we see that (3.7) holds if

$$4\pi^2\nu > \gamma_0 \frac{|\bar{k}|}{|\bar{k}|^\alpha} + (\text{const})\ \mathcal{D}_3' \left[2^{r+1} \frac{|\bar{k}|}{|\bar{k}|^\alpha} + 2^{r+2} \frac{7|\bar{k}|}{|\bar{k}|^\alpha} + \frac{|\bar{k}|^{1-r}}{|\bar{k}|^\alpha} \right] + \frac{\mathcal{G}}{\mathcal{D}_3'} \frac{1}{|\bar{k}|^\alpha}. \tag{3.8}$$

Because $\alpha > 1$ and $r > 2$, by making $\bar{k}$ large enough we can force (3.8) to hold. This shows that the solution to any Galerkin approximation stays in Ω_3 until the time t_0 and thus (3.6) holds and the proof is complete. □

Proof of Lemma 3.1. As in the proof of Theorem 1.1, we estimate separately three sums.

$$\Sigma_1 = \sum_{\substack{|l_2| \le |\frac{k}{2}| \\ l_1+l_2=k}} |a_{l_1}|\,|b_{l_2}|\frac{|k|}{|l_2|},$$

$$\Sigma_2 = \sum_{\substack{|\frac{k}{2}|<|l_2|\le 2|k| \\ l_1+l_2=k}} |a_{l_1}|\,|b_{l_2}|\frac{|k|}{|l_2|},$$

$$\Sigma_3 = \sum_{\substack{|l_2|>2|k| \\ l_1+l_2=k}} |a_{l_1}|\,|b_{l_2}|\frac{|k|}{|l_2|}.$$

Since in Σ_1, the norm of $|l_1| \ge |\frac{k}{2}|$, we can write

$$\begin{aligned}\Sigma_1 &\le \sum_{|l_2|\le|\frac{k}{2}|} |a_{l_1}|\,|b_{l_2}|\frac{|k|}{|l_2|} \le \frac{2^r(\mathcal{C})^2}{|k|^r}|k| \sum_{|l_2|\le|\frac{k}{2}|}\frac{1}{|l_2|^{r+1}} \\ &\le 2^r(\text{const})\;|k|\frac{\mathcal{C}^2}{|k|^r}, \qquad (3.9)\end{aligned}$$

where the constant is defined by the inequality

$$\sum_{l_2\in\mathbb{Z}^d\setminus 0}\frac{1}{|l_2|^{r+1}} \le \text{const}.$$

For this sum to be finite, we need $r+1>d$. For Σ_2 we have $\frac{|k|}{|l_2|}\le 2$ and hence

$$\begin{aligned}\Sigma_2 &\le 2\sum_{|\frac{k}{2}|<|l_2|\le 2|k|}\frac{\mathcal{C}^2}{|l_1|^r|l_2|^r} \le \frac{2^{r+2}(\mathcal{C})^2}{|k|^r}\sum_{|l_1|\le 3|k|}\frac{1}{|l_1|^r} \\ &\le \frac{2^{r+2}(\mathcal{C})^2}{|\bar{k}|^r}\left(\sum_{|l_1|\le 3|k|}\frac{1}{|l_1|^{2r}}\right)^{\frac12}\left(\sum_{|l_1|\le 3|k|}1\right)^{\frac12} \\ &\le 2^{r+1}(\text{const})\frac{\mathcal{C}^2(6|k|+1)}{|k|^r}. \qquad (3.10)\end{aligned}$$

Here the constant is the absolute constant defined by

$$\sum_{l_1\in\mathbb{Z}^d\setminus 0}\frac{1}{|l_1|^{2r}} \le \text{const}.$$

For this sum to be finite, we need $2r > d$. For Σ_3 we have $\frac{|k|}{|l_2|} \leq \frac{1}{2}$. Hence, we can write

$$\begin{aligned}
\Sigma_3 &\leq \frac{1}{2} \sum_{|l_2| \geq 2|\bar{k}|} \frac{\mathcal{C}^2}{|l_1|^r |l_2|^{r+1}} \\
&\leq \frac{\mathcal{C}^2}{2} \left(\sum_{\substack{|l_1| > |k| \\ l_1 \in \mathbb{Z}^d \setminus 0}} \frac{1}{|l_1|^{2r}} \right)^{\frac{1}{2}} \left(\sum_{\substack{|l_2| \geq 2|k| \\ l_2 \in \mathbb{Z}^d \setminus 0}} \frac{1}{|l_2|^{2r+2}} \right)^{\frac{1}{2}} \\
&\leq (\text{const}) \frac{|k|^{d-1-r}}{2} \frac{\mathcal{C}^2}{|k|^r} \,. \qquad (3.11)
\end{aligned}$$

Collecting together (3.9), (3.10), (3.11), we obtain the lemma. □

4. The Three-dimensional Setting

This paper is mainly concerned with presenting an elementary proof of existence and uniqueness results in the two-dimensional setting. However, these techniques can also be used to gain some insight into the three-dimensional setting. On the three torus, the Navier Stokes equations take the form

$$\begin{aligned}
&\frac{u_i}{\partial t} + \sum_{j=1,2,3} u_j \frac{\partial u_i}{\partial x_j} = \nu \Delta u_j - \frac{\partial p}{\partial x_i} + f_i \,, \quad i = 1, 2, 3 \\
&\sum_{i=1,2,3} \frac{\partial u_i}{\partial x_i} = 0 \,,
\end{aligned} \qquad (4.1)$$

where $\nu > 0$ is again the viscosity, p is the pressure, and the f_i are the components of the external, timedependent forcing. As before, we introduce the vorticity.

$$\omega(x,t) = (\omega_1(x,t), \omega_2(x,t), \omega_3(x,t)) = \left(\frac{\partial u_2}{\partial x_3} - \frac{\partial u_3}{\partial x_2}, \frac{\partial u_3}{\partial x_1} - \frac{\partial u_1}{\partial x_3}, \frac{\partial u_1}{\partial x_2} - \frac{\partial u_2}{\partial x_1} \right) .$$

The vorticity obeys the equation

$$\frac{\partial \omega_i}{\partial t} + \sum_j u_j \frac{\partial \omega_i}{\partial x_j} = \sum_j \omega_j \frac{\partial u_i}{\partial x_j} + \nu \Delta \omega_i + g_i \,, \quad i = 1, 2, 3 \qquad (4.2)$$

where the g_i are the components of curl f. Moving to Fourier space where

$$u(x,t) = \sum_{k \in \mathbb{Z}^3} u_k(t) e^{2\pi i (k,x)} \quad \text{and} \quad \omega(x,t) = \sum_{k \in \mathbb{Z}^3} \omega_k(t) e^{2\pi i (k,x)} ,$$

we obtain

$$\begin{aligned}
\frac{d\omega_k(t)}{dt} &= -2\pi i \sum_{l_1 + l_2 = k} [(u_{l_1}(t), l_2) \omega_{l_2}(t) - (\omega_{l_1}(t), l_2) u_{l_2}(t)] \\
&\quad - 4\pi^2 \nu |k|^2 \omega_k(t) + g_k(t) \,. \qquad (4.3)
\end{aligned}$$

Here the $g_k(t)$ are the Fourier components of the forcing $g(x,t)$. In addition, we can replace the Laplacian with the more general differential operator $|\nabla|^\alpha$ with $\alpha > 1$.

The incompressibility condition implies that

$$u_k(t) \perp k \tag{4.4}$$

for all $k \in \mathbb{Z}^3$. Similarly, it follows that $\omega_k(t) \perp k$, $\omega_k(t) \perp u_k(t)$, and $|\omega_k(t)| = |k||u_k(t)|$. Hence, (k, u_k, ω_k) is a right-handed orthogonal (but not orthonormal) frame.

Since $(u_{l_1}(t), l_1) = (\omega_{l_1}(t), l_1) = 0$, we can rewrite (4.3) as

$$\begin{aligned}\frac{d\omega_k(t)}{dt} &= -2\pi i \sum_{l_1+l_2=k} [(u_{l_1}(t), k)\omega_{l_2}(t) - (\omega_{l_1}(t), k)u_{l_2}(t)] \\ &\quad - 4\pi^2\nu|k|^\alpha \omega_k(t) + g_k(t)\,. \end{aligned} \tag{4.5}$$

As before, we begin by restricting our attention to a finite subset $\mathcal{Z} \subset \mathbb{Z}^3$. The finite-dimensional Galerkin system corresponding to $\mathcal{Z}$ is

$$\begin{aligned}\frac{d\omega_k(t)}{dt} &= -2\pi i \sum_{\substack{l_1+l_2=k \\ l_1,l_2\in\mathcal{Z}}} [(u_{l_1}(t), k)\omega_{l_2}(t) - (\omega_{l_1}(t), k)u_{l_2}(t)] \\ &\quad - 4\pi^2\nu|k|^\alpha \omega_k(t) + g_k(t)\,. \end{aligned} \tag{$4.5^\alpha_\mathcal{Z}$}$$

Furthermore, to simplify the arguments, we assume that the forcing $g(x,t)$ is a trigonometric polynomial which implies that all but a finite number of the g_k are identically zero. We will always analyze wave numbers above the band which is directly forced; hence, we may neglect the g_k. This is only for convenience. The forcing can be included in the same way as it was in the two-dimensional setting.

Our development is based upon the basic energy estimate (see [1, 3, 11]). It states that given any initial data such that $\sum_{k\in\mathbb{Z}^3} |u_k(0)|^2 = E_0 < \infty$ then there exists a constant E^* depending only on E_0, ν, $\sup_t |g(\cdot,t)|_{L^2}$ such that for any finite-dimensional Galerkin approximation, defined by $\mathcal{Z} \subset \mathbb{Z}^3$, we have $\sum_{k\in\mathcal{Z}} |u_k(t)|^2 < E^*$ for all $t > 0$.

When $\alpha = 2$, the system ($4.5^\alpha_\mathcal{Z}$) corresponds to the Navier–Stokes equations. Unfortunately, we are unable to prove the theorems in this setting analogous to Theorems 1.1, 1.2, and 1.3 when $\alpha = 2$. However, if we increase α, we can.

Theorem 4.1. *Consider the system ($4.5^\alpha_\mathcal{Z}$) with an $\alpha > 2.5$ and satisfying Assumption 1.2. If the initial data $\{\omega_k(0)\}$ are such that*

$$|\omega_k(0)| \le \frac{\mathcal{D}_4}{|k|^r}$$

for all $k \in \mathbb{Z}^3$ *with* $r > 1.5$, *then there exists a constant* $\mathcal{D}_4'$, *independent of* $\mathcal{Z}$, *so that*

$$|\omega_k(t)| \leq \frac{\mathcal{D}_4'}{|k|^r}$$

for all $k \in \mathbb{Z}^3$ *and* $t \geq 0$.

Theorem 4.2. *Consider the system* $(4.5^{\alpha}_{\mathcal{Z}})$ *with an* $\alpha > 2.5$ *and satisfying Assumption* 1.3. *If the initial data* $\{\omega_k(0)\}$ *are such that*

$$|\omega_k(0)| \leq \frac{\mathcal{D}_5}{|k|^r} e^{-\gamma_5 |k|}$$

for all $k \in \mathbb{Z}^3$ *with* $r > 2$, *then there exists constants* $\mathcal{D}_5' < \infty$ *and* $\gamma_5' > 0$, *both independent of* $\mathcal{Z}$, *so that*

$$|\omega_k(t)| \leq \frac{\mathcal{D}_5'}{|k|^r} e^{-\gamma_5' |k|}$$

for all $k \in \mathbb{Z}^3$ *and* $t \geq 0$.

Theorem 4.3. *Consider the system* $(4.5^{\alpha}_{\mathcal{Z}})$ *with an* $\alpha > 2.5$ *and satisfying Assumption* 1.3. *If the initial data* $\{\omega_k(0)\}$ *are such that*

$$|\omega_k(0)| \leq \frac{\mathcal{D}_6}{|k|^r}$$

for all $k \in \mathbb{Z}^3$ *with* $r > 2$, *then for any* $t_0 > 0$ *there exists constants* $\mathcal{D}_6' < \infty$ *and* $\gamma_6' > 0$, *both independent of* $\mathcal{Z}$, *so that*

$$|\omega_k(t_0)| \leq \frac{\mathcal{D}_6'}{|k|^r} e^{-\gamma_6' |k|}$$

for all $k \in \mathbb{Z}^3$.

Of these three theorems, we will only give the proof of the first. The second two will be the consequence of two more general theorems given below. They apply to all $\alpha > 1.5$ but require the additional assumption that the enstrophy of all Galerkin approximations, starting from a given set of initial data, remains uniformly bounded in time. This is not known in general. However, when $\alpha > 2.5$, Theorem 4.1 implies this bound. Hence, the two theorems below apply to $(4.5^{\alpha}_{\mathcal{Z}})$ when $\alpha > 2.5$ without any assumption on $\mathcal{E}(t)$. In light of Theorem 4.1, Theorem 4.4 and 4.5 respectively yield Theorem 4.2 and 4.3 when $\alpha > 2.5$.

Theorem 4.4. *Let* $\{u_k(t)\}$ *be a solution to* $(4.5^{\alpha}_{\mathcal{Z}})$ *with* $\alpha > 1.5$ *such that* $\sum_{\mathbb{Z}^3} |\omega_k(t)|^2 < \mathcal{E}^* < \infty$ *for all* $t > 0$. *If* $|\omega_k(0)| \leq \frac{\mathcal{D}_7}{|k|^r}$ *for some* $\mathcal{D}_7 < \infty$ *and* $r > 2$, *then for any* $t_1 > 0$ *there exists a* $\gamma_7 > 0$ *and* $\mathcal{D}_7' < \infty$ *such that*

$$|\omega_k(t_1)| \leq \frac{\mathcal{D}_7'}{|k|^r} e^{-\gamma_7 |k|}\,.$$

Theorem 4.5. *Let $\{u_k(t)\}$ be a solution to $(4.5^{\alpha}_{\mathcal{Z}})$ with $\alpha > 1.5$ such that $\sum_{\mathbb{Z}^3} |\omega_k(t)|^2 < \mathcal{E}^* < \infty$ for all $t > 0$. If for some $\mathcal{D}_8 < \infty$, $\gamma_8 > 0$, and $r > 2$, $|\omega_k(0)| \le \frac{\mathcal{D}_8}{|k|^r} e^{-\gamma_8 |k|}$, then there exists a $\gamma_8' > 0$ and $\mathcal{D}_8' < \infty$ such that for all $t > 0$*

$$|\omega_k(t)| \le \frac{\mathcal{D}_8'}{|k|^r} e^{-\gamma_8' |k|}\,.$$

The above two theorems apply to $(4.5^{\alpha}_{\mathcal{Z}})$ for $\alpha > 1.5$. In particular, this means that they cover the standard Navier–Stokes equation which corresponds to $\alpha = 2$. (One can lower the restriction on α to $\alpha > 1$ at the cost of raising the restriction on r to $r > 3$. Similarly, one lowers the restriction on r to $r > 1.5$ at the cost of making $\alpha > 2.5$.)

In proving these two theorems, it was necessary to assume that $\sum_{\mathbb{Z}^3} |\omega_k(t)|^2$ remained uniformly bounded in time. Without such an assumption, we are forced to consider only α which do not correspond to the Navier–Stokes equation. Notice that Theorem 4.1 implies that $\sum_{k\in\mathbb{Z}^3} |\omega_k(t)|^2 < \text{const} < \infty$ for all $t > 0$ and hence Theorems 4.4 and 4.5 apply showing that the solution is analytic after $t = 0$.

In proving the above results, it is again convenient to split the system $(4.5^{\alpha}_{\mathcal{Z}})$ into the equations for the real and imaginary parts of $\{u_k\}_k$ and $\{\omega_k\}_k$. Letting $u_k(t) = u_k^{(1)}(t) + iu_k^{(2)}(t)$, $\omega_k(t) = \omega_k^{(1)}(t) + i\omega_k^{(2)}(t)$, and $g_k(t) = g_k^{(1)}(t) + ig_k^{(2)}(t)$; we obtain

$$\begin{aligned}\frac{d\omega_k^{(1)}(t)}{dt} &= 2\pi \sum_{\substack{l_1+l_2=k\\ l_1,l_2\in\mathcal{Z}}} [(u_{l_1}^{(1)}(t),k)\omega_{l_2}^{(2)}(t) + (u_{l_1}^{(2)}(t),k)\omega_{l_2}^{(1)}(t) - (\omega_{l_1}^{(2)}(t),k)u_{l_2}^{(1)}(t)\\ &\quad - (\omega_{l_1}^{(1)}(t),k)u_{l_2}^{(2)}(t)] - 4\pi^2\nu|k|^{\alpha}\omega_k^{(1)}(t) + g_k^{(1)}(t)\,, \qquad (4.5^{\alpha}_{\mathcal{Z}}{}^{(1)})\\ \frac{d\omega_k^{(2)}(t)}{dt} &= -2\pi \sum_{\substack{l_1+l_2=k\\ l_1,l_2\in\mathcal{Z}}} [(u_{l_1}^{(1)}(t),k)\omega_{l_2}^{(1)}(t) - (u_{l_1}^{(2)}(t),k)\omega_{l_2}^{(2)}(t) - (\omega_{l_1}^{(1)}(t),k)u_{l_2}^{(1)}(t)\\ &\quad + (\omega_{l_1}^{(2)}(t),k)u_{l_2}^{(2)}(t)] + g_k^{(2)}(t) - 4\pi^2\nu|k|^{\alpha}\omega_k^{(2)}(t)\,. \qquad (4.5^{\alpha}_{\mathcal{Z}}{}^{(2)})\end{aligned}$$

Proof of Theorem 4.1. By the energy estimate, we know that $|u_k^{(j)}(t)| \le \sqrt{E^*}$ for all $t \ge 0$ and $j = 1, 2$. Hence, $|\omega_k^{(j)}(t)| \le |k|\sqrt{E^*}$. Fixing a K_0, set $\mathcal{D}_4'(K_0) = K_0\mathcal{D}_4$. With this choice, $|\omega_k^{(j)}(t)| \le \mathcal{D}_4'(K_0)$ for all $t \ge 0$, $j = 1, 2$, and $k \in \mathbb{Z}^3$ with $|k| \le K_0$. As before, consider the set

$$\begin{aligned}\Omega_4(K_0) = \Bigg\{ (\omega_k^{(1)}, \omega_k^{(2)})_{k\in\mathbb{Z}^3} : |\omega_k^{(1)}| \le \frac{\mathcal{D}_4'(K_0)}{|k|^r}, |\omega_k^{(2)}|\\ \le \frac{\mathcal{D}_4'(K_0)}{|k|^r} \text{ for all } k, |k| > K_0 \Bigg\}\,.\end{aligned}$$

We have to show that if K_0 is taken to be sufficiently large, the vector field points inward along $\partial\Omega_4$.

We pick a point on the boundary. For definiteness, we will again consider the case when $\omega_{\bar{k}}^{(1)} = \frac{\mathcal{D}_4'}{|\bar{k}|^r}$ and $\omega_{\bar{k}}^{(2)} \le \frac{\mathcal{D}_4'}{|\bar{k}|^r}$ for some $\bar{k}$ with $|\bar{k}| \ge K_0$ and $\omega_k^{(j)} \le \frac{\mathcal{D}_4'}{|k|^r}$ for all other k with $k \ne \bar{k}$. The theorem will be proven if we can show that there exists a K_{crit}, independent of $\mathcal{D}_4'$, so that if $|\bar{k}| \ge K_0 > K_{crit}$, then

$$\Bigg| 2\pi \sum_{\substack{l_1+l_2=\bar{k}\\ l_1,l_2\in\mathcal{Z}}} [(u_{l_1}^{(1)}(t),\bar{k})\omega_{l_2}^{(2)}(t) + (u_{l_1}^{(2)}(t),\bar{k})\omega_{l_2}^{(1)}(t)$$

$$- (\omega_{l_1}^{(2)}(t),\bar{k})u_{l_2}^{(1)}(t) - (\omega_{l_1}^{(1)}(t),\bar{k})u_{l_2}^{(2)}(t)]\Bigg| < 4\pi^2\nu|\bar{k}|^\alpha\omega_{\bar{k}}^{(1)}(t)\,.$$

Other boundary points have the same structure so we will only show the details of the calculation for this case.

We need to estimate the summation. The total sum is made of smaller sums which are dominated by sums of the form $\sum_{l_1+l_2=\bar{k}} |u_{l_1}^{(a)}||\omega_{l_2}^{(b)}||k|$ with $a, b \in \{1,2\}$. As before, we split this sum into three parts:

$$\Sigma_1 = \sum_{|l_1|\le|\frac{\bar{k}}{2}|} |u_{l_1}^{(a)}||\omega_{l_2}^{(b)}||k|\,,$$

$$\Sigma_2 = \sum_{|\frac{\bar{k}}{2}|<|l_1|\le 2|\bar{k}|} |u_{l_1}^{(a)}||\omega_{l_2}^{(b)}||k|\,,$$

$$\Sigma_3 = \sum_{2|\bar{k}|<|l_1|} |u_{l_1}^{(a)}||\omega_{l_2}^{(b)}||k|\,.$$

In Σ_1, $|l_2| \ge |\frac{\bar{k}}{2}|$ and hence

$$\Sigma_1 \le \frac{\mathcal{D}_4'}{|\bar{k}|^r} 2^r |\bar{k}| \left(\sum_{|l_1|\le|\frac{\bar{k}}{2}|} |u_{l_1}^{(a)}|^2 \right)^{\frac{1}{2}} \left(\sum_{|l_1|\le|\frac{\bar{k}}{2}|} 1 \right)^{\frac{1}{2}}$$

$$\le \frac{\mathcal{D}_4'}{|\bar{k}|^r} 2^r (\mathrm{const})\sqrt{E^*}|\bar{k}|^{\frac{5}{2}}\,.$$

The constant is defined by

$$\left(\sum_{|l_1|\le|\frac{\bar{k}}{2}|} 1 \right) \le (\mathrm{const})^2|\bar{k}|^3\,.$$

For Σ_2, we know that $|l_2| \leq 3|\bar{k}|$ and $|u_{l_1}^{(a)}| \leq \frac{\mathcal{D}_4'}{|l_1|^{r+1}}$ which gives

$$\begin{aligned}\Sigma_2 &\leq \frac{\mathcal{D}_4'}{|\bar{k}|^r} 2^r \frac{2}{|k|} |k| \sum_{|l_2| \leq 3|\bar{k}|} |\omega_{l_2}^{(b)}| \leq \frac{\mathcal{D}_4'}{|\bar{k}|^r} 2^{r+1} \sum_{|l_2| \leq 3|\bar{k}|} |l_2| |u_{l_2}^{(a)}| \\ &\leq \frac{\mathcal{D}_4'}{|\bar{k}|^r} 2^{r+1} 3|\bar{k}| \left(\sum_{|l_2| \leq 3|\bar{k}|} |u_{l_1}^{(a)}|^2 \right)^{\frac{1}{2}} \left(\sum_{|l_2| \leq 3|\bar{k}|} 1 \right)^{\frac{1}{2}} \\ &\leq \frac{\mathcal{D}_4'}{|\bar{k}|^r} 2^{r+1} 3 \text{ (const) } \sqrt{E^*} |\bar{k}|^{\frac{5}{2}} .\end{aligned}$$

Here the constant is the analogue of the constant in the estimation of Σ_1. For Σ_3, we know that $|l_2| \geq |\bar{k}|$ and thus

$$\begin{aligned}\Sigma_3 &\leq |\bar{k}| \left(\sum_{|l_1| \geq 2|\bar{k}|} |u_{l_1}|^2 \right)^{\frac{1}{2}} \left(\sum_{|l_2| \geq |\bar{k}|} |\omega_{l_2}|^2 \right)^{\frac{1}{2}} \leq |\bar{k}| \mathcal{D}_4' \sqrt{E^*} \left(\sum_{|l_2| \geq |\bar{k}|} \frac{1}{|l_2|^{2r}} \right)^{\frac{1}{2}} \\ &\leq |\bar{k}| \mathcal{D}_4' \sqrt{E^*} \frac{\text{(const)}}{|\bar{k}|^{r-\frac{3}{2}}} \leq \frac{\mathcal{D}_4'}{|\bar{k}|^r} \text{(const)} \sqrt{E^*} |\bar{k}|^{\frac{5}{2}} .\end{aligned}$$

Collecting the three estimates together we see that there is a constant, depending only on r, so that

$$\sum_{l_1 + l_2 = \bar{k}} |u_{l_1}^{(a)}| |\omega_{l_2}^{(b)}| |k| \leq \text{(const)} \frac{\mathcal{D}_4'}{|\bar{k}|^r} \sqrt{E^*} |\bar{k}|^{\frac{5}{2}} . \tag{4.6}$$

Using this estimate, we see that the condition in (2.1) will hold if

$$8\pi \text{(const)} \sqrt{E^*} |\bar{k}|^{\frac{5}{2}} \frac{\mathcal{D}_4'}{|\bar{k}|^r} < 4\pi^2 \nu |\bar{k}|^\alpha \frac{\mathcal{D}_8'}{|\bar{k}|^r} .$$

Since $\alpha > \frac{5}{2}$, this will hold for all $\bar{k}$ sufficiently large; this sets the level of K_{crit}. Notice that it does not depend on $\mathcal{D}_8'$ as was required. □

Proof of Theorem 4.4. The proof of this theorem is similar to the proof of Theorem 1.3. From the assumptions, we know that $|\omega_k(t)| \leq \sqrt{\sum_{\mathbb{Z}^3} |\omega_l(t)|^2} < \sqrt{\mathcal{E}^*}$ for all $t > 0$. We set $a_k^{(j)}(t) = u_k^{(j)} e^{\gamma_7 t |k|}$ and $b_k^{(j)}(t) = \omega_k^{(j)} e^{\gamma_7 t |k|}$ for $j = 1, 2$, where γ_7 is a constant we will set later.

Set $\mathcal{D}_7' = 2 \max(\sqrt{\mathcal{E}^*}, \mathcal{D}_7)$. Fixing a K_0, choose $\gamma_7(K_0)$ so that for all $t \in [0, t_1]$, $j \in \{1, 2\}$, and k with $|k| \leq K_0$, one has $|b_k^{(j)}(t)| \leq \frac{\mathcal{D}_7'}{|k|^r}$. Notice that by the assumption on the initial conditions, we have $|b_k^{(j)}(0)| \leq \frac{\mathcal{D}_7'}{|k|^r}$ for all k. Consider the set,

$$\Omega_7(K_0) = \left\{ (b_k^{(1)}, b_k^{(2)})_{k \in \mathbb{Z}^2 \setminus 0} \text{ with } |b_k^{(j)}| \leq \frac{\mathcal{D}_7'}{|k|^r} \text{ for } j = 1, 2 \text{ and } |k| > K_0 \right\} .$$

As before we will show that, if K_0 is chosen large enough, a point starting in Ω_7 cannot leave Ω_7 because the vector field along $\partial\Omega_7$ is pointing inward.

We pick a boundary point. For simplicity, we pick the point where $b_{\bar{k}}^{(1)} = \frac{\mathcal{D}_7'}{|k|^r}$ and all other variables satisfy the inequalities defining Ω_7. In terms of the new variables, the relevant equation of motion reads

$$\begin{aligned}\frac{db_k^{(1)}(t)}{dt} &= (\gamma_7|k| - 4\pi^2\nu|k|^\alpha)b_k^{(1)}(t)\\ &\quad - 2\pi \sum_{\substack{l_1+l_2=k\\ l_1,l_2\in\mathcal{Z}}} [(a_{l_1}^{(1)}(t),k)b_{l_2}^{(2)}(t) + (a_{l_1}^{(2)}(t),k)b_{l_2}^{(1)}(t)\\ &\quad - (b_{l_1}^{(2)}(t),k)a_{l_2}^{(1)}(t) - (b_{l_1}^{(1)}(t),k)a_{l_2}^{(2)}(t)]\frac{e^{-\gamma_7 t|l_1|}e^{-\gamma_7 t|l_2|}}{e^{-\gamma_7 t|k|}}\,.\end{aligned}$$

Since $|a_k^{(j)}(t)| = \frac{|b_k^{(j)}(t)|}{|k|}$, to insure that the vector field points inward it is sufficient to show that

$$\begin{aligned}2\pi\sum |b_{l_1}^{(1)}||b_{l_2}^{(2)}|\frac{|\bar{k}|}{|l_1|} + |b_{l_1}^{(2)}||b_{l_2}^{(1)}|\frac{|\bar{k}|}{|l_1|} + |b_{l_1}^{(2)}||b_{l_2}^{(1)}|\frac{|\bar{k}|}{|l_2|} + |b_{l_1}^{(1)}||b_{l_2}^{(2)}|\frac{|\bar{k}|}{|l_2|}\\ < (4\pi^2\nu|\bar{k}|^\alpha - \gamma_7|\bar{k}|)\frac{\mathcal{D}_7'}{|\bar{k}|^r}\,.\end{aligned}$$

Each of the terms in the above sum can be estimated with the aid of Lemma 3.1. This transforms the previous condition into

$$8\pi(\text{const})\left(2^r|\bar{k}| + 2^{r+1}(6|\bar{k}|+1)^{\frac{3}{2}} + \frac{1}{2}|\bar{k}|^{2-r}\right)\frac{(\mathcal{D}_7')^2}{|\bar{k}|^r} < (4\pi^2\nu|\bar{k}|^\alpha - \gamma_7|\bar{k}|)\frac{\mathcal{D}_7'}{|\bar{k}|^r}\,.$$

By picking K_0 large enough, we can force this condition to hold. The fact that γ_7 depends on K_0 is not a problem since it decreases as K_0 increases.

This establishes that the vector field points inward along the boundary of Ω_7 for all times in the interval $[0,t_1]$. Thus at time t_1, the trajectory is still in Ω_7. By returning to the original variables, we have the desired estimate at time t_1. □

Proof of Theorem 4.5. The proof of this theorem begins as the above theorem and then proceeds as the proof of Theorem 1.2. We change variables to $a_k^{(j)}(t) = u_k^{(j)}e^{\gamma_8|k|}$ and $b_k^{(j)}(t) = \omega_k^{(j)}e^{\gamma_8|k|}$. We use the assumption on $\sqrt{\sum_{\mathbb{Z}^3}|\omega_l(t)|^2}$ to control the lower modes. Then we use the estimates from Lemma 3.1 to control the nonlinearity. We omit the details. □

Acknowledgments

The authors thank P. Constantin, W. E, C. Fefferman, U. Frisch, G. Gallavotti, J. Mather, V. I. Judovich, F. Planchon, P. Sarnak, T. Spencer, T. Suidan, J. Vinson and V. Yakhot for useful discussions. The second author thanks NSF (grant DMS-97067994) for financial support.

References

[1] P. Constantin and C. Foiaş, *Navier–Stokes Equations*, University of Chicago Press, Chicago (1988).

[2] A. J. Chorin and J. E. Marsden, *A mathematical introduction to fluid mechanics*, Vol. 4, *Texts in Applied Mathematics*, Springer-Verlag, New York, third edition (1993).

[3] C. R. Doering and J. D. Gibbon, *Applied analysis of the Navier–Stokes equations*, Cambridge Texts in Applied Mathematics, Cambridge University Press, Cambridge (1995).

[4] C. R. Doering and E. S. Titi, *Exponential decay rate of the power spectrum for solutions of the Navier–Stokes equations*, Phys. Fluids **7** (1995) 1384–1390.

[5] C. Foiaş and R. Temam, *Gevrey class regularity for the solutions of the Navier–Stokes equations*, J. Funct. Anal. **87** (1989) 359–369.

[6] G. Gallavotti, *Ipotesi per una introduzione alla Meccanica Dei Fluidi*, Gruppo Nazionale Di Fisica Matematica, 1996.

[7] E. Hopf, *Über die Anfangswertaufgabe für die hydrodynamischen Grundgleichungen*, Math. Nachr. **4** (1951) 213–231.

[8] Heinz-Otto Kreiss, *Fourier expansions of the solutions of the Navier–Stokes equations and their exponential decay rate*, In Analyse mathématique et applications, Gauthier-Villars, Paris (1988), pp. 245–262.

[9] O. A. Ladyzhenskaya, *The Mathematical Theory of Viscous Incompressible Flow*, Gordon and Breach, New York (1969).

[10] J. Leray, *Essai sur le mouvement d'un liquide visqueux emplissant l'espace*, Acta Math. **63** (1934) 193–248.

[11] R. Temam, *Navier–Stokes equations: Theory and numerical analysis*, Vol. 2 of *Studies in Mathematics and its Applications*, North-Holland Publishing Co., Amsterdam-New York, revised edition (1979).

[12] R. Temam, *Navier–Stokes equations and nonlinear functional analysis*, Vol. 66 of *CBMS-NSF Regional Conference Series in Applied Mathematics*, Society for Industrial and Applied Mathematics (SIAM), Philadelphia, PA, second edition (1995).

[13] V. I. Yudovich, *The linearization method in hydrodynamical stability theory*, American Mathematical Society, Providence, RI, 1989. Translated from the Russian by J. R. Schulenberger.

Comments

The main result proven in this paper was obtained 10 years earlier by C. Foias and R. Temam (see [FT]). It seems that our proof is more geometric.

Recently E. Dinaburg, Dong Li, and I used the methods of this paper to extend the results to other examples of domains and boundary conditions (in preparation).

Commun. Math. Phys. 148, 601–621 (1992)

Communications in
Mathematical
Physics

Statistics of Shocks in Solutions of Inviscid Burgers Equation

Ya. G. Sinai

Landau Institute of Theoretical Physics, Moscow, Russia and Mathematics Department, Princeton University, NJ 08544, USA

Abstract. The purpose of this paper is to analyze statistical properties of discontinuities of solutions of the inviscid Burgers equation having a typical realization $b(y)$ of the Brownian motion as an initial datum. This case was proposed and studied numerically in the companion paper by She, Aurell and Frisch. The description of the statistics is given in terms of the behavior of the convex hull of the random process $w(y) = \int_0^y (b(\eta) + \eta)\, d\eta$. The Hausdorff dimension of the closed set of those y where the convex hull coincides with w is also studied.

1. General Properties of Solutions of the One-dimensional Inviscid Burgers Equation

Burgers equation is one of the most popular non-linear equations which appears in many concrete physical problems. In this paper we study some properties of solutions of the inviscid Burgers equation having as initial velocity a typical realization of the Brownian motion (as a function of the space variable). This case was proposed in a companion paper by She, Aurell, and Frisch [1] where one can find physical motivations for this case as well as many qualitative arguments and numerical results.

We start with the geometric description of the process of construction of solutions to the inviscid Burgers equation. This theory was already exposed in the pioneering works of Hopf (see [2]) and Burgers (see [3]). We present here a slightly different approach compared with [2] and [3]. The companion paper [1] also begins with this analysis. The notations in the present paper and in [1] are slightly different but it is easy to establish the correspondence between them.

We recall that the one-dimensional Burgers equation without force has the form

$$\partial_t u + u\partial_x u = \mu \partial_x^2 u, \qquad -\infty < x < \infty .$$

Y.G. Sinai, *Selecta II: Probability Theory, Statistical Mechanics, Mathematical Physics and Mathematical Fluid Dynamics*, DOI 10.1007/978-1-4419-6205-8_19, © Springer Science+Business Media, LLC 2010

Here $\mu > 0$ is the viscosity. The Hopf-Cole substitution $u = -2\mu \dfrac{\partial_x \varphi}{\varphi}$ (see [1, 3]) shows that φ satisfies the heat equation

$$\partial_t \varphi = \mu \partial_x^2 \varphi .$$

Using this fact one can write down for the solution $u = u_\mu(x,t)$ the explicit expression

$$u_\mu(x,t) = \frac{\int\limits_{-\infty}^{\infty} dy \dfrac{x-y}{t} \exp\left\{ -\dfrac{1}{2\mu} F(x,y,t) \right\}}{\int\limits_{-\infty}^{\infty} dy \exp\left\{ -\dfrac{1}{2\mu} F(x,y,t) \right\}} . \tag{1}$$

Here $F(x,y,t) = \dfrac{(x-y)^2}{2t} + \int\limits_0^y u(\eta;0)\,d\eta$ and $u(\eta;0)$ is the initial datum. The formula (1) works if $\int\limits_0^y u(\eta;0)\,d\eta = o(y^2)$ as $y \to \pm\infty$. The expression of (1) appears often for correlation function in statistical mechanics. In fact the theory of Burgers equation is closely connected with the theory analysing statistical properties of directed polymers.

Hopf in [2] and Burgers in [3] discussed the behavior of solutions in $u_\mu(x;t)$ under the limit transition $\mu \to 0$. In what follows we study the solutions $u_\mu(x;t)$ for a fixed value of t, say $t = 1$. Therefore we often omit t in our future notations. Consider

$$\begin{aligned} M(x) = \min_y F(x,y,1) &= \min_y \left[\frac{(x-y)^2}{2} + \int\limits_0^y u(\eta;0)\,d\eta \right] \\ &= \frac{x^2}{2} + \min_y \left\{ \int\limits_0^y [u(\eta;0) + \eta]\,d\eta - xy \right\} . \end{aligned}$$

Denote $w(y) = \int\limits_0^y [u(\eta;0) + \eta]\,d\eta$. The function $L_w(x) = \min_y\{w(y) - xy\}$ is the Legendre transform of w. We need the simplest properties of this transform. It will be applied below to cases where $u(\eta;0) = 0$ for $\eta < 0$ and $u(\eta;0)$ is continuous for $-\infty < \eta < \infty$. Therefore $w(y) = \dfrac{y^2}{2}$ for $y < 0$. Introduce the convex hull $C_w(y)$ of w. It is a convex function, and $C_w \le w$. Then C_w is the largest function having the last two properties.

There is also another way of describing C_w. Fix x and take a straight line having the slope x, i.e. a line given on the (y,w)-plane by the equation $w = xy + c$. Then for every x one can find such $c_0(x) = c_0$ that for all $c < c_0$ the lines $w = xy + c$ do not intersect the graph of w while for all $c > c_0$ such intersections arise. For $c = c_0$ the line $w = xy + c_0$ is tangent to the graph of w at one or several points. Put $M(x)$ to be the set of those y where the line $w = xy + c_0$ is tangent to the graph of w. Introduce $m_*(x) = \min\{y \mid y \in M(x)\}$, $m^* = \max\{y \mid y \in M(x)\}$. If $m^*(x) = m_*(x)$ then $C_w(y) = w(y)$ for $y = m_*(x) = m^*(x)$. If $m_*(x) < m^*(x)$

then $C_w(y) = xy + c_0$ for all $m_*(x) \le m^*(x)$. In other words, the graph C_w is a convex curve consisting of straight intervals and a closed set lying outside them. The derivative $F(y) = \frac{d}{dy} C_w(y)$ is in general a non-decreasing Cantor devil staircase type function which takes a constant value on each interval where C_w is linear. In these terms the Legendre transform can be written in the form

$$L_w(x) = c_0 .$$

Consider the function $G(x)$ which is the inverse function to $x = F(y)$. However it is not a well-defined object because $G(x)$ is multi-valued for those x where the set of y where $F(y) = x$ is an interval. Since we are interested in a geometric picture it is more convenient to consider $G(x)$ as a continuous curve on the plane which consists of vertical segments for those values of x, where G is discontinuous. Now we can formulate the following theorem by Hopf (see [2]).

Hopf's Theorem. *Let x be such that $M(x)$ consists of one point $y(x) = G(x)$. Then the limit $\lim_{\mu \to 0} u_\mu(x; 1) = \lim_{\mu \to 0} u_\mu(x) = u_0(x)$ exists and $u_0(x) = x - G(x)$. If $G(x)$ is an interval of positive length then there exist the limits*

$$u_0^-(x) = \lim_{x' \to -0} u_0(x') = x - m_*(x),$$

$$u_0^+(x) = \lim_{x' \to +0} u_0(x') = x - m^*(x) .$$

In both cases the limits are taken over such x' that $G(x')$ is single-valued.

We can interpret this result as follows. The function $u_0(x)$ is discontinuous for those x where $G(x)$ is multi-valued. At these points there exist the one-sided limits of $u_0(x)$ equal to $x - m_*(x)$ for the left limit and $x - m^*(x)$ for the right limit. This jump is interpreted as a shock and its size is equal to the length of the vertical segment of $G(x)$.

We shall use the following definition.

Definition 1. Cantor-type function $F(y)$ is complete if the union of intervals where F is constant is a set of full Lebesgue measure or, better to say, its complement has Lebesgue measure zero.

Let us prove now the following lemma.

Lemma 1. *If F is complete then $u_0(x)$ is differentiable* a.e. *and* $\frac{du_0(x)}{dx} = 1$ a.e.

The proof is simple. Indeed, another way to express the completeness of F is to say that the image under F^{-1} of $R^1 \setminus$ (countable set of x such that $G(x)$ is multi-valued) is a subset of R^1 of the zeroth Lebesgue measure. Put for convenience $G(x) = m^*(x)$ for all x. Then $\lim_{x' \to x} \frac{G(x') - G(x)}{x' - x} = 0$ for a.e. x with respect to the Lebesgue measure. Since $u_0(x) = x - G(x)$ for all x except the above mentioned countable set this gives the desired result.

Now we formulate the final conclusions of this section. The limit $u_0(x)$ is a discontinuous function whose discontinuities take place for those x where the equality

$x = F(y)$ holds for a segment on the y-axis of positive length. Outside this countable set $u_0(x) = x - G(x)$. The discontinuities of $u_0(x)$ are always negative, i.e. the limits from the left are bigger than the limits from the right. If the devil's stair-case $F(y)$ is complete then $u_0(x)$ is differentiable a.e. and $\frac{d}{dx}u_0(x) = 1$.

2. The Case Studied in the Companion paper [1]

The motivation for this paper was to explain some numerical results obtained by She, Aurell, and Frisch [1]. Among many cases considered by these authors there was the case of $u(x;0) = b(x)$, where $b(x)$ is a Brownian trajectory for $x \geq 0$ and $u(x;0) = 0$ for $x < 0$. According to the theory described in Sect. 1 we have to construct the random process $w(y)$, where $w(y) = \frac{y^2}{2}$ for $y < 0$ and $w(y) = \int_0^y (b(\eta) + \eta)\, d\eta$ for $y > 0$ and to study its convex hull C_w which is a non-linear and non-local functional of b. It is clear that for some $y_0 = y_0(b) < 0$ the convex hull C_w coincides with $\frac{y^2}{2}$ for $y < y_0$. Therefore $F(y) = y$ for $y < y_0$ and for such y the function F is not a devil's stair-case.

Theorem 1. *With probability* 1 *the devil's stair-case* $x = F(y)$ *is complete on the semiline* $y > 0$.

Proof. Fix $\bar{y}$ and consider the tangent line $\Gamma_{\bar{y}}$ to the graph of $w(y)$ at $y = \bar{y}$ given by the equation $w = w(\bar{y}) + (y - \bar{y})\,(b(\bar{y}) + \bar{y})$. We shall say that $\bar{y}$ is a special point for $b = \{b(y)\}$ if one can find a neighborhood U of $\bar{y}$ depending on b and such that in this neighborhood the graph of w lies above $\Gamma_{\bar{y}}$, i.e. if

$$w(y) \geq w(\bar{y}) + (y - \bar{y})(b(\bar{y}) + \bar{y}), \qquad y \in U\,.$$

We shall show that for any $\bar{y}$ the probability that it is a special point is equal to zero. Let us derive from this statement the assertion of the theorem.

Fix $Y > 0$ and consider the probability space $(C_Y, \mathscr{F}_Y, P) \times ([0, Y], \mathscr{F}, l) = (\Omega, \mathscr{D}, P)$. Here C_Y is the space of continuous functions defined on the segment $[0, Y]$, equal to zero at $y = 0$. $\mathscr{F}_Y$ is the Borel σ-algebra of the space C_Y, P is the standard Wiener measure defined on $\mathscr{F}_Y$. Further $\mathscr{F}$ is the Borel σ-algebra of the segment $[0, Y]$ and l is the normed length. Introduce the subset $A \subset \Omega$ consisting of such pairs (b, y) that y is a special point for b. It is easy to see that $A \in \mathscr{D}$. The above mentioned statement implies

$$P(A) = \int_0^Y dl(\bar{y})\, P\{\bar{y} \text{ is a special point for } b\} = 0\,,$$

and by Fubini's theorem

$$0 = P(A) = \int dP(b)\, l(\{\bar{y} \mid \bar{y} \text{ is a special point for } b\})$$

which gives $l(\{\bar{y} \mid \bar{y}$ is a special point for $b\}) = 0$ for a.e. b.

In order to prove the main statement we shall show that the probability that for some $\alpha = \alpha(b) > 0$ and all x, $0 \le x \le \alpha$,

$$w(\bar{y}+x) - w(\bar{y}) = \int_{\bar{y}}^{\bar{y}+x} (b(\eta)+\eta)\,d\eta \ge (b(\bar{y})+\bar{y})x$$

is zero. Rewrite the last inequality in the form

$$\int_0^x (b_1(\eta)+\eta)\,d\eta \ge 0, \qquad 0 \le x \le \alpha,$$

where $b_1(\eta) = b(\eta+\bar{y}) - b(\bar{y})$. It is clear that $b_1(\eta)$ has the same distribution as $b(\eta)$. Put $b_2(\eta) = b_1(\eta)+\eta$. Girsanov's theorem (see [4]) says that the probability measure corresponding to the process $b_2(\eta)$ on any finite interval of η is equivalent tot he Wiener measure.

Denote by $P_+(P_-)$ the probability with respect to the Wiener measure that there exists $\alpha_1 = \alpha_1(b) > 0$ such that $\int_0^x b(\eta)\,d\eta \ge 0$ (≤ 0) for all x, $0 \le x \le \alpha_1(b)$. By symmetry, $P_+ = P_-$. Remark now that the event whose probability we study belongs to the σ-algebra depending on the behaviour of the process in one point and therefore by the "$0-1$" law can take only the values 1 or 0. Since in our case $P_+ = P_-$ and $P_+ + P_- \le 1$ it can take only the *zeroth* value. Due to the absolute continuity of the measure corresponding to b_2 to the Wiener measure, this probability for $b_2(\eta)$ is also zero. Q.E.D.

Remark. The proof given above was shown to me by M. Yor (private communication). My original proof was more complicated.

Return now to the function F and introduce the closed set $S(b)$ of all $y > 0$ lying outside the union of intervals where F is constant. In other words, $S(b)$ consists of such $\bar{y} \in S(b)$ that the tangent line $w = (b(\bar{y})+\bar{y})(y-\bar{y}) + w(\bar{y})$ intersects the graph of the function $w(y) = \int_0^y (b(\eta)+\eta)\,d_\eta$ only at the point $(\bar{y}, w(\bar{y}))$. In the next sections we study the fractal properties of $S(b)$.

The main result of our studies is the following theorem.

Main Theorem. *With probability* 1 *the Hausdorff dimension of* $S(b)$ *is equal to* $\frac{1}{2}$.

The proof of this theorem is based upon the estimations of probabilities of small fragments of C_w which we derive in the next section.

3. Estimations of Probabilities of Small Fragments of C_w

Consider on the plane (y, w) two vertical lines $y = a_1$, $y = a_2$, $0 < a_1 < a_2$, and two strips $\Pi_1 = \{(y,w) \mid |y - a_1| \le \delta_1(a_2 - a_1)\}$, $\Delta_2 = \{(y,w) \mid |y - a_2| \le \delta_2(a_2 - a_1)\}$. In what follows $a_2 - a_1$ will tend to zero while all δ_j will remain fixed but small. Take also a straight line Γ given by the equation $w = \beta y + \beta_1 = l(y)$ and such that the point $(0, 0)$ lies above Γ. Introduce the parallelograms

$$\Pi_1 = \left\{(y, w)\big|\, |y - a_1| \le \delta_1(a_2 - a_1),\ \ |w - l(y)| \le \delta_3(a_2 - a_1)^{3/2}\right\},$$
$$\Pi_2 = \left\{(y, w)\big|\, |y - a_2| \le \delta_2(a_2 - a_1),\ \ |w - l(y)| \le \delta_3(a_2 - a_1),^{3/2}\right\}$$

We need also the segments

$$\Gamma_0 \subset \Gamma,$$
$$\Gamma_0 = \{(y, w) \mid a_1 + \delta_1(a_2 - a_1) \le y \le a_2 - \delta_2(a_2 - a_1), (y, w) \in \Gamma\},$$
$$\Gamma_{01} = \left\{(y, w) \mid a_1 - \delta_1(a_2 - a_1) \le y \le a_1 + \delta_1(a_2 - a_1),\right.$$
$$\left. w = l(y) - \frac{\delta_3}{2}(a_2 - a_1)^{3/2}\right\}$$
$$\Gamma_{02} = \left\{(y, w) \mid a_2 - \delta_2(a_2 - a_1) \le y \le a_2 + \delta_2(a_2 - a_1),\right.$$
$$\left. w = l(y) - \frac{\delta_3}{2}(a_2 - a_1)^{3/2}\right\},$$

the ray

$$\Gamma^- = \{(y, w) \mid y \le a_1 - \delta_1(a_2 - a_1), w = l_-(y)\},\ \ l_-(y) = \beta^- y + \beta_1^-,$$

whose continuation passes through the points

$$(a_1 - \delta_1(a_2 - a_2), l(a_1 - \delta_1(a_2 - a_1))) \in \Gamma$$

and

$$\left(a_2 - \delta_2(a_2 - a_1), l(a_2 - \delta_2(a_2 - a_1)) - \delta_3(a_2 - a_1)^{3/2}\right)$$

and the ray

$$\Gamma^+ = \{(y, w) \mid y \ge a_2 + \delta_2(a_2 - a_1), w = l_+(y)\},\ \ l_+(y) = \beta^+ y + \beta_1^+,$$

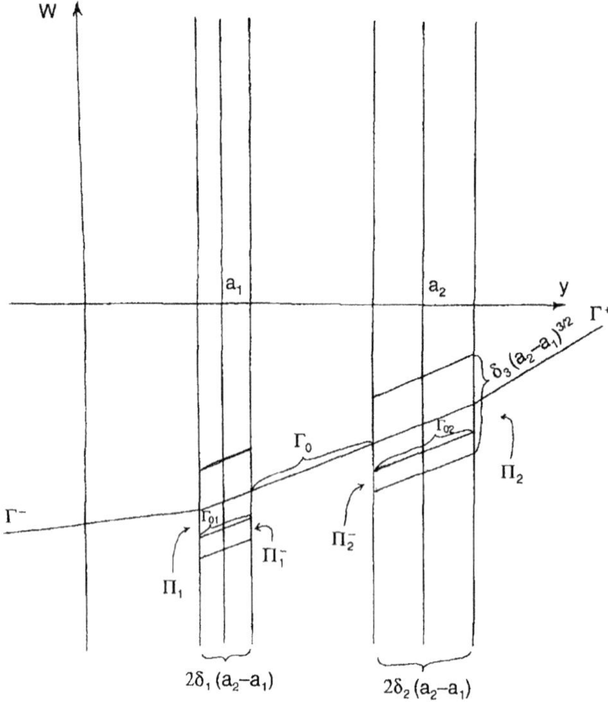

Fig. 1

whose continuation passes through the points

$$\left(a_1 + \delta_1(a_2 - a_1), l((a_1 + \delta_1(a_2 - a_1)) - \delta_3(a_2 - a_1)^{3/2}\right)$$

and

$$\left(a_2 + \delta_2(a_2 - a_1), l(a_2 + \delta_2(a_2 - a_1))\right) \in \Gamma\,.$$

Let also

$$\begin{aligned} \Pi_1^- = \{(y,w)|\, |y - a_1| \le \delta_1(a_2 - a_1), -\delta_3(a_2 - a_1)^{3/2} \\ \le w - l(y) \le -\tfrac{1}{2}\,\delta_3(a_2 - a_1)^{3/2}\}\,, \\ \Pi_2^- = \{(y,w)|\, |y - a_2| \le \delta_2(a_2 - a_1), -\delta_3(a_2 - a_1)^{3/2} \\ \le w - l(y) \le -\tfrac{1}{2}\,\delta_3(a_2 - a_1)^{3/2}\}\,, \end{aligned}$$

All these segments, rays, parallelograms, strips are drawn in Fig. 1.

Lemma 2. *If $\delta_1, \delta_2 \le$ const then any straight line passing through a point inside Π_1^- and through a point Π_2^- lies below Γ^+ and Γ^-.*

Geometrically the statement of the lemma is obvious. Remark also that the whole construction is defined as soon as Γ, Π_1, Π_2 are given. Thus they can be considered as determining parameters. Return now to the process $w(y) = \int_0^y (b(\eta) + \eta)\, d\eta$ and take another small number $\delta_4 > 0$.

Definition 2. A realization w has a right behavior (with respect to our construction) if

A$_1$) for

$$y \leq a_1 - \delta_1(a_2 - a_1)$$

the graph of w lies above Γ_-, i.e.

$$w(y) > l_-(y) \qquad \text{for all such} \quad y;$$

A$_2$)

$$l(a_1 - \delta_1(a_2 - a_1)) < w(a_1 - \delta_1(a_2 - a_1)) < l(a_1 - \delta_1(a_2 - a_1)) + \delta_3(a_2 - a_1)^{3/2};$$

$$\beta \geq w'(a_1 - \delta_1(a_2 - a_1)) = b(a_1 - \delta_1(a_2 - a_1)) + (a_1 - \delta_1(a_2 - a_1)) \geq \beta - \delta_4(a_2 - a_1)^{1/2};$$

moreover,

$$w'(a_1 - \delta_1(a_2 - a_1)) \leq \beta^-;$$

A$_3$) for all $a_1 - \delta_1(a_2 - a_1) \leq y \leq a_1 + \delta_1(a_2 - a_1)$

$$w(y) > l(y) - \delta_3(a_2 - a_1)^{3/2};$$

and there is a non-empty subset of such y that

$$w(y) < l(y) - \frac{\delta_3}{2}(a_2 - a_1)^{3/2};$$

A$_4$)

$$l(a_1 + \delta_1(a_2 - a_1)) < w(a_1 + \delta_1(a_2 - a_1)) < l(a_1 + \delta_1(a_2 - a_1)) + \delta_3(a_2 - a_1)^{3/2};$$

A$_5$) for all $a_1 + \delta_1(a_2 - a_1) \leq y \leq a_2 - \delta_2(a_2 - a_1)$

$$w(y) > l(y);$$

A$_6$)

$$l(a_2 - \delta_2(a_2 - a_1)) < w(a_2 - \delta_2(a_2 - a_1)) < l(a_2 - \delta_2(a_2 - a_1)) + \delta_3(a_2 - a_1)^{3/2};$$

A$_7$) for all $a_2 - \delta_2(a_2 - a_1) \leq y \leq a_2 + \delta_2(a_2 - a_1)$

$$w(y) > l(y) - \delta_3(a_2 - a_1)^{3/2};$$

and there is a non-empty open set of such y that

$$w(y) < l(y) - \frac{\delta_3}{2}(a_2 - a_1)^{3/2};$$

A$_8$)

$$l(a_2 + \delta_2(a_2 - a_1)) < w(a_2 + \delta_2(a_2 - a_2)) < l(a_2 + \delta_2(a_2 - a_1)) + \delta_3(a_2 - a_1)^{3/2};$$

$$\beta \leq w'(a_2 + \delta_2(a_2 - a_1)) = b(a_2 + \delta_2(a_2 - a_1)) + (a_2 + \delta_2(a_2 - a_1)) \leq \beta + \delta_4(a_2 - a_1)^{1/2};$$

moreover,

$$w'(a_2 + \delta_2(a_2 - a_1)) > \beta^+.$$

A_9) for all $y \geq a_2 + \delta_2(a_2 - a_1)$ the graph of w lies above Γ_+, i.e.

$$w(y) \geq l_+(y) \quad \text{for all such} \quad y.$$

Figure 2 shows the right behavior. Also one can easily see some symmetry in the properties (A_1)–(A_9). The reasons for our scaling will become clear from further estimations.

Theorem 2. *Let $U = U(\Gamma, \Pi_1, \Pi_2, \delta_4)$ be the event consisting of such b that w has a right behavior (see Definition 2). Then for any $b \in U$ the graph of C_w contains an interval which has endpoints inside Π_1 and Π_2.*

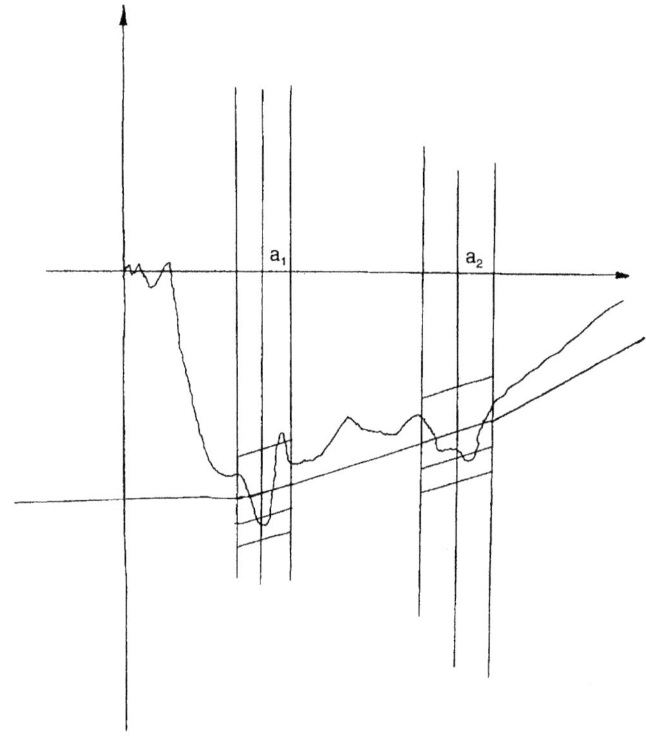

Fig. 2

Proof is simple. Consider the straight line $\Gamma_{\beta,c}$ given by the equation $w = \beta y + c$ for $c < 0$, $|c|$ is large, such that $\Gamma_{\beta,c}$ does not intersect the graph of w. Start steadily to increase c. Then there will appear such $\bar{c}$ that for all $c < \bar{c}$ there are no such intersections and $\bar{c}$ is the upper bound of c having this property. The straight line $\Gamma_{\beta,\bar{c}}$ is tangent to the graph of w at one or several points. If among these points there are points inside Π_1 as well as inside Π_2 then our statement is proven.

Suppose that all common points of the graph w and $\Gamma_{\beta,\bar{c}}$ lie inside Π_1, the case of points inside Π_2 is considered in the same way. For $\beta' > \beta$ sufficiently close to β take the analogous line $\Gamma_{\beta',\bar{c}(\beta')}$. Then all common points of w and $\gamma_{\beta',\bar{c}(\beta')}$ lie inside Π_1. One can find $\bar{\beta} > \beta$ such that for all β', $\beta < \beta' < \bar{\beta}$ we shall have the same property while for $\bar{\beta}$ the straight line $\Gamma_{\bar{\beta},c(\bar{\beta})}$ will have common points inside Π_1 and outside Π_1. From the right behavior (A_4)–(A_6) it follows easily that these points lie inside Π_1^- and Π_2^-.

Lemma 1 and (A_1)–(A_9) imply that there are no common points of w and $\Gamma_{\bar{\beta},c(\bar{\beta})}$ outside the strip $a_1 - \delta_1(a_2 - a_1) \le y \le a_2 + \delta_2(a_2 - a_1)$, Q.E.D.

One of our main estimations is given in the following theorem.

Theorem 3. *Let δ_j, $1 \le j \le 4$, be sufficiently small, and a_1, β be fixed. Then for all sufficiently small $a_2 - a_1$ the probability (with respect to the Wiener measure)*

$$P(U) \ge F(\delta_1,\delta_2,\delta_3,\delta_4,a_1)\cdot(a_2-a_1)^{1/2}Q(\delta_1,\delta_2,\beta,\beta_1,a_1,a_2),$$

where $F(\delta_1,\delta_2,\delta_3,\delta_4,a_1)$ is a positive constant and $Q(\delta_3,\delta_4,\beta,\beta_1,a_1,a_2)$ is the probability that

$$\begin{aligned}
b(a_1 - \delta_1(a_2-a_1)) &\in (-(a_1 - \delta_1(a_2-a_1))) - \delta_4(a_2-a_1)^{1/2};\\
&\quad \left(-(a_1 - \delta_1(a_2-a_1)) + \delta_4(a_2-a_1)^{1/2}\right);\\
w(a_1 - \delta_1(a_2-a_1)) &\in (l(a_1 - \delta_1(a_2-a_1))) - \delta_3(a_2-a_1)^{3/2};\\
&\quad l\left(a_1 - \delta_1(a_2-a_1)) + \delta_3(a_2-a_1)^{3/2}\right).
\end{aligned}$$

It is clear that for any compact set of values of β, β_1 the probability Q is proportional to $(a_2 - a_1)^2 \cdot \delta_3 \cdot \delta_4$.

Proof. Denote by z_1, z_2, z_3, z_4 the values of $b(a_1 - \delta_1(a_2-a_1))$, $w(a_1 - \delta_1(a_2-a_1))$, $b(a_2 + \delta_2(a_2-a_1))$, $w(a_2 + \delta_2(a_2-a_1))$ respectively, and introduce also their dimensionless values through the rescaling

$$\begin{aligned}
z_1 &= \beta + Z_1(a_2-a_1)^{1/2},\\
z_2 &= l(a_1 - \delta_1(a_2-a_1)) + Z_2(a_2-a_1)^{3/2},\\
z_3 &= \beta + Z_3(a_2-a_1)^{1/2},\\
z_4 &= l(a_2 + \delta_2(a_2-a_1)) + Z_4(a_2-a_1)^{3/2}.
\end{aligned}$$

In the case of the right behavior, i.e. $b \in U$, we have $0 < Z_1 < \delta_4$, $-\delta_3 < Z_2 < \delta_3$, $0 < Z_3 < \delta_4$, $-\delta_3 < Z_4 < \delta_3$. We need also some rescaling of y, i.e. we put $y = a_1 + (a_2 - a_1)Y$.

Using the fact that the pair $(b(y), w(y))$ is a two-dimensional Markov process we can write

$$P(U) = \int dz_1\,dz_2\,dz_3\,dz_4 \cdot p(0,0;z_1,z_2;a_1 - \delta_1(a_2-a_1)) \cdot p(z_1,z_2;z_3,z_4;$$

$$
\begin{aligned}
&a_1 - \delta_1(a_2 - a_1), a_2 + \delta_2(a_2 - a_1)) \\
&\quad \cdot P(A_1 \mid b(a_1 - \delta_1(a_2 - a_1)) = z_1, w(a_1 - \delta_1(a_2 - a_1)) = z_2)) \\
&\quad \cdot P\{A_3, A_4, A_5, A_6, A_7 \mid b(a_1 - \delta_1(a_2 - a_1)) = z_1, \\
&w(a_1 - \delta_1(a_2 - a_1) = z_2, \\
&b(a_2 + \delta_2(a_2 - a_1)) = z_3, w(a_2 + \delta_2(a_2 - a_1)) = z_4\} \\
&\quad \cdot P\{A_9 \mid b(a_2 + \delta_2(a_2 - a_1)) = z_3, w(a_2 + \delta_2(a_2 - a_1)) = z_4\}. \qquad (2)
\end{aligned}
$$

The domain of integration is determined by A_2 and A_8 and was written down in terms of the dimensionless parameters $Z_j; p(u_1, u_2; v_1, v_2; s, t)$ is the transition density of the process (b, w) from the initial state u_1, u_2 at the moment of time s to the final state v_1, v_2 at the moment t, the written probabilities describe the conditional probabilities of the corresponding properties A_j.

We study the inner factor

$$
\begin{aligned}
P\{A_3, A_4, A_5, A_6, A_7 \mid\ & b(a_1 - \delta_1(a_2 - a_1) = z_1, \\
& w(a_1 - \delta_1(a_2 - a_1)) = z_2, b(a_2 + \delta_2(a_2 - a_1)) = z_3, \\
& w(a_2 + \delta_2(a_2 - a_1)) = z_4\}.
\end{aligned}
$$

Under the described rescaling and the rescaling of the Wiener process

$$
\begin{aligned}
b(y) &= b(a_1 - \delta_1(a_2 - a_1)) + \sqrt{a_2 - a_1} B(Y), \\
w(y) &= w(a_1 - \delta_1(a_2 - a_1)) + (a_2 - a_1)^{3/2} \cdot W(y),
\end{aligned}
$$

we see that the properties (A_2)–(A_7) are expressed only in terms of the dimensionless variables and the rescaled processes B, W. Therefore for all values of z_1, z_2, z_3, z_4 under consideration

$$
\begin{aligned}
P\{A_3, A_4, A_5, A_6, A_7 \mid\ & b(a_1 - \delta_1(a_2 - a_1)) = z_1, w(a_1 - \delta_1(a_2 - a_1)) = z_2, \\
& b(a_2 + \delta_2(a_2 - a_1)) = z_3, w(a_2 + \delta_2(a_2 - a_1)) = z_4\} \\
&= F_1(Z_1, Z_2; Z_3, Z_4, \delta_1, \delta_2, \delta_3, \delta_4) > 0.
\end{aligned}
$$

The most crucial part is the estimation of

$$
P\{A_1 \mid b(a_1 - \delta_1(a_2 - a_1)) = z_1, w(a_1 - \delta_1(a_2 - a_1) = z_2\}
$$

and

$$
P\{A_9 \mid b(a_2 + \delta_2(a_2 - a_1)) = z_3, w(a_2 + \delta_2(a_2 - a_1) = z_4\}.
$$

Consider first the last probability. Now it is better to change slightly the rescaling and to consider

$$
\begin{aligned}
b(y) &= z_3 + (a_2 - a_1)^{1/2} B_1(Y - (1 + \delta_2)), \\
w(y) &= z_4 + z_3(y - (a_2 + \delta_2(a_2 - a_1))) + (a_2 - a_1)^{3/2} W_1(Y - (1 + \delta_2)), \\
& Y \geq 1 + \delta_1.
\end{aligned}
$$

The probability distribution of the processes $B_1(Y)$, $W_1(Y)$ does not depend on $(a_2 - a_1)$. Write down the equation for $\Gamma_+(y)$ in the form $l_+(y) = \beta_1^+ + \beta^+ y$, $y \geq a_2 + \delta_2(a_2 - a_1)$. Then the inequality

$$w(y) \geq \beta_1^+ + \beta^+ y, \quad y \geq a_1 + \delta_2(a_2 - a_1) \tag{3}$$

can be rewritten as follows. Since

$$\begin{aligned} w(y) &= \int_0^y (b(\eta) + \eta)\, d\eta \\ &= w(a_2 + \delta_2(a_2 - a_1)) \\ &\quad + b(a_2 + \delta_2(a_2 - a_1)) \cdot (y - (a_2 + \delta_2(a_2 - a_1))) \\ &\quad + (a_2 + \delta_2(a_2 - a_1)) \cdot (y - (a_2 + \delta_2(a_2 - a_1))) \\ &\quad + \int_0^{y-(a_2+\delta_2(a_2-a_1))} (b_1(\eta) + \eta)\, d\eta \\ &= z_4 + (z_3 + a_2 + \delta_2(a_2 - a_1))(Y - (1 + \delta_2)) \\ &\quad + \int_0^{(a_2-a_1)(Y-(1-\delta_2))} (b_1(\eta) + \eta)\, d\eta\,, \end{aligned}$$

where $b_1(\eta) = b(\eta + a_2 + \delta_2(a_2 - a_1)) - z_3$, the inequality (3) takes the form

$$\begin{aligned} & z_4 - (\beta_1^+ + \beta^+(a_2 + \delta_2(a_2 - a_1))) + (z_3 + (a_2 + \delta_2(a_2 - a_1)) - \beta^+) \\ & \cdot (a_2 - a_1) \cdot (Y - (1 + \delta_2)) \\ & + (a_2 - a_1)^{3/2} \cdot \int_0^{Y-(1+\delta_2)} (b_2(\eta) + (a_2 - a_1)^{1/2}\eta)\, d\eta > 0\,, \end{aligned} \tag{4}$$

for all Y such that $Y \geq 1 + \delta_2$, $b_2(\eta) = (a_2 - a_1)^{1/2} b_1((a_2 - a_1)^{-1}\eta)$ and has the same distribution as the initial Brownian motion. Now we see that all terms in (4) are of order $(a_2 - a_1)^{3/2}$. Indeed,

$$\begin{aligned} z_4 - (\beta_1^+ + \beta^+(a_2 + \delta_2(a_1 + a_2))) &= Z_4(a_2 - a_1)^{3/2}\,, \\ z_3 + (a_2 + \delta_2(a_2 - a_1) - \beta^+) &= z_3 + a_2 + \delta_2(a_2 - a_1) - \beta + (\beta - \beta^+) \\ &= (a_2 - a_1)^{1/2}(Z_3 + C) = (a_2 - a_1)^{1/2} Z_3^{(1)}\,, \end{aligned}$$

where C depends only on δ_1, δ_2, δ_3.

All these relations explain the reason for our rescaling. Thus we come to the dimensionless expression of (3) and (4)

$$Z_4 + Z_3^{(1)}(Y - (1 + \delta_2)) + \int_0^{Y-(1+\delta_2)} (b_2(\eta) + (a_2 - a_1)^{1/2}\eta) \geq 0\,. \tag{5}$$

It follows from (A_8) that $Z_3^{(1)} \geq 0$, $Z_4 > 0$. The probability (5) was in fact estimated in [5] and it was shown in [5] that it is not less than const$(a_2 - a_1)^{1/4}$ (one should put $\sigma = (a_2 - a_1)^{1/2}$ in Theorem 7 in [5] where const depends on δ_j, $1 \leq j \leq 4$.

The estimation of the conditional probability of (A_1), provided that $w(a_1 - \delta_1(a_2 - a_1)$, $b(a_1 - \delta_1(a_2 - a_1)$ are given, is done in a similar way (see Theorem 7′ in [5]). It is also bounded from below by const $(a_2 - a_1)^{1/4}$.

The probability Q arises from the integration over z_1, z_2 in (2). Thus the theorem is proven.

Remark. The function $F(\delta_1, \delta_2, \delta_3, \delta_4, a_1)$ shows in fact some dependence between δ_4 and δ_1, δ_2. The meaning of this dependence is quite clear. If δ_3 is relatively large and we integrate in (2) over a domain of large values of z_2 and z_3 then it becomes highly probable that $w(y)$ intersects the low side of Π_1 or Π_2 and thus the conditional probability of the right behavior inside the interval $a_1 + \delta_1(a_2 - a_1)$, $a_2 - \delta_2(a_2 - a_1)$ becomes small.

Now we are going to obtain a similar estimate from above. Assume that as above the strips Π_1, Π_2, the line Γ and two parallelograms Π_1, Π_2 are given (see Fig. 3).

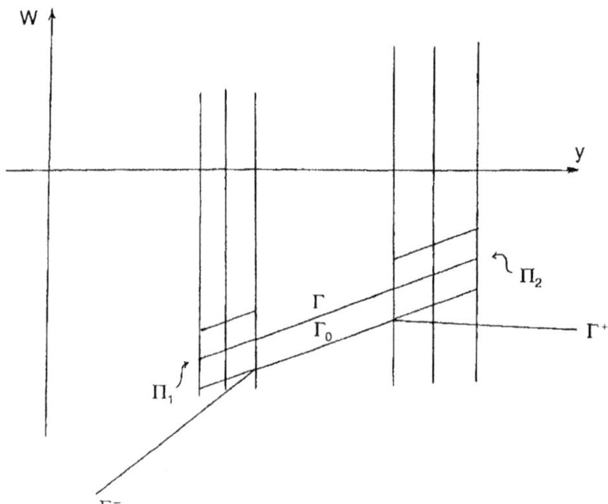

Fig. 3

Remark that for the fixed a_1, a_2, Γ the determining parameters for the whole construction are only $\delta_1, \delta_2, \delta_3$. Also we have to assume that $(0, 0)$ lies above Γ.

Theorem 4. *Suppose that a_1, a_2, Γ are given. Then for all sufficiently small $\delta_1, \delta_2, \delta_3$ the probability P that C_w has a segment whose left endpoint belongs to Π_1 while the right endpoint belongs to Π_2 satisfies the inequality*

$$P \leq F_1(\delta_1, \delta_2, \delta_3, a_1, \Gamma)(a_2 - a_1)^{1/2} \cdot (a_2 - a_1)^2 \cdot \delta_1 \delta_3$$

for all sufficiently small $(a_2 - a_1)$. *Here* $F_1(\delta_1, \delta_2, \delta_3, a_1, \Gamma)$ *is a positive constant.*

Proof. We use the same notation $l(y) = \beta y + \beta_1$ for the straight line Γ. Introduce also the ray Γ^+ passing through the points

$$\big(a_1 + \delta_1(a_2 - a_1), l(a_1 + \delta_1(a_2 - a_1)) + \delta_3(a_2 - a_1)^{3/2}\big)$$

and

$$(a_2 - \delta_2(a_2 - a_1),\ l(a_2 - \delta_2(a_2 - a_1)) - \delta_3(a_2 - a_1)^{3/2}),\quad y \geq a_2 - \delta_2(a_2 - a_1),$$

and the ray Γ^- passing through the points

$$\big(a_1 + \delta_1(a_2 - a_1),\ l(a_1 + \delta_1(a_2 - a_1)) - \delta_3(a_2 - a_1)^{3/2}\big)$$

and

$$\big(a_2 - \delta_2(a_2 - a_1),\ l(a_2 - \delta_2(a_2 - a_1)) + \delta_3(a_2 - a_1)^{3/2}\big),\quad y \leq a_1 + \delta_1(a_2 - a_1).$$

Also Γ_0 is the straight segment given by the equation

$$w = l(y) - \delta_3(a_2 - a_1)^{3/2}, a_1 + \delta_1(a_2 - a_1) \leq y \leq a_2 - \delta_2(a_2 - a_1).$$

Geometrically it is also clear that if C_w has a segment with the endpoints in Π_1 and Π_2 then w lies above Γ^+ for $y \geq a_2 - \delta_2(a_2 - a_1)$, above Γ^- for $y \leq a_1 + \delta_1(a_2 - a_1)$ and above Γ_0 for $a_1 + \delta_1(a_2 - a_1) \leq y \leq a_2 - \delta_2(a_2 - a_1)$. Now we denote $z_1', z_2', z_1'', z_2'', z_3', z_4', z_3'', z_4''$ the values of $b(a_1 - \delta_1(a_2 - a_1))$, $w(a_1 - \delta_1(a_2 - a_1))$, $b(a_1 + \delta_1(a_2 - a_1))$, $w(a_1 + \delta_1(a_2 - a_1))$, $b(a_2 - \delta_2(a_2 - a_1))$, $w(a_2 - \delta_2(a_2 - a_1))$, $b(a_2 + \delta_2(a_2 - a_1))$, $w(a_2 + \delta_2(a_2 - a_1))$, respectively. Using again the Markov property of the two-dimensional random process $b(y)$, $w(y)$ we can write

$$\begin{aligned} P \leq \int & p(0, 0; z_1', z_2'; 0, a_1 - \delta_1(a_2 - a_1)) \cdot p(z_1', z_2'; z_1'', z_2''; a_1 - \delta_1(a_2 - a_1), \\ & a_1 + \delta_1(a_2 - a_1)) \cdot p(z_1'', z_2''; z_3', z_4'; a_1 + \delta_1(a_2 - a_1), a_2 - \delta_2(a_2 - a_1)) \\ & \cdot p(z_3', z_4'; z_3'', z_4''; a_2 - \delta_2(a_2 - a_1), a_2 + \delta_2(a_2 - a_1)) \\ & \cdot P_1(z_1', z_2') \cdot P_2(z_1', z_2', z_1'', z_2'') \cdot P_3(z_1'', z_2'', z_3', z_4') \\ & \cdot P_4(z_3', z_4'; z_3'', z_4''; a_2 - \delta_2(a_2 - a_1), a_2 + \delta_2(a_2 - a_1)) \\ & \cdot P_5(z_3'', z_4'') \prod_{i=1}^{4} dz_i'\, dz_i''. \end{aligned}$$

Here $P_1(z_1', z_2')$ is the conditional probability that $w(y)$ lies above Γ^- for $y \leq a_1 - \delta_1(a_2 - a_1)$ under the conditions $b(a_1 - \delta_1(a_2 - a_1)) = z_1'$, $w(a_1 - \delta_1(a_2 - a_1)) = z_2'$; $P_2(z_1', z_2', z_1'', z_2'')$ is the conditional probability that $w(y)$ lies above the corresponding part of Γ_0 and intersects the upper side of Π_1 under the conditions $b(a_1 - \delta_1(a_2 - a_1)) = z_1'$, $w(a_1 - \delta_1(a_2 - a_1)) = z_2'$, $b(a_1 + \delta_1(a_2 - a_1)) = z_1''$,

$w(a_1 + \delta_1(a_2 - a_1)) = z_2''$, $P_3(z_1'', z_2'', z_3', z_4')$ is the conditional probability that $w(y)$ lies above Γ_0 for $a_1 + \delta_1(a_2 - a_1) \leq y \leq a_2 - \delta_2(a_2 - a_1)$ under the conditions $b(a_1 + \delta_1(a_2 - a_1)) = z_1''$, $w(a_1 + \delta_2(a_2 - a_1)) = z_2''$, $b(a_2 + \delta_2(a_2 - a_1)) = z_3'$, $w(a_2 - \delta_2(a_2 - a_1)) = z_4'$; P_4 is the analogous conditional probability with respect to Π_2 as P_2; P_5 is the conditional probability that $w(y)$ lies above Γ^+ for $y \geq a_2 + \delta_2(a_2 - a_1)$ under the conditions $b(a_2 + \delta_2(a_2 - a_1)) = z_3''$, $w(a_2 - a_1)) = z_4''$.

Introduce again the dimensionless values:

$$\begin{aligned}
z_1' &= \beta + Z_1'(a_2 - a_1)^{1/2},\\
z_2' &= l(a_1 - \delta_1(a_2 - a_1)) + Z_2'(a_2 - a_1)^{3/2},\\
z_1'' &= \beta + Z_1''(a_2 - a_1)^{1/2},\\
z_2'' &= l(a_1 + \delta_1(a_2 - a_1)) + Z_2''(a_2 - a_1)^{3/2},\\
z_3' &= \beta + Z_3'(a_2 - a_1)^{1/2},\\
z_4' &= l(a_2 - \delta_2(a_2 - a_1)) + Z_4'(a_2 - a_1)^{3/2},\\
z_3'' &= \beta + Z_3''(a_2 - a_1)^{1/2},\\
z_4'' &= l(a_2 + \delta_2(a_2 - a_1)) + Z_4''(a_2 - a_1)^{3/2}.
\end{aligned}$$

For $Z_j', Z_j'' = O(1)$, $1 \leq j \leq 4$, we can use the same arguments as above. In particular, $P_1(z_1', z_2') = O(1) \cdot (a_2 - a_1)^{1/4}$, $P_5(z_3', z_4') = O(1) \cdot (a_2 - a_1)^{1/4}$ (see Theorem 7, 7′ in [5] with $\sigma = (a_2 - a_1)^{1/2}$).

In order to estimate $P_2(z_1', z_2', z_3', z_4')$ consider two cases:

In the first case, $c \leq z_2' \leq l(a_1 - \delta_1(a_2 - a_1)) + 2\delta_3(a_2 - a_1)^{3/2}$, where c is the vertical coordinate of the intersection Γ^- and $y = a_1 - \delta_1(a_2 - a_1)$. It is easy to see that $P_2(z_1', z_2'; z_1'', z_2'')$ decays faster than exponentially as a function of Z_2'. In the second case $z_2' \geq l(a_1 - \delta_1(a_2 - a_1)) + 2\delta_3(a_2 - a_1)^{3/2}$ and the conditional probability decays faster than exponentially also as a function Z_1'. More exact estimations of the remainder terms which together with the estimate const $(a_2 - a_1)^2 \delta_1 \cdot \delta_3$ of the integral over z_1', z_2' lead to the statement of the theorem. They will be given in another publication.

Theorem 5. *Let two intervals be given* $I_1 = \{y : |y - a_1| \leq \delta_1(a_2 - a_1)\}$, $I_2 = \{y : |y - a_2| \leq \delta_2(a_2 - a_1)\}$, $a_1 \geq$ const, *and the corresponding vertical strips* $\Delta_1 = \{(y, w) \mid y \in I_2\}$, $\Delta_2 = \{(y, w) \mid y \in I_2\}$. *Then the probability* P *that* C_w *has a segment whose endpoints lie inside* Δ_1 *and* Δ_2 *respectively satisfies the inequalities*

$$F_3(\delta_1, \delta_2, a_1)(a_2 - a_1)^{1/2} \leq P \leq F_4(\delta_1, \delta_2, a_1)(a_2 - a_1)^{1/2},$$

where F_3, F_4 *are positive constants.*

Proof. The estimation from below follows easily from Theorem 3 by summation over parallelograms Π_1, Π_2. In order to get the estimation from above cover the vertical line passing through a_1 by equal intervals U_j of the length $\delta_3(a_2 - a_1)^{3/2}$

and cover the axis of angles by equal intervals Φ_s of the length $\delta_4(a_2 - a_1)^{1/2}$. In both cases the coverings are chosen so that each point is covered by at most two elements of the covering. Having the centers u_j of U_j and β_s of Φ_s we can construct the corresponding parallelograms Π_1, Π_2. If C_w has a segment with the endpoints inside Δ_1 and Δ_2 then these points lie inside at least one pair Π_1, Π_2 if δ_3 and δ_4 are sufficiently small and δ_4 is smaller than δ_3, i.e. $\delta_4 \ll \delta_3$.

The estimation of P follows from the estimation of Theorem 4 by summation over j and s.

We use the results of Theorem 3–5 in the next section for the estimation of the Hausdorff dimension of $S(b)$. However they are of more general importance because they describe some statistical properties of small shocks in solutions of the Burgers equation. One can find in [1] numerical results which are in a perfect agreement with the estimations of Theorem 4 and 5.

4. The Hausdorff Dimension of the Set $S(b)$

Take a realization

$$w(y) = \int_0^y (b(\eta) + \eta)\, d\eta \,,$$

its convex hull C_w and the closed set $S(b)$ of such $\bar{y}$ that the tangent line $w = (b(\bar{y}) + \bar{y})(y - \bar{y}) + w(\bar{y})$ intersects the graph of $w(y)$ only at the point $\bar{y}, w(\bar{y})$. In this section we study the Hausdorff dimension of $S(b) \cap [a', a'']$ for any segment $[a', a'']$, $0 < a' < a'' < \infty$.

We begin with the estimation of the fractal dimension from above, which is usually simpler. Our arguments are based upon the following lemma. Let S be a closed subset of $[a', a'']$, $O = [a', a''] \setminus S$ be its open complement. Assume that the Lebesgue measure $l(S) = 0$ and O_j are open connected components of O. Denote by N_k the number of those O_j, for which $\frac{1}{2^{k+1}} \leq l(O_j) \leq \frac{1}{2^k}$, $k = 0, 1, 2, \ldots$

Lemma 3. *If for some $c, 0 < c < 1$, and any $\delta > 0$ the numbers $N_k \leq 2^{k(c+\delta)}$ for all sufficiently large k, then the Hausdorff dimension $d(S) \leq c$.*

Proof. Fix δ and take all sufficiently large k. We define $\frac{1}{3 \cdot 2^k}$ coverings of S in the following way. Denote

$$S^{(k)} = [a', a''] \setminus O^{(k)},$$

where $O^{(k)} = \bigcup_{l(O_j) \geq \frac{1}{2^k}} O_j$. The set $S^{(k)}$ is the union of closed segments. Two neighboring open components of each segment consisting of deleted segments O_j have lengths not less than $\frac{1}{2^k}$. Cover each segment of $S^{(k)}$ by intervals of the length $\frac{1}{3 \cdot 2^k}$ in such a way that each point is covered by at most two segments. Denote the intervals of our $\frac{1}{3\cdot 2^k}$ covering by $U_s^{(k)}$. Then for any $c_1 > c+\delta$,

$$\sum_s [l(U_s^{(k)})]^{c_1} = \frac{1}{(3\cdot 2^k)^{c_1-1}} \sum_s l(U_s^{(k)})$$
$$\leq \frac{2}{(3\cdot 2^k)^{c_1-1}} \left[l(S^{(k)}) + \frac{2}{3\cdot 2^k} \sum_{j=1}^{k} N_j \right]. \tag{6}$$

The last term appears because we must take into account the lengths of parts of $U_s^{(k)}$ belonging to $O^{(k)}$. Since $l(S) = 0$,

$$l(S^{(k)}) \leq \sum_{O_j \not\subset O^{(k)}} l(O_j) \leq \sum_{p\geq k} \frac{1}{2^p} N_p \leq \frac{\text{const}}{2^{k(1-c-\delta)}}.$$

Also

$$\frac{1}{2^k}\sum_{j=1}^{k} N_j \leq \frac{\text{const}}{2^{k(1-c-\delta)}}.$$

This yields (see (6))

$$\sum_s [l(U_s^{(k)})]^{c_1} \leq \frac{\text{const}}{2^{k(c_1-c-\delta)}} \to 0$$

as $k \to \infty$. Therefore $d(S) \leq c_1$ and $d(S) \leq c$ because δ is arbitrary. The Lemma is proven.

Return to our set $S(b) \cap [a', a'']$. Lemma 4 shows that in order to estimate its Hausdorff dimension from above we have to estimate the numbers N_k.

Lemma 4. *For any $\delta > 0$ with probability 1,*

$$N_k \leq 2^{k\left(\frac{1}{2}+\delta\right)}$$

for all sufficiently large k.

Corollary. *With probability 1 the Hausdorff dimension $d(S(b)\cap[a',a'']) \leq \frac{1}{2}$.*

Corollary follows directly from Lemma 3 and 4.

Proof of Lemma 4. Fix k and choose a sufficiently large M. Decompose the segment $[a', a'']$ onto equal segments V_j of the length $\frac{1}{2^k M}$ and consider the pairs V_{j_1}, V_{j_2} such that the distance between their centers is

$$\frac{p}{M}\cdot\frac{1}{2^k}, \quad \frac{M-1}{2} \leq p \leq M.$$

Let also $C_{j_1 j_2}$ be the event consisting of such Brownian trajectories $b(\eta)$, such C_w has an interval whose endpoints have projections in V_{j_1} and V_{j_2}. If $\chi_{j_1,j_2}(b)$ is the indicator of the event $C_{j_1 j_2}$ then

$$N_k \leq \sum \chi_{C_{j_1 j_2}} .$$

It follows from Sect. 3 that the expectation $E\chi_{C_{j_1 j_2}} \leq \frac{\text{const}}{2^{\frac{k}{2}}}$ and thus

$$EN_k \leq \text{ const} \cdot M^2 \cdot 2^{\frac{k}{2}} .$$

From Chebyshev's inequality

$$P\{N_k \geq 2^{\frac{k}{2}(1+\delta)}\} \leq \text{ const}\, M^2 \cdot \frac{2^{\frac{k}{2}}}{2^{\frac{k}{2}(1+\delta)}} = \text{const}\, M^2 \cdot 2^{-\frac{k}{2}\delta} .$$

Therefore for any $\delta > 0$ the series

$$\sum P\{N_k \geq 2^{\frac{k}{2}(1+\delta)}\} < \infty .$$

In view of the Borel-Cantelli lemma for a.e. b one can find $k_0(b)$ such that for all $k \geq k_0(b)$ we shall have $N_k < 2^{\frac{k}{2}(1+\delta)}$, Q.E.D.

The estimation of the Hausdorff dimension from below is based upon Frostman's lemma. For the convenience of a reader we give its formulation adapted to our case.

Frostman's lemma (see [6]). Assume that one can find a finite measure μ concentrated on $S(b) \cap [a', a'']$ and such that for some $t > 0$,

$$\iint \frac{d\mu(y_1)\, d\mu(y_2)}{|y_1 - y_2|^t} < \infty .$$

Then the Hausdorff dimension $d(S(b) \cap [a', a'']) \geq t$.

We need some extra notations. Let O_{kj} be such components of O that

$$\frac{1}{2^k} \leq l(O_{k,j}) < \frac{1}{2^{k-1}} .$$

Also $S_j^{(k)}$ are closed components of $S(k)$. Introduce the measure μ_k which is the normed uniform measure on $S(k)$. Then for some subsequence $\{k_j\}$ the measures μ_{k_j} converge weakly to a limit which we shall denote by μ. Our purpose now is to show that for any $t < \frac{1}{2}$ the integral

$$\iint \frac{d\mu(y_1)\, d\mu(y_2}{|x - y|^t} < \infty .$$

Certainly it is sufficient to show that for any $\delta > 0$

$$\iint\limits_{|x-y|>\delta} \frac{d\mu(x)\, d\mu(y)}{|x-y|^t} \le A\,,$$

where a constant A does not depend on δ. In our case

$$\max_j \ \operatorname{diam}(S_j^{(i)}) = p_i \to 0 \quad \text{as} \quad i \to \infty\,.$$

We shall show that

$$I_m^{(i)} = \sum_{j_1<j_2} \iint\limits_{x\in S_{j_1}^{(i)}, y\in S_{j_2}^{(i)}} \frac{d\mu_m(x)\, d\mu_m(y)}{|x-y|^t} \le A\,,$$

for all sufficiently large i. (It needs some extra efforts to show that it is sufficient for our purposes.) The last sum can be written in the following way:

$$I_m^{(i)} = \sum_{p=1}^{i} \sum_j \sum_{S_{j_1}^{(i)}, S_{j_2}^{(i)} \subset S_j^{(p)}} \iint\limits_{x\in S_{j_1}^{(i)}, y\in S_{j_2}^{(i)}} \frac{d\mu_m(x)\, d\mu_m(y)}{|x-y|^t}\,.$$

Now we remark that different $S_{j_1}^{(i)}$, $S_{j_2}^{(i)}$ lying inside a segment $S_j^{(p)}$ are separated form each other by at least one interval $O_\gamma^{(p-1)}$. Therefore

$$|x-y| \ge \frac{\text{const}}{2^p} \quad \text{for } x \in S_{j_1}^{(i)},\ \ y \in S_{j_2}^{(i)}$$

and

$$I_m^{(i)} \le \text{const} \sum_{p=1}^{i} 2^{pt} \sum_j (\mu_m(S_j^{(p)}))^2\,. \tag{8}$$

We shall need the following three statements which we shall formulate as separate lemmas.

Lemma 5. *For a.e. b and any $\delta > 0$ one can find $k_0(b,\delta)$ such that*

$$N_k \ge 2^{\frac{k}{2}(1+\delta)}$$

for all $k \ge k_0(b,\delta)$.

Lemma 6. *For a.e. b and any $\beta > 0$ one can find such $i_0(b,\beta)$ that for all $i \ge i_0(b,\beta)$ the length of each $S_j^{(i)}$ is not more than $\frac{1}{2})^{i(1-\beta)}$.*

The next lemma gives a possibility to compare $\mu_k(S_j^{(i)})$ with the length $l(S_j^{(i)})$. Denote by $N_p(S_j^{(i)})$ the number of intervals $O_{p,k} \subset S_j^{(i)}$, $p > 1$, and write $N_p(S_j^{(i)}) = 2^{\frac{p}{2}} \cdot l(S_j^{(i)}) 2^{i/2(1+\varepsilon)} \gamma_p(S_j^{(i)})$, where $\varepsilon > 0$ will be chosen later.

Since $O_j^{(p)} \subset S_i^{(p-1)}$ for some i and $\cap_p \cup_{S_k^{(p)}} \subset S_j^{(i)}$ has measure zero we conclude that for each p the length of the union over k of all $S_k^{(p)} \subset S_j^{(i)}$ is equal to the sum

$$\sum_{p_1 > p} \sum_{O_{p_1,l} \subset S_j^{(i)}} l(O_{p_1,l}) \leq \text{const} \cdot l(S_j^{(i)}) \cdot 2^{\frac{i}{2}}(1+\varepsilon) \sum_{p_1 > p} 2^{-\frac{1}{2}p_1} \cdot \gamma_{p_1}(S_j^{(i)})$$

$$= \text{const}\ l(S_j^{(i)}) \cdot 2^{\frac{i}{2}(L+\varepsilon)} \cdot 2^{-\frac{p}{2}} \cdot \sum_{p_1 > p} 2^{-\frac{(p_1-p)}{2}} \gamma_{p_1}(S_j^{(i)}).$$

From Lemma 5 it follows that

$$\mu_p(S_j^{(i)}) \leq \frac{1}{\sum_j l(O_{p+1,j})} \cdot \text{const} \cdot 2^{\frac{i}{2}(1+\varepsilon)} \cdot l(S_j^{(i)}) \cdot 2^{-\frac{p}{2}} \cdot \sum_{p_1 > p} 2^{-\frac{(p_1-p)}{2}} \gamma_{p_1}(S^{(i)})$$

$$\leq \text{const} \cdot 2^{\frac{i}{2}\left(1+\frac{\varepsilon}{2}\right)} l(S_j^{(i)}) \cdot \sum_{p_1 > p} 2^{\frac{(p_1-p)}{2}} \gamma_{p_1}(S_j^{(i)}).$$

Putting this inequality in (8) and using Lemmas 5 and 6 we get

$$I_k^{(i)} \leq \text{const} \sum_{q=1}^{i} 2^{qt} \cdot 2^{\frac{1}{2}q\left(1+\frac{\varepsilon}{2}\right)} \cdot \sum_j \mu_k(S_j^{(q)}) l(S_j^{(q)}) \sum_{p_1 > p} 2^{-\frac{(p_1-p)}{2}} \gamma_{p_1}(S_j^{(q)})$$

$$\leq \text{const} \sum_{p_1 > p} 2^{-\frac{(p_1-p)}{2}} \sum_{q=1}^{t} 2^{tq} \cdot 2^{\frac{q}{2(1+\varepsilon)}} \cdot 2^{-\frac{p}{1-\beta}} \cdot \sum_j \mu_p(S_j^{(q)}) \gamma_{p_1}(S_j^{('q)}).$$

Choose β and ε so that

$$2^t \cdot 2^{\frac{2+\varepsilon}{4}} \cdot 2^{-1+\beta} = \delta < 1.$$

Then

$$I_k^{(i)} \leq \text{const} \sum_{q=1}^{i} \delta^q \sum_{p_1 > p} 2^{-\frac{(p_1-p)}{2}} \sum_j \mu_p(S_j^{(q)}) \cdot \gamma_{p_1}(S_j^{(q)}).$$

Lemma 7. *There exists such a constant B depending on ε such that for any $p, p_1 > p$,*

$$E\left(\sum_j \mu_k(S_j^{(p)}) \gamma_{k_1}(S_j^{(p)}\right) \leq B.$$

From Lemma 7 it follows that

$$E\left(\sum_{p_1>p} 2^{-\frac{(p_1-p)}{2}} \sum_j \mu_p(S_j^{(q)}) \cdot \gamma_{p_1}(S_j^{(q)})\right) \leq \text{const}\, B\,.$$

Using Fatou's lemma we can find infinite subsequences $\{l_j\}$ and $\{\bar{p}_j\}$ that

$$\lim_{s\to\infty} \sum_{q=1}^{l_s} \delta^q \sum_{p_1>\bar{p}_s} 2^{-\frac{(p_1-\bar{p}_s)}{2}} \sum_j \mu_{\bar{p}_s}(S_j^{(q)}) \cdot \gamma_{p_1}(S_j^{(q)})$$

is finite. This gives a desired result.

The proofs of Lemmata 5, 6, 7 are straightforward but lengthy. They will be published elsewhere.

Acknowledgements. I thank very much U. Frisch, Z.-S. She, and E. Aurell for many useful discussions concerning the Burgers equations and the results obtained in both papers. Also the warm hospitality of the Mathematics Department of Princeton University where the final version of this paper was finished is highly acknowledged.

References

1. She, Z.-S., Aurell, E., Frisch, U.: The inviscid Burgers equation with initial data of Brownian type Commun. Math. Phys. **148**, 623 (1992)
2. Hopf, E.: The partial differential equation $u_t + uu_x = \mu u_{xx}$, Commun. Pure Appl. Math. **3**, 201–230 (1950)
3. Burgers, J.M.: The nonlinear diffusion equation. p. 174 Dordrecht: D. Reidel 1974
4. Stroock, D.W., Varadhan, S.R.S.: Multi-dimensional diffusion processes p. 338, Berlin, Heidelberg, New York: Springer 1979
5. Sinai, Ya. G.: Distribution of some functionals of the integral of the Brownian motion. Theor. Math. Phys. (in Russian) (1992 (in press))
6. Falconer, K.: The geometry of fractal sets. Cambridge: Cambridge University Press 1985

Communicated by A. Jaffe

Comments

The problem was formulated by U. Frisch. He also did preliminary numerical studies and anticipated the answer.

Commun. Math. Phys. 224, 83 – 106 (2001)

Communications in
**Mathematical
Physics**

Gibbsian Dynamics and Ergodicity for the Stochastically Forced Navier–Stokes Equation

Weinan E[1], J. C. Mattingly[2], Ya. Sinai[3]

[1] Department of Mathematics and Program in Applied and Computational Mathematics, Princeton University, Princeton, NJ 08544, USA and School of Mathematics, Peking University, Beijing, P.R. China
[2] Department of Mathematics, Stanford University, Stanford, CA 94305, USA
[3] Department of Mathematics, Princeton University, Princeton, NJ 08544, USA and Landau Institute of Theoretical Physics, Moscow, Russia

Received: 21 November 2000 / Accepted: 9 December 2000

Dedicated to Joel L. Lebowitz, on the occasion of his 70th birthday

Abstract: We study stationary measures for the two-dimensional Navier–Stokes equation with periodic boundary condition and random forcing. We prove uniqueness of the stationary measure under the condition that all "determining modes" are forced. The main idea behind the proof is to study the Gibbsian dynamics of the low modes obtained by representing the high modes as functionals of the time-history of the low modes.

1. Introduction and Main Results

We are interested in determining conditions sufficient to insure that the stochastically-forced Navier–Stokes equation (SNS) possesses a unique stationary measure, or equivalently, that the dynamics is ergodic in the phase space. Our main result is that this holds if all the "determining modes" are forced. To prove this, we show that the dynamics of the Navier–Stokes equation can be reduced to the dynamics of the low modes, the so-called determining modes, with memory. This is the stochastic analog of results proved for the deterministic case by Foias et al. [FMRT]. We will work with the periodic boundary condition. But in principle our techniques should also apply for the more physical no-slip boundary condition.

Consider the two-dimensional Navier–Stokes equation with stochastic forcing:

$$\begin{cases} \dfrac{\partial u}{\partial t} + (u \cdot \nabla)u + \nabla p - \nu\Delta u = \dfrac{\partial W(x,t)}{\partial t} \\ \nabla \cdot u = 0 \end{cases} . \tag{1}$$

For simplicity of presentation we will take W to be of the form

$$W(x,t) = \sum_{|k|\leq N} \sigma_k w_k(t,\omega) e_k(x) m \tag{2}$$

where the w_k's are standard i.i.d complex valued Wiener process satisfying $w_{-k}(t) = \overline{w}_k(t)$, and $\sigma_k \in \mathbb{C}$, with $|\sigma_k| > 0$ and $\sigma_{-k} = \overline{\sigma}_k$, are the amplitudes of

Y.G. Sinai, *Selecta II: Probability Theory, Statistical Mechanics, Mathematical Physics and Mathematical Fluid Dynamics*, DOI 10.1007/978-1-4419-6205-8_20, © Springer Science+Business Media, LLC 2010

the forcing, $\{e_k(x) = \binom{-ik_2}{ik_1}\frac{e^{ik\cdot x}}{|k|},\ k \in \mathbb{Z}\}$ are the basis in the space of L^2 divergence-free, mean zero vector fields on $\mathbb{T}^2$, the two dimensional torus. Our techniques apply to more general cases when the higher modes are also forced, as long as $|\sigma_k|$ decays sufficiently fast as $|k| \to \infty$ or to forcing which is not diagonal in Fourier space. But we will restrict ourselves to the form in (2) for clarity.

Define $B(u, v) = -P_{div}(u \cdot \nabla)v$, $\quad \Lambda^2 u = -P_{div}\Delta u$, where P_{div} is the L^2 projection operator onto the space of divergence-free vector fields. Let $\sigma_{\max}^2 = \max\{|\sigma_k|^2 : |k| \le N\}$. $\mathcal{E}_0 = \sum_{|k|\le N} |\sigma_k|^2$ and $\mathcal{E}_1 = \sum_{|k|\le N} |k|^2|\sigma_k|^2$. Writing $u(x) = \sum_k u_k e_k(x)$, we will define $\mathbb{H}^\alpha = \left\{u = (u_k)_{k\in\mathbb{Z}^2},\ u_0 = 0,\ \sum_k |k|^{2\alpha}|u_k|^2 < \infty\right\}$ and $\mathbb{L}^2 = \mathbb{H}^0$.

We will work on a probability space $(\Omega, \mathcal{F}, \mathcal{F}_t, \mathbb{P}, \theta_t)$. We associate Ω with the canonical space generated by all $d\omega_k(t)$. $\mathcal{F}$ and $\mathcal{F}_t$ are respectively the associated global σ-algebra and filtration generated by $W(t)$. Lastly, θ_t is the shift on Ω defined by $\theta_t d\omega_k(s) = d\omega_k(s + t)$. Notice that θ_t is an ergodic group of measure-preserving transformations with respect to $\mathbb{P}$. Expectations with respect to $\mathbb{P}$ will be denoted by $\mathbb{E}$.

Projecting (1) onto $\mathbb{L}^2$, we obtain the the following system of Itô stochastic equation

$$du(x, t) + \nu\Lambda^2 u(x, t)dt = B(u, u)dt + dW(x, t). \tag{3}$$

It can be shown that (3) generates a continuous Markovian stochastic semi-flow on $\mathbb{L}^2$ defined by

$$\varphi_{s,t}^{\omega} u_0 = u(t, \omega; s, u_0). \tag{4}$$

When $s = 0$, we simply write φ_t^ω (see [Fla94, DPZ96]).

We will take the state space of (3) to be $\mathbb{L}^2$ equipped with the Borel σ-algebra. A measure $\mu(du)$ on $\mathbb{L}^2$ is stationary for the stochastic flow (3) if for all bounded continuous functions F on $\mathbb{L}^2$ and $t > 0$,

$$\int_{\mathbb{L}^2} F(u)\mu(du) = \int_{\mathbb{L}^2} \mathbb{E}F\left(\varphi_t^\omega u\right)\mu(du). \tag{5}$$

Our main result is:

Theorem 1. *There exists some absolute constant C such that if $N^2 \ge C\frac{\mathcal{E}_0}{\nu^3}$ then* (3) *has a unique stationary measure on $\mathbb{L}^2$.*

The existence of at least one stationary measure was proved in [Fla94] and [VF88]. The proof proceeds by establishing compactness for a family of empirical measures. The limiting points of these empirical measures are the stationary measures. Uniqueness has been proved under restrictive assumptions when ALL modes are forced. Flandoli and Maslowski [FM95] proved that if the σ_k's decay algebraically, i.e. if the forcing is sufficiently rough spatially, then the system has a unique stationary measure. These results were extended and refined in [Fer97]. In [Mat99], it was proven that if the viscosity was large enough the contraction induced by the Laplacian dominates and the system possesses a trivial random attractor; and hence, a unique stationary measure. We do not address convergence to the stationary measure. This and the coupling construction used to prove convergence are discussed in [Mat00]. Recently Kuksin and Shirikyan [KS] proved uniqueness of stationary measure when the Navier–Stokes equation is perturbed by a bounded degenerate kicked noise. Results similar to ours have also been obtained independently by Bricmont et al. [BKL].

Our main strategy is to reduce the dynamics of the Navier–Stokes equation to the dynamics of a finite dimensional set of low modes with memory. The reduced dynamics is no longer Markovian, but rather Gibbsian (see §2, §4). The finite dimensional Gibbsian dynamics has a non-degenerate noise, and have a unique stationary measure if the memory is short ranged.

Before proceeding further, let us observe that any given stationary measure μ can be extended to a measure on the path space, denoted by μ_p, where p stands for path or past. Consider the example of the path space $C\left((-\infty, 0], \mathbb{L}^2\right)$. Let A be a cylinder set of the type: For some $t_0, t_1, \cdots t_n, t_0 < t_1 < t_2 \cdots t_n \le 0$,

$$A = \left\{u(s) \in C\left((-\infty, 0], \mathbb{L}^2\right), u(t_i) \in A_i, i = 0, \cdots n\right\}, \tag{6}$$

where the A_i's are Borel sets of $\mathbb{L}^2$. Corresponding to A, let $B \subset \Omega \times \mathbb{L}^2$,

$$B = \{(u, \omega), u \in A_0, \varphi^{\omega}_{t_0, t_i} u \in A_i, i = 1, \cdots n\}. \tag{7}$$

We will define

$$\mu_p(A) = (\mathbb{P} \times \mu)(B), \tag{8}$$

where $(\mathbb{P} \times \mu)$ is the product measure on $\Omega \times \mathbb{L}^2$. Clearly μ_p is consistent on cylinder sets and can be extended to the natural σ-algebra using the Kolmogorov extension theorem. The natural σ-algebra is the one generated by the cylinder sets. The dynamics of the stochastic semi-flow $\{\varphi^{\omega}_t\}$ can be trivially extended to return a function from $C\left((-\infty, t], \mathbb{L}^2\right)$, given an initial function from $C\left((-\infty, 0], \mathbb{L}^2\right)$. One simply flows forward with φ from the initial condition at time 0. To avoid confusion, we will call this map ψ^{ω}_t. Symbolically, if $u(\cdot) \in C\left((-\infty, 0], \mathbb{L}^2\right)$, then $(\psi^{\omega}_t u)(s) = \varphi^{\omega}_s u(0)$ for $s \in [0, t]$ and $(\psi^{\omega}_t u)(s) = u(s)$ for $s \le 0$.

If we define the shift on trajectories by $(\theta_t v)(s) = v(s + t)$, we can define a dynamics on $C\left((-\infty, 0], \mathbb{L}^2\right)$ by $\theta_t \psi^{\omega}_t$. In other words, $\theta_t \psi^{\omega}_t u$ takes a trajectory u from $C\left((-\infty, 0], \mathbb{L}^2\right)$, extends it t units of time by flowing forward and then shifts the entire resulting trajectory back t units of time so it again lives on $C\left((-\infty, 0], \mathbb{L}^2\right)$.

It is easy to check directly that if μ is invariant then μ_p is invariant in the sense that

$$\int_{C((-\infty,0],\mathbb{L}^2)} F(u) d\mu_p(u) = \mathbb{E} \int_{C((-\infty,0],\mathbb{L}^2)} F(\theta_t \psi^{\omega}_t u) d\mu_p(u) \tag{9}$$

for all bounded functions on $C\left((-\infty, 0], \mathbb{L}^2\right)$, and $t \ge 0$.

Assume that μ and ν are two stationary measures for the stochastic flow (3), and μ_p and ν_p are respectively their induced measure on the path space $C\left((-\infty, 0], \mathbb{L}^2\right)$. It is obvious that $\mu_p = \nu_p$ implies $\mu = \nu$.

2. Reduction to the Gibbsian Dynamics

Define two subspaces of $\mathbb{L}^2$:

$$\mathbb{L}^2_\ell = \text{span}\{e_k, |k| \le N\}, \quad \mathbb{L}^2_h = \text{span}\{e_k, |k| > N\}. \tag{10}$$

We will call $\mathbb{L}^2_\ell$ the set of low modes and $\mathbb{L}^2_h$ the set of high modes. Obviously $\mathbb{L}^2 = \mathbb{L}^2_\ell \oplus \mathbb{L}^2_h$. Denote by P_ℓ and P_h the projections onto the low and high mode spaces.

Since we are concerned with stationary measures of (3), we are interested in (statistically) stationary solutions of (3) that exist for time from $-\infty$ to $+\infty$. We will show in this section that for such solutions, the high modes are completely determined by the past history of the low modes. For this purpose, we write $u(t) = \big(\ell(t), h(t)\big)$ and

$$\begin{aligned} d\ell(t) = &\Big[-\nu\Lambda^2\ell + P_\ell B(\ell,\ell)\Big]dt \\ &+\Big[P_\ell B(\ell,h) + P_\ell B(h,\ell) + P_\ell B(h,h)\Big]dt + dW(t), \end{aligned} \tag{11}$$

$$\frac{dh(t)}{dt} = \Big[-\nu\Lambda^2 h + P_h B(h,h)\Big] + P_h B(\ell,h) + P_h B(h,\ell) + P_h B(\ell,\ell). \tag{12}$$

Define the set of "nice pasts" $U \subset C\left((-\infty,0],\mathbb{L}^2\right)$ to consist of all $v:(-\infty,0]\to\mathbb{L}^2$ such that:

i) $v(t)$ is in $\mathbb{H}^2$ for all $t\le 0$.
ii) The energy averages correctly. More precisely,

$$\lim_{t\to-\infty}\frac{1}{|t|}\int_t^0 |\Lambda v(s)|^2_{\mathbb{L}^2}ds = \frac{\mathcal{E}_0}{2\nu}.$$

iii) The energy fluctuations are typical. More precisely, there exists a $T = T(v)$ such that

$$|v(t)|^2_{\mathbb{L}^2} \le \mathcal{E}_0 + \max(|t|,T)^{\frac{2}{3}}$$

for $t\le 0$. The following lemma shows that U contains almost all of the trajectories defined on the whole time interval.

Lemma 2.1. *Let μ_p be the measure on $\mathbb{C}\left((-\infty,0],\mathbb{L}^2\right)$ induced by a stationary measure μ for* (3). *Then $\mu_p\big(U\big)=1$.*

Proof of Lemma 2.1. It is proved in [Mat98] or [Fer97] that with probability one, a solution to (3) is in $\mathbb{H}^2$ for all t.

The fact that the last condition is satisfied by a set of full measure is proved in Lemma B.3. All that remains to show is ii).

From Lemma B.2 $|\Lambda v|^2_{\mathbb{L}^2}$ is in $L^1(\mu)$ for any stationary measure μ and $\int|\Lambda v|^2_{\mathbb{L}^2}\,d\mu = \frac{\mathcal{E}_0}{2\nu}$. Since the measure is invariant under shifts back in time and each ergodic component has the same average enstrophy, the ergodic theorem implies that for μ_p–almost every trajectory time average converges to the average of $|\Lambda u|^2_{\mathbb{L}^2}$ against μ. □

Given an arbitrary continuous function of time $\ell(t)$ on $\mathbb{L}^2_\ell$, we can view (12) as a closed equation with some exogenous forcing $\ell(t)$. By $\Phi_{s,t}(\ell,h_0)$, we mean the solution to (12) at time t given the initial condition h_0 at time s and the "forcing" ℓ.

Denote by $\mathcal{P}$ the set of all $\ell\in C\left((-\infty,0],\mathbb{L}^2_\ell\right)$ such that the following two conditions hold. First, $\ell = P_\ell u$ for some $u=(\ell,h)\in U$. Second, $h(t)=\Phi_{s,t}(\ell,h(s))$ for any $s<t\le 0$, where h was the matching high mode so $(\ell,h)\in U$. That is to say $h(t)$ solves (12) with low modes $\ell(t)$ and the total solution (ℓ,h) is in our space of "nice pasts". In light of Lemma 2.1 the set $\mathcal{P}$ is not empty. We now will show that this h is uniquely determined by ℓ.

Lemma 2.2. *There exists an absolute positive constant C such that if $N^2 > C\frac{\mathcal{E}_0}{\nu^3}$ then the following holds:*

If there exists two solutions $u_1(t) = \big(\ell(t), h_1(t)\big)$, $u_2(t) = \big(\ell(t), h_2(t)\big)$ corresponding to some (possibly different) realizations of the forcing and such that $u_1, u_2 \in U$, then $u_1 = u_2$, i.e. $h_1 = h_2$.

Furthermore given a solution $u(t) = \big(\ell(t), h(t)\big) \in U$, any $h_0 \in \mathbb{L}^2_h$, and $t \leq 0$, the following limit exists:

$$\lim_{t_0 \to -\infty} \Phi_{t_0,t}(\ell, h_0) = h^*$$

and $h^* = h(t)$.

Proof of Lemma 2.2. We begin with the first clam. Denote by $\rho(t) = h_1(t) - h_2(t)$. From (12) we have

$$\begin{aligned} \frac{d\rho}{dt} &= -\nu\Lambda^2\rho + P_h B(h_1, h_1) - P_h B(h_2, h_2) + P_h B(\ell, \rho) + P_h B(\rho, \ell) \\ &= -\nu\Lambda^2\rho + P_h B(\ell + h_1, \rho) + P_h B(\rho, \ell + h_2) \\ &= -\nu\Lambda^2\rho + P_h B(u_1, \rho) + P_h B(\rho, u_2). \end{aligned} \tag{13}$$

Taking the inner product with ρ, using the fact that $\langle P_h B(u_1, \rho), \rho\rangle_{\mathbb{L}^2} = 0$, gives

$$\frac{1}{2}\frac{d}{dt}|\rho|^2_{L^2} = -\nu|\Lambda\rho|^2_{L^2} + \langle P_h B(\rho, u_2), \rho\rangle_{\mathbb{L}^2}.$$

Since

$$\begin{aligned} |\langle P_h B(\rho, u_2), \rho\rangle_{\mathbb{L}^2}| &\leq \hat{C}\, |\Lambda\rho|_{\mathbb{L}^2}\, |\rho|_{\mathbb{L}^2}\, |\Lambda u_2|_{\mathbb{L}^2} \\ &\leq \frac{\nu}{2}\, |\Lambda\rho|^2_{\mathbb{L}^2} + \frac{\hat{C}^2}{2\nu}\, |\rho|^2_{\mathbb{L}^2}\, |\Lambda u_2|^2_{\mathbb{L}^2}, \end{aligned}$$

we get

$$\frac{1}{2}\frac{d}{dt}\, |\rho|^2_{\mathbb{L}^2} \leq -\frac{\nu}{2}\, |\Lambda\rho|^2_{\mathbb{L}^2} + \frac{\hat{C}^2}{2\nu}\, |\Lambda u_2|^2_{\mathbb{L}^2}\, |\rho|^2_{\mathbb{L}^2}.$$

Since ρ only contains modes with $|k| > N$, the Poincaré inequality implies

$$\frac{d}{dt}\, |\rho|^2_{\mathbb{L}^2} \leq \left(-\nu N^2 + \frac{\hat{C}^2}{\nu}\, |\Lambda u_2|^2_{\mathbb{L}^2}\right) |\rho|^2_{\mathbb{L}^2}.$$

Therefore we have, for $t_0 < t < 0$,

$$|\rho(t)|^2_{\mathbb{L}^2} \leq |\rho(t_0)|^2_{\mathbb{L}^2} \exp\left\{-\nu N^2(t - t_0) + \frac{\hat{C}^2}{\nu}\int_{t_0}^{t} |\Lambda u_2(s)|^2_{\mathbb{L}^2}\, ds\right\}. \tag{14}$$

From the third assumption on functions in U, we know that $\lim \frac{1}{t}\int_{-t}^{0} |\Lambda u_2(s)|^2_{\mathbb{L}^2}\, ds = \frac{\mathcal{E}_0}{2\nu}$. Hence for $t_0 < T_1$, where T_1 depends on t and u_2, we have

$$-\nu N^2(t - t_0) + \frac{\hat{C}^2}{\nu}\int_{t_0}^{t} |\Lambda u_2(s)|^2_{\mathbb{L}^2}\, ds \leq -\frac{\gamma}{2}(t - t_0),$$

where $\gamma = \nu N^2 - \frac{\hat{C}^2\mathcal{E}_0}{2\nu^2}$. If we set $C = \frac{\hat{C}^2}{2}$, then our assumption on N implies $\gamma > 0$.
Now using the last property of paths in U we have for any $t_0 \le T_2$,

$$\begin{aligned}|\rho(t)|^2_{\mathbb{L}^2} &\le |\rho(t_0)|^2_{\mathbb{L}^2} \exp\{-\frac{\gamma}{2}(t-t_0)\}\\ &\le 2\big[\mathcal{E}_0 + |t_0|^{\frac{2}{3}}\big]\exp\{-\frac{\gamma}{2}(t-t_0)\} \to 0\end{aligned}$$

as $t_0 \to -\infty$, where T_2 is some finite constant depending on u_1 and u_2. This completes the proof of the first part of Lemma 2.2.

To see the second part, observe that (14) only required control of $\int_{t_0}^t |\Lambda u(s)|^2_{\mathbb{L}^2}\, ds$ for one of the two solutions. If we proceed as before letting the given solution $u(t)$ play the role of u_2 and the solution to (12) starting from h_0 play the role of u_1, the we obtain the estimate

$$|\rho(t)|^2_{\mathbb{L}^2} \le |h(t_0) - h_0|^2_{\mathbb{L}^2} \exp\left\{-\nu N^2(t-t_0) + \frac{\hat{C}^2}{\nu}\int_{t_0}^t |\Lambda u(s)|^2_{\mathbb{L}^2}\, ds\right\}. \tag{15}$$

Since $u(t) = (\ell(t), h(t)) \in U$, the same reasoning as before shows that $\rho(t)$ goes to zero as $t_0 \to -\infty$. Hence the limit exists and equals $h(t)$. □

In fact the splitting into high and low modes can be accomplished even when all of the modes are forced. One replaces (12) with an Itô stochastic differential equation. This causes little complication as (13) remains a standard PDE. See [Mat98].The ideas in this section are related to the ideas of Lyapunov-Schmidt reduction and those around center and inertial manifolds. See [EFNT94] for a discussion and other references.

From now on we assume that N satisfies

$$N^2 > C\frac{\mathcal{E}_0}{\nu^3}, \tag{16}$$

where C is the constant from Lemma 2.2.

Because of Lemma (2.2), we can define a map Φ_0 which reconstructs the high modes at time zero from a given low mode trajectory stretching from zero back to $-\infty$. Before making this more precise, let us fix some notation. In general, we will use $\ell(t)$ to refer to the value of the low modes at time t and will use L^t to mean the entire trajectory from $-\infty$ to t. Hence $\ell(t) \in \mathbb{L}^2$ and $L^t \in C\big((-\infty, t], \mathbb{L}^2\big)$ and $\ell(s) = L^t(s)$ for $s \le t$. In this notation $h(0) = \Phi_0\big(L^0\big)$, where L^0 is some "low mode past" in $\mathcal{P}$ which is the projection of U to the low modes. By $\Phi_s(L^t, h(0))$ with $s \le t$, we mean the solution to (12) at time s with initial condition $h(0)$ and low mode forcing L^t. Of course $\Phi_s(L^t, h(0))$ only depends on the information in L^t between 0 and s. We can extend the definition of Φ beyond time zero by defining $\Phi_t(L^t) = \Phi_t(L^t, h(0))$, where $h(0) = \Phi_0(L^0)$.

Given the initial low mode past of $L^0 \in \mathcal{P}$, we can solve for the future of ℓ using

$$d\ell(t) = \Big[-\nu\Lambda^2\ell(t) + P_\ell B\big(\ell(t), \ell(t)\big) + G\big(\ell(t), \Phi_t(L^t)\big)\Big]\,dt + dW(t), \tag{17}$$

where

$$G(\ell, h) = P_\ell B(\ell, h) + P_\ell B(h, \ell) + P_\ell B(h, h). \tag{18}$$

Thus we have a closed formulation of the dynamics on the low modes given an initial past in $L^0 \in \mathcal{P}$. We write $L^t = \mathsf{S}_t^\omega L^0$. We reiterate that L^t is the entire trajectory from time t back to $-\infty$, whereas $\ell(t)$ is simply the value of the low modes at time t.

Except for the fact the G-term in (17) is history-dependent, (17) has the form of a standard finite dimensional stochastic ODE with non-degenerate forcing, which of course has a unique stationary measure. Our task is reduced to showing that the memory effort in (17) is not strong enough to spoil ergodicity.

Existence of the solution for memory-dependent stochastic ODEs of the type (17) was considered in the work of Ito et al. [IN].

3. Uniqueness of the Invariant Measure

3.1. Proof of the Main Theorem. Given any "nice low mode past" $L \in \mathcal{P}$, we can reconstruct the "high modes" and hence define a closed dynamics on the paths of the low modes. However, this dynamics is no longer Markovian which will produce difficulties.

Let μ be an ergodic stationary measure on $\mathbb{L}^2$ and μ_p be its extension to the path space $C\big((-\infty, 0], \mathbb{L}^2\big)$. We will also consider the restriction of μ_p to $C\big((-\infty, 0], \mathbb{L}_\ell^2\big)$, still denoted by μ_p. Lemma 2.1 says that $\mu_p(\mathcal{P}) = 1$.

Given any $L^0 \in \mathcal{P}$, let $Q_t(L^0, \cdot)$ be the measure induced on $C\big([0, t], \mathbb{L}_\ell^2\big)$ by the dynamics starting from L^0. In other words, $Q_t(L^0, \cdot)$ is the distribution of $\mathsf{S}_t^\omega L^0$ viewed as a random variable taking values in $C\big([0, t], \mathbb{L}_\ell^2\big)$. Similarly let $Q_\infty(L^0, \cdot)$ be the distribution induced on $C\big([0, \infty), \mathbb{L}_\ell^2\big)$ starting from L^0.

Consider the stochastic process defined by $\theta_t \mathsf{S}_t^\omega L^0$, where L^0 is a random variable on $\mathcal{P}$ distributed according to the invariant measure μ_p. For $t \geq 0$ it is a random process with values in $\mathcal{P}$. This is clear as all of the defining properties of U are asymptotic in t; and hence the addition of a segment of finite length does not destroy them. Since μ_p is invariant with respect to the dynamics, $\theta_t \mathsf{S}_t^\omega L^0$ is a stationary random process. Hence with probability one there exist time averages along trajectories $\theta_t \mathsf{S}_t^\omega L^0$.

Take any bounded measurable functional F from $C\big((-\infty, 0], \mathbb{L}_\ell^2\big) \to \mathbb{R}$ such that $F(L^0)$, $L^0 \in C\big((-\infty, 0], \mathbb{L}_\ell^2\big)$ depends only on a finite range of L^0. Let

$$\bar{F} = \int F(L) d\mu_p(L). \tag{19}$$

Theorem 2. *The SNS equation* (1) *has a unique stationary measure.*

The proof of Theorem 2 is based on the following two lemmas whose proofs will be given later.

Lemma 3.1. *Let L_1^0 and L_2^0 be two initial pasts in $\mathcal{P}$, such that $\ell_1(0) = \ell_2(0)$. Then $Q_\infty(L_1^0, \cdot)$ and $Q_\infty(L_2^0, \cdot)$ are equivalent.*

Recall that $\ell(\tau)$ is the solution of (16) with initial condition L.

Lemma 3.2. *For any past $L \in \mathcal{P}$ and any $t > 0$, the distribution of $\ell(t) \in \mathbb{L}_\ell^2$ conditioned at starting from L at time zero, denoted by $R_t(L, \cdot)$, satisfies the following: there exists a strictly positive function $f_{L,t} \in L^1(\mathbb{L}_\ell^2)$, such that*

$$dR_t(L, \cdot) \geq f_{L,t}(\cdot) dm(\cdot).$$

where $m(\cdot)$ is the Lebesgue measure on $\mathbb{L}_\ell^2$.

For any measure μ on $\mathbb{L}^2$ let $P_\ell \mu$ denote its projection to a measure on the low modes $\mathbb{L}^2_\ell$. Namely, $(P_\ell \mu)(B) = \mu(P_\ell^{-1}(B))$. Then we have the following direct consequence of Lemma 3.2.

Corollary 3.3. *If μ is a stationary measure then $P_\ell \mu$ has a component which is equivalent to the Lebesgue measure.*

Proof of Theorem 2. Assume that there are two different ergodic stationary measures on $\mathbb{L}^2$ called μ_1 and μ_2. They must be mutually singular. Let $\mu_{1,p}$ and $\mu_{p,2}$ be the extensions of these two measures onto the path space $\mathcal{P}$. Let L_i^0 be a random variable on $\mathcal{P}$ distributed as $\mu_{i,p}$. Since $\theta_t \mathsf{S}_t^\omega L_i^0$ is stationary with respect to $\mu_{p,i}$ we can pick a set $\mathcal{P}_i$, of full $\mu_{p,i}$-measure, such that for all $L \in \mathcal{P}_i$ One can find a functional F such as above so that $\bar{F}_1 = \int F(L) d\mu_{p,1}(L) \neq \bar{F}_2 = \int F(L) d\mu_{p,2}(L)$. This assumption will lead to a contradiction. The limit

$$\lim_{T\to\infty} \frac{1}{T} \int_0^T F(\theta_t \mathsf{S}_t^\omega L_i^o) dt = \bar{F}_i \tag{20}$$

is well defined for $\mathbb{P}$-almost every ω.

For $\ell \in \mathbb{L}^2_\ell$ define $\mathcal{P}_i(\ell) = \{L \in \mathcal{P}_i \;:\; L(0) = \ell\}$ and let $\mu_{p,i}(\,\cdot\,|\ell)$ be the conditional measure that $L(0) = \ell$. By Fubini's theorem, we know that for $P_\ell \mu_i$-almost every $\ell \in \mathbb{L}^2_\ell$ we have $\mu_{p,i}(\mathcal{P}_i(\ell) \,|\, \ell) = 1$. Hence we can find a set $A_i \subset \mathbb{L}^2_\ell$ such that $\mu_{p,i}(\mathcal{P}_i(\ell) \,|\, \ell) = 1$ for all $\ell \in A_i$ and $P_\ell \mu_i(A_i) = 1$. Define $A = A_1 \cap A_2$. Corollary 3.3 implies that $P_\ell \mu_i(A) > 0$ for $i = 1, 2$. Hence there exists some $\ell^* \in A$.

Since $\ell^* \in A_1 \cap A_2$, we know that $\mu_{p,i}(\mathcal{P}_i(\ell^*) \,|\, \ell^*) = 1$ for $i = 1, 2$. Thus there exist some $L_{*,1} \in \mathcal{P}_1(\ell^*)$ and $L_{*,2} \in \mathcal{P}_2(\ell^*)$. Notice that by construction $L_{*,1}(0) = \ell^* = L_{*,2}(0)$, and hence it follows from Lemma 3.1 that $Q_\infty(L_{*,1}, \cdot)$ and $Q_\infty(L_{*,2}, \cdot)$ are equivalent. Since $L_{*,i} \in \mathcal{P}_i(\ell^*)$, we know that we can pick $B_i \subset C\big([0,\infty), \mathbb{L}^2\big)$ such that the time average of F converges to $\bar{F}_i$ for all futures in B_i and $Q_\infty(L_{*,i}, B_i) = 1$ for $i = 1, 2$. Since the Q's are equivalent, $Q_\infty(L_{*,1}, B_1 \cap B_2) > 0$ and hence $B_1 \cap B_2$ is non-empty. This in turn implies that $\bar{F}_1 = \bar{F}_2$ which contradicts the assumption that they were not equal. □

3.2. Proofs of the lemmas. We first prove Lemma 3.1. Fix L_1^0 and L_2^0. Most of our construction will depend explicitly on them. With probability one, we can extend each of the initial pasts into the infinite future by $L_i^s = \mathsf{S}_s^\omega L_i^0$ and setting $\ell_i(s) = L_i^t(s)$ for $s \leq t$. We can also reconstruct the entire solution by using Φ_t to obtain the high modes. Set $h_i(s) = \Phi_s(L_i^s)$ and $u_i(s) = \big(\ell_i(s), h_i(s)\big)$. Fix a constant C_0 such that $|u_i(0)|^2_{\mathbb{L}^2} \leq C_0$.

We begin by constructing a set of nice future paths which will contain most trajectories. For any positive K we define

$$A_i(K) = \Big\{ f \in C\big([0,\infty), \mathbb{L}^2_\ell\big) : |v(t)|^2_{\mathbb{L}^2} + 2\nu \int_0^t |\Lambda v(s)|^2_{\mathbb{L}^2}\, ds < C_0 + \mathcal{E}_0 t + K t^{\frac{4}{5}} \quad \text{where } v(s) = f(s) + \Phi_s(f, h_i) \Big\}$$

and $A(K) = A_1(K) \cap A_2(K)$.

By Lemma A.5, we know that for any $a \in (0, 1)$ there exists a K such that

$$\mathbb{P}\Big\{\omega : \mathsf{S}_t^{\omega} L_i^0 \in A_i(K)\Big\} > 1 - \frac{a}{2} \qquad \text{for } i = 1, 2,$$

and hence

$$\mathbb{P}\Big\{\omega : \mathsf{S}_t^{\omega} L_i^0 \in A(K) \qquad \text{for } i = 1, 2\Big\} > 1 - a > 0.$$

This is just another way of saying $Q_\infty(L_i^0, A(K)) > 1 - a$.

Lemma 3.4. *Let L_1^0 and L_2^0 be two initial pasts in $\mathcal{P}$ such that $L_1^0(0) = L_2^0(0)$. Let $A(K) \subset C\big([0, \infty), \mathbb{L}_\ell^2\big)$ be as defined above. For any choice of $K > 0$, $Q_\infty(L_1^0, \cdot \cap A(K))$ is equivalent to $Q_\infty(L_2^0, \cdot \cap A(K))$.*

Proof of Lemma 3.1. Since we can choose K so that $A(K)$ has measure arbitrarily close to 1, we have that $Q_\infty(L_1^0, \cdot)$ is equivalent to $Q_\infty(L_2^0, \cdot)$. □

Proof of Lemma 3.4. We intend to use Girsanov's theorem to compare the two induced measures, $Q_\infty(L_1^0, \cdot)$ and $Q_\infty(L_2^0, \cdot)$. However we do not do so directly. To aid in our analysis, we consider the following surrogate processes y which will agree with ℓ on the set $A = A(K)$. As before, we will use $y(t)$ to denote the value of the process at time t and Y^t to be the entire trajectory up to time t.

$$\begin{aligned} dy_i(t) &= \Big[-\nu\Lambda^2 y_i(t) + P_\ell B\big(y_i(t), y_i(t)\big) \\ &\qquad + \Theta_t(Y_i^t) G\big(y_i(t), \Phi_t(Y_i^t, h_i(0))\big)\Big] dt + dW(t) \\ y_i(0) &= \ell_i(0), \end{aligned} \tag{21}$$

where

$$h_i(0) = \Phi_t(L_i^0),$$

$$\Theta_t(f) = \begin{cases} 1 & \text{if } f \in A|_{[0,t]} \\ 0 & \text{if } f \notin A|_{[0,t]} \end{cases},$$

and $A|_{[0,T]}$ is the low mode paths which agree with a path in A up to time T. Recall that $\Phi_t\big(Y_i^t, h_i(0)\big)$ is the solution to (12) with $\ell = Y$ and $h(0) = h_i(0)$.

Equation (21) is the same as (17) except for the insertion of $\Theta_t(Y_i^t)$. As long as $\Theta_s(Y_i^t) = 1$ for $s \in [0, t]$, then $y_i(s) = \ell_i(s)$ for $s \in [0, t]$.

Let $Q_\infty^y(L_1^0, \cdot)$ and $Q_\infty^y(L_2^0, \cdot)$ be the measures induced by Y_1 and Y_2 respectively. If applicable, Girsanov's theorem would imply that these measure are equivalent, that is $Q_\infty^y(L_1^0, \cdot) \sim Q_\infty^y(L_2^0, \cdot)$. For Girsanov's theorem to apply, it is sufficient that the Novikov condition holds. Namely,

$$\mathbb{E}\exp\left\{\frac{1}{2}\int_0^\infty \left|\Sigma^{-1}\Theta_t(Y_1^t) D\Big(y_1(t), \Phi_t\big(Y_1^t, h_1(0)\big), \Phi_t\big(Y_1^t, h_2(0)\big)\Big)\right|^2 dt\right\} < \infty, \tag{22}$$

where $D(g, f_1, f_2) \stackrel{\text{def}}{=} G(g, f_1) - G(g, f_2)$ and Σ is a diagonal matrix with the σ_k's on its diagonal. Here we have written the condition in terms of the y_1 process. One can also write the condition in terms of the y_2 process; the finiteness of one implies the finiteness of the other.

We will in fact show something much stronger than (22). Since $|\Sigma^{-1}| < \infty$, it would be enough to show that

$$\sup_{\omega} \int_0^{\infty} \left|\Theta_t(Y_1^t) D\left(y_1(t), \Phi_t\left(Y_1^t, h_1(0)\right), \Phi_t\left(Y_1^t, h_2(0)\right)\right)\right|^2 dt < \infty. \tag{23}$$

Putting $h_i(s) = \Phi_s(Y_1^s, h_i(0))$, $u_i(s) = \ell_i(s) + h_i(s)$, $\rho(s) = h_1(s) - h_2(s)$ and using Lemma A.4, we have

$$\left|D\left(\ell_1(s), h_1(s), h_2(s)\right)\right|_{\mathbb{L}^2}^2 \le C' \, |\rho(s)|_{\mathbb{L}^2}^2 \left[|u_1(s)|_{\mathbb{L}^2}^2 + |u_2(s)|_{\mathbb{L}^2}^2\right]. \tag{24}$$

Notice that if $\ell_i \in A|_{[0,T]}$ then for all $t \in [0, T]$,

$$\begin{aligned} |u_i(t)|_{\mathbb{L}^2}^2 &< C_0 + \mathcal{E}_0 t + K t^{\frac{4}{5}}, \\ \int_0^t |\Lambda u_i(s)|_{\mathbb{L}^2}^2 \, ds &< \frac{1}{2\nu}\left(C_0 + \mathcal{E}_0 t + K t^{\frac{4}{5}}\right), \\ |\rho(0)|_{\mathbb{L}^2}^2 = |u_1(0) - u_2(0)|_{\mathbb{L}^2}^2 &\le 2\left(|u_1(0)|_{\mathbb{L}^2}^2 + |u_2(0)|_{\mathbb{L}^2}^2\right) \le 4C_0. \end{aligned}$$

In addition, we can apply the same analysis as in Sect. 2. Starting from (14) and using the above estimates produces

$$\begin{aligned} |\rho(t)|_{\mathbb{L}^2}^2 &\le |\rho(0)|_{\mathbb{L}^2}^2 \exp\left\{-\nu N^2 t + \frac{\hat{C}^2}{\nu} \int_0^t |\Lambda u_2(s)|_{\mathbb{L}^2}^2 \, ds\right\} \\ &\le 4C_0 \exp\left\{-\nu N^2 t + \frac{\hat{C}^2}{2\nu^2}\left(C_0 + \mathcal{E}_0 t + K t^{\frac{4}{5}}\right)\right\}. \end{aligned}$$

Since by assumption $\nu N^2 > C\frac{\mathcal{E}_0}{\nu^2} = \frac{\hat{C}^2 \mathcal{E}_0}{2\nu^2}$, the second term goes to zero sufficiently fast and hence the estimate on the right-hand side of (24) decays exponentially fast. Thus,

$$\begin{aligned} \sup_{\omega} \int_0^{\infty} &\left|\Theta_t(Y_1) D\left(y_1(t), \Phi_t(Y_1^t, h_1(0)), \Phi_t(Y_1^t, h_2(0))\right)\right|^2 dt \\ &\le \sup_{f \in A} \int_0^{\infty} |D(f(r), \Phi_t(f, h_1(0)), \Phi_t(f, h_2(0)))|^2 \, dt \\ &< \text{const}(C_0) < \infty, \end{aligned}$$

which implies, $Q_\infty^y(L_1^0, \cdot) \sim Q_\infty^y(L_2^0, \cdot)$. As long as Y_i stays in A, $y_i = \ell_i$. Hence $Q_\infty^y(L_i^0, \cdot \cap A) = Q_\infty(L_i^0, \cdot \cap A)$ and finally $Q_\infty(L_1^0, \cdot \cap A) \sim Q_\infty(L_2^0, \cdot \cap A)$. □

In fact our proof provided more information than stated in Lemma 3.4. It contains some estimates uniform over a class of initial pasts which will be useful in later investigations of the convergence rate. (See [Mat00].) We state the extra information in the following corollary.

Corollary 3.5. *In the setting of the proof of Lemma 3.4, define* $\mathcal{P}' = \{L \in \mathcal{P} : |L(0) + \Phi_0(L)|_{\mathbb{L}^2} < C_0\}$. *Then there exists a constant, depending on* C_0 *and* K, *so that*

$$\sup_{L_1, L_2 \in \mathcal{P}'} \int \left|1 - \frac{dQ^y_\infty(L_1, g)}{dQ^y_\infty(L_2, g)}\right|^2 dQ^y_\infty(L_2, g) < \text{const}(C_0, K_1) < \infty.$$

We now move to the proof of Lemma 3.2. Fix $L \in \mathcal{P}$. The proof proceeds by comparing the process $\ell(t)$ to the associated Galerkin approximation living on $\mathbb{L}^2_\ell$ which we will denote by $x(t)$. The advantage is that $x(t)$ is a standard non-degenerate diffusion and hence it is Markovian and well understood.

Take $x(t)$ as the solution defined by the following stochastic ODEs:

$$\begin{aligned} dx(t) &= \left[-\nu\Lambda^2 x + P_\ell B(x, x)\right]dt + dW(t),\\ x(0) &= \ell(0). \end{aligned}$$

As in the previous section, we do not compare $x(t)$ directly to $\ell(t)$ but instead to a modified version of $\ell(t)$ which we will denote by $z(t)$. In analogy to before, we will denote the path of this process up to time t by Z^t. Before continuing let us assume without loss of generality that $|\ell(0)|_{\mathbb{L}^2} \leq C_0$ and $t \leq T$ for some positive C_0 and T. This will give our estimates some uniformity over all initial conditions inside this ball and for times $t \leq T$.

The evolution of $z(t)$ is given by

$$\begin{aligned} dz(t) &= \left[-\nu\Lambda^2 z + P_\ell B(z, z) + \Theta_t(Z^t)G\big(z, \Phi_t\left(Z^t, h_0\right)\big)\right]dt + dW,\\ z(0) &= \ell(0)\big(= L(0)\big), \end{aligned}$$

where $h_0 = \Phi_0(L)$ and G is defined in (18). As in the last section, $\Theta_t(Z^t)$ is a cut-off function. For any fixed $b_0 > 1$, we define

$$\Theta_s(Z^s) = \begin{cases} 1 & \text{if } \int_0^s |Z^s(r)|^4_{\mathbb{L}^2}\, dr < (b_0 C_0)^4 T \\ 0 & \text{otherwise} \end{cases}.$$

Here b_0 is a fixed constant to be chosen below.

For any $B \subset \mathbb{L}^2_\ell$, define

$$[B] = \left\{v \in C\big([0, t], \mathbb{L}^2_\ell\big) : v(t) \in B\right\}.$$

Then $R_t(L(0), B) = Q_t(L, [B])$.

Letting $Q^x_t(L, \cdot)$ and $Q^z_t(L, \cdot)$ be the two measures induced on $C\big([0, t], \mathbb{L}^2_\ell\big)$ by the dynamics of x and z respectively. Lemma 3.2 will be a consequence of the following two lemmas.

Lemma 3.6. *Fix any* $b_0 > 1$. *(The constant used in defining the* z *process.) Then the following holds: For any* $L \in \mathcal{P}$ *and* $t \geq 0$, $Q^x_t(L(0), \cdot)$ *is equivalent to* $Q^z_t(L, \cdot)$.

Lemma 3.7. *For any* b_0 *the following holds: For any* $L \in \mathcal{P}$ *and* $t \geq 0$, *there exists a positive function* $g(\cdot)$ *so that* $Q^x_t(L(0), [B] \cap A) \geq \int_B g(y)dm(y)$, *where* $m(\cdot)$ *is the Lebesgue measure.*

We now use these two lemmas to prove Lemma 3.2.

Proof of Lemma 3.2. Observe that by construction as long as the trajectories stay in A, $x(t) = \ell(t)$. Hence using Lemma 3.7, we have

$$R_t(L, B) = Q_t(L, [B]) \geq Q_t(L, [B] \cap A) = Q_t^z(L, [B] \cap A),$$

$$Q_t^x(L(0), [B] \cap A) \geq \int_B g(L(0), y) dm(y),$$

where $g(L(0), y)$ is a positive function in y. Since Lemma 3.6 says that $Q_t^z(\ell, \cdot \cap A)$ is equivalent to $Q_t^x(L(0), \cdot \cap A)$, we know that $R_t(L(0), B)$ is also bounded from below by a positive measure equivalent to the Lebesgue measure. □

We now turn to Lemma 3.6. Our construction gives some measure of uniform control which is useful for estimating the rate the system converges to the stationary measure. (See [Mat00].) We state these more precise estimates in the following corollary.

Corollary 3.8. *Fix a $C_0 > 0$ and define $\mathcal{P}' = \{L \in \mathcal{P} : |L(0) + \Phi_0(L)|_{\mathbb{L}^2} < C_0\}$. Then for any $\alpha \in (0, 1)$ there exists a $b_0 > 0$ (the constant used to define A) so that:*

$$\inf_{t \in [0,T]} \inf_{L \in \mathcal{P}'} \mathbb{P}\{\mathsf{S}_t^\omega L \in A\} > 1 - a,$$

$$\sup_{L \in \mathcal{P}'} \int \left| 1 - \frac{dQ_t^z(L, g)}{dQ_t^x(L, g)} \right|^2 dQ_t^x(L, g) < K(C_0, t)$$

for $t \in [0, T]$, where K is a constant depending on C_0 and t such that for each C_0, $K \to 0$ as $t \to 0$.

Proof of Lemma 3.6 and Corollary 3.8. Girsanov's theorem would imply the result if the Novikov condition

$$\mathbb{E} \exp\left\{ \frac{1}{2} \int_0^t |\Theta_s(Z^s)|^2 \left| G\big(z(s), \Phi_s(Z^s, h_0)\big) \right|_{\mathbb{L}^2}^2 ds \right\} < \infty$$

holds. As in the proof of Lemma 3.4, we will prove the stronger condition

$$\sup_{z(\cdot) \in A} \int_0^t \left| G\big(z(s), \Phi_s(Z^s, h_0)\big) \right|_{\mathbb{L}^2}^2 ds < \infty.$$

Using Lemma A.4, we obtain the following estimate on G:

$$\left| G\big(z(s), \Phi_s(Z^s, h_0)\big) \right|_{\mathbb{L}^2}^2 \leq C' \big[|z(s)|_{\mathbb{L}^2}^2 \, |h(s)|_{\mathbb{L}^2}^2 + |h(s)|_{\mathbb{L}^2}^4 \big],$$

where $h(s) = \Phi_s(Z^s, h_0))$. By Lemma C.1 we know that if z is in A then $\sup_{s \in [0,t]} |h(t)|_{\mathbb{L}^2}$ is less than some C_1, where C_1 depends on $|h_0|_{\mathbb{L}^2}$ and the b_0, C_0 and T used to define A. Hence for any $z \in A$, we have

$$\begin{aligned}
\int_0^t \left| G\big(z(s), \Phi_s(Z^s, h_0)\big) \right|_{\mathbb{L}^2}^2 ds &\leq C' \int_0^t \big[|z(s)|_{\mathbb{L}^2}^2 \, |h(s)|_{\mathbb{L}^2}^2 + |h(s)|_{\mathbb{L}^2}^4 \big] ds \\
&\leq C' \left(\int_0^t |z(s)|_{\mathbb{L}^2}^4 \, ds \right)^{\frac{1}{2}} \left(\int_0^t |h(s)|_{\mathbb{L}^2}^4 \, ds \right)^{\frac{1}{2}} + C' C_1^4 t \\
&\leq C' (b_0 C_0)^2 T^{\frac{1}{2}} C_1^2 t^{\frac{1}{2}} + C' C_1^4 t.
\end{aligned}$$

Hence Novikov's condition holds and the lemma is proven. □

Proof of Lemma 3.7. The basic idea is as follows. Some of the paths which satisfy the condition defining A can be described by requiring that some norm of the paths be less than some fixed $f_k^*(t)$ at time t. Such a condition has the advantage that it corresponds to fixing a zero boundary condition along the boundary of some region for the associated Fokker-Planck equation. Since the diffusion is nondegenerate this process has a positive density on the interior of this region. By carefully picking f_k^* we can have the region contain sets arbitrarily far away from the origin. We now make this precise.

Fix a $L \in \mathcal{P}$, and a $t > 0$. For $k = 0, 1, 2, \ldots$ define the disk D_k by

$$D_k = \{f \in \mathbb{L}^2_\ell : |f|^4_{\mathbb{L}^2} \in [2^k, 2^{k+1})\}$$

and let $\bar{D}_k$ be the closure of D_k. We will construct $g(\cdot) = \sum g_k(\cdot)\mathbf{1}_{D_k}$, where g_k is strictly positive on $\bar{D}_k$ and zero outside of $\bar{D}_k$.

Let f_k^* be a non-decreasing, positive, real-vaued C^∞ function f_k^* such that $f_k^*(s) = (C_0^4 + \alpha_k)^{\frac{1}{4}}$ for $s \in [0, (1-\alpha_k)t - \varepsilon]$ and $f_k^*(s) = (100 \cdot 2^{k+1})^{\frac{1}{4}}$ for $s \in [(1-\alpha_k)t, t]$ and linearly interpolates in $[(1-\alpha_k)t - \varepsilon, (1-\alpha_k)t]$. α_k is some number in $(0, 1)$ chosen so that $\int_0^t (f_k^*(r))^4 dr < (b_0 C_0)^4 T$. This is possible as long as $b_0 > 1$ and $t \le T$.

Now define the subset H_k of $C([0, t], \mathbb{L}^2_\ell)$ by

$$H_k = \left\{ f \in C\big([0, t], \mathbb{L}^2_\ell\big) : \sup_{s \in [0,t]} |f(s)|_{\mathbb{L}^2} \le f_k^*(s) \right\}.$$

By the choice of f_k^* it is clear that $H_k \subset A$, where A is the same set used in the definition of z.

Now consider the process $x_k'(t)$ which follows the same equation as $x(t)$ except that it is killed whenever the trajectory leaves H_k. Another way of saying this is $x_k'(t)$ is the process $x(t)$ conditioned on staying in H_k. The transition density of this process $g_k'(s, \ell(0), y)$ is the solution to the Kolmogorov equation with the same generator as x but with zero boundary conditions along the boundary of H_k. Since the generator is elliptic, we know that $g_k'(t, \ell(0), y)$ is strictly positive everywhere in the interior of H_k. Since the trace of H_k at time t strictly contains D_k, we know that $g_k'(t, \ell(0), y)$ is strictly positive for $y \in \bar{D}_k$. Also by construction it is clear that $Q_t^x(\ell(0), H_k) > 0$ for all k. Let $a_k = Q_t^x(\ell(0), H_k)$ and set $g_k(\cdot) = a_k g_k'(t, \ell(0), \cdot)\mathbf{1}_{D_k}(\cdot)$.

All that remains is to verify that this choice of g_k constructs a g with the desired minorization property since it is clearly everywhere positive. Without loss of generality it is enough to show it for a B contained in some arbitrary D_k. Then

$$\begin{aligned} Q_t^x(\ell(0), [B] \cap A) &\ge Q_t^x(\ell(0), [B] \cap H_k) \ge \mathbb{P}_{\ell(0)}\{x \in [B] \;\&\; x \in H_k\} \\ &\ge \mathbb{P}_{\ell(0)}\{x \in [B] \mid x \in H_k\}\mathbb{P}_{\ell(0)}\{x \in H_k\} \\ &\ge a_k \int_B g_k'(t, \ell(0), y)dm(y) = \int_B g_k(y)dm(y). \end{aligned}$$ □

4. Stationary Measures and Thermodynamical Formalism

In this section we make a few general heuristic remarks about the methodology behind our approach.

The starting point of our construction is rewriting the original Navier–Stokes equation with random forcing as a finite-dimensional system of ordinary stochastic differential equations whose drift coefficients depends on the whole past:

$$d\ell = [-\nu\Lambda^2\ell + P_\ell B(\ell,\ell) + G(\ell,\Phi_t(L^t))]dt + dW. \tag{25}$$

From (25)

$$dW = d\ell - [-\nu\Lambda^2\ell + P_\ell B(\ell,\ell) + G(\ell,\Phi_t(L^t))]dt. \tag{26}$$

The measure corresponding to all $dw_k(t), k \in \mathcal{Z}_\nu, -\infty < t < \infty$ can be symbolically written as

$$\int \exp\left\{-\frac{1}{2}\sum_{k\in\mathcal{Z}_\nu}\frac{1}{|\sigma_k|^2}\int_{-\infty}^{\infty}\left|\frac{dw_k(t)}{dt}\right|^2 dt\right\} \quad \prod_k dw_k(t).$$

Here $\mathcal{Z}_\nu$ is the set of modes that are forced. The substitution of the expression for dw_k from (26) gives

$$\exp\left\{\int_{-\infty}^{\infty}\mathcal{L}_1(\ell(t))dt + \int_{-\infty}^{\infty}\mathcal{L}_2(\ell(t))dt \right.$$
$$\left. -\frac{1}{2}\sum_{k\in\mathcal{Z}_\nu}\frac{1}{|\sigma_k|^2}\int_{-\infty}^{\infty}\left|\frac{d\ell_k(t)}{dt}\right|^2 dt\right\} \quad \prod_k d\ell_k(t),$$

where

$$\mathcal{L}_1(\ell(t)) = -\frac{1}{2}\left(-\nu\Lambda^2\ell + P_\ell B(\ell,\ell) + G(\ell,\Phi_t(L^t))\right)^2,$$

$$\int_{-\infty}^{\infty}\mathcal{L}_2(\ell(t))dt = \int_{-\infty}^{\infty}\sum_{k\in\mathcal{Z}_\nu}\frac{1}{|\sigma_k|^2}\left(-\nu\Lambda^2\ell + P_\ell B(\ell,\ell) + G(\ell,\Phi_t(L^t))\right)_k d\ell_k(t).$$

The factor $\exp\left\{-\frac{1}{2}\sum_{k\in\mathcal{Z}_\nu}\frac{1}{|\sigma_k|^2}\int_{-\infty}^{\infty}\left|\frac{d\ell_k(t)}{dt}\right|^2 dt\right\} \quad \prod_k d\ell_k(t)$ can be considered as the differential of a "free measure" which in our case is a finite-dimensional white noise.

The "Lagrangians" $\mathcal{L}_1$, $\mathcal{L}_2$ describe the non-local interaction of $\ell(t)$ with the past. The whole expression shows that the stationary measure for the SNS system is actually a Gibbs state constructed with the help of Lagrangians $\mathcal{L}_1$, $\mathcal{L}_2$.

The estimations of the growth of $\mathcal{L}_1$, $\mathcal{L}_2$ as a function of the growth of $|\ell_k(s)|_{\mathbb{L}^2}$, $s \to -\infty$ show the class of realizations for which the conditional distributions can be defined. Therefore we have a weaker form of the Gibbs state. R. L. Dobrushin in his last papers and talks stressed the importance of this class of probability distributions. Since we are dealing all the time with probability distributions, the free energy of our Gibbs state is zero. It would be interesting to develop a general theory of existence and uniqueness of Gibbs states for general Lagrangians $\mathcal{L}_1$, $\mathcal{L}_2$ so that our result becomes a particular case of a more general statement.

5. Conclusion

When analyzing the ergodic properties of an infinite dimensional stochastic process, one of the most delicate aspects is often finding the correct topology in which to work. One of the principle advantages of the approach presented in this paper is that it evades this difficulty. We trade an infinite dimensional diffusion process for a finite dimensional Itô process with memory.

We have tried to present the simplest case of our theory, so that the exposition would be unencumbered. In fact the proofs contained in this work have proved a more general theorem than originally stated. Consider forcing defined by

$$W(x,t)=\sum_{k\in\mathcal{Z}}\sigma_k w_k(t,\omega)e_k(x),$$

where $\mathcal{Z}$ is some finite subset of $\mathbb{Z}^2$ such that $(0,0)\notin\mathcal{Z}$ and $k\in\mathcal{Z}$ if and only if $\sigma_k>0$. If we define

$$\mathbb{L}^2_\ell=\mathrm{span}\{e_k,k\in\mathcal{Z}\},\quad \mathbb{L}^2_h=\mathrm{span}\{e_k,k\notin\mathcal{Z}\}$$

and

$$N_-=\sup\big\{N:k\in\mathcal{Z}\text{ for all }k\text{ with }0<|k|\le N\big\}.$$

With these definitions all of the previous lemmas and theorems hold with the role of N replaced by N_-. In particular, if $N_-^2>C\frac{\mathcal{E}_0}{\nu^3}$ the system has a unique invariant measure.

This formulation emphasizes the nature of our principle assumption. By requiring that all of the low modes are forced, we are essentially requiring that the reduced Gibbsian dynamics are elliptic in nature. Some steps towards dealing with a hypo-elliptic setting have been made. In [EMatt], finite dimensional truncations of the two dimensional SNS equation were studied and shown to be ergodic under minimal assumptions. In [EM], a reaction diffusion equation was studied under degenerate forcing.

Our arguments can be easily extended to the case where the forcing of the kth mode has the form $f_k+\sigma_k dw_k(t)$, f_k is a constant, $f_k=0$ and $\sigma_k=0$ for $k\notin\mathcal{Z}$ or the case when the forcing is not diagonal in Fourier space.

Our approach can also be extended in several other different directions. We can consider the case when the high modes are also forced. As long as the forcing of the high modes decays sufficiently fast, our argument still applies with almost no change. The Wiener process in the forcing can be replaced by other diffusion processes such as the Ornstein-Uhlenbeck process. Dissipative PDEs such as the Cahn-Hilliard equation and the Ginzburg-Landau equations can also be studied using the same method. Finally, exponential convergence of empirical distributions to the stationary distribution can be proved.

A. Energy Estimates

In this Appendix, we prove a number of estimates controlling the evolution of the energy and enstrophy. Estimates for higher Sobolev norms are also possible, see [Mat98] for examples. In all cases, they are analogous to the standard results in the deterministic setting. Here we do not limit ourselves to forcing with only finitely many active modes. We will characterize the forcing in terms of the $\mathcal{E}_l$ defined by $\mathcal{E}_l\overset{\text{def}}{=}\sum|k|^{2l}|\sigma_k|^2$. We begin with the basic energy and enstrophy estimates in the stochastic setting.

Lemma A.1. *For any $p > 1$, we have*

$$\mathbb{E}\,|u(t)|_{\mathbb{L}^2}^{2p} + 2p\nu \int_0^t \mathbb{E}\,|\Lambda u(s)|_{\mathbb{L}^2}^2\,|u(s)|_{\mathbb{L}^2}^{2(p-1)}\,ds \le \mathbb{E}\,|u(0)|_{\mathbb{L}^2}^{2p} + C_0 \int_0^t \mathbb{E}\,|u(s)|_{\mathbb{L}^2}^{2(p-1)}\,ds,$$

$$\mathbb{E}\,|\Lambda u(t)|_{\mathbb{L}^2}^{2p} + 2p\nu \int_0^t \mathbb{E}\left|\Lambda^2 u(s)\right|_{\mathbb{L}^2}^2 |\Lambda u(s)|_{\mathbb{L}^2}^{2(p-1)}\,ds \le \mathbb{E}\,|\Lambda u(0)|_{\mathbb{L}^2}^{2p} + C_1 \int_0^t \mathbb{E}\,|\Lambda u(s)|_{\mathbb{L}^2}^{2(p-1)}\,ds.$$

Here $C_i = p\mathcal{E}_i + 2p(p-1)\sigma_{\max}^2$ and $\sigma_{\max}^2 = \sup|\sigma_k|^2$. In the case $p = 1$, we have the equalities

$$\mathbb{E}\,|u(t)|_{\mathbb{L}^2}^2 + 2\nu \int_0^t \mathbb{E}\,|\Lambda u(s)|_{\mathbb{L}^2}^2 = \mathbb{E}\,|u(0)|_{\mathbb{L}^2}^2 + \mathcal{E}_0 t, \tag{27}$$

$$\mathbb{E}\,|\Lambda u(t)|_{\mathbb{L}^2}^2 + 2\nu \int_0^t \mathbb{E}\left|\Lambda^2 u(s)\right|_{\mathbb{L}^2}^2 = \mathbb{E}\,|\Lambda u(0)|_{\mathbb{L}^2}^2 + \mathcal{E}_1 t. \tag{28}$$

Proof. We begin by fixing a positive integer M and considering the Galerkin approximation defined by $u^{(M)}(t) = \sum_{|k|\le M} u_k^{(M)}(t)e_k$. $u^{(M)}(t)$ satisfies an equation of exactly the same form as the full solution except the nonlinearity has been projected to those terms of order less than or equal to M. We will also need $\mathcal{E}_l^M \stackrel{\text{def}}{=} \sum_{|k|\le M} |k|^{2l}|\sigma_k|^2$. Our estimates will be independent of the order of approximation M. For simplicity, we will sometimes neglect the superscript M.

Applying Itô's formula to the map $\{u_k\} \mapsto \left(\sum |u_k|^2\right)^p$ produces,

$$d\,|u(t)|_{\mathbb{L}^2}^{2p} = 2p\,|u(t)|_{\mathbb{L}^2}^{2(p-1)}\left[-\nu\,|\Lambda u(t)|_{\mathbb{L}^2}^2\,dt + \langle u(t), dW\rangle_{\mathbb{L}^2}\right] + 2p(p-1)\,|u(t)|_{\mathbb{L}^2}^{2(p-2)}\Big(\sum_k |u_k(t)|^2|\sigma_k|^2\Big)dt + p\,|u(t)|_{\mathbb{L}^2}^{2(p-1)}\,\mathcal{E}_0^M\,dt \tag{29}$$

for the energy moments and

$$d\,|\Lambda u(t)|_{\mathbb{L}^2}^{2p} = 2p\,|\Lambda u(t)|_{\mathbb{L}^2}^{2(p-1)}\left[-\nu\left|\Lambda^2 u(t)\right|_{\mathbb{L}^2}^2 dt + \langle \Lambda^2 u(t), dW\rangle_{\mathbb{L}^2}\right] + 2p(p-1)\,|\Lambda u(t)|_{\mathbb{L}^2}^{2(p-2)}\Big(\sum_k |k|^2|\sigma_k|^2|u_k(t)|^2\Big)dt + p\,|\Lambda u(t)|_{\mathbb{L}^2}^{2(p-1)}\,\mathcal{E}_1^M\,dt \tag{30}$$

for the enstrophy moments.

Here $\langle \Lambda^\alpha u(t), dW(t)\rangle_{\mathbb{L}^2}$ is shorthand for $\sum |k|^\alpha u_k(t)\sigma_k dw_k(t)$. In the first, we have used the fact that $\langle B(u,u), u\rangle_{\mathbb{L}^2} = 0$ and in the second the fact that $\langle B(u,u), \Lambda^2 u\rangle_{\mathbb{L}^2} = 0$. Since, on the torus, the structure of the energy and the enstrophy equations are the same we will continue giving all of the details for analysis of the enstrophy equation.

The analysis for the energy equation proceeds analogously, see [Mat99, Mat98]. For a fixed $H > 0$, we introduce the stopping time

$$T = \inf\{t \geq 0 : \left|\Lambda^2 u(t)\right|_{\mathbb{L}^2}^2 \geq H^2\}.$$

Denoting by M_t the local martingale term in (30), we define the stopped martingale M_t^T by

$$M_t^T = \int_0^t 2p\,|\Lambda u(s \wedge T)|_{\mathbb{L}^2}^{2(p-1)} \langle \Lambda^2 u(s \wedge T), dW(s)\rangle_{\mathbb{L}^2}.$$

M_t^T has the advantage that its quadratic variation, denoted by $[M^T, M^T]_t$, is clearly finite.

$$\begin{aligned}
[M^T, M^T]_t &\leq 2p\sigma_{\max}^2 \int_0^t \left|\Lambda^2 u(s \wedge T)\right|_{\mathbb{L}^2}^{2p} ds \\
&\leq 2p\sigma_{\max}^2 \int_0^t \left|\Lambda^2 u(s \wedge T)\right|_{\mathbb{L}^2}^{2p} ds \leq 2p\sigma_{\max}^2 H^{2p} t < \infty.
\end{aligned}$$

Because $\mathbb{E}[M^T, M^T]_t < \infty$ we know that $\mathbb{E}M_t^T = 0$. And because $t \wedge T$ is a bounded stopping time the Optional Stopping Time Lemma says that $\mathbb{E}M_{t\wedge T}^T = 0$. Since $M_{t\wedge T} = M_{t\wedge T}^T$, we have

$$\mathbb{E}\,|\Lambda u(t \wedge T)|_{\mathbb{L}^2}^2 + 2\nu\mathbb{E}\int_0^{t\wedge T} \left|\Lambda^2 u(s)\right|_{\mathbb{L}^2}^2 ds = \mathbb{E}\,|\Lambda u(0)|_{\mathbb{L}^2}^2 + \mathcal{E}_1^M \mathbb{E}(t \wedge T),$$

and when $p > 1$,

$$\begin{aligned}
&\mathbb{E}\,|\Lambda u(t \wedge T)|_{\mathbb{L}^2}^{2p} + 2p\nu\mathbb{E}\int_0^{t\wedge T} |\Lambda u(t)|_{\mathbb{L}^2}^{2(p-1)} \left|\Lambda^2 u(s)\right|_{\mathbb{L}^2}^2 ds \\
&\quad = \mathbb{E}\,|\Lambda u(0)|_{\mathbb{L}^2}^{2p} + \mathbb{E}\int_0^{t\wedge T} 2p(p-1)\,|\Lambda u(s)|_{\mathbb{L}^2}^{2(p-2)} \Big(\sum_k |k|^2 |\sigma_k|^2 |u_k(s)|^2\Big) \\
&\qquad\qquad + p\,|\Lambda u(s)|_{\mathbb{L}^2}^{2(p-1)}\, \mathcal{E}_1^M\, ds.
\end{aligned}$$

Hence

$$\begin{aligned}
&\mathbb{E}\,|\Lambda u(t \wedge T)|_{\mathbb{L}^2}^{2p} + 2p\nu\mathbb{E}\int_0^{t\wedge T} |\Lambda u(t)|_{\mathbb{L}^2}^{2(p-1)} \left|\Lambda^2 u(s)\right|_{\mathbb{L}^2}^2 ds \\
&\qquad \leq \mathbb{E}\,|\Lambda u(0)|_{\mathbb{L}^2}^{2p} + \left[2p(p-1)\sigma_{\max}^2 + p\mathcal{E}_1^M\right]\mathbb{E}\int_0^{t\wedge T} |\Lambda u(s)|_{\mathbb{L}^2}^{2(p-1)} ds.
\end{aligned}$$

Since $u(t)$ is continuous in time, $T \to \infty$ as $H \to \infty$ and hence $T \wedge t \to t$. Thus we obtain

$$\mathbb{E}\,|\Lambda u(t)|_{\mathbb{L}^2}^2 + 2\nu\mathbb{E}\int_0^t \left|\Lambda^2 u(s)\right|_{\mathbb{L}^2}^2 ds = \mathbb{E}\,|\Lambda u(0)|_{\mathbb{L}^2}^2 + \mathcal{E}_1^M t,$$

$$\mathbb{E}\,|\Lambda u(t)|_{\mathbb{L}^2}^{2p} + 2p\nu\mathbb{E}\int_0^t |\Lambda u(t)|_{\mathbb{L}^2}^{2(p-1)} \left|\Lambda^2 u(s)\right|_{\mathbb{L}^2}^2 ds$$
$$\leq \mathbb{E}\,|\Lambda u(0)|_{\mathbb{L}^2}^{2p} \left[2p(p-1)\sigma_{\max}^2 + p\mathcal{E}_1^M\right]\mathbb{E}\int_0^t |\Lambda u(s)|_{\mathbb{L}^2}^{2(p-1)} ds.$$

Recall that we have been calculating with an Mth order Galerkin approximation. For the $p = 1$ equation, the right hand side converges to the desired right hand side. With this bound on $\mathbb{E}\,|\Lambda u(t)|_{\mathbb{L}^2}^2$ in hand we can take the $M \to \infty$ limit of the $p = 2$ equation. Analogously, once we have taken the limit in the pth equation we have the dominating bound needed to take the limit in the $p + 1$ equation. □

In our setting, the Poincaré inequality reads $|\Lambda f|_{\mathbb{L}^2}^2 > |f|_{\mathbb{L}^2}^2$ and $\left|\Lambda^2 f\right|_{\mathbb{L}^2}^2 > |\Lambda f|_{\mathbb{L}^2}^2$. This allows us to close the above inequalities. After applying Gronwall's inequality, we obtain the following estimates which are uniform in time.

Corollary A.2.

$$\mathbb{E}\,|u(t)|_{\mathbb{L}^2}^2 \leq e^{-2\nu t}\mathbb{E}\,|u(0)|_{\mathbb{L}^2}^2 + \frac{\mathcal{E}_0}{2\nu}\left(1 - e^{-2\nu t}\right),$$
$$\mathbb{E}\,|\Lambda u(t)|_{\mathbb{L}^2}^2 \leq e^{-2\nu t}\mathbb{E}\,|\Lambda u(0)|_{\mathbb{L}^2}^2 + \frac{\mathcal{E}_1}{2\nu}\left(1 - e^{-2\nu t}\right).$$

For any $p > 1$,

$$\mathbb{E}\,|u(t)|_{\mathbb{L}^2}^{2p} \leq e^{-2\nu t}\mathbb{E}\,|u(0)|_{\mathbb{L}^2}^{2p} + C_0\int_0^t e^{-2\nu(t-s)}\mathbb{E}\,|u(s)|_{\mathbb{L}^2}^{2(p-1)} ds,$$
$$\mathbb{E}\,|\Lambda u(t)|_{\mathbb{L}^2}^{2p} \leq e^{-2\nu t}\mathbb{E}\,|\Lambda u(0)|_{\mathbb{L}^2}^{2p} + C_1\int_0^t e^{-2\nu(t-s)}\mathbb{E}\,|\Lambda u(s)|_{\mathbb{L}^2}^{2(p-1)} ds.$$

We use standard estimates in the tri-linear term $\langle B(u, v), w\rangle_{\mathbb{L}^2}$ specialized to our two dimensional setting. Its proof can be found in [CF88] for example.

Lemma A.3. *Let* α, β, γ *be positive real numbers such that* $\alpha + \beta + \gamma \geq 1$ *and* $(\alpha, \beta, \gamma) \neq (0, 0, 1)$, *or* $(0, 1, 0)$, *or* $(1, 0, 0)$,

$$|\langle B(u, v), w\rangle_{\mathbb{L}^2}| \leq C\left|\Lambda^\alpha u\right|_{\mathbb{L}^2}\left|\Lambda^{\beta+1} v\right|_{\mathbb{L}^2}\left|\Lambda^\gamma w\right|_{\mathbb{L}^2}.$$

Using this lemma we prove the following estimate specialized to the two dimensional setting with periodic boundary conditions.

Lemma A.4. *Let* $\{e_k, k \in \mathbb{Z}^2\}$ *be a basis for* $\mathbb{L}^2$. *Consider a splitting of* $\mathbb{L}^2 = \mathbb{L}_\ell^2 + \mathbb{L}_h^2$. *Let* N^+ *be in* $\sup\{|k| : \exists\, e_k$ *with* $e_k \in \mathbb{L}_\ell^2\}$ *and* P_ℓ *be the projector onto* $\mathbb{L}_\ell^2$. *If* $u, v \in \mathbb{L}^2$ *then*

$$|P_\ell B(u, v)| \leq C(N^+)^3\,|u|_{\mathbb{L}^2}\,|v|_{\mathbb{L}^2}.$$

Proof of Lemma A.4. In the periodic setting, P_ℓ, P_{div}, and $(-\Delta)^s$ all are simply Fourier multipliers and hence commute with one other. Recall that $B(u, v) = P_{div}(u \cdot \nabla)v$ and hence,

$$\begin{aligned} |P_\ell B(u, v)| &= \sup_{\substack{w \in \mathbb{L}^2 \\ |w|=1}} |\langle P_\ell B(u, v), w\rangle_{\mathbb{L}^2}| = \sup_{\substack{w \in \mathbb{L}^2 \\ |w|=1}} |\langle B(u, v), P_\ell w\rangle_{\mathbb{L}^2}| \\ &= \sup_{\substack{w \in \mathbb{L}^2 \\ |w|=1}} |\langle B(u, P_\ell w), v\rangle_{\mathbb{L}^2}| \leq C\, |u|_{\mathbb{L}^2}\, |v|_{\mathbb{L}^2} \sup_{\substack{w \in \mathbb{L}^2 \\ |w|=1}} \left|\Lambda^3 P_\ell w\right|_{\mathbb{L}^2} \\ &\leq C(N^+)^3\, |u|_{\mathbb{L}^2}\, |v|_{\mathbb{L}^2} \sup_{\substack{w \in \mathbb{L}^2 \\ |w|=1}} |w|_{\mathbb{L}^2} \leq C(N^+)^3\, |u|_{\mathbb{L}^2}\, |v|_{\mathbb{L}^2}. \qquad \square \end{aligned}$$

Lemma A.5. *Fix any $\delta > \frac{1}{2}$, $a \in (0, 1)$ and $C_1 > 0$. Let $u(t) = \varphi_t^\omega u_0$. There exists a $K_1 > 0$ such that whenever $|u_0|^2_{\mathbb{L}^2} < C_0$,*

$$\mathbb{P}\left\{|u(t)|^2_{\mathbb{L}^2} + 2\nu \int_0^t |\Lambda u(s)|^2_{\mathbb{L}^2}\, ds \leq C_0 + \mathcal{E}_0 t + K_1(t+1)^\delta \textit{ for all } t \geq 0\right\} \geq 1 - a.$$

Proof of Lemma A.5. The energy equation reads

$$|u(t)|^2_{\mathbb{L}^2} + 2\nu \int_0^t |\Lambda u(s)|^2_{\mathbb{L}^2}\, ds = |u_0|^2_{\mathbb{L}^2} + \mathcal{E}_0 t + \int_0^t \langle u(s), dW(s)\rangle_{\mathbb{L}^2}.$$

Since $|u_0|^2_{\mathbb{L}^2} < C_0$, all we need to show is that

$$\mathbb{P}\left\{M_t \leq K_1(t+1)^\delta \text{ for } t \geq 0\right\} \geq 1 - a$$

for K_1 large enough, where $M_t = \int_0^t \langle u(s), dW(s)\rangle_{\mathbb{L}^2}$. The quadratic variation $[M, M]_t$ can be calculated and one sees that

$$[M, M]_t \leq \sigma^2_{\max} \int_0^t |u(s)|^2_{\mathbb{L}^2},$$

and hence

$$([M, M]_t)^p \leq \sigma^{2p}_{\max} \left(\int_0^t |u(s)|^2_{\mathbb{L}^2}\right)^p \leq \sigma^{2p}_{\max} t^{p-1} \int_0^t |u(s)|^{2p}_{\mathbb{L}^2}\, ds.$$

From Corollary A.2, we know that if $|u(0)|^2_{\mathbb{L}^2} < C_0$, then there exists a constant $C_p(C_0)$ so that $\mathbb{E}\,|u(t)|^{2p}_{\mathbb{L}^2} \leq C_p$ for all $t \geq 0$ and $p \geq 1$.

Now define the events

$$A_k = \left\{\sup_{s \in [0,k]} |M_s| > K_1 k^\delta\right\}.$$

By the Doob–Kolmogorov martingale inequality we have

$$\mathbb{P}\{A_k\} \le \frac{\mathbb{E}([M,M]_t)^p}{K_1^{2p}k^{2p\delta}} \le \frac{\sigma_{\max}^{2p}C_p}{K_1^{2p}}\frac{k^p}{k^{2p\delta}}.$$

Lastly observe that

$$\mathbb{P}\{M_t \le K_1(t+1)^\delta\} \ge 1 - \mathbb{P}\Big\{\bigcup_k A_k\Big\} \ge 1 - \sum_k \mathbb{P}\{A_k\}.$$

By the previous estimate on $\mathbb{P}\{A_k\}$, for any $\delta > \frac{1}{2}$ we see that the sum is finite for p sufficiently large. Specifically, we need $\delta > \frac{1}{2}(1+\frac{1}{p})$. Lastly, the sum can be made as small as we want by increasing K_1. □

B. Properties of Stationary Measures

We now establish a number of properties, derived from the dynamics, which any stationary measure must possess.

Lemma B.1. *For any stationary measure all energy moments are finite. In fact for any $p \ge 1$ there exist a constant $C_p < \infty$ such that*

$$\int_{\mathbb{L}^2} |u|_{\mathbb{L}^2}^{2p}\, d\mu(u) < C_p$$

for all stationary measures μ. In particular $C_1 = \frac{\mathcal{E}_0}{2\nu}$.

Proof. We will consider the case when $p = 1$. The other cases follow by the same method. For any $\epsilon > 0$ there exists a b_ϵ such that $\mu\{u \in \mathbb{L}^2 : |u|_{\mathbb{L}^2}^2 \le b_\epsilon\} > 1-\epsilon$. Let B_ϵ denote $\{u \in \mathbb{L}^2 : |u|_{\mathbb{L}^2}^2 \le b_\epsilon\}$. For any $H > 0$ and $t > 0$, we have

$$\begin{aligned}\int_{\mathbb{L}^2}\Big(|u|_{\mathbb{L}^2}^2 \wedge H\Big)\,d\mu(u) &= \int_{\mathbb{L}^2}\mathbb{E}\Big(|\varphi_{0,t}^{\omega}u|_{\mathbb{L}^2}^2 \wedge H\Big)\,d\mu(u)\\ &\le H\epsilon + \int_{B_\epsilon}\mathbb{E}\Big(|\varphi_{0,t}^{\omega}u|_{\mathbb{L}^2}^2 \wedge H\Big)\,d\mu(u)\\ &\le H\epsilon + \int_{B_\epsilon}\mathbb{E}\Big(|\varphi_{0,t}^{\omega}u|_{\mathbb{L}^2}^2\Big)\,d\mu(u).\end{aligned}$$

Applying the first bound in Corollary A.2 gives

$$\int_{\mathbb{L}^2}\Big(|u|_{\mathbb{L}^2}^2 \wedge H\Big)\,d\mu(u) \le H\epsilon + \frac{\mathcal{E}_0}{2\nu} + e^{-2\nu t}\Big(b_\epsilon - \frac{\mathcal{E}_0}{2\nu}\Big).$$

Taking the limit as $t \to \infty$ and then observing that ϵ was arbitrary, we obtain

$$\int_{\mathbb{L}^2}\Big(|u|_{\mathbb{L}^2}^2 \wedge H\Big)\,d\mu(u) = \int_U\Big(|u|_{\mathbb{L}^2}^2 \wedge H\Big)\,d\mu(u) \le \frac{\mathcal{E}_0}{2\nu}.$$

Taking $H \to \infty$ gives that the energy of any stationary measure is bounded by $\frac{\mathcal{E}_0}{2\nu}$. The argument for higher moments of the energy is the same □

Lemma B.2. *For any stationary measure* μ,

$$\int_{\mathbb{L}^2} |\Lambda u|^2_{\mathbb{L}^2}\, d\mu(u) = \frac{\mathcal{E}_0}{2\nu}.$$

In addition if the forcing is such that $\mathcal{E}_1 < \infty$ *then*

$$\int_{\mathbb{L}^2} \left|\Lambda^2 u\right|^2_{\mathbb{L}^2} d\mu(u) = \frac{\mathcal{E}_1}{2\nu} \quad \text{and} \quad \int_{\mathbb{L}^2} |\Lambda u|^{2p}_{\mathbb{L}^2}\, d\mu(u) < C_1(p) < \infty$$

for all $p \geq 1$.

Proof. Using Eq. (27), we have that for any initial condition $u_0 \in \mathbb{L}^2$,

$$\mathbb{E}\left|\varphi_{0,t}u_0\right|^2_{\mathbb{L}^2} + 2\nu \int_0^t \mathbb{E}\left|\Lambda\varphi_{0,s}u_0\right|^2_{\mathbb{L}^2} ds = |u_0|^2_{\mathbb{L}^2} + \mathcal{E}_0 t.$$

Here we have switched the time integral and the expectation by the Fubini–Tonelli theorem because the integrand is non-negative. We know from Lemma B.1 that any stationary measure has finite energy moments. Hence averaging with respect to the stationary measure gives

$$\begin{aligned}\int_{\mathbb{L}^2} \mathbb{E}\left|\varphi_{0,t}u_0\right|^2_{\mathbb{L}^2} d\mu(u_0) + 2\nu \int_{\mathbb{L}^2}\int_0^t \mathbb{E}\left|\Lambda\varphi_{0,s}u_0\right|^2_{\mathbb{L}^2} ds\, d\mu(u_0)\\ = \int_{\mathbb{L}^2} |u_0|^2_{\mathbb{L}^2}\, d\mu(u_0) + \mathcal{E}_0 t.\end{aligned}$$

Because μ was a stationary measure, we have that

$$\int_{\mathbb{L}^2} \mathbb{E}\left|\varphi_{0,t}u_0\right|^2_{\mathbb{L}^2} d\mu(u_0) = \int_{\mathbb{L}^2} |u_0|^2_{\mathbb{L}^2}\, d\mu(u_0)$$

and

$$\int_{\mathbb{L}^2}\int_0^t \mathbb{E}\left|\Lambda\varphi_{0,s}u_0\right|^2_{\mathbb{L}^2} ds = t\int_{\mathbb{L}^2} |\Lambda u_0|^2_{\mathbb{L}^2}\, d\mu(u_0).$$

Hence $2\nu \int_{\mathbb{L}^2} |\Lambda u_0|^2_{\mathbb{L}^2}\, d\mu(u_0) = \mathcal{E}_0$, concluding the proof of the first claim.

We now turn to the enstrophy moments. By the first part of this lemma, we know that there exist a $U \subset \mathbb{H}^1$ such that $\mu(U) = 1$. We now can proceed just as in Lemma B.1 to prove that all of the enstrophy moments are finite.

To find the expected value of the $\mathbb{H}^2$ norm we use Eq. (28). Then we proceed exactly as we did to obtain the expected value of the enstrophy (the $\mathbb{H}^1$ norm). □

Lemma B.3. *Let* μ_p *be the measure induced on* $C\big((-\infty, 0], \mathbb{L}^2_\ell\big)$ *by any given stationary measure* μ. *Fix any* $K_0 > 0$ *and* $\delta > \frac{1}{2}$. *Then for* μ_p*-almost every trajectory in* $C\big((-\infty, 0], \mathbb{L}^2_\ell\big)$, $v(s)$, *there exists a constant* T *such that for* $s \leq 0$,

$$|v(s)|^2_{\mathbb{L}^2} \leq \mathcal{E}_0 + K_0 \min(T, |s|)^\delta.$$

Proof. The basic energy estimate, derived from (29), reads:

$$|v(t)|^2_{\mathbb{L}^2} = |v(t_0)|^2_{\mathbb{L}^2} + \mathcal{E}_0(t-t_0) - 2\nu \int_{t_0}^{t} |\Lambda v(s)|^2_{\mathbb{L}^2}\, ds + \int_{t_0}^{t} \langle v(s), dW(s)\rangle_{\mathbb{L}^2},$$

for any $t_0 < t \le 0$. There is no problem writing the integration against the Wiener path in the above integral. Our stochastic PDE had pathwise defined solutions. Therefore if we know the initial condition $v(t_0)$ and the trajectory of $v(s)$ for $s \in [t_0, t]$ the increments of the Wiener process on the interval $[t_0, t]$ are uniquely defined.

For any $k \ge 1$, the above estimate implies

$$\sup_{s\in[-k,-k+1]} |v(s)|^2_{\mathbb{L}^2} \le |v(-k)|^2_{\mathbb{L}^2} + \mathcal{E}_0 + \sup_{s\in[-k,-k+1]} F_k(s),$$

where $F_k(s) = -2\nu \int_{-k}^{s} |\Lambda v(r)|^2_{\mathbb{L}^2}\, dr + M_k(s)$ and $M_k(s) = \int_{-k}^{s} \langle v(r), dW(r)\rangle_{\mathbb{L}^2}$.

Now define

$$A_k = \left\{ v(s) : \sup_{s\in[-k,-k+1]} |v(s)|^2_{\mathbb{L}^2} \le \mathcal{E}_0 + K_0|k-1|^{\delta} \right\}$$

and $U_T = \cap_{k>T} A_k$. Since the U_T are an increasing collection of sets it will be sufficient to prove that the $\lim_{T\to\infty} \mu_p(U_T) = 1$. This is the same as showing that $\lim_{T\to\infty} \mu_p(U_T^c) = 0$. Now since $\mu_p(U_T^c) \le \sum_{k>T} \mu_p(A_k^c)$, we need only to show that $\sum_{k>0} \mu_p(A_k^c) < \infty$:

$$\mu_p(A_k^c) \le \mu_p \left\{ v(s) : |v(-k)|^2_{\mathbb{L}^2} \ge \frac{K_0}{2}|k-1|^{\delta} \right\}$$
$$+ \mu_p \left\{ v(s) : \sup_{s\in[-k,-k+1]} F_k(s) \ge \frac{K_0}{2}|k-1|^{\delta} \right\},$$

The first term is the most straightforward. Lemma B.2 implies that the second moment of the energy is uniformly bounded by some constant C_2. Hence Chebyshev's inequality produces

$$\mu_p \left\{ v(s) : |v(-k)|^2_{\mathbb{L}^2} \ge \frac{K_0}{2}|k-1|^{\delta} \right\} \le \frac{4}{K_0^2|k-1|^{2\delta}} \mathbb{E}\, |v(-k)|^4_{\mathbb{L}^2} \le \frac{4C}{K_0^2|k-1|^{2\delta}}$$

which is summable as long as $\delta > \frac{1}{2}$.

The second term proceeds in the same way but with Chebyshev's inequality replaced by the exponential martingale estimate. The exponential martingale inequality controls the size of a martingale minus something proportional to its quadratic variation (see [RY94, Mao97] for example). The details are given in the following.

The key observation is that we can control $F_k(s)$ by controlling $M_k(s) - \alpha[M_k, M_k](s)$, where $[M_k, M_k](s)$ is the quadratic variation of the martingale $M_k(s)$ and α is a constant we will choose presently. First notice that with probability one,

$$[M_k, M_k](s) = \int_{-k}^{s} \sum_{l} |\sigma_l|^2 |v_l(r)|^2 dr \le \sigma^2_{\max} \int_{-k}^{s} |v(r)|^2_{\mathbb{L}^2}\, dr$$
$$\le \sigma^2_{\max} \int_{-k}^{s} |\Lambda v(r)|^2_{\mathbb{L}^2}\, dr$$

and hence

$$F_k(s) \leq M_k(s) - \frac{2\nu}{\sigma_{\max}^2}[M_k, M_k](s)$$

almost surely. In this setting, the exponential martingale inequality states that for positive α and β,

$$\mathbb{P}\left\{\sup_{s\in[-k,0]} M_k(s) - \frac{\alpha}{2}[M_k, M_k](s) > \beta\right\} \leq e^{-\alpha\beta}.$$

Taking $\alpha = \frac{4\nu}{\sigma_{\max}^2}$ we find

$$\mu_p\left\{v(s) : \sup_{s\in[-k,-k+1]} F_k(s) \geq \frac{K_0}{2}|k-1|^\delta\right\} \leq \exp\left(-\frac{2\nu K_0}{\sigma_{\max}^2}|k-1|^\delta\right).$$

Since this is summable for any $\delta > 0$, the proof is complete. □

C. Control of High Modes by Low Modes

Lemma C.1. *If $h(t)$ is the solution to (12) with some low mode forcing $\ell \in C([0,t], \mathbb{L}^2_\ell)$, then $\sup_{s\in[0,t]} |h(s)|_{\mathbb{L}^2}$ is bounded by a constant depending on $|h(0)|_{\mathbb{L}^2}$ and $\int_0^t |\ell|^4_{\mathbb{L}^2}\, ds$.*

Proof. Taking the inner product of (12) with h produces

$$\frac{1}{2}\frac{d}{dt}|h(t)|^2_{\mathbb{L}^2} = -\nu\,|\Lambda h|^2_{\mathbb{L}^2} + \langle P_h B(h,\ell), h\rangle_{\mathbb{L}^2} + \langle P_h B(\ell,\ell), h\rangle_{\mathbb{L}^2}$$

because $\langle P_h B(\ell,h), h\rangle_{\mathbb{L}^2} = \langle P_h B(h,h), h\rangle_{\mathbb{L}^2} = 0$. Next using Lemma A.3 produces,

$$\begin{aligned}\frac{1}{2}\frac{d}{dt}|h(t)|^2_{\mathbb{L}^2} &\leq -\nu\,|\Lambda h|^2_{\mathbb{L}^2} + C\,|\Lambda h|_{\mathbb{L}^2}\,|h|_{\mathbb{L}^2}\,|\Lambda\ell|_{\mathbb{L}^2} + C\,|\Lambda h|_{\mathbb{L}^2}\,|\Lambda\ell|^2_{\mathbb{L}^2}\\ &\leq \frac{C}{2\nu}\,|h|^2_{\mathbb{L}^2}\,|\Lambda\ell|^2_{\mathbb{L}^2} + \frac{C}{2\nu}\,|\Lambda\ell|^4_{\mathbb{L}^2}\end{aligned}$$

Since $\ell \in \mathbb{L}^2_\ell$ we have $|\Lambda\ell|_{\mathbb{L}^2} \leq (N^+)\,|\ell|_{\mathbb{L}^2}$, where $N^+ = \sup\{|k| : \exists\, e_k \text{ with } e_k \in \mathbb{L}^2_\ell\}$, and hence after applying Gronwall's Lemma we have

$$\begin{aligned}|h(t)|^2_{\mathbb{L}^2} \leq{}& C_1\,|h(0)|^2_{\mathbb{L}^2}\exp\left(a_1\int_0^t |\ell|^2_{\mathbb{L}^2}\,ds\right)\\ &+ C_2\left(\int_0^t |\ell|^4_{\mathbb{L}^2}\,ds\right)\exp\left(a_1\int_0^t |\ell|^2_{\mathbb{L}^2}\,ds\right).\end{aligned}$$

Since by Hölder inequality,

$$\int_0^t |\ell|^2_{\mathbb{L}^2}\,ds \leq t\int_0^t |\ell|^4_{\mathbb{L}^2}\,ds,$$

the proof is complete. □

Acknowledgements. The authors would like to thank Gérard Ben Arous, Amir Dembo, Perci Diaconis, Yitzhak Katznelson, Di Liu, George Papanicolaou and Andrew Stuart for useful discussions. The work of the first author is partially supported by a Presidential Faculty Fellowship from the NSF. The work of the second author is partially supported by NSF grant DMS-9971087. The work of the third author is partially supported by NSF grant DMS-9706794 and RFFI grant 99-01-00314.

References

[BKL] Bricmont, J., Kupiainen, A., and Lefevere, R.: Preprint

[CDF97] Crauel, H., Debussche, A., and Flandoli, F.: Random attractors. J. Dynam. Diff. Eqs. **9** no. 2, 307–341 (1997)

[CF88] Constantin, P. and Foiaş, C.: *Navier–Stokes equations*. Chicago: University of Chicago Press, 1988

[EMatt] E, W. and Mattingly, J.C.: Ergodicity for the Navier–Stokes Equation with Degenerate Random Forcing: Finite Dimensional Approximation. Submitted

[EM] Eckmann, J.P., and Hairer, M.: Uniqueness of the invariant measure for a stochastic PDE driven by degenerate noise. Preprint

[EFNT94] Eden, A., Foias, C., Nicolaenko, B., and Temam, R.: *Exponential attractors for dissipative evolution equations*. Research in Applied Mathematics, New York: John Wiley and Sons and Masson, 1994

[Fer97] Ferrario, B.: Ergodic results for stochastic Navier–Stokes equation. Stochastics and Stochastics Rep. **60**, no. 3–4, 271–288 (1997)

[Fla94] Flandoli, F.: Dissipativity and invariant measures for stochastic Navier–Stokes equations. NoDEA **1**, 403–426 (1994)

[FM95] Flandoli, F. and Maslowski, B.: Ergodicity of the 2-D Navier–Stokes equation under random perturbations. Commun. Math. Phys. **171**, 119–141 (1995)

[FMRT] Foias, C., Manley, O., Rosa, R., Temam, R.: Navier–Stokes Equations and Turbulence. To be published

[IN] Ito, K., Nisio, M.: On stationary solutions of a stochastic differential equation. J. Math. Kyoto Univ. **4**, 1–75 (1964)

[KS] Kuksin, S. and Shirikyan, A.: Stochastic Dissipative PDE's and Gibbs Measures. Commun. Math. Phys. **213**, 291–330 (2000)

[Mao97] Mao, X.: *Stochastic differential equations and their applications*. Horwood Series in Mathematics & Applications, Chichester: Horwood Publishing Limited, 1997

[Mat98] Mattingly, J.C.: The stochastically forced Navier–Stokes equations: Energy estimates and phase space contraction. Ph.D. thesis, Princeton University, 1998

[Mat99] Mattingly, J.C.: Ergodicity of 2D Navier–Stokes equations with random forcing and large viscosity. Commun. Math. Phys. **206** no. 2, 273–288 (1999)

[Mat00] Mattingly, J.C.: Exponential convergence for the stochastically forced Navier–Stokes equations and other partially dissipative dynamics. Submitted

[RY94] Revuz, D. and Yor, M.: *Continuous martingales and Brownian motion*. Second ed., Grundlehren der Mathematischen Wissenschaften, Vol. **293**, Berlin: Springer-Verlag, 1994

[Str82] Stroock, D.W.: *Lectures on topics in stochastic differential equations*. Bombay: Tata Institute of Fundamental Research, 1982, with notes by Satyajit Karmakar

[SV79] Stroock, D.W. and Varadhan, S.R.S.: *Multidimensional diffusion processes*. Berlin: Springer-Verlag, 1979

[VF88] Vishik, M. and Fursikov, A.: *Mathematical problems of statistical hydrodynamics*. Dordrect: Kluwer Academic Publishers, 1988

Communicated by G. Gallavotti

Comments

It is generally believed that Navier – Stokes system with random forcing is a good model for turbulence. Since in the two-dimensional case strong existence and uniqueness results are known, it was natural to try to analyze the two-dimensional Navier – Stokes system with periodic boundary conditions as the simplest case. The basic result shows that the dynamics can be described as finite-dimensional with memory which is a typical model of statistical mechanics. This approach allows us to prove the uniqueness of the stationary measure. Similar results were obtained in the paper by J. Bricmont, A. Kupiainen, and R. Lefevere (see [BKL]) and by S. Kuksin and A. Shirikyan (see [KS]). Recently, M. Hairer and J. Mattingly significantly extended the results by allowing the action of random forcing to act only on a few modes (see [HM]).

Blow Ups of Complex Solutions of the $3\mathcal{D}$-Navier-Stokes System and Renormalization Group Method

by

Dong Li[1] and Ya. G. Sinai[1 2]

Abstract: We consider complex-valued solutions of the three-dimensional Navier-Stokes system without external forcing on R^3. We show that there exists an open set in the space of 10-parameter families of initial conditions such that for each family from this set there are values of parameters for which the solution develops blow up in finite time.

Keywords: Navier-Stokes system, renormalization group theory, fixed point, the linearization near the fixed point, spectrum of the linearized group, Hermite polynomials

[1]Program in Applied and Computational Mathematics, Princeton University, Princeton, New Jersey, U.S.A.

[1]Mathematics Department, Princeton University, Princeton, New Jersey, U.S.A. &

[2]Landau Institute of Theoretical Physics, Moscow, Russia

Y.G. Sinai, *Selecta II: Probability Theory, Statistical Mechanics, Mathematical Physics and Mathematical Fluid Dynamics*, DOI 10.1007/978-1-4419-6205-8_21,

§1. Introduction.

There are many phenomena in nature which can be considered as some manifestation of blow ups, like hurricanes, tornadoes, sandstorms, etc. If we believe that Navier-Stokes system describes well enough the motions of real gases and fluids under normal conditions, then it gives some reasons to expect that blow ups in solutions of this system also exist.

We consider in this paper the $3\mathcal{D}$-Navier-Stokes system for incompressible fluids moving without external forcing on R^3 with viscosity equal to 1. After Fourier transform it becomes a non-local, non-linear equation for a non-known function $v(k,t)$ with values in C^3, $k \in R^3$, $t > 0$. The incompressibility condition takes the form* $\langle v(k,t),\, k\rangle = 0$ and

$$v(k,t) = \exp\{-t|k|^2\}\, v(k,0) + i \int_0^t \exp\{-(t-s)|k|^2\}\, ds \cdot \int_{R^3} \langle\, v(k-k', s),\, k\rangle \cdot P_k\, v(k',s)\, dk' \tag{1}$$

In this expression $v(k,0)$ is an initial condition and P_k is the orthogonal projection to the subspace orthogonal to k, i.e. $P_k v = v - \frac{\langle v,k\rangle \cdot k}{\langle k,k\rangle}$. The formula (1) shows that the Navier-Stokes system is genuinely infinite-dimensional dynamical system: the value $v(k,t)$ is determined by the integration over all "degrees of freedom" and previous moments of time.

The problem of blow ups in solutions of the Navier-Stokes System(NSS) appeared after classical works of J. Leray (see [Le 1]) where he proved the existence of the weak solutions of NSS. O. Ladyzenskaya proved the existence of strong solutions of 2-dim NSS in bounded domains (see [La 1]). Many important contributions to the modern understanding of the 2-dim fluid dynamics were done by C. Foias and R. Temam (see [FT]), V. Yudovich (see [Y1]), Giga ([G1]) and others. However, the situation with the 3-dim NSS remained unclear. The Clay mathematical institute announced the problem of existence of strong solutions of the 3-dimensional NSS as one of the most important problem in mathematics of the XXI-century (see [Cl]).

In this paper we omit the condition that $v(k,t)$ is the Fourier transform of a real-valued vector field in the x-space and consider (1) in the space of all possible complex-valued functions with values in C^3. In this situation the energy inequality does not hold. Detailed

*Since $k \in R^3$, $v(k,t) \in C^3$, the order in the inner product is important.

assumptions concerning the initial condition $v(k,0)$ will be discussed later (see §7). In all cases $v(k,0)$ will be bounded functions whose support is a neighborhood of some point $(0,0,k^{(0)})$. The incompressibility condition implies that the components $v_1(k,0)$,$v_2(k,0)$ of $v(k,0)$ are arbitrary functions of k while $v_3(k,0)$ can be found from the incompressibility condition $\langle v,k\rangle = 0$.

Various methods (see, for example, [K], [C], [S1]) allow to prove in such cases the existence and uniqueness of classical solutions of (1) on finite intervals of time. For these solutions (see, for example [S2])

$$|v(k,t)| \leq \text{const} \exp\{-\text{const}\sqrt{t}\cdot|k|\}, \; 0 \leq t \leq t_0 . \tag{2}$$

Presumably, $v(k,t)$ has an asymptotics of this type but this requires more work. According to a conventional wisdom, possible blow ups are connected with the violation of (2).

In this paper we fix t and consider one-parameter families of initial conditions $v_A(k,t) = Av(k,0)$. We show that for some special $v(k,0)$ one can find critical values $A_{cr} = A_{cr}(t)$ such that the solution $v_{A_{cr}}(k,s)$ blows up at t so that for $t' < t$ both the energy and the enstrophy are finite while at $t' = t$ they both become infinite. Even more, for $t' < t$ the solution decays exponentially outside some region depending on t. As $t' \uparrow t$ this region expands to an unbounded domain in R^3.

Our main approach is based on the renormalization group method which is so useful in probability theory, statistical physics and the theory of dynamical systems. It is rather difficult to give the exact formulation of our result in the introduction because it uses some notions, parameters, etc., which will appear in the later sections. Loosely speaking, we show that in ℓ-parameter families of initial conditions, for $\ell = 10$, one can find values of parameters for which the solutions develop blow ups of the type we already described. The meaning of ℓ is explained in §4, §5, §6.

We thank C. Fefferman, W.E, K. Khanin and V. Yakhot for many useful discussions. A big part of the text was prepared during the visit of the second author of the Mathematics Department of California Institute of Technology and we thank the Department for its very warm hospitality. We also thank G. Pecht for her excellent typing of the manuscript. The financial support from NSF Grant DMS 0600996 given to the second author is highly acknowledged.

§2. Power Series for Solutions of the 3$\mathcal{D}$-Navier-Stokes-Systems and Preliminary Changes of Variables

Our general approach is based upon the method of power series which were introduced in [S1], [S2]. We write down the solution of (1) in the form:

$$v_A(k,t) = \exp\{-t|k|^2\} \cdot A\,v(k,0) + \int_0^t \exp\{-(t-s)|k|^2\} \cdot \sum_{p>1} A^p\,h_p(k,s)\,ds \tag{3}$$

The substitution of (3) into (1) gives the system of recurrent equations connecting the functions h_p:

$$h_1(k,s) = \exp\{-s|k|^2\}\,v(k,0), \tag{4}$$

$$h_2(k,s) = i\int_{R^3} \langle v(k-k',0)\,,k\rangle\,P_k\,v(k',0) \cdot \exp\{-s|k-k'|^2 - s|k'|^2\}\,d^3k', \tag{5}$$

$$h_p(k,s) = i\int_0^s ds_2 \int_{R^3} \langle\, v(k-k',0),k\,\rangle\; P_k h_{p-1}(k',s_2) \cdot$$
$$\exp\{-s|k-k'|^2 - (s-s_2)|k'|^2\}\,d^3k' + i\sum_{\substack{p_1+p_2=p\\ p_1,p_2>1}} \int_0^s ds_1 \int_0^s ds_2 \int_{R^3} \langle\, h_{p_1}(k-k',s_1),k\,\rangle \cdot$$
$$P_k h_{p_2}(k',s_2) \cdot \exp\{-(s-s_1)|k-k'|^2 - (s-s_2)|k'|^2\}\,d^3k' +$$
$$i\int_0^s ds_1 \int_{R^3} \langle\, h_{p-1}(k-k',s_1),\,k\,\rangle\; P_k\,v(k',0) \cdot \exp\{-(s-s_1)|k-k'|^2 - s|k'|^2\}\,d^3k'. \tag{6}$$

Clearly, $h_p(k,s) \perp k$ for every $p \ge 1$, $k \in R^3$.

It follows from the results of [S2] that the series (3) converges for sufficiently small s and gives a classical solution of (1). Make the following change of variables which simplifies (4), (5), (6). Put $\tilde{k} = k\sqrt{s}$, $\tilde{k}' = k'\sqrt{s}$, introduce relative times $\tilde{s}_1, \tilde{s}_2, s_1 = \tilde{s}_1 s, s_2 = \tilde{s}_2 s$ and denote $g_r(\tilde{k},s) = h_r\left(\frac{\tilde{k}}{\sqrt{s}}, s\right), r \ge 1$. Then

$$g_1(\tilde{k},s) = \exp\{-|\tilde{k}|^2\} \cdot v\left(\frac{\tilde{k}}{\sqrt{s}},0\right), \tag{4'}$$

$$g_2(\tilde{k}, s) = h_2\left(\frac{\tilde{k}}{\sqrt{s}}, s\right) = \frac{i}{s^2}\int_{R^3} \langle v\left(\frac{\tilde{k}-\tilde{k}'}{\sqrt{s}}, 0\right), \tilde{k}\,\rangle \cdot$$

$$P_{\tilde{k}} v\left(\frac{\tilde{k}'}{\sqrt{s}}, 0\right) \exp\{-|\tilde{k}-\tilde{k}'|^2 - |\tilde{k}'|^2\}\, d^3\tilde{k}', \tag{5'}$$

$$g_p(\tilde{k}, s) = \frac{i}{s}\int_0^1 d\tilde{s}_2 \int_{R^3} \langle v\left(\frac{\tilde{k}-\tilde{k}'}{\sqrt{s}}, 0\right), \tilde{k}\,\rangle \cdot P_{\tilde{k}} g_{p-1}(\tilde{k}\sqrt{\tilde{s}_2}\,, \tilde{s}_2 s)$$

$$\exp\{-|\tilde{k}-\tilde{k}'|^2 - (1-\tilde{s}_2)|\tilde{k}'|^2\} d^3\tilde{k}' +$$

$$+\, i \sum_{\substack{p_1+p_2=p\\ p_1>1,\, p_2>1}} \int_0^1 d\tilde{s}_1 \int_0^1 d\tilde{s}_2 \int_{R^3} \langle\, g_{p_1}((\tilde{k}-\tilde{k}')\,\sqrt{\tilde{s}_1}, \tilde{s}_1 s), \tilde{k}\,\rangle \cdot$$

$$P_{\tilde{k}} g_{p_2}(\tilde{k}'\sqrt{\tilde{s}_2}, \tilde{s}_2 s)\, \exp\{-(1-\tilde{s}_1)\,|\tilde{k}-\tilde{k}'|^2 - (1-\tilde{s}_2)|\tilde{k}'|^2\}\, d^3\tilde{k}'$$

$$+\, \frac{i}{s}\int_0^1 d\tilde{s}_1 \int_{R^3} \langle\, g_{p-1}((\tilde{k}-\tilde{k}'))\,\sqrt{\tilde{s}_1}\,,\, \tilde{s}_1 s), \tilde{k}\rangle P_{\tilde{k}}\, v\left(\frac{\tilde{k}'}{\sqrt{s}}, 0\right) \cdot$$

$$\exp\{-(1-\tilde{s}_1)|\tilde{k}-\tilde{k}'|^2 - |\tilde{k}'|^2\}\, d^3\tilde{k}' \tag{6'}$$

The function $g_2(\tilde{k}, s)$ has a singularity at $s = 0$ even in the case of functions with compact support: for small s its values are of order $\frac{1}{\sqrt{s}}$. This singularity is integrable and all $g_p(k, s)$, $p > 2$, are bounded. The singularity is connected with our choice of the coordinates $\tilde{k}, \tilde{k}'$.

The formulas (4)-(6) or (4′)-(6′) resemble convolutions in probability theory. For example, if $C = \mathsf{supp}\, v(k, 0)$ then $\mathsf{supp}\, h_p = \underbrace{C + C + \cdots + C}_{p \text{ times}}$. Therefore it is natural to expect that

h_p and g_p satisfy some form of the limit theorem of probability theory. This question will be discussed in more detail in the next sections.

Make another change of variables. Assume that we have some p. The terms in (6′) with $p_1 \le p^{1/2}$ and $p_2 \le p^{1/2}$ will be called boundary terms. They will be treated as remainder terms and will be estimated later. Suppose that we have some number $\tilde{k}^{(0)}$ which later will be assumed to be sufficiently large. Introduce the vector $\widetilde{\mathcal{K}}^{(r)} = (0, 0, r\tilde{k}^{(0)})$. These will be the points near which all g_r will be concentrated, $p^{1/2} \le r \le p - p^{1/2}$. We write $\tilde{k} = \widetilde{\mathcal{K}}^{(r)} + \sqrt{r} \cdot Y, Y \in R^3$. Thus instead of $\tilde{k}$ we have the new variable $Y = (Y_1, Y_2, Y_3)$ which typically will take values $O(1)$. Put $\tilde{\kappa}^{(0)} = (0, 0, \tilde{k}^{(0)})$.

In all integrals over $\tilde{s}_1, \tilde{s}_2$ in (6′) make another change of variables $1 - \tilde{s}_j = \frac{\theta_j}{p_j^2}$, $j = 1, 2$. Instead of the variable of integration $\tilde{k}'$ introduce Y' where $\tilde{k}' = \widetilde{\mathcal{K}}^{(p_2)} + \sqrt{p}Y'$. We write $\tilde{g}_r(Y, s) = g_r(\widetilde{\mathcal{K}}^{(r)} + \sqrt{r}\,Y, s)$, $\gamma = \frac{p_1}{p}$, $\frac{p_2}{p} = 1 - \gamma$. Then from (6′)

$$\tilde{g}_p(Y, s) = g_p(\widetilde{\mathcal{K}}^{(p)} + \sqrt{p}\,Y, s) = p^{5/2} \Bigg[i \sum_{\substack{p_1 \cdot p_2 > \sqrt{p} \\ p_1 + p_2 = p}} \int_0^{p_1^2} d\theta_1 \int_0^{p_2^2} d\theta_2 \cdot \frac{1}{p_1^2 \cdot p_2^2} \cdot$$
$$\int_{R^3} \left\langle \tilde{g}_{p_1}\left(\frac{Y - Y'}{\sqrt{\gamma}}, \left(1 - \frac{\theta_1}{p_1^2}\right) s\right), \tilde{\kappa}^{(0)} + \frac{Y}{\sqrt{p}} \right\rangle \cdot P_{\tilde{\kappa}^{(0)} + \frac{Y}{\sqrt{p}}}\, \tilde{g}_{p_2}\left(\frac{Y'}{\sqrt{1-\gamma}}, \left(1 - \frac{\theta_2}{p_2^2}\right) s\right) \cdot$$
$$\exp\left\{ -\theta_1 \left| \tilde{\kappa}^{(0)} + \frac{Y - Y'}{\sqrt{p} \cdot \gamma} \right|^2 - \theta_2 \left| \tilde{\kappa}^{(0)} + \frac{Y'}{\sqrt{p}\,(1 - \gamma)} \right|^2 \right\} d^3 Y' \Bigg]. \tag{7}$$

This is the main recurrent relation which we shall study in the next sections. It is of some importance that in front of (7) we have the factor $p^{5/2}$ and inside the sum the factor $\frac{1}{p_1^2} \cdot \frac{1}{p_2^2}$. Both are connected with the new scaling inherent to the Navier-Stokes system.

§3. The Renormalization Group Equation

As $p \longrightarrow \infty$ the recurrent equation (7) takes some limiting form which will be derived in this section. All remainders which appear in this way are listed and estimated in §8.

The main contribution to (7) comes from p_1, p_2 of order p. If $Y, Y' = O(1)$ then $\frac{Y - Y'}{\sqrt{p}}$, $\frac{Y'}{\sqrt{p}}$ are small compared to $\tilde{\kappa}^{(0)} = (0, 0, \tilde{k}^{(0)})$. Therefore the Gaussian term in (7) can be replaced by $\exp\{-(\theta_1+\theta_2)|\tilde{k}^{(0)}|^2\}$, $\tilde{s}_1$ and $\tilde{s}_2$ can be replaced by 1 and the integrations over θ_1, θ_2 and Y' can be done separately. Thus instead of (7) we get a simpler recurrent relation:

$$\tilde{g}_p(Y,s) = \frac{i}{|\tilde{k}^{(0)}|^4} p^{5/2} \sum_{\substack{p_1\cdot p_2 > p^{1/2} \\ p_1+p_2=p}} \frac{1}{p_1^2\cdot p_2^2} \cdot \int_{R^3} \left\langle \tilde{g}_{p_1}\left(\frac{Y-Y'}{\sqrt{\gamma}}, s\right), \tilde{\kappa}^{(0)} + \frac{Y}{\sqrt{p}} \right\rangle \cdot$$

$$\cdot P_{\tilde{\kappa}^{(0)}+\frac{Y}{\sqrt{p}}}\, \tilde{g}_{p_2}\left(\frac{Y'}{\sqrt{(1-\gamma)}}, s\right) d^3Y' . \tag{8}$$

In view of incompressibility

$$\begin{aligned}
&\left\langle \tilde{g}_{p_1}\left(\tfrac{Y-Y'}{\sqrt{\gamma}}, s\right), \tilde{\kappa}^{(0)} + \tfrac{Y}{\sqrt{p}} \right\rangle = \\
&= \tfrac{1}{p_1} \left\langle g_{p_1}\left(\kappa^{(0)}\cdot p_1 + \tfrac{Y-Y'}{\sqrt{\gamma}}\cdot\sqrt{p_1}, s\right), \kappa^{(0)}p_1 + Y\cdot\gamma\sqrt{p} \right\rangle \\
&= \tfrac{1}{p_1} < g_{p_1}\left(\kappa^{(0)}p_1 + \tfrac{Y-Y'}{\sqrt{\gamma}}\cdot\sqrt{p_1}, s\right), \kappa^{(0)}p_1 + \tfrac{Y-Y'}{\sqrt{\gamma}}\sqrt{p_1} > + \\
&+ \tfrac{1}{p_1} < g_{p_1}\left(\kappa^{(0)}p_1 + \tfrac{Y-Y'}{\sqrt{\gamma}}\cdot\sqrt{p_1}, s\right), Y\gamma\cdot\sqrt{p} - (Y-Y')\sqrt{p} >= \\
&= \tfrac{1}{\sqrt{p_1}}\cdot < \tilde{g}_{p_1}\left(\tfrac{Y-Y'}{\sqrt{\gamma}}, s\right), \tfrac{Y-Y'}{\sqrt{\gamma}} > \cdot(\gamma-1) + \\
&+ \tfrac{1}{\sqrt{p_2}} < \tilde{g}_{p_1}\left(\tfrac{Y-Y'}{\sqrt{\gamma}}, s\right), Y'\sqrt{(1-\gamma)} >
\end{aligned} \tag{9}$$

Write $\tilde{g}_p$ in the form

$$\tilde{g}_p(Y,s) = (G_1^{(p)}(Y,s), G_2^{(p)}(Y,s), \frac{1}{\sqrt{p}} F^{(p)}(Y,s)) . \tag{10}$$

Since $\tilde{k} = \tilde{\kappa}^{(0)}\cdot p + Y\sqrt{p}$, the incompressibility implies

$$< g_r(\tilde{k},s), \tilde{k} > = < g_r(\tilde{k},s), \frac{\tilde{k}}{r} > = 0 \tag{11}$$

and for $Y = O(1)$

$$\frac{Y_1}{\sqrt{r}}\cdot G_1^{(r)}(Y,s) + \frac{Y_2}{\sqrt{r}} G_2^{(r)}(Y,s) + \frac{\tilde{k}^{(r)}}{\sqrt{r}}\cdot F^{(r)}(Y,s) = O\left(\frac{1}{r}\right) . \tag{12}$$

In our approximation we replace (12) by

$$Y_1 G_1^{(r)}(Y,s) + Y_2 G_2^{(r)}(Y,s) + F^{(r)}(Y,s) = 0\,. \tag{13}$$

Thus for given Y_1, Y_2, Y_3 the component F_r can be expressed through $G_1^{(r)}, G_2^{(r)}$. This remains to be true even if we do not neglect the rhs of (13). Return back to (9). From (13)

$$\begin{aligned}
&< \tilde{g}_{p_1}\left(\tfrac{Y-Y'}{\sqrt{\gamma}}, s\right), \tilde{\kappa}^{(0)} + \tfrac{Y}{\sqrt{p}} >= \tfrac{1}{\sqrt{p}}\Big[\tfrac{\gamma-1}{\sqrt{\gamma}} < \tilde{g}_{p_1}\left(\tfrac{Y-Y'}{\sqrt{\gamma}}, s\right), \tfrac{Y-Y'}{\sqrt{\gamma}} > \\
&+ \sqrt{1-\gamma} < \tilde{g}_{p_1}\left(\tfrac{Y-Y'}{\sqrt{\gamma}}, s\right), \tfrac{Y'}{\sqrt{1-\gamma}} > \Big] = \\
&= \tfrac{1}{\sqrt{p}}\Big[\tfrac{\gamma-1}{\sqrt{\gamma}}\Big(\tfrac{Y_1-Y_1'}{\sqrt{\gamma}} G_1^{(p_1)}\left(\tfrac{Y-Y'}{\sqrt{\gamma}}, s\right) + \tfrac{Y_2-Y_2'}{\sqrt{\gamma}} G_2^{(p_1)}\left(\tfrac{Y-Y'}{\sqrt{\gamma}}, s\right) \\
&+ \tfrac{Y_3-Y_3'}{\sqrt{\gamma}} \cdot \tfrac{1}{\sqrt{p_1}} \cdot F^{(p_1)}\left(\tfrac{Y-Y'}{\sqrt{\gamma}}, s\right) + \sqrt{(1-\gamma)}\Big(\tfrac{Y_1'}{\sqrt{1-\gamma}} G_1^{(p_1)}\left(\tfrac{Y-Y'}{\sqrt{\gamma}}, s\right) \\
&\qquad + \tfrac{Y_2'}{\sqrt{1-\gamma}} G_2^{(p_1)}\left(\tfrac{Y-Y'}{\sqrt{\gamma}}, s\right) + \tfrac{1}{\sqrt{p_2}}\tfrac{Y_3'}{\sqrt{1-\gamma}} F^{(p_1)}\left(\tfrac{Y-Y'}{\sqrt{\gamma}}, s\right)\Big)\Big]\,.
\end{aligned} \tag{14}$$

In our approximation the inner product in (14) can be replaced by

$$\begin{aligned}
&\tfrac{1}{\sqrt{p}}\Big[\tfrac{\gamma-1}{\sqrt{\gamma}}\Big(\tfrac{Y_1-Y_1'}{\sqrt{\gamma}} G_1^{(p_1)}\left(\tfrac{Y-Y'}{\sqrt{\gamma}}, s\right) + \tfrac{Y_2-Y_2'}{\sqrt{\gamma}} G_2^{(p_1)}\left(\tfrac{Y-Y'}{\sqrt{\gamma}}, s\right)\Big) \\
&\qquad + \sqrt{1-\gamma}\left(\frac{Y_1'}{\sqrt{1-\gamma}} G_1^{(p_1)}\left(\frac{Y-Y'}{\sqrt{\gamma}}, s\right) + \frac{Y_2}{\sqrt{1-\gamma}} G_2^{(p_1)}\left(\frac{Y-Y'}{\sqrt{\gamma}} s\right)\right)\Big]
\end{aligned} \tag{15}$$

According to the definition of the projector

$$P_{\tilde{\kappa}^{(0)}+\frac{Y}{\sqrt{p}}}\, \tilde{g}_{p_2}\left(\tfrac{Y'}{\sqrt{1-\gamma}}, s\right) = \tilde{g}_{p_2}\left(\tfrac{Y'}{\sqrt{1-\gamma}}, s\right)$$

$$-\frac{< \tilde{g}_{p_2}\left(\frac{Y'}{\sqrt{1-\gamma}}, s\right), \tilde{\kappa}^{(0)} + \frac{Y}{\sqrt{p}} > \cdot (\tilde{\kappa}^{(0)} + \frac{Y}{\sqrt{p}})}{< \tilde{\kappa}^{(0)} + \frac{Y}{\sqrt{p}}, \tilde{\kappa}^{(0)} + \frac{Y}{\sqrt{p}} >} = \tilde{g}_{p_2}\left(\frac{Y'}{\sqrt{1-\gamma}}, s\right) + O\left(\frac{1}{\sqrt{p_2}}\right). \tag{16}$$

This shows that in the main order of magnitude the projector is the identity operator and we come to a simpler recurrent relation instead of (8):

$$\tilde{g}_p(Y,s) = i\,\lambda_1 \cdot \sum_{\substack{p_1 \cdot p_2 > p^{1/2} \\ p_1+p_2=p}} \frac{p^2}{p_1^2 \cdot p_2^2} \int_{R^3} \left[\frac{\gamma-1}{\sqrt{\gamma}} \left(\frac{Y_1 - Y_1'}{\sqrt{\gamma}} \cdot \right.\right.$$

$$\left. G_1^{(p_1)} \left(\frac{Y-Y'}{\sqrt{\gamma}}, s \right) + \frac{Y_2 - Y_2'}{\sqrt{\gamma}} G_2^{(p_1)} \left(\frac{Y-Y'}{\sqrt{\gamma}}, s \right) \right) + \sqrt{1-\gamma}$$

$$\left. \left(\frac{Y_1'}{\sqrt{1-\gamma}} G_1^{(p_1)} \left(\frac{Y-Y'}{\sqrt{\gamma}}, s \right) + \frac{Y_2'}{\sqrt{1-\gamma}} G_2^{(p_1)} \left(\frac{Y-Y'}{\sqrt{\gamma}}, s \right) \right) \right] \cdot \tilde{g}_{p_2} \left(\frac{Y'}{\sqrt{1-\gamma}}, s \right) d^3Y'. \tag{17}$$

The main assumption which we shall check below in the next sections concerns the asymptotic form of $\tilde{g}_p(Y,s)$ as $p \longrightarrow \infty$: for some interval $S^{(p)} = [S_-^{(p)}, S_+^{(p)}]$ on the time axis and some Λ, positive $\sigma^{(1)}$, $\sigma^{(2)}$ and for all $r < p$

$$\tilde{g}_r(Y,s) = \Lambda^{r-1} r \cdot \frac{\sigma^{(1)}}{2\pi} \exp\left\{ -\frac{\sigma^{(1)}}{2} \left(|Y_1|^2 + |Y_2^2| \right) \right\} \cdot \sqrt{\frac{\sigma^{(2)}}{2\pi}} \exp\left\{ -\frac{\sigma^{(2)}}{2} |Y_3|^2 \right\} \cdot$$

$$\left(H_1(Y_1, Y_2, Y_3) + \delta_1^{(r)}(Y,s),\, H_2(Y_1, Y_2, Y_3) + \delta_2^{(r)}(Y,s),\, \delta_3^{(r)}(Y,s) \right) \tag{18}$$

where $\delta_j^{(r)}(Y,s) \longrightarrow 0$ as $r \longrightarrow \infty$, $j = 1,2,3$. Later we shall explain in more detail in what sense the convergence to zero takes place. The substitution of (18) into (17) gives

$$\tilde{g}_p(Y,s) = \frac{i}{|\tilde{k}^{(p)}|^4} \cdot p \cdot \Lambda^{p-2} \cdot$$

$$\sum_{\gamma = \frac{p_1}{p}} \frac{1}{p} \cdot \gamma^{\frac{1}{2}} (1-\gamma)^{\frac{1}{2}} \cdot \int_{R^3} \left[\frac{\gamma-1}{\sqrt{\gamma}} \cdot \left(\frac{Y_1 - Y_1'}{\sqrt{\gamma}} \cdot H_1 \left(\frac{Y-Y'}{\sqrt{\gamma}} \right) + \right.\right.$$

$$\left. + \frac{Y_2 - Y_2'}{\sqrt{\gamma}} H_2 \left(\frac{Y-Y'}{\sqrt{\gamma}} \right) \right) + \sqrt{1-\gamma} \left(\frac{Y_1'}{\sqrt{1-\gamma}} H_1 \left(\frac{Y-Y'}{\sqrt{\gamma}} \right) + \right.$$

$$\left.\left. + \frac{Y_2'}{\sqrt{1-\gamma}} H_2 \left(\frac{Y-Y'}{\sqrt{\gamma}} \right) \right) \right] H \left(\frac{Y'}{\sqrt{1-\gamma}} \right) \cdot$$

$$\frac{\sigma^{(1)}}{2\pi\gamma} \cdot \exp\left\{ -\frac{\sigma^{(1)}}{2} \left(\frac{|Y_1 - Y_1'|^2 + |Y_2 - Y_2'|^2}{\gamma} \right) \right\} \cdot$$

$$\frac{\sigma^{(1)}}{2\pi(1-\gamma)} \exp\left\{-\frac{\sigma^{(1)}}{2}\frac{|Y_1'|^2+|Y_2'|^2}{1-\gamma}\right\} \cdot$$

$$\sqrt{\frac{\sigma^{(2)}}{2\pi\gamma}} \exp\left\{-\frac{\sigma^{(2)}}{2}\frac{|Y_3-Y_3'|^2}{\gamma}\right\} \cdot \sqrt{\frac{\sigma^{(2)}}{2\pi(1-\gamma)}} \exp\left\{-\frac{\sigma^{(2)}}{2}\frac{|Y_3'|^2}{1-\gamma}\right\} d^3Y'. \tag{19}$$

Here

$$H\left(\frac{Y'}{\sqrt{1-\gamma}}\right) = \left(H_1\left(\frac{Y_1'}{\sqrt{1-\gamma}}, \frac{Y_2'}{\sqrt{1-\gamma}}, \frac{Y_3'}{\sqrt{1-\gamma}}\right),\right.$$

$$\left. H_2\left(\frac{Y_1'}{\sqrt{1-\gamma}}, \frac{Y_2'}{\sqrt{1-\gamma}}, \frac{Y_3'}{\sqrt{1-\gamma}}\right), 0\right).$$

We do not mention explicitly the dependence of H on s.

The last sum looks like a Riemannian integral sum whose limit takes the form as $p \longrightarrow \infty$:

$$\Lambda \exp\left\{-\tfrac{\sigma^{(1)}}{2}(|Y_1|^2+|Y_2|^2)\right\} \cdot \frac{\sigma^{(1)}}{2\pi} \cdot \exp\left\{-\frac{\sigma^{(2)}|Y_3|^2}{2}\right\} \sqrt{\frac{\sigma^{(2)}}{2\pi}}\, H(Y)$$

$$= \frac{i}{|\tilde{k}^{(0)}|^4} \int_0^1 \gamma^{\frac{1}{2}}(1-\gamma)^{\frac{1}{2}} d\gamma \int_{R^3} \frac{\sigma^{(1)}}{2\pi\gamma} \exp\left\{-\frac{\sigma^{(1)}(|Y_1-Y_1'|^2+|Y_2-Y_2'|^2)}{2\gamma}\right\}$$

$$\frac{\sigma^{(1)}}{2\pi(1-\gamma)} \exp\left\{-\frac{\sigma^{(1)}(|Y_1'|^2+|Y_2'|^2)}{2(1-\gamma)}\right\} \cdot \sqrt{\frac{\sigma^{(2)}}{2\pi\gamma}} \exp\left\{-\frac{\sigma^{(2)}|Y_3-Y_3'|^2}{2\gamma}\right\} \cdot$$

$$\sqrt{\frac{\sigma^{(2)}}{2\pi(1-\gamma)}} \exp\left\{-\frac{\sigma^{(2)}|Y_3'|^2}{2(1-\gamma)}\right\} \left[-\frac{\gamma-1}{\sqrt{\gamma}}\left(\frac{Y_1-Y_1'}{\sqrt{\gamma}} H_1\left(\frac{Y-Y'}{\sqrt{\gamma}}\right) + \frac{Y_2-Y_2'}{\sqrt{\gamma}} H_2\left(\frac{Y-Y'}{\sqrt{\gamma}}\right)\right)\right.$$

$$\left. + \gamma^{\frac{1}{2}}(1-\gamma)\left(\frac{Y_1'}{\sqrt{1-\gamma}} H_1\left(\frac{Y-Y'}{\sqrt{\gamma}}\right) + \frac{Y_2'}{\sqrt{1-\gamma}} H_2\left(\frac{Y-Y'}{\sqrt{1-\gamma}}\right)\right)\right] \cdot H\left(\frac{Y'}{\sqrt{1-\gamma}}\right) d^3Y'. \tag{20}$$

The integral over Y_3 is the usual convolution. Therefore we can look for functions H_1, H_2 depending only on Y_1, Y_2, i.e. $H_1(Y) = H_1(Y_1, Y_2)$, $H_2(Y) = H_2(Y_1, Y_2)$. Write down the equation for H_1, H_2 which does not contain Y_3:

$$\exp\left\{-\tfrac{\sigma^{(1)}}{2}|Y|^2\right\} \cdot \frac{\sigma^{(1)}}{2\pi} \cdot H(Y) = \int_0^1 d\gamma \int\limits_{R^2} \frac{\sigma^{(1)}}{2\pi\gamma} \cdot \exp\left\{-\frac{\sigma^{(1)}|Y-Y'|^2}{2\gamma}\right\} \cdot \frac{\sigma^{(1)}}{2\pi(1-\gamma)} \cdot$$

$$\exp\left\{-\tfrac{\sigma^{(1)}}{2(1-\gamma)} \cdot |Y'|^2\right\} \left[-(1-\gamma)^{3/2}\left(\tfrac{Y_1-Y_1'}{\sqrt{\gamma}} \cdot H_1\left(\tfrac{Y-Y'}{\sqrt{\gamma}}\right) + \tfrac{Y_2-Y_2'}{\sqrt{\gamma}} H_2\left(\tfrac{Y-Y'}{\sqrt{\gamma}}\right)\right)\right.$$

$$\left. + \gamma^{\frac{1}{2}}(1-\gamma)\left(\frac{Y_1'}{\sqrt{1-\gamma}} H_1\left(\frac{Y-Y'}{\sqrt{\gamma}}\right) + \frac{Y_2'}{\sqrt{1-\gamma}} H_2\left(\frac{Y-Y'}{\sqrt{\gamma}}\right)\right)\right] \cdot H\left(\frac{Y'}{\sqrt{1-\gamma}}\right) d^2Y'. \tag{21}$$

Here $Y = (Y_1, Y_2)$, $Y' = (Y_1', Y_2')$, $H(Y) = (H_1(Y_1, Y_2), H_2(Y_1, Y_2)$. This is our main equation for the fixed point of the renormalization group which we shall analyze in the next section (see also §7).

§4. Analysis of the Equation (21)

The solutions to the equation (21) have a natural scaling with respect to the parameter $\sigma = \sigma^{(1)}$. Namely, if we solve the equation (21) for $\sigma = 1$ and let the corresponding solution be $H(Y)$, then the general solution for arbitrary σ is given by the formula

$$H_\sigma(Y) = \sqrt{\sigma}\, H(\sqrt{\sigma}Y). \tag{22}$$

This is analogous to the usual scaling of the Gaussian fixed point in probability theory. Thus, it is enough to consider the equation (21) for $\sigma = 1$. We shall show that there exists a three-parameter family of solutions to the equation (21) for $\sigma = 1$. The equation (21) takes a simpler form if we use expansions over Hermite polynomials. All necessary facts about Hermite polynomials are collected in the Appendix 1. For $H(Y_1, Y_2) = (H_1(Y_1, Y_2), H_2(Y_1, Y_2))$, we write

$$H_j(Y_1, Y_2) = \sum_{m_1, m_2 \geq 0} h^{(j)}(m_1, m_2)\, He_{m_1}(Y_1)\, He_{m_2}(Y_2), \quad j = 1, 2 \tag{23}$$

where $He_m(z)$ are the Hermite polynomials of degree m with respect to the Gaussian density $\frac{1}{\sqrt{2\pi}} \exp\left\{-\frac{1}{2}z^2\right\}$. We have (see (42)):

$$zHe_m(z) = He_{m+1}(z) + mHe_{m-1}(z)\,,\ m > 0 \tag{24}$$

and

$$He_0(z) = 1, \qquad zHe_0(z) = z = He_1(z).$$

Also we use the formula (see (43))

$$\int_{\mathbb{R}^1} He_{m_1}\left(\frac{Y-Y'}{\sqrt{\gamma}}\right)\frac{1}{\sqrt{2\pi}}\exp\left\{-\frac{|Y-Y'|^2}{2\gamma}\right\} He_{m_2}\left(\frac{Y'}{\sqrt{1-\gamma}}\right)\frac{1}{\sqrt{2\pi}}$$

$$\exp\left\{-\frac{|Y'|^2}{2(1-\gamma)}\right\}dY' = \gamma^{\frac{m_1+1}{2}}(1-\gamma)^{\frac{m_2+1}{2}} He_{m_1+m_2}(Y)\frac{1}{\sqrt{2\pi}}\exp\left\{-\frac{|Y|^2}{2}\right\}. \tag{25}$$

Substituting (23) into (21) and using (24), (25), we come to the system of equations for the coefficients $h(m_1, m_2)$ which is equivalent to (21):

$$h^{(j)}(m_1, m_2) = \sum_{\substack{m_1'+m_1''=m_1\\ m_2'+m_2''=m_2}} J^{(1)}_{m'm''}\cdot\left\{(B_1h^{(1)})(m_1', m_2') + (B_2h^{(2)})(m_1', m_2')\right\} h^{(j)}(m_1'', m_2'')$$

$$+J^{(2)}_{m'm''}\cdot\left\{h^{(1)}(m_1', m_2')\,(B_1h^{(j)})(m_1'', m_2'') + h^{(2)}(m_1', m_2')\,(B_2h^{(j)})(m_1'', m_2'')\right\} \tag{26}$$

where $m' = m_1' + m_2'$, $m'' = m_1'' + m_2''$ and

$$\begin{cases} J^{(1)}_{m'm''} = -\displaystyle\int_0^1 \gamma^{\frac{m'}{2}}(1-\gamma)^{\frac{m''+3}{2}}d\gamma \\ J^{(2)}_{m'm''} = \displaystyle\int_0^1 \gamma^{\frac{m'+1}{2}}(1-\gamma)^{\frac{m''+2}{2}}d\gamma \end{cases} \tag{27}$$

$$(B_1\,h^{(j)})\,(m_1', m_2') = h^{(j)}\,(m_1'-1, m_2') + (m_1'+1)\,h^{(j)}(m_1'+1, m_2')$$

$$(B_2\,h^{(j)})\,(m_1', m_2') = h^{(j)}\,(m_1', m_2'-1) + (m_2'+1)\,h^{(j)}\,(m_1', m_2'+1)$$

To simplify the system (26), we shall look for solutions with $h^{(j)}(0,0) = 0$, $j = 1,2$. Below we sometimes write $h^{(j)}(m_1, m_2)$ as $h^{(j)}_{m_1m_2}$ or $h^{(j)}_{m_1,m_2}$ when there is no confusion. Similar conventions will be applied to $J^{(j)}(m_1, m_2)$. For $m_1 + m_2 = 1$, we have

$$\begin{cases} h_{10}^{(1)} & = & J_{01}^{(1)} \cdot (h_{10}^{(1)} + h_{01}^{(2)}) \cdot h_{10}^{(1)} + J_{10}^{(2)} \cdot (h_{10}^{(1)} h_{10}^{(1)} + h_{10}^{(2)} h_{01}^{(1)}) \\ h_{01}^{(1)} & = & J_{01}^{(1)} \cdot (h_{10}^{(1)} + h_{01}^{(2)}) \cdot h_{01}^{(1)} + J_{10}^{(2)} \cdot (h_{01}^{(1)} h_{10}^{(1)} + h_{01}^{(2)} h_{01}^{(1)}) \\ h_{10}^{(2)} & = & J_{01}^{(1)} \cdot (h_{10}^{(1)} + h_{01}^{(2)}) \cdot h_{10}^{(2)} + J_{10}^{(2)} \cdot (h_{10}^{(1)} h_{10}^{(2)} + h_{10}^{(2)} h_{01}^{(2)}) \\ h_{01}^{(2)} & = & J_{01}^{(1)} \cdot (h_{10}^{(1)} + h_{01}^{(2)}) \cdot h_{01}^{(2)} + J_{10}^{(2)} \cdot (h_{01}^{(1)} h_{10}^{(2)} + h_{01}^{(2)} h_{01}^{(2)}) \end{cases}$$

where $J_{01}^{(1)} = -1/3$ and $J_{10}^{(2)} = 1/6$. There are two cases:

Case 1. $h_{10}^{(1)} + h_{01}^{(2)} = -6$. In this case $(h_{10}^{(1)}, h_{01}^{(1)}, h_{10}^{(2)}, h_{01}^{(2)})$ only needs to satisfy:

$$(h_{10}^{(1)} + 3)^2 = 9 - h_{01}^{(1)} h_{10}^{(2)}$$

This is a two parameter family of solutions.

Case 2. $h_{10}^{(1)} + h_{01}^{(2)} \neq -6$. In this case $(h_{10}^{(1)}, h_{01}^{(1)}, h_{10}^{(2)}, h_{01}^{(2)})$ can be uniquely determined and we have $h_{10}^{(1)} = h_{01}^{(2)} = -2$, $h_{01}^{(1)} = h_{10}^{(2)} = 0$.

For the rest of this paper we shall consider only the case 2 for which $h_{10}^{(1)} = h_{01}^{(2)} = -2$, $h_{01}^{(1)} = h_{10}^{(2)} = 0$. Let us write down the recurrent relations for $m_1 + m_2 = 2$, $j = 1, 2$:

$$\begin{cases} h_{20}^{(j)} & = & -(2J_{20}^{(2)} + 4J_{02}^{(1)} + 4J_{11}^{(2)}) h_{20}^{(j)} + 2J_{11}^{(1)} \cdot h_{10}^{(j)} \cdot h_{20}^{(1)} + h_{10}^{(j)} \cdot J_{11}^{(1)} \cdot h_{11}^{(2)} \\ h_{11}^{(j)} & = & -(2J_{20}^{(2)} + 4J_{02}^{(1)} + 4J_{11}^{(2)}) h_{11}^{(j)} + J_{11}^{(1)} h_{01}^{(j)} \cdot (2h_{20}^{(1)} + h_{11}^{(2)}) + J_{11}^{(1)} \cdot h_{10}^{(j)} \cdot (h_{11}^{(1)} + 2h_{02}^{(2)}) \\ h_{02}^{(j)} & = & -(2J_{20}^{(2)} + 4J_{02}^{(1)} + 4J_{11}^{(2)}) h_{02}^{(j)} + 2J_{11}^{(1)} \cdot h_{01}^{(j)} \cdot h_{02}^{(2)} + h_{01}^{(j)} \cdot J_{11}^{(1)} \cdot h_{11}^{(1)} \end{cases}$$

It is not difficult to check that the only solution to the above system is $h_{20}^{(j)} = h_{02}^{(j)} = h_{11}^{(j)} = 0$. Solving the recurrent relations for $m_1 + m_2 = 3$ gives us:

$$\begin{cases} h_{03}^{(1)} = h_{30}^{(2)} = 0 \\ h_{12}^{(1)} = h_{03}^{(2)} \\ h_{21}^{(1)} = h_{12}^{(2)} \\ h_{30}^{(1)} = h_{21}^{(2)} \end{cases}$$

This shows that $(h_{12}^{(1)}, h_{21}^{(1)}, h_{30}^{(1)})$ can be considered as free parameters. For any $p \geq 4$, the recurrent relations for $m_1 + m_2 = p$ form a linear system of equations for the variables

$\{h^{(j)}_{m_1,p-m_1}\}^p_{m_1=0}$ with coefficients depending on $h^{(j)}_{01}$ and $h^{(j)}_{10}$ only. In principle, they can be solved and an explicit expression for the solutions can be found. We emphasize here that if the free parameters take real values then the whole solution is also real.

It is not difficult to check that for any values of $(h^{(1)}_{12}, h^{(1)}_{21}, h^{(1)}_{30})$, one can find all $h^{(j)}_{m_1,m_2}$ $(m_1+m_2 \geq 4)$ by using (26). The solution we obtain is formal in the sense that it satisfies (26) but h_{m_1,m_2} with $m_1+m_2=p$ may not decay as $p \longrightarrow \infty$. We are now ready to formulate the theorem concerning the existence of formal solutions to (26).

Theorem 4.1. *For any values of $(h^{(1)}_{12}, h^{(1)}_{21}, h^{(1)}_{30})$, there exists a unique formal solution to the recurrent equation (26).*

Thus, theorem 4.1 claims the existence of a three-parameter family of solutions of (21) parameterized by $h^{(1)}_{12}$, $h^{(1)}_{21}$ and $h^{(1)}_{30}$. It turns out that if $h^{(1)}_{12}$, $h^{(1)}_{21}$ and $h^{(1)}_{30}$ are sufficiently small, then $h^{(j)}_{m_1,m_2}$ decay as $m_1+m_2=p$ tends to infinity. Let us say that $h^{(j)}_{m_1,m_2}$ has degree d if $m_1+m_2=d$. For each $d \geq 4$, introduce the vector $h^{(d)} = (h^{(1)}_{0,d}, h^{(1)}_{1,d-1}, \ldots, h^{(1)}_{d,0}, h^{(2)}_{0,d}, \ldots, h^{(2)}_{d,0})^T$. The vector $h^{(d)}$ contains all terms of degree d. By the recurrent relation (26)

$$C^{(d)}\, h^{(d)} = b^{(d)} \tag{28}$$

where the vector $b^{(d)}$ contains terms of degree $\leq d-1$. Also $C^{(d)} \in \mathbb{R}^{(2d+2)\times(2d+2)}$ is a matrix:

$$C^{(d)}_{k\ell} = \begin{cases} 1 - \dfrac{16d-16+32k}{(d+1)(d+3)(d+5)}, & \text{if } 1 \leq k = l \leq d+1 \\[2ex] 1 - \dfrac{80d+80-32k}{(d+1)(d+3)(d+5)}, & \text{if } d+2 \leq k = l \leq 2d+2 \\[2ex] -\dfrac{32(d-k+2)}{(d+1)(d+3)(d+5)}, & \text{if } 2 \leq k \leq d+1, l = d+k \\[2ex] -\dfrac{32(k-d-1)}{(d+1)(d+3)(d+5)}, & \text{if } d+2 \leq k \leq 2d+1, l = k-d \\[2ex] 0, & \text{all other cases} \end{cases}$$

It is easy to check that if $d \geq 4$, then $C^{(d)}$ is nonsingular and as $d \longrightarrow \infty$, $C^{(d)}$ converges to the identity matrix. This observation immediately implies the following lemma:

Lemma 4.2. *Let $(C^{(d)})^{-1}$ be the inverse matrix of $C^{(d)}$ for $d \geq 4$. There exists an absolute constant $C_1 > 0$ such that for all $d \geq 4$*

$$\| (C^{(d)})^{-1} \| \leq C_1 .$$

We are now ready to derive an estimate which gives the decay of solutions of the recurrent relation (26).

Theorem 4.3. *If $|h_{12}^{(1)}| \leq \delta$, $|h_{21}^{(1)}| \leq \delta$, $|h_{30}^{(1)}| \leq \delta$ and δ is sufficiently small, then for some $C_2 > 0$, $0 < \rho < \frac{1}{4}$, we have*

$$|h_{m_1,m_2}^{(j)}| \leq C_2 \frac{\rho^{m_1+m_2}}{\Gamma\left(\frac{m_1+m_2+7}{2}\right)} \qquad \forall\, m_1 \geq 0,\ m_2 \geq 0,\ j = 1, 2.$$

Proof. We begin by noting that $h_{m_1 m_2}^{(j)} = 0$ if $m_1 + m_2$ is even. This can be easily proven by using the recurrent relation (26) and the fact that $h_{00}^{(j)} = 0$ and $h_{m_1,m_2}^{(j)} = 0$ for $m_1 + m_2 = 2$. Let $0 < \rho_1 < 1$, ρ_1 will be chosen sufficiently small. We shall use induction on $m_1 + m_2$ where $m_1 + m_2$ is odd. According to the induction hypothesis

$$|h_{m_1,m_2}^{(j)}| \leq \frac{\rho_1^{m_1+m_2+2}}{\Gamma\left(\frac{m_1+m_2+7}{2}\right)} g(m_1 + m_2) \tag{29}$$

for every $3 \leq m_1 + m_2 \leq d - 2$ where $d \geq L$ is an odd number and L will be chosen later to be sufficiently large. Also g is a function to be specified later. We shall comment on the choice of L and verify the induction hypothesis for $3 \leq m_1 + m_2 \leq L$ later. Let us show that the same inequality holds for $m_1 + m_2 = d$. Without any loss of generality, let us consider $j = 1$. The case $j = 2$ is similar. Fix m_1 and let $b_{m_1}^{(d)}$ be the $(m_1 + 1)^{\text{th}}$ component of the vector $b^{(d)}$ in the equation (28). We now estimate $b_{m_1}^{(d)}$ using the induction hypothesis (29) and the equation (26):

$$\begin{aligned}
|b_{m_1}^{(d)}| \le & \sum_{m'=2}^{d-3} \left|J^{(1)}_{m',m''}\right| \cdot 2 \cdot \frac{\rho_1^{m'+3}}{\Gamma\left(\frac{m'+8}{2}\right)} \cdot \frac{\rho_1^{m''+2}}{\Gamma\left(\frac{m''+7}{2}\right)} \cdot (m'+1)g(m'+1)g(m'') \\
& + \sum_{m'=4}^{d-3} \left|J^{(1)}_{m',m''}\right| \cdot 2 \cdot \frac{\rho_1^{m'+1}}{\Gamma\left(\frac{m'+6}{2}\right)} \cdot \frac{\rho_1^{m''+2}}{\Gamma\left(\frac{m''+7}{2}\right)} \cdot (m'+1) \cdot g(m'-1)g(m'') \\
& + \sum_{m'=3}^{d-2} \left|J^{(2)}_{m',m''}\right| \cdot 2 \cdot \frac{\rho_1^{m'+2}}{\Gamma\left(\frac{m'+7}{2}\right)} \cdot \frac{\rho_1^{m''+3}}{\Gamma\left(\frac{m''+8}{2}\right)} \cdot (m'+1) \cdot (m''+1) \cdot g(m')g(m''+1) \\
& + \sum_{m'=3}^{d-4} \left|J^{(2)}_{m',m''}\right| \cdot 2 \cdot \frac{\rho_1^{m'+2}}{\Gamma\left(\frac{m'+7}{2}\right)} \cdot \frac{\rho_1^{m''+1}}{\Gamma\left(\frac{m''+6}{2}\right)} \cdot (m'+1) \cdot g(m')g(m''-1) \\
& + 12\left(\left|J^{(1)}_{2,d-2}\right| + \left|J^{(1)}_{d-1,1}\right| + \left|J^{(2)}_{d-2,2}\right| + \left|J^{(2)}_{1,d-1}\right|\right) \cdot \frac{\rho_1^{d}}{\Gamma\left(\frac{d+5}{2}\right)} \cdot g(d-2)
\end{aligned}$$

The last term in the rhs of the above inequality comes from the case where $h_{m_1'm_2'}$ or $h_{m_1''m_2''}$ is of degree one since the induction hypothesis holds only for $3 \le m_1 + m_2 \le d-2$. Also in the estimation of the first four terms we use the fact that for fixed (m', m_1), there are at most $\min\{m'+1, m''+1\}$ tuples of (m', m_1'', m_2', m_2'') such that $m_1' + m_1'' = m_1$, $m_2' + m_2'' = m_2$, $m_1' + m_2' = m'$ and $m_1'' + m_2'' = m''$. By (27), we have

$$\left|J^{(1)}_{m',m''}\right| = \frac{\Gamma\left(\frac{m'+2}{2}\right)\Gamma\left(\frac{m''+5}{2}\right)}{\Gamma\left(\frac{m'+m''+7}{2}\right)}$$

$$\left|J^{(2)}_{m',m''}\right| = \frac{\Gamma\left(\frac{m'+3}{2}\right)\Gamma\left(\frac{m''+4}{2}\right)}{\Gamma\left(\frac{m'+m''+7}{2}\right)}$$

and for some constant $C_3 > 0$

$$\left|J^{(1)}_{2,d-2}\right| + \left|J^{(1)}_{d-1,1}\right| + \left|J^{(2)}_{d-2,2}\right| + \left|J^{(2)}_{1,d-1}\right| \le \frac{C_3}{d^2}$$

Therefore

$$
\begin{aligned}
|b_{m_1}^{(d)}| \leq & \frac{2\rho_1^{d+5}}{\Gamma\left(\frac{d+7}{2}\right)} \cdot \sum_{m'=2}^{d-3} \frac{\Gamma\left(\frac{m'+2}{2}\right)\cdot(m'+1)}{\Gamma\left(\frac{m'+8}{2}\right)} \cdot \frac{\Gamma\left(\frac{m''+5}{2}\right)}{\Gamma\left(\frac{m''+7}{2}\right)} \cdot g(m'+1)g(m'') \\
& + \frac{2\rho_1^{d+3}}{\Gamma\left(\frac{d+7}{2}\right)} \cdot \sum_{m'=4}^{d-3} \frac{\Gamma\left(\frac{m'+2}{2}\right)\cdot(m'+1)}{\Gamma\left(\frac{m'+6}{2}\right)} \cdot \frac{\Gamma\left(\frac{m''+5}{2}\right)}{\Gamma\left(\frac{m''+7}{2}\right)} \cdot g(m'-1)g(m'') \\
& + \frac{2\rho_1^{d+5}}{\Gamma\left(\frac{d+7}{2}\right)} \cdot \sum_{m'=3}^{d-2} \frac{\Gamma\left(\frac{m'+3}{2}\right)\cdot(m'+1)}{\Gamma\left(\frac{m'+7}{2}\right)} \cdot \frac{\Gamma\left(\frac{m''+4}{2}\right)\cdot(m''+1)}{\Gamma\left(\frac{m''+8}{2}\right)} \cdot g(m')g(m''+1) \\
& + \frac{2\rho_1^{d+3}}{\Gamma\left(\frac{d+7}{2}\right)} \cdot \sum_{m'=3}^{d-4} \frac{\Gamma\left(\frac{m'+3}{2}\right)\cdot(m'+1)}{\Gamma\left(\frac{m'+7}{2}\right)} \cdot \frac{\Gamma\left(\frac{m''+4}{2}\right)}{\Gamma\left(\frac{m''+6}{2}\right)} \cdot g(m')g(m''-1) \\
& + \frac{\rho_1^{d+2}}{\Gamma\left(\frac{d+7}{2}\right)} \cdot \frac{C_3}{d^2} \cdot \frac{\Gamma\left(\frac{d+7}{2}\right)}{\Gamma\left(\frac{d+5}{2}\right)} \cdot \frac{12}{\rho_1^2} \cdot g(d-2) \\
\leq & \frac{\rho_1^{d+2}}{\Gamma\left(\frac{d+7}{2}\right)} \cdot \rho_1 \cdot C_4 \cdot \left(\sum_{m'=2}^{d-3} g(m'+1)g(m'') + \sum_{m'=4}^{d-3} g(m'-1)g(m'') \right. \\
& \left. \sum_{m'=3}^{d-2} g(m')g(m''+1) + \sum_{m'=3}^{d-4} g(m')g(m''-1) \right) + \frac{\rho_1^{d+2}}{\Gamma\left(\frac{d+7}{2}\right)} \cdot C_5 \cdot \frac{g(d-2)}{d\cdot\rho_1}
\end{aligned}
$$

where C_4, C_5 are some constants. Now we specify the choice of the function g. Let $g(m)$ be such that $g_1 = \alpha$ and

$$
g(m) = \sum_{p=1}^{m-1} g(p)g(m-p) \qquad \text{for } m > 1
$$

By the method of formal power series it is not difficult to show that

$$
g(m) = \frac{1}{2} \cdot \frac{(2m-1)!!}{m!} \cdot (2\alpha)^m
$$

Clearly, we have $const \leq \frac{g(m+1)}{g(m)} \leq const$, and this immediately gives us

$$
\begin{aligned}
|b_{m_1}^{(d)}| \leq & \frac{\rho_1^{d+2}}{\Gamma\left(\frac{d+7}{2}\right)} \cdot C_6 \cdot \sum_{m'=1}^{d} g(m')g(d-m') + \frac{\rho_1^{d+2}}{\Gamma\left(\frac{d+7}{2}\right)} \cdot \frac{C_6}{d\cdot\rho_1} g(d) \\
\leq & \frac{\rho_1^{d+2}}{\Gamma\left(\frac{d+7}{2}\right)} \cdot g(d) \cdot \left(C_6\rho_1 + \frac{C_6}{d\cdot\rho_1} \right)
\end{aligned}
$$

where $C_6 > 0$ is some constant. Now by Lemma 4.2, we obtain that

$$
|h_{m_1 m_2}| \leq \frac{\rho_1^{d+2}}{\Gamma\left(\frac{d+7}{2}\right)} g(d) \cdot C_1 \cdot \left(C_6\rho_1 + \frac{C_6}{d\cdot\rho_1} \right)
$$

Choose ρ_1 so small that $C_1C_6\rho_1 < \frac{1}{2}$ and $\rho_1 \cdot 4\alpha < \frac{1}{4}$. Then take L so large that $\frac{C_1C_6}{\rho_1 L} < \frac{1}{2}$. This clearly implies

$$|h_{m_1,m_2}| \leq \frac{\rho_1^{d+2}}{\Gamma\left(\frac{d+7}{2}\right)} g(d)$$

We now justify the induction hypothesis (29). Recall that our free parameters are $h_{12}^{(1)}$, $h_{21}^{(1)}$ and $h_{30}^{(1)}$. It is easy to check that if we set $h_{12}^{(1)} = h_{21}^{(1)} = h_{30}^{(1)} = 0$, then $h_{m_1 m_2} = 0$ for any $m_1 + m_2 \geq 2$. Since L is fixed, and $0 < |h_{12}^{(1)}| < \delta$, $0 < |h_{21}^{(1)}| < \delta$, $0 < |h_{30}^{(1)}| < \delta$ with sufficiently small δ, then the induction hypothesis is satisfied. A simple estimate on g gives that

$$g(m) \leq (4\alpha)^m$$

Thus the theorem is proven if one takes $\rho = 4\alpha\rho_1$.

As it is stated our solutions of (20) are determined by five parameters $\sigma^{(1)}, h_{12}^{(1)}, h_{21}^{(1)}, h_{30}^{(1)}, \sigma^{(2)}$. However, it turns out that these parameters are not independent and σ_1 can be expressed through ($h_{12}^{(1)}, h_{21}^{(1)}, h_{30}^{(1)}$). Namely, let $G^{\sigma^{(1)}, h_{12}^{(1)}, h_{21}^{(1)}, h_{30}^{(1)}, \sigma^{(2)}}(Y)$ be the solution of (20). Then

$$G^{(\sigma^{(1)}, h_{12}^{(1)}, h_{21}^{(1)}, h_{30}^{(1)}, \sigma^{(2)})}(Y) = G^{(1, \sigma^{(1)}(h_{12}^{(1)} - 1) + 1\,,\, \sigma^{(1)} h_{21}^{(1)},\, \sigma^{(1)}(h_{30}^{(1)} - 1) + 1,\, \sigma^{(2)})}(Y).$$

This equality is proved at the end of §6. We formulate now the final result concerning the existence of solutions of (21).

Theorem 4.2. *Let $\sigma^{(1)} > 0$, $\sigma^{(2)} > 0$ and $h_{12}^{(1)}, h_{21}^{(1)}, h_{30}^{(1)}$ be sufficiently small. Then there exists a solution of (20) which has the following form*

$$G^{(\sigma^{(1)}, h_{12}^{(1)}, h_{21}^{(1)}, h_{30}^{(1)}, \sigma^{(2)})}(Y_1, Y_2, Y_3) = \exp\left\{-\frac{\sigma^{(1)}}{2}\left(|Y_1|^2 + |Y_2|^2\right)\right\} \cdot$$

$$\cdot \frac{\sigma^{(1)}}{2\pi} \cdot \exp\left\{-\frac{\sigma^{(2)}}{2}|Y_3|^2\right\} \sqrt{\frac{\sigma^{(2)}}{2\pi}} \cdot \sqrt{\sigma^{(1)}}\, H^{(h_{12}^{(1)}, h_{21}^{(1)}, h_{30}^{(1)})}\left(\sqrt{\sigma^{(1)}}\, Y_1, \sqrt{\sigma^{(1)}}\, Y_2\right).$$

Here $H^{(h_{12}^{(1)}, h_{21}^{(1)}, h_{30}^{(1)})}$ is the solution of (21) with the given $h_{12}^{(1)}, h_{21}^{(1)}, h_{30}^{(1)}$.

As it was already mentioned the parameters σ_1, $h_{12}^{(1)}$, $h_{21}^{(1)}$, $h_{30}^{(1)}$, σ_2 are not independent and actually the set of solutions depends on four independent parameters (see Lemma 6.2).

From the estimate in Theorem 4.3 and from known asymptotic formulas for the Hermite polynomials it follows that the series giving $H^{(h_{12}^{(1)}, h_{21}^{(1)}, h_{30}^{(1)})}$ converges for every $Y = (Y_1, Y_2)$. Better estimates are also easily available.

§5. The Linearization Near Fixed Point

Denote $h_{12}^{(1)} = x^{(1)}$, $h_{21}^{(1)} = x^{(2)}$, $h_{30}^{(1)} = x^{(3)}$. Our fixed points have the following form

$$
\begin{aligned}
G^{(\sigma^{(1)},x^{(1)},x^{(2)},x^{(3)},\sigma^{(2)})} &= \frac{\sigma^{(1)}}{2\pi}\exp\left\{-\frac{\sigma^{(1)}(Y_1^2+Y_2^2)}{2}\right\}\cdot \\
&\cdot\sqrt{\frac{\sigma^{(2)}}{2\pi}}\exp\left\{-\frac{\sigma^{(2)}Y_3^2}{2}\right\}\left(H_1^{(\sigma^{(1)},x^{(1)},x^{(2)},x^{(3)})}(Y_1,Y_2),\ H_2^{(\sigma^{(1)},x^{(1)},x^{(2)},x^{(3)})}(Y_1,Y_2),\ 0\right)
\end{aligned}
\tag{30}
$$

Recall that $H^{(\sigma^{(1)},x^{(1)},x^{(2)},x^{(3)})} = \sqrt{\sigma^{(1)}}H^{(1,x^{(1)},x^{(2)},x^{(3)})}(\sqrt{\sigma^{(1)}}Y_1,\sqrt{\sigma^{(1)}}Y_2)$ and $H^{(1,x^{(1)},x^{(2)},x^{(3)})}$ are described in §4.

The strategy of the proof of the main result is based on the method of renormalization group. At the p-th step of our procedure, we consider an interval on the time axis $S^{(p)} = \left[S_-^{(p)}, S_+^{(p)}\right]$ such that $S^{(p+1)} \subseteq S^{(p)}$. From our estimates it will follow that $\bigcap\limits_p S^{(p)} = [S_-, S_+]$ is an interval of positive length. We want to find conditions under which $\tilde{g}_r(Y,s)$, $s \in S^{(p)}$, have a representation

$$
\begin{aligned}
\tilde{g}_r(Y,s) &= \Lambda^{r-1}r\frac{\sigma^{(1)}}{2\pi}\exp\left\{-\frac{\sigma^{(1)}(Y_1^2+Y_2^2)}{2}\right\}\sqrt{\frac{\sigma^{(2)}}{2\pi}}\exp\left\{-\frac{\sigma^{(2)}Y_3^2}{2}\right\}\cdot \\
&\left(H_1^{(\sigma^{(1)},x^{(1)},x^{(2)},x^{(3)})}(Y)+\delta_1^{(r)}(Y,s),\ H_2^{(\sigma^{(1)},x^{(1)},x^{(2)},x^{(3)})}(Y)+\delta_2^{(r)}(Y,s),\ \delta_3^{(r)}(Y,s)\right)
\end{aligned}
$$

where $\delta_1^{(r)}$, $\delta_2^{(r)}$, $\delta_3^{(r)}$ tend to zero as $r\to\infty$. The renormalization is based on the crucial observation (see above) that for large p, the sum over p_1 is a Riemannian integral sum for an integral over γ changing from 0 to 1. Let us write

$$
\begin{aligned}
\tilde{g}_r(Y,s)\,\Lambda^{-r+1}(r^{-1}\exp\left\{\frac{\sigma^{(1)}(Y_1^2+Y_2^2)}{2}+\frac{\sigma^{(2)}Y_3^2}{2}\right\}\left(\frac{2\pi}{\sigma^{(1)}}\right)\left(\frac{2\pi}{\sigma^{(2)}}\right)^{\frac{1}{2}} &= \\
= H^{(\sigma^{(1)},x^{(1)},x^{(2)},x^{(3)})}(Y_1,Y_2) + \delta^{(r)}(\gamma,Y,s)&
\end{aligned}
\tag{31}
$$

where $\delta^{(r)}(\gamma,Y,s) = \left\{\delta_j^{(r)}(\gamma,Y,s),\ 1\le j\le 3\right\} = \delta^{(p)}(\gamma,Y,s)$, $\gamma = \dfrac{r}{p}$. It is natural to consider the set of functions $\{\delta^{(p)}(\gamma,Y,s)\}$ as a small perturbation of our fixed point (30). Recall that the third component of $H^{(\sigma^{(1)},x^{(1)},x^{(2)},x^{(3)})}$ is zero because of incompressibility and $\delta_3^{(p)}$ can be found from the incompressibility condition. Clearly,

$$
\delta^{(p+1)}(\gamma,Y,s) = \delta^{(p)}\left(\frac{p+1}{p}\gamma,Y,s\right),\quad \gamma\le\frac{p}{p+1}.
$$

The formula for $\delta^{(p+1)}(1, Y, s)$ follows from (21):

$$\exp\left\{-\frac{\sigma^{(1)}}{2}(|Y_1|^2 + |Y_2|^2) - \frac{\sigma^{(2)}}{2}|Y_3|^2\right\} \cdot \frac{\sigma^{(1)}}{2\pi}\sqrt{\frac{\sigma^{(2)}}{2\pi}} \cdot \delta_j^{(p+1)}(1, Y, s)$$

$$= \int_0^1 d\gamma \int_{\mathbb{R}^3} \frac{\sigma^{(1)}}{2\pi\gamma} \cdot \sqrt{\frac{\sigma^{(1)}}{2\pi\gamma}} \cdot \frac{\sigma^{(2)}}{2\pi(1-\gamma)} \cdot \sqrt{\frac{\sigma^{(2)}}{2\pi(1-\gamma)}} \cdot$$

$$\exp\left\{-\frac{\sigma^{(1)}(|Y_1 - Y_1'|^2 + |Y_2 - Y_2'|^2)}{2\pi\gamma} - \frac{\sigma^{(2)}|Y_3 - Y_3'|^2}{2\pi\gamma} - \frac{\sigma^{(1)}(|Y_1'|^2 + |Y_2'|^2)}{2\pi(1-\gamma)} - \right.$$

$$\left. - \frac{\sigma^{(2)}|Y_3'|^2}{2\pi(1-\gamma)}\right\} \left\{\left[-(1-\gamma)^{\frac{3}{2}}\left(\frac{Y_1 - Y_1'}{\sqrt{\gamma}} H_1\left(\frac{Y - Y'}{\sqrt{\gamma}}\right) + \frac{Y_2 - Y_2'}{\sqrt{\gamma}} H_2\left(\frac{Y - Y'}{\sqrt{\gamma}}\right)\right)\right.\right.$$

$$+ \gamma^{\frac{1}{2}}(1-\gamma)\left(\frac{Y_1'}{\sqrt{1-\gamma}} H_1\left(\frac{Y - Y'}{\sqrt{\gamma}}\right) + \frac{Y_2'}{\sqrt{1-\gamma}} H_2\left(\frac{Y - Y'}{\sqrt{\gamma}}\right)\right] \delta_j^{(p+1)}\left(1-\gamma, \frac{Y'}{\sqrt{1-\gamma}}, s\right)$$

$$+ \left[-\left(1-\gamma\right)^{\frac{3}{2}}\left(\frac{Y_1 - Y_1'}{\sqrt{\gamma}} \delta_1^{(p+1)}\left(\gamma, \frac{Y - Y'}{\sqrt{\gamma}}, s\right) + \frac{Y_2 - Y_2'}{\sqrt{\gamma}} \delta_2^{(p+1)}\left(\gamma, \frac{Y - Y'}{\sqrt{\gamma}}, s\right)\right)\right.$$

$$\left. + \gamma^{\frac{1}{2}}(1-\gamma)\left(\frac{Y_1'}{\sqrt{1-\gamma}} \delta_1^{(p+1)}\left(\gamma, \frac{Y - Y'}{\sqrt{\gamma}}, s\right) + \frac{Y_2'}{\sqrt{1-\gamma}} \delta_2^{(p+1)}\left(\gamma, \frac{Y - Y'}{\sqrt{\gamma}}, s\right)\right)\right)\Bigg]$$

$$\left. H_j\left(\frac{Y'}{\sqrt{1-\gamma}}, s\right)\right\} d^3Y', \qquad j = 1, 2 \tag{32}$$

We did not include in the last expression terms which are quadratic in δ because in this section we consider only the linearized map.

Another way to introduce the semi-group of linearized maps is the following. Take $\theta > 0$ which later will tend to zero. Denote $\gamma_j = (1+\theta)^{-j}$, $j = 0, 1, 2, \ldots$ Our semigroup will act on the space Δ of functions $\delta(\gamma, Y)$ with values in C^3 such that

1. for each γ, $0 \le \gamma \le 1$, the function $\delta(\gamma, Y)$ belongs to the Hilbert space $L^2 = L^2(R^3)$ of square-integrable functions with respect to the weight $\left(\frac{\sigma^{(1)}}{2\pi}\right)^{\frac{3}{2}} \exp\{-\frac{\sigma^{(1)}Y^2}{2}\}$, $Y = (Y_1, Y_2, Y_3)$;

2. As a function of γ it is a continuous curve in this Hilbert space and $\max_{0\leq\gamma\leq 1}\|\delta(\gamma,Y)\|_{L^2} < \infty$.

Define the linearized map L_θ corresponding to θ as follows:

1. for $\gamma_{j+1}\leq\gamma\leq\gamma_j$, $j=1,2,\dots$

$$L_\theta(\delta(\gamma,Y)) = \delta(\gamma(1+\theta),Y);$$

2. for $\dfrac{1}{1+\theta}\leq\gamma\leq 1$ the function $L_\theta(\delta(\gamma,Y))$ is given by the formula

$$L_\theta(\delta(\gamma,Y)) = \delta_{p_1}(1,Y,s)$$

where p_1 is found from the relation $\dfrac{p_1}{p}=\gamma$.

In other words at $\gamma=1$ we use (32) to find the new $\delta^{(p+1)}(1,Y,s)$. After that we apply 1.

It is easy to see that there exist the limits $\lim_{\substack{\theta\to 0\\ n\theta\to t}} L_\theta^n = A^t$ and the operators A^t constitute a semi-group. For $\gamma<1$, $t>0$ such that $\gamma e^t<1$

$$A^t\delta(\gamma,Y) = \delta(\gamma e^t,Y)$$

Let $\mathcal{A}$ be the infinitesimal generator of the semi-group A^t. In §6 we study in more detail the spectrum and eigenfunctions of $\mathcal{A}$.

Lemma 5.1. *The eigenfunctions of the group A^t have the form*

$$\delta(\gamma,Y) = \gamma^\alpha\tilde{\Phi}_\alpha(Y)$$

where $\tilde{\Phi}_\alpha$ is a function with values in C^3 satisfying (32).

In more detail, if we take $\delta(\gamma,Y)=\gamma^\alpha\tilde{\Phi}_\alpha(Y)$ and substitute it into the rhs of (32) we get in the lhs $\delta^{(p+1)}(Y)=\tilde{\Phi}_\alpha(Y)$.

Proof. If $\delta(\gamma, Y)$ is an eigenfunction then from the formula for A^t

$$A^t\delta(\gamma, Y) = \delta(\gamma e^{-t}, Y) = e^{-\alpha t}\delta(\gamma, Y)$$

Let $\gamma \to 1$. Then

$$\delta(e^{-t}, Y) = e^{-\alpha t}\Phi(Y) = \gamma^{\alpha}\Phi(Y)$$

Lemma is proven.

The space Δ is spanned by the eigenfunctions of $\{A^t\}$ in the sense that for any $h \in \Delta$ we have the expansion

$$h(\gamma, Y) = \sum_{\alpha \in \mathrm{spec}\mathcal{A}} C^{(\alpha)}\gamma^{\alpha}\Phi_{\alpha}(Y)$$

The coefficients $C^{(\alpha)}$ are found with the help of the eigenfunctions of the conjugate system $\{(A^*)^t\}$. The form of the conjugate semi-group and its eigenfunctions can be investigated using the described above discrete approximation. We do not dwell more on this.

§6. The Spectrum of the Group of Linearized Maps

In this section we show that the solutions of (21) studied in §4 have $l^{(u)} = 4$ unstable eigenvalues and $l^{(n)} = 6$ neutral eigenvalues. Therefore in the renormalization group approach we consider 10– parameter families of initial conditions (see below).

As was already mentioned, in the limit $p \longrightarrow \infty$ the linearized maps generate a semi-group of operators acting in the space Δ of functions $f^{(j)}(\gamma, Y)$, $0 \leq \gamma \leq 1$, $Y \in \mathbb{R}^3$, $j = 1, 2$ which are continuous as functions of γ in the Hilbert space L^2. At $\gamma = 1$, the functions $f^{(j)}(\gamma, Y)$ satisfy the boundary condition which follows from (32):

$$\exp\left\{-\frac{\sigma^{(1)}}{2}(|Y_1|^2 + |Y_2|^2) - \frac{\sigma^{(2)}}{2}|Y_3|^2\right\} \cdot \frac{\sigma^{(1)}}{2\pi}\sqrt{\frac{\sigma^{(2)}}{2\pi}} \cdot f^{(j)}(1, Y)$$

$$= \int_0^1 d\gamma \int_{\mathbb{R}^3} \frac{\sigma^{(1)}}{2\pi\gamma} \cdot \sqrt{\frac{\sigma^{(2)}}{2\pi\gamma}} \cdot \frac{\sigma^{(1)}}{2\pi(1-\gamma)} \cdot \sqrt{\frac{\sigma^{(2)}}{2\pi(1-\gamma)}} \cdot$$

$$\exp\left\{-\frac{\sigma^{(1)}(|Y_1-Y_1'|^2+|Y_2-Y_2'|^2)}{2\pi\gamma}-\frac{\sigma^{(2)}|Y_3-Y_3'|^2}{2\pi\gamma}-\frac{\sigma^{(1)}(|Y_1'|^2+|Y_2'|^2)}{2\pi(1-\gamma)}-\right.$$

$$\left.-\frac{\sigma^{(2)}|Y_3'|^2}{2\pi(1-\gamma)}\right\}\left\{\left[-(1-\gamma)^{\frac{3}{2}}\left(\frac{Y_1-Y_1'}{\sqrt{\gamma}}H_1\left(\frac{Y-Y'}{\sqrt{\gamma}}\right)+\frac{Y_2-Y_2'}{\sqrt{\gamma}}H_2\left(\frac{Y-Y'}{\sqrt{\gamma}}\right)\right)\right.\right.$$

$$\left.+\gamma^{\frac{1}{2}}(1-\gamma)\left(\frac{Y_1'}{\sqrt{1-\gamma}}H_1\left(\frac{Y-Y'}{\sqrt{\gamma}}\right)+\frac{Y_2'}{\sqrt{1-\gamma}}H_2\left(\frac{Y-Y'}{\sqrt{\gamma}}\right)\right]f^{(j)}\left(1-\gamma,\frac{Y'}{\sqrt{1-\gamma}}\right)$$

$$+\left[-\left(1-\gamma\right)^{\frac{3}{2}}\left(\frac{Y_1-Y_1'}{\sqrt{\gamma}}f^{(1)}\left(\gamma,\frac{Y-Y'}{\sqrt{\gamma}}\right)+\frac{Y_2-Y_2'}{\sqrt{\gamma}}f^{(2)}\left(\gamma,\frac{Y-Y'}{\sqrt{\gamma}}\right)\right)\right.$$

$$\left.+\gamma^{\frac{1}{2}}(1-\gamma)\left(\frac{Y_1'}{\sqrt{1-\gamma}}f^{(1)}\left(\gamma,\frac{Y-Y'}{\sqrt{\gamma}}\right)+\frac{Y_2'}{\sqrt{1-\gamma}}f^{(2)}\left(\gamma,\frac{Y-Y'}{\sqrt{\gamma}}\right)\right)\right)\right]$$

$$\left.H_j\left(\frac{Y'}{\sqrt{1-\gamma}}\right)\right\}d^3Y',\qquad j=1,2 \tag{33}$$

Denote by $\mathcal{R}_p$ the linear operator which transforms $\{\delta^{(p)}(\gamma,Y,s)\}$ into $\{\delta^{(p+1)}(\gamma,Y,s)\}$. Here s is a parameter which plays no role in this section. As it was explained in §5, for each t there exists the limit $\lim_{p\longrightarrow\infty}\mathcal{R}_p^{tp}=A^t$ so that the operators A^t constitute a semi-group having an infinitesimal generator $\mathcal{A}=\lim_{t\downarrow 0}\frac{A^t-I}{t}$. In our case $\mathcal{A}\delta(\gamma,Y,s)=\gamma\frac{\partial\delta(\gamma,Y,s)}{\partial\gamma}$, $0<\gamma<1$ and for $\gamma=1$ the function $\delta(1,Y,s)$ satisfies the boundary condition (33) in which $f(\gamma,Y)=\delta^{(p+1)}(1,Y,s)$.

If α is an eigenvalue of $\mathcal{A}$, then the corresponding eigenfunction has the form $\gamma^\alpha\Phi_{\alpha,\sigma^{(1)},\sigma^{(2)}}(Y)$ (see Lemma 5.1), where $\Phi_{\alpha,\sigma^{(1)},\sigma^{(2)}}(Y)$ satisfies the equation (33) with $f(\gamma,Y)=\gamma^\alpha\Phi_{\alpha,\sigma^{(1)},\sigma^{(2)}}(Y)$. If $\Re(\alpha)>0$ ($\Re(\alpha)=0$) then the corresponding eigenvalue is called unstable (neutral). All other eigenvalues are called stable. The subspaces generated by unstable, neutral, stable eigenvalues are denoted by $\Gamma^{(u)}$, $\Gamma^{(n)}$, $\Gamma^{(s)}$ respectively.

As before, for $\Phi^{(j)}_{\alpha,\sigma^{(1)},\sigma^{(2)}}(Y)$ the following scaling relation with respect to $\sigma^{(1)},\sigma^{(2)}$ is valid:

$$\Phi^{(j)}_{\alpha,\sigma^{(1)},\sigma^{(2)}}(Y)\propto\Phi^{(j)}_{\alpha,1,1}(\sqrt{\sigma^{(1)}}Y_1,\sqrt{\sigma^{(1)}}Y_2,\sqrt{\sigma^{(2)}}Y_3)$$

Therefore it is enough to consider the above equation (33) for $\sigma^{(1)}=\sigma^{(2)}=1$. We again use the expansion over *Hermite* polynomials:

$$\Phi^{(j)}_{\alpha,1,1}(Y) = \Phi^{(j)}_{\alpha}(Y) = \sum_{m_1,m_2,m_3} f^{(j)}_{\alpha}(m_1,m_2,m_3)\, He_{m_1}(Y_1)\, He_{m_2}(Y_2) He_{m_3}(Y_3)$$

Here j takes values $1,2,3$. Since in m_3 it is the usual convolution and H does not depend on Y_3, it is enough to look for solutions of (33) having the form $f_{m_1,m_2}\delta_{m_3}$. Put $\beta = \alpha + \frac{m_3}{2}$ and $f^{(j)}_{\beta}(m_1,m_2) = f^{(j)}_{\alpha}(m_1,m_2)\delta_{m_3}$. We come to the linear system of recurrent relations

$$\begin{aligned} f^{(j)}_{\beta}(m_1,m_2) = & \sum_{\substack{m_1'+m_1''=m_1\\ m_2'+m_2''=m_2}} J^{(1)}_{m',m''+2\beta}\left((B_1h^{(1)})(m_1',m_2') + (B_2h^{(2)})(m_1',m_2')\right) f^{(j)}_{\beta}(m_1'',m_2'') \\ & + J^{(2)}_{m',m''+2\beta}\cdot h^{(1)}(m_1',m_2')\cdot(B_1f^{(j)}_{\beta})(m_1'',m_2'') \\ & + J^{(2)}_{m',m''+2\beta}\cdot h^{(2)}(m_1',m_2')\cdot(B_2f^{(j)}_{\beta})(m_1'',m_2'') \\ & + J^{(1)}_{m'+2\beta,m''}\cdot\Big((B_1f^{(1)}_{\beta})(m_1',m_2') + (B_2f^{(2)}_{\beta})(m_1',m_2')\Big)h^{(j)}(m_1'',m_2'') \qquad (34) \\ & + J^{(2)}_{m'+2\beta,m''}\cdot f^{(1)}_{\beta}(m_1',m_2')\cdot(B_1h^{(j)})(m_1'',m_2'') \\ & + J^{(2)}_{m'+2\beta,m''}\cdot f^{(2)}_{\beta}(m_1',m_2')\cdot(B_2h^{(j)})(m_1'',m_2'') \end{aligned}$$

Introduce the vector

$$f^{(d)}_{\beta} = \left(f^{(1)}_{\beta}(0,d),\ f^{(1)}_{\beta}(1,d-1),\dots,f^{(1)}_{\beta}(d,0)\ f^{(2)}_{\beta}(0,d),\ f^{(2)}_{\beta}(1,d-1),\dots,f^{(2)}_{\beta}(d,0)\right)^T$$

The vector $f^{(d)}_{\beta}$ contains all terms of degree d. The recurrent relation (34) can be written as

$$C^{(d)}_{\beta}\, f^{(d)}_{\beta} = b^{(d)}_{\beta}$$

where the vector $b^{(d)}_{\beta}$ contains terms of degree $\le d-1$. Also $C^{(d)}_{\beta} \in \mathbb{R}^{2(d+1)\times 2(d+1)}$ is a matrix. Let $C^{(d)}_{\beta}(k,\ell)$ be its (k,ℓ)-entry. Then

$$C_\beta^{(d)}(k,\ell) = \begin{cases} 1 - \dfrac{16d + 32\beta - 16 + 32k}{(d+2\beta+1)(d+2\beta+3)(d+2\beta+5)}, & \text{if } 1 \le k = \ell \le d+1 \\ 1 - \dfrac{80d + 160\beta + 80 - 32k}{(d+2\beta+1)(d+2\beta+3)(d+2\beta+5)}, & \text{if } d+2 \le k = \ell \le 2d+2 \\ -\dfrac{32(d+2\beta-k+2)}{(d+2\beta+1)(d+2\beta+3)(d+2\beta+5)}, & \text{if } 2 \le k \le d+1, \ell = d+k \\ -\dfrac{32(k-d-2\beta-1)}{(d+2\beta+1)(d+2\beta+3)(d+2\beta+5)}, & \text{if } d+2 \le k \le 2d+1, \ell = k-d \\ 0, & \text{all other cases} \end{cases}$$

Note that $d + 2\Re(\beta) > -1$.

Lemma 6.1. *Assume $\Re(\beta) \ge 0$. There exists an integer $d_* > 0$, independent of β, such that for all $d \ge d_*$, the matrix $C_\beta^{(d)}$ is invertible.*

Proof. As d tends to infinity, $C_\beta^{(d)}$ tends to the identity matrix if $\Re(\beta) \ge 0$. A simple estimate on the diagonal and off-diagonal entries shows that for all β such that $\Re(\beta) \ge 0$ and sufficiently large d, the matrix $C_\beta^{(d)}$ becomes diagonally dominant. This implies the existence of the needed d_* and its independence of β.

A similar statement holds if we assume that $\Re(\beta) \ge -A$ for any given $A \le 0$. We formulate it as the following lemma.

Lemma 6.1′. *For any $A \ge 0$, there exists an integer $d_*(A) > 0$ which depends only on A, such that for all $d \ge d_*(A)$ and all β with $\Re(\beta) \ge -A$, the matrix $C_\beta^{(d)}$ is invertible.*

By Lemma 6.1, to find all eigen-values of $\mathcal{A}$ it amounts to solve the equation $\det(C_\beta^{(d)}) = 0$. The eigenvalue α is then found from the relation $\beta = \alpha + \frac{m_3}{2}$. Let

$$a_1 = \left(1 - \frac{16}{(d+2\beta+3)(d+2\beta+5)}\right) \Big/ \left(\frac{32}{(d+2\beta+1)(d+2\beta+3)(d+2\beta+5)}\right)$$

Then a_1 is the eigen-value of the matrix $\tilde{C}^{(d)} \in \mathbb{R}^{2(d+1)\times 2(d+1)}$ given by:

$$\tilde{C}^{(d)}(k,\ell) = \begin{cases} k-1, & \text{if } 1 \le k = \ell \le d+1 \\ 2d+2-k, & \text{if } d+2 \le k = \ell \le 2d+2 \\ d+2-k, & \text{if } 2 \le k \le d+1, \ell = d+k \\ k-d-1 & \text{if } d+2 \le k \le 2d+1, \ell = k-d \\ 0, & \text{all other cases} \end{cases}$$

It is not difficult to find that the eigen-values of $\tilde{C}^{(d)}$ are 0 and $d+1$ with algebraic multiplicity $d+2$ and d respectively. Solve the equations $a_1 = 0$ or $a_1 = d+1$ and use the condition $d + 2\Re(\beta) > -1$. The possible values of β are then given by

$$\beta = \frac{3-d}{2} \quad \text{or} \quad \frac{\sqrt{17}-4-d}{2}, \quad d = 1, 2, 3, \cdots$$

This fact immediately gives the following lemma.

Lemma 6.2. *Let $(\tilde{C}_\beta^{(d)})^{-1}$ be the inverse matrix of $\tilde{C}_\beta^{(d)}$ for $d \ge d^*(\beta)$, where $d^*(\beta) = 3 - 2\beta$ or $\sqrt{17} - 4 - 2\beta$ is an integer. Then there exists an absolute constant $C_2 > 0$ such that for all $d \ge d^*(\beta)$*

$$\| (\tilde{C}_\beta^{(d)})^{-1} \| \le C_2 .$$

We now state our theorem about the properties of the solutions to the recurrent relation (34).

Theorem 6.3. *The only possible values of β for which (34) have nonzero solutions $f_\beta^{(j)}(m_1, m_2)$ is given by:*

$$\beta = \frac{3-m}{2} \quad \text{or} \quad \frac{\sqrt{17}-4-m}{2}, \quad m = 1, 2, 3, \cdots$$

The corresponding solutions $f_\beta^{(j)}(m_1, m_2)$ have the following property:

a) $\beta = (\sqrt{17} - 4 - m)/2$. In this case $f_\beta^{(j)}(m_1, m_2) = 0$ for any $0 \le m_1 + m_2 < m$. For $d = m$, we have

$$f_\beta^{(1)}(r, d-r) = -(d-r+1) f_\beta^{(2)}(r-1, d-r+1), \quad r = 1, 2, \cdots, d$$

$f_\beta^{(1)}(0,d)$, $f_\beta^{(2)}(d,0)$ are free parameters. $f_\beta^{(j)}(m_1,m_2)$ for $m_1+m_2 \geq m+1$ are uniquely determined if the values of the $m+2$ free parameters $f_\beta^{(1)}(r,m-r), r=0,1,\cdots,m$, and $f_\beta^{(2)}(m,0)$ are specified.

b) $\beta=(3-m)/2$. In this case $f_\beta^{(j)}(m_1,m_2)=0$ for any $0 \leq m_1+m_2 < m$. For $d=m$, we have $f_\beta^{(1)}(0,d)=f_\beta^{(2)}(d,0)=0$, and

$$f_\beta^{(1)}(r,d-r)=f_\beta^{(2)}(r-1,d-r+1), \quad r=1,\cdots,d$$

are free parameters. $f_\beta^{(j)}(m_1,m_2)$ for $m_1+m_2 \geq m+1$ are uniquely determined if the values of the m free parameters $f_\beta^{(1)}(r,m-r), r=1,\cdots,m$ are specified.

In both case a) and b), the solutions $f_\beta^{(j)}(m_1,m_2)$ is zero for $m_1+m_2=m+1,m+3,\ldots$. Since $f_\beta^{(j)}$ depends linearly on the free parameters, we have for some $C_3>0$, $0<\rho<\frac{1}{4000}$

$$\left|f_\beta^{(j)}(m_1,m_2)\right| \leq C_3 \frac{\rho^{m_1+m_2+2\beta}}{\Gamma\left(\frac{m_1+m_2+2\beta+3}{2}\right)}, \quad \forall m_1 \geq 0,\ m_2 \geq 0,\ j=1,2.$$

Proof. Property a) and b) are straightforward computations. From recurrent relation (34), by parity it is obvious that $f_\beta^{(j)}(m_1,m_2)=0$ for $m_1+m_2=m+1$. An easy induction shows that $f_\beta^{(j)}(m_1,m_2)=0$ for $m_1+m_2=m+3,m+5,\ldots$ We now prove the decay estimate. The strategy of the proof is the same as in theorem 4.3. From the proof of theorem 4.3, it is clear that by choosing the parameters $(x^{(1)},x^{(2)},x^{(3)})$ sufficiently small, we have

$$|h_{m_1,m_2}^{(j)}| \leq \frac{\rho^{m_1+m_2+2}}{\Gamma\left(\frac{m_1+m_2+7}{2}\right)} \quad \forall m_1 \geq 0,\ m_2 \geq 0, m_1+m_2 \geq 3,\ j=1,2.$$

Our induction hypothesis for $f_\beta^{(j)}(m_1,m_2)$ is

$$\left|f_\beta^{(j)}(m_1,m_2)\right| \leq \frac{\rho^{m_1+m_2+2\beta}}{\Gamma\left(\frac{m_1+m_2+2\beta+3}{2}\right)} \quad \forall m \leq m_1+m_2 < d,\ j=1,2.$$

where $d \geq L$ and $d-m$ is an even number (note that $f_\beta^{(j)}(m_1,m_2)=0$ for $m_1+m_2=m+1,m+3,\ldots$). We assume that L is a sufficiently large number and will verify the induction assumption for $m \le d \le L$ later. Now for $m_1 + m_2 = d$, by lemma 6.2, we have

$$
\begin{aligned}
\left|f_\beta^{(j)}(m_1,m_2)\right| \le & C_2 \cdot \sum_{m'=2}^{d-m}(m'+1)\cdot\left|J^{(1)}_{m',m''+2\beta}\right|\cdot 2(m'+1)\cdot\frac{\rho^{m'+3}}{\Gamma\left(\frac{m'+8}{2}\right)}\cdot\frac{\rho^{m''+2\beta}}{\Gamma\left(\frac{m''+2\beta+3}{2}\right)}\\
&+C_2\cdot\sum_{m'=4}^{d-m}(m'+1)\cdot\left|J^{(1)}_{m',m''+2\beta}\right|\cdot 2\cdot\frac{\rho^{m'+1}}{\Gamma\left(\frac{m'+6}{2}\right)}\cdot\frac{\rho^{m''+2\beta}}{\Gamma\left(\frac{m''+2\beta+3}{2}\right)}\\
&+C_2\cdot\left|J^{(1)}_{2,d-2+2\beta}\right|\cdot 4\cdot\frac{\rho^{d-2+2\beta}}{\Gamma\left(\frac{d+2\beta+1}{2}\right)}\\
&+C_2\cdot\sum_{m'=3}^{d-m+1}(m'+1)\cdot\left|J^{(2)}_{m',m''+2\beta}\right|\cdot 2\cdot\frac{\rho^{m'+2}}{\Gamma\left(\frac{m'+7}{2}\right)}\cdot(m''+1)\cdot\frac{\rho^{m''+2\beta+1}}{\Gamma\left(\frac{m''+2\beta+4}{2}\right)}\\
&+C_2\cdot\sum_{m'=3}^{d-m-1}(m'+1)\cdot\left|J^{(2)}_{m',m''+2\beta}\right|\cdot 2\cdot\frac{\rho^{m'+2}}{\Gamma\left(\frac{m'+7}{2}\right)}\cdot\frac{\rho^{m''+2\beta-1}}{\Gamma\left(\frac{m''+2\beta+2}{2}\right)}\\
&+C_2\cdot\left|J^{(2)}_{1,d-1+2\beta}\right|\cdot 4\cdot\frac{\rho^{d-1+2\beta}}{\Gamma\left(\frac{d+2\beta+2}{2}\right)}\\
&+C_2\cdot\sum_{m'=m-1}^{d-3}(m''+1)\cdot\left|J^{(1)}_{m'+2\beta,m''}\right|\cdot 2\cdot(m'+1)\cdot\frac{\rho^{m'+2\beta+1}}{\Gamma\left(\frac{m'+2\beta+4}{2}\right)}\cdot\frac{\rho^{m''+2}}{\Gamma\left(\frac{m''+7}{2}\right)}\\
&+C_2\cdot\sum_{m'=m+1}^{d-3}(m''+1)\cdot\left|J^{(1)}_{m'+2\beta,m''}\right|\cdot 2\cdot\frac{\rho^{m'+2\beta-1}}{\Gamma\left(\frac{m'+2\beta+3}{2}\right)}\cdot\frac{\rho^{m''+2}}{\Gamma\left(\frac{m''+7}{2}\right)}\\
&+C_2\cdot\left|J^{(1)}_{d-1+2\beta,1}\right|\cdot 4\cdot\frac{\rho^{d-2+2\beta}}{\Gamma\left(\frac{d+2\beta+1}{2}\right)}\\
&+C_2\cdot\sum_{m'=m}^{d-2}(m''+1)\cdot\left|J^{(2)}_{m'+2\beta,m''}\right|\cdot 2\cdot\frac{\rho^{m'+2\beta}}{\Gamma\left(\frac{m'+2\beta+3}{2}\right)}\cdot\frac{\rho^{m''+3}}{\Gamma\left(\frac{m''+8}{2}\right)}\cdot(m''+1)\\
&+C_2\cdot\sum_{m'=m}^{d-4}(m''+1)\cdot\left|J^{(2)}_{m'+2\beta,m''}\right|\cdot 2\cdot\frac{\rho^{m'+2\beta}}{\Gamma\left(\frac{m'+2\beta+3}{2}\right)}\cdot\frac{\rho^{m''+1}}{\Gamma\left(\frac{m''+6}{2}\right)}\\
&+C_2\cdot\left|J^{(2)}_{d-2+2\beta,2}\right|\cdot 4\cdot\frac{\rho^{d-2+2\beta}}{\Gamma\left(\frac{d+2\beta+1}{2}\right)}
\end{aligned}
$$

$$\leq C_2 \cdot \frac{\rho^{d+2\beta}}{\Gamma\left(\frac{d+2\beta+3}{2}\right)} \cdot \left(\sum_{m'=2}^{d-m} \frac{8\rho^3}{d+2\beta+5} + \sum_{m'=4}^{d-m} \frac{8\rho}{d+2\beta+5} + \frac{8}{\rho^2 \cdot (d+2\beta+5)} \right.$$
$$+ \sum_{m'=3}^{d-m+1} \frac{8\rho^3}{d+2\beta+5} \cdot \frac{2(m''+1)}{d+2\beta+3} + \sum_{m'=3}^{d-m-1} \frac{8\rho}{d+2\beta+5} + \frac{8}{\rho(d+2\beta+5)}$$
$$+ \sum_{m'=m-1}^{d-3} \frac{8\rho^3}{d+2\beta+5} \cdot \frac{2(m'+1)}{d+2\beta+3} + + \sum_{m'=m+1}^{d-3} \frac{8\rho}{d+2\beta+5} + \frac{16}{\rho^2(d+2\beta+5)}$$
$$\left. + \sum_{m'=m}^{d-2} \frac{16\rho^3}{d+2\beta+5} + \sum_{m'=m}^{d-4} \frac{8\rho}{d+2\beta+5} + \frac{16}{\rho^2(d+2\beta+5)} \right)$$
$$\leq C_2 \cdot \frac{\rho^{d+2\beta}}{\Gamma\left(\frac{d+2\beta+3}{2}\right)} \cdot 2000\rho \leq \frac{\rho^{d+2\beta}}{\Gamma\left(\frac{d+2\beta+3}{2}\right)}$$

where in the second last inequality above we have used the fact that $d \geq L$ and L is sufficiently large such that $d/(d+2\beta+3) \leq 2$. It is clear that it suffices for us to take $L = 2m$. To check the inductive assumption for $m \leq d \leq 2m$, we recall that $f_\beta^{(j)}(m_1, m_2)$ depends linearly on several free parameters. If we let them be sufficiently small, then it is clear that the inductive assumption is satisfied for $m \leq d \leq 2m$. Our theorem is proved.

We now formulate our main theorem about the spectrum of the linearized operator.

Theorem 6.4. *The spectrum of the operator $\mathcal{A}$ consists of the following eigen-values*

$$\mathsf{spec}\,(\mathcal{A}) = \left\{1, \frac{1}{2}, 0, \lambda_m^{(1)}, \lambda_m^{(2)}, m \geq 1\right\}.$$

where $\lambda_m^{(1)} = -\frac{m}{2}$, $\lambda_m^{(2)} = \frac{\sqrt{17}-4-m}{2}$, $m \geq 1$.

The first eigen-values have multiplicities $\nu_1 = 1$, $\nu_{\frac{1}{2}} = 3$, $\nu_0 = 6$. The eigen-values $\lambda_m^{(1)}$, $\lambda_m^{(2)}$ correspond to the stable part of the spectrum and also have finite multiplicities given by: $\nu_{\lambda_m^{(1)}} = \frac{(m+3)(m+4)}{2}$, $\nu_{\lambda_m^{(2)}} = \frac{m(m+5)}{2}$.

For each $\alpha \in \mathsf{spec}\,(\mathcal{A})$, the eigenfunctions $f_\alpha^{(j)}(m_1, m_2, m_3)$ have the following property:

a) $f_\alpha^{(j)}(m_1, m_2, m_3)$ is compactly supported in the m_3 variable, i.e., there exists an integer $m_3^* = m_3^*(\alpha)$ such that

$$f_\alpha^{(j)}(m_1, m_2, m_3) = 0 \quad \text{if } m_3 > m_3^*$$

b) $f_\alpha^{(j)}(m_1, m_2, m_3)$ *decays faster than exponentially, more precisely, there exist constants* $C_3 = C_3(\alpha) > 0$ *and* $0 < \rho < \frac{1}{4000}$, *such that*

$$\left|f_\alpha^{(j)}(m_1, m_2, m_3)\right| \le C_3 \frac{\rho^{m_1+m_2+m_3+2\alpha}}{\Gamma\left(\frac{m_1+m_2+m_3+2\alpha+3}{2}\right)}, \qquad \forall\ m_1, m_2, m_3 \ge 0$$

The system of eigenfunctions is complete in the following sense. Let $\Gamma^{(s)}$ *be the stable linear subspace of* Δ *generated by all eigenfunctions with* $\Re(\lambda) < 0$, $\Gamma^{(u)}$ *be the unstable subspace generated by all eigenfunctions with eigenvalues* $\lambda > 0$, *and* $\Gamma^{(n)}$ *be the neutral subspace generated by all eigenfunctions with eigenvalue* $\lambda = 0$. *Then* $\dim \Gamma^{(u)} = 4$, $\dim \Gamma^{(n)} = 6$ *and*

$$\Delta = \Gamma^{(u)} + \Gamma^{(n)} + \Gamma^{(s)}.$$

Proof. By Lemma 6.1, we only need to examine β for which $\det(C_\beta^{(d)}) = 0$. From previous arguments, we have that for $d \ge 1$, $\beta = -\frac{d-3}{2}$ or $\frac{\sqrt{17}-4-d}{2}$. We discuss the spectrum separately in the following three cases.

1° unstable spectrum: $\alpha = 1, 1/2$.

a) $\alpha = 1$. Since $\beta = \alpha + \frac{m_3}{2}$, the only possibility is that $\beta = 1$, $d = 1$ and $m_3 = 0$. The eigenspace is one-dimensional with $f_{000}^{(1)} = f_{000}^{(2)} = f_{010}^{(1)} = f_{100}^{(2)} = 0$, $f_{100}^{(1)} = f_{010}^{(2)}$ is a free parameter and the remaining part of all higher degree terms ($f_{m_1,m_2,0}^{(j)}$ with $m_1 + m_2 \ge 2$) is uniquely determined once we specify $f_{100}^{(1)}$.

b) $\alpha = 1/2$. Possible cases are $m_3 = 0$, $\beta = 1/2$, $d = 0, 2$ or $m_3 = 1$, $\beta = 1$, $d = 1$. In the first case we have $f_{m_1,m_2,0}^{(j)} = 0$ for $m_1 + m_2 \le 1$, $f_{110}^{(1)} = f_{020}^{(2)}$, $f_{200}^{(1)} = f_{110}^{(2)}$ are two free parameters, all other terms of higher degee ($f_{m_1,m_2,0}^{(j)}$ with $m_1 + m_2 \ge 3$) are uniquely determined once we specify the above four parameters. In the second case we have $f_{001}^{(1)} = f_{001}^{(2)} = f_{011}^{(1)} = f_{101}^{(2)} = 0$, $f_{101}^{(1)} = f_{011}^{(2)}$ is a free parameter and the remaining part of all higher degree terms ($f_{m_1,m_2,1}^{(j)}$ with $m_1 + m_2 \ge 2$) is uniquely determined once we specify $f_{101}^{(1)}$. Putting two cases together, we see that the dimension of the eigenspace is 3.

This gives $\dim \Gamma^{(u)} = 4$.

2° neutral spectrum: Here we have $\alpha = 0$, and three cases.

a) $m_3 = 2$. Then $\beta = 1$. The eigenspace is one-dimensional with $f^{(1)}_{002} = f^{(2)}_{002} = f^{(1)}_{012} = f^{(2)}_{102} = 0$, $f^{(1)}_{102} = f^{(2)}_{012}$ is a free parameter and the remaining part of all higher degree terms ($f^{(j)}_{m_1,m_2,2}$ with $m_1 + m_2 \geq 2$) is uniquely determined once we specify $f^{(1)}_{102}$. This eigenvector is connected with $\frac{\partial}{\partial \sigma^{(2)}}$ which corresponds to the variation of the parameter $\sigma^{(2)}$ of the fixed point.

b) $m_3 = 1$. Then $\beta = 1/2$. We have $f^{(j)}_{m_1,m_2,1} = 0$ for $m_1 + m_2 \leq 1$, $f^{(1)}_{111} = f^{(2)}_{021}$, $f^{(1)}_{201} = f^{(2)}_{111}$ are two free parameters, all other terms of higher degee ($f^{(j)}_{m_1,m_2,1}$ with $m_1 + m_2 \geq 3$) are uniquely determined once we specify the above two parameters. Clearly the eigenspace is two-dimensional. This eigenspace does not correspond to any change of parameters of the fixed point.

c) $m_3 = 0$. Then $\beta = 0$. We have $f^{(j)}_{m_1,m_2,0} = 0$ for $m_1 + m_2 \leq 2$, $f^{(1)}_{030} = f^{(2)}_{300} = 0$, $f^{(1)}_{120} = f^{(2)}_{030}$, $f^{(1)}_{210} = f^{(2)}_{120}$, $f^{(1)}_{300} = f^{(2)}_{210}$ are three free parameters. All other terms of higher degee ($f^{(j)}_{m_1,m_2,0}$ with $m_1 + m_2 \geq 4$) are uniquely determined once we specify the above three parameters. This eigenspace corresponds to $(\frac{\partial}{\partial x^{(1)}}, \frac{\partial}{\partial x^{(2)}}, \frac{\partial}{\partial x^{(3)}})$.

Putting all three cases together, we see that $\dim \Gamma^{(n)} = 6$.

3° stable spectrum: $\Re(\alpha) < 0$.

There are two cases.

Case 1: $\alpha = -\frac{m}{2}$, $m \geq 1$. Recall that $\beta = \alpha + \frac{m_3}{2}$, and m_3 satisfies $0 \leq m_3 \leq m + 2$. By theorem 6.3, for each such β, the number of free parameters is $3 - 2\beta$. Then the total multiplicity ν_α is given by

$$\nu_\alpha = \sum_{m_3=0}^{m+2} 3 - (-m + m_3) = \frac{(m+3)(m+4)}{2}$$

Case 2: $\alpha = \frac{\sqrt{17}-4-m}{2}$, $m \geq 1$. $\beta = \alpha + \frac{m_3}{2}$, and m_3 satisfies $0 \leq m_3 \leq m-1$. By theorem 6.3, we have

$$\nu_\alpha = \sum_{m_3=0}^{m-1} (m - m_3 + 2) = \frac{m(m+5)}{2}$$

It follows easily that the eigenfunctions $f_\alpha^{(j)}(m_1, m_2, m_3)$ is compactly supported in the m_3 variable. By theorem 6.3, the decay estimate on $f_\alpha^{(j)}(m_1, m_2, m_3)$ is obvious.

It turns out that the eigenvector corresponding to $\frac{\partial}{\partial \sigma^{(1)}}$ is in the eigenspace spanned by the eigenvectors $(\frac{\partial}{\partial x^{(1)}}, \frac{\partial}{\partial x^{(2)}}, \frac{\partial}{\partial x^{(3)}})$. More precisely we have the following:

Lemma 6.3. *Let* $t_1 = x^{(1)} - 1$, $t_2 = x^{(2)}$, $t_3 = x^{(3)} - 1$. *Then*

$$\tilde{G}^{(\sigma^{(1)}, t_1, t_2, t_3, \sigma^{(2)})}(Y) = G^{(\sigma^{(1)}, x^{(1)}, x^{(2)}, x^{(3)}, \sigma^{(2)})}(Y) \tag{34}$$

where $G^{(\sigma^{(1)}, x^{(1)}, x^{(2)}, x^{(3)}, \sigma^{(2)})}$ *is defined in (30). The function* $\tilde{G}$ *satisfies the following scaling relation:*

$$\tilde{G}^{(\sigma^{(1)}, t_1, t_2, t_3, \sigma^{(2)})}(Y) = \tilde{G}^{(1, \sigma^{(1)}t_1, \sigma^{(1)}t_2, \sigma^{(1)}t_3, \sigma^{(2)})}(Y) \tag{35}$$

Proof. Let $f^{(j),0}_{m_1,m_2,0}$ correspond to the eigenvector $\frac{\partial}{\partial \sigma^{(1)}}$, then a simple calculation shows that

$$f^{(j),0}_{m_1,m_2,0} = (m_1 + m_2 - 1)\, h^{(j)}_{m_1 m_2} + h^{(j)}_{m_1-2,m_2} + h^{(j)}_{m_1,m_2-2}\,.$$

If $f^{(j),1}_{m_1,m_2,0}$, $f^{(j),2}_{m_1,m_2,0}$ and $f^{(j),3}_{m_1,m_2,0}$ correspond to the eigenvectors $\frac{\partial}{\partial x^{(1)}}$, $\frac{\partial}{\partial x^{(1)}}$, and $\frac{\partial}{\partial x^{(3)}}$ respectively, then clearly we have

$$f^{(j),0}_{m_1,m_2,0} = \left(x^{(1)} - 1\right) f^{(j),1}_{m_1,m_2,0} + x^{(2)} f^{(j),2}_{m_1,m_2,0} + \left(x^{(3)} - 1\right) f^{(j),3}_{m_1,m_2,0}$$

This immediately gives

$$\left[\sigma^{(1)} \frac{\partial}{\partial \sigma^{(1)}} - \left(x^{(1)} - 1\right) \frac{\partial}{\partial x^{(1)}} - x^{(2)} \frac{\partial}{\partial x^{(2)}} - \left(x^{(3)} - 1\right) \frac{\partial}{\partial x^{(3)}}\right] G^{(\sigma^{(1)}, x^{(1)}, x^{(2)}, x^{(3)}, \sigma^{(2)})}(Y) = 0\,.$$

Regarding this as a transport equation in the variables $(\sigma^{(1)}, t_1, t_2, t_3)$, we can easily find that $\tilde{G}$ satisfies the scaling (35). Lemma is proved.

This lemma actually shows in what sense the parameters $\sigma^{(1)}, x^{(1)}, x^{(2)}, x^{(3)}$ are dependent.

As was shown in §4, we have the five-parameter family of fixed points $G^{(\sigma^{(1)},x^{(1)},x^{(2)},x^{(3)},\sigma^{(2)})}$. We use the notation $\pi = (\sigma^{(1)}, x^{(1)}, x^{(2)}, x^{(3)}, \sigma^{(2)})$ and write $G^{(\pi)}$ instead of $G^{(\sigma^{(1)},x^{(1)},x^{(2)},x^{(3)},\sigma^{(2)})}$. The spectrum of the linearization of the equation for the fixed point does not depend on π (see §5) and has $\ell^{(u)} = 4$ unstable eigenvectors $\Phi_j^{(u)}(Y_1, Y_2, Y_3)$, $1 \le j \le \ell^{(u)} = 4$ and $\ell^{(n)} = 6$ neutral eigenvectors $\Phi_{j'}^{(n)}(Y_1, Y_2, Y_3)$, $1 \le j' \le \ell^{(n)} = 6$.

§7. The Choice of Initial Conditions and the Initial Part of the Inductive Procedure

The equation (21) for the fixed point which was derived in §3 is non-typical from the point of view of the renormalization group theory because it contains the integration over γ, $0 \le \gamma \le 1$. On the other hand, since we consider the Cauchy problem for (1) we are given only the initial condition $v(k, 0)$ which produces through the recurrent relations (4), (5), (6) or (4′), (5′), (6′) the whole set of functions $h_r(k, s)$ or $g_r(\tilde{k}, s)$. For large p and $r \le p$ they can be considered as depending on $\gamma = \frac{r}{p}$ and our procedure is organized in such a way that for γ which are away from zero $\tilde{g}_r$ are close to their limits. Therefore the initial part of our process should be discussed in more detail. This is done in this section.

We take $k^{(0)}$ which will be assumed to be sufficiently large, introduce the neighborhood

$$A_1 = \{k : \left|k - \kappa^{(0)}\right| \le D_1 \sqrt{k^{(0)} ln k^{(0)}}\}$$

where $\kappa^{(0)} = (0, 0, k^{(0)})$ and D_1 is also sufficiently large. Our initial conditions will be zero outside A_1. Inside A_1 they have the form

$$v(k, 0) = \frac{1}{2\pi} \exp\left\{-\frac{Y_1^2 + Y_2^2}{2}\right\} \left(H^{(0)}(Y_1, Y_2) + \sum_{j=1}^{4} b_j^{(u)} \Phi_j^{(u)}(Y_1, Y_2, Y_3) + \right.$$
$$\left. \sum_{j'=1}^{6} b_{j'}^{(n)} \Phi_{j'}^{(n)}(Y_1, Y_2, Y_3) + \Phi(Y_1, Y_2, Y_3; b^{(u)}, b^{(n)}) \right) \frac{1}{\sqrt{2\pi}} \exp\left\{-\frac{Y_3^2}{2}\right\}$$

In this expression $k = k^{(0)} + \sqrt{k^{(0)}}Y$, $H^{(0)}(Y_1, Y_2) = (H_1^{(0)}(Y_1, Y_2), H_2^{(0)}(Y_1, Y_2), 0)$ is the fixed point of our renormalization group (see §4) corresponding to the parameters $\sigma_1^{(1)} = \sigma_1^{(2)} = 1$, $x_1 = x_2 = x_3 = 0$. Also $\Phi_j^{(u)}$, $\Phi_{j'}^{(n)}$ are unstable and neutral eigen-functions for $H^{(0)}$ described in §6, $b_j^{(u)}$ and $b_{j'}^{(n)}$ are our main parameters, $-\rho_1 \le b_j^{(u)}, b_{j'}^{(n)} \le \rho_1$ where ρ_1 is another constant which depends on $k^{(0)}$. Its value will also be specified later. Each function $\Phi(Y_1, Y_2, Y_3; b^{(u)}, b^{(n)})$, $b^{(u)} = \{b_j^{(u)}\}$, $b^{(n)} = \{b_{j'}^{(n)}\}$ is small in the sense that they satisfy

$$\sup_{Y,b} \left|\Phi(Y_1, Y_2, Y_3; b^{(u)}, b^{(n)})\right| \le D_2,$$

$$\sup \left\|\Phi(Y_1, Y_2, Y_3; \bar{b}^{(u)}, \bar{b}^{(n)}) - \Phi(Y_1, Y_2, Y_3; b^{(u)}, b^{(n)})\right\| \le D_2(\left|\bar{b}^{(u)} - b^{(u)}\right| + \left|\bar{b}^{(n)} - b^{(n)}\right|).$$

Due to the presence of $b^{(u)}$, $b^{(n)}$, we have $l = l^{(u)} + l^{(n)} = 10$-parameter families of initial conditions, due to the presence of Φ we have an open set in the space of such families.

Let

$$A_r = \{k : |k - r\kappa^{(0)}| \le D_1\sqrt{rk^{(0)} \ln rk^{(0)}}\}$$

and the variable Y be such that $k = r\kappa^{(0)} + \sqrt{rk^{(0)}}Y$. Assume that for $r < p$, $|Y| \le D_1\sqrt{\ln rk^{(0)}}$

$$h_r(r\kappa^{(0)} + \sqrt{rk^{(0)}}Y, s) = Z_p(s)\Lambda_p^{r-1}(s)r\tilde{g}_r(Y, s)$$

and

$$\begin{aligned}\tilde{g}_r(Y, s) = & \frac{1}{2\pi}\exp\left\{-\frac{Y_1^2 + Y_2^2}{2}\right\}\frac{1}{\sqrt{2\pi}}\exp\left\{-\frac{Y_3^2}{2}\right\} \cdot r \\ & \cdot \Big(H_1^{(0)}(Y_1, Y_2) + \delta_1^{(r)}(Y_1, Y_2, Y_3), H_2^{(0)}(Y_1, Y_2) + \delta_2^{(r)}(Y_1, Y_2, Y_3), \\ & \qquad \frac{1}{\sqrt{rk^{(0)}}}(F^{(r)}(Y_1, Y_2) + \delta_3^{(r)}(Y_1, Y_2, Y_3))\Big)\end{aligned}$$

where in view of incompressibility

$$H_1^{(0)}Y_1 + H_2^{(0)}Y_2 + F^{(r)} = 0 \tag{36}$$

We shall derive a system of recurrent relations for $Z_p(s)$ and $\Lambda_p(s)$ for $p < p_0$. All $\delta_j^{(r)}$ will be considered as remainder terms.

Outside A_r we assume that

$$|h_r(rk^{(0)} + \sqrt{rk^{(0)}}Y, s)| \leq \frac{1}{(rk^{(0)})^{\lambda_1}}$$

where λ_1 is another constant which depends on C_1.

Returning back to (6) take the term with some p_1, p_2, $p_1 + p_2 = p$ and introduce the new variable of integration Y' where $k' = p_2 k^{(0)} + \sqrt{pk^{(0)}}Y'$. Introduce also the variables θ_1, θ_2, $0 \leq \theta_1 \leq (p_1 k^{(0)})^2$, $0 \leq \theta_2 \leq (p_2 k^{(0)})^2$ where $s_1 = s\left(1 - \frac{\theta_1}{(p_1 k^{(0)})^2}\right)$, $s_2 = s\left(1 - \frac{\theta_2}{(p_2 k^{(0)})^2}\right)$.

Then from (6)

$$
\begin{aligned}
&h_p(p\kappa^{(0)}+\sqrt{pk^{(0)}}Y,s)=Z_{p+1}(s)\Lambda_{p+1}^p(s)p\tilde g_p(Y,s)\\
&=(pk^{(0)})^{\frac52}i\int_0^{((p-1)k^{(0)})^2}d\theta_2\int_{\mathbb R^3}\exp\left\{-\theta_2|\kappa^{(0,0)}+\frac{Y'}{\sqrt{(p-1)k^{(0)}}}|^2\right\}\cdot\\
&\qquad Z_p(s(1-\frac{\theta_2}{((p-1)k^{(0)})^2})\cdot\Lambda_p^{p-1}(s(1-\frac{\theta_2}{((p-1)k^{(0)})^2}))\cdot(p-1)\cdot Z_p(s)\Lambda_p(s)\\
&\left\langle\tilde g_1((Y-Y')\sqrt{pk^{(0)}},0),\kappa^{(0,0)}+\frac{Y}{\sqrt{pk^{(0)}}}\right\rangle P_{\kappa^{(0,0)}+\frac{Y}{\sqrt{pk^{(0)}}}}\tilde g_{p-1}\left(Y'\sqrt{\frac{p}{p-1}},s(1-\frac{\theta_2}{((p-1)k^{(0)})^2})\right)d^3Y'+\\
&+ip\sum_{\substack{p_1+p_2=p\\p_1,p_2>1}}\frac1p\frac{(pk^{(0)})^{\frac52}p_1p_2}{(p_1k^{(0)})^2(p_2k^{(0)})^2}\int_0^{(p_1k^{(0)})^2}d\theta_1\int_0^{(p_2k^{(0)})^2}d\theta_2\int_{\mathbb R^3}\left(\frac1{2\pi}\right)^{\frac32}\exp\left\{-\frac{(Y_1-Y_1')^2+(Y_2-Y_2')^2+(Y_3-Y_3')^2}{2\gamma}\right\}\\
&\left\langle\tilde g_{p_1}\left(\frac{Y-Y'}{\sqrt\gamma},s\left(1-\frac{\theta_1}{(p_1k^{(0)})^2}\right)\right),\kappa^{(0,0)}+\frac{Y}{\sqrt{pk^{(0)}}}\right\rangle P_{\kappa^{(0,0)}+\frac{Y}{\sqrt{pk^{(0)}}}}\tilde g_{p_2}\left(\frac{Y'}{\sqrt{1-\gamma}},s\left(1-\frac{\theta_2}{(p_2k^{(0)})^2}\right)\right)\\
&\qquad Z_p(s(1-\frac{\theta_1}{(p_1k^{(0)})^2}))\cdot\Lambda_p^{p_1-1}(s(1-\frac{\theta_1}{(p_1k^{(0)})^2}))\cdot Z_p(s(1-\frac{\theta_2}{(p_2k^{(0)})^2}))\cdot\Lambda_p^{p_2-1}(s(1-\frac{\theta_2}{(p_2k^{(0)})^2}))\cdot\\
&\left(\frac1{2\pi}\right)^{\frac32}\exp\left\{-\frac{(Y_1')^2+(Y_2')^2+(Y_3')^2}{2(1-\gamma)}\right\}\exp\left\{-\theta_1|\kappa^{(0,0)}+\frac{Y-Y'}{\gamma\sqrt{pk^{(0)}}}|^2\right\}\exp\left\{-\theta_2|\kappa^{(0,0)}+\frac{Y'}{(1-\gamma)\sqrt{pk^{(0)}}}|^2\right\}+\\
&+\frac{i(pk^{(0)})^{\frac52}(p-1)}{((p-1)k^{(0)})^2}\int_0^{((p-1)k^{(0)})^2}d\theta_1\int_{\mathbb R^3}\exp\left\{-\theta_1|\kappa^{(0,0)}+\frac{Y-Y'}{\sqrt{(p-1)k^{(0)}}}|^2\right\}\\
&\qquad Z_p(s(1-\frac{\theta_1}{((p-1)k^{(0)})^2})\cdot\Lambda_p^{p-1}(s(1-\frac{\theta_1}{((p-1)k^{(0)})^2}))\cdot Z_p(s)\Lambda_p(s)\\
&\left\langle\tilde g_{p-1}((Y-Y')\sqrt{\frac{p}{p-1}},s(1-\frac{\theta_1}{((p-1)k^{(0)})^2}),\kappa^{(0,0)}+\frac{Y-Y'}{\sqrt{pk^{(0)}}}\right\rangle P_{\kappa^{(0,0)}+\frac{Y}{\sqrt{pk^{(0)}}}}\tilde g_1\left(Y'\sqrt{p},s)\right)d^3Y'
\end{aligned}\tag{37}
$$

Here $\gamma=\dfrac{p_1}{p}$ and $\kappa^{(0,0)}=(0,0,1)$. Now we shall modify (37) for $p_1>1$, $p_2>1$ similar to what we did in §3. Later we discuss the terms with $p_1=1$ and $p_2=1$. The modification consists of four steps.

Step 1. All terms $s\left(1-\frac{\theta_1}{(p_1k^{(0)})^2}\right),s\left(1-\frac{\theta_2}{(p_2k^{(0)})^2}\right)$ are replaced by s.

Step 2. Write

$$\frac{(pk^{(0)})^{\frac{5}{2}} p_1 p_2}{(p_1 k^{(0)})^2 (p_2 k^{(0)})^2} = \frac{(pk^{(0)})^{\frac{1}{2}}}{(k^{(0)})^2 \gamma (1-\gamma)}$$

Step 3. Consider the inner product

$$(pk^{(0)})^{\frac{1}{2}} \left\langle \tilde{g}_{p_1} \left(\frac{Y - Y'}{\sqrt{\gamma}}, s \right), \kappa^{(0,0)} + \frac{Y}{\sqrt{pk^{(0)}}} \right\rangle$$

Up to remainders and from (36) it equals to

$$\left(\frac{1}{2\pi}\right)^{\frac{3}{2}} \exp\left\{ -\frac{(Y_1 - Y_1')^2 + (Y_2 - Y_2')^2 + (Y_3 - Y_3')^2}{2\gamma} \right\}$$

$$\left[H_1^{(0)} \left(\frac{Y_1 - Y_1'}{\sqrt{\gamma}}, \frac{Y_2 - Y_2'}{\sqrt{\gamma}} \right) Y_1 + H_2^{(0)} \left(\frac{Y_1 - Y_1'}{\sqrt{\gamma}}, \frac{Y_2 - Y_2'}{\sqrt{\gamma}} \right) Y_2 + \frac{1}{\sqrt{\gamma}} F^{(p_1)} \left(\frac{Y - Y'}{\sqrt{\gamma}}, s \right) \right] =$$

$$= \left(\frac{1}{2\pi}\right)^{\frac{3}{2}} \exp\left\{ -\frac{(Y_1 - Y_1')^2 + (Y_2 - Y_2')^2 + (Y_3 - Y_3')^2}{2\gamma} \right\}$$

$$\left[H_1^{(0)} \left(\frac{Y_1 - Y_1'}{\sqrt{\gamma}}, \frac{Y_2 - Y_2'}{\sqrt{\gamma}} \right) Y_1 + H_2^{(0)} \left(\frac{Y_1 - Y_1'}{\sqrt{\gamma}}, \frac{Y_2 - Y_2'}{\sqrt{\gamma}} \right) Y_2 - \right.$$

$$\left. - H_1^{(0)} \left(\frac{Y_1 - Y_1'}{\sqrt{\gamma}}, \frac{Y_2 - Y_2'}{\sqrt{\gamma}} \right) \frac{Y_1 - Y_1'}{\gamma} - H_2^{(0)} \left(\frac{Y_1 - Y_1'}{\sqrt{\gamma}}, \frac{Y_2 - Y_2'}{\sqrt{\gamma}} \right) \frac{Y_2 - Y_2'}{\gamma} \right] =$$

$$= \left(\frac{1}{2\pi}\right)^{\frac{3}{2}} \exp\left\{ -\frac{(Y_1 - Y_1')^2 + (Y_2 - Y_2')^2 + (Y_3 - Y_3')^2}{2\gamma} \right\}$$

$$\left\{ -\frac{\gamma - 1}{\sqrt{\gamma}} \left[H_1^{(0)} \left(\frac{Y_1 - Y_1'}{\sqrt{\gamma}}, \frac{Y_2 - Y_2'}{\sqrt{\gamma}} \right) \frac{Y_1 - Y_1'}{\sqrt{\gamma}} + H_2^{(0)} \left(\frac{Y_1 - Y_1'}{\sqrt{\gamma}}, \frac{Y_2 - Y_2'}{\sqrt{\gamma}} \right) \frac{Y_2 - Y_2'}{\sqrt{\gamma}} \right] + \right.$$

$$\left. + \sqrt{1 - \gamma} \left[H_1^{(0)} \left(\frac{Y_1 - Y_1'}{\sqrt{\gamma}}, \frac{Y_2 - Y_2'}{\sqrt{\gamma}} \right) \frac{Y_1'}{\sqrt{1 - \gamma}} + H_2^{(0)} \left(\frac{Y_1 - Y_1'}{\sqrt{\gamma}}, \frac{Y_2 - Y_2'}{\sqrt{\gamma}} \right) \frac{Y_2'}{\sqrt{1 - \gamma}} \right] \right\}$$

Let us stress again that $H_j^{(0)}(Y, s)$ depend only on Y_1, Y_2 and s. With respect to Y_3 we have the usual convolution.

Step 4. Replace the projection operator by the identity operator. It is not the reduction to the Burgers system because the incompressibility condition is preserved.

Now we shall modify the first and the last terms in (37). For the first one we can write

$$\frac{(pk^{(0)})^{\frac{5}{2}}(p-1)}{((p-1)k^{(0)})^2}\int\limits_0^{((p-1)k^{(0)})^2} d\theta_2 \int\limits_{\mathbb{R}^3} \exp\left\{-\theta_2|\kappa^{(0,0)}+\frac{Y'}{\sqrt{(p-1)k^{(0)}}}|^2\right\}\cdot$$

$$\cdot\exp\left\{-s|\kappa^{(0)}+(Y-Y')\sqrt{pk^{(0)}}|^2\right\}\left\langle v(\kappa^{(0)}+(Y-Y')\sqrt{pk^{(0)}},0),\kappa^{(0,0)}+\frac{Y}{\sqrt{pk^{(0)}}}\right\rangle\cdot \tag{38}$$

$$\cdot P_{\kappa^{(0)}+\frac{Y}{\sqrt{pk^{(0)}}}}\tilde{g}_{p-1}\left(Y'\sqrt{\frac{p}{p-1}},s(1-\frac{\theta_2}{((p-1)k^{(0)})^2})\right)d^3Y'$$

The factor $(p-1)$ comes from the inductive assumption concerning h_{p-1}. As before, we replace $\exp\left\{-\theta_2|\kappa^{(0,0)}+\frac{Y'}{\sqrt{(p-1)k^{(0)}}}|^2\right\}$ by $\exp\{-\theta_2\}$, $P_{\kappa^{(0)}+\frac{Y}{\sqrt{pk^{(0)}}}}$ by the identity operator and $\tilde{g}_{p-1}\left(Y'\sqrt{\frac{p}{p-1}},s(1-\frac{\theta_2}{((p-1)k^{(0)})^2})\right)$ by $\tilde{g}_{p-1}(Y'\sqrt{\frac{p}{p-1}},s)$. All corrections are included in the remainder terms.

For the Gaussian term in $v(\kappa^{(0)}+(Y-Y')\sqrt{pk^{(0)}},0)$ we can write $\frac{1}{(2\pi)^{\frac{3}{2}}}\exp\left\{\frac{|Y-Y'|^2p}{2}\right\}$. This shows that typically $Y-Y'=O(\frac{1}{\sqrt{p}})$. For the third component $F^{(1)}$ of $v(\kappa^{(0)}+(Y-Y')\sqrt{pk^{(0)}},0)$ using the incompressibility condition we can write

$$F^{(1)}(\kappa^{(0)}+(Y-Y')\sqrt{pk^{(0)}},0)=$$

$$-\frac{1}{\sqrt{k^{(0)}}}\left((Y_1-Y_1')\sqrt{p}H_1^{(0)}((Y-Y')\sqrt{p})+(Y_2-Y_2')\sqrt{p}H_1^{(0)}((Y-Y')\sqrt{p})+O(\frac{1}{\sqrt{k^{(0)}}})\right)\cdot$$

$$\cdot\exp\left\{-\frac{p|Y-Y'|^2}{2}\right\}$$

For the inner product in (38)

$$\sqrt{pk^{(0)}}\left\langle v(\kappa^{(0)}+(Y-Y')\sqrt{pk^{(0)}},0),\kappa^{(0,0)}+\frac{Y}{\sqrt{pk^{(0)}}}\right\rangle=\exp\left\{-\frac{p|Y-Y'|^2}{2}\right\}\cdot$$

$$\cdot\Big[H_1^{(0)}((Y-Y')\sqrt{p})Y_1+H_1^{(0)}((Y-Y')\sqrt{p})Y_2-$$

$$-\sqrt{p}\left((Y_1-Y_1')\sqrt{p}H_1^{(0)}((Y-Y')\sqrt{p})+(Y_2-Y_2')\sqrt{p}H_1^{(0)}((Y-Y')\sqrt{p})\right)+O(\frac{1}{\sqrt{k^{(0)}}})\Big]$$

The expression in the square brackets grows as $\sqrt{p}$ and therefore

$$\sqrt{pk^{(0)}}\left\langle v(\kappa^{(0)}+(Y-Y')\sqrt{pk^{(0)}},0),\kappa^{(0,0)}+\frac{Y}{\sqrt{pk^{(0)}}}\right\rangle=$$

can be replaced by

$$-\sqrt{p}\left[(Y_1-Y_1')\sqrt{p}H_1^{(0)}((Y-Y')\sqrt{p})+(Y_2-Y_2')\sqrt{p}H_1^{(0)}((Y-Y')\sqrt{p})-\right.$$
$$\left.-\frac{1}{\sqrt{p}}\left(H_1^{(0)}((Y-Y')\sqrt{p})Y_1+H_1^{(0)}((Y-Y')\sqrt{p})Y_2\right)\right]$$

Further,

$$\exp\{-s|\kappa^{(0)}+(Y-Y')\sqrt{pk^{(0)}}|^2\}=\exp\{-s|k^{(0)}|^2\}\cdot$$
$$\cdot\exp\{-2sk^{(0)}\langle\kappa^{(0,0)},(Y-Y')\sqrt{p}\sqrt{k^{(0)}}\rangle\}\exp\{-s|Y-Y'|^2pk^{(0)}\}$$

The first factor takes values $O(1)$, the others can be written as $1+O(\frac{1}{\sqrt{k^{(0)}}})$. The main order of magnitude of (38) takes the form

$$p\exp\{-s(k^{(0)})^2\}\frac{(p-1)}{p}\frac{1}{p}\left[-\int_{\mathbb{R}^3}\left[(Y_1-Y_1')\sqrt{p}H_1^{(0)}((Y-Y')\sqrt{p})+(Y_2-Y_2')\sqrt{p}H_1^{(0)}((Y-Y')\sqrt{p})\right]+\right.$$
$$+\frac{1}{\sqrt{p}}\left[H_1^{(0)}((Y-Y')\sqrt{p})Y_1+H_1^{(0)}((Y-Y')\sqrt{p})Y_2\right]\left(\frac{p}{2\pi}\right)^{\frac{3}{2}}\exp\left\{-\frac{|Y-Y'|^2p}{2}\right\}\cdot$$
$$\left.\cdot\left(\frac{1}{2\pi}\right)^{\frac{3}{2}}\exp\left\{-\frac{|Y'|^2p}{2(p-1)}\right\}H^{(0)}\left(Y'\sqrt{\frac{p}{p-1}}\right)d^3Y'\right]$$

A similar expression can be written for the last term in (37). Remark that due to our choice of the interval $S^{(1)}$ the product $s(k^{(0)})^2=O(1)$.

Now we derive the recurrent formula for $Z_p(s)$ and $\Lambda_p(s)$. Since our special fixed point $H^{(0)}$ is a Hermite polynomial of first degree, the convolution of $H^{(0)}$ over Y' (see (37)) gives us simply the product of $H^{(0)}$ and the Gaussian term and a polynomial in γ. The function $H^{(0)}$ and the Gaussian term can then be taken out of the summation in γ and this gives us the following recurrent system for $Z_p(s)$ and $\Lambda_p(s)$:

$$\begin{aligned}&Z_{p+1}(s)\Lambda_{p+1}^p(s)\\&=\sum_{p_1+p_2=p}\frac{1}{p}\cdot\frac{i}{(k^{(0)})^2}\cdot(6\gamma^2-10\gamma+4)\cdot Z_p(s)^2\cdot\Lambda_p^p(s)\cdot(1-e^{-s(p_1k^{(0)})^2})\cdot(1-e^{-s(p_2k^{(0)})^2})\end{aligned}\tag{39}$$

where the factor $(6\gamma^2 - 10\gamma + 4)$ comes from the convolution of $H^{(0)}$ with itself. Now if we take $Z_p(s) = -i(k^{(0)})^2$ and write $\frac{\Lambda_{p+1}(s)}{\Lambda_p(s)} = 1 + \frac{\xi_{p+1}}{p^2}$, then we have

$$\left(1 + \frac{\xi_{p+1}}{p^2}\right)^p = \sum_\gamma \frac{1}{p} \cdot (6\gamma^2 - 10\gamma + 4) \cdot (1 - e^{-s(p_1 k^{(0)})^2})(1 - e^{-s(p_2 k^{(0)})^2})$$

then it is not difficult to see that there exists bounded ξ_{p+1} (with an bound independent of p) such that the equality holds. It is an elementary fact that the limit

$$\Lambda(s) = \lim_{p\to\infty} \Lambda_{p+1}(s) = \Lambda_1 \prod_{k=1}^{\infty} \left(1 + \frac{\xi_{k+1}}{k^2}\right)^k$$

exists.

Now we discuss the behavior of all remainders for $p < \left(k^{(0)}\right)^{\lambda_2}$.

By $\Phi_j^{(u)}$, $\Phi_{j'}^{(n)}$ we denote the eigen-vectors of the linearized renormalization group corresponding to the fixed point $H^{(0)}$. For each p we make the following inductive assumption for $\delta^{(r)}(Y, s)$, $r < p$:

$$\delta^{(r)}(Y,s) = \sum_{j=1}^{4} \left(b_{j,r}^{(u)} + \beta_{j,r}^{(u)}\right)\Phi_j^{(u)} + \sum_{j'=1}^{6} \left(b_{j',r}^{(n)} + \beta_{j',r}^{(n)}\right)\Phi_{j'}^{(n)} + \Phi_r^{(st)}, \quad \gamma = \frac{r}{p-1}$$

where $b_{j,r}^{(u)} = (p-1)^{\alpha_j} b_j^{(u)} \gamma^{\alpha_j^{(u)}}$, $b_{j',r}^{(n)} = b_{j'}^{(n)}$, $\Phi_r^{(st)}$ is a function which belongs to the stable subspace of the linearized renormalization group, $\gamma = \frac{r}{p-1}$.

As we go from $p-1$ to p, the variable $\gamma = \frac{r}{p-1}$ is replaced by $\gamma' = \frac{r}{p} = \gamma \cdot \frac{p-1}{p}$. Therefore

$$\begin{aligned}\left(b_{j,r}^{(u)} + \beta_{j,r}^{(u)}\right)\gamma^{\alpha_j}\Phi_j^{(u)} &= \left((p-1)^{\alpha_j} b_j^{(u)} + \beta_{j,r}^{(u)}\right) \cdot \left(\frac{p}{p-1}\right)^{\alpha_j} \cdot (\gamma')^{\alpha_j}\Phi_j^{(u)} \\ &= \left(p^{\alpha_j} b_j^{(u)} + \left(\frac{p}{p-1}\right)^{\alpha_j} \cdot \beta_{j,r}^{(u)}\right) \cdot (\gamma')^{\alpha_j}\Phi_j^{(u)}.\end{aligned}$$

In the same way for the neutral eigen-functions we have

$$\left(b_{j'}^{(n)} + \beta_{j',r}^{(n)}\right)\Phi_{j'}^{(n)}$$

because $\alpha_{j'} = 0$. In the same way one can transform $\Phi^{(st)}$.The coefficients $\beta_{j,r}^{(u)}$, $\beta_{j',r}^{(n)}$ are small compared to the first term . An important conclusion is that the projections to the unstable directions increase, projections to the neutral directions remain the same and projections to the stable directions decrease. As was already said, in the case of unstable and neutral directions the term containing $b_j^{(u)}$ or $b_{j'}^{(n)}$ is the main term.

Now we discuss the form of $\delta^{(p)}(Y,s)$. It is the sum of three types of terms.

a_1). The term which depends linearly on all $\delta^{(r)}(Y,s)$. Especially important is the part which contains all $p^{\alpha_j} b_j^{(u)} (\gamma')^{\alpha_j} \Phi_j^{(u)}$, $b_{j'}^{(n)} \Phi_{j'}^{(n)}$. If we were to have and be in the limiting regime $H^{(0)}$ then the integral will give $p^{\alpha_j} b_j^{(u)} \left(1+\frac{1}{p}\right)^{\alpha_j} \Phi_j^{(u)} = (p+1)^{\alpha_j} b_j^{(u)} \Phi_j^{(u)}$ since $\gamma' = 1$. However, $H^{(r)}$ are slightly different from $H^{(0)}$. Therefor we shall have a small correction which is included in all $\beta_{j,p}^{(u)}$, $\beta_{j',p}^{(n)}$ and in $\Phi_p^{(st)}$. We denote it as $\beta_{pj1}^{(u)}$, $\beta_{pj'1}^{(n)}$, $\Phi_{p,1}^{(st)}$. The we have terms which are linear functions of all $\beta_{j,r}^{(u)}$, $\beta_{j',r}^{(n)}$ and $\Phi_r^{(st)}$. They will give us $\beta_{pj2}^{(u)}$, $\beta_{pj'2}^{(n)}$, $\Phi_{p,2}^{(st)}$.

a_2). The term which is the sum of all quadratic expressions depending on $\delta^{(p_1)}$, $\delta^{(p_2)}$. We expand it using our basis of $\Phi_j^{(u)}$, $\Phi_{j'}^{(n)}$ and all stable eigen-vectors.

a_3). The term which contains all corrections which arise during the four steps described above. We also expand it in the same way as in a_2)).

The sum of all terms gives $\beta_{p,j}^{(u)}$, $\beta_{p,j'}^{(n)}$, Φ_p^{st}.

We use this procedure till $p = p_0 = (k^{(0)})^{\lambda_2}$. The procedure for $p > p_0$ is discussed in §9.

§8. The List of Remainders and Their Estimates

In the beginning of §7 we described 10-parameter families of initial conditions which we consider in this paper. We mentioned above that for each p we have an interval $S^{(p)} = \left[S_-^{(p)}, S_+^{(p)}\right]$ on the time axis. Actually these intervals are changing when $p = p_n = (1+\epsilon)^n$ where $\epsilon > 0$ is a constant. Therefore we shall write $S^{(n)} = \left[S_-^{(p_n)}, S_+^{(p_n)}\right]$ and hope that there will be no confusion.

In this and the next section we consider $p > (k^{(0)})^{\lambda_2}$. Each function $\tilde{g}_r(Y,s)$, $3 \le r < p$, has the following representation:

in the domain $|Y| \le C_1\sqrt{\ln rk^{(0)}}$, $Y = (Y_1, Y_2, Y_3) \in \mathcal{R}^3$

$$\tilde{g}_r(Y,s) = \Lambda^{r-1} \cdot r \cdot \frac{\sigma^{(1)}}{2\pi} \exp\left\{\frac{\sigma^{(1)}}{2}\left(|Y_1|^2 + \right.\right.$$

$$\left.\left. + |Y_2|^2\right)\right\} \cdot \sqrt{\frac{\sigma^{(2)}}{2\pi}} \exp\left\{-\frac{\sigma^{(2)}}{2}|Y_3|^2\right\} \cdot \left(H^{(0)}(Y_1,Y_2) + \delta^{(r)}(Y,s)\right);$$

in the domain $|Y| > C_1\sqrt{\ln(rk^{(0)})}$:

$$\begin{aligned} &\frac{\sigma^{(1)}}{2\pi} \exp\left\{-\frac{\sigma^{(1)}}{2}\left(|Y_1|^2 + |Y_2|^2\right)\right\} \cdot \sqrt{\frac{\sigma^{(2)}}{2\pi}} \cdot \\ &\quad \cdot \exp\left\{-\frac{\sigma^{(2)}}{2}|Y_3|^2\right\} \cdot \left|H^{(0)}(Y_1,Y_2) + \delta^{(r)}(Y,s)\right| \\ &\le \Lambda^{r-1} \cdot r \cdot \frac{1}{r^{\lambda_3 - 1}} \end{aligned}$$

for some constant $\lambda_3 > 0$. We use the formula (7) to get $\tilde{g}^{(p)}(Y,s)$. New remainders appear in one of the following ways.

Type I. The recurrent relation (7) does not coincide with the equation for the fixed point and actually is some perturbation of this equation. The difference produces some remainders which tend to zero as $p \to \infty$.

Type II. For the limiting equation all eigen-vectors in the linear approximation are multiplied by some constant. In the equation (7) it is no longer true and the difference generate some remainders. (see also §9).

Type III. The remainders which are quadratic functions of all previous remainders.

§8A. The Remainders of Type I.

We call the domain A the set $\{|Y| \leq D_1\sqrt{\ln(rk^{(0)})}\}$ and the domain B the set $\{|Y| > D_1\sqrt{\ln(rk^{(0)})}\}$. The estimates will be done separately in each domain. We include the first, the second and the last two terms in (7) in the remainders. We shall estimate only the first one, the others are estimated in the same way.

Domain A: We have

$$\beta_p^{(1)}(Y,s) = (p+1)^{\frac{5}{2}} \cdot \frac{i}{sp^2} \cdot \int_0^{p^2} d\theta_2 \int_{\mathbb{R}^3} < v\left(\left(k^{(0)} + \frac{Y-Y'}{\sqrt{s}}\sqrt{p+1}\,,\,0\right);b\right),$$

$$\sqrt{s}\,k^{(0)} + \frac{Y}{\sqrt{p+1}} > P_{\sqrt{s}\,k^{(0)}+\frac{Y}{\sqrt{p+1}}}\,\tilde{g}_p\left(Y',\left(1-\frac{\theta_2}{p^2}\right)s\right)$$

$$\exp\left\{-\left|\sqrt{s}\,k^{(0)} + (Y-Y')\sqrt{p+1}\right|^2 - \frac{\theta_2}{p^2}\left|\sqrt{s}\,k^{(0)}\,p + Y'\sqrt{p+1}\right|^2\right\}d^3Y'$$

Here b means the collection of all parameters of $v(k;0)$. The the main contribution to the integral comes from $Y-Y' = O\left(\frac{1}{\sqrt{p+1}}\right)$. In this domain in the main order of magnitude

$$\langle v(k^{(0)} + \frac{Y-Y'}{\sqrt{s}}\sqrt{p+1},\,0\,;\,b),\,\sqrt{s}\,k^{(0)}\rangle = O(1)$$

Assuming that $v(k^{(0)} + \frac{Y-Y'}{\sqrt{s}}\sqrt{p+1},\ 0;\ b)$ is differentiable `w.r.t` the first three variables we see that the inner product

$$\langle v(k^{(0)} + \frac{Y-Y'}{\sqrt{s}}\sqrt{p+1},\,0\,;\,\alpha)\,,\,\sqrt{s}\,k^{(0)} + \frac{Y}{\sqrt{p+1}}\rangle$$

is of order $O(1)$. For $\tilde{g}_p$ we can write using our inductive assumptions

$$\tilde{g}_p\left(Y',\left(1-\frac{\theta_2}{p^2}\right)s\right) = \Lambda^{p-1}\cdot p\cdot\frac{\sigma^{(1)}}{2\pi}\cdot\sqrt{\frac{\sigma^{(2)}}{2\pi}}\cdot\exp\left\{-\frac{\sigma^{(1)}\left(|Y_1|^2 + Y_2|^2\right)}{2}\right\}$$

$$\cdot\exp\left\{-\frac{\sigma^{(2)}\left(|Y_3|^2\right)}{2}\right\}\cdot\mathcal{H}^{(p)}\left(Y',\left(1-\frac{\theta_2}{p^2}\right)s\right).$$

Also

$$\exp\left\{-\frac{\theta_2}{p^2}\left|\sqrt{s}\,k^{(0)}\,p + Y'\,\sqrt{p+1}\right|^2\right\} = \exp\left\{-\theta_2\left|\sqrt{s}\,k^{(0)} + \frac{Y'\sqrt{p+1}}{p}\right|^2\right\}$$

and in the main order of magnitude the integration over θ_2 does not depend on Y'. Thus we can write

$$|\beta_p^{(1)}(Y,s)| \leq \Lambda^{(p-2)} \cdot p \cdot \exp\left\{-\frac{\sigma^{(1)}}{2}\left(|Y_1|^2 + |Y_2|^2\right)\right\}$$
$$\cdot \exp\left\{-\frac{\sigma^{(2)}}{2}|Y_3|^2\right\} \cdot \frac{D_4}{p} \tag{40}$$

Here and later various constants whose exact values play no role in the arguments will be denoted by the letter D with indices. Since in the expression for $\tilde{g}_{p+1}$ we have the factors $\Lambda^p \cdot (p+1) \cdot \exp\left\{-\frac{\sigma^{(1)}}{2}\left(|Y_1|^2 + |Y_2|^2\right\} \cdot \frac{\sigma^{(1)}}{2\pi}\sqrt{\frac{\sigma^{(2)}(s)}{2\pi}} \cdot \exp\left\{-\frac{\sigma^{(2)}}{2}|Y_3|^2\right\}\right.$, the estimate (40) shows that $|\beta_p^{(1)}(Y,s)|$ is relatively smaller than $\tilde{g}_{p+1}$ with an order $O(\frac{1}{p})$. This is good enough for our purposes. We did not discuss the errors which follow from the fact that the expressions in the previous formulas depend on θ_2.

Domain B: The smallness of $\beta_p^{(1)}(Y,s)$ in this case follows easily from several inequalities and arguments.

1°: $|Y| \leq D_4\sqrt{pk^{(0)}}$ because $|k| \leq D_5 pk^{(0)}$.

2°: $|Y - Y'| \leq D_6\sqrt{k^{(0)}}$ because $v(k,0;b)$ has a compact support.

3°: If $|Y - Y'| \leq \dfrac{2s_+}{\sqrt{p}}$ then

$$\exp\left\{-\left|\sqrt{s}\,k^{(0)} + (Y - Y')\,\sqrt{p+1}\right|^2\right\} \leq 1$$

If $|Y - Y'| \geq \dfrac{2s_+}{\sqrt{p}}$ then

$$\exp\left\{-\left|\sqrt{s}\,k^{(0)} + (Y - Y')\sqrt{p+1}|^2\right\} \leq \exp\left\{-\frac{s_+}{4}\left|Y - Y'|^2\right\}\right. .$$

4°: If $|Y'| \geq D_7 \sqrt{p}$ then

$$\exp\left\{-\frac{\theta_2}{p^2}\left|\sqrt{s}\,k^{(0)}p + Y'\sqrt{p+1}\right|^2\right\} \leq \exp\{-C_8\theta_2\}$$

5°: If $|Y'| \leq D_7 \sqrt{p}$ then

$$\exp\left\{-\frac{\theta_2}{p^2}\left|\sqrt{s}\,k^{(0)}\,p + Y'\sqrt{p+1}\right|^2\right\} \leq 1\,.$$

6°: We have

$$\exp\left\{-\frac{\sigma^{(1)}}{2}\left(|Y_1'|^2 + |Y_2'|^2\right) - \frac{\sigma^{(2)}}{2}|Y_3'|^2\right\}$$

$$= \exp\left\{-\frac{\sigma^{(1)}}{2}\left(|Y_1 - (Y_1 - Y_1')|^2 + |Y_2 - (Y_2 - Y_2')|^2\right)\right.$$

$$\left. - \frac{\sigma^{(2)}}{2}|Y_3 - (Y_3 - Y_3')|^2\right\}$$

$$= \exp\left\{-\frac{\sigma^{(1)}}{2}\left(|Y_1|^2 + |Y_2|^2\right) - \frac{\sigma^{(2)}}{2}\left(|Y_3|^2\right)\right\}$$

$$\cdot \exp\left\{\sigma^{(1)}\left(Y_1(Y_1 - Y_1') + Y_2(Y_2 - Y_2')\right) + \sigma^{(2)}\,Y_3(Y_3 - Y_3')\right.$$

$$\left. - \frac{\sigma^{(1)}(s)}{2}\left(|Y_1 - Y_1'|^2 + |Y_2 - Y_2'|^2\right) - \frac{\sigma^{(2)}}{2}|Y_3 - Y_3'|^2\right\}.$$

If $|Y - Y'| \leq \dfrac{2s_+}{\sqrt{p}}$ then

$$\exp\left\{\sigma^{(1)}\left(Y_1(Y_1 - Y_1') + Y_2(Y_2 - Y_2')\right) + \sigma^{(2)}(s)\,Y_3(Y_3 - Y_3')\right.$$

$$\left. - \frac{\sigma^{(1)}}{2}\left(|Y_1 - Y_1'|^2 + |Y_2 - Y_2'|^2\right) - \frac{\sigma^{(2)}}{2}|Y_3 - Y_3'|^2\right\} \leq C_8\,.$$

If $|Y - Y'| > \frac{2s_+}{\sqrt{p}}$ then we have an integral of the function which is the product of some Gaussian factor and $|\mathcal{H}^{(p)}(Y)|$. Direct estimate shows as before that in this case

$$|\beta_p^{(1)}(Y,s)| \le \Lambda^{(p-1)} \cdot p \cdot e^{-\frac{\sigma^{(1)}}{2}(|Y_1|^2 + Y_2|^2)} \cdot e^{-\frac{\sigma^{(2)}}{2}|Y_3|^2} \cdot \frac{D_8}{p^{\frac{3}{2}}}$$

which is also good for us.

In the same way one can estimate terms with relatively small p_1 and $p - p_1$ (i.e., $p_1 \le \sqrt{p}$ or $p_1 \ge p - \sqrt{p}$. The remainders will be of order $\frac{1}{\sqrt{p_1}} \cdot \frac{1}{p}$. The next set of remainders comes from splitting the integration over θ and Y' (see (7) and beginning of §3). We may assume that $p_1 > \sqrt{p}$ or $p_1 < p - \sqrt{p}$ because other terms were estimated before. Put

$$\tilde{\tilde{g}}_{p+1}(Y,s) = i\,(p+1)^{\frac{5}{2}} \sum_{\substack{p_1+p_2=p+1\\ p_1, p_2 > \sqrt{p}}} \int_0^{p_1^2} d\theta_1 \int_0^{p_2^2} d\theta_2 \cdot \frac{1}{p_1^2 p_2^2}$$

$$\int_{\mathbb{R}^3} \langle \tilde{g}_{p_1}\left((Y - Y')\frac{(1-\frac{\theta_1}{p_1^2})^{\frac{1}{2}}}{\sqrt{\gamma}}, \left(1 - \frac{\theta_1}{p_1^2}\right)s\right), \sqrt{s}\,k^{(0)} + \frac{Y}{\sqrt{p+1}}\rangle$$

$$P_{\sqrt{s}\,k^{(0)} + \frac{Y}{\sqrt{p+1}}}\tilde{g}_{p_2}\left(\frac{Y'(1-\frac{\theta_2}{p_2^2})^{\frac{1}{2}}}{\sqrt{(1-\gamma)}}, \left(1 - \frac{\theta_2}{p_2^2}\right)s\right)$$

$$\cdot \exp\left\{-\theta_1 |\sqrt{s}\,k^{(0)} + \frac{Y - Y'}{\sqrt{p+1}\cdot\gamma}|^2 - \theta_2|\sqrt{s}\,k^{(0)} + \frac{Y - Y'}{\sqrt{p+1}\gamma}|^2\right\}.$$

Using the inductive assumption we can rewrite the last expression as follows:

$$\tilde{\tilde{g}}_{p+1}(Y,s) = i\,(p+1) \sum_{\substack{p_1+p_2=p+1\\ p_1, p_2 > \sqrt{p}}} \int_0^{p_1^2} d\theta_1 \int_0^{p_2^2} d\theta_2$$

$$\Lambda^{p_1-1} \cdot \Lambda^{p_2-1} \cdot \frac{1}{\gamma(1-\gamma)} \cdot \frac{1}{p+1} \exp\left\{-\frac{\sigma^{(1)}}{2}\frac{|Y_1 - Y_1'|^2 + |Y_2 - Y_2'|^2}{\gamma}\right.$$

$$- \frac{\sigma^{(2)}}{2} \frac{|Y_3 - Y_3'|^2}{\gamma} - \frac{\sigma^{(1)}}{2} \frac{|Y_1'|^2 + |Y_2'|^2}{(1-\gamma)}$$

$$- \frac{\sigma^{(1)}}{2} \cdot \frac{|Y_3'|^2}{1-\gamma} \Bigg\} \cdot p^{\frac{1}{2}} < \mathcal{H}^{(p_1)} \left(Y - Y', s \left(1 - \frac{\theta_1}{p_1^2} \right) \right),$$

$$\sqrt{s}\, k^{(0)} + \frac{Y}{\sqrt{p}} > \cdot P_{\sqrt{s}\, k^{(0)} + \frac{Y}{\sqrt{p}}} \mathcal{H}^{(p_2)} \left(Y', s \left(1 - \frac{\theta_2}{p_2^2} \right) \right) d^3 Y'.$$

As was explained before, due to incompressibility in the Domain A , the inner product

$$\langle \mathcal{H}^{(p_1)} \left(Y - Y'; s \left(1 - \frac{\theta_1}{p_1^2} \right) \right), \sqrt{s}\, k^{(0)} + \frac{Y}{\sqrt{p}} \rangle$$

takes values $O(\frac{1}{\sqrt{p}})$ because the first two components of the vector $\sqrt{s}\, k^{(0)} + \frac{Y}{\sqrt{p}}$ are of order $O(\frac{1}{\sqrt{p}})$. Therefore the product

$$\sqrt{p}\, \langle \mathcal{H}^{(p_1)} \left(Y - Y', s \left(1 - \frac{\theta_1}{p_1^2} \right) \right), \sqrt{s}\, k^{(0)} + \frac{Y}{\sqrt{p}} \rangle$$

takes values of order $O(1)$.

The remainder can be written in the following form:

$$\beta_p^{(2)}(Y, s) = i \sum_{\substack{p_1 + p_2 = p+1 \\ p_1, p_2 > \sqrt{p}}} \frac{1}{\gamma(1-\gamma)} \cdot \frac{1}{p} \cdot \int_0^{p_1^2} d\theta_1 \int_0^{p_2^2} d\theta_2$$

$$\cdot \Lambda^{p_1 - 1} \cdot \Lambda^{p_2 - 1} \cdot \frac{1}{\Lambda^p} \cdot \int_{\mathbb{R}^3} \exp \left\{ - \frac{\sigma^{(1)} \left(|Y_1 - Y_1'|^2 + |Y_2 - Y_2'|^2 \right)}{2\gamma} \right.$$

$$\left. - \frac{\sigma^{(2)}}{2\gamma} \cdot \frac{|Y_3 - Y_3'|^2}{2\gamma} - \frac{\sigma^{(1)} \left(|Y_1'|^2 + Y_2'|^2 \right)}{2(1-\gamma)} - \frac{\sigma^{(2)} |Y_3'|^2}{2(1-\gamma)} \right\} < \mathcal{H}^{(p_1)} \left(Y - Y', s \left(1 - \frac{\theta_1}{p_1^2} \right) \right),$$

$$\sqrt{s}\,k^{(0)} + \frac{Y}{\sqrt{p}} > \cdot P_{\sqrt{s}\,k^{(0)} + \frac{Y}{\sqrt{p}}} \mathcal{H}^{(p_2)}\left(Y', \left(1 - \frac{\theta_2}{p_2^2}\right) s\right) \cdot$$

$$\cdot \exp\left\{ -\theta_1 |\sqrt{s}\,k^{(0)} + \frac{Y - Y'}{\sqrt{p}\gamma}|^2 - \theta_2 |\sqrt{s}\,k^{(0)} + \frac{Y'}{\sqrt{p}\,(1-\gamma)}|^2 \right\} \cdot$$

$$-i \sum_{\substack{p_1 + p_2 = p+1 \\ p_1, p_2 > 1}} \frac{1}{\gamma(1-\gamma)} \cdot \frac{1}{p} \cdot \int_0^{p_1^2} \exp\{-\theta_1 s\}\, d\theta_1 \int_0^{p_2^2} \exp\{-\theta_2 s\}\, d\theta_2$$

$$\int_{\mathbb{R}^3} \exp\left\{ - \frac{\sigma^{(1)}\left(|Y_1 - Y_1'|^2 + |Y_2 - Y_2'|^2\right)}{2\gamma} - \frac{\sigma^{(2)}(|Y_3 - Y_3'|^2)}{2\gamma} \right.$$

$$\left. - \frac{\sigma^{(1)}\left(|Y_1'|^2 + Y_2'|^2\right)}{2(1-\gamma)} - \frac{\sigma^{(2)}\,|Y_3'|^2}{2(1-\gamma)} \right\}$$

$$\cdot p^{\frac{1}{2}} \cdot \langle \mathcal{H}^{(p_1)}(Y - Y'), \sqrt{s}\,k^{(0)} + \frac{Y}{\sqrt{p}} \rangle\, P_{\sqrt{s}\,k^{(0)} + \frac{Y}{\sqrt{p}}} \mathcal{H}^{(p_2)}(Y', s)\, d^3Y' .$$

We did not include the factor $\Lambda^{p-1} \cdot p$ because it is a part of the inductive assumption. This remainder is estimated in the following way.

First we consider

$$R_1 = \left(\left| \sqrt{s}\,k^{(0)} + \frac{Y - Y'}{\sqrt{p}\gamma} \right|^2 - s \right) + \left(\left| \sqrt{s}\,k^{(0)} + \frac{Y'}{\sqrt{p}\,(1-\gamma)} \right|^2 - s \right)$$

As before, consider the domain where

$$|Y - Y'| \le D_9 \sqrt{\ln(pk^{(0)})},\ |Y'| \le D_{10} \sqrt{\ln(pk^{(0)})} .$$

We write

$$R_1 = \frac{|Y-Y'|^2}{p\cdot\gamma_1^2} + \frac{|Y'|^2}{p\cdot\gamma_2^2} + C_{11}\left(\frac{|Y-Y'|}{\sqrt{p}\,\gamma|} + \frac{|Y'|}{\sqrt{p}(1-\gamma)}\right).$$

In the Domain A

$$|R_1| \le \frac{C_{12}\ln(pk^{(0)})}{pk^{(0)}}.$$

Therefore

$$R_2 = \exp\left\{-\theta_1\left|\sqrt{s}\,k^{(0)} + \frac{Y-Y'}{\sqrt{p}\,\gamma}\right|^2 - \theta_2\left|\sqrt{s}\,k^{(0)} + \frac{Y'}{\sqrt{p}\gamma^2}\right|^2\right\}$$
$$- \exp\{-\theta_1 s\}\cdot\exp\{-\theta_2 s\}$$

$$= \exp\{-(\theta_1+\theta_2)s)\}\cdot\left[\exp\left\{-\theta_1\left(\left|\sqrt{s}\,k^{(0)} + \frac{Y-Y'}{\sqrt{p}\gamma}\right|^2 - s\right)\right.\right.$$
$$\left.\left.\cdot\exp\left\{-\theta_2\left(\left|\sqrt{s}\,k^{(0)} + \frac{Y'}{\sqrt{p}(1-\gamma)}\right|^2 - s\right)\right)\right\} - 1\right]$$

and in the Domain A

$$|R_2| \le \exp\{-(\theta_1+\theta_2)s\}\left(\frac{\theta_1\cdot C_{13}}{\sqrt{p}\gamma} + \frac{\theta_2\ln p}{\sqrt{p}(1-\gamma)}\right).$$

This shows that in the Domain A we can replace the exponent

$$\exp\left\{-\theta_1|\sqrt{s}\,k^{(0)} + \frac{Y-Y'}{\sqrt{p}\gamma}|^2 - \theta_2|\sqrt{s}\,k^{(0)} + \frac{Y'}{\sqrt{p}(1-\gamma)}|^2\right\}$$

by $\exp\{-(\theta_1+\theta_2)s(k^{(0)})^2\}$ and the remainder will be not more than $\frac{D_{14}\ln p}{\sqrt{p}}$. This is enough for our purposes. In the Domain B the estimates are similar because again the main contribution to the integral comes from $|Y-Y'|\le D_9\sqrt{\ln p}$, $|Y'|\le D_{10}\sqrt{\ln p}$. In other words, in the Domain B we can replace the product of the Gaussian factors and $\mathcal{H}^{(p)}$ by

$$\exp\left\{-\frac{1}{2}\sigma^{(1)}\left(|Y_1-Y_1'|^2 + |Y_2-Y_2'|^2\right) - \frac{1}{2}\sigma^{(2)}|Y_3-Y_3'|^2\right.$$

$$- \frac{1}{2}\sigma^{(1)}\left(|Y_1'|^2 + |Y_2'|^2\right) - \frac{1}{2}\sigma^{(2)}(|Y_3'|^2)\Big\} .$$

This is also enough for our purpose.

The next remainder of Type I comes from the difference between the sum over γ and the corresponding integral. The remainder $\beta_p^{(3)}(Y, s)$ is the difference between the sum

$$i \sum_{\substack{p_1+p_2=p+1 \\ p_1, p_2 > \sqrt{p}}} \sqrt{\gamma}\sqrt{(1-\gamma)} \cdot \frac{1}{p} \cdot \int_{\mathbb{R}^3} \exp\Big\{ - \frac{\sigma^{(1)}\left(|Y_1 - Y_1'|^2 + |Y_2 - Y_2'|^2\right)}{2\gamma}$$

$$- \frac{\sigma^{(2)}(|Y_3 - Y_3'|)^2}{2\gamma} - \frac{\sigma^{(1)}(|Y_1'|^2 + |Y_2'|^2)}{2(1-\gamma)} - \frac{\sigma^{(2)}|Y_3'|^2}{2(1-\gamma)}\Big\}$$

$$\cdot \left(\frac{1}{2\pi\gamma}\right)^{\frac{3}{2}} \cdot \left(\frac{1}{2\pi(1-\gamma)}\right)^{\frac{3}{2}} \cdot p^{\frac{1}{2}} \cdot \langle \mathcal{H}^{(p_1)}((Y-Y')), \sqrt{s}\,k^{(0)} + \frac{Y}{\sqrt{p}}\rangle$$

$$P_{\sqrt{s}\,k^{(0)} + \frac{Y}{\sqrt{p}}} \mathcal{H}^{(p_2)}(Y', s)\, d^3Y'$$

and the corresponding integral over γ from 0 to 1. It is easy to check that this difference is not more than $\frac{C_{14}}{\sqrt{p}}$.

§8B. The Remainders of Type II and III

All remainders of Type II appear because we use the sums (over p_1) instead of the integrals . The functions $\mathcal{H}\left(\frac{Y-Y'}{\sqrt{\gamma}}\right)$ are defined for all γ. We use a linear interpolation to define $\delta(\gamma, Y, s)$ for all γ. From our inductive assumptions it follows that $|\delta_p(\gamma, Y, s)| \le \frac{C_{16}}{\sqrt{p}}$. Therefore, the remainders which follow from the difference between the sum and the integral also satisfy this estimate.

It remains to consider quadratic expressions of $\delta_p(\gamma, Y, s)$. The Gaussian density is present in all these expressions. Therefore, all the remainders are not more than $\frac{C_{17}}{p}$.

§9. Final Steps in the Proof of the Main Result

In this section we consider our procedure for $p > p_0$. Introduce the sequence p_n, $p_n = (1+\epsilon)p_{n-1} = (1+\epsilon)^n p_0$, where $\epsilon > 0$ is small (see below). These are the values of p when we make the renormalization of our parameters. For $p \neq p_n$, no renormalization is done.

In §7 we explained the choice of our fixed point $H^{(0)}$. The corresponding eigen-functions are denoted by $\Phi_j^{(u)}$ and $\Phi_{j'}^{(n)}$. Also we have eigen-functions of the stable part of the spectrum. Consider p, $p_m < p < p_{m+1}$. By induction we assume that we have an interval on the time axis $\left[S_-^{(m)}, S_+^{(m)}\right]$ and $s \in \left[S_-^{(m)}, S_+^{(m)}\right]$, $r < p$, so that

$$\begin{aligned}\tilde{g}_r(Y,s) =& \Lambda^{r-1} \cdot r \cdot (H^{(0)}(Y) + \delta^{(r)}(Y,s)) \cdot \\ & \cdot \frac{\sigma^{(1)}}{2\pi} \exp\left\{-\frac{\sigma^{(1)}(Y_1^2+Y_2^2)}{2}\right\} \cdot \sqrt{\frac{\sigma^{(2)}}{2\pi}} \exp\left\{-\frac{\sigma^{(2)}Y_3^2}{2}\right\}\end{aligned}$$

If $\gamma = \frac{r}{p-1}$ then

$$\delta^{(r)}(Y,s) = \sum_{j=1}^{4}\left(b_{j,p}^{(u)} + \beta_{j,r}^{(u)}\right)\gamma^{\alpha_j^{(u)}}\Phi_j^{(u)}(Y) + \sum_{j'=1}^{6}\left(b_{j',p}^{(n)} + \beta_{j',r}^{(n)}\right)\Phi_j^{(n)}(Y) + \Phi_r^{(st)}(Y,\gamma).$$

here $\beta_{j,r}^{(u)}, \beta_{j',r}^{(n)}$ are small corrections to the main terms $b_{j,p}^{(u)}$, $b_{j',p}^{(n)}$, $\Phi_r^{(st)}$ can be written as a series w.r.t. the stable eigen-functions. (see Appendix II).

At one step of our procedure $p-1$ is replaced by p, γ is replaced by $\gamma' = \gamma \cdot \frac{p-1}{p}$ and $\gamma^{\alpha_j^{(u)}}$ is replaced by $\left(1+\frac{1}{p-1}\right)^{\alpha_j^{(u)}} \cdot (\gamma')^{\alpha_j^{(u)}}$, $b_{j,p}^{(u)} + \bar{\beta}_{j,r}^{(u)}$ is replaced by $(\bar{b}_{j,p}^{(u)} + \beta_{j,r}^{(u)})\left(1+\frac{1}{p-1}\right)^{\alpha_j^{(u)}}$. During the whole interval $p_m < p < p_{m+1}$ the variable $b_{j,p_m}^{(u)}$ acquires the factor

$$\prod_{p_m<p<p_{m+1}}\left(1+\frac{1}{p-1}\right)^{\alpha_j^{(u)}} \approx e^{(1+\epsilon)\alpha_j^{(u)}}.$$

A similar statement holds for the stable part of the spectrum. The neutral part remains the same since $\alpha_{j'}^{(n)} = 0$.

Now we shall discuss $\delta^{(p)}(Y,s)$ using (7). As in §7 $\delta^{(p)}(Y,s)$ consists of three parts.

Part I. In all $\delta^{(r)}$ the main term is the one which contains our basic parameters $b_j^{(u)}$, $b_{j'}^{(n)}$. We consider terms in (7) which are linear in $b_j^{(u)}$, $b_{j'}^{(n)}$. As it follows from the definition of the linearized group and its spectrum we get $\left(1+\frac{1}{p}\right)^{\alpha_j^{(u)}} b_{j,p}^{(u)}$. For the neutral part we get 1 because $\alpha_{j'}^{(n)}=0$. We put $b_{j,p+1}^{(u)}=b_{j,p}^{(u)}\cdot\left(1+\frac{1}{p}\right)^{\alpha_j^{(u)}} b_{j,p}^{(u)}$, $b_{j',p+1}^{(n)}=b_{j',p}^{(n)}$. The stable part is transformed accordingly.

Part II. The term which is the sum of quadratic functions of all $\delta^{(r)}$. We expand it using the basis of our functions $\Phi_j^{(u)}$, $\Phi_{j'}^{(n)}$ and the functions from the stable part of the spectrum. The result is included in $\beta_{j,p}^{(u)}$, $\beta_{j',p}^{(n)}$ and the stable function $\Phi_p^{(st)}(Y,s)$.

Part III. All remainders which arise because the formulas for finite p are different from the limiting formulas. These remainders were estimated in §6. The result is written as a series w.r.t. our basis and the corresponding terms are included in $\beta_{j,p}^{(u)}$, $\beta_{j',p}^{(n)}$ and the stable part of the spectrum.

Finally we have

$$b_{j,p+1}^{(u)}=b_{j,p}^{(u)}\left(1+\frac{1}{p}\right)^{\alpha_j^{(u)}}, \quad b_{j,p+1}^{(n)}=b_{j,p}^{(n)}$$

and the formulas for $\beta_{j,p}^{(u)}$, $\beta_{j',p}^{(n)}$ and $\Phi_p^{(st)}(Y,s)$. This works for $p<p_{m+1}$. If $p=p_{m+1}$, then we introduce new variables (rescaling!)

$$b_{j,p_{m+1}}^{(u)}=b_{j,p_{m+1}-1}^{(u)}\left(1+\frac{1}{p_{m+1}}\right)^{\alpha_j^{(u)}}+\beta_{j,p_{m+1}}^{(u)},$$

$$b_{j',p_{m+1}}^{(n)}=b_{j',p_{m+1}-1}^{(n)}+\beta_{j',p_{m+1}}^{(n)}.$$

It is our other inductive assumption that

$$-\rho_1^m\le b_{j,p_m}^{(u)}\le\rho_1^m,\quad -\rho_1^m\le b_{j,p_m}^{(n)}\le\rho_1^m$$

where $0<\rho_1<1$ but ρ_1 is sufficiently close to 1.

Let $\Delta_{m+1}^{(m+1)}=\left[-\rho_1^{m+1},\rho_1^{m+1}\right]$ and $\Delta_m^{(m+1)}=\left\{(b_{j,p_m}^{(u)},b_{j',p_m}^{(n)}):\ -\rho_1^{m+1}\le b_{j,p_{m+1}}^{(u)},\ b_{j',p_{m+1}}^{(n)}\le\rho_1^{m+1}\right\}$. It follows easily from the estimates of $\beta_{j,p_{m+1}}^{(u)}$, $\beta_{j',p_{m+1}}^{(n)}$ that $\Delta_m^{(m+1)}\subseteq\Delta_m^{(m)}$. If $\Delta_0^{(m)}=\left\{(b_j^{(u)},b_{j'}^{(n)}):\ (b_{j,m}^{(u)},b_{j',m}^{(n)})\in\Delta_m^{(m)}\right\}$, then $\Delta_0^{(m)}$ is a decreasing sequence of closed sets. The intersection $\bigcap_m\Delta_0^{(m)}$ gives us the values of parameters for which $\delta^{(p)}\to\infty$ as $p\to\infty$.

We make also some shortening of the time interval $S^{(m)}$. In the formulas for $\delta^{(r)}$ there are several remainders which appear when we replace in all expressions s' and s'' by s. We estimate these remainders using the fact that our functions satisfy the Lipschitz condition and the Lipschitz constants and the maxima of their values decay as some power of p. We choose the interval $S^{(m+1)} \subset S^{(m)}$ so that when we consider $s \in S^{(m+1)}$ these remainders do not violate the basic inclusion $\Delta_m^{(m+1)} \subset \Delta_m^{(m)}$. It is easy to see $S^{(m+1)}$ can be chosen so that $S^{(m)} \setminus S^{(m+1)}$ consists of two intervals whose lengths decay exponentially. Therefore $\bigcap_m S^{(m)}$ is an interval of positive length.

The transformation $(b_{j,p_{m+1}}^{(u)}, b_{j',p_{m+1}}^{(n)}) \to (b_{j,p_m}^{(u)}, b_{j',p_m}^{(n)})$ is given by smooth functions and is sufficiently close to the identity map. The last step in the renormalization is the replacement in all $\delta^{(r)}$, $r < p_{m+1}$ the variables $b_{j,p_m}^{(u)}$, $b_{j',p_m}^{(n)}$ by their expressions through $b_{j,p_{m+1}}^{(u)}$, $b_{j',p_{m+1}}^{(n)}$. The form of $\delta^{(r)}$ in new variables is the same as before.

The Choice of Constants

The main constants which are used in the construction are the following:

1. $k^{(0)}$ which determines the position of the domain where $v(k,0)$ is concentrated.
2. D_1 is the constant which determines the size of the neighborhood where $v(k,0)$ is concentrated.
3. ρ_1 determines the size of the neighborhood where the main parameters $b_j^{(u)}$, $b_{j'}^{(n)}$ vary.
4. D_2 is the constant which determines the possible size of perturbations $\Phi^{(st)}$ in the form of $v(k,0)$.
5. λ_1 is the power which gives the estimation of the decay of h_r in the domain B.
6. λ_2 is the parameter which determines the size of the first part of the procedure.
7. ϵ determines the values of p where we make the renormalization.

The value of $k^{(0)}$ should be sufficiently large. All estimate of the remainders which appear during the first half of the procedure are less than $\frac{const}{(k^{(0)})^{\frac{1}{2}}}$. They should be so small that the estimates of all $\beta_{jr}^{(u)}$, $\beta_{j'r}^{(n)}$ are much smaller than ρ_1. On the other hand, ρ_1 should be small but not too small. It should be small in order to make the quadratic part of our formulas smaller than the linear part. However ρ cannot be too small in order that we could choose the next interval $[-\rho^{m+1}, \rho^{m+1}]$. This can be achieved by the choice of $k^{(0)}$. The parameter λ_2 should be small. In this case the estimates of all corrections are easier. However, after λ_2 is chosen the value of $k^{(0)}$ can be taken sufficiently large depending on λ_2. The parameter λ_1 can be arbitrarily large in order to make the perturbation arbitrarily small. The value of D_1 determines the estimates in the domain B which decay as $\frac{1}{\left(k^{(0)}\right)^{\lambda_1}}$. We choose D_1 so that $\lambda_1 > \frac{1}{2}$. The value of ϵ is chosen small so that we can write with a good precision the action of the linearized renormalization group.

§10. Critical Value of Parameters and Behavior of Solutions near the Singularity Point

We return back to the first formulas:

$$v_A(k,t) = \exp\{-t|k|^2\}\, A\cdot v(k,0) + \int_0^t \exp\{-(t-s)|k|^2\}\cdot \sum_{p>1} A^p h_p(k,s)\, ds$$

or

$$v_A(k,t) = \exp\{-t|k|^2\}\, A\cdot v(k,0) + \int_0^t \exp\{-(t-s)|k|^2\}\cdot \sum_{p>1} A^p g_p(k\sqrt{s},s)\, ds\,. \quad (41)$$

Our construction gives us the interval $S = \bigcap_n S^{(n)}$ on the time axis such that for each $t \in S$ we can find the values of parameters $b_j^{(u)} = b_j^{(u)}(t)$, $1 \le j \le 4$ and $b_{j'}^{(n)} = b_{j'}^{(n)}(t)$, $1 \le j' \le 6$ such that we have the representation (31) with $\delta^{(r)} \to 0$ as $r \to \infty$. It is easy to see that $A_{cr}(t) = \Lambda^{-1}(t)$. If so then $A^p h_p(k,t)$ is concentrated in the domain with the center at $\frac{\kappa^{(0)}p}{\sqrt{t}}$ having the size $O(\sqrt{p})$ and there it takes values $O(p)$. This immediately implies that at t the energy is infinite.

Consider $t' < t$. It is important to investigate the behavior of $E(t')$ and the enstrophy $\Omega(t')$ of the same solution with $A = A_{cr}(t)$ when t' is close to t. Denote $\Delta t = t - t'$. It follows easily from the proof of the main result that $\Lambda(t')/_{\Lambda(t)} = (1 - C_1\Delta t + O(\Delta t))$ for some constant C_1. Since $A_{cr}^p \cdot (\Lambda(t'))^p = A_{cr}^p \cdot (\Lambda(t))^p \cdot \left(\Lambda(t')/_{\Lambda(t)}\right)^p = (1 - C\Delta t + o(\Delta t))^p$. It is clear that the terms in (41) are close to each other for $p \le O\left(\frac{ln(\Delta t)^{-1}}{\Delta t}\right)$. For $p >> \frac{ln(\Delta t)^{-1}}{\Delta t}$ the product $A_{cr}^p(\Lambda(t'))^p$ tends exponentially to zero and dominates other terms of the expansion. Therefore it is enough to consider $|k| \leq O\left(\frac{ln(\Delta t)^{-1}}{\Delta t}\right)$ and in this domain the solution grows as $|k|^{\frac{3}{2}}$. The factor $|k|^{\frac{1}{2}}$ appears because for any k the values for which the terms in (41) give the essential contribution to the solution belonging to an interval of the size $O(\sqrt{|k|}) = O(\sqrt{p})$. From this argument it follows easily that $E(t') \sim \left(\frac{ln(\Delta t)^{-1}}{\Delta t}\right)^6$ and $\Omega(t') \sim \left(\frac{ln(\Delta t)^{-1}}{\Delta t}\right)^8$.

Appendix I. Hermite Polynomials and their basic properties

Take $\sigma > 0$ and write

$$He_n^{(\sigma)}(x) = (-1)^n e^{\frac{\sigma x^2}{2}} \frac{d^n}{dx^n} e^{-\frac{\sigma x^2}{2}}, \quad n \geq 0.$$

It is clear that $He_n^{(\sigma)}(x) = \sigma^n x^n + \cdots$, where dots mean terms of smaller degree. We shall call $He_n^{(\sigma)}$ the n-th Hermite polynomial. It is clear that $He_0^{(\sigma)}(x) = 1$, $He_1^{(\sigma)}(x) = \sigma x$, $He_2^{(\sigma)}(x) = \sigma^2 x^2 - \sigma$ and so on. In general, $He_n^{(\sigma)}(x) = \sigma^{\frac{n}{2}} He_n^{(1)}(\sqrt{\sigma}x)$. It is easy to check that

$$\sigma x He_n^{(\sigma)}(x) = He_{n+1}^{(\sigma)}(x) + \sigma n He_{n-1}^{(\sigma)}(x). \tag{42}$$

The Fourier transform of $He_m^{(\sigma)}(x) e^{-\frac{\sigma x^2}{2}} \sqrt{\frac{\sigma}{2\pi}}$ is $(i\lambda)^m e^{-\frac{\lambda^2}{2\lambda}}$. This implies the formula for convolution:

$$\int_{\mathbb{R}^1} He_{m_1}^{(\sigma)}(x-y) e^{-\frac{\sigma(x-y)^2}{2}} \sqrt{\frac{\sigma}{2\pi}} \cdot He_{m_2}^{(\sigma)}(y) e^{-\frac{\sigma y^2}{2}} \sqrt{\frac{\sigma}{2\pi}} dy = He_{m_1+m_2}^{(\sigma)}(x) e^{-\frac{\sigma x^2}{2}} \sqrt{\frac{\sigma}{2\pi}} \tag{43}$$

Take positive $\gamma_1, \gamma_2, \gamma_1 + \gamma_2 = 1$ and consider the convolution of $He_{m_1}^{(\sigma)}(\frac{x}{\sqrt{\gamma_1}}) e^{-\frac{\sigma x^2}{2\gamma_1}} \cdot \sqrt{\frac{\sigma}{2\pi\gamma_1}}$ and $He_{m_2}^{(\sigma)}(\frac{x}{\sqrt{\gamma_2}}) e^{-\frac{\sigma x^2}{2\gamma_2}} \cdot \sqrt{\frac{\sigma}{2\pi\gamma_2}}$. Their Fourier transforms are $(i\lambda\sqrt{\gamma_1})^{m_1} e^{-\frac{\lambda^2\gamma_1}{2\sigma}}$ and $(i\lambda\sqrt{\gamma_2})^{m_2} e^{-\frac{\lambda^2\gamma_2}{2\sigma}}$ respectively. The product of these two functions is $\gamma_1^{\frac{m_1}{2}} \gamma_2^{\frac{m_2}{2}} (i\lambda)^{m_1+m_2} e^{-\frac{\lambda^2}{2\sigma}}$. Therefore the convolution is $\gamma_1^{\frac{m_1}{2}} \cdot \gamma_2^{\frac{m_2}{2}} He_{m_1+m_2}^{(\sigma)}(x) e^{-\frac{\sigma x^2}{2}}$.

References

[C] M. Cannone. Harmonic Analysis Tools for Solving the Incompressile Navier-Stokes Equations. Handbook of Mathematical Fluid Dynamics, vol. 3, 2002.

[Cl] Clay Mathematical Institute. The Millennium Prize Problems, 2006.

[F-T] C. Foias and R. Temam. Gevrey Classes of Regularity for the Solutions of the Navier-Stokes Equations. J. of Funct. Anal. 87, 1989, 359-369.

[G] Y. Giga, T. Miyakawa. Navier-Stokes Flow in R^3 with Measures as Initial Vorticity and Morrey spaces. Commu. Partial Differential Equations, 14, 1989, 577-618.

[K] T. Kato. Strong L^p-solution of the Navier-Stokes Equation in R^m, with Applications to Weak Solutions. Math. Zeitschrift, 187, 1984, 471-480.

[La] O. Ladyzenskaya. The mathematical theory of viscous incompressible flow. New York: Gordon and Breach Science Publishers, 1969.

[Le] J. Leray. Étude de diverses équations intégrales non linéaires et de quelques problémes que pose l'hydrodynamique. J. Math. Pures Appl. 12, 1993, 1-82

[Si 1] Ya. G. Sinai. Power Series for Solutions of the Navier-Stokes System on R^3. Journal of Stat. Physics, vol. 121, No. 516, 2005, 779-804.

[Si 2] Ya. G. Sinai. Diagrammatic Approach to the $3D$-Navier-Stokes System. Russian Math. Surveys, vol. 60, No.5, 2005, 47-70.

[Y] V.I. Yudovich. The Linearization Method in Hydrodynamical Stability Theory. Trans. Math. Mon. Amer. Math. Soc. Providence, RI 74(1984).

March 31, 2007:gpp

Comments

Many experts in PDE and fluid dynamics tried to prove the existence of strong solutions for the $3D$-Navier – Stokes system. So far all these attempts have been unsuccessful. We tried to prove an opposite result by constructing solutions which exist only during a finite interval of time and at the end of this interval blows up. We considered complex-valued solutions which do not satisfy the energy inequality. It is a big simplification compared with the usual case. We used the renormalization group method in the form similar to the one used by M. Feigenbaum in his papers on the universality of period-doubling bifurcations in one-parameter families of one-dimensional maps. The results show that blow ups can appear in ten-parameter families of initial conditions. Later we generalized our technique to the case of Burgers systems of any dimension and some one-dimensional systems (see [LS1, LS2]).

Journal of Statistical Physics, Vol. 64, Nos. 1/2, 1991

Two Results Concerning Asymptotic Behavior of Solutions of the Burgers Equation with Force

Ya. G. Sinai[1]

Received February 26, 1991

We consider the Burgers equation with an external force. For the case of the force periodic in space and time we prove the existence of a solution periodic in space and time which is the limit of a wide class of solutions as $t \to \infty$. If the force is the product of a periodic function of x and white noise in time, we prove the existence of an invariant distribution concentrated on the space of space-periodic functions which is the limit of a wide class of distributions as $t \to \infty$.

KEY WORDS: Burgers equation; white noise; local central limit theorem; partition function.

1. FORMULATION OF THE RESULTS

We consider the one-dimensional Burgers equation with force having the form

$$u_t + u \cdot u_x = \mu u_{xx} + F'(x)\, B(t), \qquad -\infty < x < \infty \tag{1}$$

Here $F(x)$ is a C^1-periodic function of period x_0. The assumptions concerning B will be formulated later. The initial data $u(x;0)$ are derivatives $u(x;0) = v'(x)$, where $v(x)$ are typical realizations of some random process. The probability distribution corresponding to v is denoted by P_0. It is defined on the natural σ-algebra of subsets of the space V of absolutely continuous functions $v(x)$. We assume that:

1. There exists a constant C_0 such that with P_0-probability 1

$$|v(x)| \leqslant C_0$$

for all x.

[1] Landau Institute of Theoretical Physics, Moscow, USSR, and Princeton University, Princeton, New Jersey.

Y.G. Sinai, *Selecta II: Probability Theory, Statistical Mechanics, Mathematical Physics and Mathematical Fluid Dynamics*, DOI 10.1007/978-1-4419-6205-8_22, © Springer Science+Business Media, LLC 2010

2. $A = E\{\exp[-v(x)]\}$ does not depend on x.
3. There exists γ, $0<\gamma<1/2$, such that for P_0-almost every $v(x)$,

$$\lim_{n\to\infty} \sup_{\substack{a\in[0,x_0]\\ |m|\leqslant n, m\in \mathbb{Z}^1}} \frac{1}{2[n^\gamma]+1}\left|\sum_{|k-m|\leqslant[n^\gamma]} e^{-v(a+kx_0)} - A\right| = 0$$

It is easy to give concrete examples of P_0 for which condition 3 is true.

Theorem 1. Let B be a continuous periodic function of period τ_0. There exists a solution $u^{(0)}(x, t)$ of (1) periodic in x with period x_0 and periodic in time with period τ_0 such that for P_0-a.e. v

$$\lim_{t\to\infty} [u(x, t) - u^{(0)}(x, t)] = 0$$

for any x, $-\infty < x < \infty$.

Remarks. 1. Our method of proof also gives an explicit expression for $u^{(0)}(x, t)$.

2. The theorem remains true if the force in the Burgers equation takes the form $\partial F(x, t)/dx$, where $F(x, t)$ is a function periodic in space with period x_0 and periodic in time with period τ_0.

3. The theorem remains true for bounded functions v such that $v'(x)\to 0$ as $x\to\pm\infty$, $v(x)\to$ const as $x\to\pm\infty$.

4. The convergence in Theorem 1 is pointwise. After giving the proof, we discuss stronger statements concerning the character of convergence.

In Theorem 2 we assume that $B(t)$ is a white noise. This means that for any t_1, t_2, $t_1<t_2$, a random variable $b(t_1, t_2) = \int_{t_1}^{t_2} B(\tau)\, d\tau$ is defined such that:

(a$_1$) $b(t_1, t_2)$ has the Gaussian distribution with mean value equal to zero and dispersion $Eb^2(t_1, t_2) = \sigma(t_2 - t_1)$ for some $\sigma > 0$.

(a$_2$) For nonoverlapping intervals (t_1', t_2') and (t_1'', t_2'') the random variables $b(t_1', t_2')$ and $b(t_1'', t_2'')$ are independent.

Denote by $M((t_1, t_2)$ the least σ-algebra generated by all $b(t_1', t_2')$, where $t_1 < t_1' < t_2' < t_2$, and let $\{T^t\}$ be the measure-preserving flow on the space of all random variables $b(t_1', t_2')$, where each T^t transforms $M(t_1, t_2)$ to $M(t_1 - t, t_2 - t)$ and

$$(T^t b)(t_1', t_2') = b(t_1 - t', t_2 - t')$$

for any t_1', t_2', $t_1 + t < t_1' < t_2' < t_2 + t$.

Assume that P_0 satisfies the same conditions as in Theorem 1.

Theorem 2. Let P_t be the natural probability distribution on the space of solutions $u(x, t)$ of (1) induced by P_0, $0<t<\infty$. Then P_t converges weakly as $t \to \infty$ to some probability distribution Q which does not depend on P_0 and is concentrated on the space of functions periodic in x with period x_0.

The proof of Theorem 1 is given in Section 2. In Section 3 we expound the proof of Theorem 2. The actual statement which we show is the following. For any $\varepsilon<0$ we find $t_0(\varepsilon)$, a set $C \in M(0, t_0(\varepsilon))$, and a functional $H_x(\{B(t_1, t_2)\}, 0 \leqslant t_1, t_2 \leqslant t_0(\varepsilon))$ defined on C and such that $\text{Prob}(C) \geqslant 1-\varepsilon$ and if $T^{-t} b \in C$, then

$$|u(x, t)-H_x(b(t_1, t_2), t-t_0(\varepsilon) \leqslant t_1, t_2 \leqslant t)| \leqslant \varepsilon$$

In other words, for increasing t, the solution $u(x, t)$ becomes a functional of the realization of white noise $B(\tau)$, $0<\tau<t$, with "short memory." This memory can be estimated in a more precise way. The functional H_x depends periodically on x. Theorems 1 and 2 are valid also for the multi-dimensional Burgers equation. Only small modifications in the proofs are needed.

2. PROOF OF THEOREM 1

After the appropriate rescaling of x and μ we may assume that the period $\tau_0=1$. We use the Cole–Hopf substitution $u=-2\mu(\varphi_x/\varphi)$,[1,2] and get for φ the equation

$$\varphi_t=\mu \varphi_{xx}-\frac{1}{2\mu} F(x) B(t) \varphi \tag{2}$$

The Feynman–Kac formula[3] makes it possible to write down φ as a functional integral. Namely, denote by $\Pi_{W_1, W_2}^{(t_1, t_2)}$ the corresponding Wiener measure on the space of continuous functions $w(\tau)$, $t_2 \leqslant \tau \leqslant t_1$, such that $w(t_1)=w_1$, $w(t_2)=w_2$. Then

$$\begin{aligned} \varphi(x; t)= & \int_{-\infty}^{\infty} dy\, \{\exp[-v(y)]\} \\ & \times \int \left\{\exp\left[\int_0^t F(W(\tau)) B(\tau)\, d\tau\right]\right\} d\Pi_{x, y}^{(t, 0)}(W) \end{aligned} \tag{3}$$

Put $t=t_0$, $t-j=t_j$, $j \geqslant 1$, and $j \in \mathbb{Z}^1$, and find r such that $t_{r+2}<0<t_{r+1}$. Fix the numbers $a_1, a_2, \ldots, a_r, a_{r+1}$, $a_j \in [0, x_0)$ and rewrite (3) as follows:

$$\varphi(x;t)=\int_0^{x_0}\cdots\int_0^{x_0} da_1\,da_2\cdots da_r\,da_{r+1}\sum_{\substack{m_1,m_2,\ldots,m_{r+1}\\ m_j\in\mathbb{Z}^1}} K_{t_1}(x,a_1+mx_0)$$
$$\times\prod_{j=2}^{r} K_{t_j}(a_{j-1}+m_jx_0,a_j+m_jx_0)$$
$$\times K_f(a_r+m_rx_0,a_{r+1}+m_{r+1}x_0)\,e^{-v(a_{r+1}+m_{r+1}x_0)} \tag{4}$$

Here

$$K_{t_j}(W_1,W_2)=\int\left\{\exp\left[\int_{t_{j-1}}^{t_j} F(W(\tau))\,B(\tau)\,dt\right]\right\}d\Pi_{(W_1,W_2)}^{(t_j,t_{j-1})}(W)$$
$$K_f(W_1,W_2)=\int\left\{\exp\left[\int_0^{t_r} F(W(\tau))\,B(\tau)\,dt\right]\right\}d\Pi_{(W_1,W_2)}^{(t_r,0)}(W)$$

The periodicity of F in space and that of B in time imply the following relations:

1. $K_{t_j}(W_1,W_2)=K_{t_j}(W_1+mx_0,W_2+mx_0),\qquad 2\leqslant j\leqslant r$

for all $m\in\mathbb{Z}^1$.

2. $K_{t_2}(W_1,W_2)=K_{t_3}(W_1,W_2)=\cdots K_{t_r}(W_1,W_2)=K(W_1,W_2)$

The functions $K_{t_1}(w_1,w_2)$, $K_f(w_1,w_2)$ depend on the fractional part $\{t\}$ and thus are periodic in time with period 1. Introduce the sums

$$Z_{t_1}(x;a_1)=\sum_{m\in\mathbb{Z}^1} K_{t_1}(x,a_1+mx_0)$$
$$Z_{t_j}(a_{j-1},a_j)=Z_{t_j}(a_{j-1},a_j)$$
$$=\sum_{m\in\mathbb{Z}^1} K_{t_j}(a_{j-1},a_j+mx_0),\qquad 2\leqslant j\leqslant r$$
$$Z_f(a_r,a_{r+1})=\sum_{m\in\mathbb{Z}^1} K_f(a_r,a_{r+1}+mx_0)$$

and the probabilities

$$p_{t_1}(x,a_1+m_1x_0)$$
$$=Z_{t_1}^{-1}(x,a_1)\,K_{t_1}(x,a_1+m_1x_0)$$
$$p_{t_j}(a_{j-1}+m_{j-1}x_0,a_j+m_jx_0)$$
$$=Z_{t_j}^{-1}(a_{j-1},a_j)\,K_{t_j}(a_{j-1}+m_{j-1}x_0,a_j+m_jx_0),\qquad 2\leqslant j\leqslant r$$
$$p_f(a_r+m_rx_0;a_{r+1}+m_{r+1}x_0)$$
$$=Z_f^{-1}(a_r,a_{r+1})\,K_f(a_r+m_rx_0;a_{r+1}+m_{r+1}x_0)$$

Consider the sequence of independent random variables $\xi_1, \xi_2, \ldots, \xi_{r+1}$, where each variable ξ_j takes values $m \in \mathbb{Z}^1$, and ξ_1 has the distribution with the probabilities $p_1(x, a_1 + mx_0)$, while ξ_j, $2 \leqslant j \leqslant r$, have the distribution with the probabilities $p_{t_j}(a_{j-1}, a_j + mx_0) = p(a_{j-1}, a_j + mx_0)$, which depends only on $\{t\}$ but not on j, and ξ_{r+1} has the distribution with the probabilities $p_f(a_r, a_{r+1} + mx_0)$, which also depend only on $\{t\}$. Then the sum in (4) can be rewritten as follows:

$$
\begin{aligned}
\sum &= \sum (a_1, a_2, \ldots, a_{r+1}) \\
&= \sum_{m_1, m_2, \ldots, m_{r+1}} K_{t_1}(x, a_1 + m_1 x_0) \\
&\quad \times \prod_{j=2}^{r} K_{t_1}(a_{j-1} + m_{j-1} x_0, a_j + m_j x_0) \\
&\quad \times K_f(a_r + m_r x_0, a_{r+1} + m_{r+1} x_0)\, e^{-v(a_{r+1} + m_{r+1} x_0)} \\
&= Z_{t_1}(x; a_1)\, Z(a_1, a_2) \cdots Z(a_{r-1}, a_r)\, Z_f(a_r, a_{r+1}) \\
&\quad \times \sum_{m_1, m_2, \ldots, m_{r+1}} p_{t_1}(x, a_1 + m_1 x_0) \\
&\quad \times \prod_{j=2}^{r} p_j(a_{j-1} + m_{j-1} x_0, a_j + m_j x_0) \\
&\quad \times p_f(a_r + m_r x_0, a_{r+1} + m_{r+1} x_0)\, e^{-v(a_{r+1} + m_{r+1} x_0)} \\
&= Z_{t_1}(x; a_1) \prod_{j=2}^{r} Z_j(a_{j-1}, a_j)\, Z_f(a_r, a_{r+1}) \\
&\quad \times \sum_{n_1, n_{r+1}} p_{t_1}(x, a_1 + n_1 x_0) \prod_{j=2}^{r} p_j(a_{j-1}, a_j + n_j x_0) \\
&\quad \times p_f(a_r, a_{r+1} + n_{r+1} x_0)\, e^{-v(a_{r+1} + (n_1 + \cdots + n_{r+1}) x_0)} \\
&= Z_{t_1}(x; a_1) \prod_{j=2}^{r} Z(a_{j-1}, a_j)\, Z_f(a_r, a_{r+1}) \\
&\quad \times E_\xi e^{-v(a_{r+1} + (\xi_1 + \cdots + \xi_{r+1}) x_0)}
\end{aligned} \tag{5}
$$

where E_ξ is the expectation with respect to the joint distribution of the random variables ξ_j, $1 \leqslant j \leqslant r+1$. Put $\mu(a_1) = E\xi_1$, $\mu(a_{j-1}, a_j) = E\xi_j$ for $1 \leqslant j \leqslant r+1$, $d(a_1) = D(\xi_1) = E_\xi(\xi_1 - \mu(a_1))^2$, $d(a_{j-1}, a_j) = E_\xi(\xi_j - \mu(a_{j-1}, a_j))^2$, $M = \mu(a_1) + \sum_{j=2}^{r+1} \mu(a_{j-1}, a_j)$, and $D = d(a_1) + \sum_{j=2}^{r+1} d(a_{j-1}, a_j)$.

Lemma 1. Under the conditions of Theorem 1, the sequence of random variables $\xi_1, \xi_2, \ldots, \xi_{r+1}$ satisfies the central local limit theorem of probability theory in the form

$$\sup_{a_1, \ldots, a_{r+1}} \left| P_\xi(\xi_1 + \cdots + \xi_{r+1} = m) - \frac{1}{(2\pi D)^{1/2}} e^{-(m-M)^2/2D} \right| \leqslant \varepsilon_t \tag{6}$$

where ε_t tends to zero as $t \to \infty$.

The statement of Lemma 1 means that the convergence to zero of the difference in (6) is uniform in m_1 and $a_1, a_2, \ldots, a_{r+1}$. Lemma 1 can be easily proven by standard methods of probability theory (see, e.g., ref. 4). We omit the proof. During the proof one must keep in mind the boundedness of $|F|$ and $|B|$.

Consider in more detail the expectation

$$E = E_\xi e^{-v_0(a_{r+1} + (\xi_1 + \cdots + \xi_{r+1}) x_0)}$$

In view of Lemma 1 and property 1 of P_0, it is equal to

$$E = \sum_m e^{-v_0(a_{r+1} + m x_0)} P_\xi\{\xi_1 + \cdots + \xi_{r+1} = m\}$$

$$= \sum_m e^{-v_0(a_{r+1} + m x_0)} \frac{1}{(2\pi D)^{1/2}} e^{-(m-M)^2/2D} + e^{c_0} \delta_t$$

where $\delta_t \to 0$ as $t \to \infty$. Using property 3 of P_0, we easily get

$$E = A + \delta_t^{(1)}$$

where $\delta_t^{(1)} \to 0$ as $t \to \infty$ uniformly in $a_1, \ldots, a_{r+1}$. Thus,

$$\varphi(x; t) \sim A \int \cdots \int da_1\, da_2 \cdots da_r\, da_{r+1}\, Z_{t_1}(x; a_1)$$

$$\times \prod_{j=2}^{r} Z(a_{j-1}, a_j)\, Z_f(a_r, a_{r+1}) \tag{7}$$

The expression (7) can be studied with the methods of statistical mechanics. Consider $Z(a', a'')$ as a transfer matrix of a one-dimensional system and find its positive eigenvector $e(a)$ and the corresponding positive eigenvalue λ:

$$\int e(a')\, Z(a', a'')\, da' = \lambda e(a'')$$

Introduce the Markov transition operator π with the transition probabilities

$$\pi(a', a'') = \frac{Z(a', a'')\, e(a')}{\lambda e(a'')}$$

giving the density of the transition $a'' \to a'$. Then (7) can be rewritten as

$$\begin{aligned}\varphi(x; t) &\sim A\lambda^r \int \cdots \int da_1 \cdots da_r\, da_{r+1}\, Z_{t_1}(x; a_1)[e(a_1)]^{-1}\, \pi(a_2, a_3)\\ &\quad \times \cdots \pi(a_{r-1}, a_r)\, Z(a_r, a_{r+1})\, e(a_r)\\ &= A\lambda^r \int Z_{t_1}(x; a_1)[e(a_1)]^{-1}\, \pi^{(r)}(a_1, a_r)\\ &\quad \times Z(a_r, a_{r+1})\, e(a_r)\, da_1\, da_r\, da_{r+1}\end{aligned}$$

The ergodic theorem for the Markov chain generated by the operator π shows that $\pi^{(r)}(a_1, a_r)$ asymptotically does not depend on a_r and is exponentially close to the stationary distribution of this chain. Denote this distribution by $\pi(a_1)$. It is well known that it has the form $e(a_1)\, e^*(a_1)$, where $e^*(a_1)$ is the positive eigenvector of the adjoint operator, i.e.,

$$\int Z(a', a'')\, e^*(a'')\, da'' = \lambda e^*(a')$$

Thus

$$\varphi(x, t) \sim A \cdot A_1 \lambda^r \int Z_{t_1}(x; a_1)\, e^*(a_1)\, da_1$$

where $A_1 = \iint Z(a_r, a_{r+1})\, e(a_r)\, da_r\, da_{r+1}$.

Taking the derivative of the rhs of (4) with respect to x and making the same analysis, we find a similar expression for φ_x:

$$\varphi_x(x, t) \sim AA_1\lambda^r \int \frac{\partial}{\partial x} Z_{t_1}(x; a_1)\, e(a_1)\, da_1$$

Finally we get that for $t \to \infty$

$$u = -2\mu \frac{\varphi_x(x; t)}{\varphi(x; t)} \sim 2\mu \frac{\int [\partial Z_{t_1}(x; a_1)/\partial x]\, e(a_1)\, da_1}{\int Z_{t_1}(x; a_1)\, e(a_1)\, da_1} \tag{8}$$

The rhs of (8) is a solution of (1) periodic in space and time and (8) gives the assertion of Theorem 1.

It is clear that the properties of smoothness of

$$u^{(0)}(x; t) = -2\mu \frac{\int [\partial Z_{t_1}(x; a_1)/\partial x]\, e(a_1)\, da_1}{\int Z_{t_1}(x; a_1)\, e(a_1)\, da_1}$$

depend on the smoothness of F. In particular, if $F(x) \in C^k(S^1)$, then $u^{(0)}(x; t) \in C^{k-1}(S^1)$ for any t and one can prove easily the convergence of $(\partial^i/\partial x^i)\, u(x; t)$, $i \leqslant k-1$, to $(\partial^i/\partial x^i)\, u^{(0)}(x; t)$. Also, the convergence in Theorem 1 is uniform in x on any compact subset of R^1. Certainly in general it cannot be uniform on the whole line, because of fluctuations of v.

3. PROOF OF THEOREM 2

Again we use the Cole–Hopf substitution, which now gives the expression for φ in the form

$$\varphi(x; t) = \int_{-\infty}^{\infty} dy\, \{\exp[-v(y)]\} \times \left\{\exp\left[\int_0^t F(w(\tau))\, db(\tau)\right]\right\} d\Pi_{(x,\, y)}^{(t,0)}(W) \tag{9}$$

Here $\int_0^t F(w(\tau))\, db(\tau)$ is a stochastic integral, and $B(\tau) = db(\tau)/d\tau$ is white noise. It is worthwhile to stress that $\{w(\tau)\}$ and $\{b(\tau, 0)\}$ are statistically independent Brownian motions. Therefore $\varphi(x; t)$ is random because of the randomness of b. We proceed in the same way as before. Take an integer $r = r(t)$ for which $r/t \to 1$ as $t \to \infty$ and divide the interval $(0, t)$ into r equal parts. Denote the points of the division by $t = t_0 > t_1 > \cdots > t_r = 0$, and rewrite (9) as follows:

$$\varphi(x; t) = \int \cdots \int da_1 \cdots da_r \sum_{\substack{m_1, \ldots, m_r \\ m_j \in \mathbb{Z}^1}} K_1(x, a_1 + m_1 x_0) \times K_j(a_{j-1} + m_{j-1} x_0, a_j + m_j x_0) \exp[-v(a_r + m_r x_0)] \tag{10}$$

where

$$K_j(a', a'') = \int \left\{\exp\left[\int_{t_j}^{t_{j-1}} F(w(\tau))\, db(\tau)\right]\right\} d\Pi_{a', a''}^{(t_{j-1}, t_j)}(W)$$

In the case of white noise the operators $K_j(a', a'')$ are random and statistically independent in a natural sense for different j. The periodicity of F in x implies

$$K_j(a', a'') = K_j(a' + m x_0, a'' + m x_0), \qquad m \in \mathbb{Z}^1$$

This gives again a possibility to reduce the summation $\sum_{m_1,\ldots,m_r}$ to a problem concerning independent differently distributed random variables. Namely, introduce the partition functions

$$Z_1(x, a_1) = \sum_m K_1(x, a_1 + mx_0)$$

$$Z_j(a_{j-1}, a_j) = \sum_{m \in \mathbb{Z}^1} K_j(a_{j-1}, a_j + mx_0)$$

and the corresponding probabilities

$$p_1(x, a_1 + m_1 x_0) = Z_1^{-1}(x, a_1)\, K(x, a_1 + m_1 x_0)$$

$$\begin{aligned} p_j(a_{j-1} + m_{j-1} x_0, a_j + m_j x_0) &= Z_j^{-1}(a_{j-1}, a_j)\, K_j(a_{j-1} + m_{j-1} x_0, a_j + m_j x_0) \\ &= Z_j^{-1}(a_{j-1}, a_j)\, K_j(a_{j-1}, a_j + (m_j - m_{j-1})\, x_0) \end{aligned}$$

Then (10) takes the form

$$\begin{aligned} \varphi(x; t) = \int da_1\, da_2 \cdots da_r\, Z_1(x, a_1)\, Z_2(a_1, a_2) \cdots Z_r(a_{r-1}, a_r) \\ \times \sum_{n_1, n_2, \ldots, n_r \in \mathbb{Z}^1} p_1(x, a_1 + n_1 x_0) \\ \times p_2(a_1, a_2 + n_2 x_0) \cdots p_r(a_{r-1}, a_r + n_r x_0) \\ \times \exp\{-r[a_r + (n_1 + \cdots + n_r)\, x_0]\} \end{aligned}$$

Let $\xi_1, \ldots, \xi_r$ be r independent integer-valued random variables where each ξ_j has the distribution $p_j(a_{j-1}, a_j + mx_0)$, $a_0 = x$. We can write

$$\begin{aligned} \sum_{n_1, n_2, \ldots, n_r \in \mathbb{Z}^1} p_1(x, a_1 + n_1 x_0)\, p_2(a_1, a_2 + n_2 x_0) \cdots p_r(a_{r-1}, a_r + n_r x_0) \\ \times \exp\{-r[a_r + (n_1 + \cdots + n_r)\, x_0]\} \\ = E_\xi \exp\{-r[a_r + (\xi_1 + \cdots + \xi_r)\, x_0]\} \qquad (11) \end{aligned}$$

Again as in Section 2 we encounter two problems. The first one concerns the validity of the local central limit theorem of probability theory, while the second one consists of the possibility of replacing the average (11) by its mathematical expectation A. Since the distribution P_0 has the properties 1–3 (see Section 1), the second problem is simple because the local central limit theorem and the stability of the averages (see property 3) of the distribution P_0 show that (11) is equivalent to A as $t \to \infty$.

In order to study the local central limit theorem, introduce

$$\mu_1(a_1) = E\xi_1, \qquad \mu_j(a_{j-1}, a_j) = E\xi_j, \qquad d_1(a_1) = D(\xi_1)$$

$$d_j(a_{j-1}, a_j) = D(\xi_j), \qquad 2 \leqslant j \leqslant r, \qquad \mathcal{M}_r = \mu_1(a_1) + \sum_{j=2}^{r} \mu_j(a_{j-1}, a_j)$$

$$D_r = d(a) + \sum_{j=2}^{r} d_j(a_{j-1}, a_j)$$

Certainly, $\mathcal{M}_r$ and D_r are andom variables, since they are functions of b.

Let $t \to \infty$. Consider the probability $P_b(t)$ (with respect to b) that the random variables $\xi_1, \xi_2, \ldots, \xi_r$ satisfy the local central limit theorem (lclt) in the form described in the Lemma 1.

Lemma 3. $P_b(t) \to 1$ as $t \to \infty$.

The proof of the lemma is simple and we shall describe only the main steps. It uses characteristic functions. It is easy to show that there exists a finite covering of S^1 by arcs $C_0, C_1, C_2, \ldots, C_s$, $p > 0$, $\delta > 0$, such that C_0 is a symmetric neighborhood of 1 and for any C_j, $1 \leqslant j \leqslant s$, the probability (with respect to B) that the characteristic function has on C_j the absolute value less than $1 - \delta$ is greater than p. This gives easily an exponential estimation for the characteristic function of the sum $\sum_{j-1}^{r} \xi_j$ outside a small neighborhood of 1. The rest follows the traditional way of proving the local central limit theorem.[4]

Thus, under the conditions of Theorem 1 and for those b for which the lclt is true we can write again

$$\varphi(x; t) \sim A \int da_1\, da_2 \cdots da_r\, Z_1(x, a_1) \prod_{j=2}^{r} Z_j(a_{j-1}, a_j) \tag{12}$$

Now $Z_j(a_{j-1}, a_j)$ are b-independent random variables. The analysis of (12) can be done again with the help of the theory of non-homogeneous Markov chains.

Namely, consider the conditional probabilities

$$\pi_1(a_1/a_2) = \frac{Z_2(a_1, a_2)}{\int Z_2(a_1, a_2)\, da_1}$$

$$\pi_j(a_j/a_{j+1}) = \frac{\int Z_2(a_1, a_2) \cdots Z_j(a_{j-1}, a_j)\, Z_{j+1}(a_j, a_{j+1})\, da_1 \cdots da_{j-1}}{\int Z_2(a_1, a_2) \cdots Z_j(a_{j-1}, a_j)\, Z_{j+1}(a_j, a_{j+1})\, da_1 \cdots da_j}$$

We can use them to rewrite the rhs of (12) in another way:

$$\varphi(x; t) \sim A\, \Xi_r \int Z_1(x, a_1)\, \pi_1(a_1 \mid a_2)\, \pi_2(a_2 \mid a_3) \cdots$$
$$\times\, \pi_{r-1}(a_{r-1} \mid a_r)\, \pi_r(a_r)\, da_1 \cdots da_r$$

where

$$\Xi_r = \int da_1 \cdots da_r\, Z_2(a_1, a_2) \cdots Z_r(a_{r-1}, a_r)$$

plays the role of partition function. For the derivative $\varphi_x(x; t)$ we have a similar expression:

$$\varphi_x(x; t) \sim A\Xi_r \int \frac{\partial Z_1}{\partial x}(x, a_1)\, \pi_1(a_1 \mid a_2)\, \pi_2(a_2 \mid a_3) \cdots$$
$$\times\, \pi_{r-1}(a_{r-1} \mid a_r)\, \pi(a_r)\, da_1 \cdots da_r$$

Therefore this yields for the solution $u(x; t)$ of the Burgers equation

$$u(x, t) = -2\mu \frac{\varphi_x(x; t)}{\varphi(x; t)} \sim \frac{-2\mu \int (\partial Z_1/\partial x)(x, a_1)\, \pi_1(a_1 \mid a_2) \cdots}{\int Z_1(x, a_1)\, \pi_1(a_1 \mid a_2) \cdots}$$
$$\times \frac{\cdots \pi_{r-1}(a_{r-1} \mid a_r)\, \pi(a_r)\, da_1 \cdots da_r}{\cdots \pi_{r-1}(a_{r-1} \mid a_r)\, \pi(a_r)\, da_1 \cdots da_r}$$
$$= \frac{\int (\partial Z_1/\partial x)(x, a_1)\, \pi_1(a_1 \mid a_k)\, \pi_k(a_{k+1} \mid a_k) \cdots}{\int Z_1(x, a_1)\, \pi_1(a_1 \mid a_k)\, \pi_k(a_{k+1} \mid a_k) \cdots}$$
$$\times \frac{\pi_{r-1}(a_{r-1} \mid a_r)\, \pi(a_r)\, da_1\, da_k \cdots da_r}{\pi_{r-1}(a_{r-1} \mid a_r)\, \pi(a_r)\, da_1\, da_k \cdots da_r}$$

for any k. Here $\pi(a_1 \mid a_k)$ is the conditional density corresponding to the joint probability density

$$\frac{Z_2(a_1, a_2) \cdots Z_r(a_{r-1}, a_r)\, da_1\, da_2 \cdots da_r}{\Xi}$$

Now we remark that for large k the conditional distribution $\pi(a_1 \mid a_k)$ becomes almost independent of $a_k, a_{k+1}, \ldots, a_r$ and thus independent of $B(\tau)$, $0 \leqslant \tau \leqslant t_k$. This follows easily from the ergodic theorem for Markov chains. To be more precise, let us formulate the following lemma.

Lemma 4. There exist positive constants $\rho < 1$ and $C_2 < \infty$ and events $S_k \in B(0, k)$, $k = 1, 2, \ldots$, $P_b(S_k) > 1 - C_2 \rho^k$ and a functional

$H_x^{(k)}(b(t_1, t_2),\ 0 \leqslant t_1,\ t_2 \leqslant k)$ defined on $B(0, k)$ such that if $(b(t_1, t_2), t-k \leqslant t_1 < t_2 \leqslant t) \in S_k$, then

$$|u(x, t) - H_x^{(k)}(b(t_1, t_2), t-k \leqslant t_1 < t_2 \leqslant t)| \leqslant C_2 \rho^k$$

The functional $H_x^{(k)}$ is a periodic function of x of period x_0.

The proof of the lemma goes as follows. The transition densities $\pi_j(a_{j-1} \mid a_j)$ are bigger than some constant $\sigma > 0$ with a positive probability. It is easy to show that with the probability not less than $1 - C_2 \rho^k$ the number of such j is bigger then βk for some $\beta > 0$. Then the conditional distribution $\pi(a_1 \mid a_k)$ does not depend on k. The periodicity of $H_x^{(k)}$ on x follows easily from the expressions for $\varphi(x; t)$.

ACKNOWLEDGMENTS

This paper was done mainly during my stay at the Mathematics Department of Princeton University. I would like to thank the Chairman of the Department, Prof. S. Kochen, for warm hospitality. I also thank M. Aizenmann, J. Lebowitz, A. Majda, T. Spencer, H. Spohn, and D. Szasz for useful discussions. I would especially like to thank V. Yachot, who explained to me the importance and the beauty of the Burgers equation and with whom I several times discussed the properties of its solutions.

REFERENCES

1. E. Hopf, *Commun. Pure Appl. Math.* 3:201 (1950).
2. J. M. Burgers, *The Nonlinear Diffusion Equation* (Reidel, Dordrecht, 1974).
3. B. Simon, *Functional Integration and Quantum Physics* (Academic Press, New York, 1979).
4. B. V. Gnedenko and A. N. Kolmogorov, *Limit Theorems for Sums of Independent Random Variables* (FizMatGiz, Moscow, 1949).

Comments

The first theorem shows the difference between the Burgers equation and the Navier – Stokes system. In the case of Navier – Stokes system, complicated invariant limiting sets like strange attractors are possible while in the case of a Burgers system these sets are simpler. The second theorem uses some arguments taken from statistical mechanics.

References

[BKL] J. Bricmont, A. Kupiainen, R. Lefevere, Ergodicity of the 2*D* Navier – Stokes equations with random forcing, *Comm. Math. Phys.*, **224** (2001), no. 1, 65–81.

[EKMS] Weinan E, K. Khanin, A. Mazel, Ya. G. Sinai, Probability distribution functions for the random forced Burgers equations, *Phys. Rev. Lett.*, **78** (1997), 1904–1907.

[FT] C. Foias, R. Temam, Gevrey class regularity for the solutions of the Navier – Stokes equations, *J. Funct. Anal.*, **87** (1989), no. 2, 359–369.

[HM] M. Hairer, J. Mattingly, Ergodicity of the 2*D* Navier – Stokes equations with degenerate stochastic forcing, *Ann. Math. (2)*, **164** (2006), no. 3, 993–1032.

[KS] S. Kuksin, A. Shirikyan, Stochastic dissipative PDE's and Gibbs measures, *Comm. Math. Phys.*, **213** (2000), 291–330.

[LS1] D. Li, Ya. G. Sinai, Complex singularities of the Burgers system and renormalization group method, in: *Current Developments in Mathematics*, 2006, Int. Press, Somerville, MA, 2008, pp. 181–210.

[LS2] D. Li, Ya. G. Sinai, Complex singularities of solutions of some 1D hydrodynamic models, *Phys. D*, **237** (2008), no. 14–17, 1945–1950.

Printed in Great Britain
by Amazon

18322698R00305